Lecture Notes in Economics and Mathematical Systems 633

For further volumes:
http://www.springer.com/series/300

Kurt Marti • Yuri Ermoliev
Marek Makowski
Editors

Coping with Uncertainty

Robust Solutions

Springer

Editors
Prof. Dr. Kurt Marti
Federal Armed Forces University Munich
Aero-Space Engineering and Technology
Werner-Heisenberg-Weg 39
85577 Neubiberg
Germany
kurt.marti@unibw-muenchen.de

Prof. Dr. Yuri Ermoliev
International Institute for Applied
Systems Analysis (IIASA)
Schloßplatz 1
2361 Laxenburg
Austria
ermoliev@iiasa.ac.at

Dr. Marek Makowski
International Institute for Applied
Systems Analysis (IIASA)
Schloßplatz 1
2361 Laxenburg
Austria
marek@iiasa.ac.at

ISSN 0075-8442
ISBN 978-3-642-03734-4 e-ISBN 978-3-642-03735-1
DOI 10.1007/978-3-642-03735-1
Springer Heidelberg Dordrecht London New York

Library of Congress Control Number: 2009934495

Cover design: SPi Publisher Services

Printed on acid-free paper

Springer is part of Springer Science+Business Media (www.springer.com)

Preface

The aim of the series of workshops on "Coping with Uncertainty" (*CwU*) organized at IIASA, Laxenburg, Austria, has been to provide researchers and practitioners from different areas with an interdisciplinary forum for discussing various ways of dealing with uncertainties in diverse areas, including environmental and social sciences, economics, policy-making, management, and engineering. The workshops proved to be successful, especially in cross-disciplinary sharing methods, ideas, and open problems.

Science-based support for addressing the on-going global changes needs solutions for fundamentally new scientific problems, which in turn require new concepts and tools. A key issue concerns a vast variety of practically irreducible uncertainties, including potential extreme events of high multidimensional consequences, which challenge traditional models, and thus require new concepts and analytical tools. This type of uncertainty critically dominates, e.g., the climate change debates. In short, the dilemma is concerned with enormous costs versus massive uncertainties of potentially extreme impacts. Traditional scientific approaches usually rely on real observations and experiments. Yet no sufficient observations exist for new problems, and "pure" experiments, and learning by doing may be very expensive, dangerous, or simply impossible. In addition, the available historical observations are often contaminated by "experimentator", i.e., past actions, and policies. The complexity of new problems does not allow us to achieve enough certainty just by increasing the resolution of models or by bringing in more links. They require explicit treatment of uncertainties using "synthetic" information composed of available "hard" data from historical observations, the results of possible experiments, and scientific facts as well as "soft" data from experts' opinions, scenarios, stakeholders, and public opinion. As a result of all these factors, our assessment will always have poor estimates. Therefore, the role of science-based support for addressing the new problems increasingly changes from the traditional "deterministic predictions" analysis to the design of strategies that are robust against the involved uncertainties and risks.

This volume contains contributions based on selected presentation at the *CwU*2007 workshop. The workshop aimed at contributing to a better understanding between practitioners dealing with the safety of complex systems under uncertainty, and scientists working on either corresponding modeling approaches, or on methods that can be adapted for improving the understanding and management of uncertainty. The focus of the CwU 2007 was on novel approaches to supporting robust decision-making and design, especially when uncertainty is irreducible, consequences might be enormous, and the decision process involves stakeholders with diverse interests. Presentations dealt with open problems in this field, limitations of known approaches, novel methods and techniques, or lessons from applications of various approaches. In particular, contributions on the following issues were presented:

- Modeling different types of uncertainty (probabilistic and non-probabilistic)
- The formulation of appropriate deterministic substitute problems for different types of uncertainty
- Robustness of efficient solutions with respect to inherent uncertainties
- Simulation tools (for optimal decision/design under uncertainty)
- Safety and security of humans, environment, and vital infrastructure facing catastrophe risks
- Lessons that can be learned from designing and operating highly reliable systems
- Downscaling and discounting methods for handling spatial and temporal scales
- Benefits and costs of (partial) postponing decisions (aimed at reducing uncertainties)
- Open problems in the adequate treatment of uncertainties
- Concrete applications in economics, finance, engineering, energy, population, air quality, climate change, ecology, forestry, and other environmental problems

The workshop was organized at IIASA in December 2007, jointly by:

* IIASA – International Institute for Applied Systems Analysis, Laxenburg, Austria
* Federal Armed Forces University Munich, Germany

The scientific Program Committee included: Yuri Ermoliev, IIASA, Laxenburg (A); Leen Hordijk, IIASA, Laxenburg (A); Marek Makowski, IIASA, Laxenburg (A); Kurt Marti, Federal Armed Forces University Munich (D); Gerhard I. Schuëller, University of Innsbruck (A).

The organizers gratefully acknowledge the support of:

- GAMM – International Association of Applied Mathematics and Mechanics, and
- IIASA – International Institute for Applied Systems Analysis

Their generous support enabled the participation of many researchers who otherwise could not have attended the Workshop.

This volume contains chapters based on selected presentations at the CwU 2009 and an introductory short summary of the key issues related to the robust solutions. The chapters are organized into the following four parts:

1. *Modeling of uncertainty* discusses descriptions of uncertainties of different types (probabilities, theory of evidence and possibility, imprecise probabilities, fuzzy sets and variables).

2. *Robust solutions under uncertainty* presents new approaches to discounting applied to evaluation of investments for catastrophic risk management, and to cost-effective and environmentally-safe emission trading under uncertainties, as well as modern quantitative modeling methodologies for analysis of network risks and design of robust networks under uncertainty.
3. *Analysis and optimization of technical systems and structures under uncertainty* deals with state estimation of dynamical systems in case of uncertainties of initial conditions and dynamic parameters described by means of certain ellipsoids, and with the derivation of stochastic linear programs for the reliability-based optimization of plane frames under stochastic uncertainty with respect to external loadings and material parameters.
4. *Analysis and optimization of economic and engineering systems under uncertainty* discusses the variability of the atmospheric deposition of nitrogen in the sea, the treatment of risks and uncertainties in planning agricultural production allocation and expansion, the uncertainty in greenhouse gas emission estimates, consequences of the weather forecasts for the optimal control of agricultural production, and the estimation error in retrieving carbon dioxide column abundances obtained from the GOSAT satellite.

We express our gratitude to all referees, and we thank all authors for the timely delivery of the final version of their contributions to this volume. Furthermore, we thank Ms Elisabeth Lößl of the Federal Armed Forces University Munich for her support in the preparation of this volume. Finally, we thank Springer-Verlag for including the Proceedings in the Springer Lecture Notes Series "LNEMS".

Munich, Laxenburg
June 2009

Kurt Marti
Yuri Ermoliev
Marek Makowski

Contents

Contributors

Jerzy Bartnicki Norwegian Meteorological Institute, P.O. Box 43, Blindern, 0313 Oslo, Norway, jerzy.bartnicki@met.no

F.L. Chernousko Institute for Problems in Mechanics, Russian Academy of Sciences, pr. Vernadskogo, 101-1, 119526 Moscow, Russia, chern@ipmnet.ru

T. Doeswijk Systems and Control Group, Wageningen University, P.O. Box 17, 6700 AA Wageningen, The Netherlands, timo.doeswijk@wur.nl

Y. Ermoliev International Institute for Applied Systems Analysis, Schlossplatz 1, 2361 Laxenburg, Austria, ermoliev@iiasa.ac.at

T. Ermolieva International Institute Applied Systems Analysis, Shlossplatz 1, 2361 Laxenburg, Austria, ermol@iiasa.ac.at

G. Fischer International Institute Applied Systems Analysis, Shlossplatz 1, 2361 Laxenburg, Austria, fisher@iiasa.ac.at

A. Gaivoronski Department of Industrial Economics and Technology Management, Norwegian University of Science and Technology, Alfred Getz vei 1, 7491 Trondheim, Norway, Alexei.Gaivoronski@iot.ntnu.no

Olgierd Hryniewicz Systems Research Institute, Newelska 6, 01-447 Warsaw, Poland, hryniewi@ibspan.waw.pl

Matthias Jonas International Institute for Applied Systems Analysis, Schlossplatz 1, 2361 Laxenburg, Austria, jonas@iiasa.ac.at

Koji Kanefuji The Institute of Statistical Mathematics, Tokyo, Japan, kanefuji@ism.ac.jp

K.J. Keesman Systems and Control Group, Wageningen University, P.O. Box 17, 6700 AA Wageningen, The Netherlands, karel.keesman@wur.nl

Edith Kovács Department of Mathematics, ÁVF College of Management of Budapest, Villányi út 11-13, 1114 Budapest, Hungary, kovacs.edith@avf.hu

Daniel Lieberman ICF International, 1725 Eye St. NW., Suite 1000, Washington, DC 20006, USA (currently at Chevron Corporation, 6001 Bollinger Canyon Rd, K2339, San Ramon, CA 94583, USA), dlieberman@chevron.com

M. Makowski International Institute for Applied Systems Analysis, Schlossplatz 1, 2361 Laxenburg, Austria, marek@iiasa.ac.at

Gregg Marland International Institute for Applied Systems Analysis, Schlossplatz 1, 2361 Laxenburg, Austria, marland@iiasa.ac.at

K. Marti Aero-Space Engineering and Technology, Federal Armed Forces University Munich, 85577 Neubiberg, Germany, kurt.marti@unibw-muenchen.de

Yukio Matsumoto Association of International Research Initiatives for Environmental Studies, Tokyo, Japan, y-matsu@airies.or.jp

Zbigniew Nahorski Systems Research Institute of the Polish Academy of Sciences, ul. Newelska 6, 01-447 Warsaw, Poland, zbigniew.nahorski@ibspan.waw.pl

Sten Nilsson International Institute for Applied Systems Analysis, Schlossplatz 1, 2361 Laxenburg, Austria, nilsson@iiasa.ac.at

L. Sun Institute for Applied Systems Analysis, Laxenburg, Austria, sun@iiasa.ac.at

Tamás Szántai Institute of Mathematics, Budapest University of Technology and Economics, Műegyetem rkp. 3, 1111 Budapest, Hungary, szantai@math.bme.hu

Mitsuhiro Tomosada The Institute of Statistical Mathematics, Tokyo, Japan, tomosada@y7.dion.ne.jp

Hiroe Tsubaki The Institute of Statistical Mathematics, Tokyo, Japan, tsubaki@ism.ac.jp

Thomas White Canadian Forest Service, Pacific Forestry Centre, 506 West Burnside Road, Victoria, BC, Canada V8Z 1M5, thos_white@hotmail.com

Tatsuya Yokota National Institute for Environmental Studies, Ibaraki, Japan, yoko@nies.go.jp

Chapter 1
General Remarks on Robust Solutions

Y. Ermoliev, M. Makowski, and K. Marti

We summarize here the background and key concepts related to robust solutions in the context of supporting decision-making for problems characterized by deep uncertainties, which also were in the focus of the previous workshops on *Coping with uncertainty*, see, e.g., [3]. Although such problems are fundamentally different from statistical decision models, yet basic ideas of robust statistics are applicable to methods supporting robust decision-making under uncertainty. The main new issues are concerned with a proper representation of uncertainty, and its interactions with decisions. In particular, a key issue is the sensitivity of robust decisions with respect to low probability catastrophic events, that are of critical importance for analyzing global change problems. Robust decisions for problems exposed to extreme catastrophic events are essentially different from over-simplified decisions that ignore such events. Specifically, a proper treatment of extreme/rare events requires new paradigms of rational decisions, new performance indicators, and new spatio-temporal dimensions of heterogeneous interdependencies including network externalities and risks. This, in particular, needs new approaches to downscaling, upscaling and discounting.

Global change processes, in particular climate change, involve inherently unpredictable complex interactions between natural and human-created systems therefore proper modeling of these processes must rely on adequate treatment of uncertainties, and their effects on human's decisions. Traditional natural science models are based on relations whose validity is estimated from repetitive experiments and observations. If experiments do not affect the underlying relations, then repetitive observations allow to derive them by using the statistical decision theory. Unfortunately, human-created processes do not follow fixed relations. Elements of these processes change their dimensions and structure. For example, introduction of new

Y. Ermoliev (✉) and M. Makowski
International Institute for Applied Systems Analysis, Schlossplatz 1, 2361 Laxenburg, Austria,
e-mail: ermoliev@iiasa.ac.at, marek@iiasa.ac.at

K. Marti
Aero-Space Engineering and Technology, Federal Armed Forces University Munich, Neubiberg, Munich, Germany,
e-mail: kurt.marti@unibw-muenchen.de

K. Marti et al. (eds.), *Coping with Uncertainty*, Lecture Notes in Economics and Mathematical Systems 633, DOI 10.1007/978-3-642-03735-1_1,

technologies may increase or reduce uncertainties, risks, critical thresholds and discontinuities. Exact identification of global climate change processes is impossible because such processes are non-stationary, have delayed responses, and human or natural actions may have catastrophic irreversible consequences.

Under inherent uncertainty of heterogeneous processes the role of integrated models rests on the ability to guide comparative analysis of rational decisions. Although exact evaluations are impossible, the preference structure among decisions can be a stable basis for a relative ranking of alternatives in order to design robust policies, which must be in a sense optimal against all relevant uncertainties. It is commonly known that finding out (without exact measurement) which of two parcels is the heavier is much easier than evaluating weight of each parcel.

The term "robust" was introduced into statistics in 1953 by Box [2] and acquired recognition after the publication of a path-breaking paper by Huber [5]. As Huber admits, researchers had long been concerned with the sensitivity of standard estimation procedures to "bad" observations (outliers), and the word "robust" was loaded with many, sometimes inconsistent connotations, frequently for the simple reason of conferring respectability on it. Appeal for robustness [4] probably dates back to prehistory of statistics. A distant outlier in observations ruins the least square analysis, therefore rejection of outliers is a sort of robust statistical procedure. The discussion about the rejection of outliers is at least as old as the 1777 publications of Daniel Bernoulli [1]. The mean is not robust to outliers, whereas the median is robust. Therefore, switching from the mean to the median for long-tailed data increases robustness. This is also equivalent to switching from quadratic (least square) smooth optimization to non-smooth optimization principles.

According to Huber [5], "...any statistical procedure...should be robust in the sense that small deviations from the model assumptions should impair the performance only slightly". This concept of robustness corresponds to standard mathematical ideas of continuity and stability: when disturbances become small, the performance of the perturbed model also deviates slightly. In other words, a robust procedure is in a certain sense optimal with respect to all uncertainties from a neighborhood of the model. Huber introduced rigorous notions of robustness based on probabilistic minimax approach and Choquet capacity (imprecise probabilities), which lead to specific non-smooth stochastic optimization models. By using appropriate neighborhoods of probability measures (e.g., with respect to ϵ-contaminated probabilities, Levy distance, or Kolmogorov distance), he derived robust estimators optimizing the worst that can happen over the neighborhood of the model with respect to a certain performance indicator. Neighborhoods of probability measures can also be characterized by Choquet capacities, i.e., functions which define sets of probability measures by taking all probabilities which lie bellow (or above) a capacity (point-wise).

These basic ideas of robust statistics as well as the infinitesimal robustness introduced by Hampel [4] are also used for more general decision problems under uncertainty. In particular, the infinitesimal approach is based on the fact that many statistics and solutions of general decision models can be considered as functionals in the space of probability measures. The robustness information is then provided by the inference functions, roughly speaking, a derivative (in the space of probability measures) of a statistic or a performance indicator at an underlying distribution. There are also important concepts of Bayesian and non-Bayesian robustness, where we need not only robustness against deviations from the given parametric model, but also against uncertainties of the prior distributions.

The word "robust" has become fashionable in statistical decision theory and other disciplines dealing with data analysis and decisions [4], in particular, for dealing with questions such as: Which data are of critical importance and should be examined with a special care? What methods provide the greatest safety? How safe are results of a model that is known only approximately?

As the concept of statistical robustness is in a sense similar to the problem of local stability of dynamic systems, the robustness in deterministic control theory was introduced as an additional requirement on the stability of optimal trajectories. In other words, additional constraints were introduced in the form of a stability criterion. Optimization theory provides tools for analyzing and solving various decision making problems. For deterministic models robustness is defined similar to probabilistic minimax robustness in statistics: to optimize the worst that can happen to performance indicators over solutions x that satisfy feasibility constraints for all admissible values of uncertainty $\omega \in \Omega$. The set Ω is often characterized by a finite number of scenarios or simple sets such as intervals or ellipsoidal uncertainty which, in a sense, attempt to substitute for normal probability distributions in a simple but inconsistent with statistical analysis manner. It is clear that this type of deterministic worst-case robustness leads to extremely conservative decisions.

After the word robust become fashionable in statistics, it is being used in many senses, e.g., "quantitative robustness", "Π-robustness", "B-robustness", see, e.g., [4]. The situation becomes more complex for general decision problems under uncertainty dealing with quite different decision situations which may have a vast variety of different facets of robustness. Therefore, in order to avoid dangerous confusions, the term "robust" must be precisely defined in every specific context. This is similar with other notions whose meaning depends on the context, e.g., fairness, efficiency, optimality. For example, a decision is optimal only with respect to precisely specified conditions.

Statistical decision theory deals with situations in which a model of uncertainty and corresponding optimal solution are defined by a sampling model characterized by a probability measure P with an unknown vector of "true parameters" x^*. Vector x^* defines a desirable optimal solution, its performance can be observed from the sampling model, and the problem is to recover x^* from these data. Potential estimates of x^* define feasible solutions x of a statistical decision problem. It is essential that x does not affect the sampling model P so that the optimality and robustness of solutions can be evaluated by a distance from x^* by using its observable performance.

The general problems of decision making under uncertainty deal with fundamentally different situations. The model of uncertainty, feasible solutions, and performance of the optimal solution are not given and all of these have to be characterized from the context of the decision making situation, e.g., socio-economic, technological, environmental, and risk considerations. As there is no information on true optimal performance, robustness cannot be also characterized by a distance from an observable true optimal performance. Therefore, the general decision problems may have rather different facets of robustness. In particular, probabilistic minimax solutions may seem too pessimistic or too optimistic for coping with potentially catastrophic events. Therefore, other concepts of robustness are required. Such concepts may be based on, e.g., expected utility maximization, stochastic optimization, stochastic minimax models.

In the presence of uncertainty, any related decision x results in multiple outcomes such as costs, benefits, damages, and risks, as well as indicators of fairness, equity, and environmental impacts. The outcomes depend not only on decisions x but also on uncertainty characterized by $\omega \in \Omega$, where Ω, denotes a set of admissible scenarios.

Scenario analysis is often used as a straightforward approach to find a decision that is "optimal" with respect to all scenarios by attempting to solve the decision problem for all possible scenarios. Unfortunately, a given decision x for different scenarios ω may have rather contradictory outcomes which do not really tell us which decision is reasonably good for all scenarios.

In 1738 mathematician Daniel Bernoulli (see discussion in [7]) introduced the concept of expected utility maximization as a rule for choosing decisions under multiple outcomes. It is assumed that all outcomes $g_i(x, \omega)$, $i = 1, \ldots, K$ can be summarized in a single measure of preferability, e.g., a monetary payoff denoted by: $q(x, \omega) = Q(g_1, \ldots, g_K)$. The standard expected utility model suggests to choose a decision x that maximizes an expected utility function

$$U(x) = Eu(q(x, \omega)) = \int u(q(x, \omega))P(d\omega),$$

i.e., in a sense, for all $\omega \in \Omega$, where $u(\cdot)$ is a utility associated with an aggregate outcome $q(x, \omega)$. The shape of u defines attitudes to risks. This model presupposes that one can rank the alternative scenarios ω according to weights – objective or subjective probability measure P.

The use of a probability measure as a degree of belief was formalized by Ramsey [6]. Savage published [7] a thorough treatment of expected utility maximization based on subjective probability as a degree of belief. As a result of this work the use of probability measure became a standard approach for modeling uncertainty by using "hard" observations and "soft" public and expert opinions in a consistent way within a single model. Although a decision maximizes the expected utility, in a sense, for all scenarios, still it cannot be considered to be a robust solution. The shortcomings of the expected utility model are well known. Generally, it is practically impossible to find a utility function that enables satisfactory aggregation of various attributes in one preferability measure, including attitudes to different risks, the distributional aspects of gains and losses, the rights of future generations, and responsibilities for environmental protection.

For complex problems it is natural that different performance indicators should be used to evaluate robustness of the integrated system in the same way as we use indicators of health (e.g., temperature and blood pressure for humans). An expected utility model is a specific case of stochastic optimization (STO) models that use various performance indicators $f_i(x, \omega)$, $i = 1, \ldots, m$, one of which can be the expected utility (disutility). These indicators depend on outcomes $g_k(x, \omega)$, $k = 1, \ldots, K$, on x and ω, i.e.,

$$f_i(x, \omega) := q_i(g_1, \ldots, g_K, x, \omega).$$

A rather general STO problem is formulated as optimization (maximization or minimization) of the expectation function

$$F_0(x) = Ef_0(x, \omega) = \int f_0(x, \omega) P(d\theta)$$

subject to constraints

$$F_i(x) = Ef_i(x, \omega) = \int f_i(x, \omega) P(d\theta) \geq 0, \quad i = 1, \ldots, m.$$

The choice of proper indicators $f_i(x, \omega)$ and outcomes $g_k(x, \omega)$, $k = 1, \ldots, K$, is essential for the robustness of x. Globally or regionally aggregated outcomes are less uncertain but they may not reveal potentially dramatic heterogeneities induced by global changes on individuals, governments, and the environment. For instance, an aggregate income or growth indicators may not reveal an alarming gap between poor and rich regions, which may cause future instabilities.

By choosing appropriate outcomes $g_k(x, \omega)$ and functions $f_i(x, \omega)$, STO models allowing a natural and flexible way to represent various risks, abrupt changes, spatio-temporal heterogeneities, equity constraints and the sequential resolution of uncertainty in time. Often, under proper robustness requirements, $f_i(x, \omega)$ are analytically intractable, non-smooth, and even discontinuous functions, and probability measure P is chosen from a feasible set, thus is imprecise; moreover, P often depends on x, which is essential for modeling endogenous extreme events and catastrophic risks, and non-Bayesian robustness. It is often practically impossible to identify uniquely subjective (and objective) probability as a degree of beliefs. Most people cannot clearly distinguish between probability ranging roughly from 0.3 to 0.7. Decision analysis often has to rely on imprecise statements, for example, that event e_1 is more probable than event e_2 or that the probability p_1, p_2 of event e_1 or of event e_2 is greater than 50% and less than 90%. Therefore, feasible sets of probabilities may be represented by inequalities such as $p_1 \geq p_2, 0.5 \leq p_1 + p_2 \leq 0.9$. As in robust statistics, the robust solutions of general decision models can be derived by using worst-case (for a given decision) probability distribution from the feasible sets of distributions satisfying constraints of STO model. The applicability of the infinitesimal approaches is facilitated by the fact that solutions of STO models are functionals of underlying probability measures.

In contrast to the expected utility based approaches, the STO models support integrated solutions composed of anticipative ex-ante and adaptive ex-post decisions, which in turn allows to model flexible decision processes with adaptive adjustments of anticipative decisions when new information becomes available. The ability of STO models to incorporate both anticipative ex-ante and adaptive ex-post decisions induces non-smooth criteria and risk aversion among ex-ante decisions that implicitly depends on input data and practically cannot be characterized by an exogenous utility function. In particular, even in the simplest linear model (see [3] and chapters of Part II of this volume) the co-existence of ex-ante and ex-post decisions induces quantile-type risk measures. Therefore, this approach allows to substitute mean values by median or/and other quantiles without destroying convexity (concavity) of models.

A fundamentally new aspect of robustness for general decision problems is the sensitivity (or discontinuity) of robust solutions with respect to extreme (low probability and high consequences) events affecting large territories and communities, see [3]. Although potential extreme events with large-scale catastrophic impacts dominate the global change policy discussions there is no agreement on the corresponding measures of the changes and the impacts. For example, a commonly used measure is the projected global mean temperature change, which falls within the difference between the average temperature of cities and their surrounding rural areas, and cannot be reliable indicator of climate change. Actually, global climate change impacts can be properly evaluated only in terms of local temperature variability and related extreme events, in particular, heat waves, floods, droughts, wind-storms, diseases, and sea-level rise.

Unfortunately, ignorance of extreme events explicitly or implicitly through inadequate treatments of their impacts can be dramatically misleading. The latter is especially dangerous since it provides an illusion of truly comprehensive analysis. A 500-year disaster, e.g., an extreme flood that occurs on average once in 500-years is often considered as irrelevant for many future generations, but actually it may occur next year and destabilize economy for many years. However, existing extremal value theory deals primarily with independent variables quantifiable by a single number, e.g., money. Therefore, the use of this theory may also be misleading, because catastrophes are definitely not quantifiable events in this sense. They have different patterns, spatial and temporal dimensions and induce heterogeneity of losses and gains, which exclude the use of aggregate space-less characteristics. Globally, an average resident may even benefit from some climate change scenarios, while some regions would be simply wiped out.

Ignorance of spatio-temporal details leads to models and solutions which are insensitive to large-scale extreme events (outliers), therefore formally can be viewed as robust solutions demonstrating insignificance of threats related to on-going global changes, say, climate change. The approaches presented in [3] and in this volume allow to properly address issues of robustness of solutions for problems exposed to catastrophic events. In particular, probabilistic maximin (minimax) robustness may not be sufficient for this. A more general approach is the combination of the probabilistic and so-called stochastic maximin approach maximizing

$$\min_{p \in P} E \min_{z \in Z} f_0(x, y, z, \xi)$$

under constraints of STO model, where x are decision variables, and scenario ω is defined by the vector of variables y, z, and ξ: $\omega = (y, z, \xi)$. The components of ω are: $y \in Y$, variables representing uncertainty ranked by an objective or subjective probability measure from P; $z \in Z$, variables representing potential extreme random scenarios, as in the extremal value theory; ξ are variables ranked by a fixed probability measure as in the basic STO models. The concept of stopping time allows to focus the analysis on the least-probable and the most destructive extreme events. As shown in [3], there are strong connections between the stopping time concept and the stochastic maximin.

In the absence of sufficient information, models play a key role in comparative analysis of alternative solutions for designing robust policies. Any policy analysis focuses attention on situations where processes can be changed by decisions that should be selected in the best possible manner. In reality, a proper structure of models, e.g., sets of proper decision are also uncertain and they can be specified through a dialogue of users with models. Specification of such models include use of quantiles, thresholds, and stopping times, which in turn require specific nonsmooth stochastic optimization methods. Effective analysis of such models requires development of specific fast adaptive Monte Carlo optimization procedures.

References

1. Bernoulli, D.: Dijudicatio maxime probabilis plurium observationum discrepantium atque verisimililima inductio inde formanda. Acta Acad. Sci. Petropolit **1**, 3–33 (1777). English translation by Allen, C.G.: Biometrika, **48**, 3–13 (1961)
2. Box, G.: Non-normality and tests on variances. Biometrika **40**, 318–335 (1953)
3. Ermoliev, Y., Hordijk, L.: Facets of robust decisions. In: Marti, K., Ermoliev, Y., Makowski, M., Pflug, G. (eds.) Coping with Uncertainty: Modeling and Policy Issues, pp. 3–28. Springer, Berlin, Heidelberg, New York (2006)
4. Hampel, F., Ronchetti, E., Rousseeuw, P., Stahel, W.: Robust Statistics: The Approach Based on Influence Functions. Wiley, New York (1986)
5. Huber, P.: Robust Statistics. Wiley, New York (1981)
6. Ramsey, F.: Truth and probability. In: Braithwaite, R. (ed.) The Foundations of Mathematics and Other Logical Essays, pp. 156–198. Kegan, Paul, Trench, Trubner & Co., London (1931). Originally published in 1926
7. Savage, L.: The Foundations of Statistics, 2nd edn. Dover, New York (1972). Originally published in 1954

Part I
Modeling of Uncertainty and Probabilistic Issues

Chapter 2
On Joint Modelling of Random Uncertainty and Fuzzy Imprecision

Olgierd Hryniewicz

Abstract The paper deals with the problem of the mathematical description of uncertainties of different type. It has been demonstrated by many authors that the theory of probability is not always suitable for the description of uncertainty related to vagueness. We briefly present some of the most promising theories which have been recently proposed for coping with this problem. Then, we concentrate our attention on the application of fuzzy random variables which seem to be very useful for the joint modelling of random uncertainty and fuzzy imprecision, and for the statistical analysis of imprecise data. The application of the statistical methodology for fuzzy data, called fuzzy statistics, is illustrated with two practical examples, typical for the problems of systems analysis. First example is devoted to the problem of the estimation of greenhouse gases inventories. In the second example, typical for the problems of making decisions using small amount of available data, we show how fuzzy approach can be used for the improvement of sequential statistical tests.

2.1 Introduction

Coping with uncertainty is an important problem in many areas of science, but in systems analysis and decision sciences it becomes a really crucial one. In both these branches of science uncertainty is always present, as systems analysts and decision makers have never full information about past, current and future "states of the world". Thus, their information about consequences of proposed by them actions, which is necessary, e.g. for finding optimal decisions, is rarely precise and fully reliable. Moreover, they usually cannot present descriptions of processes of their interest with accuracy which is typical, e.g. for physics, chemistry or astronomy. Therefore, they need to use a formal language that could be used for sufficiently precise description of uncertain events, actions, etc. For many decades it appeared

---

O. Hryniewicz
Systems Research Institute, Newelska 6, 01-447 Warsaw, Poland,
e-mail: hryniewi@ibspan.waw.pl

K. Marti et al. (eds.), *Coping with Uncertainty*, Lecture Notes in Economics and Mathematical Systems 633, DOI 10.1007/978-3-642-03735-1_2,

to the majority of scientists that the theory of probability and mathematical statistics is the only methodology that should be used for the formal description of uncertainty. However, during last few decades many scientists working in such areas like psychology, economic sciences, quantum physics, artificial intelligence, etc. have raised questions about possible inadequacy of the classical (Kolmogorov's) theory of probability when applied for solving their particular problems.

This situation is far from being unexpected. If we look at any good dictionary of, e.g. English we can find that the word "probable", which definitely describes uncertainty, has many synonyms and other related words. For example, such words like "possible", "plausible", and expressions like "likely to be true", "hopeful", "to be expected" (and their antonyms) are used for the description of a state on uncertainty. One may expect that they are used for expressing slightly different types of uncertainty, and are not fully exchangeable with the word "probable". The differences between their meanings raised doubts among philosophers and mathematicians about the role of the classical theory of probability as the sole mathematical language for the description of uncertainty.

Basic problems with the applicability of the classical theory of probability inspired mathematicians who proposed other, not necessarily equivalent, theories of probability. Some of them are described in a classical book of Fine [24]. Other doubts were raised by the founder of the theory of fuzzy sets L.A. Zadeh who claimed that the classical theory of probability cannot describe uncertainty related to innate imprecision of many notions and ideas expressed in a plain human language. In his seminal paper [95] Zadeh proposed to use the formalism of fuzzy sets as the formal language of the theory of possibility. Another criticism of the classical probability came from economists and psychologists. The Nobel Prize winner in economics H. Simon in his book [77] noticed that people do not make their decisions according to the principle of expected utility which is based on the classical theory of probability. Other doubts were raised by other Nobel Prize winners Tversky and Kahneman who noticed in their works (see, for example, [83]) that probabilities evaluated by humans are not necessarily additive, as it is assumed in the classical theory of probability. An interesting description of theoretical and practical problems with the applicability of the classical theory of probability can be found in the paper by Bordley [5] who also noticed that this theory is in a certain sense incompatible with the quantum physics.

The existence of many extensions and modifications of the classical theory of probability creates problems for systems analysts and decision makers who are expected to model systems in presence of uncertainties of different types. In the second section of this paper we present a very brief description of several generalizations of classical probability. This has to be done in order to set borders between those areas where classical probability is still the best (and probably the only) mathematical model of uncertainty and the areas where its generalizations are needed. We claim that in the majority of practical cases the combination of classical probability and Zadeh's theory of possibility is sufficient for the description of complex systems and making decisions. We describe this methodology in the third section of the paper. Examples of the application of fuzzy random models and fuzzy statistics are given in the fourth section of the paper. We show that in case of information that has both random and imprecise nature some additional indices, like possibility and necessity measures, are indispensable for a correct description of decision making problems.

2.2 Generalizations of Classical Probability and Their Applications in Decision Making

2.2.1 Measures of Uncertainty and Criteria of Their Evaluation

Close analysis of the human perception of uncertainty reveals that this concept does not have one, unanimously approved, interpretation. Zimmermann [96] notes that any definition of uncertainty has to be to some extent arbitrary and subjective. He proposes the following one:

> Uncertainty implies that in a certain situation a person does not dispose about information which quantitatively and qualitatively is appropriate to describe, prescribe or predict deterministically and numerically a system, its behavior or other characteristica.

This definition is definitely technology-oriented. One may say that everything what prevents us to describe reality in a deterministic way may be considered as a facet of uncertainty. Thus, there exist many different causes of uncertainty. Zimmermann [96] lists the following: lack of information (quantitative or qualitative), abundance of information, conflicting evidence, ambiguity, measurement, and belief. It is not surprising that he does not believe that the general theory of uncertainty which is able to describe these completely different sources of uncertainty exists, and appropriate mathematical models should be context-dependent. They could be formulated either as different generalizations of probability, or may be formulated in another way, such as Pawlak's rough set theory (see, e.g. [67]) or convex modeling proposed by Ben-Haim and Elishakoff [3].

If we look at different theories of uncertainty we can notice that they can be divided into two general groups: those based on methods of mathematical logics (such as the rough sets theory) or those based on the theory of probability and its generalizations. In this paper we restrict our interest only to the second one. This restriction arises from a fact that in the majority of practical cases the classical theory of probability is sufficient for the mathematical description of uncertainty. This popularity of classical probabilistic models of uncertainty makes many specialists to believe that classical probability is the only mathematical theory that is sufficient for the formal description of uncertainty. Insufficiency of this approach was noticed only recently, mainly by specialists in decision-making or expert systems. Peter Walley, who is the one of the most prominent persons representing that group of scientists, presents the list of pertinent mathematical models, in order of their generality [88] (in parentheses, there are given the most important, according to Walley, references, and indicated by him typical areas of application):

- Possibility measures and necessity measures ([18, 95], vague judgments of uncertainty in natural language)
- Belief functions and plausibility functions ([12, 76], multivalued mappings and non-specific information)
- Choquet capacities of order 2 ([8, 13, 45], some types of statistical neighborhood in robustness studies, and various economic applications)
- Coherent upper and lower probabilities ([45, 56, 78], personal betting rates, and upper and lower bounds for probabilities)
- Coherent upper and lower previsions ([86, 87, 92], buying and selling prices for gambles, upper and lower bounds for expectations, and envelopes of expert opinions)
- Sets of probability measures ([4, 36, 58], partial information about an unknown probability measure, and robust statistical models)
- Sets of desirable gambles ([86,91,93], preference judgments in decision making)
- Partial preference orderings ([34, 86], preference judgments in decision making)

Walley [88] also notices:

- Partial comparative probability orderings ([25, 47, 49], qualitative judgements of uncertainty)

In order to evaluate and compare all competing theories of uncertainty (including classical probability) Walley [87] proposes to take into account the following criteria:

(a) Interpretation
(b) Imprecision
(c) Calculus
(d) Consistency
(e) Assessment
(f) Computation

The proposed measure of uncertainty has to be sufficiently easy to understand by its users. For example, conclusions inferred from the application of the theory should be clear enough to be useful for making decisions. It should be able to model partial or complete ignorance, reflected, for example, in imprecision of statements of natural language. There should be rules for merging uncertainties, updating, and using them in inferential processes. There should be methods for the evaluation of coherence of all assessments formulated using the theory and its assumptions. A useful theory of uncertainty should provide guidance how to make assessments about uncertain events and handle imprecise judgments of different types. Finally, it should be computationally feasible. More comprehensive interpretation of the criteria presented above and their practical and theoretical importance can be found in [87].

All these requirements may have different importance in different applications, and none of the existing theories and measures of uncertainty fulfills them sufficiently well. For example, the classical theory of probability does not meet sufficiently well criteria (b) and (e), and for this reason philosophers, mathematicians, economists, psychologists, and specialists in expert systems have been making a lot of efforts in order to introduce more general, and more useful in specific applications, theories of uncertainty. In the following subsections of this section we present very brief description of some of these theories. This presentation is needed, in our opinion, for the understanding of limits which still exist if we try to cope with uncertainty inherent in the analysis of complex systems.

2.2.2 Probability

The theory of probability is the best known, and the most frequently used in practice, theory of uncertainty. Many specialists claim that it is the only consistent theory that describes all types of uncertainty, both *aleatoric* describing randomness observed in repeated experiments and *epistemic* describing uncertain information provided by human beings (e.g. experts). It has a well established mathematical formulation, and efficient methods for processing information described using probabilistic concepts. However, despite its four hundred years lasting history its fundamentals are still subject to different interpretations and controversies. In general, there exist two different interpretations of the classical probability: an objective "frequentist" interpretation, based on the analysis of empirical observations of series of events, and subjective "Bayesian" approach, based of subjective assessment of probabilities of events. It is interesting that even in the 1960s the second approach was dismissed as "non-scientific" by the majority of statisticians, and not present in nearly all popular textbooks. On the other hand, the supporters of the Bayesian approach presented in books of Savage [74] and de Finetti [26, 27] pointed out apparent incoherences inherent for the frequentist approach (see, for example, an excellent monograph by Lindley [59]).

The basics of the theory of probability are well described in all textbooks on probability and statistics. Therefore, there is no need to present them in details in this paper. However, we are going to point out those assumptions of this theory which are criticized by some authors who see them as main reasons of discrepancies between theory and practice of coping with uncertainties.

According to Kolmogorov's theory of probability there exists a sample space Ω consisting of *disjoint* elements, called elementary states, or simply states. These states need not be necessarily observable. Then, Kolmogorov postulates a Borel-field set B consisting of some, but not necessarily all, subsets of Ω. The elements of B are called events and represent observable outcomes of actual or hypothetical experiments. Probabilities are assigned only to elements of B, so they are not assigned to those states of Ω that do not belong to B. One can argue, however, that those events which cannot be included in an appropriate σ-algebra are irrelevant and may be treated as Lebesgue null sets. A counter-argument to this point of view may be the following: in presence of partial information (or partial ignorance) we may not be entitled to make such claims. The consequence of these assumptions is far-reaching, and it does not depend upon the interpretation of probability. It means that every event can be *precisely* described using the elements of Ω. Another consequence refers to the feature which Walley [87] calls "Bayes dogma of precision". According to this feature of the theory of probability, every uncertain entity (physical object, value of a parameter, etc.) can always be described by a *precisely* defined probability distribution. Some supporters of this assumption even question the reality of situations when the only available information about uncertainty can be presented in form of an interval (see quotations presented in the paper [23] which summarizes discussions on the problems related to the treatment of epistemic uncertainty). When probabilities are assessed by frequencies of observed precisely defined events there are no fundamental problems with the accuracy of their evaluation. However, when they are assessed subjectively (and we have to remember that according to the followers of the Bayesian probability and statistics it is the only *coherent* way of doing this) it is assumed that they may be interpreted as precisely defined fair betting rates. The behavioral interpretation of probabilities in terms of fair betting rates was originally introduced by de Finetti (see [27]) who has shown that betting in favor of an event A against its complement A^C will not lead to sure loss only if betting odds are $P(A)$ to $1 - P(A)$, where $P(A)$ is equal to the probability of event A. Moreover, from the postulate of fair betting rates and some additional coherence requirements one can derive that probability is nonnegative, normalized, and finitely-additive set function. In addition, Savage [74] proved that the theory of subjective probabilities constitutes the basis for an axiomatized and coherent theory of decision-making.

Despite of all these unquestionable advantages, empirical observations show however, that in the presence of partial or full ignorance about events of interest such precise assessments of probability cannot be made. Moreover, the actual behaviour of decision-makers differs from that prescribed by the theory based on classical precise probabilities. Bordley [5] argues that this apparent inconsistency is due to the fact that real decision-makers take into account incompleteness of their information. This leads us (and many other researches working in the area of decision sciences) to the conclusion that in case of imprecisely defined states and events it is not possible to obtain precise values of their probabilities, and hence to make precise prescriptions in decision-making processes. A counterargument to this opinion presented by rather dogmatic followers of the classical Bayesian approach to probability is the following: their theory shall be considered as the normative one, and all the differences between the theory and the actual human behaviour are always due to human weakness and shall be overcome by using more precise measurements and precise problem formulation. In many practical cases this is definitely true. However, even in principle this standpoint can be questioned using the results from quantum physics. Bordley [5] shows that as the consequence of the Heisenberg Uncertainty Principle some events cannot be precisely observed, and in such a case precise probability statements are in principle impossible. Therefore, generalizations of classical probability are necessary if we want to deal with imprecisely defined events and with partial information about probabilities of their occurrence.

It has to be stressed, however, that once the probability distributions describing uncertain phenomena have been defined, the mechanisms of the theory of probability are sufficient for further analysis of uncertain phenomena. For this reason probabilistic methods, such as Bayesian methods, are generally used in practice. Basic principles of the application of the Bayesian methods are described in classical textbooks by Raiffa and Schleifer [69] and De Groot [11]. Examples of particular applications are presented in numerous publications from virtually all areas of science. The only point of criticism formulated against this approach stems from the fact that according to some scientists the existing imprecise or incomplete information does not allow us to propose those precisely defined probability distributions without making some additional assumptions. Other approaches proposed in order to cope with this problem are presented in the next sections of this paper.

2.2.3 Dempster–Shafer Theory of Evidence and Possibility Theory

The notion of *possibility* attracted attention of philosophers, economists, logicians, etc. Dubois and Prade [20] notice that first attempts to formalize the concept of possibility were made in the late 1940s by the economist Shackle [75] who proposed a calculus of "potential surprise" as the base for decision-making. The works of many authors, who have noticed the deficiency of the theory of probability in dealing with many practical problems have led to more or less independent formalizations of two similar theories of uncertainty: Dempster–Shafer theory of evidence and possibility theory.

The concept of possibility can be interpreted in different ways. For example, it can be understood as an objective notion or as an epistemic and subjective one. Zadeh [95] understands possibility as objective feasibility; an objective measure of physical easiness to achieve a certain goal. By his famous example of a possibilistic statement, "it is possible for Hans to eat six eggs for breakfast", he shows an exemplary information which is difficult, or even hardly possible, to be formalized using theory of probability. This type of interpretation of possibility is closely related to the idea of preference. Alternatives that are more easily achieved (more feasible) are usually more preferred. This relation has been described in details in the paper by Dubois, Fargier, and Prade [14]. Second interpretation of possibility is an epistemic one, and is given in terms of plausibility. An event is fully plausible when its occurrence does not create any surprise. This type of interpretation has subjectivistic nature and means that possibility may represent consistency of the observed event with available knowledge. Possibility, understood as plausibility of an event, may also have an objectivistic interpretation, and can be evaluated from the observations of upper bounds of frequency of its occurrence [17]. There also exists a deontologic interpretation of possibility (something is possible when it is allowed by law), but satisfactory formal description of this type of possibility has not been proposed yet. All those interpretations, which in the majority of practical cases are hardly exclusive, lead to the same mathematical description based on fuzzy sets and fuzzy logic (see the book by Dubois and Prade [18] for more information).

The basic notion of the possibility theory is a *possibility distribution function* $\pi(\omega)$, defined on the possibility space Ω (a frame of discernment) which may not be the same as the sample space defined in the theory of probability. The value of $0 \leq \pi(\omega) \leq 1$ represents the measure of possibility of an element ω from the set Ω. It is usually assumed that $sup\{\pi(\omega) : \omega \in \Omega\} = 1$. A possibility measure of a subset A of Ω is defined as $\pi(A) = sup\{\pi(\omega) : \omega \in A\}$. There exist many versions (extensions) of the possibility theory, but in all of them the axiom of finite additivity, characteristic for the probability theory, has been replaced by the axiom of *maxitivity*. Let A and B be two events, and $\Pi(A)$ and $\Pi(B)$ be, respectively, their possibilities. Then,

$$\Pi(A \cup B) = max(\Pi(A), \Pi(B)). \tag{2.1}$$

In his seminal paper [95] Lotfi A. Zadeh proposed to use the formalism of the fuzzy sets theory as the mathematical formalism of the possibility theory. According to this proposal the possibility distribution function that assigns measures of possibility (understood according to the assumed interpretation of this notion) to elements of a certain set (or equivalently, to values of a certain numerical variable) may be interpreted as the membership function assigned to that set. This interpretation allows to use a well developed formal mechanism of the fuzzy sets theory in many different applications. The book by Dubois and Prade [18] describes the links between these two theories, and presents methods for the calculation of numerical values of possibility and necessity measures, both typical for the possibility theory. Moreover, Dubois and Prade [18] pointed out possible links between possibility and probability. According to this interpretation possibility distributions may be regarded as upper envelopes for families of probability distributions. The mutual relation between these two major theories of uncertainty have been later explained and clarified using the Dempster–Shafer theory of evidence. The recent results published in papers of Walley and de Cooman [89], and de Cooman [9] show that measures of possibility are the special case of imprecise probabilities, and thus have a well defined behavioural interpretation.

The original motivation for the development of the possibility theory was to describe imprecise notions or imprecise pieces of information given as statements of a natural language such as, e.g. "costs are high", "time to failure is about 5 hours", etc. As a matter of fact, the founders of the possibility theory saw this theory as fundamentally different from the probability theory. They considered the possibility theory as the formalism for the description of uncertain events or uncertain (partial) information in cases where the probability theory failed to provide satisfactory description. On the other hand, the theory of evidence (also known as the theory of belief functions) proposed originally by Dempster [12] and further developed by Shafer [76] aimed at the generalization of the probability theory for dealing with such problems. The basic assumptions of the Dempster–Shafer theory of evidence look similar to the basic assumptions of the theory of probability. It is assumed that there exists a certain possibility space Ω, but probability measures, called in this theory "probability mass assignments", in contrast to the probability theory, are defined on its whole power set 2^{Ω}. This allows to assign probability to an event formed by a set algebra of the elements of the possibility space Ω which in the case of this particular event are indistinguishable, as this is typical for imprecisely described notions.

In the Dempster–Shafer theory of evidence uncertainty is measured using belief functions. A belief function *Bel*, defined on all subsets of the possibility space Ω, is written in the form

$$Bel(A) = \sum_{B \subseteq A} m(B), \tag{2.2}$$

where m is a probability mass assignment function on all subsets of Ω, such that $m(\emptyset) = 0$, $m(B) \geq 0$ for all $B \subseteq \Omega$ and $\sum_{B \subseteq \Omega} m(B) = 1$. In the Dempster–Shafer theory of evidence there exists also a notion of plausibility which is conjugate to the notion of belief. The conjugate function to the belief function is called the plausibility function *Pl*, and is defined by

$$Pl(A) = 1 - Bel(A^C) = \sum_{B \cap A \neq \emptyset} m(B), \tag{2.3}$$

where A^C is the complement of the set A. It has been shown that there exists a close relationship between the Dempster–Shafer theory of evidence and the possibility theory. When all elements, say $A_1, A_2, \ldots, A_n$, of the set Ω form a nested set (such that $A_1 \subseteq A_2 \subseteq \ldots \subseteq A_n$) then there exists a direct relationship between the mass probability assignments of the Dempster–Shafer theory and the possibility distribution defined as a membership function of a certain fuzzy set (see [19] for more general results). Thus, the possibility theory (for one of its possible interpretations) may be regarded as a special case of the more general Dempster–Shafer theory of evidence. This relationship was used by several authors, who proposed methods for making probability – possibility transformations. Unfortunately, a unique one-to-one transformation between probability and possibility does not exist. Thus, any transformation of this type can lead to loss of information (especially from less precise, but containing more information about the event of interest, possibility to more precise probability), and transformation methods proposed by some authors differ in the methodology used to decrease that loss. One of these methods, proposed by Klir [29,48] is based on the principle of information invariance. The other approach, based on the optimization of information content, has been proposed by Dubois et al. [21]. More information on the problem of probability – possibility transformation can be also found in [20].

Despite their very close relations, the possibility theory and the Dempster–Shafer theory of evidence are used in different areas of application. The Dempster–Shafer theory is mainly used in building computer expert systems or, in a more general setting, in computerized decision support systems. It is not used, however, in data analysis and in those instances of decision-making processes where statistical data (precise or imprecise) have to be merged with subjective information. In this particular domain of application the possibility theory is used in practice much more frequently. We will discuss those applications of the possibility theory in the next section of this paper.

2.2.4 Imprecise Probabilities and Their Generalizations

Possibility theory and its main tool, the possibility distribution, have been found very useful for the formal description of imprecise information. However, claims – expressed, for example, by Zadeh – that it is also useful for the formalization of imprecise descriptions of probabilities (in statements like "event A is much more probable than event B") have been questioned by Walley [86]. In one of his toy examples Walley considers the following information about possible outcomes (win(W), draw (D) or loss(L)) of a football game [86, 87]:

(a) Probably *not* W.
(b) W is more probable than D.
(c) D is more probable than L.

Walley convincingly explains that information of this type cannot be used, without making some arbitrary assumptions, for the precise evaluation of the probability of, e.g. the win $P(W)$. The only consistent evaluation can be done in terms of imprecise probabilities: lower probability $\underline{P}(W)$ and upper probability $\bar{P}(W)$.

Different concepts, such as *interval probabilities* or *non-additive probabilities*, introduced independently by many authors, are covered by the framework of imprecise probabilities. The most comprehensive theory of imprecise probabilities, defined as lower and upper probabilities, was proposed by Walley [86] who introduced the notion of coherent lower (upper) probability. Coherent lower probability may be interpreted as a lower envelope of a set of probability measures that fulfills certain coherence requirements. The similar interpretation exists for the upper probability. The behavioural interpretation of lower probabilities was proposed by Walley [86]. This interpretation is based on the generalization of a similar interpretation of subjective probabilities introduced by de Finetti. According to Walley (see also [87]) the lower (or upper) probability of an event A can be interpreted by specifying acceptable betting rates for betting on (or against) A. If the betting odds on A are x to $1 - x$ one will bet *on* A if $x \leq \underline{P}(A)$ and *against* A if $x \geq \bar{P}(A)$. The choice is not determined if x is between $\underline{P}(A)$ and $\bar{P}(A)$. The basic properties of lower and upper probabilities can be summarized as follows:

(a) $\underline{P}(\emptyset) = \bar{P}(\emptyset) = 0$.
(b) $\underline{P}(\Omega) = \bar{P}(\Omega) = 1$.
(c) $\bar{P}(A) = 1 - \underline{P}(A^C)$.
(d) $0 \leq \underline{P}(A) \leq \bar{P}(A) \leq 1$.
(e) $\underline{P}(A) = \underline{P}(B) \leq \underline{P}(A \cup B) \leq \underline{P}(A) + \bar{P}(B) \leq \bar{P}(A \cup B) \leq \bar{P}(A) + \bar{P}(B)$, for disjoint events A and B.
(f) $\underline{P}(A \cup B) + \underline{P}(A \cap B) \geq \underline{P}(A) + \underline{P}(B)$, for all events A and B.

It is worth to note, that precise probabilities are the special case of lower and upper probabilities such that $\underline{P}(A) = \bar{P}(A)$. Moreover, possibility and necessity measures of the possibility theory and belief and plausibility measures of the Dempster–Shafer theory of evidence are lower and upper probabilities. On the other hand, as it was noted by Walley (see, e.g. [87] and [88]) not all lower and upper probabilities can be interpreted as possibility measures or measures of the Dempster–Shafer theory.

In certain problems of decision-making lower and upper probabilities are not sufficient for dealing with imprecise information. Their further generalization was proposed by Walley (see [86,88]) in his theory of lower and upper previsions. These measures of uncertainty are introduced in terms of *gambles*. A gamble X, defined on a non-empty space Ω of the outcomes of an experiment represents an uncertain reward $X(\omega)$ if the outcome of the experiment is $\omega \in \Omega$. Gambles are expressed in units of some linear utility scale. Below, we present the formal definition of lower and upper previsions, as it was given in [88].

Definition 2.1 (Walley [88]). A bounded mapping from Ω to $\mathbb{R}$ (the real numbers) is called a *gamble*. Let $\mathcal{K}$ be a nonempty set of gambles. A mapping $\underline{P} : \mathcal{K} \rightarrow \mathbb{R}$ is called a *lower prevision* or *lower expectation*. A lower prevision is said to be *coherent* when it is the lowest envelope of some set of linear expectations, i.e. when there is a nonempty set of probability measures, $\mathcal{M}$, such that $\underline{P}(X) = min\{E_P(X) : P \in \mathcal{M}\}$ for all $X \in \mathcal{K}$, where $E_P(X)$ denotes the expectation of X with respect to P. The conjugate upper prevision is determined by $\bar{P}(X) = -\underline{P}(-X) = max\{E_P(X) : P \in \mathcal{M}\}$.

The lower prevision $\underline{P}(X)$ can be interpreted as a supremum acceptable buying price for X. Indicators of events are particular cases of gambles. In such a case, lower and upper previsions coincide with lower and upper probabilities. More information about lower (and upper) previsions can be found in the cited above papers of Walley or in the survey paper by Miranda [61].

The lower (upper) previsions seem to be general enough for the description of subjectively perceived uncertainty. Classical (Bayesian) probabilities, measures of possibility, and Dempster–Shafer measures of evidence can be interpreted as special cases of lower (upper) previsions. From a theoretical point of view this theory is sufficiently well developed. However, there exist some basic problems which require further investigations. For example, the problem of updating the values of imprecise probabilities when new pieces of information are available (i.e. the problem of conditioning) still needs some investigations, as a single generalization of the Bayes updating rule has not been proposed yet. The existing problems with updating procedures are related, for example, to the problems of dealing with observations whose prior probabilities are equal to zero. Other problems arise in relation to concepts of independence or conditional independence. There exist also problems with modeling (in terms of uncertainty) the concepts of preference and weak preference (for example, lower previsions cannot distinguish preference from weak preference). All these problems (for more information, see [88]) motivate researchers to look for more general mathematical models of uncertainty. Some of these models have been indicated in the first subsection of this section. It is interesting that the need to develop more general models of uncertainty has been also recognized in the community of classical Bayesians. The concept of robust Bayesian inference (see, e.g. the paper by Berger [4]) seems to be closely related to the problems presented in this paper. Moreover, some concepts of frequency-based robust statistics (the notion of ϵ-contamination) can be interpreted using the language of the theory of imprecise probabilities and its generalizations.

2.3 Fuzzy Random Variables and Fuzzy Statistics

The brief description of the existing main theories of uncertainty presented in the preceding section shows that neither of them is fully sufficient to cope with real problems where statistical data are both random and imprecise. The attempts to propose such a theory resulted with the introduction of the notion of a fuzzy random variable. This notion has been defined by many authors. Historically, the first widely accepted definition was proposed by Kwakernaak [54, 55]. Kruse [52] proposed an interpretation of this notion, and according to this interpretation a fuzzy random variable $\widetilde{Z}$ may be considered as a perception of an unknown usual random variable $Z : \Omega \to \mathcal{R}$, called an original of $\widetilde{Z}$. Below, we present another, slightly modified, version of this definition presented in Grzegorzewski [37].

A fuzzy set $\widetilde{A}$ described by an upper semicontinuous membership function $\mu_A : \mathcal{R} \to [0, 1]$ is a fuzzy number when it is:

(a) Normal, i.e. there exists a $t_0 \in \mathcal{R}$ such that $\mu(t_0) = 1$.
(b) Fuzzy convex, i.e. $\mu(\lambda s + (1 - \lambda)t) \geq min\{\mu(s), \mu(t)\}$ for $s, t \in \mathcal{R}$ and $\lambda \in [0, 1]$.
(c) The set $\{u_0\} = cl\{t \in \mathcal{R} : \mu(t) > 0\}$ is a compact set.

The operator in (c) denoted by cl is the closure operator (see, e.g. Dubois and Prade [15] for a more detailed description). A space of all fuzzy numbers will be denoted by $\mathcal{FN}(\mathcal{R})$. We have $\mathcal{FN}(\mathcal{R}) \subset \mathcal{F}(\mathcal{R})$, where $\mathcal{F}(\mathcal{R})$ denotes a space of all fuzzy sets on the real line. Fuzzy numbers are completely defined by their α-cuts. The α-cut, $\alpha \in (0, 1]$, of a fuzzy number $\widetilde{A}$ with its membership function μ_A is a closed crisp set defined as

$$X_\alpha = \{t \in \mathcal{R} : \mu_X(t) \geq \alpha\}. \tag{2.4}$$

In order to describe α-cuts let us use the following notation: $A_\alpha = [A_\alpha^L, A_\alpha^U]$, where

$$A_\alpha^L = \inf\{t \in \mathcal{R} : \mu_A(t) \geq \alpha\}, \tag{2.5}$$
$$A_\alpha^U = \sup\{t \in \mathcal{R} : \mu_A(t) \geq \alpha\}. \tag{2.6}$$

Definition 2.2 (Grzegorzewski [37]). Let $(\Omega, \mathcal{A}, P)$ be a probability space, where Ω is a set of all possible outcomes of the random experiment, $\mathcal{A}$ is a σ-algebra of subsets of Ω, and P is a probability measure.

A mapping $\widetilde{X} : \Omega \to \mathcal{FN}(\mathcal{R})$, where $\mathcal{FN}(\mathcal{R})$ is the space of all fuzzy numbers, is called a fuzzy random variable if it satisfies the following properties:

1. $\{X_\alpha(\omega) : \alpha \in [0,1]\}$ is a set representation of $\widetilde{X}(\omega)$ for all $\omega \in \Omega$.
2. For each $\alpha \in (0,1]$ both $X_\alpha^L = X_\alpha^L(\omega) = \inf X_\alpha(\omega)$ and $X_\alpha^U = X_\alpha^U(\omega) = \sup X_\alpha(\omega)$, are usual real-valued random variables on $(\Omega, \mathcal{A}, P)$.

There exists also another popular definition of a fuzzy random variable proposed by Puri and Ralescu [68] and based on the notion of set-valued mapping and random sets. Below, we present this definition in a form given in [32].

Definition 2.3 (Gil et al. [32]). Let $\mathcal{FN}(\mathcal{R})$ be the space of all fuzzy numbers. Given a probability space $(\Omega, \mathcal{A}, P)$, a mapping $\widetilde{X} : \Omega \to \mathcal{FN}(\mathbb{R}^p)$ is said to be a *fuzzy random variable* (also called *fuzzy random set*) if for all $\alpha \in (0,1]$ the set-valued mappings $X_\alpha : \Omega \to \mathcal{K}(\mathbb{R}^p)$, where $\mathcal{K}$ is the class of the non-empty subsets of $\mathbb{R}^p$, defined so that for all $\omega \in \Omega$ $X_\alpha(\omega) = (X(\omega))_\alpha$, are random sets (that is, Borel-measurable mapping with the Borel σ-field generated by the topology associated with the Haussdorf metric on $\mathcal{K}(\mathbb{R}^p)$).

Fuzzy random variables may be used to model random and imprecise measurements. First statistical methods for the analysis of such imprecise fuzzy data were developed in the 1980s. Kruse and Meyer [53] proposed a general methodology for dealing with fuzzy random data. In case of their methodology fuzzy random data are described by fuzzy random variables defined according to Definition 2.2. This assumption has very important practical consequences. First of all it means that there exists an underlying non-fuzzy probability distribution that governs the origins of the observed imprecise fuzzy data. The parameters of this distribution have non-fuzzy values, but because of the fuzziness of observed data they cannot be precisely estimated. Their fuzziness comes directly from the fuzziness of statistical data and disappears when statistical data are precise. Therefore, fuzzy statistical methods developed according to the methodology proposed by Kruse and Meyer shall be regarded as straightforward generalization of classical (non-fuzzy) statistical methods. Using the aforementioned methodology Kruse and Meyer [53] proposed methods for the construction of estimators and confidence intervals for the parameters of the probability distributions of fuzzy random variables. According to the interpretation of fuzzy randomness proposed in [53] the observed values of estimators of the parameters of the probability distributions of fuzzy random variables are fuzzy. For this reason these estimators are called fuzzy, despite the fact that their fuzziness is only the reflection of the fuzziness of data, and not of the fuzziness of the estimation procedure itself. The same methodology may be applied to the estimators of the limits of confidence intervals that are also represented by fuzzy numbers. The most important practical consequence of the adoption of the Kruse and Meyer's methodology is that all relevant formulae for fuzzy estimators and other fuzzy statistics can be obtained by fuzzification of well known formulae of classical non-fuzzy statistics. Relevant formulae are obtained by straightforward application of the Zadeh's extension principle (for its definition see the book by Dubois and Prade[15] or any other textbook on fuzzy sets). As this operation is purely technical, we are not going to present this problem in details. The appropriate formulae can be found in papers devoted to the estimation of parameters of concrete probability distribution when available statistical data are fuzzy. Below, we present only a simple illustrative example.

Let us consider the problem of the estimation of the parameter θ of the exponential distribution defined by the density function $f(t) = (1/\theta)exp(-t/\theta), t > 0$. The maximum likelihood estimator of this parameter is given as a simple arithmetic average of the observed values in the sample, i.e.

$$\theta^{\star} = \sum_{i=1}^{n} t_i/n, \tag{2.7}$$

where $t_i, i = 1, \dots, n$ are the values observed in the sample. Suppose now, that instead of crisp values $t_i, i = 1, \dots, n$ we observe fuzzy values described by their respective α-cuts $[t_{i,L}^{\alpha}, t_{i,U}^{\alpha}], i = 1, \dots, n, \alpha \in (0, 1]$. Now, we can apply the Zadeh's extension principle and fuzzify (2.7) arriving at the fuzzy version of $\theta^{\star}$ described by the set of respective α-cuts:

$$\left[\sum_{i=1}^{n} t_{i,L}^{\alpha}/n, \sum_{i=1}^{n} t_{i,R}^{\alpha}/n\right] \tag{2.8}$$

The same approach may be applied in all cases when there exist explicit formulae for computation of point estimators and confidence intervals in the case of crisp data. When these formulae do not exist (for example, when maximum likelihood estimators are obtained as numerical solutions of nonlinear equations) the application of the Zadeh's extension principle requires the solution of complex mathematical programming problems.

Despite the fact that the generalization of well known statistical methods to the fuzzy case is relatively straightforward, the construction of fuzzy statistical tests and making statistical decisions is far from being trivial. Fuzzy statistical tests may be developed for testing both non-fuzzy (precise) and fuzzy (imprecise) statistical hypotheses, and for fuzzy (imprecise) and non-fuzzy (precise) statistical data. For example, statistical methods for testing fuzzy hypotheses have been considered in the papers by Saade and Schwarzlander [73], Saade [72], Watanabe and Imaizumi [90], Arnold [1, 2] Taheri and Behboodian [80], and Grzegorzewski and Hryniewicz [39]. When the data are also fuzzy, interesting solutions have been proposed in the papers by Arnold [1], Casals et al. [6], Kruse and Meyer [53], Saade [72], Saade and Schwarzlander [73], Son et al. [79], Watanabe and Imaizumi [90], Römer and Kandel [70], and Montenegro et al. [62]. Grzegorzewski [37] has proposed a unified approach for testing statistical hypotheses with vague data which is a direct generalization of the classical approach. Below, we present his definition of the fuzzy statistical test.

Let $\widetilde{Z}_1, \dots, \widetilde{Z}_n$ denote a fuzzy sample, i.e. a sample consisting of fuzzy random variables representing fuzzy perception of the usual random sample $Z_1, \dots, Z_n$

from the population described by the probability distribution P_Θ, and let δ be a given number from the interval (0, 1). Grzegorzewski [37] has defined a fuzzy test as follows:

Definition 2.4 (Grzegorzewski [37]). A function

$$\varphi : (\mathcal{FN}(\mathcal{R}))^n \to \mathcal{F}(\{0,1\}), \tag{2.9}$$

where $\mathcal{F}(\{0,1\})$ is the set of possible decisions, is called a fuzzy test for the hypothesis H, on the significance level δ, if

$$\sup_{\alpha\in[0,1]} P\left\{\omega \in \Omega : \varphi_\alpha\left(\widetilde{Z}_1(\omega),\ldots,\widetilde{Z}_n(\omega)\right) \subseteq \{0\}|H\right\} \le \delta$$

where φ_α is the α-level set (α-cut) of $\varphi\left(\widetilde{Z}_1,\ldots,\widetilde{Z}_n\right)$.

When we test statistical hypotheses about the values of the parameters of probability distributions we utilize a well known equivalence between the set of values of the considered probability distribution parameter for which the null hypothesis is accepted and a certain confidence interval for this parameter. The same equivalence exists in the case of statistical tests with fuzzy data.

When statistical data are precise (crisp), for testing the null hypothesis $H:\theta \le \theta_0$, and the alternative hypothesis $K:\theta > \theta_0$ we use a one-to-one correspondence between the acceptance region for this test on the significance level δ and the one-sided confidence interval $[\underline{\pi}_l, +\infty)$ for the parameter θ on the confidence level $1-\delta$, where $\underline{\pi}_l = \underline{\pi}_l(Z_1,\ldots,Z_n;\delta)$. This equivalence was utilized by Kruse and Meyer [53] who introduced the notion of a fuzzy confidence interval for the unknown parameter θ, when the data are fuzzy. In the considered case, a fuzzy equivalent of $[\underline{\pi}_l, +\infty)$ can be defined by the following α-cuts (for all $\alpha \in (0,1]$):

$$\begin{aligned}\underline{\Pi}_\alpha^L &= \underline{\Pi}_\alpha^L(\widetilde{Z}_1,\ldots,\widetilde{Z}_n;\delta)\\ &= \inf\Big\{t \in \mathcal{R} : \forall i \in \{1,\ldots n\}\exists z_i \in \left(\widetilde{Z}_i\right)_\alpha \\ &\qquad \text{such that } \underline{\pi}_l(z_1,\ldots,z_n;\delta) \le t\Big\}\end{aligned} \tag{2.10}$$

Similarly, we can define a fuzzy equivalent of the one-sided confidence interval $(-\infty, \overline{\pi}_u]$, as given in Grzegorzewski [37]:

$$\begin{aligned}\overline{\Pi}_\alpha^U &= \overline{\Pi}_\alpha^U(\widetilde{Z}_1,\ldots,\widetilde{Z}_n;\delta)\\ &= \sup\Big\{t \in \mathcal{R} : \forall i \in \{1,\ldots n\}\exists z_i \in \left(\widetilde{Z}_i\right)_\alpha \\ &\qquad \text{such that } \overline{\pi}_u(z_1,\ldots,z_n;\delta) \ge t\Big\}\end{aligned} \tag{2.11}$$

where $\overline{\pi}_u(z_1,\ldots,z_n;\delta) = \underline{\pi}_l(z_1,\ldots,z_n;1-\delta)$.

The notion of the one-sided fuzzy interval can be used to define a fuzzy test. In the considered case of one-sided statistical hypothesis, a function $\varphi : (\mathcal{FN}(\mathcal{R}))^n \rightarrow \mathcal{F}(\{0, 1\})$ with the following α-cuts:

$$\varphi_\alpha(\widetilde{Z}_1, \ldots, \widetilde{Z}_n) = \begin{cases} \{1\} & if\ \theta_0 \in (\underline{\Pi}_\alpha \backslash (\neg\underline{\Pi})_\alpha), \\ \{0\} & if\ \theta_0 \in ((\neg\underline{\Pi})_\alpha \backslash \underline{\Pi}_\alpha), \\ \{0, 1\} & if\ \theta_0 \in (\underline{\Pi}_\alpha \bigcap (\neg\underline{\Pi})_\alpha), \\ \emptyset & if\ \theta_0 \notin (\underline{\Pi}_\alpha \bigcup (\neg\underline{\Pi})_\alpha) \end{cases} \tag{2.12}$$

is a fuzzy test for $H : \theta \leq \theta_0$, against $K : \theta > \theta_0$, on the significance level δ (Grzegorzewski [37]). In a similar way, we can define fuzzy tests for testing other one-sided hypotheses such as $H : \theta \geq \theta_0$, against $K : \theta < \theta_0$, and for testing two-sided hypotheses about θ.

It is worthy to note that in certain cases the application of the fuzzy test defined above does not lead to a clearly indicated decision. This feature is far from being unexpected because, unless we make some additional assumptions, we should not expect precise answers to questions presented in a form of statistical hypotheses, if we infer these answers from the analysis of imprecise data. Let us note, however, that we face the similar situation when we use classical statistical methods. In that case a decision cannot be made without setting in advance an appropriate significance level of test. In case of fuzzy statistical data the knowledge of the significance level is not enough, so we have to use some additional indicators that would be helpful in making decisions. There exist several approaches that are suitable for solving this problem. One of these approaches which is formulated in the language of the possibility theory has been proposed by Hryniewicz [43].

In order to present a possibilistic approach to the problem of statistical testing when both data and statistical hypotheses are imprecise let us consider a fuzzy equivalent of testing the hypothesis $H : \vartheta \leq \vartheta_H$ when we observe a random sample $(X_1, \ldots, X_n)$. In case of precisely formulated hypotheses and precise statistical data we can use a well known method (for reference, see, e.g. the book of Lehmann [57]) and calculate a one-sided confidence interval on a confidence level $1 - \delta$ from the formula $[\pi_L(X_1, \ldots, X_n; 1 - \delta), \infty)$. We reject the null hypothesis on the significance level δ if the observed value of $\pi_L\,(X_1, \ldots, X_n; 1 - \delta)$ is larger than ϑ_H, i.e. when the inequality $\vartheta_H < \pi_L\,(x_1, \ldots, x_n; 1 - \delta)$ holds. Similarly, we reject the hypothesis $H : \vartheta \geq \vartheta_H$ on the significance level δ when the inequality $\vartheta_H > \pi_U\,(x_1, \ldots, x_n; 1 - \delta)$ holds, where $\pi_U\,(x_1, \ldots, x_n; 1 - \delta)$ is the observed value of the upper limit of the one-sided confidence interval $(-\infty, \pi_U\,(X_1, \ldots, X_n; 1 - \delta)]$ on a confidence level $1 - \delta$. When we test the hypothesis $H : \vartheta = \vartheta_H$ on the significance level δ we reject it if either $\vartheta_H < \pi_L\,(x_1, \ldots, x_n; 1 - \delta/2)$ or $\vartheta_H > \pi_U\,(x_1, \ldots, x_n; 1 - \delta/2)$ holds, where $\pi_L\,(x_1, \ldots, x_n; 1 - \delta/2)$ is the observed in the sample value of the lower limit of the two-sided confidence interval $\pi_L\,(X_1, \ldots, X_n; 1 - \delta/2)$ on a confidence level $1 - \delta$, and the observed value of its upper limit $\pi_U\,(X_1, \ldots, X_n; 1 - \delta/2)$ is given by $\pi_U\,(x_1, \ldots, x_n; 1 - \delta/2)$. Thus, when we test a hypothesis about the value of the parameter ϑ we find a respective confidence interval, and compare it to the hypothetical value.

Dubois et al. [22] proposed to use statistical confidence intervals of parameters of probability distributions for the construction of possibility distributions of these parameters in a fully objective way. According to their approach, the family of two-sided confidence intervals

$$[\pi_L(x_1, \ldots, x_n; 1 - \delta/2), \pi_U(x_1, \ldots, x_n; 1 - \delta/2)], \delta \in (0, 1) \tag{2.13}$$

forms the *possibility distribution* $\tilde{\vartheta}$ of the estimated value of the unknown parameter ϑ. In a similar way it is possible to construct one-sided possibility distributions based on one-sided nested confidence intervals. Hryniewicz [43] proposed to compare this possibility distribution with a hypothetical value of the tested parameter. For this purpose he proposed to use the necessity of strict dominance measure introduced by Dubois and Prade [16] for measuring the necessity of the strict dominance relation $\tilde{A} \succ \tilde{B}$, where $\tilde{A}$ and $\tilde{B}$ are fuzzy sets. This measure, called the *Necessity of Strict Dominance* index (NSD), is defined as

$$NSD = Ness\left(\tilde{A} \succ \tilde{B}\right) = 1 - \sup_{x,y;x \leq y} \min\{\mu_A(x), \mu_B(y)\}. \tag{2.14}$$

Hryniewicz [43] has shown that in the classical case of precise statistical data and precisely defined statistical hypotheses the value of the NSD index is equal to the p-value of the test.

In case of fuzzy data the confidence intervals used for the construction of the possibility distribution of the estimated parameter θ can be replaced by their fuzzy equivalents, calculated according to the methodology proposed by Kruse and Meyer [53]. In his paper Hryniewicz [43] assumes that the value of the significance level of the corresponding statistical test δ is equal to the possibility degree α that defines the respective α-cut of the possibility distribution of $\tilde{\vartheta}$. He also assumes that in the possibilistic analysis of statistical tests on the significance level δ we should take into account only those possible values of the fuzzy sample whose possibility is not smaller than δ. Thus, the α-cuts of the membership function $\mu_F(\vartheta)$ denoted by $\left[\mu_{F,L}^{(\alpha)}, \mu_{F,U}^{(\alpha)}\right]$ are equivalent to the α-cuts of the respective fuzzy confidence intervals on a confidence level $1 - \alpha$.

In order to consider the most general case let us also assume that the hypothetical value of the tested parameter may be also imprecisely defined by a fuzzy number $\tilde{\theta}_H$ described by the membership function $\mu_H(\theta)$. Possibilistic evaluation of the results of statistical fuzzy test consists now in the comparison of the possibility distribution of the estimated parameter θ, and the possibility distribution of the hypothetical value of this parameter. Let us illustrate this procedure by assuming that our fuzzy hypothesis is given as $H : \theta \leq \widetilde{\theta_H}$. In the case of crisp data we compare the lower limit of the one-sided confidence interval on a given confidence level $1 - \delta$ with the respective α-cut of the membership function that describes $\widetilde{\theta_H}$. In the possibilistic framework described in [43] it means that we compare the possibility distribution of $\tilde{\vartheta}_L$ based on the left-hand sides of the confidence intervals with the fuzzy value of $\widetilde{\theta}_H$. This distribution is defined by the membership function

$$\mu_{F,L}^{(\alpha)} = \inf_{\gamma \geq \alpha} \Pi_L^{\gamma}(\widetilde{X}_1, \ldots, \widetilde{X}_n; 1 - \delta), \tag{2.15}$$

where

$$\begin{aligned} \Pi_L^{\gamma} &= \Pi_L^{\alpha}(\widetilde{X}_1, \ldots, \widetilde{X}_n; 1 - \delta) \\ &= \inf \Big\{ t \in \mathcal{R} : \forall i \in \{1, \ldots n\} \exists x_i \in \big(\widetilde{X}_i\big)_{\gamma} \\ &\qquad \text{such that } \pi_L(x_1, \ldots, x_n; 1 - \delta) \leq t \Big\}, \end{aligned} \tag{2.16}$$

and $\pi_L\,(x_1, \ldots, x_n; 1 - \delta)$ is the left-hand side limit of the classical confidence interval. In a similar way we can define a fuzzy equivalent of the upper limit of the one-sided confidence interval $(-\infty, \pi_U\,(X_1, \ldots, X_n; 1 - \delta)]$.

In the presence of fuzzy data we have to compare the possibility distribution $\tilde{\vartheta}_{F,L}$ of the estimated value of the unknown parameter θ represented by its α-cuts given by (2.15) with the fuzzy value of $\widetilde{\theta}_H$. In such a case we have to find the intersection point of the membership function $\mu_{F,L}(\vartheta)$ and the left-hand side of $\mu_H(\theta)$. The NSD index of the relation $\tilde{\vartheta}_{F,L} \succ \tilde{\vartheta}_H$ is equal to one minus the ordinate of this point, i.e.

$$Ness\left(\tilde{\vartheta}_{F,L} \succ \widetilde{\theta}_H\right) = 1 - \sup \min\left(\mu_{F,L}(\vartheta), \mu_H(\theta)\right). \tag{2.17}$$

The NSD index defined by (2.17) can be regarded as the generalization of the observed test size p (also known as *p-value* or *significance*) for the case of imprecisely defined statistical hypotheses and vague statistical data. In exactly the same way we can find the NSD index for other one-sided and two-sided statistical hypotheses.

Statistical analysis of fuzzy random data can be also done in the Bayesian framework. First results presenting the Bayesian decision analysis for imprecise data were given in papers by Casals et al. [6, 7], and Gil [30]. In these papers the authors described fuzzy observations using the notion of the fuzzy information system by Zadeh [95] and Tanaka et al. [82]. As this approach seems to be not very effective in practical applications, other approaches have been proposed, for example, by Viertl [84], Frühwirth-Schnatter [28], and Taheri and Behboodian [81]. Comprehensive Bayesian model comprising fuzzy data, fuzzy hypotheses, and fuzzy utility function has been proposed in the paper by Hryniewicz [42].

When we interpret fuzzy random variables according to the definition proposed by Puri and Ralescu (see Definition 2.3 above) the statistical analysis of fuzzy data is unfortunately not so simple. The reasons for this difficulty stem from the fact that in that case the underlying classical probability distribution does not exist anymore. For example, it is difficult to formulate fuzzy equivalents of the Central Limit Theorem, as the concept of asymptotic normal distribution cannot be directly applied. Thus, the statistical procedures have to be constructed on a different theoretical basis. Some authors, see [32] for an overview of the problem, define statistical tests in terms of distances, in different metrics, between the observed fuzzy value of a test statistics and the hypothetical value of a certain characteristic of the fuzzy random variable, e.g. its fuzzy expected value. Therefore, the resulting statistics, such as a variance of the fuzzy random quantity, are real-valued in contrast to fuzzy statistics arriving from the model by Kwakernaak and Kruse. From among few papers devoted to this problem one can mention the papers by Körner [50], Körner and Näther [51], and Montenegro et al. [62]. The problem of lacking underlying probability distribution can be overcome by using a bootstrap methodology, as it has been recently proposed, in the papers by Gil et al. [33], González-Rodríguez et al. [35], and Montenegro et al. [63]. In the case of the Bayesian analysis of fuzzy random variables interesting results have been proposed in the paper by Gil and López-Díaz [31].

2.4 Applications of Fuzzy Statistics in Systems Analysis

Systems analysis is oriented on solving complex problems where precise mathematical models are used for a simplified (and sometimes even oversimplified) description of reality. The main problem of every researcher who has to apply the methods of systems analysis in real applications is related to coping with uncertainties of different kinds. What is important, not all of these uncertainties can be described by well developed methods like theory of probability and mathematical statistics. The methodology of fuzzy statistics, presented in the previous section, gives possibility to describe phenomena where probabilistic randomness is merged with possibilistic imprecision (fuzziness). In this section we present two applications of this methodology which seem to be useful in solving real problems. The character of this paper does not allow us to present too many details that may be necessary to fully understand these applications. The details will be presented in forthcoming papers dedicated to particular problems.

2.4.1 Example 1: Verification of the Kyoto Protocol

First, let us consider the problem of the verification of commitments agreed in the Annex I to Kyoto Protocol. The Parties who accept the Kyoto Protocol agreed to reduce the national emissions of greenhouse gases by a specified percentage. The main problem with the verification of these commitments stems from a fact that these emissions cannot be directly measured. Therefore, according to the IPCC Guidance document [46], the total emission X is estimated as a sum of emissions of every type of activity, evaluated indirectly using certain measures describing those activities. For example, the emission from electric coal power plants is evaluated using the knowledge about the amount of burned coal. A simple, but commonly used in practice, model has a linear form

$$x_i = \sum_{j=1}^{m} x_{ij} = \sum_{j=1}^{m} c_{ij} a_{ij}, i = 0, 1, \ldots, \tag{2.18}$$

where a_{ij} is the j-th activity measure for the i-th year, and c_{ij} is the emission factor for the i-th year that enables us to calculate greenhouse gas emission knowing the activity measure for the j-th activity. At a national scale the values of activity measures a_{ij} are definitely uncertain, and the level of associated uncertainty strongly depends on the type of activity. Similarly, the values of emission factors c_{ij} may be highly uncertain, as in many cases they are evaluated using experts opinions. More detailed information on this problem can be found in papers by Winiwarter [94], and by Rypdal and Winiwarter [71].

Suppose now, that for each activity we observe a time series, which consists of $n+1$ yearly observations: $(a_{0j}, a_{1j}, \ldots, a_{nj}), j = 1, \ldots, m$. For sake of simplicity we assume that these are realizations of stochastic processes, and the uncertainty related to their values are purely random. The nature of uncertainty assigned to the associated emission factors $(c_{0j}, c_{1j}, \ldots, c_{nj}), j = 1, \ldots, m$, is hardly easy for precise evaluation. This uncertainty contains undoubtedly a random factor (for example, for an electric coal power plant the emission rate varies randomly with randomly varying quality of burned coal), but may also contain another factor, related to imprecise opinions of experts. For this reason specialists assume that the values of emission factors, and more generally, the values of emissions from different activities, should be evaluated in terms of intervals of possible values. We claim, however, that even in case of very vague information about the values of c_{ij} this assumption is too restrictive. It seems to us that the representation of this information using possibility distributions is much more informative. Therefore, we assume that each value of the emission factors is given as a *fuzzy number* represented by a set of its α-cuts: $\left[c_{ij,L}^{\alpha}, c_{ij,R}^{\alpha}\right], \alpha \in (0, 1]$. Therefore, the evaluated yearly emission is a realization of a *fuzzy random variable* defined as

$$\tilde{X}_i = \sum_{j=1}^{m} \tilde{X}_{ij} = \sum_{j=1}^{m} \tilde{c}_{ij} A_{ij}, i = 0, 1, \ldots, \tag{2.19}$$

where A_{ij} is a random value of the j-th activity in the i-th year.

Suppose now, that having the observed realizations of fuzzy random variables $\tilde{x}_{ij}, j = 1, \ldots, m, j = 0, 1, \ldots, n$ we want to predict the total emission for the commitment year k. Let $\left(a_{kj,L}^{\gamma}, a_{kj,R}^{\gamma}\right)$ be the prediction confidence interval on the confidence level γ for the amount of the j-th activity in the commitment year k. Note, that for $\gamma = 0$ this interval shrinks to the point-wise forecast for the value of x_{kj}. From the theory presented in the previous section we know that *fuzzy confidence interval* for $\tilde{X}_{kj}$ on the confidence level γ is represented by a set of α-cuts (nested α-level intervals): $\left(a_{kj,L}^{\gamma} * c_{ij,L}^{\alpha}, a_{kj,R}^{\gamma} * c_{ij,R}^{\alpha}\right)$. The construction of the fuzzy prediction interval for the total emission X_k in the commitment year k is a difficult task and, in general, may require the application of Monte Carlo methods. However, when the forecasts for a_{kj} are described by normal distributions this construction is straightforward.

According to the Annex I to Kyoto Protocol a country fulfills its commitment if in the commitment year k its emission does not exceed the value $y_k = \rho x_0$, where x_0 is the emission in the base year, and ρ is a coefficient agreed upon in the Kyoto Protocol. Our analysis of the verification problem we begin from the simplest case when the value of y_k is given, and the activity measures in the commitment year $a_{kj}, j = 1, \ldots, m$ have been already established. Thus, we have to verify if $\tilde{x}_k \leq y_k$. From the theory of fuzzy sets we know that the unique method for the verification of this inequality does not exist. However, for this purpose we can use the NSD index defined by (2.14) for the relation $y_k \succ \tilde{x}_k$. By simple calculations we can show that this possibility measure is greater than zero if the α-cut of $\tilde{x}_k$ at the level $\alpha = 1$ is located to the left of y_k, i.e. if the inequality $x^1_{k,R} \leq y_k$ holds. Thus, we have

$$NSD = 1 - \alpha_\star \tag{2.20}$$

where

$$\alpha_\star : x^{\alpha_\star}_{k,R} = y_k. \tag{2.21}$$

From these equations we can also see that $NSD = 1$ if the whole support of $\tilde{x}_k$ is located to the left of y_k. In a more general case, when the emission in the base year is given as a fuzzy number (due to the fuzziness of the emission factors), the commitment may be verified by the calculation of the NSD index for the relation $\tilde{y}_k \succ \tilde{x}_k$. In such a case, for the computation of NSD we use (2.20), but the value of $\alpha_\star$ is now calculated from the following expression:

$$\alpha_\star : x^{\alpha_\star}_{k,R} = y^{\alpha_\star}_{k,L}. \tag{2.22}$$

In the most general case we may want to know in advance, after the evaluations of emissions in years numbered from 0 (base year) to n have been observed, if the commitment in the year k is likely to be fulfilled. We can solve this problem by the verification of a statistical hypothesis that the expected value of the predicted emission (estimated from fuzzy random data) is lower than a given fuzzy valued limit. Thus, we have the problem described in the previous section. The NSD index that describes the necessity that this hypothesis is true can be calculated from (2.15)–(2.17).

2.4.2 *Example 2: Sequential Testing of a Hypothesis About the Mean Value in the Normal Distribution*

As a possible application of the fuzzy statistical methodology in the systems analysis we may consider the problem of testing a hypothesis about the mean value of a random variable described by a normal distribution when sampling costs are high, and we are forced to observe as few sample items as possible. This situation often happens when we have to control costs of a large project, which consists of many individually assessed partial costs. If the number of partial costs is relatively large, we can assume – following the Central Limit Theorem – that the observed total cost, say X, is distributed according to the normal distribution $N(\mu, \sigma)$. Suppose now that we are interested in keeping the total costs constant for a certain period of time at a level μ_0. A simple, and the most effective statistical test for verification of the statistical hypothesis $H_0 : \mu = \mu_0$ against the alternative $H_1 : \mu = \mu_1$ is the sequential probability ratio test proposed originally by Wald (see the book by Lehmann [57] for more information) for the case of the known value of the standard deviation σ. The test statistic, based on a sample $(x_1 \ldots x_n)$ is, in that case, a simple sum of re-scaled observations

$$S_n = \sum_{i=1}^{n} (x_i/\sigma). \tag{2.23}$$

Let α be the probability of the type-I error (probability of an erroneous rejection of H_0), and β be the probability of the type-II error (probability of an erroneous acceptance of H_1). We accept the null hypothesis H_0 if

$$S_n \geq \frac{A}{\mu_1 - \mu_0} + \frac{n}{2}(\mu_1 + \mu_0), \tag{2.24}$$

where $A = \beta/(1 - \alpha)$. We reject H_0 in favour of H_1 if

$$S_n \leq \frac{B}{\mu_1 - \mu_0} + \frac{n}{2}(\mu_1 + \mu_0), \tag{2.25}$$

where $B = (1 - \beta)/\alpha$. If neither of these inequality holds, we have to increase the sample size by one, and repeat the same procedure.

In the sequential test described above we have assumed that the value of σ is known. In practice, we never know this value in advance, but when the amount of historical data is large enough we can estimate σ, and take this estimated value as the known one. However, when the available amount of data is scarce, as it is usually the case in the analysis of large systems, we cannot proceed this way. A possible way out is to use a procedure proposed by Hryniewicz [44] for the analysis of reliability data.

Having some historical data we can estimate the value of σ. When the available sample size is equal to m, and the estimated value is given by $\sigma_m^\star$, we can calculate the two-sided confidence interval for the unknown value of σ:

$$\left(\sigma_L(\gamma) = \sqrt{\frac{(m-1)(\sigma_m^\star)^2}{\chi^2_{m-1}((1+\gamma)/2)}}, \sigma_R(\gamma) = \sqrt{\frac{(m-1)(\sigma_m^\star)^2}{\chi^2_{m-1}((1-\gamma)/2)}}\right), \tag{2.26}$$

where $\chi^2_{m-1}(\gamma)$ is the quantile of order γ of the chi-square distribution with $m-1$ degrees of freedom. The confidence intervals defined by (2.26) can be used for the construction of the *possibility distribution* of σ defined by its δ-cuts (we use here the symbol of δ because the symbol α, traditionally used in this context, has been used for the description of the type-I error of the sequential test). The left-hand side limits of this possibility distribution are given by

$$\mu_L^{\delta}(\sigma) = \begin{cases} \sigma_L(1-\delta) & \text{if } \delta \geq \delta_0 \\ \sigma_L(1-\delta_0) & \text{if } \delta < \delta_0 \end{cases}, \tag{2.27}$$

where δ_0 is a small number close to 0 (e.g. 0.01). Similarly, the right-hand limits are given by

$$\mu_R^{\delta}(\sigma) = \begin{cases} \sigma_R(1-\delta) & \text{if } \delta \geq \delta_0 \\ \sigma_R(1-\delta_0) & \text{if } \delta < \delta_0 \end{cases}. \tag{2.28}$$

If we assume that σ is a fuzzy number defined by this possibility distribution we immediately find that the test statistic S_n becomes also fuzzy. Thus, we cannot directly verify if inequalities (2.24) and (2.25) are fulfilled. However, we may assume that they are fulfilled if the NSD index for respective fuzzy relations is greater than a prescribed value.

In this second example of the application of fuzzy statistics we haven't used any subjective imprecise fuzzy information. The fuzziness has been introduced in a purely objective way using some historical statistical data. We can also generalize this problem by allowing imprecise hypotheses about the values of μ_0 and μ_1. The methodology for dealing with such a problem is the same, but the necessary calculations become much more complicated.

2.5 Conclusions

In the paper we have presented different methods used for mathematical modeling of uncertainty. We have concentrated on modeling of epistemic uncertainty, i.e. uncertainty evaluated subjectively by human beings like experts which reflects lack of full information about phenomena of interest. We have presented basic information about the most important (among many others) theories: theory of probability, which is the basic theory of uncertainty, and alternative theories of uncertainty like Dempster–Shafer theory of evidence, theory of imprecise probabilities, and theory of possibility. Considerable space has been devoted to the theory of fuzzy random variables which links classical probability with fuzzy sets and theory of possibility. This choice stems from the fact that, as for now, this is the only theory which provides mathematical foundations for comprehensive analysis of data representing aleatoric randomness and epistemic imprecision.

It has to be stressed once more that the theory of probability (together with classical mathematical statistics) seems to be the only reasonable methodology for dealing with randomness. However, its ability to model other types of uncertainty has been questioned by many scholars. Alternative theories of uncertainty, like those mentioned above, have been developed to cope with problems where classical probability provides questionable solutions. Some of these theories, like theory of imprecise probabilities, may be regarded as generalizations of the classical probability, some others, like theory of possibility, were introduced as completely different theories. It has to be noted, however, that recently published theoretical results reveal many links between all these theories making them rather complementary than competitive.

Alternative theories of uncertainty are relatively very young in comparison to the theory of probability. Therefore, they are not as matured as the theory of probability, and there exist questions for which unequivocal answers have not been proposed yet. The current situation is well illustrated by an experiment proposed in the paper by Oberkampf et al. [66]. In this experiment prominent representatives of different methodologies were asked to provide numerical evaluation of outputs of systems described by equations with uncertain parameters. Different types of sources of uncertainty were represented in this experiment, allowing for using different types of models, and different methods for aggregation of uncertainty. The results of this experiment, described in [23], show that the usage of different approaches resulted in different evaluations. It is interesting, however, that prominent representatives of the probabilistic approach, who were invited to this experiment, either refused to provide their evaluation, arguing that the input information was unrealistically imprecise, or provided results obtained after the introduction of additional assumptions which had not been stated in the original description of this experiment.

The results of the experiment mentioned above, and analyses presented in overview papers, like the paper by Helton et al. [40] or by Möller and Beer [65], reveal the most important challenges specialists in dealing with uncertainty are confronted with. First, in case of the analysis of complex systems and structures the analysts who apply alternative approaches for coping with uncertainty face very serious computational problems. They have to use methods, such as interval computations, which are computationally much more demanding in comparison to methods used in case of traditional probabilistic models. Second, Monte Carlo methodology for simulation of systems has to be extended in order to cope with information which is not random by nature. For example, the existing methodology for simulation of fuzzy random variables is still under development. Third, for some alternative theories of uncertainty unequivocally accepted methods of propagation of uncertainty have not been proposed yet. And finally, some very general theories of uncertainty still need methods for coping with statistical data.

The challenges mentioned above are less visible when we apply classical theory of probability. The price for this favorable situation is sometimes high. We have to introduce additional assumptions, which in presence of partial information (or partial ignorance) cannot be verified. Therefore, there is still a need for finding more appropriate methods for coping with uncertainty. The methodology of fuzzy random variables, described in numerous recently published papers, and books like the books by Viertl [85] or by Möller and Beer [64], in our opinion seems to be, as for now, the most useful in practice.

References

1. Arnold, B.F.: Statistical tests optimally meeting certain fuzzy requirements on the power function and on the sample size. Fuzzy Sets Syst. **75**, 365–372 (1995)
2. Arnold, B.F.: An approach to fuzzy hypothesis testing. Metrika **44**, 119–126 (1996)
3. Ben-Haim, Y., Elishakoff, I.: Convex Models of Uncertainty in Applied Mechanics. Elsevier, Amsterdam (1985)
4. Berger, J.O.: An overview of robust Bayesian analysis (with discussion). Test **3**, 5–124 (1994)
5. Bordley, R.F.: Reformulating decision theory using fuzzy set theory and Shafer's theory of evidence. Fuzzy Sets Syst. **139**, 243–266 (2003)
6. Casals, R., Gil, M.A., Gil, P.: The fuzzy decision problem: an approach to the problem of testing statistical hypotheses with fuzzy information. Eur. J. Oper. Res. **27**, 371–382 (1986)
7. Casals, M.R., Gil, M.A., Gil, P.: On the use of Zadeh's probabilistic definition for testing statistical hypotheses from fuzzy information. Fuzzy Sets Syst. **20**, 175–190 (1986)
8. Choquet, G.: Theory of capacities. Annales de l'Institut Fourier **5**, 131–295 (1953–1954)
9. de Cooman, G.: A behavioural model for vague probability assessments. Fuzzy Sets Syst. **154**, 305–358 (2005)
10. Cutello, V., Montero, J.: An extension of the axioms of utility theory based on fuzzy rationality measures. In: Fodor, J., De Baets, B., Perny, P. (eds.) Preference and Decisions under Incomplete Knowledge, pp. 33–50. Physica, Heidelberg (1999)
11. De Groot, M.H.: Optimal Statistical Decisions. McGraw Hill, New York (1970)
12. Dempster, A.P.: Upper and lower probabilities induced by multivalued mapping. Ann. Math. Stat. **38**, 325–339 (1967)
13. Denneberg, D.: Non-Additive Measure and Integral. Kluwer, Dordrecht (1994)
14. Dubois, D., Fargier, H., Prade, H.: Possibility theory in constraint satisfaction problems: Handling priority, preference and uncertainty. Appl. Intell. **6**, 287–309 (1996)
15. Dubois, D., Prade, H.: Fuzzy Sets and Systems: Theory and Applications. Academic, New York (1980)
16. Dubois, D., Prade, H.: Ranking fuzzy numbers in the setting of possibility theory. Inform. Sci. **30**, 184–244 (1983)
17. Dubois, D., Prade, H.: Fuzzy sets and statistical data. Eur. J. Oper. Res. **25**, 345–356 (1986)
18. Dubois, D., Prade, H.: Possibility Theory. Plenum, New York (1988)
19. Dubois, D., Prade, H.: When upper probabilities are possibility measures. Fuzzy Sets Syst. **49**, 65–74 (1992)
20. Dubois, D., Prade, H.: Quantitative possibility theory and its probabilistic connections. In: Grzegorzewski, P., Hryniewicz, O., Gil, M.A. (eds.) Soft Methods in Probability, Statistics and Data Analysis, pp. 3–26. Physica, Heidelberg (2002)
21. Dubois, D., Prade, H., Sandri, S.: On possibility/probability transformations. In: Lowen, R., Roubens, M. (eds.) Fuzzy Logic. State of the Art, pp. 103–112. Kluwer, Dordrecht (1993)
22. Dubois, D., Foulloy, L., Mauris, G., Prade, H.: Probability–possibility transformations, triangular fuzzy-sets and probabilistic inequalities. In: Proc. of the Ninth International Conference IPMU, Annecy, 1077–1083 (2002)
23. Ferson, S., Joslyn, C.A., Helton, J.C., Oberkampf, W.L., Sentz, K.: Summary from the epistemic uncertainty workshop: consensus amid diversity. Reliab. Eng. Syst. Safety **85**, 355–369 (2004)
24. Fine, T.L.: Theories of Probability. Academic, New York (1973)
25. Fine, T.L.: An argument for comparative probability. In: Butts, R.E., Hintikka, J. (eds.) Basic Problems in Methodology and Linguistics, pp. 105–119. Reidel, Dordrecht (1977)
26. de Finetti, B.: Probability, Induction and Statistics. Wiley, Chichester (1972)
27. de Finetti, B.: Theory of Probability, vols. 1 and 2. Wiley, Chichester (1974–1975). [English translation of a book published earlier in Italian]
28. Frühwirth-Schatter, S.: Fuzzy Bayesian inference. Fuzzy Sets Syst. **60**, 41–58 (1993)
29. Geer, J.F., Klir, G.J.: A mathematical analysis of information – preserving transformations between probabilistic and possibilistic formulation of uncertainty. Int. J. General Syst. **20**, 143–176 (1992)

30. Gil, M.A.: Probabilistic–possibilistic approach to some statistical problems with fuzzy experimental observations. In: Kacprzyk, J., Fedrizzi, M. (eds.) Combining Fuzzy Imprecision with Probabilistic Uncertainty in Decision Making, pp. 286–306. Springer, Berlin (1988)
31. Gil, M.A., López-Díaz, M.: Fundamentals of Bayesian analyses of decision problems with fuzzy-valued utilities. Int. J. Approx. Reason. **15**, 203–224 (1996)
32. Gil, M.A., López-Díaz, M., Ralescu, D.A.: Overwiev on the development of fuzzy random variables. Fuzzy Sets Syst. **157**, 2546–2557 (2006)
33. Gil, M.A., Montenegro, M., González-Rodríguez, G., Colubi, A., Casals, M.R.: Bootstrap approach to the multi-sample test of means with imprecise data. Comput. Stat. Data Anal. **51**, 148–162 (2006)
34. Giron, F.J., Rios, S.: Quasi-Bayesian behaviour: a more realistic approach to decision making? (with discussion). In: Bernardo, J.M., DeGroot, M.H., Lindley, D.V., Smith, A.F.M. (eds.) Bayesian Statistics I, pp. 17–38 and 49–66. Valencia University Press, Valencia (1980)
35. González-Rodríguez, G., Montenegro, M., Colubi, A., Gil, M.A.: Bootstrap techniques and fuzzy random variables: synergy in hypothesis testing with fuzzy data. Fuzzy Sets Syst. **157**, 2608–2613 (2006)
36. Good, I.J.: Subjective probability as the measure of a non-measurable set. In: Nagel, E., Supes, P., Tarski, A. (eds.) Logic, Methodology and Philosophy of Science, pp. 319–329. Stanford University press, Stanford CA (1962)
37. Grzegorzewski, P.: Testing statistical hypotheses with vague data. Fuzzy Sets Syst. **112**, 501–510 (2000)
38. Grzegorzewski, P., Hryniewicz, O.: Testing hypotheses in fuzzy environment. Mathware Soft Comput. **4**, 203–217 (1997)
39. Grzegorzewski, P., Hryniewicz, O.: Soft methods in hypotheses testing. In: Ruan, D., Kacprzyk, J., Fedrizzi, M. (eds.) Soft Computing for Risk Evaluation and Management. Applications in Technology, Environment and Finance, pp. 55–72. Physica, Heidelberg (2001)
40. Helton, J.C., Johnson, J.D., Oberkampf, W.L: An exploration of alternative approaches to the representation of uncertainty in model predictions. Reliab. Eng. Syst. Safety **85**, 39–71 (2004)
41. Hryniewicz, O.: Possibilistic interpretation of the results of statistical tests. In: Proc. of the Eight International Conference IPMU, Madrid, vol. I, 215–219 (2000)
42. Hryniewicz, O.: Possibilistic approach to the Bayes statistical decisions. In: Grzegorzewski, P., Hryniewicz, O., Gil, M.A. (eds.) Soft Methods in Probability, Statistics and Data Analysis, pp. 207–218. Physica, Heidelberg (2002)
43. Hryniewicz, O.: Possibilistic decisions and fuzzy statistical tests. Fuzzy Sets Syst. **157**, 2665–2673 (2006)
44. Hryniewicz, O.: Testing reliability with imprecise requirements and life data. In: Guedes Soares, C., Zio, E. (eds.) Safety and Reliabilty for Managing Risk, pp. 889–896. Taylor & Francis, London (2006)
45. Huber, P.J.: Robust Statistics. Wiley, New York (1981)
46. IPCC Good Practice Guidance and Uncertainty Management in National Greenhouse Gas Inventories. IPCC National Greenhouse Gas Inventories Programme, Technical Support Unit, Hayama, Japan (2000). http://www.ipcc-nggip.iges.or.jp/public/gp/english/
47. Keynes, J.M.: A Treatise on Probability. Macmillan, London (1921)
48. Klir, G.J.: A principle of uncertainty and information invariance. Int. J. General Syst. **17**, 249–275 (1990)
49. Koopman, B.O.: The bases of probability. Bull. Am. Math. Soc. **46**, 763–774 (1940)
50. Körner, R.: An asymptotic α-test for the expectation of fuzzy random variables. J. Stat. Plan. Infer. **83**, 331–346 (2000)
51. Körner, R., Näther, W.: Linear regression with random fuzzy variables: extended classical estimates, best linear estimates, least square estimates. Inform. Sci. **109**, 95–118 (1998)
52. Kruse, R.: The strong law of large numbers for fuzzy random variables. Inform. Sci. **28**, 233–241 (1982)
53. Kruse, R., Meyer, K.D.: Statistics with Vague Data. Riedel, Dordrecht (1987)
54. Kwakernaak, H.: Fuzzy random variables, part I: definitions and theorems. Inform. Sci. **15**, 1–15 (1978)

55. Kwakernaak, H.: Fuzzy random variables, Part II: algorithms and examples for the discrete case. Inform. Sci. **17**, 253–278 (1979)
56. Kyburg, H.E.: Probability and the Logic of Rational Belief. Wesleyan University Press, Middletown, CO (1961)
57. Lehmann, E.L.: Testing Statistical Hypotheses, 2nd edn. Wiley, New York (1986)
58. Levi, I.: The Enterprise of Knowledge. MIT, London (1980)
59. Lindley, D.V.: Bayesian Statistics, A Review. SIAM, Philadelphia (1971)
60. Lindley, D.V.: Making Decisions, 2nd rev. edn. Wiley, New York (1991)
61. Miranda, E.: A survey of the theory of coherent lower previsions. Int. J. Approx. Reason. **48**, 628–658 (2008)
62. Montenegro, M., Casals, M.R., Lubiano, M.A., Gil, M.A.: Two-sample hypothesis tests of means of a fuzzy random variable. Inform. Sci. **133**, 89–100 (2001)
63. Montenegro, M., Colubi, A., Casals, M.R., Gil, M.A.: Asymptotic and bootstrap techniques for testing the expected value of a fuzzy random variable. Metrika **59**, 31–49 (2004)
64. Möller, B., Beer, M.: Fuzzy randomness – uncertainty in civil engineering and computational mechanics. Springer, Berlin (2004)
65. Möller, B., Beer, M.: Engineering computation under uncertainty – capabilities of non-traditional models. Comput. Struct. **86**, 1024–1041 (2008)
66. Oberkampf, W.L., Helton, J.C., Joslyn, C.A., Wojtkiewicz, S.F., Ferson, S.: Challenge problems: uncertainty in systems response given uncertain parameters. Reliab. Eng. Syst. Safety **85**, 11–19 (2004)
67. Pawlak, Z.: Rough sets. Fuzzy Sets Syst. **17**, 99–102 (1985)
68. Puri, M.L., Ralescu, D.A.: Fuzzy random variables. J. Math. Anal. Appl. **114**, 409–422 (1986)
69. Raiffa, H., Schleifer, R.: Applied Statistical Decision Theory. MIT, Cambridge (1961)
70. Römer, Ch., Kandel, A.: Statistical tests for fuzzy data. Fuzzy Sets Syst. **72**, 1–26 (1995)
71. Rypdal, K., Winiwarter, W.: Uncertainty in greenhouse gas emission inventories – evaluation, comparability and implications. Environ. Sci. Policy **4**, 104–116 (2001)
72. Saade, J.: Extension of fuzzy hypothesis testing with hybrid data. Fuzzy Sets Syst. **63**, 57–71 (1994)
73. Saade, J., Schwarzlander, H.: Fuzzy hypothesis testing with hybrid data. Fuzzy Sets Syst. **35**, 197–212 (1990)
74. Savage, L.J.: The Foundation of Statistics. Wiley, New York (1954)
75. Shackle, G.L.S.: Decision, Order and Time in Human Affairs, 2nd edn. Cambridge University Press, Cambridge (1961)
76. Shafer, G.: A Mathematical Theory of Evidence. Princeton University Press, Princeton, NJ (1976)
77. Simon, H.: Models of Man. Wiley, New York (1955)
78. Smith, C.A.B.: Consistency in statistical inference and decision (with discussion). J. R. Stat. Soc. B **36**, 1–37 (1974)
79. Son, J.Ch., Song, I., Kim, H.Y.: A fuzzy decision problem based on the generalized Neyman–Pearson criterion. Fuzzy Sets Syst. **47**, 65–75 (1992)
80. Taheri, S.M., Behboodian, J.: Neyman–Pearson lemma for fuzzy hypotheses testing. Metrika **49**, 3–17 (1999)
81. Taheri, S.M., Behboodian, J.: A Bayesian approach to fuzzy hypotheses testing. Fuzzy Sets Syst. **123**, 39–48 (2001)
82. Tanaka, H., Okuda, T., Asai, K.: Fuzzy information and decisions in statistical model. In: Gupta, M.M., et al. (eds.) Advances in Fuzzy Sets Theory and Applications, pp. 303–320. North-Holland, Amsterdam (1979)
83. Tversky, A., Kahneman, D.: Extensional versus intuitive reasoning: the conjunctive fallacy in probability judgements. Psychol. Rev. **91**, 293–315 (1983)
84. Viertl, R.: Is it necessary to develop a fuzzy Bayesian inference. In: Viertl, R. (ed.) Probability and Bayesian Statistics, pp. 471–475. Plenum, New York (1987)
85. Viertl, R.: Statistical methods for non-precise data. CRC, Boca Raton (1996)
86. Walley, P.: Statistical Reasoning with Imprecise Probabilities. Chapman & Hill, London (1991)

87. Walley, P.: Measures of uncertainty in expert systems. Artif. Intell **83**, 1–58 (1996)
88. Walley, P.: Towards a unified theory of imprecise probability. Int. J. Approx. Reason. **24**, 125–148 (2000)
89. Walley, P., de Cooman, G.: A behavioral model for linguistic uncertainty. Inform. Sci. **134**, 1–37 (2001)
90. Watanabe, N., Imaizumi, T.: A fuzzy statistical test of fuzzy hypotheses. Fuzzy Sets Syst. **53**, 167–178 (1993)
91. Williams, P.M.: Coherence, strict preference and zero probabilities. In: Proc. of the Fifth Intern. Cong. of Logic, Methodology and Philosophy of Science (1975), vol. VI, pp. 29–30.
92. Williams, P.M.: Indeterminate probabilities. In: Przelecki, M., Szaniawski, M., Wojcicki, R. (eds.) Formal Methods in the Methodology of Empirical Sciences, pp. 229–246. Riedel, Dordrecht (1976)
93. Wilson, N., Moral, S.: A logical view of probability. In: Cohn, A. (ed.) Proc. of the Eleventh Europ. Conf. on Artificial Intelligence, pp. 386–390. Wiley, London (1994)
94. Winiwarter, W.: National greenhouse gas inventories: understanding uncertainties versus potential for improving reliability. Water, Air Soil Pollut. Focus **7**, 443–450 (2004)
95. Zadeh, L.A.: Fuzzy sets as a basis for a theory of possibility. Fuzzy Sets Syst. **1**, 3–28 (1978)
96. Zimmermann, H.-J.: An application-oriented view of modeling uncertainty. Fuzzy Sets Syst. **122**, 190–198 (2000)

Chapter 3
On the Approximation of a Discrete Multivariate Probability Distribution Using the New Concept of t-Cherry Junction Tree

Edith Kovács and Tamás Szántai

Abstract Most everyday reasoning and decision making is based on uncertain premises. The premises or attributes, which we must take into consideration, are random variables, so that we often have to deal with a high dimensional discrete multivariate random vector. We are going to construct an approximation of a high dimensional probability distribution that is based on the dependence structure between the random variables and on a special clustering of the graph describing this structure. Our method uses just one-, two- and three-dimensional marginal probability distributions. We give a formula that expresses how well the constructed approximation fits to the real probability distribution. We then prove that every time there exists a probability distribution constructed this way, that fits to reality at least as well as the approximation constructed from the Chow–Liu dependence tree. In the last part we give some examples that show how efficient is our approximation in application areas like pattern recognition and feature selection.

3.1 Introduction

The goal of our paper is to approximate a high dimensional joint probability distribution using two and three-dimensional marginal distributions, only.

The idea of such kind of approximations was given by Chow and Liu [7]. In their work they construct a first order tree taking into account the mutual information gains of all pairs of random variables. They proved that their approximation is optimal in the sense of Kullback–Leibler divergence [12, 15].

In order to give an approximation that uses lower dimensional marginal probability distributions, there are many algorithms developed. Most of them first construct

Edith Kovács (✉)
Department of Mathematics, ÁVF College of Management of Budapest, Villányi út 11-13, 1114 Budapest, Hungary
e-mail: kovacs.edith@avf.hu

Tamás Szántai
Institute of Mathematics, Budapest University of Technology and Economics, Műegyetem rkp. 3, 1111 Budapest, Hungary

K. Marti et al. (eds.), *Coping with Uncertainty*, Lecture Notes in Economics and Mathematical Systems 633, DOI 10.1007/978-3-642-03735-1_3,

a Bayesian network (directed acyclic graph) see [1] and then they obtain from this in several steps a junction tree. See [11] for a good overview.

Many algorithms were developed for obtaining a Bayesian Network. These algorithms start with the construction of the Chow–Liu tree and then this graph is transformed, by adding edges and then delete the superfluous edges using conditional independence tests. The number of conditional independence tests is diminished by searching the minimal d-separating set [2, 6].

After the graphical structure is determined a number of quantitative operations have to be performed on it (see [13] and [9]).

In our paper we suppose to be known just the three-dimensional marginals of the joint probability distribution (indeed that implies that the second and first order marginals are known, too). From this information we first construct a graphical model, and then a junction tree, named t-cherry-junction tree.

The construction of our graphical model is inspired by the graphical structure, named cherry tree, and t-cherry tree introduced by Bukszár and Prékopa in [3].

We emphasize that our method does not use the construction of the Bayesian network first, we are going to use just the third order marginals and the information contents related to them, and to certain pairs of random variables involved.

The paper is organized as follows:

- The second section contains the theory, formulas, and connections between them that are used in the third section.
- The third section contains the introduction of the concept of the t-cherry-junction tree, the formula that gives the Kullback–Leibler divergence associated to the approximation introduced, and a proof that the approximation associated to the t-cherry-junction tree is at least as good as Chow–Liu's approximation is.
- The last section contains some applications of the approximation introduced in order to exhibit some advantages of this approach.

3.2 Preliminaries

3.2.1 Notations

Let $\mathbf{X} = (X_1, X_2, \ldots, X_n)^T$ be an n-dimensional random vector with the joint probability distribution

$$P\left(X_1 = x_{i_1}^1, \ldots, X_n = x_{i_n}^n\right), i_1 = 1, \ldots, m_1, \ldots, i_n = 1, \ldots, m_n.$$

For this we will use also the abbreviation

$$P(\mathbf{X}) = P(X_1, X_2, \ldots, X_n).$$

This shorter form will be applied in sums and products, too. So we will write

$$\sum_{\mathbf{X}} P(\mathbf{X}) = \sum_{i_1=1}^{m_1} \cdots \sum_{i_n=1}^{m_n} P(X_1 = x_{i_1}^1, \ldots, X_n = x_{i_n}^n)\,(=1) \tag{3.1}$$

and for example if $H = \{j, k, l\}$ is a three element subset of the index (vertex) set $\{1, 2, \ldots, n\}$ then $\mathbf{X}_H = (X_j, X_k, X_l)^T$ and

$$\sum_{\mathbf{X}_H} P(\mathbf{X}_H) = \sum_{i_j=1}^{m_j} \sum_{i_k=1}^{m_k} \sum_{i_l=1}^{m_l} P(X_j = x_{i_j}^j, X_k = x_{i_k}^k, X_l = x_{i_l}^l)\,(=1) \tag{3.2}$$

and

$$\prod_{\mathbf{X}_H} P(\mathbf{X}_H) = \prod_{i_j=1}^{m_j} \prod_{i_k=1}^{m_k} \prod_{i_l=1}^{m_l} P(X_j = x_{i_j}^j, X_k = x_{i_k}^k, X_l = x_{i_l}^l) \tag{3.3}$$

$P_a(\mathbf{X})$ will denote the approximating joint probability distribution of $P(\mathbf{X})$.

3.2.2 Cherry Tree and t-Cherry Tree

The cherry tree is a graph structure introduced by Bukszár and Prékopa (see [3]). A generalization of this concept, called hyper cherry tree can be found in [4] by Bukszár and Szántai. Let be given a nonempty set of vertices V.

Definition 3.1. The graph defined recursively by the following steps is called *cherry tree*:

1. Two vertices connected by an edge is the smallest cherry tree.
2. By connecting a new vertex of V, by two new edges to two existing vertices of a cherry tree, one obtains a new cherry tree.
3. Each cherry tree can be obtained from (1) by the successive application of (2).

Remark 3.1. A cherry tree is an undirected graph. We emphasize that it is not a tree.

Definition 3.2. We call *cherry*, a triplet of vertices formed from two existing vertices and a new one connected with them in step (2) of Definition 3.1.

For the cherry we will use the notation introduced in [4] by Bukszár and Szántai: $(\{l, m\}, k)$, where l and m are the existing vertices, k is the newly connected vertex and $\{k, l\}, \{k, m\}$ are the new edges.

Remark 3.2. From a set of n vertices we obtain a cherry tree with $n - 2$ cherries.

Remark 3.3. We denote by $\mathcal{H}$ the set of all cherries of the cherry tree and by ϵ the set of edges of the cherry tree.

Remark 3.4. A cherry tree is characterized by a set V of vertices, the set $\mathcal{H}$ of cherries and the set ϵ of edges. We denote a cherry-tree by $\Delta = (V, \mathcal{H}, \epsilon)$.

In our paper we need the concept of the t-cherry tree introduced by Bukszár and Prékopa in [3]. To get the concept of t-cherry tree one has to apply the more restrictive Step (2′) in Definition 3.1 instead of Step (2):

(2′) By connecting a new vertex of V by two new edges to two *connected* vertices of a cherry tree, one obtains a new cherry tree.

Definition 3.3. The graph defined recursively by (1), (2′) and (3) is called *t-cherry tree.*

Remark 3.5. A pair of adjacent vertices from the t-cherry tree may be used several times, for connecting new vertices to them.

3.2.3 Junction Tree

The junction tree is a very prominent and widely used structure for inference in graphical models (see [5] and [12]).

Let $X = \{X_1, \ldots, X_n\}$ be a set of random variables defined over the same probability field and $\mathbf{X} = (X_1, \ldots, X_n)^T$ an n-dimensional random vector.

Definition 3.4. A tree with the following properties is called junction tree over X:

1. To each node of the tree, a subset of X called cluster and the marginal probability distribution of these variables is associated.
2. To each edge connecting two clusters of the tree, the subset of X given by the intersection of the connected clusters and the marginal probability distribution of these variables is associated.
3. If two clusters contain a random variable, than all clusters on the path between these two clusters contain this random variable (running intersection property).
4. The union of all clusters is X.

Notations:

- $\mathcal{C}$ – the set of clusters
- C – a cluster
- $\mathbf{X}_C$ – the random vector with the random variables of C as components
- $P(\mathbf{X}_C)$ – the joint probability distribution of $\mathbf{X}_C$
- $\mathcal{S}$ – the set of separators
- S – a separator
- $\mathbf{X}_S$ – the random vector with the random variables of S as components
- $P(\mathbf{X}_S)$ – the joint probability distribution of $\mathbf{X}_S$

The junction tree provides a joint probability distribution of $\mathbf{X}$:

$$P(\mathbf{X}) = \frac{\prod\limits_{C \in \mathcal{C}} P(\mathbf{X}_C)}{\prod\limits_{S \in \mathcal{S}} P(\mathbf{X}_S)^{(\nu_S - 1)}},$$

where ν_S is the number of those clusters which contain all of the variables involved in S.

3.3 t-Cherry-Junction Tree

3.3.1 Construction of a t-Cherry-Junction Tree

Let $X = \{X_1, \ldots, X_n\}$ be a set of random variables defined over the same probability field and denote by $\mathbf{X} = (X_1, \ldots, X_n)^T$ the corresponding n-dimensional random vector. Together with the construction of the t-cherry tree over the set of vertices $V = \{1, \ldots, n\}$ we can construct a t-cherry-junction tree in the following way.

Algorithm 1. *Construction of a t-cherry-junction tree.*

1. The first cherry of the t-cherry tree identifies the first cluster of the t-cherry junction tree. The vertices of the first cherry give the indices of the random variables belonging to the first cluster.
2. Similar way each new cherry of the t-cherry tree identifies a new cluster of the t-cherry-junction tree. The separators contain pairs of random variables with indices corresponding to the connected vertices used in Step (2′) of Definition 3.3.

Theorem 3.1. *The t-cherry-junction tree constructed by Algorithm 1 is a junction tree.*

Proof. We can check the statements of Definition 3.4:

- The first statement of definition is obvious.
- The separator that connects two clusters contains the intersection of the two clusters. This follows from the construction of the separator sets.
- Let us suppose that there exists a variable X_m which belongs to two clusters on a path, so that on this path exist a cluster that does not contain X_m. This implies that on this past exist two neighboring cluster so that one of them contains X_m and the other one does not contain X_m. This means that in the t-cherry tree there are two cherries connected so that one of them contains the vertex m the other one does not contain the vertex m. According to the point (2) of Definition 3.1, m is a new vertex connected, but since X_m belongs already to another cluster, the vertex m belongs to another cherry, too. That is a contradiction, because that means that m is not a new vertex.

- The union of all sets associated to the clusters is X, because of the union of set of indices is V.

Remark 3.6. To each cluster $\{X_l, X_m, X_k\}$ a three-dimensional joint probability distribution $P(X_l, X_m, X_k)$ can be associated. To each separator $\{X_l, X_m\}$ a two-dimensional joint probability distribution $P(X_l, X_m)$ can be associated.

The joint probability distribution associated to the t-cherry-junction tree is a joint probability distribution over the variables $X_1, \ldots, X_n$, given by

$$P(\mathbf{X}) = \frac{\prod\limits_{\{X_l,X_m,X_k\}\in\mathcal{C}} P(X_l, X_m, X_k)}{\prod\limits_{\{X_l,X_m\}\in\mathcal{S}} P(X_l, X_m)^{\nu_{lm}-1}}, \tag{3.4}$$

where $\nu_{lm} = \#\{\{X_l, X_m\} \subset C \,|\, C \in \mathcal{C}\}$.

3.3.2 The Approximation of the Joint Distribution Over X by the Distribution Associated to a *t*-Cherry-Junction Tree

Chow and Liu introduced a method to approximate optimally an n-dimensional, discrete joint probability distribution by the one- and two-dimensional probability distributions using first-order dependence tree. It is shown that the procedure presented in their paper yields an approximation with minimum difference of information, in the sense of Kullback–Leibler divergence.

In this part, we first give a formula for the Kullback–Leibler divergence between the approximated distribution associated to the t-cherry-junction tree and the real joint probability distribution; we then conclude what we have to take into account to minimize the divergence. For the proof of this theorem we need a lemma:

Lemma 3.1. *In a t-cherry-junction tree for each variable* $X_m \in X$*:*

$$\#\{\{X_l, X_m\} \mid (\{X_l, X_m\}, X_k) \in \mathcal{C}\} = \#\{(\{X_l, X_m\}, X_k) \mid (\{X_l, X_m\}, X_k) \in \mathcal{C}\} - 1$$

Proof. Let denote $t = \#\{\{X_l, X_m\} \mid (\{X_l, X_m\}, X_k) \in \mathcal{C}\}$ for a given $X_m \in X$.

- Case $t = 0$.
  The statement is a consequence on one hand of Definition 3.4 that is the union of all sets associated to the nodes (clusters) is X, so every vertex from X, have to appear at least in one cluster. On the other hand, if X_m would be contained in two clusters than there must exist one separator set containing this vertex (point 3) in Definition 3.4), but we supposed $t = 0$.
- Case $t > 0$.
  If two clusters contain a variable X_m, than all clusters from the path between the two clusters contain X_m (running intersection property). From this results that

the clusters containing X_m are the nodes of a connected graph, and this graph is a first order tree. If this graph contains $t+1$ clusters, than it contains t separator sets (definition of a tree in which we have separators instead of edges and clusters instead of nodes). From the definition of the junction tree (Definition 3.4) results that every separator set connecting two clusters contains the variables from the intersection of the clusters. So there are t separator sets that contain the variable X_m.

The goodness of the approximation of a probability distribution can be quantified by the Kullback–Leibler divergence between the real and the approximating probability distributions (see for example [8] or [15]). The Kullback–Leibler divergence expresses somehow the distance between two probability distributions. As smaller value it has as better the approximation is.

Theorem 3.2. *If we denote by $P_a(\mathbf{X})$ the approximating joint probability distribution associated to the t-cherry-junction tree [see formula (3.4)], then the Kullback–Leibler divergence between the real and the approximating joint probability distribution is given by:*

$$\begin{aligned} I(P(\mathbf{X}), P_a(\mathbf{X})) = -H(\mathbf{X}) - \sum_{(X_l, X_m, X_k)\in\mathcal{C}} I(X_l, X_m, X_k) \\ + \sum_{\{X_l, X_m\}\in\mathcal{S}} (\nu_{lm}-1) I(X_l, X_m) + \sum_{k=1}^{n} H(X_k). \end{aligned} \tag{3.5}$$

Proof.

$$\begin{aligned} I(P(\mathbf{X}), P_a(\mathbf{X})) &= \sum_{\mathbf{X}} P(\mathbf{X}) \log_2 \frac{P(\mathbf{X})}{P_a(\mathbf{X})} \\ &= \sum_{\mathbf{X}} P(\mathbf{X}) \log_2 P(\mathbf{X}) - \sum_{\mathbf{X}} P(\mathbf{X}) \log_2 P_a(\mathbf{X}) \\ &= -H(\mathbf{X}) - \sum_{\mathbf{X}} P(\mathbf{X}) \log_2 \frac{\prod_{\{X_l, X_m, X_k\}\in\mathcal{C}} P(X_l, X_m, X_k)}{\prod_{\{X_l, X_m\}\in\mathcal{S}} P(X_l, X_m)^{\nu_{lm}-1}} \\ &= -H(\mathbf{X}) - \sum_{\mathbf{X}} P(\mathbf{X}) \Bigg[\log_2 \prod_{\{X_l, X_m, X_k\}\in\mathcal{C}} P(X_l, X_m, X_k) \\ &\qquad - \log_2 \prod_{\{X_l, X_m\}\in\mathcal{S}} P(X_l, X_m)^{\nu_{lm}-1} \Bigg] \\ &= -H(\mathbf{X}) - \sum_{\mathbf{X}} P(\mathbf{X}) \log_2 \prod_{\{X_l, X_m, X_k\}\in\mathcal{C}} P(X_l, X_m, X_k) \\ &\quad + \sum_{\mathbf{X}} P(\mathbf{X}) \log_2 \prod_{\{X^l, X^m\}\in\mathcal{S}} P(X_l, X_m)^{\nu_{lm}-1} \end{aligned}$$

From Definition 3.4 follows that the union of the clusters of the junction tree is the set X. From the Lemma 3.1 we know that each vertex appears once more in clusters than in separator sets. So by adding and subtracting the sum

$$\sum_{\mathbf{X}} P(\mathbf{X}) \log_2 \left[\prod_{\{X_l, X_m, X_k\} \in \mathcal{C}} P(X_l) P(X_m) P(X_k) \right]$$

we obtain the following:

$$\begin{aligned}
&I(P(\mathbf{X}), P_a(\mathbf{X})) \\
&\quad = -H(\mathbf{X}) - \sum_{\mathbf{X}} P(\mathbf{X}) \log_2 \frac{\prod_{\{X_l, X_m, X_k\} \in \mathcal{C}} P(X_l, X_m, X_k)}{\prod_{\{X_l, X_m, X_k\} \in \mathcal{C}} P(X_l) P(X_m) P(X_k)} \\
&\qquad + \sum_{\mathbf{X}} P(\mathbf{X}) \log_2 \frac{\prod_{\{X_l, X_m\} \in \mathcal{S}} P(X_l, X_m)^{\nu_{lm}-1}}{\prod_{\{X_l, X_m\} \in \mathcal{S}} [P(X_l) P(X_m)]^{\nu_{lm}-1}} - \sum_{\mathbf{X}} P(\mathbf{X}) \log_2 \prod_{X_k \in X} P(X_k) \\
&\quad = -H(\mathbf{X}) - \sum_{\mathbf{X}} P(\mathbf{X}) \sum_{\{X_l, X_m, X_k\} \in \mathcal{C}} \log_2 \frac{P(X_l, X_m, X_k)}{P(X_l) P(X_m) P(X_k)} \\
&\qquad + \sum_{\mathbf{X}} P(\mathbf{X}) \sum_{\{X_l, X_m\} \in \mathcal{S}} (\nu_{lm} - 1) \log_2 \frac{P(X_l, X_m)}{P(X_l) P(X_m)} + \sum_{X_k \in X} H(X_k).
\end{aligned}$$

Since (X_l, X_m, X_k) and (X_l, X_m) are components of the random vector $\mathbf{X}$, we have the relations:

$$\begin{aligned}
&\sum_{\mathbf{X}} P(\mathbf{X}) \sum_{\{X_l, X_m, X_k\} \in \mathcal{C}} \log_2 \frac{P(X_l, X_m, X_k)}{P(X_l) P(X_m) P(X_k)} \\
&\quad = \sum_{\{X_l, X_m, X_k\} \in \mathcal{C}} \sum_{(X_l, X_m, X_k)^T} P(X_l, X_m, X_k) \log_2 \frac{P(X_l, X_m, X_k)}{P(X_l) P(X_m) P(X_k)}
\end{aligned}$$

and

$$\begin{aligned}
&\sum_{\mathbf{X}} P(\mathbf{X}) \sum_{\{X_l, X_m\} \in \mathcal{S}} (\nu_{lm} - 1) \log_2 \frac{P(X_l, X_m)}{P(X_l) P(X_m)} \\
&\quad = \sum_{\{X_l, X_m\} \in \mathcal{S}} \sum_{(X_l, X_m)^T} (\nu_{lm} - 1) P(X_l, X_m) \log_2 \frac{P(X_l, X_m)}{P(X_l) P(X_m)}.
\end{aligned}$$

Taking into account these relations and applying the notion of the mutual information content for two and three variables:

$$I(X_l, X_m) = \sum_{(X_l,X_m)^T} P(X_l, X_m) \log_2 \frac{P(X_l, X_m)}{P(X_l)P(X_m)},$$

$$I(X_l, X_m, X_k) = \sum_{(X_l,X_m,X_k)^T} P(X_l, X_m, X_k) \log_2 \frac{P(X_l, X_m, X_k)}{P(X_l)P(X_m)P(X_k)}$$

we obtain (3.5) and the statement of the theorem has been proved.

Observation 1. We can observe that for minimizing the Kullback–Leibler divergence between the real probability distribution and the approximation obtained from the *t*-cherry-junction tree we have to maximize the difference between the sum of the mutual divergence of the clusters and the sum of the mutual divergence of the separators denoted by S:

$$S = \sum_{\{X_l,X_m,X_k\}\in\mathcal{C}} I(X_l, X_m, X_k) - \sum_{\{X_l,X_m\}\in\mathcal{S}} (\nu_{lm} - 1) I(X_l, X_m)$$

Observation 2. If we wish to compare two approximations of a joint probability distribution associated to two different *t*-cherry-junction trees, we have just to:

- Sum the information contents of the clusters
- Sum the information contents of the separators
- Make the difference between them
- Claim a *t*-junction-cherry tree be better than another one, if it produces greater value of S

3.3.3 The Relation Between the Approximations Associated to the First-Order Dependence Tree and *t*-Cherry-Junction Tree

A natural question is: can the *t*-cherry-junction tree give better approximation than the first order tree given by Chow and Liu does? Let us remind that Chow and Liu introduced a method for finding an optimal first order tree that minimizes the Kullback–Leibler divergence.

If $pa(j)$ denotes the parent node of j, the joint probability distribution associated to the Chow–Liu first order dependence tree is given as follows:

$$P_{Ch-L}(\mathbf{X}) = \prod_{i=1}^{n} P(X_{m_i} | X_{pa(m_i)}),$$

where $\{m_1, \dots, m_n\}$ is a permutation of the numbers $1, 2, \dots, n$ and if $pa(j)$ is the empty set, then by definition $P(X_j | X_{pa(j)}) = P(X_j)$.

In the following Algorithm 2 we will show how one can construct a t-cherry-junction tree from a Chow–Liu first order dependence tree. After this in Theorem 3.3 we prove that the t-cherry-junction tree constructed this way gives at least as good approximation of the real probability distribution as the Chow–Liu first order dependence tree does.

Algorithm 2. *The construction of a t-cherry-junction tree from a Chow–Liu first order dependence tree.*

Let us regard the spanning tree behind the Chow–Liu first order dependence tree. It is sufficient to give an algorithm for constructing a t-cherry tree from this spanning tree. Then by Algorithm 1 we can assign a t-cherry-junction tree to this t-cherry tree:

1. The first cherry of the t-cherry tree let be defined by any three vertices of the spanning tree which are connected by two edges.
2. We add a new cherry to the t-cherry tree by taking a new vertex of the spanning tree adjacent to the so far constructed t-cherry tree.
3. We repeat step (2) till all vertices from the spanning tree become included in the t-cherry tree.

Theorem 3.3. *If $P_{Ch-L}(\mathbf{X}) = \prod_{i=1}^{n} P(X_{m_i}|X_{pa(m_i)})$ is the approximation associated to the Chow–Liu first order dependence tree there always exists a t-cherry-junction tree with associated probability distribution $P_{t-ch}(\mathbf{X})$ that approximates $P(\mathbf{X})$ at least as well as $P_{Ch-L}(\mathbf{X})$ does.*

Proof. We construct the t-cherry-junction tree from the Chow–Liu first order dependence tree using Algorithm 2.

The Kullback–Leibler divergence formally looks like it is given in (3.5).

In the case of the approximation obtained by the Chow–Liu method, the Kullback–Leibler divergence is given as follows (see [7]):

$$I(P(\mathbf{X}), P_{Ch-L}(\mathbf{X})) = -H(\mathbf{X}) - \sum_{i=1}^{n} I(X_{m_i}, X_{pa(m_i)}) + \sum_{i=1}^{n} H(X_i). \tag{3.6}$$

Because the first and last terms of the Kullback–Leibler divergences are the same in the case of the two approximations (3.5) and (3.6), we denote the sums that we have to compare by S_{Ch-L} in the case of Chow–Liu's approximation and S_{t-ch} in the case of t-cherry-junction tree approximation:

$$S_{Ch-L} = \sum_{X_{m_i} \in X} I(X_{m_i}, X_{pa(m_i)}) \tag{3.7}$$

and

$$S_{t-ch} = \sum_{\{X_l, X_m, X_k\} \in C} I(X_l, X_m, X_k) - \sum_{\{X_l, X_m\} \in \mathcal{S}} (\nu_{lm} - 1) I(X_l, X_m) \tag{3.8}$$

In the case of formula (3.7) we can apply the formula

$$I(X,Y) = H(X) - H(X|Y)$$

(see [8], Formula (2.43) on p. 16), while in the case of formula (3.8) we can apply

$$I(X,Y,Z) = H(Z) + I(X,Y) - H(Z|X,Y)$$

which easily can be derived from formulae given in book [8].

So from (3.7) we get

$$\begin{aligned} S_{Ch-L} &= \sum_{X_{m_i} \in X} \left[H(X_{m_i}) - H(X_{m_i}|X_{pa(m_i)}) \right] \\ &= \sum_{X_{m_i} \in X} H(X_{m_i}) - \sum_{X_{m_i} \in X} H(X_{m_i}|X_{pa(m_i)}) \end{aligned} \tag{3.9}$$

and from (3.8) we get

$$\begin{aligned} S_{t-ch} &= \sum_{\{X_l,X_m,X_k\} \in \mathcal{C}} \left[H(X_k) + I(X_l,X_m) - H(X_k|X_l,X_m) \right] \\ &\quad - \sum_{\{X_l,X_m\} \in \mathcal{S}} (\nu_{lm} - 1) I(X_l,X_m) \\ &= \sum_{X_k \in X} H(X_k) - \sum_{\{X_l,X_m,X_k\} \in \mathcal{C}} H(X_k|X_l,X_m) \end{aligned} \tag{3.10}$$

From (3.9) and (3.10) we conclude that we have to compare only

$$\sum_{X_{m_i} \in X} H(X_{m_i}|X_{pa(m_i)})$$

and

$$\sum_{\{X_l,X_m,X_k\} \in \mathcal{C}} H(X_k|X_l,X_m)$$

To each X_k corresponds an X_{m_i} because these both are the vertices of X and each of them appears exactly once. The edge connecting X_{m_i} and $X_{pa(m_i)}$ is contained in the t-cherry tree as it was constructed according to Step (2) in the proof of Theorem 3.3. From this we can conclude that for each X_k contained in a cluster $\{X_l, X_m, X_k\}$ one of X_l and X_m is $X_{pa(k)}$. Now we can apply the inequality (see book [8], Formula (2.130) on p. 36):

$$H(X_k|X_i,X_j) \leq H(X_k|X_i)$$

with equality just in case X_k is conditionally independent of X_j. From this it is obvious that $S_{t-ch} \geq S_{Ch-L}$, and indeed the relation between the Kullback–Leibler divergences of the two approximations is:

$$I(P(\mathbf{X}), P_{t-ch}(\mathbf{X})) \leq I(P(\mathbf{X}), P_{Ch-L}(\mathbf{X})).$$

This proves the statement of the theorem.

Observation 1. If the underlying dependence structure is a first order dependence tree than we got equality between the two divergences. (Because of the conditional independences that take place between the unlinked vertices of the first order dependence tree).

3.4 Some Practical Results of Our Approximation and Discussions

This section consists of three parts. In the first part we consider two different approximations of a given eight-dimensional discrete joint probability distribution:

- The approximation given by the Chow–Liu method
- The approximation corresponding to the *t*-cherry tree constructed by the algorithm given in the proof of Theorem 3.3

In the second part of this section we use the approximations to make some prediction, when the values taken by two out of eight random variables are known. From this information we are going to recognize the most probable values of the remaining six random variables. In the third part we use the influence diagram underlying the t-cherry-junction tree to make a feature selection.

Two Approximations of an Eight-Dimensional Discrete Probability Distribution

First we construct an eight-dimensional discrete probability distribution in the following way. The one-dimensional marginal distributions are generated randomly. Then the so called North-West corner algorithm was applied to determine an initial feasible solution of an eight-dimensional transportation problem, where the quantities to be transported were the marginal probability values. This way the "transported probabilities" are concentrated on 44 different directions, i.e. the constructed eight-dimensional discrete probability distribution is concentrated on 44 eight-dimensional vector instead of the possible 2,116,800. An other advantage is that this way we can get as high as possible positive correlations between the components of the random vector **X**. For having also high negative correlations, we combined the North-West corners with South-West corners.

Now we suppose that only the two and three-dimensional marginals of the eight-dimensional joint probability distribution are known. Using these we are going to construct a probability distribution associated to the *t*-cherry-junction tree. First we construct the Chow–Liu spanning tree and then transform the correspondent Chow–Liu dependence tree in a *t*-cherry junction tree using the Algorithm 2.

Table 3.1 The mutual information gains in decreasing order for the construction of the Chow–Liu spanning tree

Index pair	Information gain
58	**1.660755**
18	**1.644658**
28	**1.641684**
25	1.611500
15	1.605311
56	**1.593626**
17	**1.558543**
47	**1.556713**
16	1.436039
14	1.423012
68	1.421927
26	1.416225
78	1.367303
13	**1.334538**

Table 3.2 The three variable mutual information contents ordered in decreasing way. The boldfaced ones are used in the *t*-cherry-junction tree

Index triplet	Information content
158	**3.615752**
156	**3.431799**
258	**3.414463**
168	3.384689
568	3.357803
128	3.322943
125	3.321265
256	3.278311
147	**3.276440**
178	**3.270665**
157	3.197210
578	3.167661
268	3.151382
148	3.137220
135	**3.120396**

The first step is to calculate the mutual information contents of every pair and triplet of random variables. We then order them in descending way.

The Chow–Liu tree is constructed by a greedy algorithm from the two-dimensional information gains. In Table 3.1 the ordered mutual information gains are given. The mutual information gains used by Chow–Liu's method are in boldface. In Fig. 3.1 the Chow–Liu tree can be seen.

The joint probability distribution associated to the constructed Chow–Liu dependence tree is:

Table 3.3 The two variable mutual information contents ordered in decreasing way. The boldfaced ones are used in the t-cherry-junction tree

Index pair	Information content
58	**1.660755**
18	**1.644658**
28	1.641684
25	1.611500
15	**1.605311**
56	1.593626
17	**1.558543**

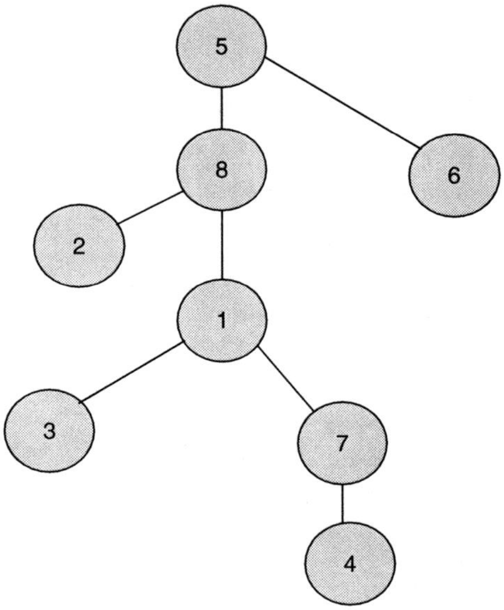

Fig. 3.1 The Chow–Liu spanning tree

$$
\begin{aligned}
&P_{Ch-L}(\mathbf{X})\\
&\quad = P(X_5)P(X_8|X_5)P(X_2|X_8)P(X_6|X_5)P(X_1|X_8)P(X_3|X_1)P(X_7|X_1)P(X_4|X_7)\\
&\quad = \frac{P(X_5X_8)P(X_2X_8)P(X_5X_6)P(X_1X_8)P(X_1X_3)P(X_1X_7)P(X_4X_7)}{P(X_8)P(X_5)P(X_8)P(X_1)P(X_1)P(X_7)}.
\end{aligned}
$$

The Kullback–Leibler divergence corresponding to the divergence between the real probability distribution and the approximation given by Chow–Liu method can be calculated as follows:

$$
\begin{aligned}
&I(P(\mathbf{X}), P_{Ch-L}(\mathbf{X}))\\
&\quad = \sum_{i=1}^{8} H(X_i) - H(\mathbf{X}) - \big[I(X_5, X_8) + I(X_2, X_8) + I(X_5, X_6) + I(X_1, X_8)\\
&\qquad + I(X_1, X_3) + I(X_1, X_7) + I(X_4, X_7)\big]\\
&\quad = 16.908113 - 4.634363 - 10.990524 = 1.283226.
\end{aligned}
$$

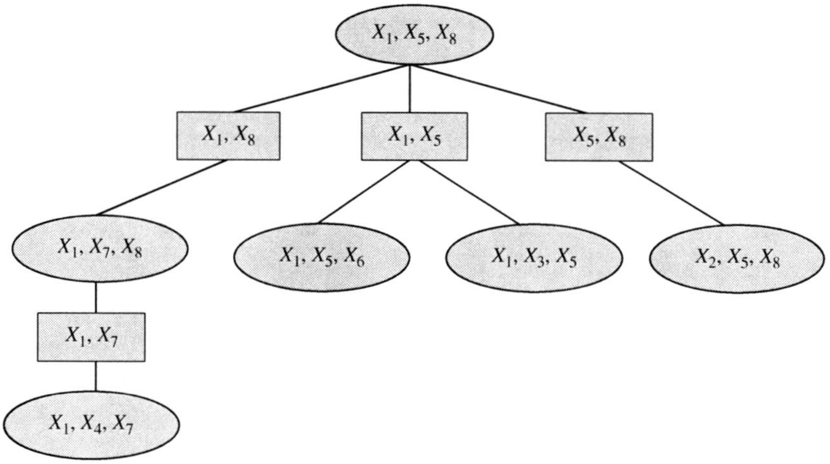

Fig. 3.2 The t-cherry-junction tree

The t-cherry-junction tree constructed by Algorithm 2 can be seen in Fig. 3.2. The three and the two variable mutual information contents corresponding to the clusters and separators of the t-cherry-junction tree are given in bold face in Tables 3.2 and 3.3

The joint probability distribution associated to the constructed t-cherry-junction tree is:

$$P_{t-ch}(\mathbf{X}) = \frac{P(X_1X_5X_8)P(X_1X_7X_8)P(X_1X_5X_6)P(X_1X_3X_5)P(X_2X_5X_8)P(X_1X_4X_7)}{P(X_1X_8)P(X_1X_5)P(X_1X_5)P(X_5X_8)P(X_1X_7)}.$$

The Kullback–Leibler divergence between the real probability distribution and the approximation associated to the t-cherry-junction tree is:

$$\begin{aligned} I(P(\mathbf{X}), P_{t-ch}(\mathbf{X})) &= \sum_{i=1}^{8} H(X_i) - H(\mathbf{X}) - \big[\, I(X_1, X_5, X_8) + I(X_1, X_7, X_8) + I(X_1, X_5, X_6) \\ &\quad + I(X_1, X_3, X_5) + I(X_2, X_5, X_8) + I(X_1, X_4, X_7) - I(X_1, X_8) \\ &\quad - I(X_1, X_5) - I(X_1, X_5) - I(X_5, X_8) - I(X_1, X_7)\big] \\ &= 16.908113 - 4.634363 - 20.129510 + 8.074578 = 0.218818. \end{aligned}$$

The real eight-dimensional distribution has 44 different vectors with probabilities different from 0. The Chow–Liu approximation has 453 vectors; the t-cherry-junction tree approximation has 93 vectors with probabilities different from zero. From the two Kullback–Leibler divergences calculated above we can observe easily that the approximation associated to the t-cherry-junction tree is much better ("closer" to the reality) than the approximation constructed from the Chow–Liu tree.

Table 3.4 Comparison of the predicted values with the real ones in the case of the two different approximations

The most probable vectors predicted by Ch–L approx.	$P(\mathbf{X})$	$P_a(\mathbf{X})$	The most probable vectors in the reality	$P(\mathbf{X})$
7 5 1 7 1 2 7 1	0.003487	0.003443	7 5 1 7 1 2 7 1	0.003487
6 5 1 7 1 2 7 3	0.006802	0.005149	6 5 1 7 1 2 7 3	0.006802
3 3 3 **4** 4 5 **3** 5	0.000000	0.001696	3 3 3 5 4 5 5 5	0.009853
2 2 3 4 5 5 3 7	0.011576	0.003718	2 2 3 4 5 5 3 7	0.011576
4 4 3 6 **2** **3** 6 4	0.000000	0.004076	4 4 3 6 3 4 6 4	0.012450
3 2 3 4 4 5 3 6	0.019070	0.008043	3 2 3 4 4 5 3 6	0.019070
8 5 1 7 1 2 7 1	0.019399	0.019155	8 5 1 7 1 2 7 1	0.019399
3 2 3 4 **5** 5 3 7	0.010207	0.006777	3 2 3 4 4 5 3 7	0.025312
7 5 1 7 1 2 7 2	0.027852	0.027503	7 5 1 7 1 2 7 2	0.027852
9 5 1 7 1 2 8 1	0.034906	0.037489	9 5 1 7 1 2 8 1	0.034906
4 3 3 6 4 **5** 6 5	0.000000	0.033668	4 3 3 6 4 4 6 5	0.044913
5 4 3 7 **2** **3** 7 4	0.000000	0.030836	5 4 3 7 3 4 7 4	0.049737
1 2 4 **1** 5 5 **2** 7	0.027323	0.033325	1 2 4 2 5 5 3 7	0.054170
6 5 **3** 7 1 2 7 2	0.000000	0.001522	6 5 1 7 1 2 7 2	0.128273
6 4 **2** 7 **3** 3 7 4	0.050302	0.009993	6 4 1 7 2 3 7 4	0.152472
The most probable vectors predicted by t–ch approx.	$P(\mathbf{X})$	$P_a(\mathbf{X})$	The most probable vectors in the reality	$P(\mathbf{X})$
7 5 1 7 1 2 7 1	0.003487	0.003487	7 5 1 7 1 2 7 1	0.003487
6 5 1 7 1 2 7 3	0.006802	0.006802	6 5 1 7 1 2 7 3	0.006802
3 3 3 5 4 5 5 5	0.009853	0.009497	3 3 3 5 4 5 5 5	0.009853
2 2 3 4 5 5 3 7	0.011576	0.009733	2 2 3 4 5 5 3 7	0.011576
4 4 3 6 3 4 6 4	0.012450	0.009917	4 4 3 6 3 4 6 4	0.012450
3 2 3 **5** 4 5 **4** 6	0.012682	0.018480	3 2 3 4 4 5 3 6	0.019070
8 5 1 7 1 2 7 1	0.019399	0.019399	8 5 1 7 1 2 7 1	0.019399
3 2 3 4 4 5 3 7	0.025312	0.020359	3 2 3 4 4 5 3 7	0.025312
7 5 1 7 1 2 7 2	0.027852	0.027852	7 5 1 7 1 2 7 2	0.027852
9 5 1 7 1 2 8 1	0.034906	0.034906	9 5 1 7 1 2 8 1	0.034906
4 3 3 6 4 4 6 5	0.044913	0.037797	4 3 3 6 4 4 6 5	0.044913
5 4 3 7 3 4 7 4	0.049737	0.058277	5 4 3 7 3 4 7 4	0.049737
1 2 4 2 5 5 3 7	0.054170	0.044752	1 2 4 2 5 5 3 7	0.054170
6 5 1 7 1 2 7 2	0.128273	0.128273	6 5 1 7 1 2 7 2	0.128273
6 4 1 7 2 3 7 4	0.152472	0.154778	6 4 1 7 2 3 7 4	0.152472

Application for Pattern Recognition

In this part we are testing our approximations for the following pattern recognition problem. We suppose that the values of X_1 and X_8 are known. For these given values we want to predict the most probable values of the other six random variables. In Table 3.4 one can see these predictions made with the help of the approximation.

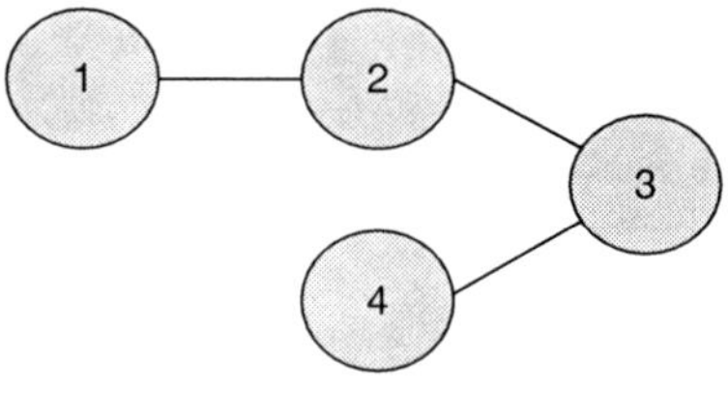

Fig. 3.3 A possible dependence diagram of four random variables

In the left side of the table the predicted values of the random variables $X_2, \dots, X_7$ are given for the two different approximations. In the right side of the table the most probable values of the same random variables are given according to the real probability distribution (the same in the case of both approximations). The rows of the table are in ascending order according to the real probabilities. As the wrong predicted values are typed in boldface one easily can find them for each approximation. Let us observe that as long as the number of the wrong predicted values is 13 in the case of the Chow–Liu approximation, the same number equals only 2 in the case of the new t-cherry-junction tree approximation.

Feature Selection: Forecasting the Values of a Random Variable Which Depends on Many Others

The main idea of feature selection is to choose a subset of input random variables by eliminating features with little or no predictive information. In supervised learning the feature selection is useful when the main goal is to find feature subset that produces higher classification accuracy.

In practice many times we have a lot of attributes (random variables) that depend more or less on each other. The problem is how to select a few of them to make a good forecast of the variable we are interested in. The pairwise mutual information contents are not sufficient to make such a decision. To highlight this let us consider the following example. If we have four random variables with the relations between their pairwise mutual information contents: $I(X_2, X_3) > I(X_1, X_3) > I(X_3, X_4)$, and want to take into account only two random variables to forecast the values of X_3, we would use X_2 and X_1. But if we have the dependence diagram (see Fig. 3.3) we should decide in an other way. As X_1 influences X_3 only through X_2 we should rather use X_2 and X_4 for the forecast of the values of X_3. To solve such problems we give a method that uses the t-cherry-junction tree. If we are interested in forecasting a random variable X_i, we have to select from the cherry tree the clusters that contain X_i. We obtain in this way a sub junction tree. This results immediately from the properties of the junction tree. We have to take into consideration, just the variables that belong to this sub junction tree.

If in our earlier numerical example we are interested in the forecast of X_8 we select from the t-cherry-junction tree the clusters that contain X_8. From Fig. 3.2 these are (X_1, X_5, X_8), (X_1, X_7, X_8) and (X_2, X_5, X_8). Now we can conclude that for our purpose it is important to know the joint probability distribution of the random vector $(X_1, X_2, X_5, X_7, X_8)^T$. This can be obtained as a marginal distribution of the distribution associated to the cherry tree, which can be also expressed as:

$$P(X_1, X_2, X_5, X_7, X_8) = \frac{P(X_1, X_5, X_8)P(X_1, X_7, X_8)P(X_2, X_5, X_8)}{P(X_1, X_8)P(X_5, X_8)}.$$

Now to test our method we calculate the value of the conditional entropy $H(X_8 \mid X_1, X_2, X_5, X_7) = 0.214946$ from the *real* probability distribution. If we choose another random vector $(X_1, X_2, X_5, X_7, X_8)^T$, then we get $H(X_8 \mid X_1, X_2, X_5, X_6) = 0.235572$.

It is interesting to see that $I(X_1, X_2, X_5, X_6, X_8) = 7.210594$ is greater than $I(X_1, X_2, X_5, X_7, X_8) = 7.087809$.

Acknowledgements This work was partly supported by the grant No. T047340 of the Hungarian National Grant Office (OTKA).

References

1. Acid, S., Campos, L.M.: BENEDICT: An algorithm for learning probabilistic belief networks. In: Sixth International Conference IPMU 1996, 979–984 (1996)
2. Acid, S., Campos, L.M.: An algorithm for finding minimum d-separating sets in belief networks. In: Proceedings of the 12th Conference on Uncertainty in Artificial Intelligence, pp. 3–10. Morgan Kaufmann, San Mateo (1996)
3. Bukszár, J., Prékopa, A.: Probability bounds with cherry trees. Math. Oper. Res. **26**, 174–192 (2001)
4. Bukszár, J., Szántai, T.: Probability bounds given by hypercherry trees. Optim. Methods Software **17**, 409–422 (2002)
5. Castillo, E., Gutierrez, J., Hadi, A.: Expert Systems and Probabilistic Network Models. Springer, Berlin (1997)
6. Cheng, J., Bell, D.A., Liu, W.: An algorithm for Bayesian belief network construction from data. In: Proceedings of AI&StAT'97, 83–90 (1997)
7. Chow, C.K., Liu, C.N.: Approximating discrete probability distribution with dependence trees. IEEE Trans. Inform. Theory **14**, 462–467 (1968)
8. Cover, T.M., Thomas, J.A.: Elements of Information Theory. Wiley, New York (1991)
9. Cowell, R.G., Dawid, A.Ph., Lauritzen, S.L., Spiegelhalter, D.J.: Probabilistic Networks and Expert Systems. Statistics for Engineering and Information Science. Springer, Berlin (1999)
10. Csiszar, I.: I-divergence geometry of probability distributions and minimization problems. Ann. Probab. **3**, 146–158 (1975)
11. Huang, C., Darwiche, A.: Inference in belief networks: A procedural guide. Int. J. Approx. Reason. **15**(3), 225–263 (1996)
12. Hutter, F., Ng, B., Dearden, R.: Incremental thin junction trees for dynamic Bayesian networks. Technical report, TR-AIDA-04-01, Intellectics Group, Darmstadt University of Technology, Germany, 2004. Preliminary version at http://www.fhutter.de/itjt.pdf
13. Jensen, F.V., Lauritzen, S.L., Olesen, K.: Bayesian updating in casual probabilistic networks by local computations. Comput. Stat. Q. **4** 269–282 (1990)
14. Jensen, F.V., Nielsen, T.D.: Bayesian networks and decision graphs, 2nd edn. Information Science and Statistics. Springer, New York (2007)
15. Kullback, S.: Information Theory and Statistics. Wiley, New York (1959)

Part II
Robust Solutions Under Uncertainty

Chapter 4
Induced Discounting and Risk Management

T. Ermolieva, Y. Ermoliev, G. Fischer, and M. Makowski

Abstract The goal of this paper is to specify and summarize new approaches to discounting proposed in our catastrophic risk management studies. The main issue is concerned with justification of investments, which may turn into benefits over long and uncertain time horizon. For example, how can we justify mitigation efforts for expected 300-year flood that can occur also next year. The discounting is supposed to impose time preferences to resolve this issue, but this view may be dramatically misleading. We show that any discounted infinite horizon sum of values can be equivalently replaced by undiscounted sum of the same values with random finite time horizon. The expected duration of this stopping time horizon for standard discount rates obtained from capital markets does not exceed a few decades and therefore such rates may significantly underestimate the net benefits of long-term decisions. The alternative undiscounted random stopping time criterion allows to induce social discounting focusing on arrival times of the main concern (stopping time) events rather than horizons of market interests.
In general, induced discount rates are conditional on the degree of social commitment to mitigate risk. Random stopping time events affect these rates, which alter the optimal mitigation efforts that, in turn, change events. This endogeneity of the induced discounting restricts exact evaluations necessary for using traditional deterministic methods and it calls for stochastic optimisation methods. The paper provides insights in the nature of discounting that are critically important for developing robust long-term risk management strategies.

4.1 Introduction

The implication of uncertainties and risks for justifying long-term investments is a controversial issue. How can we justify investments, which may possibly turn into benefits over long and uncertain time horizons in the future? This is a key issue for catastrophic risk management. For example, how can we justify investments in

T. Ermolieva (✉), Y. Ermoliev, G. Fischer, and M. Makowski
International Institute Applied Systems Analysis, Shlossplatz 1, 2361 Laxenburg, Austria,
e-mail: ermol@iiasa.ac.at, ermoliev@iiasa.ac.at, fisher@iiasa.ac.at, marek@iiasa.ac.at

K. Marti et al. (eds.), *Coping with Uncertainty*, Lecture Notes in Economics and Mathematical Systems 633, DOI 10.1007/978-3-642-03735-1_4,

climate change mitigations, say, in flood defense systems to cope with foreseen extreme 1,000-, 500-, 250-, and 100-floods? The lack of proper evaluations for dealing with extreme events dramatically contributes to increasing losses from human-made and natural disasters [20]. The analysis of floods that occurred in the summer of 2002 across central Europe [14] shows that the potential areas of vulnerability to extreme floods have multiplied as a consequence of failed development planning. Underestimation and ignorance of low probability/high consequence events have led to the growth of buildings and industrial land and sizable value accumulation in flood prone areas without proper attention being paid to flood mitigations. A challenge is that an endogenously created catastrophe,[1] say a 300-year flood, has never occurred before in a given region. Therefore, purely adaptive policies relying on historical observations provide no awareness of the "unknown" risk although, a 300-year flood may occur next year. For example, the 2002 floods in Austria, Germany and the Czech Republic were classified (in different regions) as 1,000-, 500-, 250-, and 100-year events [14].

A key issue is development of policies with proper long-term perspectives. The traditional discounting is supposed to impose necessary time preferences, but this view may be dramatically misleading. There are several possibilities for choosing discount rates (see, for example, the discussion in [2, 19, 24, 29]). The traditional approach is to use the rates obtained in capital markets. The geometric or exponential discount factor $d_t = (1+r)^{-t} = e^{-ln(1+r)t} \approx e^{-rt}$ (for small r) is usually connected with a constant rate r of returns from capital markets. Since returns in capital markets are linked to assets with a lifespan of a few decades, this choice may completely reduce the impacts that investments have beyond these intervals (Sect. 4.2). Another serious problem [21, 31] arises from the use of the expected value $r = E\xi$ and the discount factor $(1+r)^{-t}$. It implies additional significant reduction of future values in contrast to the expected discount factor $E(1+\xi)^{-t}$, because $E(1+\xi)^{-t} >> (1+r)^{-t}$. These issues are discussed in Sects. 4.2 and 4.3.

An appropriate interest rate is especially difficult to define when decisions involve time horizons beyond the interests of the current generation. If future generations are not present in the market, e.g., long-term environmental damages are not included in production costs, the market interest rates do not reflect the preferences of future generations. According to Arrow et al. [2] "the observed market rates of interest refer to how individuals are willing to trade off consumption over their own life. These may or may not bear close correspondence to how a society is willing to trade off consumption across generations".

Debates on proper discount rates for long-term problems have a long-standing history [2, 29]. Ramsey [26] argued that applying a positive discount rate r to discount values across generations is unethical. Koopmans [17], contrary to Ramsey, argued that zero discount rate r would imply an unacceptably low level of current consumption. The use of so-called social discount rates produces two effects [2].

[1] As a consequence of inappropriate policies.

The "prescriptive" approach tends to generate relatively low discount rates and thus favors mitigation measures and the wellbeing of future generations. The "descriptive" approach tends to generate higher discount rates and thus favors less spending on mitigation and the wellbeing of the current generation.

The constant discount rate has only limited justification [3, 12, 24, 29]. As a compromise between "prescriptive" and "descriptive" approaches, Cline [4] argues for a declining discount rate: 5% for the first 30 years, and 1.5% later. There have been proposals for other schedules and attempts to justify the shape of proper decline. Papers [21, 31] show that uncertainty about r produces a certainty-equivalent discount rate, which will generally be declining with time. Weitzman [31] proposed to model discount rates by a number of exogenous time dependent scenarios. He argued for rates of 3–4% for the first 25 years, 2% for the next 50 years, 1% for the period 75–300 years and 0 beyond 300 years. Newel and Pizer [21] analyzed the uncertainty of historical interest rates by using data on the US market rate for long-term government bonds. They proposed a different declining discount rate justified by a random walk model. Chichilinsky [3] proposed a new concept for long-term discounting with a declining discount rate by attaching some weight on the present and the future consumption. All these papers aim to derive an appropriate exogenous social discount rate.

Sections 4.2 and 4.3 develop a different approach for social discounting. It is shown that any discounted sum, so-called net present value (NPV) criterion, $\sum_{t=0}^{\infty} d_t V_t$ of expected values $V_t = Ev_t$ for random variables (r.v.) $v_t, t = 0, 1, \ldots, d_t = (1 + r_t)^{-t}$ under constant and declining discount rates r_t equals the average undiscounted (in the agreement with Ramsey's concerns) random sum $E \sum_{t=0}^{\tau} v_t$ with a random stopping time τ defined by the given discounting d_t. Therefore, discount rates can be associated with the occurrences of "stopping time" random events determining a finite "internal" discount-related horizon $[0, \tau]$. The expected duration of τ and its standard deviation σ under modest market interest rates of 3.5% is approximately 30 years, which may have no correspondence with expected, say, 300-year extreme events and $\sigma \approx 300$. Conversely, it is shown that any stopping time random event induces a discounting. A set of mutually exclusive stopping time random events, e.g., 1,000-, 500-, 250-, and 100-year floods, induces discounting with time-declining discount rates. This case corresponds also to the discounting with uncertain discount rates r. In particular, a single stopping time random event with the standard geometric probability distribution induces the standard discounting with constant discount rate r and $d_t = (1 + r)^{-t}$.

The effects of catastrophes on the stream of values $v_t, t = 0, 1, \ldots,$ differ from the effects of market uncertainties. Section 4.4 indicates that catastrophic events pose new challenges. They often create so-called endogenous, unknown (with the lack and even absence of adequate observations) and interdependent risks, which may potentially affect large territories and communities and, on the other hand, are dramatically affected by risk management decisions. As a consequence, catastrophic risks generally make it impossible to use traditional economic and insurance models [1, 3, 5, 8, 9, 16]. The concept of undiscounted random stopping time criteria allows to induce social discounting that focuses on arrivals of catastrophic events rather then the lifetime of market products. Since risk management decisions affect the occurrence of disasters in time and space, the induced discounting may depend on spatio-temporal distributions of extreme events and feasible sets of decisions, i.e., it can be viewed as a spatio-temporal discounting. The implicit dependence of the stopping time discounting on random events and decisions calls for the use of stochastic optimization methods, which allows also to address the variability (Remark 4.2) of discounted criteria by using random value $\sum_{t=0}^{\tau} v_t$ even for deterministic $v_t, t = 0, 1, \ldots$. Section 4.5 establishes connections of stopping time discounting with dynamic versions of CVaR (Conditional Value-at-Risk) risk measures. Section 4.6 illustrates how misperception of induced discounting may provoke catastrophes. Section 4.7 provides concluding remarks.

4.2 Standard and Stopping Time Induced Discounting

This section illustrated the main idea by using the standard geometric discounting. The choice of discount rate as a prevailing interest rate within a time horizon of existing financial markets is well established [18]. Uncertainties, especially related to extreme events, challenge the possibility of markets to offer proper rates for longer time horizons. The following simple Proposition 4.1 and Remark 4.2 clarify the main concerns.

The traditional financial approaches [18] often use the so-called net present value (NPV) criteria to justify investments. An investment is defined as an expected cash flow stream $V_0, V_1, \ldots, V_T, V_t = Ev_t$, over a time horizon $T \leq \infty$. Assume that r is a constant prevailing market interest rate, then alternative investments are compared by $V = V_0 + d_1 V_1 + \ldots + d_T V_T$, where $d_t = d^t, d = (1+r)^{-1}, t = 0, 1, \ldots, T$, is the discount factor and V denotes NPV.

It is usually assumed that a long-term investment activity has an infinitely long time horizon, i.e.,

$$V = \sum_{t=0}^{\infty} d_t V_t. \tag{4.1}$$

The stream of values $V_t, t = 0, 1, \ldots,$ can represent an expected cash flow stream of a long-term investment activity. In economic growth models and integrated assessment models [19, 22, 29] the value V_t represents utility $U(x^t)$ of an infinitely living representative agent, or welfare $V_t = \sum_{i=1}^{n} a_i u_i(x_i^t)$ of a society with representative agents $i = \overline{1, n}$, utilities u_i, consumptions x_i^t and welfare weights α_i. Natural selection theory treats (4.1) as Darwinian fitness [28], where discount factors d_t are associated with hazard rates of an environment (Example 4.2).

The infinite time horizon in (4.1) creates an illusion of truly long-term analysis. Proposition 4.1 shows that in fact deterministic evaluation (4.1) accounts only for values V_t from a finite random horizon $[0, \tau]$ defined by a random stopping time τ with the discount-related geometric probability distribution $P[\tau \geq t] = d_t$.

Proposition 4.1. *Consider a discounted sum (4.1) with* $d_t = d^t, d = (1+r)^{-1}$, $r > 0$. *Let* $q = d, p = 1 - q$, *and* τ *be a random variable with the geometric probability distribution* $P[\tau = t] = pq^t, t = 0, 1, \ldots$. *Then* $d_t = P[\tau \geq t]$ *and*

$$\sum_{t=0}^{\infty} d^t V_t = \sum_{t=0}^{\infty} P[\tau \geq t] V_t = E \sum_{t=0}^{\tau} V_t. \tag{4.2}$$

Conversely, for any stopping time τ *with a geometric probability distribution*

$$E \sum_{t=0}^{\tau} V_t = \sum_{t=0}^{\infty} d_t V_t, d_t = P[\tau \geq t].$$

Proof. We have $P[\tau \geq t] = \sum_{k=t}^{\infty} pq^k = pq^t(1-q)^{-1} = q^t = d_t$. Conversely,

$$E \sum_{t=0}^{\tau} V_t = \sum_{t=0}^{\infty} P[\tau = t] \sum_{k=0}^{t} = V_k = \sum_{t=0}^{\infty} pq^t \sum_{k=0}^{t} V_k$$
$$= \sum_{t=0}^{\infty} \left(\sum_{k=t}^{\infty} pq^k \right) V_t = \sum_{t=0}^{\infty} d_t V_t.$$

That is, any discounted deterministic sum (4.1) equals to the average undiscounted random sum $\sum_{t=0}^{\tau} V_t$ of the same values V_t. In other words, the discount factor $d_t = d^t$ induces an "internal" discount-relate time horizon $[0, \tau]$ with the geometrically distributed τ. Conversely, any geometrically distributed τ and the criterion $E \sum_{t=0}^{\tau} V_t$ induces the geometric discounting in the sum $\sum_{t=0}^{\infty} d_t V_t$.

Remark 4.1. (Random stopping time horizon). We can consider$[0, \tau]$ being a random stopping time horizon associated with the first occurrence of a "killing", i.e., a catastrophic stopping time event. The probability that this event occurs at $t = 0, 1, \ldots$ is p and pq^t is the probability that this event occurs first time at t, i.e., τ has a geometric probability distribution. Since $p = 1 - d, d = (1+r)^{-1}$, then the expected duration of τ, $E\tau = 1/p = 1 + 1/r$. Therefore, for the interest rate of 3.5%, $r = 0.035$, the expected duration is $E\tau \approx 30$ years, i.e., this rate orients the policy analysis on an expected 30-year time horizon. The standard deviation $\sigma = \sqrt{q/p}$, i.e., it equals approximately 30 years The bias in favor of the present in discounting with the rate of 3.5% is easily illustrated [24]. For a project with long-run benefits or costs, 1 Euro of benefits or costs in years 50, 100, and 200, has a present value respectively of 0.18, 0.003, and practically 0 Euros. Definitely, this rate may have no correspondence to how society has to deal with a 300-year flood, i.e., a flood with the expected arrival time equal to 300 years. Therefore, in the risk management τ can be associated with the arrival of potential catastrophic events rather than with horizons of market interests. The induced social discounting $d_t = P[\tau \geq t]$ in this case would have proper long-term perspectives dependent on spatio-temporal patterns of catastrophes and risk management decisions (see Proposition 4.3 and Sect. 4.4). The discount rate r can be viewed also as a killing (hazard) rate [15] which makes the life expectancy of an otherwise infinitely living representative agent or society equal to $1 + 1/r$ years. Yet, depending on a concrete situation, stopping time τ can be also associated with the arrival time of a reward.

Remark 4.2. (Variability of NPV). Disadvantages of this standard criterion (4.1) are well known [18]. In particular, the NPV critically depends on the prevailing interest rate which may not be easily defined in practice. In addition, the NPV does not reveal the temporal variability of cash flow streams. Two alternative streams may easily have the same NPV despite the fact that in one of them all the cash is clustered within a few periods, but in another it is spread out evenly over time. This type of temporal heterogeneity is critically important for dealing with catastrophic losses which occur suddenly as a "spike" in time and space [9].

The criterion $E\sum_{t=0}^{\tau} V_t, V_t = Ev_t$ has visible advantages. In particular, it allows to address distributional aspects and robust strategies [6] by analyzing the random variable $\sum_{t=0}^{\tau} V_t$ (even for deterministic $v_t = V_t$), e.g., its quantiles defined as maximal $y = y_\delta$ satisfying safety constraint

$$P\left[\sum_{t=0}^{\tau} v_t \geq y\right] \geq \delta.$$

Equivalently, y_δ maximizes the concave function (see discussion in [6], p. 16)

$$y + \delta^{-1} E min\left\{0, \sum_{t=0}^{\tau} v_t - y\right\}.$$

The optimal value of this function defines the so-called CVaR (Conditional Value-at-Risk) risk measure [27].

Therefore, if variables v_t depend on some decisions x (as in Sect. 4.4), then the maximization of function

$$F(x) = \left[y + \delta^{-l} E min\left\{0, \sum_{t=0}^{\tau} v_t - y\right\}\right].$$

allows easy control of highly nonlinear (even for linear in x function v_t) the safety constraints (quantiles of $\sum_{t=0}^{\tau} v_t$) in an optimal manner defined by a function $F(x)$ that is adjusted to CVaR risk measure (see also Sect. 4.5).

Remark 4.3. (Shock testing). The sensitivity of models w.r.t. "shocks" (extreme scenarios, events, stresses) is often assessed by introducing them into discounted criteria [22, 29]. From Proposition 4.1 it follows that this may lead to serious miscalculations. Let us consider criterion (4.1) with discount factors, $d_t = d^t, d = (1+r)^{-1}$ and assume that a "shock" arrives at a random time moment $\theta \in \{0, 1, \ldots\}$ with probability $P[\theta = t] = \pi\gamma^t, \gamma = 1 - \pi = (1+\rho)^{-1}$. Then the expected value, $E\sum_{t=0}^{\theta} d_t V_t = \sum_{t=0}^{\infty} d^t \gamma^t V_t = E\sum_{t=0}^{\tau} \gamma^t V_t = E\sum_{t=0}^{min(\tau,\theta)} V_t$, where $P[\tau = t] = pq^t$ with $q = d, p = 1 - q$. Therefore, the stopping time of the "shocked" evaluation $E\sum_{t=0}^{\theta} d^t V_t$ is defined by $min(\tau, \theta)$. The discount rate of this evaluation is $(1+r)^{-1} \cdot (1+p)^{-1} = (1 + r + \rho + r\rho)^{-1}$, i.e., the shocked evaluation increases the rate of the original discounting and, hence, the bias in favor of the present.

Example 4.1. (Catastrophic Risk Management). The implications of Proposition 4.1 for long-term policy analysis are rather straightforward. Let us consider some important cases. It is realistic to assume [24] that the cash flow stream, typical for investment in a new nuclear plant, has the following average time horizons. Without a disaster the first six years of the stream reflect the costs of construction and commissioning followed by 40-years of operating life when the plant is producing positive cash flows and, finally, a 70-year period of expenditure on decommissioning. The flat discount rate of 5%, as Remark 4.1 shows, orients the analysis on a 20-year time horizon. It is clear that a lower discount rate places more weight on distant costs and benefits. For example, the explicit treatment of a potential 200-year disaster would require at least the discount rate of 0.5% instead of 5%. A related example is investments in climate change mitigations to cope with potential climate change related extreme events. Definitely, a rate of 3.5%, as often used in integrated assessment models [29], can easily illustrate that climate change does not matter. A shock testing of these models reduces even further their internal stopping time horizon.

Example 4.2. (Darwinian fitness). Ramsey [26] had introduced discounting, first of all, as a mathematical device ensuring the convergence of infinite horizon cumulative values. Its various explanations supported by empirical studies were proposed afterwards suggesting that humans and animals place less weights on the future then on the present [28]. A reason is that future rewards run more risk of disappearing. Hence, they should be discounted, where the discount rate is the hazard rate. For example, evidence from selection experiments indicates the existence of a trade-off between short-term and long-term fertility, i.e., the existence of life-history strategy that discounts the future. In other words, natural selection puts a premium on immediate reproductivity. Accordingly, an animal can be treated as a rational optimizer maximizing its Darwinian fitness, that can be taken to be equivalent to maximizing the expected number of offsprings. In a simple case, fitness is defined [28] then as integral $F = \int_0^\infty m(t)s(t)dt$, where $m(t)$ is the expected rate of reproductive output at age t if the animal survives to that age, and $s(t)dt$ is the probability of surviving to age t. It is highly unlikely that an animal is able to learn discount factors (probability density $s(t)$) in order to maximize the Darwinian fitness. The equivalent distribution free stopping time criterion requires observations of only lifetime intervals τ, which can be easily used for adaptive adjustments of life-history strategies.

4.3 Time Declining Discount Rates

This section extends Proposition 4.1 to general time declining discount rates. It also shows that a time declining discount rate can be associated even with a set of mutually exclusive geometrically distributed extreme (stopping time) events. This rate is determined in a sense by the least probable event.

Let us consider now a stream of random variables (r.v.) $v_0, v_1, \dots$ affected by a set of random events including potential catastrophic events. Formally, we can think of v_t as a function $v_t(\omega)$ defined on a probability space $\{\Omega, P\}$ with the set Ω of related random events and the probability measure P on Ω. We assume that v_t does not depend on the "future", i.e., we assume that $\{\Omega, P\}$ is adapted to a sequence of increasing σ -algebras $A_0 \subseteq A_1 \subseteq \dots$ (subsets of events from Ω, which occur before $t = 0, 1, \dots$), such that v_t is measurable (defined on) w.r.t. A_t. In what follows, all random variables are assumed to be defined on $\{\Omega, P\}$.

Let $\sigma_{k,t} = \sigma(v_k, \dots, v_t)$ be the σ-algebra generated by $v_k, \dots, v_t$. Consider a stopping time τ, which we define as a r. v. $\tau \in \{0, 1, \dots\}$, such that event, $\{\tau \le t\}, t = 0, 1, \dots$ does not depend on values $v_{t+1}, v_{t+2}, \dots$, i.e., $\sigma_{t+1,\infty}$.

Proposition 4.2. *Consider a discounted sum* $\sum_{t=0}^{\infty} d_t V_t, d_t = (1 + r_t)^{-t}$, *where* r_t *is an increasing positive sequence,* $V_t = Ev_t$. *Then there is a stopping time* τ *such that* $P[\tau \ge t] = d_t$ *and*

$$\sum_{t=0}^{\infty} d_t V_t = \sum_{t=0}^{\infty} P[\tau \ge t] \, Ev_t = E \sum_{t=0}^{\tau} v_t. \tag{4.3}$$

Conversely, let $E \mid v_t \mid$ *is uniformly bounded. Then, for any stopping time* τ

$$E \sum_{t=0}^{\tau} v_t = \sum_{t=0}^{\infty} d_t V_t, d_t = P[\tau \ge t],$$

where V_t *is conditional expectation:*

$$V_t = E[v_t \mid \tau \ge t]$$

Proof. Consider such any r.v. $\tau, \tau \in \{0, 1, \dots\}$ that $\{\tau \le t\}$ does not dependent on values $v_0, \dots, v_{t-1}$ and $P[\tau = t] = d_t - d_{t+1}, t = 0, 1, 2, \dots$. Clearly, $P[\tau \ge 0] = d_0 - d_1 + d_1 - d_2 + \dots = d_0 = 1$, $P[\tau \ge t] = d_t$ and

$$\sum_{t=0}^{\infty} d_t V_t = \sum_{t=0}^{\infty} P[\tau \ge t] V_t.$$

Let now $f_t := \sum_{k=0}^{t} v_k$. From the rearrangement known as the Kolmogorov–Prohorov's theorem it follows that

$$Ef_\tau = \sum_{t=0}^{\infty} E[f_t; \tau = t] = \sum_{t=0}^{\infty} \sum_{k=0}^{t} E[v_k; \tau = t] = \sum_{k=0}^{\infty} E[v_k; \tau \ge k]$$
$$= \sum_{k=0}^{\infty} P[\tau \ge k] V_k.$$

where $V_k = E[v_k \mid \tau \geq k]$ and $E[v_t; A]$, denotes unconditional expectation $E[v_t I_A]$, I_A is the indicator function of event A. The last assertion follows from the identity $\{\tau \geq t\} = \{\tau > t-1\}$, i.e., from the independence of $\{\tau \geq t\}$ on $\sigma_{t,\infty}$. The change in the order of sums is possible due to the uniform boundness of $E \mid v_t \mid$.

Corollary 4.1. *If $v_0, v_1, \ldots$ are independent r.v. or $\{\tau \geq t\}, t = 0, 1, 2, \ldots,$ does not depend on $v_0, v_1, \ldots, v_{t-1}$,then V_t in both cases of Proposition 4.2 is unconditional expectation $V_t = Ev_t$. If $v_0, v_1, \ldots$ are independent identically distributed r.v., then the Wald's identity follows from Proposition 4.1:*

$$E \sum_{t=0}^{\tau} v_t = Ev_0 E_\tau.$$

Proof. It follows from the following rearrangements:

$$\sum_{t=0}^{\infty} P[\tau \geq t] = \sum_{k=0}^{\infty} \sum_{t=k}^{\infty} P[\tau = t] = \sum_{t=0}^{\infty} t P[\tau = t] = E\tau.$$

Example 4.3. (Expected catastrophic losses). Assume that a catastrophic event may occur at $t = 0, 1, 2, \ldots$ with probability p. It is usually defined as $(1/p)$-year event, say a 100-year flood. Define τ as the arrival time of the first catastrophe and let $v_t = 0, 0 \leq t \leq \tau - 1$, $v_\tau = L_\tau$, where L_τ is conditional expected losses given that the event occurs at τ. Since $l_t \neq 0$ only for, $t = \tau$, then the expected (unconditional) losses at τ are:

$$Ev_\tau = pL_0 + qpL_1 + q^2 pL_2 + \ldots = \sum_{t=0}^{\infty} q^t V_t = \sum_{t=0}^{\infty} P[\tau \geq t] V_t,$$

where $V_t = pL_t$.

The next proposition shows that a set of even geometrically distributed events can induce discounting with time declining discount rates. Let us assume that there is a set of mutually exclusive events (see also Sect. 4.4) of "magnitude" $i = 1, \ldots, n$. The probability of scenario i is $\theta_i, \sum_{i=1}^{n} \theta_i = 1$ and, conditional on this scenario, the event i occurs for the first time at τ_i with the probability $P[\tau_i = t] = p_i q_i^t, q_i = 1 - p_i, t = 0, 1, \ldots$. Thus, the occurrence of events at t is characterized by a mixed geometric distribution $\sum_{i=1}^{n} \theta_i p_i q_i^t$. Let τ be the arrival time of a first event. Then $d_t = P[\tau \geq t] = \sum_{i=1}^{n} \theta_i P[\tau_i \geq t]$. Since $P(\tau_i \geq t) = p_i q_i^t + p_i q_i^{t+1} + \ldots = q_i^t$, then evaluation (4.1) takes the form

$$V = \sum_{t=0}^{\infty} d_t V_t, d_t = \sum_{i=1}^{n} \theta_i q_i^t. \tag{4.4}$$

This equation essentially modifies the standard geometric discounting. Nevertheless, the induced discount factors d_t for large t tend to be defined by the smallest discount rate of the least probable event. The following proposition is similar to the conclusion in [31].

Proposition 4.3. *Discount factor* $d_t = \sum_{i=1}^{n} \theta_i q_i^t$ *in (4.4) is determined for* $t \to \infty$ *by the standard geometric discount factor* $q_i^t *$ *associated with the least probable event* $i^\bullet$,

$$p_{i*} = \min_i \ p_i \ : d_t / q_{i*}^t \to \theta_{i*} \text{ for } t \to \infty.$$

Proof. $d_t = q_{i*}^t \sum_{i=1}^{n} \theta_i \chi_i(t)$, where $\chi_i(t) = (q_i/q_i*)^t$. From $p_{i*} < p_i, p_i = 1 - q_i$, it follows that $\chi_i(t) \to 0, t \to \infty$, for $i \neq i^\bullet$ and $\chi_{i\bullet}(t) = 1$. Hence, $d_t/q_{i*}^t \to \theta_{i*}$ for $t \to \infty$.

Remark 4.4. (Finite time horizon T). Propositions 4.1, 4.2, 4.3 hold true also for a finite time horizon $T < \infty$ after substituting probabilities $P[\tau = t]$, $P[\tau \geq t]$ by conditional probabilities $P[\tau = t \mid \tau \leq T]$ and $P[\tau \geq t \mid \tau \leq T]$.

Remark 4.5. (Distribution free approach). Propositions 4.1, 4.2 provide two alternative approaches for discounting: standard discounted criterion of the left-hand side of (4.1), (4.2) with an exogenous discounting, or undiscounted criterion of the right hand side with τ defined by random arrival time of stopping time events. Proposition 4.3 shows that the corresponding induced discounting $d_t = P[\tau \geq t]$ can be a complex implicit function of spatio-temporal patterns of events. The next section illustrates, that τ may depend also on various decisions. All these make it rather difficult to evaluate exact risk profiles $P[\tau \geq t]$ and exogenous discount factors d_t. Therefore, this would require the use of the distribution-free random stopping time criterion and STO methods rather then the standard distribution-based discounted criterion and deterministic optimization methods.

4.4 Endogenous Discounting

This section summarizes typical motivations for developing spatio-temporal catastrophic risk management models with rather natural versions of the stopping time concepts. A typical model may include often the following loop and the potential for positive feedbacks, branching and disequilibrium:

1. Stopping time induces discounting in the form of dynamic risk profiles $d_t = P[\tau \geq t]$.
2. The discounting affects optimal mitigation efforts.
3. Mitigation efforts affect the stopping time τ, risk profiles $P[\tau \geq t]$ and the discounting d_t (return to point 1).

This means that the stopping time criterion induces endogenous spatio-temporal endogenous discounting.

Example 4.4. (Evaluation of a Flood Management Program). Consider a simple version of the catastrophic flood management model developed for the Upper Tisza river region [9]. The spatio-temporal structure of this model was motivated by the following reasons.

Throughout the world, the losses from floods and other natural disasters are mainly absorbed by the immediate victims and their governments [13]. The insurance industry and its premium payers also absorb a portion of catastrophic losses, but even in the wealthy countries this share is relatively small. With increasing losses from floods, governments are concerned with escalating costs for flood prevention, flood response, compensation to victims, and public infrastructure repair. As a new policy, many officials would like to increase the responsibility of individuals and local governments for flood risks and losses [25], but this is possible only through location-specific analysis of risk exposures and potential losses, the mutual interdependencies of these losses, and the sensitivities of the losses to new risk management strategies.

This is a methodologically challenging task requiring at least the development of spatio-temporal catastrophe models [5, 8, 9, 30]. Although rich data usually exist on aggregate levels, the sufficient location specific data are not available, especially data relevant to new policies. Moreover, catastrophes affect large territories and communities producing mutually dependent losses with analytically intractable multidimensional probability distributions dependent also on various decisions. This critically distinguishes the arising problems from a standard risk management situations, e.g., the well-known asset-liability management. The standard methods, in particular, the existing extreme event theory, are not applicable to rational management of catastrophic risks. The new GIS-based catastrophe models [9,30] are needed to simulate the occurrence of potential extreme events and the samples of mutually dependent catastrophic losses for which no or very few historic observations exist.

In general, a catastrophe model represents the study region by grids, e.g., a relatively small pilot Upper Tisza region is represented by $1{,}500 \times 1{,}500$ grids [9]. Depending on the purpose of the study, these grids are aggregated into a much smaller number of cells (locations, compartments) $j = 1, 2, \dots, m$. These cells may correspond to a collection of households at a certain site, a collection of grids with similar land-use characteristics, or an administrative district or grid with a segment of gas pipeline. The choice of cells provides a desirable representation of losses. Accordingly, cells are characterized by their content, in general, not necessarily in monetary units. Values can be measured in real terms, without using an aggregate dollar value. The content of cells is characterized by the vulnerability curves calculating random damages to crops, buildings, infrastructure, etc., under a simulated catastrophic scenario.

Catastrophic floods which are simulated by the catastrophe model, affect at random different cells and produce mutually dependent random losses L^t_j, $j = 1, \dots, m$, from a catastrophic event at time t. These losses can be modified by various decisions. Some of the decisions reduce losses, say a dike, whereas others spread them on a regional, national, and international level, e.g., insurance contracts. If $x = (x_1, x_2, \dots, x_n)$ is the vector of the decision variables, then L^t_j is a random function $L^t_j(x)$.

Flood occurrences in the region are modeled according to specified probabilistic scenarios of catastrophic rainfalls and the reliability of dikes. There are three dikes allocated along the region's river branch. Each of them may break after the occurrence at a random time of a 100-, 150-, 500-, and 1,000-year rainfall characterized by the so-called up-stream discharge curves calculating the amount of discharged water to the river branch per unit of time. In fact, the discharge curves upscale the information about complex rainfall and run-off processes affected by land-use and land-transformation policies. This brings considerable uncertainty in the definition of a $1/p$-year flood, $p = 1/100$, 1/150, 1/500,1/1,000. Therefore, a 100-year discharge curve may represent, in fact, a set of floods with different frequencies p, say, $1/150 \leq p \leq 1/100$. In addition to the interval, the uncertainty about p can be given by a prior distribution. Therefore, a single discharge curve, in general, corresponds to a set of $1/p$-year floods, where p is characterized by a prior probability distribution. For example, it can be characterized by a finite number of probabilistic scenarios $p_1, \ldots, p_n$ with prior probabilities $\theta_1, \ldots, \theta_n$ as in Proposition 4.3.

The stopping time can be defined differently, depending on the purpose of the policy analysis. A catastrophic flood in our example occurs due to the break of one of the three dikes. These events are considered as mutually exclusive events, since the break of a dike in the pilot region releases the "pressure" on other dikes. Therefore, the stopping time τ can be defined as the first time moment of a dike break. In this case, the probability or induced discount factor $d_t = P[\tau \geq t]$ is an implicit function of t, probabilities $\theta_i, p_i, i = 1, \ldots, n$, and the probability of a dike break. The situation is complicated further by the deterioration of dikes in time and/or by inappropriate maintenance of the flood protection system (see also Sect. 4.6), e.g., modifications to the dikes, the removal of some of them, and building new retention areas and reservoirs. Besides these structural decisions, the stopping time τ can be affected by other decisions, e.g., land use policies. Accordingly, depending on goals, the definition of stopping time τ can be further modified. For example, let us assume that the region [10] participates in the flood management program through payments to a mutual catastrophe fund, which has to support a flood protection system and compensates losses to victims. To enforce the participation in the program, the government provides only partial coverages of losses. The stability of this program critically depends on the insolvency of the fund that may require a new definition of τ. Let β be a fixed investment rate enabling the support of the system of dikes on a certain safety level and ξ be a random time of a first catastrophic flood. Denote by L_j^ξ random losses at location j, $j = \overline{1, m}$, at time $t = \xi$ and by π_j the premium rate paid by location j to the mutual catastrophe fund. Then, its accumulated risk reserve at time ξ together with a fixed partial compensation of losses $\chi \sum_j L_j^\xi$ by the government is $R_\xi = \xi \sum_j \pi_j + \chi \sum_j L_j^\xi - \sum_j \varphi_j L_j^\xi - \beta\xi$, where $0 \leq \varphi_j \leq 1$, is the portion of losses compensated by the fund at location j. Let us also assume that the functioning of the flood management program is considered as a long-term activity assuming that growth and aging processes compensate each other. Then, the insolvency of the fund is associated with the event:

$$\xi \sum_j \pi_j + \chi \sum_j L_j^\xi - \sum_j \varphi_j L_j^\xi - \beta\xi < 0. \tag{4.5}$$

Inequality (4.5) defines extreme random events affected by various feasible decisions x including components $(\pi_j, \varphi_j, \chi, b_j, \beta, j = \overline{1,m}$. The likelihood of event (4.5) determines the vulnerability of the program. It is more natural now to define the stopping time τ as the first time when event (4.5) occurs. In this case τ would depend on all components of vector x and the induced discounting would focus on time horizons associated with the occurrence of the event (4.5).

4.5 Dynamic Risk Profiles and CVaR Risk Measure

Example 4.5 illustrates that the probability distributions $P[\tau \geq t]$, $t = 0, 1, \ldots,$ itself represent key safety characteristics of catastrophic risk management programs. Induced discounting $d_t = P[\tau \geq t]$ then "controls" these risk profiles implicitly through their contributions to discounted goals of programs. Another possibility as this section shows is to impose explicitly safety constraints of the type $P[\tau \geq t] \geq \gamma_t$, for some safety levels $\gamma_t, t = 0, 1, \ldots$. In this case resulting robust strategies would directly control the safety constraints.

Example 4.5. (Safety constraints). The occurrence of disasters is often associated with the likelihood of some processes abruptly passing "vital" thresholds. This is a typical situation for insurance, where the risk process is defined similar to (4.5) by flows of premiums and claims whereas thresholds are defined by insolvency constraints. A similar situation arises in the control of environmental targets and in the design of disaster management programs [5,8,9]. Assume that there is a random process R_t and the threshold is defined by a random ρ_t. In spatial modeling, R_t and ρ_t can be large-dimensional vectors reflecting the overall situation in different locations of a region. Let us define the stopping time τ as the first time moment t when R_t is below ρ_t. By introducing appropriate risk management decisions x it is often possible to affect R_t and ρ_t in order to ensure the safety constraints $P[R_t \geq \rho_t] \geq \gamma$, for some safety level γ, or $\gamma_t, t = 0, 1, 2, \ldots$.

The use of this type safety constraints is a rather standard approach for coping with risks in the insurance, finance, and nuclear industries. For example, the safety regulations of nuclear plants assume that the violation of safety constraints may occur only once in 10^7 years, i.e., $\gamma = 1 - 10^{-7}$. It is remarkable that the use of stopping time criterion as in the right-hand side of (4.2) has strong connections with the dynamic safety constraints and dynamic versions of static CVaR risk measures [27]. Let us illustrate this by using the simplest version of climate change stabilization models discussed in [23].

Assume that $R_t = \sum_{k=0}^{t} x_k$, where decision variables

$$x_k \geq 0, k = 0, 1, \ldots, t, \qquad t \leq T < \infty.$$

We can consider x_k to be a CO_2 emission reduction at the beginning of period k. At time t the target value on total emission reduction R_t in period t is given as a random variable ρ_t. It is assumed that the exact value of ρ_t may be revealed at a random period τ, $P[\tau \geq t] = d_t$. The decision path $x = (x_0, x_1, \dots, x_T)$ has to be chosen ex-ante in period $t = 0$ to mitigate climate change impacts associated with the case $R_\tau < \rho_\tau$. Consider the loss function associated with emission mitigation strategy x and given τ:

$$V(x) = E\sum_{t=0}^{\tau}[c_t x_t + b_t max\{0, \rho_t - R_t\} I_{t=\tau}], \tag{4.6}$$

where deterministic coefficients c_t can be viewed as marginal costs, and b_t as risk factors.

This can be written (Example 4.3) as

$$V(x) = \sum_{t=0}^{T} d_t \left[c_t x_t + b_t Emax \left\{ 0, \rho_t - \sum_{k=0}^{t} x_k \right\} \right].$$

Assume that $V(x)$ is a continuously differentiable function, e.g., a component of random vector $\rho = (\rho_0, \rho_1, \dots, \rho_T)$ has a continuous density function. Also, assume for now that there exists a positive optimal solution $x^* = (x_0^*, x_1^*, \dots, x_T^*)$, $x_t^* > 0$, minimizing $V(x)$ subject to $x_t \geq 0, t = 0, 1, \dots, T$. Then, from the optimality condition for stochastic minimax problems (see discussions in [6], p. 16) it follows that for $x = x^*$,

$$V_{x_t} = c_t - \sum_{k=t}^{T} b_k P\left[\sum_{s=0}^{k} x_s \leq \rho_k\right] = 0, t = 0, 1, \dots, T.$$

From this it follows sequentially for $t = T, T-1, \dots, 0$,

$$P\left[\sum_{k=0}^{T} x_k \leq \rho T\right] = c_T/b_T, P\left[\sum_{k=0}^{t} x_k \leq \rho_t\right] = (c_t - c_{t+1})/b_t, t = 0, 1, \dots, T-1. \tag{4.7}$$

Since $Emax\{0, \rho_t - R_t\} = E\rho_t I_{\rho_t \geq R_t} - R_t P[\rho_t \geq R_t]$, then from (4.7) it follows that $V(x^*) = Ep b_\tau I_{\rho_\tau \geq R_t}$, which can be viewed as a dynamic CVaR (Conditional-Value-at-Risk) risk measure. Equations (4.7) can be used to control dynamic risk profiles, say, profiles with a given safety level γ as in Example 4.5:

$$1 - \gamma = c_T/b_T = (c_t - c_{t+1})/b_t t = 0, 1, \dots, T-1,$$

by appropriate choice of risk factors b_t similar to stationary CVaR risk measures. In this case the minimization of (4.6) controls safety constraints (4.7) with given safety level γ, i.e.,

$$P\left[\sum_{k=0}^{t} x_k \leq \rho_k\right] = 1 - \gamma, \qquad t = 0, 1, \dots, T. \tag{4.8}$$

This is a remarkable result, since the safety constraints, as a rule, are non-convex and even discontinuous, whereas the minimization of function (4.6) is often a convex problem for important practical cases.

Equations (4.7) are derived so far from the existence of the positive optimal solution x^*. The following proposition clarifies this assumption.

Proposition 4.4. *The existence of positive optimal solution follows from* $c_T/d_T < 1, (c_t - c_{t+1})/d_t < 1, t = 0, 1, \ldots, T-1$, *and the monotonicity of quantiles* $\beta_t, \beta_0 < \beta_1 < \ldots < \beta_T$ *defined by equations*

$$P[\beta_T \le \rho_T] = c_T/d_T, P[\beta_t \le \rho_t] = (c_t - c_{t+1})/d_t, t = 0, 1, \ldots, T-1$$

Proof. Indeed, the first requirement guarantees that $x_0^* > 0, \sum_{k=0}^{t} x_0^* > 0$, $t = 1, 2, \ldots, T$. From the second requirement it follows that $x_0^* + x_1^* > x_0^*$, $i.e., x_1^* > 0$, and so on.

Let us note that in general cases outlined in Example 4.5, process R_t is given by stochastic equations $R_{t+1} - R_t = g(t, x_t), t = 0, 1, \ldots, T-1$, where $g(t, x_t)$ is a random function. In this case (4.7), (4.8) would have a form of conditional expectation rather then quantiles. It is even easy to see for $g(t, x_t) = a_t x_t$, where a_t are random variables. In rather general cases a minimization problem (4.6) can be solved by distribution-free stochastic optimization methods proposed in [5,7–9], i.e., methods which don't use (in general) exact probability distributions.

Remark 4.6. (Robust decision). The stopping time τ in model (4.6) is not associated with the violation of safety constraint (4.8). In catastrophic risk management the model (4.6) is usually considered as an auxiliary submodel. For example, if random ρ_t are affected by a set of decisions y with a cost function $F(y)$, then the minimization of function $V(x) + F(y)$ yields robust decision minimizing total costs under safety constraints (4.8) and a dynamic version of the CVaR risk measure.

4.6 Intertemporal Inconsistency

The time consistency of discounting means that the evaluation of an investment project today ($t = 0$), will have the same discount factor as the evaluation of the same project after any time interval $[0, T]$ in the future. In other words, despite delayed implementation of the project we always found ourselves in the same environment. Only geometric or exponential discounting, $d_t = d^t = e^{ln(d)t} = e^{-\lambda t}$, where $\lambda = -ln(d)$, defines a homogeneous time consistent preference:

$$\sum_{t=0}^{\infty} d^t V_t = V_0 + dV_1 + \ldots + d^{T-1} V_{T-1} + d^T [V_T + dV_{T+1} + \ldots]$$

This is also connected with the geometric probability distribution of the discount-related stopping time τ in (4.2): if $P[\tau \geq t] = d^t, 0 < d < 1$, then

$$P[\tau = t] = d^t - d^{t+1} = (1-d)d^t, t = 0, 1, \ldots.$$

In other words, the consistency is the direct consequence of the well-known "memoryless" feature of geometric and exponential probability distributions: for any $t \geq 0, s \geq 0$,

$$P[\tau = t + s \mid \tau \geq t] = d^{t+s}(1-d)/d^t = d^s(1-d)$$

Hence, independently of waiting time t, the probability of the stopping time occurrence at $t + s$ is the same as at the initial time moment $t = 0$.

For other discount factors with time-dependent rates, their time inconsistency requires appropriate adjustments of discount factors for projects undertaken later rather than earlier. The misperception of this inconsistency may provoke increasing vulnerability and catastrophic losses. Let us consider typical scenarios of such developments. Section 4.4 shows that the adequate perception of proper discounting is a challenging task requiring models that allow the explicit evaluation of related risk profiles. This section, in fact, illustrates that the design of such models has to be considered as a key mitigation measure to cope with increasing vulnerability.

A number of authors distinguish between various types of so-called "imperfect altruism" resulting in the lack of social commitment to mitigate risks. For example, there were alluded definitions of a naive, a sophisticated and a committed (ideal) society. The main differences between these three societies and how they provoke catastrophes are summarized in [11] by using a simplified flood management model outlined in Sect. 4.4. This model has the fixed 100-year horizon T in which three societies, the naive, the sophisticated, and the committed, live and plan for coping with the catastrophic losses that may occur due to break of a dyke from 150-year flood with time consistent geometric probability distribution. They are able to mitigate the reliability of dikes and losses by paying fair premiums to the catastrophe fund. But, depending on their perception of risk profiles or induced discounting, the results are dramatically different.

The current generation of *The Naive Society* is aware of a possible catastrophe. It maximizes the (identical for all generations) value function taking into account the potential need to save for paying premiums. Unfortunately, it has a misleading view on the catastrophe, namely, if the catastrophe has not occurred in the later generation the society believes that it will not occur within the current generation with the same probability. Thus, it relies on geometric probability distribution and fails to take into account the time inconsistency induced by increasing the probability of a dike break due to aging processes. Therefore, the first generation of the society postpones the implementation of decisions, i.e., the naive society puts also its preferences on consumption as the first priority consuming at a higher rate than it actually plans.

For the next generation the time is shifted forward by 20 years, and the second generation, similar to the first, plans but does not implement saving actions essential for the catastrophe fund to function. The risk profiles, time preferences, premiums, and the actions are not adjusted towards the real escalating risks. In a similar way, behave the next generations. The plans are never implemented and the view on a catastrophe remains time invariant despite dramatic increase of risk.

The Sophisticated Society implies a correct understanding of the time-inconsistent discounting induced by the deteriorating system of dikes. But this society, similar to the naive planners, also evaluates present consumption to be much higher than the future one. This leads to postponing the decisions. Due to these delays, the risk burden is increasingly shifted to the next generation, calculated premiums become higher and higher. If a catastrophe occurs, this society will also be not prepared to cope with losses as catastrophe management is not functioning.

The "pathologies" of these societies can be explained by their misperception of risks, and, the lack of committed actions.

The Committed Society is similar to that of the sophisticated society. In contrast though, this society is able to implement decisions because its calculations demonstrate that the delays in actions may dramatically affect individuals and the growth of societies as a whole. Individuals could be better off if their consumption options were limited and their choices constrained by anticipating risks. As a direct consequence of the committed actions, the premiums that the society pays for coping with catastrophes in 100 years time are much lower than those of the sophisticated society.

4.7 Concluding Remarks

The proposed new approach to discounting is based on undiscounted stopping-time criterion which is equivalent to the standard discounted criterion in the case of market-related discount factors. In general, the stopping time criterion induces the discounting that depends on spatio-temporal patterns of catastrophes and various relevant decisions. More formally, this paper demonstrates that discount factors $d_t, t = 0, 1, \dots$ can be associated with the occurrence of an extreme ("killing") "stopping time" event at random time τ with probability $P[\tau \geq t] = d_t$. Consequently, the infinite discounted sum $\sum_{t=0}^{\infty} d_t V_t$, $V_t = E v_t$, is replaced by the undiscounted expectation $E \sum_{t=0}^{\tau} v_t$ within the finite interval $[0, \tau]$. The use of the stopping time criterion $E \sum_{t=0}^{\tau} V_t$ induces the standard discounting in the case when τ is associated with the lifetime of market products. In dealing with catastrophic risks, the stopping time τ can be associated with the arrival time of potential catastrophic events. The use of random criterion $\sum_{t=0}^{\tau} v_t$ allows to address the variability of valuations even in the case of deterministic flows $V_0, V_1, \dots$. In this case, it is often important to substitute the expected value of random sum $\sum_{t=0}^{\tau} v_t$ by its quantiles. Mitigation efforts affect the occurrence of extreme events and, thus, they affect discounting, which in turn affects mitigations. This endogeneity of discounting restricts exact evaluations of d_t and the consequent use of deterministic methods and it calls for specific stochastic optimization methods.

Acknowledgements The authors thankfully acknowledge collaboration with several IIASA colleagues who have in various ways helped us in the research that motivated this report. We cannot mention all such contributions. Therefore we explicitly thank only Sten Nilsson and Michael Obersteiner of the Forestry Program for their valuable input to the earlier joint publications.

References

1. Arrow, K.: The theory of risk bearing small and great risks. J. Risk Uncertain. **12**, 103–111 (1996)
2. Arrow, K., Cline, W., Maeler, K., Munasinghe, M., Squitieri, R., Stigliz, J.: Intemporal equity, discounting, and economic efficiency. In: Bruce, J., Lee, H., Haites, E. (eds.) Climate Change 1995: Economic and Social Dimensions, pp. 125–144. Cambridge University Press, Cambridge, (1996)
3. Chichilinsky, G.: What is sustainable development? Land Econ. **73**, 123–135 (1997)
4. Cline, W.: Discounting for the very long term. In: Portney, P., Weyant, J. (eds.) Discounting and Intergenerational Equity, Resources for the Future. RFF, Washington DC (1999)
5. Ermoliev, Y., Ermolieva, T., MacDonald, G., Norkin, V.: Stochastic optimization of insurance portfolios for managing exposure to catastrophic risks. Ann. Oper. Res. **99**, 207–225 (2000)
6. Ermoliev, Y., Hordijk, L.: Facets of robust decisions. In: Marti et al. (eds.) Coping with Uncertainty: Modeling and Policy Issues, *Lecture Notes in Economics and Mathematical Systems*, vol. 581. Springer, Berlin, Heidelberg, New York (2006). ISBN 978-3-540-35258-7
7. Ermoliev, Y., Norkin, V.: Stochastic optimization of risk functions via parametric smoothing. In: Marti, K., Ermoliev, Y., Pflug, G. (eds.) Dynamic Stochastic Optimization, pp. 225–249. Springer, Berlin (2003)
8. Ermolieva, T.: The design of optimal insurance decisions in the presence of catastrophic risks. Interim Report IR-97-068, International Institute for Applied Systems Analysis, Laxenburg, Austria (1997)
9. Ermolieva, T., Ermoliev, Y.: Catastrophic risk management: Flood and seismic risks case studies. In: Wallace, W.Z.S. (ed.) Applications of Stochastic Programming, MPS-SIAM Series on Optimization. SIAM, Philadelphia (2006)
10. Ermolieva, T., Ermoliev, Y., Fischer, G., Galambos, I.: The role of financial instruments in integrated catastrophic flood management. Multinational Fin. J. **7**(3&4), 207–230 (2003)
11. Ermolieva, T., Ermoliev, Y., Hepburn, C., Nilsson, S., Obersteiner, M.: Induced discounting and its implications to catastrophic risk management. Interim Report IR-03-029, International Institute for Applied Systems Analysis, Laxenburg, Austria (2003)
12. Frederick, S., Loewenstein, G., O'Donoghue, T.: Time discounting and time preference: Critical review. J. Econ. Lit. **XL**, 351–401 (2002)
13. Froot, K.: The Limited Financing of Catastrophe Risk: and Overview. Harvard Business School and National Bureau of Economic Research, Harvard (1997)
14. Grollman, T., Simon, S.: Floods: Harbingers of climate change. Risk Transf. **1/3**, 1–9 (2003)
15. Haurie, A.: Integrated assesment modeling for global climate change: An infinite horizon optimization viewpoint. Environ. Model. Assess. **8**, 117–132 (2003)
16. Heal, G., Kriström, B.: Uncertainty and climate change. Environ. Resour. Econ. **22**, 3–39 (2002)
17. Koopmans, T.: On the Concept of Optimal Economic Growth Econometric Approach to Development. Rand Mc Nally, Chicago (1966)
18. Luenberger, D.: Investment Science. Oxford University Press, Oxford (1998)
19. Manne, A.: Equity, efficiency and discouning. In: Portney, P., Weyant, J. (eds.) Discounting and Intergenerational Effects, Resources for the Future, pp. 391–394. RFF, Washington DC (1999)
20. Munich Re: Climate change and increase in loss trend persistence. Press Release, Munich Re (15 March 1999). http://www.munichre.com/default_e.aspx

21. Newel, R., Pizer, W.: Discounting the distance future: How much do uncertain rates increase valuations? Economics Technical Series, Pew Center on Global Climate Change (2000). http//: www.pewclimate.org
22. Nordhaus, W., Boyer, J.: Warming the World: Economic Models of Global Warming. MIT, Cambridge, MA (2001)
23. O'Neill, B., Ermoliev, Y., Ermolieva, T.: Endogenous risks and learning in climate change decision analysis. In: Marti et al. (eds.) Coping with Uncertainty: Modeling and Policy Issues, *Lecture Notes in Economics and Mathematical Systems*, vol. 581. Springer, Berlin, Heidelberg, New York (2006). ISBN 978-3-540-35258-7
24. OXERA: A social time preference rate for use in long-term discounting, pp. 1–74. OXERA, Oxford (2002)
25. Proposal, P.: Flood risk management policy in the upper tisza basin: A system analytical approach. Tech. rep., International Institute for Applied Systems Analysis, Laxenburg, Austria (1997)
26. Ramsey, F.: A mathematical theory of savings. Econ. J. **138**, 543–559 (1928)
27. Rockafellar, T., Ursayev, T.: Optimization of conditional value at risk. J. Risk **2/3**, 21–41 (2000)
28. Sozou, P., Seymour, R.: Augumented discounting: Interaction between ageing and time-preference behavious. Proc. R. Soc. London B **270**, 1047–1053 (2003)
29. Toth, F.: Intergenerational equity and discounting. Integrated Assess. **1**, 127–136 (2000)
30. Walker, G.: Current developments in catastrophe modelling. In: Britton, N., Oliver, J. (eds.) Financial Risks Management for Natural Catastrophes, pp. 17–35. Griffith University, Brisbane, Australia (1997)
31. Weitzman, M.: Keep on discounting. But .... In: Portney, P., Weyant, J. (eds.) Discounting and Intergenerational Equity, Resources for the Future, chap. 3. RFF, Washington DC (1999)

Chapter 5
Cost Effective and Environmentally Safe Emission Trading Under Uncertainty

T. Ermolieva, Y. Ermoliev, G. Fischer, M. Jonas, and M. Makowski

Abstract The aim of this paper is to analyze robust cost-effective and environmentally safe carbon emission trading schemes under uncertainties of emissions and costs, and asymmetric information of participants. The proposed model allows to control explicitly the safety of Kyoto (or other) targets by taking long-term perspectives on emission trading. The dynamics of this scheme is driven by bilateral trades with different endogenous disequilibrium prices between mutually beneficial trades, but finally the system converges to cost-effective and environmentally safe global equilibrium. The safety constraints work as a discounting mechanism that discounts the reported emissions to detectable undershooting levels. This, in turn, provides incentives for participants to reduce uncertainties. The model shows that uncertainties and short term market perspectives may easily prevent price-based trading to be environmentally safe and cost-effective scheme. The desirable equilibrium emerges only under proper price-formation mechanisms. The role of the proposed computerized multi-agent trading system is central for dealing with long-term perspectives, irreversibility and lock-in equilibriums of trades. This system can be viewed as a device for decentralized collective regulation of trades based on unified approaches to modeling of uncertainty, calculation of costs and trading rules.

5.1 Introduction

The public property of large scale pollution makes it impossible to organize complete environmental markets with private demand for and private supply of pollution control [2, 3, 7, 15, 18, 20, 22]. Yet, the idea of carbon trading markets is becoming increasingly popular for global climate change control. At the same time, the existence of various exogenous and endogenous inherent uncertainties raises serious concerns regarding the ability of carbon trading markets to

T. Ermolieva (✉), Y. Ermoliev, G. Fischer, M. Jonas, and M. Makowski
Institute for Applied Systems Analysis, Laxenburg, Austria

K. Marti et al. (eds.), *Coping with Uncertainty*, Lecture Notes in Economics and Mathematical Systems 633, DOI 10.1007/978-3-642-03735-1_5,

fulfill the main purpose of the climate change control without creating world-wide irreversible socio-economic and environmental disruptions. Definitely, that interests of profit oriented markets may contradict the main concerns of the Kyoto agreements [29].

Considered in this paper is bilateral exchange of emission rights. It is assumed that parties with high emission reduction costs buy emissions from parties with low emission reduction costs within prescribed targets. In other words, parties can be engaged in a mutually beneficial bilateral emission exchange process [8] driven by cost minimizing and environmentally safe bilateral trades without the need for a market. This approach is close to important ideas on decentralized non-monetary exchange [21, 28].

In contrast, carbon trading markets, which become increasingly popular in recent years, are more similar to stock markets. Parties hold a number of permits to emit a specific amount of emissions. Parties that cannot to keep their emissions at the given level (called cap) must buy permits on the market at a prevailing market price.

Thus, in the bilateral emission trading scheme the exchange of emission rights is driven by the abatement costs and safety constraints, while in the carbon trading markets the exchange of emissions is driven by prevailing market prices. Such price signals with potential bubbles created by speculators may have no connections with minimization of abatement costs and achieving environmental safety constraints.

There are two principle approaches to control pollution: centralized cooperative command-and-control methods and decentralized market simulating schemes. If there was a social planner (central agency or regulator) fully informed about emissions and abatement cost functions of all parties, the primal problem of finding emission levels that meet given environmental standards in a cost-effective way would be a straightforward task. This could be done by dealing with nonconvex cost functions typically encountered in long-term evaluations involving new technologies with increasing returns. However, without such a planner, the primal model has to be solved in a decentralized manner.

The aim of this paper is to develop an integrated approach for designing cost-effective and environmentally safe decentralized emission trading schemes robust with respect to uncertainties of emissions, costs and asymmetric information of parties. The bilateral emission trading scheme of Sect. 5.5 corresponds to a decentralized solution of the primal model, whereas schemes of Sect. 5.7 simulate decentralized price-based market's solutions. The cost-effectiveness and environmental safety of latter solutions critically depend on proper price signals, which usually reflect instantaneous market situations rather than long-term costs and environmental constraints of the dual model. The complexity of the primal model is a vital issue for the existence of proper prices.

There is a number of uncertainties, affecting outcomes of examined model. First of all, emissions of Green-House Gases (GHGs) are not directly observable. A comprehensive discussion of related issues can be found in the volume by [1], and in [12]. In general, emissions can be estimated with information on the GHG-emitting activities by applying specific conversion factors and from atmospheric measurements using inversion models. The accuracy of the reported emissions depends on the quality of the monitoring system in each specific country and on the accuracy of the conversion factors used [26]. As emissions of GHGs cannot be observed perfectly, the uncertainty can be misused by concealing unreported emissions. A central issue becomes a trade-off between reductions of emissions and uncertainties. For example, carbon prices in the European Union crashed and caused instabilities in late April 2005 after the Czech Republic, Estonia, France, the Netherlands and Sweden reported lower than anticipated emissions [4, 27].

Apart from emissions, another essential uncertainties are those related to the emission amounts and reduction costs. Parties have incentives to keep this information private and the specific costs may remain unknown to the other parties. Besides, they may vary according to unknown market conditions. They are also subject to both industry wide and firm specific shocks.

The novelty of this paper is in integrated analysis of emission trading schemes under various types of natural and human related uncertainties. Section 5.2 illustrates the need for proper treatment of uncertainties by using available historical observations of CO_2 emissions. It shows that the use of uncertainty intervals as practiced by International Panel for Climate Change (IPCC) can leave out of consideration an essential mass of potential emissions. Therefore, Sect. 5.3 introduces a simple and realistic stochastic model allowing to represent both, human related uncertainties and uncertainties associated with the natural variability of emissions. The model focuses on proper representation of potentially controversial experts' judgments and path-dependencies of emissions. As a result, this allows us to introduce safety constraints and undershooting mechanism to control the robustness of emission targets during trading process.

Section 5.4 introduces a basic model allowing to analyze different trading schemes. In particular, the model shows that the trade equilibrium under uncertainty is significantly affected by uncertainty. This emphasizes the need for proposed integrated modeling of uncertainties, safety constraints and emission trading schemes.

The dynamic bilateral trading scheme of Sect. 5.5 can be viewed as a stochastic decomposition procedure. The trade at each step takes place towards minimization of safety-adjusted costs of meeting parties. This generates disequilibrium random prices which are endogenously driven towards the cost-effective and environmentally safe equilibrium price. This section analyses also difficulties involved in setting up such an equilibrium price in monetary trading schemes. Standard market models usually imply (e.g., by an arbitrage free type of assumption) that markets operate under equilibrium prices. Section 5.6 outlines a computerized Multi-Agent Decentralized Trading System dealing with the irreversibility of emission trades. Section 5.7 analyses path-dependencies of myopic trading schemes relying on instantaneous markets situations. It shows that short-term market perspectives preclude achieving desirable long-term emission reduction goals. Section 5.8 concludes and outlines important numerical results. The Appendix provides a proof of the convergence. It also discusses stable core solutions of bilateral emission trading scheme.

5.2 Uncertainties and Trends of Carbon Fluxes

This section illustrates a general need for proper representation of emission uncertainties. Next section addresses these issues in a more specific context of emission trading.

Uncertainties of emissions are often represented by means of intervals. In reality, emissions may have different likelihoods within these intervals, i.e., rather general skewed probability distributions. In this case, the use of uncertainty intervals can leave out of consideration essential patterns of emission changes as in Figs. 5.2 and 5.3.

Figures 5.1, 5.2, and 5.3 illustrate trends and natural variability of net carbon fluxes on the global scale http://lgmacweb.env.uea.ac.uk/lequere/co2/carbon.budget.htm. The global carbon budget is composed of the fossil fuel emissions, the emissions stemming from land use, the ocean uptake, and the terrestrial uptake estimated as a residual of all the sources minus the ocean uptake and atmosphere increase.

Figures 5.2 and 5.3 show the dynamics of changes in emissions and emissions uncertainties. The histogram in Fig. 5.2 is skewed to the left. In the next study period, Fig. 5.3, the situation changes: more values are concentrated on the right hand side. Between these two periods, the system turns from sink to source of CO_2. Definitely, it is impossible to characterize these changes only by uncertainty intervals.

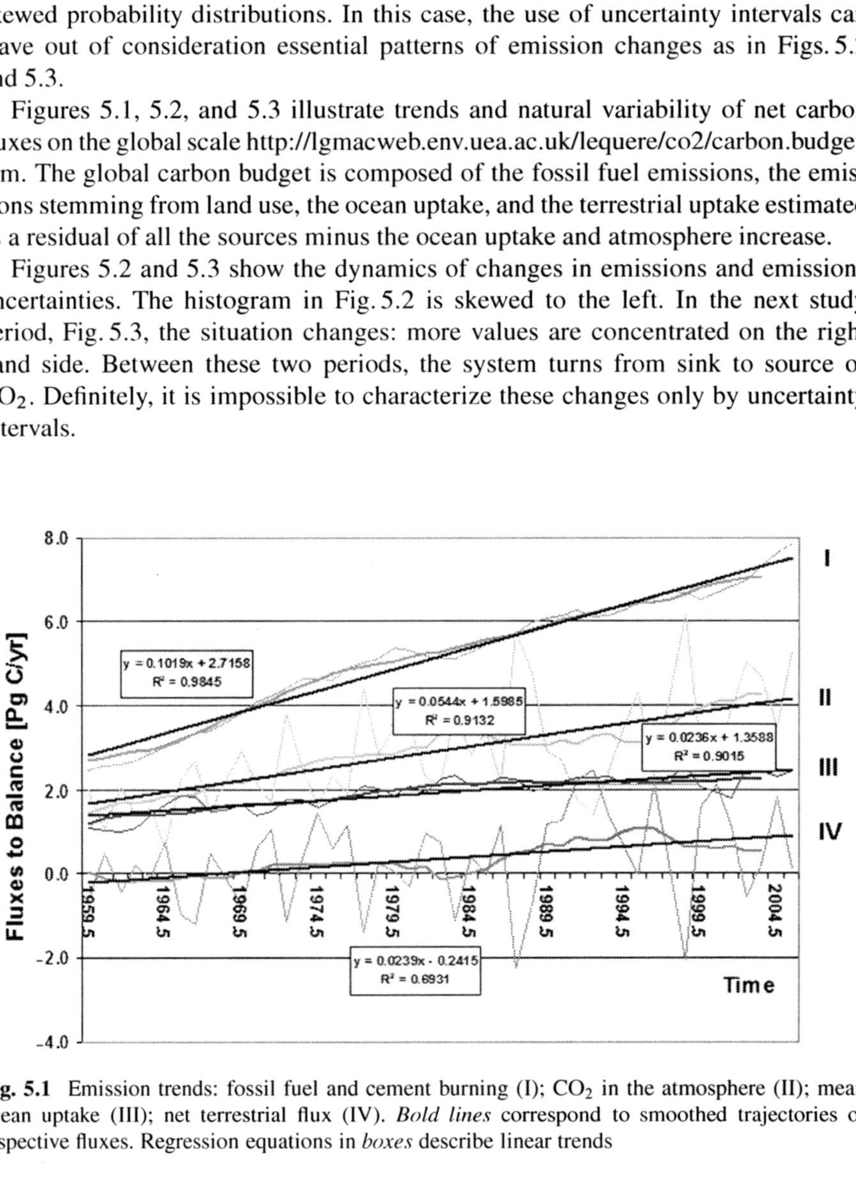

Fig. 5.1 Emission trends: fossil fuel and cement burning (I); CO_2 in the atmosphere (II); mean ocean uptake (III); net terrestrial flux (IV). *Bold lines* correspond to smoothed trajectories of respective fluxes. Regression equations in *boxes* describe linear trends

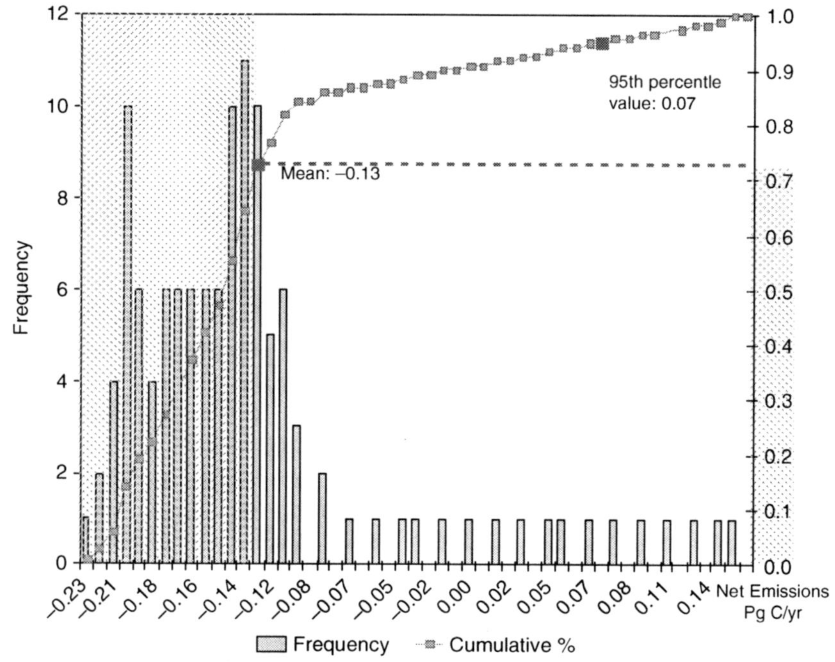

Fig. 5.2 Global CO_2 net terrestrial uptake, 1960–1970

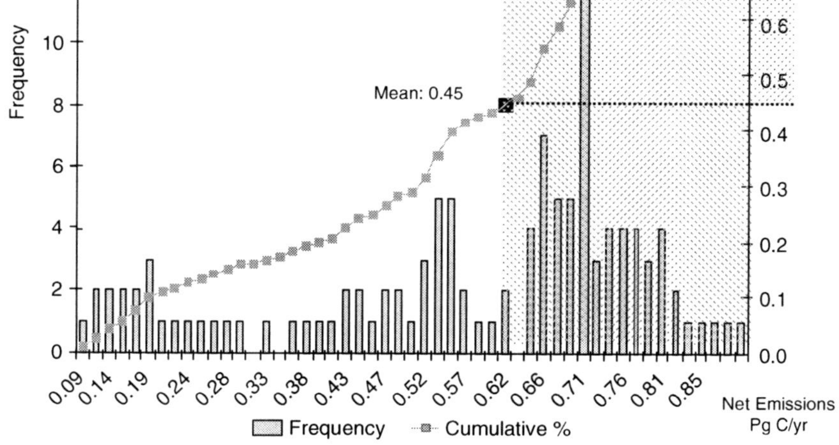

Fig. 5.3 Global CO_2 net terrestrial uptake, 1985–1995

5.3 Detectability of Emission Changes

A simple way to introduce the detectability of emission changes can be based on a straightforward representation of emission trends and uncertainties by equally-sided intervals as in Fig. 5.4 (see [1, 19]). The main idea is illustrated in Fig. 5.4. Let us assume that uncertainty of emission e_1 in the base year t_1 is characterized by equally sided interval $[e_1 - \epsilon, e_1 + \epsilon]$. The uncertainty of reported emission e_2 in the commitment year t_2 ($t_1 < t_2$) is characterized by the same type of interval $[e_2 - \epsilon, e_2 + \epsilon]$. We assume that $e_1 > e_2$, although the case $e_1 < e_2$ is also possible, e.g., as a result of emission trading. The detectability of emission changes requires that the change in net carbon emissions $\Delta e = e_1 - e_2$ at time t_2 is greater than the uncertainty in the reported net carbon emissions at time t_2.

Under the non-restrictive assumption that the first-order linear approximations for emissions (as in Fig. 5.1), $e(t)$ and uncertainties $\epsilon(t)$ trends are applicable for $t_1 \leq t \leq t_2$, the detection time t^* is defined as the first time moment when net emission change Δe outstrips the uncertainty interval. In a sense, this is a worst-case evaluation. As in Figs. 5.2 and 5.3, considerable mass of real emissions can be concentrated in a much smaller subinterval. For example, the uncertainty interval of random variable with a normal probability distribution is $(-\infty, \infty)$, whereas practically entire probability mass may be concentrated within $[-1, 1]$ interval. Therefore, by using stochastic uncertainty models it is possible to derive with high probability a more optimistic t^*. This is the main issue of stochastic models discussed in [16, 17]. An overview of different approaches can be found in [1]. The goal of the stochastic models is to rank the trading parties by a safety indicator showing the percentage of detectable emission changes within a given time interval.

Let us consider a rather general stochastic model for a representation of a controversial data about uncertainties. We assume that the uncertainty of emissions e_1,

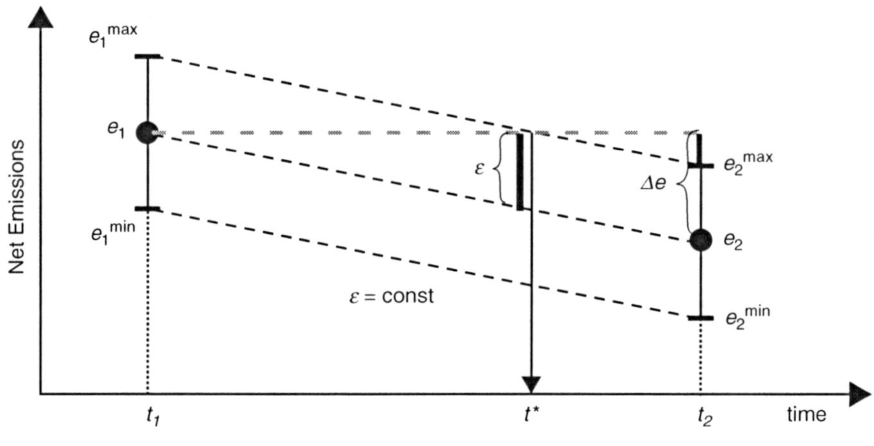

Fig. 5.4 Simplified illustration of detection time t^*

e_2 is characterized by a set of, in general, disconnected intervals. For example, Figs. 5.2 and 5.3 may suggest to represent uncertainty by a number of subintervals characterized by simple, say, uniform (conditional) probability distributions.

The following simple example illustrates the main idea of the model.

Example 5.1 (Controversial experts). Experts judgments are used in situations with the lack or even absence of real observations. Assume that two experts, Ex.1 and Ex.2, characterize the uncertainty of e_1 by overlapping intervals $[1, 8]$, $[5, 10]$. Then overall uncertainty of e_1 can be characterized by intervals $[1, 5]$, $[5, 8]$, $[8, 10]$ with likelihoods $1/4, 2/4, 1/4$ derived from the "voting" of experts: $(1, 0)$, $(1, 1)$, $(0, 1)$, i.e., for interval $[1, 5]$ votes only Ex.1, both experts vote for interval $[5, 8]$, and only Ex.2 votes for $[8, 10]$. In general, experts may characterize uncertainty by disconnected intervals. For example, Ex.1 may insist on equally probable intervals $[1, 3]$, $[5, 8]$ conditional on implementation of different technologies. Uncertainty of e_2 can be characterized in a similar manner.

Consider now a general model. For simplicity of notation we omit the index i of parties. Assume that (specific for each party i) uncertain emission e_1 is characterized by intervals $[e_1^{\min}, e_1^1]$, $[e_1^2, e_1^3]$, $[e_1^4, e_1^5], \ldots, [e_1^R, e_1^{\max}]$ with probabilities p_r, $\sum_r p_r = 1$. These intervals can be derived from real observations, experts opinions, and scenarios of future developments. In addition, likelihoods of emissions within an interval can be characterized by a conditional on r distribution, say, uniform, normal or the degenerated distribution concentrated in the middle of this interval as in Fig. 5.4. In a similar manner, emissions e_2 are characterized by intervals $[e_2^{\min}, e_2^1]$, $[e_2^2, e_2^3]$, $[e_2^4, e_2^5], \ldots, [e_2^L, e_2^{\max}]$ with some conditional on r and l distributions. Path-dependencies between emissions e_1 and e_2 are induced by the following stochastic model. An interval r at the base year t_1 is selected with the probability p_r, emission level e_1 is sampled from the conditional distribution in this interval; an interval l of a trend from r to l is selected with probability q_{rl}, $\sum_l q_{rl} = 1$, and finally, the end point e_2 of the random linear path (e_1, e_2) is sampled from the distribution in interval l conditional on r and l. Let us denote the obtained linear random path by $e(t, \omega)$, $t_1 \leq t \leq t_2$, where ω denotes the pair of points (e_1, e_2).

Linear paths $e(t, \omega)$ create the uncertainty ranges at t_2. For example, if $e(t_1, \omega)$ belongs to interval r, then the uncertainty of e_2 is defined on the basis of only feasible transitions from interval r to random intervals l with positive q_{rl}, $q_{rl} > 0$.

The proposed stochastic model allows to introduce path-dependencies of emissions subject to some essential conditions, say, the implementation of new emission reduction technologies or monitoring equipment. This may simplify the detection of emission changes. Exact detection is in general a difficult task because the resolution of all involved uncertainties may be prohibitively costly. Yet, it is possible to define likelihoods of changes. For example, it is possible to find a minimal time t_2 such that emission changes are detected during $[t_1, t_2]$ with a specified likelihood. It is also possible to find the likelihood of the changes within given interval $[t_1, t_2]$ that is used in the next section.

Remark 5.1 (Modifications of model). The proposed stochastic model can be further generalized or simplified subject to available data. Straight lines of linear emission paths can be theoretically substituted by more general stochastic paths (processes) or scenario trees, although the proposed linear stochastic trends allow simple calculations. If path-dependencies of emissions are not essential, then the model directly deals only with uncertainties of e_2. In more general situations, uncertainties are also characterized by a set of probability distributions, i.e., there is a set of feasible p_r, q_{rl}. There exist different approaches to deal with arising "uncertainty-of-uncertainty" issues, in particular, the use of non-Bayesian worst-case distributions [6].

Comparative analysis of deterministic and stochastic simple detection models can be found at http://www.iiasa.ac.at/Research/FOR/unc-prep.html,FOR/vt-concept.html, and in [10].

5.4 Trade Equilibrium Under Uncertainty

GHG's control policy as other environmental policies have to be designed in such a way that they are environmentally safe and cost-effective. The models proposed in this section provide a basis for designing rather different decentralized emission trading schemes.

The models reflect the following key features. The participants (countries, companies or other emitting entities) are given a right to emit a specific amount for which they obtain an equivalent number of allowances (emission permits). Such amounts are called the "cap" (Kyoto or other targets). If participants emit more than the corresponding cap reduced by the amount of uncertainty (undershooting level) ensuring that the actual emission does not exceed the cap with a given safety (likelihood) level, they are required to reduce uncertainty or/and to buy additional credits from the parties which emit less than their cap. The transfer of permits is called "trading". Standard deterministic models belong to a specific class of the proposed models. Since they ignore uncertainty, actual emissions may considerably overshoot allowed targets.

Let us briefly consider a deterministic model with uncertainty intervals proposed in [13, 14], that will be further extended to include stochastic safety constraints. The decision problem of each party can be separated in two interdependent subproblems. Firstly, for a given amount of permits, each party solves individual problem deciding whether to spend resources on abating emissions or investing in uncertainty reduction to satisfy emission targets. This problem does not require the information from any other party. Secondly, the party needs to decide whether or not to exchange permits with other parties. This decision problem involves the cost functions of other parties. In the model this information is private and therefore the methodology of decentralized optimization [8, 10] is required.

For the individual optimization problem, we define the least costs $f_i(y_i)$ for party i to comply with imposed targets for a given amount of permits y_i and the target K_i as the minimum of emission reduction costs $c_i(x_i)$ and costs of uncertainty reduction (e.g., by investments in monitoring) $d_i(u_i)$:

$$f_i(y_i) = \min_{u_i, x_i} [c_i(x_i) + d_i(u_i)], \tag{5.1}$$

$$x_i + u_i \leq K_i + y_i, x_i \geq 0, u_i \geq 0, \tag{5.2}$$

for all i, where x_i is the reported emissions at source i, u_i is its uncertainty, and y_i is the amount of emission permits acquired by source i (y_i is negative if i is a net supplier of permits). Therefore, constraint (5.2) requires that the reported emission x_i undershoots the target K_i by the level of uncertainty u_i. Similar concept of undershooting is also used in [24]. There are also suggestions [23] to represent uncertainty by a fraction of x_i. Example 5.2 shows that this case can be reduced to the case of additive uncertainty as in (5.2). Let us also note, that the model can be formulated in terms of emission reductions that require only slight changes of terminology.

The second optimization problem with asymmetric information involves finding the permit vector $y = (y_1, \ldots, y_n)$ or distribution of permits minimizing unknown total or social cost function

$$F(y) = \sum_{i=1}^{n} f_i(y_i) \tag{5.3}$$

subject to

$$\sum_{i=1}^{n} y_i = 0. \tag{5.4}$$

Suppose that the cost functions $c_i(x_i)$ and $d_i(u_i)$ are positive, decreasing, convex in x_i and u_i respectively and continuously differentiable. Therefore, $f_i(y_i)$ is also convex, positive, decreasing and differentiable. Then, from the Lagrangian minimization a trade equilibrium can be defined as the vector $y = (y_1, \ldots, y_n)$ satisfying the following equations:

$$f_i'(y_i) = -\lambda, \qquad \sum_{i=1}^{n} y_i = 0. \tag{5.5}$$

The condition (5.5) states that the marginal value of a permit shall in equilibrium be equal to a specific unknown level (price) λ same for all parties. It is clear that at the equilibrium vector y^* the constraints (5.2) will hold with equality, i.e.,

$$f_i(y) = \max_{x_i}[c_i(x_i) + d_i(K_i + y_i - x_i)] = \max_{u_i}[c_i(K_i + y_i - u_i) + d_i(u_i)].$$

Therefore from (5.1), (5.2) it follows that at the equilibrium $y_i = y_i^*$, $\lambda = \lambda^*$, $x_i = x_i^*$, $u_i = u_i^*$:

$$c_i'(x_i) = d_i'(u_i) = -\lambda, \qquad \sum_{i=1}^{n} y_i = 0, \tag{5.6}$$

where (x_i^*, u_i^*) is the solution of the subproblem (5.1), (5.2) for (y^*, λ^*), $y^* = (y_1^*, \ldots, y_n^*)$ satisfying (5.5). This equation states that in the cost-effective and environmentally safe equilibrium, the marginal cost of holding emissions down to x_i^* will be equal to the marginal cost of holding uncertainty down to u_i^*. It shows that the explicit introduction of uncertainty u_i and the safety constraints (5.2) into emission trading schemes may significantly affect the equilibrium and, hence, the design of proper emission trading schemes. In particular, it means that equilibrium market prices λ^* must satisfy (5.5), (5.6). In other words, if λ^* is known and f_i are convex functions, then λ^* decentralizes joint model (5.3), (5.4) into individual solutions of (5.5). For non-convex function $F(y)$, (5.5) are not sufficient to find an equilibrium solution of the model.

However, there is no social planner that knows the cost functions of all parties. Therefore, even in the convex case the optimal value of $F(y)$ and λ^* cannot be resolved by solving (5.5). The scheme of bilateral trade presented in Sect. 5.5 allows to compute the equilibrium x_i^*, u_i^*, λ_i^* without revealing private information on functions f_i.

Remark 5.2 (Long-term perspectives, detectability and undershooting). The basic model can be easily extended to a dynamic version. In this article we do not consider it explicitly. Instead, we introduce below long-term perspectives by explicit treatment of future uncertainties and dynamic trading processes. The environmental constraint (5.2) assumes that the known emissions plus the uncertainty of emissions undershoot the emission target. This corresponds exactly to the detectability concept in Fig. 5.4. Constraint (5.2) discounts, in a sense, the reported emissions to levels undershooting emission targets. As (5.6) show, this provides incentives for the uncertainty reduction.

In the stochastic model of Sect. 5.3 uncertainty of emissions by party i at the commitment year t_2 is characterized by a random variable $e(t_2, \omega_i)$. A reported emission x_i provides additional information that modifies $e(t_2, \omega)$. For example, if it is known for sure that x_i belongs to an interval l_i, then distribution of $e_i(t_2, \omega_i)$ is induced only by feasible transitions from initial intervals r_i to l_i with corresponding probability distributions. We can also say that reported emission x_i transforms $e_i(t_2, \omega_i)$ into a random variable $\epsilon_i(x_i, \omega)$. Therefore, (5.2) of the deterministic model has to be understood now in a probabilistic sense as the following safety constraint. Let us define the uncertainty of reported emission x_i as $\xi_i(x_i, \omega) = \epsilon_i(t_2, x_i, \omega) - x_i$. Then the safety constraint can be written as probabilistic version of the deterministic constraint (5.2):

$$P[x_i + \xi_i(x_i, \omega) \le K_i + y_i] \ge Q_i, \tag{5.7}$$

for all parties i, where Q_i is a safety level ensuring that the probability of all potential emission paths to x_i satisfying the emission target K_i exceeds Q_i. Thus the interval uncertainty u_i is substituted by a random variable $\xi_i(x_i, \omega)$ dependent, in general, on x_i. In reality, the uncertainty characterized by ξ_i can be reduced by improvements of monitoring systems. Let us introduce the variable u_i to control ξ_i

within the desirable safety level Q_i. If $z_i(x_i)$ is the minimal z such that

$$P[\xi_i(x_i, \omega_i) \leq z] \geq Q_i,$$

then the safety constraint (5.7) can be substituted by the following equivalent constraint

$$x_i + u_i \leq K_i + y_i, u_i \leq z_i(x_i), \tag{5.8}$$

Remark 5.3 (Risk-based undershooting). Equation (5.8) shows that the stochastic model induces risk-based upper bounds on uncertainty intervals. Therefore, it allows, e.g., in cases illustrated by Figs. 5.3 and 5.4 to introduce but risk-based undershooting of emission targets defined by "critical" quantile $z_i(x_i)$, which is less conservative than standard interval uncertainty.

For the simplicity of notation, let us denote now by ω the vector of all uncertain parameters affecting cost functions and emissions, i.e., some components of ω such as market prices affect only cost functions, whereas other components affect emissions. In other words, suppose that all uncertain variables are defined on a probability space with a set of scenarios (events) ω. For random cost functions $c_i(x_i, \omega)$, $d_i(u_i, \omega)$, we can redefine functions $f_i(y_i)$ in (5.1) as

$$f_i(y_i) = \min_{x_i u_i} E[c_i(x_i, \omega) + d_i(u_i, \omega)], \tag{5.9}$$

where the minimization in (5.9) is subject to constraint (5.8). In this model ex-ante decisions x_i, u_i take a long-term perspective: they have to be optimal against all potential future scenarios ω and threats regulated by safety constraints. Uncertainties of cost functions c_i, d_i may be due to unknown in advance market performance, production shocks, and technological uncertainties.

Example 5.2 (Linear equivalent). Often, $\xi_i(x_i, \omega)$ is represented as $\xi_i(x_i, \omega) = \gamma_i x_i + \epsilon_i$, where $0 \leq \gamma_i \leq 1$, and ϵ_i is a random variable. In particular, uncertainty u_i in (5.2) can be given as $u_i = \gamma_i x_i$. The uncertainty in these cases can be controlled by γ_i in the following manner. Let $\epsilon_i(Q)$ be the minimal z such that $P[\epsilon_i \leq z] \geq Q_i$, e.g., $\epsilon_i(Q) = 0$ for constraints (5.2). Then constraint (5.8) is reduced to linear constraint

$$x_i + u_i \leq K_i - \epsilon_i(Q_i) + y_i, u_i \leq x_i.$$

After solving individual subproblem subject to this constraint, the optimal γ_i can be found as $\gamma = u_i/x_i$.

Remark 5.4 (Nonconvexity). Safety constraints (5.7) are well known in financial applications as the Value-at-Risk indicator. Similar constraints are typical for safety regulation of insurance companies, nuclear power plants, and catastrophic risk management [9]. Unfortunately, due to these constraints $f_i(y_i)$ and $F(x)$ may not be a convex function. In order to ensure convexity and/or robustness of decisions x_i, u_i

under rare extreme events it is possible to modify slightly cost functions c_i, d_i based on the "Conditional Value at Risk" [25] indicators as it is used in [9] and in [6] for general stochastic optimization problems.

Additional nonconvexities of functions $f_i(y_i)$ may be due to increasing returns of cost functions $c_i(\cdot)$, $d_i(\cdot)$, with respect to new emission abatement technologies with increasing returns. Nonconvexities generate the so-called duality gap between solutions of the basic model (5.3), (5.4) and its dual model (see Sect. 5.6) precluding the price-based dual (market) schemes to achieve cost-effective and environmentally safe solutions.

5.5 Dynamic Bilateral Trading Processes

The overall goal of the parties participating in the emission trading is to jointly achieve emission targets by redistributing the emission permits y_i, i.e., to find a vector y that would minimize social costs of all parties (5.3) under safety constraints (5.8), where cost functions $f_i(y_i)$ are defined according to (5.9). It is assumed that a party i knows its expected cost function $f_i(y_i)$, but the expected cost function $F(y)$ is unknown.

The basic feature of the trading scheme is similar to the procedure in [8] for convex function $F(y)$: two parties meet (e.g., picked at random) and, if possible, exchange emission permits in a mutually beneficial way. A new pair is picked and the procedure is repeated. The Appendix provides the proof that this dynamic process will lead the parties to an equilibrium despite the information of each party's cost is private and $F(y)$ is not necessarily a convex function.

The following simple equations illustrate that the bilateral exchange of emissions is beneficial for both parties. Let $y^k = (y_1^k, \ldots, y_n^k)$ be the vector of emission permits after k trades. Consider two parties i and j at step k with permits y_i^k and y_j^k. An exchange of permits between them leads to a new distribution of permits $y^{k+1} = (y_1^{k+1}, \ldots, y_n^{k+1})$, $y_l^{k+1} = y_l^k$ for $l \neq i, j$. According to (5.5), if there exist any two parties i and j having different marginal costs on emission reduction $f_i'(y_i^k) \neq f_j'(y_j^k)$, then the permit vector $y^k = (y_1^k, \ldots, y_n^k)$ is not cost efficient. Without loss of generality, assume that $f_i'(y_i^k) - f_j'(y_j^k) < 0$. Constraint (5.4) requires that the feasible exchange in permits has to be such that $y_i^{k+1} + y_j^{k+1} = y_i^k + y_j^k$.

If we take $y_i^{k+1} = y_i^k + \Delta_k$ and $y_j^{k+1} = y_j^k - \Delta_k$, $\Delta_k > 0$, then the new feasible distribution of permits reduces the total costs of parties $f_i(y_i^k) + f_j(y_j^k)$ and hence the total cost $F(y^k)$:

$$
\begin{aligned}
F(y^{k+1}) - F(y^k) &= f_i(y_i^{k+1}) + f_j(y_j^{k+1}) - f_i(y_i^k) - f_j(y_j^k) \\
&= \Delta_k(f_i'(y_i^k) - f_j'(y_j^k)) + o(\Delta_k) < 0,
\end{aligned}
$$

for a small Δ_k. We also have

$$f_i(y_i^{k+1}) - f_i(y_i^k) < f_j(y_j^k) - f_j(y_j^{k+1}). \tag{5.10}$$

i.e., the new distribution of permits reduces costs of j more than increases cost of i. Hence j is able to compensate i for the increased costs in a mutually beneficial way.

Let us summarize the trade scheme more precisely. We assume that after picking up (say at random) a pair of parties i, j these parties are able to find y_i^{k+1}, y_j^{k+1} minimizing

$$f_i(y_i) + f_j(y_j) \tag{5.11}$$

subject to constraints $y_i + y_j = y_i^k + y_j^k$, $y_i \geq 0$, $y_j \geq 0$.

This problem is solved by parties i and j only. For continuously differentiable functions $f_i(y_i)$, $f_j(y_j)$, a party j that decreases emission permit by $\Delta_k > 0$ may negotiate with i such a level Δ_k that equalizes marginal costs, i.e., $f_i^{'}(y_i^k - \Delta_k) = f_j^{'}(y_j^k + \Delta_k) = \lambda_k$, where λ_k can be viewed as an equilibrium price (in general stochastic) at step k. Let us note that price process λ_k is driven endogenously by cost-minimizing decisions of meeting parties, what is fundamentally different from standard models of financial markets with exogenously given price processes.

The sequential bilateral trades can go on as long as there are two parties with different marginal costs. The bilateral exchange of emissions equalizes marginal costs which define an intermediate "local" equilibrium price λ_k. During the process, marginal costs and prices will differ between the sequential trades, but finally the trading system converges to an equilibrium with marginal costs of all parties equal to equilibrium price as in (5.5).

It is important to compare the outlined bilateral trading scheme with a basic price-based scheme. A cost-effective and environmentally safe price signal is a solution of the dual model to the basic primal model (5.3)–(5.4). It involves finding the price λ maximizing the following concave and, in general, non-differentiable function

$$\phi(\lambda) = \min_y \sum_{i=1}^n (f_i(y_i) + \lambda y_i).$$

A price signal λ decentralizes the solution of internal minimization problem into individual subproblems: find solutions $y_i(\lambda)$ minimizing functions $f_i(y_i) + \lambda y_i$. In general, solutions $y_i(\lambda)$ do not satisfy the balance equation, i.e., $\sum_{i=1}^n y_i(\lambda) \neq 0$, therefore the price λ has to be adjusted towards the desirable balance. The common idea is to change current λ_k at time $k = 0, 1, \ldots$ proportionally to the imbalance, i.e., $\phi^{'}(\lambda)$ for continuously differentiable $\phi(\lambda)$:

$$\lambda_{k+1} = \lambda_k + \rho_k \sum_{i=1}^n y_i(\lambda_k)$$

with a small step-size ρ_k. From the convergence results of quasi-gradient methods (see, e.g., discussion in [5]) it follows that with $\rho_k = const/k$, the sequence λ_k

converges to a price $\bar{\lambda}$ maximizing $\phi(\lambda)$. If $y_i(\lambda)$ are unique solutions (for any λ and $i = \overline{1,n}$)), then $\phi(\lambda)$ is continuously differentiable function and $\bar{\lambda}$ fulfills the balance equation ($\sum_{i=1}^{n} y_i(\bar{\lambda}) = 0$) independently of the convexity $f_i(y_i)$. Otherwise, additional coordinated search is required to select $\overline{y_i}$ from the set of solutions $y_i(\bar{\lambda})$ in order to guarantee (achieve) the balance $\sum_{i=1}^{n} y_i(\bar{\lambda}) = 0$. The coordination of parties is also required for tracking values of imbalances $\sum_{i=1}^{n} y_i(\lambda_k)$ for adjusting prices λ_k, $k = 0, 1, \ldots$. Fundamental difficulties arise in the case of markets uncertainties (see Sect. 5.7) and duality gap, i.e., when $\max \phi(\lambda) < \min_y \{F(y), \sum_{i=1}^{n} y_i = 0\}$ for nonconvex function $F(y)$.

5.6 Computerized Multi-agent Decentralized Trading System

The proposed perfect market system implies that trades being bilateral, sequential (dynamic) and random do not impair the cost savings even if parties only have information on their own cost. However, there are essential obstacles that can inhibit real markets from perfect functioning according to proposed procedure. In a perfect market, a party that has sold permits in an early stage of the trading process would be able to cancel its earlier transaction. In the real emission trading market, this type of counter-actions may be impossible due to irreversibility of decisions: investments may already have been made, and these investment costs are largely sunk costs. This is the fundamental obstacle involved in the design of cost-minimizing and environmentally safe emission trading markets.

Price-based trading schemes have additional inherent obstacles. Designing environmentally safe and cost-effective price-based emission trading markets is equivalent to solving of the dual model asking for the same full information as the solution of the primary model. Additional critical limitation is the duality gap which occurs in nonconvex cases and uncertainty of market prices. The available computer technology and numerically stable optimization procedures allow to organize computerized (say, web-based) multi-agent decentralized trading system to resolve these issues.

One can imagine a distributed computer network that connects computers of parties with the computer of a central agency. The party in an anonymous manner stores information on its specific cost functions, and other characteristics of the underlying optimization model (5.8), (5.9) including specific probability distributions. The central agency stores information on the emission detection model. The computer of the central agency generates a pair of parties i, j and in an anonymous manner negotiates with computers of these partners a proper Δ_k that solves the subproblem (5.11). This can be easily organized without revealing private information of the parties. The process is repeated until equilibrium levels have been reached. This procedure allows to discover equilibrium solution that can then be implemented in reality. The information about the equilibrium price λ^* allows also to identify so-called core solution defining stable coalition of parties (see the Appendix). A network of interconnected computers is essential for a rapid, smooth and robust functioning of emission trading market. There would also be a clear separation between a first stage, in which provisional bids are made between computers of parties and reconstructing is allowed, and a second stage, when contracts have been concluded and investments in emission control are implemented.

It is well known [2] that generally the market does not generate desirable outcomes if market prices fail to reflect socio-economic and environmental impacts. In this case it is typically necessary to establish negotiation processes between involved parties to determine desirable collective solutions. From this perspective, the proposed trading system can be viewed as a device for collective negotiations and decision-making in the presence of inherent uncertainties and irreversibilities.

5.7 Myopic Market Processes

The basic model (5.3), (5.4), (5.8), (5.9) takes long-term perspectives on emission permit trading. Parties use expectations and safety constraints in order to achieve cost-effective and environmentally safe outcomes robust against future developments. The resulting trading scheme is similar to non-monetary exchange economy [28] important for environmental control. There are no demand and supply functions. Instead, the safety constraints enforce parties to invest in emission and uncertainty reductions and consequently act as supplier of mutually beneficial emission permits until a global equilibrium emerges.

The situation becomes dramatically different in the case of price-based schemes under markets uncertainties affecting cost functions c_i, d_i of parties. The short term market perspectives orient parties on instantaneous information about prices and costs. At time interval k parties observe market-related components of uncertainty ω_k and thus know their instantaneous cost functions $c_i(x_i, \omega_k)$, $d_i(x_i, \omega_k)$. Based on this information, parties calculate cost functions

$$f_i(y_i, \omega_k) = \min_{u_i, x_i}[c_i(x_i, \omega_k) + d_i(u_i, \omega_k)], \tag{5.12}$$

subject to the safety constraints (5.8) conditional on observable uncertainties. They minimize then

$$\sum_{i=1}^{n} f_i(y_i, \omega_k) \tag{5.13}$$

subject to

$$\sum_{i=1}^{n} y_i = 0$$

by using observed price signals π_k, $k = 1, 2, \ldots$, which separates joint model (5.13) into independent individual minimization of cost functions

$$f_i(y_i, \omega_k) + \pi_k y_i$$

by parties $i = 1, \ldots, n$, where $\pi_k y_i$ is the cost of buying ($y_i > 0$) or selling ($y_i < 0$) emission permits y_i. If π_k coincides with an equilibrium spot price, then solutions $y_i(\omega_k)$, $i = 1, \ldots, n$, of these individual models may coincide (Sect. 5.5) with the solution of joint model (5.13). In particular, they may satisfy the balance equation $\sum_{i=1}^{n} y_i = 0$. Otherwise, market prices π_k may cause disruptions of the balance and crashes of prices similar to the European carbon prices in April 2005. The following example illustrates a typical situation.

Example 5.3 (Market's uncertainty). Suppose that $f_i(y_i)$ are known deterministic functions, i.e., only prices are random. At time interval k there is a favorable situation for the exchange of emission permits for some parties i and j, e.g., $f_i'(y_i^k) \neq f_j'(y_j^k)$. Instead, high market price π_k, $\pi_k > \left|f_j'(y_j^k)\right| > \left|f_i'(y_i^k)\right|$, forces both parties to reduce emissions in the excess of targets in order to sell surpluses on the market. Disequilibrium price π_k creates an oversupply of emission permits that pushes the market price π_{k+1} towards 0. This may prevent to sell reduced emissions which turned to be of higher marginal costs with respect to new prices.

The myopic model (5.12)–(5.13) yields decisions $x_i(k, \omega_k)$, $u_i(k, \omega_k)$, $y_i(k, \omega_k)$, $k = 1, 2, \ldots$. Such decisions depend on case-specific realizations of the random variable ω_k, therefore are not robust. At time interval $k + 1$ new observation ω_{k+1} may contradict ω_k requiring significant revisions of these decisions, which may be impossible due to their irreversibility. In order to achieve a convergence, the parties must adopt a precautionary incremental behavior with respect to arriving new information. Let us consider first this type of trading scheme for the basic model (5.3)–(5.4), (5.9).

The bilateral dynamic trading process of Sect. 5.5 has deep roots in the structure of so-called stochastic gradients in the linear subspace defined by (5.4). Namely, it is easy to prove that the vector

$$g(y) = (f_1'(y_1) - \frac{1}{n}\sum_{j=1}^{n} f_j'(y_j), \ldots, f_n'(y_n) - \frac{1}{n}\sum_{j=1}^{n} f_j'(y_j))$$

is the projection of $gradF(y) = (f_1'(y_1), \ldots, f_n'(y_n))$ in this subspace. A stochastic gradient then can be defined as the following. Pick up at random a pair (i, j) and define stochastic vector

$$\xi(y) = \frac{(n-1)}{2}(0, \ldots, 0, f_i'(y_i) - f_j'(y_j), 0, \ldots, 0, f_j'(y_j) - f_i'(y_i), 0, \ldots, 0),$$

i.e., $\xi(y) = (0, \ldots, 0)$ for $i = j$. This vector is a stochastic gradient of $F(y)$ [5], i.e., the conditional expectation $E[\xi(y)|y] = g(y)$, assuming that pairs (i, j) of distinct parties are chosen with equal probability $1/n(n-1)$. Therefore, instead of complete minimization of function (5.11) at step k, parties can move from y^k in the random direction $\xi(y^k)$ with a small step size α_k. This type of stochastic decentralized optimization processes are important in cases when functions $f_i(y_i)$

are not calculated exactly, e.g., they are affected by unknown random variables ω as in (5.12). It is easy to check that

$$F(y^k - \alpha_k \xi(y^k)) = F(y^k) - \alpha_k [f'_{i_k}(y^k_{i_k}) - f'_{j_k}(y^k_{j_k})]^2 + o(\alpha_k), \qquad (5.14)$$

where (i_k, j_k) is picked up at step k random pair (i, j). Therefore, sequential change of emissions defined by equations

$$y^{k+1} = y^k - \alpha_k \xi(y^k) \qquad (5.15)$$

produces monotonically decreasing for small α_k (contrary to standard stochastic gradient methods [5] random sequence $\{y^k\}$. The convergence analysis of this scheme is similar to the proof of the Theorem in the Appendix. It is also possible to derive from (5.14), that the scheme of Sect. 5.5 is in fact equivalent to the procedure (5.15) with the full step-size α_k equalizing marginal costs,

$$f'_i(y_i^{k+1}) = f'_j(y_j^{k+1}).$$

The basic adjustments in (5.15) are again pair-wise, but random encounters. Other encounters are also possible assuming that each party meets every other party.

This type procedure is also applicable in the case when parties use only observable random functions $c_i(x, \omega)$, $d_i(u, \omega)$. Suppose that instead of myopic decisions $y_i(k, \omega_k)$ parties make precautionary incremental and adaptive adjustments of vector y. We can define stochastic vector $\xi(y, \omega)$ similar to vector $\xi(y)$ as

$$\xi(y, \omega) = \frac{(n-1)}{2}(0, \ldots, 0, f'_i(y_i, \omega) - f'_j(y_j, \omega), \ldots, f'_j(y_j, \omega)$$
$$-f'_i(y_i, \omega), 0, \ldots, 0)$$

and proceed with changes of emission permits y^k according to procedure (5.15) with $\xi(y^k$ substituted by $\xi(y^k, \omega_k)$. Under standard assumptions the conditional expectation $E[\xi(y, \omega)|y] = g(y)$, where $g(y)$ is the projection of $grad G(y)$,

$$G(y) = \sum_{i=1}^{n} Ef_i(y, \omega). \qquad (5.16)$$

The convergence of the trading scheme (5.15) with the vector $\xi(y^k, \omega_k)$ to a solution minimizing $G(y)$ subject to $\sum_{i=1}^{n} y_i = 0$ can be derived from general results on the convergence with probability 1 of stochastic quasigradient methods [5]. In particular, it requires the proper step-size multipliers α_k, e.g., $\alpha_k = const/k$ is applicable. This requirement presumes no knowledge of underlying data. Yet, it suffices to stabilize exchanges y^k.

Remark 5.5 (Short-term market decisions). Method (5.15) with vector $\xi(y^k, \omega_k)$ leads to an array of bilateral trading schemes with a variety of trading rules. Yet, it is important to note a significant difference between functions $G(y)$ and $F(y)$ defined by (5.16) and (5.9), respectively; function $G(y)$ focuses on the variability of short-term market decisions $x_i(k,\omega)$, $u_i(k,\omega)$, whereas $F(y)$ focuses on forward-looking decisions x_i, u_i, y_i robust against all future eventualities ω. Thus, in contrast to trading scheme of Sect. 5.6, the minimization of function (5.16) takes short-term market perspectives on emission and uncertainty reductions driven by random observations ω_k of market situations. It also treats the safety constraints conditionally on myopic decisions $x_i(k,\omega)$, $u_i(k,\omega)$, that prevents to achieve robust cost-effective and environmentally safe outcomes specified by (5.3), (5.4), (5.7), (5.9). Such outcomes are guaranteed only through replacement of myopic overacting decisions $x_i(k,\omega)$, $u_i(k,\omega)$ by incremental adjustments of x_i, u_i similar to adjustments of y^k.

5.8 Concluding Remarks

The paper analyzes cost effective and environmentally safe carbon trading schemes explicitly incorporating different types of exogenous and endogenous uncertainties on emissions and the abatement costs. The feasibility of decentralized market pollution control mechanisms is usually discussed under strong assumptions that all actions are made simultaneously at known equilibrium prices, what implies existence of perfectly informed social planner. The examined dynamic bilateral trading schemes are not based on price signals, and the emerging emission prices implicitly depend on the costs functions and the safety constraints on environmental targets. With the safety constraints, the parties set the level of their exposure toward uncertainties and risks. The safety constraints discount the reported emissions to undershooting detectable levels. This type of undershooting concept or discounting should become a key element in a robust regulation of emission trades together with unified approaches to modeling emission uncertainties and cost functions. The paper shows that myopic price-based trading schemes are not able to achieve cost-effective and environmentally safe solutions. The irreversibility of trades calls for the use of the proposed computerized emission trading system providing, in a sense, collective decentralized regulation of trades. The procedures provide a constructive and easy approach for designing robust emission trading schemes. All decisions are fully decentralized, individual contrary to cost effective and environmentally safe price mechanisms requiring additional coordination to stabilize leading otherwise to nowhere trading processes. Our approach is close to important ideas on decentralized non-monetary exchange. Bilateral trading scheme with deterministic interval uncertainty has been applied [14] for the fossil fuel related carbon emissions of the major Parties of the Kyoto Protocol. Numerical findings indicate that the compliance costs increase significantly for USA, Japan and the European Union, if uncertainty of the emission levels is considered. However, although the Central and Eastern Europe, Russia, and Ukraine have larger uncertainties in emission levels, their net costs may decrease as they can sell emission reductions at a higher price. Additional simple calculations according to Remark 5.3 show that stochastic uncertainty in the emission levels reduces the compliance costs of parties.

Appendix

The convergence of trading scheme in Sect. 5.5 was proved in [8] for convex functions $f_i(y_i), i = \overline{1,n}$. The following proof covers the case of nonconvex functions.

Proof of Convergence

Theorem (Convergence to an equilibrium). Let $f_i(y_i) \geq 0$ be continuously differentiable functions and let Y^* be the set of equilibriums equalizing marginal values f_i' as in (5.5), $F(Y^*) = F(y), y \in Y^*$, y^k is defined as in Sect. 5.5. Then either

1. $y^k \in Y^*$ after a finite number of steps, or
2. The sequence $\{F(y^k)\}$ converges to its equilibrium value from the set $F(y^*)$ and all cluster points of $\{y^k\}$ belong to Y^*, or
3. If Y^* contains only a single point y^*, then $\{y^k\}$ converges to this point.

Proof. The sequence $\{F(y^k)\}$, $k = 1, \ldots$, is monotonically decreasing, $F(y^k) \geq 0$. Therefore, there exist a limit $\overline{F} = \lim_k F(y^k)$. Let us prove that $\overline{F} \in F(Y^*)$. Suppose there exists a convergent subsequence y^{k_s}, $y^{k_s} \rightarrow \overline{y}$, $s \rightarrow \infty$ and $\overline{y} \notin Y^*$. Therefore, there exist i, j such that $f_i'(\overline{y}_i) \;/\!\!= f_j'(\overline{y}_j)$. It means that $\lim F(y^{k_s+1}) < F(\overline{y})$, what contradicts the convergence of $F(y^k)$, i.e., $\overline{F} \in F(Y^*)$, and all cluster points of the bounded sequence y^k belong to Y^*. Hence, if Y^* is a singleton, then y^k converges to y^*.

A Core Solution

From (5.10) it follows that at each step k cooperating parties i, j can redistribute joint cost $f_i(y_i^{k+1}) + f_j(y_j^{k+1}) = \phi_i^{k+1} + \phi_j^{k+1}$, $\phi_i^{k+1} < f_i(y_i^k)$, $\phi_j^{k+1} < f_j(y_j^k)$. Therefore at the equilibrium $y^* = (y_1^*, \ldots, y_n^*)$ parties will deal actually with payments $\phi_i^* < f_i(y_i^0)$ such that $\sum_{i=1}^n \phi_i^* = \sum_{i=1}^n f_i(y_i^*) \; : \; = F_I$ where $I = 1, \ldots, n$. From this equation follows the Pareto efficiency of $\phi^* = (\phi_i^*)_{i=1,\ldots,n}$. An important question is whether the grand coalition I of parties is stable, i.e., $\sum_{I \in C} \phi_i^* \leq F_c$ for any other coalition $C \subseteq I$. Accordingly, a distribution of payments ϕ^* is a core solution if it satisfies these two equation. The bilateral trading procedure allows to find the equilibrium price λ^*. If function $F(y)$ is convex, then the payment distribution $\phi_i^* = f_i(y_i^*) + \lambda^* y_i^*$ is a core solution. If the function $F(y)$ is globally Lipschitz continuous, then the core solution remains the same (see discussion in [11]).

References

1. Accounting for climate change. In: Lieberman, D., Jonas, M., Nahorski, Z., Nilsson, S. (eds.) Uncertainty in Greenhouse Gas Inventories Verification, Compliance, and Trading. Springer, Berlin (2007)

2. Baumol, W., Oates, W.: The use of standards and prices for protection of the environment. Swedish J. Econ. **73**, 42–54 (1971)
3. EEA: Application of the emissions trading directive by EU member states. Tech. Rep. 2, European Environment Agency (EEA), Denmark (2006). http://reports.eea.europa.eu/technical_report_2006_2/en/technicalreport_2_2006.pdf
4. Energy Business Review: Volatility the only certainty in EU carbon market. Energy Business Review (2006). http://www.energy-business-review.com/article_feature.asp?guid=FD09D7CA-3EFC-4229-BA86-1D968025DF5B
5. Ermoliev, Y.: Stochastic quasigradient methods. In: Floudas, C., Pardalos, P. (eds.) Encyclopedia of Optimization. Springer, New York (2009). ISBN 978-0-387-74758-3
6. Ermoliev, Y., Hordijk, L.: Facets of robust decisions. In: Marti, K., Ermoliev, Y., Makowski, M., Pflug, G. (eds.) Coping with Uncertainty: Modeling and Policy Issues, pp. 3–28. Springer, Berlin, Heidelberg, New York (2006)
7. Ermoliev, Y., Klaasen, G., Nentjes, A.: The design of cost effective ambient charges under incomplete information and risk. In: van Ierland, E., Gorka, K. (eds.) Economics of Atmospheric Pollution, *NATO-ASI Series, Partnership Sub-Series, 2. Environment*, vol. 14, pp. 123–151. Springer, Berlin (1996)
8. Ermoliev, Y., Michalevich, M., Nentjes, A.: Markets for tradeable emissions and ambient permits: A dynamic approach. Environ. Resour. Econ. **15**(1), 39–56 (2000)
9. Ermolieva, T., Ermoliev, Y.: Catastrophic risk management: Flood and seismic risks case studies. In: Wallace, S., Ziemba, W. (eds.) Applications of Stochastic Programming, MPS-SIAM Series on Optimization. SIAM, Philadelphia (2006)
10. Ermolieva, T., Jonas, M., Makowski, M.: The difference between deterministic and probabilistic detection of emission changes: Toward the use of the probabilistic verification time concept. In: Proc. of the 2nd International Workshop on Uncertainty in Greenhouse Gas Inventories, IIASA. Systems Research Institute of the Polish Academy of Sciences (2007)
11. Evstigneev, I., Flaam, S.: Sharing nonconvex cost. J. Global Optim. **20**, 257–271 (2001)
12. Gillenwater, M., Sussman, F., Cohen, J.: Practical policy applications of uncertainty analysis for national greenhouse gas inventories. Water Air Soil Pollut. **4–5**(7), 451–474 (2007)
13. Godal, O.: Simulating the carbon permit market with imperfect observations of emissions: Approaching equilibrium through sequential bilateral trade. Technical Report IR-00-060, International Institute for Applied Systems Analysis, Laxenburg, Austria (2000)
14. Godal, O., Ermoliev, Y., Klaassen, G., Obersteiner, M.: Carbon trading with imperfectly observable emissions. Environ. Resour. Econ. **2**(25), 151–169 (2003). Available also as IIASA Reprint RP-03-08
15. Grubb, M., Neuhoff, K.: Allocation and competitiveness in the eu emissions trading scheme: policy overview. Climate Policy **6**(1), 7–30 (2006)
16. Hudz, H.: Verification times underlying the kyoto protocol: Consideration of risk. Interim Report IR-02-066, International Institute for Applied Systems Analysis, Laxenburg, Austria (2002). http://www.iiasa.ac.at/Publications/Documents/IR-02-066.pdf
17. Hudz, H., Jonas, M., Ermolieva, T., Bun, R., Ermoliev, Y., Nilsson, S.: Verification times underlying the kyoto protocol: Consideration of risk. VT Concept, International Institute for Applied Systems Analysis, Laxenburg, Austria (2003). http://www.iiasa.ac.at/Research/FOR/vt_concept.html
18. IETA: Emission trading master agreement for the eu scheme. In: Version 2.1 as of 13 June. International Emission Trading Association (IETA), Geneva, Switzerland (2005). http://www.ieta.org/ieta/www/pages/getfile.php?docID=1001
19. Jonas, M., Nilsson, S., Obersteiner, M., Gluck, M., Ermoliev, Y.: Verification times underlying the kyoto protocol: Global benchmark calculations. Interim Report IR 99-062, International Institute for Applied Systems Analysis, Laxenburg, Austria (1999). http://www.iiasa.ac.at/Publications/Documents/IR-99-062.pdf
20. Klaassen, G., Nentjes, A.: Creating markets for air pollution control in Europe and the USA. Environ. Resour. Econ. **10**, 125–146 (1997)
21. Madden, P.: Efficient sequences of non-monetary exchange. Rev. Econ. Stud. **42**(4), 581–596 (1975)

22. Montgomery, D.: Markets in licences and efficient pollution control programs. J. Econ. Theor. **5**, 395–418 (1972)
23. Nahorski, Z., Horabik, J.: Greenhouse gas emission permit trading with different uncertainties in emission sources. J. Energy Eng. **2**(134), 47–52 (2008)
24. Nahorski, Z., Horabik, J., Jonas, M.: Compliance and emissions trading under the kyoto protocol: Rules for uncertain inventories. Water Air Soil Pollut. **7**(4-5), 539–558 (2007)
25. Rockafellar, T., Ursayev, T.: Optimization of conditional value at risk. J. Risk **2/3**, 21–41 (2000)
26. Rypdal, K., Winiwarter, W.: Uncertainties in greenhouse gas emission inventories evaluation, comparability and implications. Environ. Sci. Policy **4**(2–3), 107–116 (2001)
27. Schleicher, S., Kettner, C., Köppl, A., Thenius, G.: Stringency and distribution in the eu emissions trading scheme? The 2005 evidence. Working paper, Fondazione Eni Enrico Mattei (FEEM), Milano, Italy (2007). http://www.feem.it/NR/rdonlyres/6E9369C4-82F8-4115-AF93-1D4AF586CBF9/2262/22 07.pdf
28. Starr, R.: Decentralized nonmonetary trade. Econometrica **44**(5), 1087–1089 (1976)
29. UNFCCC: Kyoto protocol to the united nationas framework convention on climate change. (1997). http://www.unfccc.de

Chapter 6
Robust Design of Networks Under Risks

Y. Ermoliev, A. Gaivoronski, and M. Makowski

Abstract Study of network risks allows to develop insights into the methods of building robust networks, which are also critical elements of infrastructures that are of a paramount importance for the modern society. In this paper we show how the modern quantitative modeling methodologies can be employed for analysis of network risks and for design of robust networks under uncertainty. The approach is illustrated by an important problem arising in the process of building the information infrastructure for the advanced mobile data services.

We show how the portfolio theory developed in the modern finance can be used for design of robust provision network. Next, the modeling frameworks of Bayesian nets and Markov fields are used for the study of several problems fundamental for the process of service adoption such as the sensitivity of networks, the direction of improvements, and the propagation of participants' attitudes on social networks.

6.1 Introduction

This paper is dedicated to a study of the network risks which are the key issues defining the robustness of infrastructures. There are similarities between network risks and catastrophic risks: both have interdependencies in space and time. An appropriate analysis of these two classes of risks requires adaptation, integration, extension and further development of methodologies for quantitative modeling of uncertainty and risks. Such methodologies have emerged during recent decades in diverse fields such as economics and finance, optimization, simulation of stochastic and multiagent systems. For a more detailed treatment we have selected two of

A. Gaivoronski (✉)
Department of Industrial Economics and Technology Management, Norwegian University of Science and Technology, Alfred Getz vei 1, 7491 Trondheim, Norway,
e-mail: Alexei.Gaivoronski@iot.ntnu.no

Y. Ermoliev and M. Makowski
International Institute for Applied Systems Analysis, Schlossplatz 1,2361 Laxenburg, Austria,
e-mail: ermoliev@iiasa.ac.at, marek@iiasa.ac.at

K. Marti et al. (eds.), *Coping with Uncertainty*, Lecture Notes in Economics and Mathematical Systems 633, DOI 10.1007/978-3-642-03735-1_6,

such methodologies: portfolio theory of finance and Bayesian networks, both coupled with optimization approach show how these methodologies can be extended and applied for the study of network risks with the emphasis on information infrastructure.

Besides clarifying the methodological issues, we also aim at creation of integrated modeling decision support environment for analysis of network risks. Such environment enables identification and evaluation of critical bottlenecks inherent in important infrastructures seen as specialized networks and allow to give advice to planning and regulating bodies on robust design and improvement of these infrastructures.

More specifically, we look at the risk adapted performance networks composed of nodes and links of different levels of complexity. The risk adjusted performance of each node can be improved by selecting appropriate control parameters. In addition, performance of each node is affected by uncertainties. These network elements are designed from the point of view of local tradeoff between local performance and risk. The risk can be exogenous to the network as well as endogenous, generated by inappropriate functioning of other network elements. It is important that the overall performance of the network is also affected by the risk on the global level. This risk is understood as a possibility that the global performance can differ, sometimes drastically, from the expected network performance. The key issue in the designing of robust networks is to assure that a local level of risk/performance tradeoff results in a desirable risk/performance tradeoff on the global level. This is one of the central issues which we aim to clarify.

Here we briefly discuss three examples of infrastructure which can be described as risk/performance networks.

1. *Energy infrastructure.* Electric power infrastructure can be described as a network with several levels of hierarchy where the nodes correspond to production, distribution and consumption facilities while the arcs represent transmission lines. On the local level each node has production and consumption targets subject to uncertainty and risk manifested as the equipment failures, local demand variations, local weather patterns important for hydro and wind generation. On the global level this infrastructure should meet electric power demand of consumers and industries subject to uncertainties and risk of disruptions, prices for fuel and energy, weather, societal attitudes towards certain generation technologies and climate change.

Earlier such performance/risk tradeoff was much easier to achieve because the public utilities managed generation and transport in almost each country. Now the electric power industry is deregulated or being deregulated in almost all developed countries. The industry is now composed of many independent actors which decide their production plans according to the market conditions. Besides, the new types of actors have entered the field, like energy contract traders and speculators, and energy exchanges. Yet they have to act in concert if this infrastructure is to fulfill in a robust manner the energy needs of society at large. This is a critical issue as the power shortages in California and price surges in Norway have shown.

What are robust risk management methods which will mitigate these new risks which result from market forces and individual actors' behavior? What is the robust way to assure that the local decisions on risk/performance tradeoff which every actor takes will transform into optimal or even acceptable tradeoff on the global level? What lessons developing countries can learn from the experience of developed countries in this respect? These are the questions which our paper aims to answer.

2. *Gas transport and distribution infrastructure.* Similar issues of networked risk/performance management arise in other types of infrastructure. For example, developing the gas transport and consumption infrastructure in Europe largely follows deregulation patterns of electric power infrastructure according to EU directives, and the same is true for the railroad transport.

3. *Information and communication network infrastructure.* It can be described as a superposition of several layers of hierarchical networks each one consisting of nodes connected with links. There are also mappings connecting different layers. The network nodes are represented by heterogeneous devices like routers, switches, cross-connects, etc. Each of these devices is equipped with control structures that govern communication flows through the network; the control is composed of communication protocols, routing tables, call admission rules, etc. The control parameters are tuned largely independently in order to meet performance targets of each node. Uncertainty on the node level comes from highly variable communication flows, but also from actions of adjacent nodes. There are also risks of equipment failures, congestion, malicious attacks, link failures which threaten the performance targets.

Each of the nodes is built to achieve admissible tradeoff between performance, costs and risk on the local level. The entire network, however, should satisfy various global performance targets, like a satisfaction of communication and information demand with quality of service guarantees for the user population. Besides, the operation of this infrastructure should be economically attractive for industrial actors which own its different parts. Global sources of risks and uncertainty include both external component (changing usage patterns and global malicious disturbances) and internal component (connected with conflicting interests of different actors).

How this local tradeoff between performance and risk at the node level affects the global tradeoff on the infrastructure level and vice versa? What are the economically sound principles for further robust development and operation of this infrastructure under inherent risk and uncertainty? Where are the bottlenecks which threaten its global performance? These are the questions which require the new methodology of the network risk analysis and robust risk management to which this paper aims to contribute.

Methods for taking optimal decisions under uncertainty and related issues of risk management have been at the center of methodological development in the couple of last decades, and recently they have met also a considerable and rising industrial interest. One should mention stochastic programming which is a hot topic in operations research community now and it has become an important modeling tool in quantitative finance, energy, telecommunications and other industrial fields. Understanding importance of risk management in finance resulted in the development of several risk management paradigms and industrial standards which are now being gradually adopted also in other industrial branches. However, these and other methods mainly consider control of relatively simple systems facing external sources of risk and uncertainty. The real challenge is to look at the network of such systems and study the effects of risk and uncertainty on the network overall performance.

In this paper we look how such general considerations of network risks are adapted to the case of information infrastructure consisting of several components that are in different stages of development. We consider development and deployment of such a component characterized by increasing an importance: namely advanced mobile data services. We show how the related risks can be analyzed using and further extending for this case the quantitative risk modeling methodologies. In particular, the portfolio theory developed in finance is extended for analysis and design of service provision networks (Sects. 6.2–6.6). Next, Bayesian nets and Markov fields models are used in order to predict and analyze the sensitivity of networks, the directions of improvements, and the service adoption patterns, all of which depend on complex interplay of attitudes of different groups of population. Finally, we discuss development of two components of advanced methodological toolbox for analysis of network risk.

6.2 Cooperative Provision of Advanced Mobile Data Services

Design of advanced mobile data services to be carried on the 3G networks and the networks of next generations is the hot topic in telecommunication industry and research. This is because the business success of such services provision will define the business success of the mobile operators and other relevant industrial actors in the near to medium future. In this respect considerable attention is given to design and development of service provision platforms which support a set of tools and basic services that facilitate development, deployment and customization of specialized services by service providers and even non-professional end users.

Deployment and operation of service provision platforms and provision of individual services requires collaboration of different industrial actors contributing their individual capabilities and expertise to the common goal. One can think about fixed network operators, mobile operators, providers of different information content, internet providers, software developers and other actors that join forces to provide a successful service. Provision of a service involves assuming different roles and industrial actors can combine such roles. All this gives a rich picture of service provision environment where a multitude of actors cooperate and compete in order to deliver to customers a wide range of services in a profitable manner.

Understandably, the main research and development effort so far has been concentrated on technological and engineering aspects enabling provision of advanced mobile data services. The history of information technology testifies, however, that the possession of the best technological solution is not necessarily enough to assure the business success of an enterprise. A very important and sometimes neglected aspect is the design and evaluation of appropriate business model to support the service provision. Business models for provision of a service requiring a single actor are pretty well understood, both organizationally and economically. This is the case, for example, of the provisioning of the traditional voice service over a fixed network. When an actor evaluates the economic feasibility of entering the provision of such service, he can employ quantitative tools developed by investment science, like estimation of the Net Present Value of such project [21]. Usually an actor should choose between several service provisioning projects, each providing return on investment and generating the risky cash flows. Then the portfolio theory [24] suggests the way to balance between return and risk and select the best portfolio of projects taking into account the actor's risk attitudes. The adequate risk management is especially important in a highly volatile telecommunication environment and the industrial standards in this respect are starting to emerge, originating from the financial industry [3]. Industrial projects in high-tech industries are often characterized by considerable uncertainty and at the same time carry different flexibilities. The real options approach [29] allows to take these flexibilities into account while making evaluation of the profitability of the project. Stochastic programming [2, 10, 13, 20] provide the optimization models for adequate treatment of uncertainty in the planning of service provision.

Business models for cooperative service provision involving different constellations of actors are studied to a much lesser extent. The understanding of their importance has lead to some qualitative analysis methods, e.g., [11, 19]; however, quantitative analysis at a level available for the single actor case remains a challenge. The methods mentioned above are developed to be used by a single actor engaged in selection and risk management of his portfolio of industrial projects. The influence of the other actors is present only implicitly on the stage of estimation of the future cash flows. This is not enough for an adequate analysis of collaborative service provision. Suppose, for example, that a service provider delivers a service to a population of users and receives a revenue for this delivery. If a service is composed of modules and enablers provided by different actors then this service provider has to decide on the revenue division between the actors which will make it attractive to them to participate in the service provision. The provider should explicitly incorporate in the evaluation of profitability of his project such revenue sharing decision together with a concept of what is attractive to other actors. Here we contribute to the adaptation and further development of the methods of evaluation and risk management of business models and industrial projects for the case of the collaborative service provision. We look at the actors engaging in a service provision as making a decision about the composition of their portfolio of services to which they are going to contribute. They do this independently following the risk management framework of the portfolio theory. The pricing and revenue sharing schemes induce each actor to contribute the right amount of provision capacity to its participation in the service provision. We develop a two tier modeling framework which supports selection of pricing and revenue sharing in an optimal way. This is done by utilizing the approach of stochastic optimization with bilevel structure [1] and combining it with the portfolio theory.

6.3 Simplified Model of the Service Portfolio

In this section we develop a quantitative description of the service provisioning model involving several actors having as the background the environment presented in the previous section.

6.3.1 Description of Services

A composition of a service can be quite complex, especially if we take into account that various components can be services themselves and subject to further disaggregation. For the purposes of clarity we start from a simple description which still possess the main features of the provision environment important for business modeling. Namely, two levels of the service composition will be considered here as shown in the example illustrated in Fig. 6.1.

In this case the service environment is composed of two types of services. The first type is comprised from services with structure and provision we consider in some detail. They can be provided in the context of a service platform and therefore they are referred to as *platform* services. There will be also *3rd party* services whose structure is of no concern to our modeling purposes. They are present in the model for the purpose of the adequate modeling of the environment in which the provisioning of the platform services happens. Let us now consider the model of provisioning of platform services.

The main building blocks of the platform services are service *enablers* indexed by $i = 1 : N$ and *services* indexed by $j = 1 : M$. Enablers are measured in units relevant for their description, like bandwidth, content volume, etc. The relation between enablers and services is described by coefficients λ_{ij} which measure the amount of enabler i necessary for provision of the unit amount of service j. Thus, a service j can be described by vector

$$\lambda_j = (\lambda_{1j}, \ldots, \lambda_{Nj}) \tag{6.1}$$

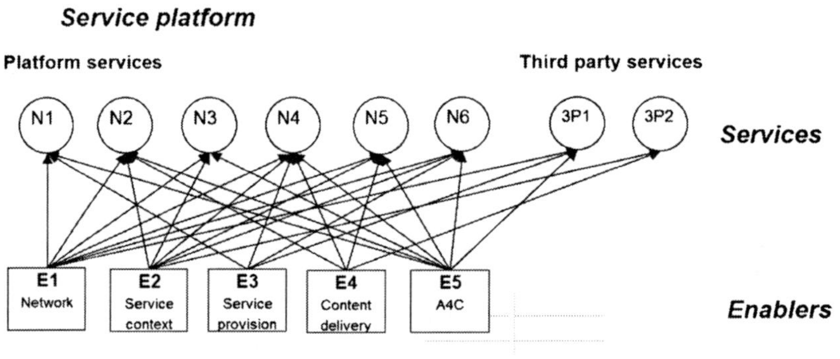

Fig. 6.1 Service provision for business persons on the move

This description is obtained from analysis of the usage scenarios described in the Sect. 6.2. A service j generates a revenue v_j per unit of service. This quantity depends on the service pricing which in its turn depends on the user behavior and market structure. For a moment let us assume that v_j is a random variable with a known distribution; in Sect. 6.4.3 we describe this revenue in more detail. The distribution can be derived from the expert estimates and from simulation models which would explore the structure of user preferences and market features. The random variables v_j can be correlated due to the service substitution, macroeconomic phenomena and other causes.

Services can be provided by different constellations of actors. In this paper we consider a constellation where the actors are the enterprises having the capability to provide service enablers assuming different *roles*, indexed by $k = 1 : K$. Actors may choose to join forces to provide a service. Contribution of a given actor consists of taking responsibility for provision of one or more enablers of the service. Sometimes these actors are referred to as *enabler providers*. There is also an actor providing the service aggregation functionality and organizing the overall service delivery to the end users; this actor is referred to as a *service provider*; it can provide the whole bundle of platform services and decides which services are includes in this bundle. Often it collects the revenue from the end users and distributes it among the enabler providers.

Example 6.1. Service provision for business persons on the move (see Fig. 6.1). This is a simplified yet realistic example of service provision which was developed on the basis of the results of the EU project SPICE and the NFR project ISIS. The terminal service is a smart mobile phone used by a business person on the move.

We consider the services which run on the service platform and the third party services which partially compete with them, being accessible from the same terminal. We present a simplified example composed of just six native services, two third party services, and five enablers (in actual applications there are hundreds of services and dozens of enablers, distributed in several service platforms) available in this service platform. However, services in this platform correspond well to the business offer of a typical service provider. Service bundles have been defined in accordance with the market segments, corresponding customer classes, user behavior, requirements, and various subscription schemes. More specifically, we consider the following services.

Native services of the platform:

1. N1 – Messaging
2. N2 – Audio conferencing
3. N3 – Video conferencing
4. N4 – Location based services
5. N5 – News
6. N6 – Point of Interest service

Third party services:

1. 3P1 – Third party Information service
2. 3P2 – Third party News service

The following business actors collaborate in providing the mobile service bundle to the users:

1. E1 – *Network provider* – providing the network access.
2. E2 – *Context provider* – service context retrieval and management.
3. E3 – *Service provider* – responsible for service provision.
4. E4 – *Content provider* – content retrieval and management.
5. E5 – *Provider of A4C* (authentication, authorization, auditing, accounting and charging) enabler. This actor often coincides with the service provider, but one can envisage also the cases when it is a distinct actor.

Besides, there are one or more providers of the third party services which are in partial competition with the platform services.

The objective of an enabler provider is to select a portfolio of services to which this actor will make a contribution. This decision is made on the grounds of balance between projected profit from enabler provision balanced against the risk of variations in demand and service acceptance among the prospective users of services. In order to quantify such decision process it is necessary to use a simplified profit model for an actor.

It is assumed that the revenue v_j generated by a unit of service j is distributed among the actors participating in the service creation. There can be different schemes for such subdivision. It is assumed here that this distribution is performed using a vector of revenue shares

$$\gamma_j = \left(\gamma_{1j}, \ldots, \gamma_{Nj}\right), \; \gamma = (\gamma_{11}, \ldots, \gamma_{N1}, \ldots, \gamma_{1M}, \ldots, \gamma_{NM})$$

such that an actor which contributes with the enabler i receives the revenue $\gamma_{ij} v_j$. Determination of these revenue sharing coefficients is one of the objectives of the design of the business model for service provision.

Besides platform services the actors can supply enablers also to the third party services. The structure of these services is not specified and it is assumed that they are fully described by the revenue v_{ij} generated by provision of the unit of enabler i to third party service j, $j = M + 1, \ldots, \bar{M}$.

This example will be treated in more detail in Sect. 6.6.

6.3.2 *Profit Model of an Actor*

Let us consider the situation when all the actors have already developed the capacities for provision of enablers. Thus, the investment process necessary for creation and expansion of these capacities is not discussed here; however, it is considered at the later stages. For this reason at this stage it is enough to consider only the variable costs due to the operation of capacities and provision of enablers. Alternatively, one can assume that the cost structure includes both the operational and discounted portion of the investment costs for enabler development, recalculated down to the enabler and the service instances.

For further formulation of the actor's profit model let us introduce the following notations:

c_{ik} – Unit provision costs for enabler i by actor k
W_{ik} – Provision capability of enabler i of actor k
x_{ijk} – The portion of provision capability for enabler i of actor k dedicated to participation in provision of service j

Now the revenue of actor k obtained from contribution to provision of the platform service j can be expressed as follows. The quantity $x_{ijk} W_{ik}$ is the volume of provision of enabler i dedicated by actor k to service j. Assuming that the required quantity of other enablers is available, this results in the volume of service j in which the actor k participates to be $x_{ijk} W_{ik}/\lambda_{ij}$. The total revenue from this service is equal to $v_j x_{ijk} W_{ki}/\lambda_{ij}$ and the part of the revenue which goes to actor k is $v_j x_{ijk} W_{ik} \gamma_{ij}/\lambda_{ij}$. For the third party service the revenue will be $v_{ij} x_{ijk} W_{ik}$.

The total costs incurred by actor k for the provision of enabler i to service j is equal to $x_{ijk} c_{ik} W_{ik}$.

In order to simplify the following discussion let us assume now that the actor k participates in the provision of service j by contributing only one enabler $i = i(k, j)$ or assuming only one role. Taking the profit π_k to be the difference between the revenue and costs, the profit of the actor k can be expressed as follows:

$$\pi_k = \sum_{j=1}^{M} \left(v_j x_{ijk} W_{ik} \frac{\gamma_{ij}}{\lambda_{ij}} - x_{ijk} c_{ik} W_{ik} \right) + \sum_{j=M+1}^{\bar{M}} \left(v_{ij} x_{ijk} W_{ik} - x_{ijk} c_{ik} W_{ik} \right)$$

$$= \sum_{j=1}^{M} x_{ijk} W_{ik} c_{ik} \left(\frac{v_j \gamma_{ij}}{c_{ik} \lambda_{ij}} - 1 \right) + \sum_{j=M+1}^{\bar{M}} x_{ijk} W_{ik} c_{ik} \left(\frac{v_{ij}}{c_{ik}} - 1 \right)$$

In the expression above index i depends on the values of indices j and k. Now let us assume that the actor k assumes only one role which consists in the provision of enabler i to different services which require this enabler. Thus, we consider a generic actor whose role is to provide enabler i to different services. Then we can simplify notations by taking $x_{ijk} = x_{ij}$, $W_{ik} = W_i$, $c_{ik} = c_i$, $\pi_k = \pi_i$. In this case the profit is defined by:

$$\pi_i = W_i c_i \left(\sum_{j=1}^{M} x_{ij} \left(\frac{v_j \gamma_{ij}}{c_i \lambda_{ij}} - 1 \right) + \sum_{j=M+1}^{\bar{M}} x_{ij} \left(\frac{v_{ij}}{c_i} - 1 \right) \right).$$

Dividing the profit by the total costs $W_i c_i$ we obtain the return r_i on investment by a generic actor which assumes the role of provision of enabler i to services which require this enabler:

$$r_i = \sum_{j=1}^{M} x_{ij} \left(\frac{v_j \gamma_{ij}}{c_i \lambda_{ij}} - 1 \right) + \sum_{j=M+1}^{\bar{M}} x_{ij} \left(\frac{v_{ij}}{c_i} - 1 \right). \qquad (6.2)$$

6.3.3 Service Portfolio: Financial Perspective

The profit representation (6.2) allows us to look at the enabler provision from the point of view of the financial portfolio theory [24]. The actor with the role to provide the enabler i has to choose from all possible available services requiring this enabler the set of services which provide this enabler. In other words, he has to select his service portfolio. This portfolio is defined by shares x_{ij} of his provision capability,

$$x_i = \left(x_{i1}, \ldots, x_{i\bar{M}} \right)$$

Return coefficients associated with the participation in each platform service are expressed as

$$r_{ij} = \frac{v_j \gamma_{ij}}{c_i \lambda_{ij}} - 1, \; j = 1 : M \qquad (6.3)$$

and for the third party services these coefficients are

$$r_{ij} = \frac{v_{ij}}{c_i} - 1, \; j = M + 1 : \bar{M}. \qquad (6.4)$$

These coefficients depend on the random variables, in particular the revenue per unit of service v_j and the revenue per component provision v_{ij}. Randomness here is due to the uncertainty in demand and the user acceptance of service. Also the enabler provision costs c_i and enabler shares λ_{ij} are random variables due to the uncertainty inherent in the service usage patterns and the evolution of costs. Besides, the costs c_i are often estimates of the provision costs of enabler provider i made by another actor. Such estimates are inherently imprecise and therefore described by random variables similarly to how it was done in [1]. The expected return coefficients are

$$\mu_{ij} = \gamma_{ij} \mathbb{E} \frac{v_j}{c_i \lambda_{ij}} - 1, \; j = 1 : M, \; \mu_{ij} = \mathbb{E} \frac{v_{ij}}{c_i} - 1, \; j = M + 1 : \bar{M} \qquad (6.5)$$

and expected return $\bar{r}_i(x_i)$ of the service portfolio is

$$\bar{r}_i(x_i) = \sum_{j=1}^{\bar{M}} \mu_{ij} x_{ij} = \sum_{j=1}^{M} x_{ij} \left(\gamma_{ij} \mathbb{E} \frac{v_j}{c_i \lambda_{ij}} - 1 \right) + \sum_{j=M+1}^{\bar{M}} x_{ij} \left(\mathbb{E} \frac{v_{ij}}{c_i} - 1 \right) \qquad (6.6)$$

However, the realized return can differ substantially from the expected return due to the uncertainty discussed above. This introduces the risk $R(x_i)$ for an actor which assumes the enabler provision role. The financial theory traditionally measures this risk as the variance of the portfolio return [24]. Recently several different risk measures were introduced, in particular Value at Risk (VaR) and its many modifications. The VaR has attained the level of industrial standard in the financial risk management [3]. In this section the variance and the standard deviation of the return are used as the risk measure. It is not clear yet what is the best risk measure to consider in the context of portfolio problems related to the collaborative service provision. Besides the classical choice of standard deviation which we follow here, the promising candidates are the *VaR* mentioned above [14] and the Conditional Value at Risk [15]. The identification of the most relevant risk measure in this context is outside the scope of this paper; it will be pursued in our subsequent research. What is important here is the consideration of the risk measures allow an actor to estimate the probability and size of his future losses. Thus, we take

$$R(x_i) = \text{StDev}(r_i(x_i)) = \text{StDev}\left(\sum_{j=1}^{\bar{M}} r_{ij}x_{ij}\right) \tag{6.7}$$

where the return coefficients r_{ij} are taken from (6.3), (6.4).

Portfolio theory looks at the portfolio selection as the trade-off between risk and return. Its application to our problem of service portfolio consists of the following steps.

1. *Construction of the efficient frontier.* An average return target η is given. The risk of service portfolio is minimized with constraint on this return target. The risk minimization problem is defined by:

$$\min_x \ \text{StDev}^2\left(\sum_{j=1}^{\bar{M}} r_{ij}x_{ij}\right) \tag{6.8}$$

$$\sum_{j=1}^{\bar{M}} \mu_{ij}x_{ij} = \eta \tag{6.9}$$

$$\sum_{j=1}^{\bar{M}} x_{ij} = 1, \ x_{ij} \geq 0 \tag{6.10}$$

Solution of this problem for all admissible values of target return η provides the set of service portfolios which are a reasonable candidates for selection by the actor provides the enabler j. They constitute the *efficient frontier* of the set of all possible service portfolios. This concept is illustrated in Fig. 6.2.

Each service portfolio x can be characterized by the pair (risk, return) defined by (6.7) and (6.6) respectively. Therefore it can be represented as a point in the risk-return space depicted in Fig. 6.2. The set of such points for all possible portfolios describes all existing relations between risk and return and is called *the feasible set.* Which of possible service portfolios an actor should choose? It depends on the

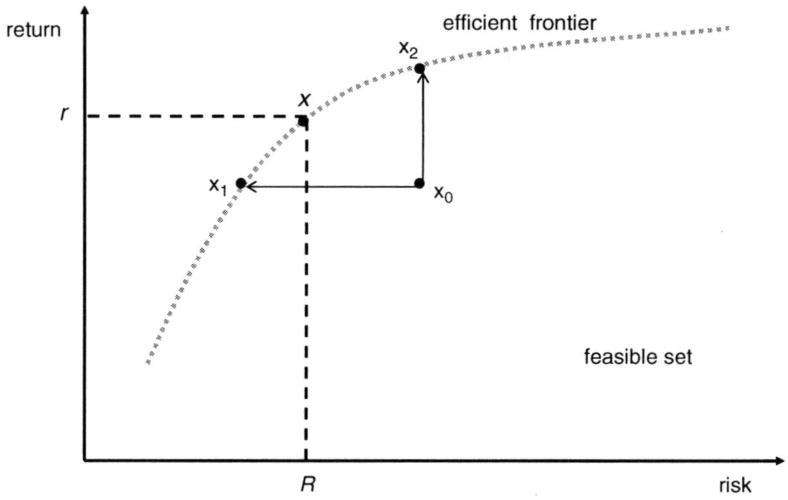

Fig. 6.2 Selection of service portfolio

objectives which an actor pursues. Here we assume that an actor's decision depends on return and risk only. Namely, an actor seeks the highest possible return among equally risky alternatives and the lowest possible risk among equally profitable alternatives. This is a simplification, because in reality the actors can be driven by other considerations, like increase of market share, revenue, regulatory constraints, etc. However, the consideration of only risk and return provides with a reasonable starting point for analysis of business models. More complex cases can be taken into account in a similar manner by introducing additional constraints on the feasible set or by modifying the concept of performance. For example, suppose that an actor has three objectives: the return and the market share to maximize and the risk to minimize. Then these three criteria will be used for specification of an achievement function used in the multi-criteria problem analysis, see, e.g., [30]. Such an analysis supports examination of efficient solutions having different trade-offs between the criteria values, and help the users in finding such trade-offs the correspond best to his/her preferences.

The example illustrated in Fig. 6.2 show that some service portfolios should be preferred to others. For example, let us consider portfolio x_0 to which corresponds the point in risk-return space inside the feasible set shown in Fig. 6.2. It is clear that the portfolio x_2 should be preferred to x_0, because x_2 has the same risk as portfolio x_0 but it has larger return. Similarly, portfolio x_1 also should be preferred to x_0 because it provides the same return with smaller risk. In other words, portfolio x_0 is dominated by both portfolios x_1 and x_2 and therefore it is not rational to select it when only these two criteria are considered. The actor whose decisions are guided by risk and return should consider only nondominated portfolios which constitute the *efficient frontier*, depicted by dotted curve in Fig. 6.2. Such an efficient frontier can be computed by solving the problem (6.8)–(6.10) for different values of η. It can be considered as a set of efficient solutions of the two objective optimization problem: minimizing risk and maximizing return under constraints (6.10). The choice of a solution from the efficient set depends on the preferences the user has between risk and return, and it is easy for experienced managers.

The above outlined approach to the multicriteria analysis is suitable for problems with two criteria, i.e., for which it is possible to compute and analyze the *efficient frontier.* For more than two criteria one should use a suitable method for multi-criteria analysis. However, one has to note that the most popular approach to multi-criteria analysis believed to be reliable and intuitive, namely the aggregation of criteria through a weighted sum is actually neither reliable nor intuitive. Limitations of the weighted-sum approach are discussed, e.g., in [22, 23]. Therefore, for analysis of problems with more than two criteria one should consider a truly multicriteria analysis approach, see, e.g., [30], and a modular tool (such as ISAAP described in [18]) that supports interactive analysis of trade-offs between conflicting objectives.

2. *Selection of the target service portfolio.* The previous step resulted in the selection of a much smaller set of efficient service portfolios from the set of all possible service portfolios. These portfolios form the efficient frontier in the risk-return space. An actor selects its target service portfolio from this efficient set by choosing the trade-off between risk and return. One way to achieve this trade-off is to consider the largest risk an actor is willing to take. Suppose that the value of such risk is R (see Fig. 6.2). Then the actor should choose the portfolio x on efficient frontier with this value of risk. Suppose that this service portfolio yields return η. No other portfolio yields better return without increasing the risk. If an actor is not satisfied with return η this means that he should increase his risk tolerance or look for opportunities to participate in the service provision not yet described in this model. Or, such actor should seek more advantageous revenue sharing scheme.

From these considerations it is clear that all important opportunities of participation in service provision should be included in the model. For example, suppose that an actor assumes the role of the content provision and can contribute a content to the advanced mobile data service, and at the same time the same content can be contributed to, say, traditional newspapers. Both opportunities should be included in the model with the traditional service being modeled as a third party service.

6.4 Modeling of Collaborative Service Provision

In the previous section we highlighted the importance of having an adequate forecast of the cash flows generated by services in order to quantitatively evaluate the economic future of the service, and business models which support the service provision. Due to the inherent uncertainty of the user response and technological development any such forecast should be given in terms of random variables which assign probabilities to different scenarios of user response and possible evolution of other uncertain parameters. Forecasts should take into account the mutual influence of services which result in correlation between cash flows generated by different services.

Such a model allows to consider providers of different service enablers as actors which independently select the service portfolios having their targets described in terms of return on investment and risk tolerance. However, a service can become a reality only if the participation in its provision is consistent with these individual targets. This means that all actors which cover the roles indispensable for provision of a particular service should have the corresponding services in their efficient service portfolio. In other words, the service portfolios of the relevant actors should be *compatible*. There are several items which affect the risk/return characteristics of a service portfolio and decide whether a particular service is offered. One is the cash flow generated by a service j, another is the revenue sharing scheme γ_j. Besides, the enabler provision capacities, industrial risk/return standards, market prices, all play a role in making service portfolios compatible. In this section we characterize the properties which facilitate the service portfolio compatibility and develop a model for selection of the revenue sharing scheme.

6.4.1 Service Provision Capacities

According to (6.1) a platform service j is described by vector λ_j of the service enablers. Let us denote by I_j the set of enablers which are present in the service description in nonzero quantities:

$$I_j = \{i : \lambda_{ij} > 0\}$$

For each enabler $i \in I_j$ an actor willing to take the role of provision of this enabler should be found. This means that the position j in the service portfolio of an actor that provides enabler $i \in I_j$ should be positive: $x_{ij} > 0$. The value of this position allows to estimate the enabler provision capacity which an actor should possess. Indeed, x_{ij} is a fraction of provision capacity which an actor is going to dedicate to provision of enabler i to service j. Therefore, λ_{ij}/x_{ij} is the capacity necessary to provide a unit of service j. Suppose that $B_j^{\min}$ is the minimal volume of provision of service j which makes such provision viable, and B_j is the target volume of service provision for a generic constellation of actors which is going to provide this service. Then we have the following constraints on the service provision capacities of actors:

$$W_i x_{ij} \geq \lambda_{ij} B_j^{\min}, \ i \in I_j \tag{6.11}$$

if the provision of service j will be viable at all, and

$$W_i x_{ij} \geq \lambda_{ij} B_j, \ i \in I_j \tag{6.12}$$

if only one actor with provision capability of enabler i is desirable in the constellation which provides service j. These constraints can help to make decisions about the nature of the actors which should be encouraged to participate in the provision of different services. For example, some enablers of some services will be provided by established actors with large provision capacity. In such cases the share x_{ij} of capacity dedicated to service j can be small. In other cases the service enablers will be provided by startups with relatively small capacity. In such cases the share x_{ij} should be smaller or equal to 1. These shares implicitly depend on the revenue sharing scheme γ_j through the solution of problem (6.8)–(6.10) and in the latter case it may be beneficial on the initial stages of service penetration to encourage startups by appropriate adjustment of the revenue sharing scheme.

The constraints (6.11)–(6.12) can be also looked at as the constraints on the composition of service portfolio. Suppose that $W_i^{\max}$ is the maximal desirable component provision capacity which an actor providing enabler i should possess. Then the smallest share x_{ij} dedicated to service j should be

$$x_{ij} \geq \lambda_{ij} \frac{B_j^{\min}}{W_i^{\max}}, \quad i \in I_j \tag{6.13}$$

6.4.2 *Risk/Return Industrial Expectations*

Provision of advanced mobile data services involves different actors coming from different backgrounds and industries. There are many startups, but there are also established actors from other industrial branches. For example, the content provision where the same content can be provided to newspapers, internet and mobile terminals. Such actors have preferences for tradeoffs between admissible and/or desirable returns on investment and the corresponding risks. Often such preferences are influenced by industrial standards and expectations inherited from their previous activities. One way to express such expectations is to include all generic projects, in which an actor can be involved in its traditional business, as services in the set of all third party services considered in the model. This is especially useful approach, if the revenues from the traditional activities influence and are influenced by the revenues from the mobile services under consideration. Another possibility is to account for these expectations explicitly. This can be done by introducing the relation between the expected return $\bar{r}_i(x_i)$ and risk [26] $R(x_i)$ from (6.6), (6.7) as follows:

$$\bar{r}_i(x_i) \geq a_i + b_i R(x_i) \tag{6.14}$$

where a_i is the return on investment associated with traditional activity while $b_i R(x_i)$ is the risk premium associated with the participation in provision of the advanced mobile data services. The coefficients a_i and b_i will depend also on individual characteristics of an actor, e.g., its size, market position.

Additionally an actor has the risk tolerance expressed in terms of the upper bound on risk which it is willing to take irrespective of return:

$$R(x_i) \leq \bar{R} \tag{6.15}$$

The upper bound on admissible risk $\bar{R}$ also depends on the characteristics of a particular actor. To put it simply, this is the maximal loss an actor can afford during the time period under consideration.

6.4.3 Pricing

The revenue per unit of service v_j together with the service composition λ_j and the revenue sharing scheme γ_j defines the unit price p_i of enabler i:

$$p_i = \frac{v_j \gamma_{ij}}{\lambda_{ij}}$$

This is a random variable since the revenue is random. Therefore the expected price $\bar{p}_i = \mathbb{E} p_i$ is:

$$\bar{p}_i = \frac{\gamma_{ij}}{\lambda_{ij}} \mathbb{E} v_j$$

An actor providing the enabler i may have the target p_i^* for the price of its product, and the tolerances Δ^+ and Δ^- within which it is willing to accept a price. These targets can result from the market prices in established industries, internal market studies, or internal cost estimates. This leads to the following constraint

$$p_i^* - \Delta^- \leq \frac{\gamma_{ij}}{\lambda_{ij}} \mathbb{E} v_j \leq p_i^* + \Delta^+ \tag{6.16}$$

This constraint should be taken into account while considering the revenue sharing schemes.

6.4.4 Revenue Sharing Schemes

Now let us look at the problem of selecting the revenue sharing coefficients γ_j which would be compatible with the concerted provision of a platform service. Summarizing the discussion presented in Sects. 6.3, 6.4.1 and 6.4.2 we obtain that the actor supplying enabler i selects a portfolio of services $x_i = (x_{i1}, \ldots, x_{i\bar{M}})$ by solving the following problem

$$\max_{x_i} \bar{r}_i(x_i, \gamma_j) \tag{6.17}$$

subject to constraints

$$\sum_{j=1}^{\bar{M}} x_{ij} = 1, \; x_{ij} \geq 0 \tag{6.18}$$

$$\bar{r}_i(x_i, \gamma_j) \geq a_i + b_i R(x_i, \gamma_j) \tag{6.19}$$

$$R(x_i, \gamma_j) \leq \bar{R} \tag{6.20}$$

where $\bar{r}_i(x_i, \gamma_j)$ is the expected return of the actor on his expenditure and $R(x_i, \gamma_j)$ is the risk defined by (6.6) and (6.7), respectively. We emphasize here the dependence of risk and return on the revenue sharing scheme γ_j. Solution of this problem provides service portfolios $x_i(\gamma_j)$ for all generic actors providing enabler i for the platform services $j = 1 : M$. These service portfolios depend on the revenue sharing schemes γ_j. Let us now concentrate on a j-th service. In order that a provision of this service becomes possible it is necessary that all actors providing the necessary enablers to this service include it in their service portfolios in the desirable proportions. This is denoted by:

$$x_i(\gamma_j) \in X_j \text{ for all } i \in I_j \tag{6.21}$$

where the set X_j is defined, for example, by constraints (6.13). The feasible set of the revenue sharing coefficients is defined by the set X_j and if these conditions are not satisfied then the service is not available.

Suppose now that the enabler number 1 of service j is a service aggregation enabler which is provided by an actor which bears overall responsibility for the functioning of service and receives the revenue stream from the end users. Its responsibility includes also the division of the revenue stream between the participating actors and thus also a selection of the revenue sharing coefficients γ_j. It selects these coefficients in such a way that the constraints (6.21) are satisfied. Between all such revenue sharing coefficients selects those which maximize its return. This can be formulated as the following optimization problem.

$$\max_{\gamma_j} \bar{r}_1(x_1(\gamma_j), \gamma_j) \tag{6.22}$$

subject to constraints

$$x_i(\gamma_j) \in X_j \text{ for all } i \in I_j \tag{6.23}$$

$$\gamma_j \in \Gamma_j \tag{6.24}$$

where the set Γ_j is defined, for example, by constraints (6.16). Even simpler, this actor may wish to maximize its revenue share

$$\max_{\gamma_j} \gamma_{1j} \tag{6.25}$$

under constraints (6.23)–(6.24). Observe that the feasible set of this optimization problem depends on the solution of the other actor's optimization problems (6.17)–(6.20).

6.5 Properties of the Models and Implementation Issues

In Sect. 6.4.4 we presented two models for strategic assessment of collaborative provision of mobile data services. These models possess quite complicated structure, although we made a few simplifying assumptions. They can be considered as a special type of stochastic optimization problems with bilevel structure [1], where the lower level is composed of the problems of individual component providers (6.17)–(6.20) while the upper level contains the problem of service provider (6.22)–(6.24). Stochasticity comes from the uncertainty inherent in the characteristics of advanced data services and the user response to them. So far we have adopted a simple treatment of uncertainty by substituting the random variables by their expected values, and, when the special structure of the problem allowed, the covariance matrix. This can be viewed as a special type of the deterministic equivalent of the stochastic programming problems, a technique widely used in stochastic programming (see [2] for more discussion on different types of the deterministic equivalents). A more detailed description of the uncertainty can be introduced in these models following the approach presented in [1]. Moreover, different bilevel optimization problems have drawn considerable attention recently, see [5, 6, 25, 28].

Such problems provide quite a challenge to the state-of-the-art numerical optimization procedures. While many theoretical issues are understood reasonably well, the solution techniques have not yet reached the off-the shelf commercial applicators available for linear and some nonlinear programming problems. The main challenge here is that the upper level problems can be highly nonlinear and nonconvex with multiple local minima. Therefore a substantial work is needed for exploiting the structure of such problems. Still, our aim here is to describe a set of decision support tools for evaluation of business models, where the computational complexities should be hidden from the end user. We have found that this aim can be achieved by a combination of customized implementation with the use of general purpose mathematical modeling systems and commercial software. The general architecture of the system under development is shown in Fig. 6.3.

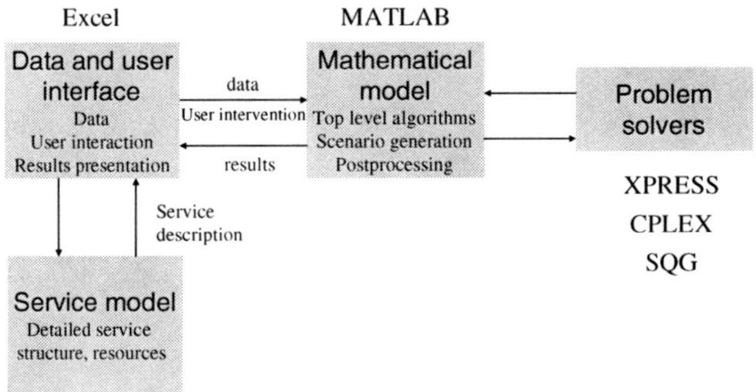

Fig. 6.3 Architecture of decision support system for evaluation of business models of service provision

The system is composed of four components: data and user interface, a set of service models, a set of mathematical models, and a library of solvers.

Data and user interface is implemented in Excel due to its familiarity to potential users. Its purpose is to provide an easy tool for storing and changing the data describing the service and customer properties, for presentation of results of business modeling and for providing the capability to the system user to ask what-if questions pertaining to different scenarios. For example, the efficient frontier illustrated in Fig. 6.2 is presented to the user through this component.

Service model provides the capability to perform detailed modeling of advanced data services. It is implemented in a specialized modeling language which has the features necessary for describing communication sequences. This model provides the aggregated description of services composition λ_j from (6.1).

Mathematical model implements the quantitative description of the business decision process of collaborative service provision from the previous sections. It imports data from data interface and implements the top level structures and algorithms necessary for representation and solution of models (6.17)–(6.20), (6.22)–(6.24). The custom algorithms for analysis and solution of these models are also implemented in MATLAB. This component is also responsible for calling external software for solving subproblems where standard approaches and commercial software are available. For example (6.8)–(6.10) is a quadratic programming problem which can be solved by many solvers, e.g., by the CPLEX and MATLAB optimization toolboxes.

Library of solvers contains solvers for linear and nonlinear programming problems and some a specialized solvers for stochastic programming problems, e.g., SQG [12].

The system depicted in Fig. 6.3 is now in advanced stages of development, in particular the service model component and some mathematical models of service provisioning were implemented in MATLAB [1]. The next section describes some of the results of a case study performed using this system.

6.6 Case Study

This case study deals with analysis of the service provider centric business model for provision of the platform bundle of services to a business person on the move who uses a smart mobile phone to access the corresponding service offer. The setting of this case study is described in Example 6.1 introduced in Sect. 6.3. The prototype of the decision support system implementing models presented in this paper was used for the analysis of this case study; it includes the models described in Sects. 6.3 and 6.4.

Considerable data preparation effort was made for this case study. First of all, we have developed the service composition matrix, showing which enabler participates in which services. This relation between different enablers and services is shown in Fig. 6.1. We have obtained also an average estimate of the service usage (in service instances) in a period of interest, and prices per service instance. We have estimated these data by averaging various service composition and business scenarios. On the basis of technical and economic analysis we have obtained the cost estimates and the correlation matrix showing the correlation between the usage of services and the variance of service usage.

Suppose that the service provider uses our DSS for performing the feasibility study for provision of this bundle of business services similarly to the discussion at the end of Sect. 6.3.3. There are many different what-if questions of interest to the prospective service provider to which this DSS can provide answers. Let us provide an example of this analysis. Suppose that a manager of the service provider feels that the success of the whole enterprise depends critically on the quality and offer of specialized content which can be obtained for his services by engaging prospective content providers (enabler E4 from Fig. 6.1). He wants to get an insight into the properties of the content providers which may be interested in collaboration with the provider and in the chances that his service offer in this respect will stand against the competition of the third party services. One way to do this is to look how the service portfolios and risk/profit preferences of prospective partners depend on correlation and relative pricing of his offer against the offer of competition. Figures 6.4 and 6.5 provide examples of answers which our DSS can deliver.

Figures 6.4 and 6.5 show how the characteristics and attitudes of the content providers towards the service platform depend on the alternatives the competitors can offer to them. Figure 6.4 shows risk/profit efficient frontiers similar to the frontier presented in Fig. 6.2 while Fig. 6.5 depicts the percentage share of the content provision capability of the content providers dedicated to the service platform. In other words, the Fig. 6.5 shows the market share of the service platform in the market for this specific type of content provision dependent on the risk tolerance of the content providers. The competing offer is described by the average price per unit of

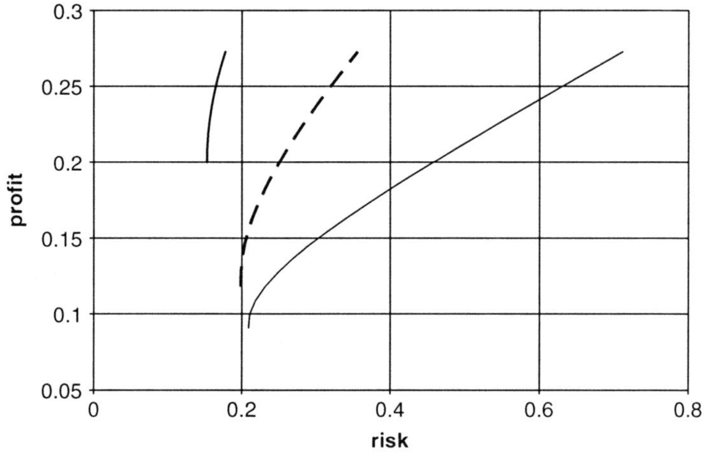

Fig. 6.4 Dependence of risk/return preferences of content providers on the strength of competition

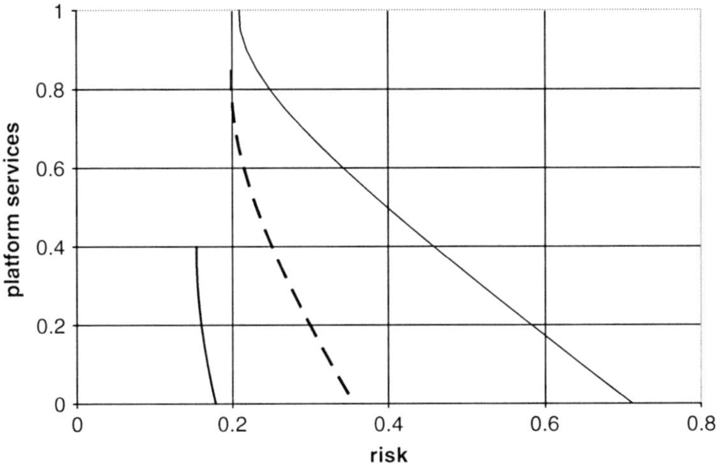

Fig. 6.5 Share of platform services in the service portfolios of content providers

content and by how the actual price can differ from the average price dependent on the future market conditions, as measured by the price variance.

The figures present three scenarios. In all scenarios the competition tries to undercut the service platform by offering about 15% higher average price to content providers for their services. The three scenarios differ by how strong the competition is, namely its capability to maintain the price consistently higher under the changing market conditions. In scenario 1 shown by the thick solid line the competition is strong and has its price variance about two times smaller compared with the platform offer. In scenario 2 depicted by the thick dashed line the competition is about as strong as the platform offer and has the similar price variance. In scenario 3 shown with thin solid line the competition is weaker than the platform and has about twice higher variability of its offer to content providers than the platform.

The results show that in scenario 1 with the strong competition only economically weak content providers with small tolerance towards losses are interested in the collaboration with the platform. Often this corresponds to small firms or even individuals who can not sustain large losses. For such entities participation in the platform means additional security and insurance against losses in the case the strong competing offer prove to be deceitful. Even then, the interest of such firms drops sharply when their risk tolerances grow even by a small amount.

Scenario 3 corresponds to the opposite case when the platform faces aggressive but economically relatively weak competition. Its weakness manifests itself in a large variability of the price offer to the content providers despite the 15% higher average price. In this case the market share of the platform services is much higher and the platform manages to attract also strong actors with higher capacities to sustain losses. Also the market share drops slower with the increase of the loss tolerance of the agents. Scenario 2 corresponds to the intermediate case when the competition is about as strong as the platform and has about the same capability to maintain its price offer to the service providers.

Similar patterns arise when the variability of the revenue steam of content providers is not due to the changes in the unit price of the content but due to the variability of the usage frequencies of this content. Having these predictions, the platform service provider can now realistically weight its strengths and weaknesses, invest more effort into the market research and decide under which market conditions, with what kind of partners, and with what kind of competition it can successfully operate the platform.

6.7 Dynamics of Attitudes

In the previous sections we have analyzed the network risks which arise in the process of collaborative provision of advanced mobile data services which form an important part of information infrastructure. In this section we discuss another aspect of the same situation, namely the process of service adoption by a heterogeneous population of users. The process of service adoption is of fundamental importance to the successful development of information infrastructure because by its very nature the value of the infrastructure for a given user grows with the amount of users already covered by its different components. Adopting the language of microeconomics one can say that the elements of information infrastructure exhibit strong externalities. Modeling these externalities and related risks requires tools and approaches for quantitative modeling of attitudes. In this and subsequent sections we consider a possible methodology based on Bayesian nets.

We formulate a stochastic, dynamic model of attitude formation that takes special account of individual interactions and networks governing intrinsic dynamics of attitudes. The model also accounts explicitly for various external factors such as new information, stimulus, events, actions or some sort of social pressure. If different sets of external factors are activated at different times, the system may show more or less complex dynamics, in particular, it may lead to different alternative attractors. We distinguish two types of influences: (1) interactions among individuals, and (2) the influence of external factors.

According to our model different individuals may receive different information. Information with subjective judgments is transformed through chains of communications to other individuals. Attitudes change in a probabilistic manner depending on attitude of other individuals and information about the external factors. Individuals are socially linked by relationships mediated through a series of intertwining interactions and resulting in a highly diverse social network. This complexity can be modeled by Bayesian networks or more general Markov fields. This notion is a natural generalization of Markov chains to dynamic and spatial processes, whose domains are not necessarily linearly ordered.

6.7.1 Simplified Model: Direct and Indirect Interdependencies

In order to demonstrate the dynamics of attitude change, we begin with representing the public in groups with similar attitudes. Empirical attitude research structures the sampled population into cohorts, possibly by age, sex, income, profession, ethnicity, geographic location, political party affiliation, etc. We assume that population may be divided into "similar-attitude" groups such that an individual has a higher probability of sharing the same attitudes with others in the similar-attitude group than with individuals in other groups. Individuals communicate mainly with individuals within their group, but also with individuals in other groups. Figure 6.6 shows a simplified illustration of possible interactions between five similar-attitude groups that individuals of each group have the same attitudes. The groups are represented by nodes, and interactions are represented by arrows. Thus there is a link from group 3 to group 1, meaning that group 3 has an influence (positive or negative) on group 1. There are also links from group 1 to group 2 and from group 2 to 3. The example can be generalized to N groups, where the arrow from node i to node j indicates a link from i to j.

The direct links between nodes can be represented by the adjacency matrix shown in Table 6.1. An element of this $N \times N$ matrix indicates the position and possibly the strength of the direct dependency links from i to j.

In addition to the direct dependencies, individuals are also indirectly influenced by one another by chains of communications. If A is the adjacency matrix, then an element (i, j) of the matrix $A^2 = A \times A$ represents the number of sequential dependency paths of length 2 involving an intermediate group from i to j; in general, A^l indicates the number of sequential dependency paths of length l with $l - 1$ intermediate groups from i to j. Thus, the entries of the matrix

$$A + A^2 + \ldots + A^l$$

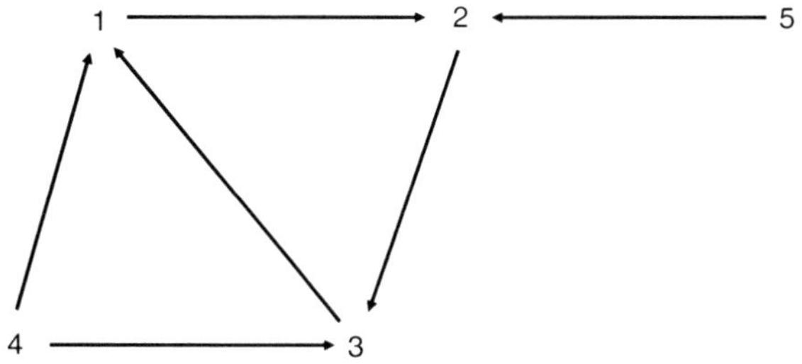

Fig. 6.6 Graph of direct relationships

Table 6.1 The incidence matrix of the graph from Fig. 6.6

Nodes	1	2	3	4	5
1	0	1	0	0	0
2	0	0	1	0	0
3	1	0	0	0	0
4	1	0	1	0	0
5	0	1	0	0	0

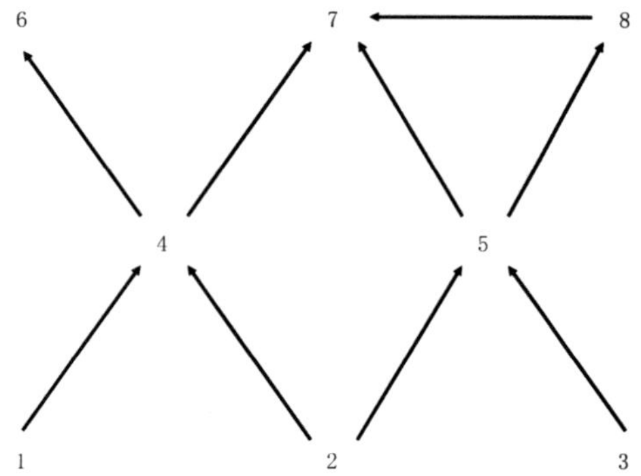

Fig. 6.7 Acyclic graph

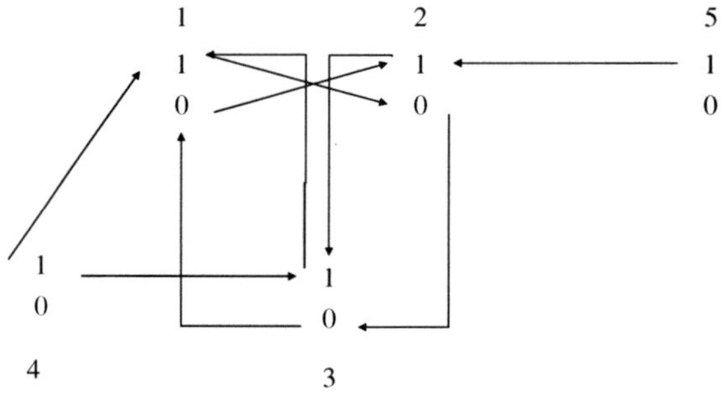

Fig. 6.8 Limit cycle attractor: waves of opinions

represent the number of all possible direct and indirect paths of length smaller or equal to l.

The graph on Fig. 6.6 has a cycle between nodes $1, 2, 3$. The graph on Fig. 6.7 is acyclic which represents hierarchical structure of opinion formation.

Consider now a simple situation of how the links between individuals may affect their attitudes. Suppose that links between five groups of individuals are represented in Fig. 6.6. Figure 6.8 indicates that groups $1, 2, 3$ have two possible (no–yes) attitudes $j = 0, 1$. This is represented by subnodes 0 or 1 inside of each node $i = 1, 2, 3$. We can consider $0 - 1$ states of these nodes. In this example we assume the deterministic nature of interactions. Therefore assume the state $j = 1$ is settled down at nodes $i = 1, 2, 3$ at the initial time interval $t = 0$. In a general case at time $t = 0$ the state $j = 1$ is accepted only by a fraction of a group with a certain probability.

Interdependencies between groups may change their attitudes. Arrows in the graph of Fig. 6.9 indicate that individuals of the group 1 are influenced by group 3 in the sense that it expresses solidarity with group 1 taking the same opinion. Group 2 is antagonistic (in opposition) to group 1. Group 3 has the solidarity with group 2.

It easy to see a cyclic change of responses in time. When individuals of group 2 learn about the attitude of group 1 (state of node $i = 1$), the group 2 changes the attitude to the opposite. Since the initial state of all nodes $j = 1$, the next state of

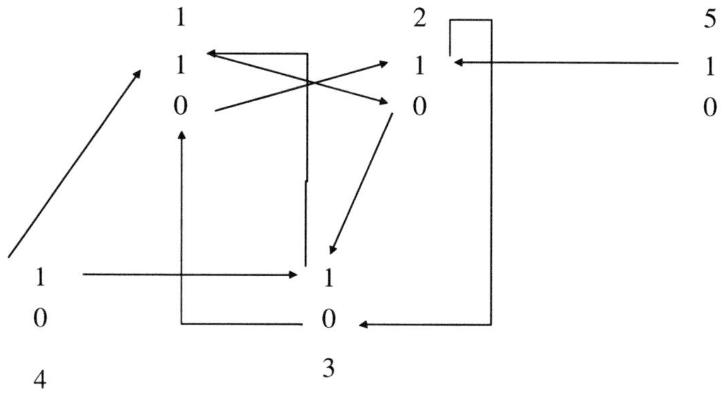

Fig. 6.9 Fixed point attractors

node 2 is $j = 0$ which triggers changes of states at nodes 3, 1 to $j = 1$. These changes lead again to state $j = 1$ at nodes $1, 2, 3$, and so on.

Now suppose that group 3 is antagonistic to the group 2 (see Fig. 6.9), If group 2 learns first the response of group 1 it changes the attitude 1 to 0 and so on until attitudes reach values $(1, 0, 1)$ for groups $1, 2, 3$, respectively. If group 3 learns first the attitude of group 2, the attitudes are settled down at states $(0, 1, 0)$. Hence, the behavior of the attitudes may display two attractors: $(1, 0, 1)$ and $(0, 1, 0)$.

These simple examples suggest that there may be waves of attitudes. Any opinion survey at a particular time may not represent opinion at a later time. These examples also demonstrate that delays in the learning of attitudes may change the pattern of the overall dynamics towards different attractors. In the model of the next section we take more general point of view on driving forces of attitude changes. It is assumed that members of a group may react differently at attitudes of other groups. They also may "hesitate" to react as the opposition or the solidarity.

6.7.2 Model Formulation

In the examples of the previous section individuals of each group have the same attitude. The individuals of a group also react in the same manner at "signals" (attitudes) from other groups. In this section we undertake a more general view. It is assumed that only a fraction p_j^{it} of members, the group i have attitude j at time t. Thus in the previous examples there may be fractions p_0^{it}, p_1^{it} of each group $i = 1, 2, 3$ with attitudes 0 and 1 at time t. Of course, $p_0^{it} + p_1^{it} = 1$.

Individuals of a group communicate with different individuals of other groups. Therefore, their attitudes are influenced by random samples of information from adjacent (neighbor) groups. Individuals may form their attitudes at a particular time interval t, on the basis of various rules besides just "solidarity" or "opposition" principles. For example, they may follow the majority attitude from a sampled opinions. In general, such behavior induces a conditional probability for an individual of a group i to take an attitude from given set of possible attitudes. This probability is conditioned on their current attitudes, the attitudes of adjacent groups and some exogenous variables, or external factors.

Let us now formulate the model precisely. To do so, we must represent the driving forces of attitude changes or dependencies between groups as well as dependencies of these relations on external factors. We opted for probabilistic description based on conditional distributions.

The model distinguishes N groups $i = 1, \dots, N$ of individuals. The number of the relevant groups depends on the issue under consideration and on the level of detail represented in the model.

Individuals display different attitudes to the given issue, ranging from hostile to very favorable. We assume that there is a finite number M of possible attitudes. The attitudes of individuals of group i are described by the random variable ζ_i which takes values from 1 to M. In other words, we assume that individuals from group i statistically follow the same pattern of attitude formation given by the distribution of ζ_i. In this sense, we can say that individuals of group i share approximately the same view. The attitudes of the population are described by random vector $\zeta = (\zeta_1, \dots, \zeta_N)$. A fixed value of this vector is denoted by z and the set of all possible attitudes by Z. Let us denote by p_j^{it} the probability that a member of group i assumes the attitude j at time t:

$$p_j^{it} = \mathcal{P}(\zeta_i^t = j)$$

Naturally,

$$\sum_{j=1}^{M} p_j^{it} = 1, \quad p_j^{it} \geq 0$$

Interactions between individuals are represented as a graph similar to that described in Sect. 6.7.1. Groups $i = 1, \dots, N$ correspond to nodes of the graph and direct links between individuals of groups are represented by arrows between nodes. Thus there are two sets: nodes $V = 1, \dots, N$ and the set of arrows (directed arcs) U. Let us denote this as $G = (V, U)$. If nodes i, j belong to V, $i, j \in V$ and there is an arrow from i to j, $(i, j) \in U$ then i is an adjacent to j node. Define as V_j the set of all nodes adjacent to the j node. Let z_{V_j} be a subvector of the vector of attitudes indexed by V_j. For example, in the case of dependencies indicated by graph in Fig. 6.6 we have $V_1 = \{1, 3, 4\}$, $V_2 = \{1, 2, 5\}$, $V_3 = \{2, 3, 4\}$, $V_4 = \{\emptyset\}$, $V_5 = \{\emptyset\}$ where $\emptyset$ is the symbol of the empty set. Then $z_{V_1} = (z_1, z_3, z_4)$, $z_{V_2} = (z_1, z_2, z_5)$.

Individuals change their attitudes depending on current attitudes of their own and adjacent groups, and on a vector x of exogenous external factors or variables. The attitude formation is described by a conditional distribution $H^i(z_i | z_{V_i}, x)$ for each individual of group i: the probability for an individual of group i to have an attitude $z_i = 1, 2, \dots, M$ when attitudes of groups V_i are Z_{V_i} and external factors have values x.

We assume that function H^i can be derived on the basis of appropriate questionnaires answered by a representative member of individuals from each group, such as: "What is your attitude (from 1 to M) if attitudes of the adjacent groups are z_{V_i} and the environment is $x = (x_1, \ldots, x_n)$?". Functions H^i may also depend on time interval t, but we skip it in order to simplify the notation. Functions H^i define the dynamics of the attitude change according to the following relation:

$$p_j^{i,t+1} = \sum_{z_{V_i} \in Z} H^i(\zeta_i^{t+1} = j | \zeta_{V_i}^t = z_{V_i}; x^t) \mathcal{P}(\zeta_{V_i}^t = z_{V_i}) \qquad (6.26)$$

The groups i with $V_i = i$ can be identified with "leaders", which influence opinions of other individuals but are not influenced themselves.

To define completely the dynamics of the system described by (6.26) it is necessary to fix initial attitudes distributions for $t = 0$. In case when the corresponding graph of direct influences is acyclic, it is enough to define these distributions for nodes i such that $V_i = i$.

Equation (6.26) together with initial distributions allows us in principle to calculate p_j^{it} for any $t \geq 0$. Of course, for complex graphs it is practically impossible to derive analytical formulas for p_j^{it} as functions of external factors x. The existence of analytical expressions for all p_j^{it} provides an easy tool to analyze implications of changes in x. The next section is devoted to the analysis of attitudes changes in the case when such a possibility does not exist. The approach is based on the stochastic version of (6.26) dealing directly with random variables ζ_i^t, $i = 1, \ldots, N$ by using the Monte-Carlo simulation techniques. Instead of H^i, $i = 1, \ldots, N$ the approach allows also to use myopic rules for generating random changes of ζ^t, i.e., the approach allows to analyze cases when functions H^i are given implicitly. One such important case arises in the situations when individuals form their attitudes by asking acquaintances from adjacent groups and use some simple rules based on majority or minority of sampled attitudes. Let us now formulate some problems that are important in this context.

Problem 1. Evaluation of attitudes

The objective here is to predict attitudes of various population groups. As we have seen in the previous section, the attitudes of different groups change in intricate ways and are subject to changes in external factors, as well as direct and indirect dependencies. Direct dependencies involve relatively few adjacent groups, while indirect dependencies and external factors may involve all or almost all population groups. Thus, in Fig. 6.1 group 2 is affected directly only by groups 1, 5. But there exists a path from 3 to 2, and from 4 to 2. Therefore, indirectly individuals of group 2 are affected by all groups. The direct dependencies are much easier to study experimentally through surveys and questionnaires. Suppose that we managed to study the direct dependencies between attitudes of different adjacent population groups. The problem is to predict the public attitudes as the result of complex direct and indirect interactions by using the information about direct dependencies. As was outlined in the previous section, this problem involves the calculation of all possible direct and indirect paths by using the adjacency matrix (Table 6.1). Formally the problem is formulated as follows. Given conditional distributions $H^i(z_i|z_{V_i};x)$ for $i = 1 : N$ and the values of x^t for $t = 0, 1, \dots, T-1$ find distributions $\mathcal{P}(\zeta_i^T)$, $i = 1 : N$ of random variables ζ_i^T, i.e., the public attitudes at time T.

Problem 2. Response interpretation

This problem deals with the interpretation (identification) of public response to a mixture of events and efforts which influence the public. This interpretation is made on the basis of our knowledge which has two components. The first component is the a priori information on direct dependencies deduced from responses in the past. The second component consists of new direct observations of attitudes for some groups to a given new issue. This type of knowledge can be called a posteriori knowledge. The response interpretation deals with the following questions:

- Suppose that we have direct observations on the attitudes of only some selected groups, or we have observations of aggregated response from several groups. How can we recover attitudes of unobserved groups?
- How to use the newly acquired a posteriori knowledge to update our knowledge about direct dependencies between groups?
- Often a public response is the result of mixture of different, sometimes conflicting events and efforts. What is the contribution of each single event to the attitude dynamics?

Formally these problems can be formulated as follows. Let us consider only first problem. Denote by $V_{\mathcal{E}}$ the set of observed groups. Given conditional distributions $H^i(z_i|z_{V_i};x)$ for $i = 1 : N$, the values of x^t and distributions $\mathcal{P}(\zeta_i^t)$, $i \in V_{\mathcal{E}}$ for $t = 0, 1, \dots, T-1$ find distributions $\mathcal{P}(\zeta_i^t)$ for $i \notin V_{\mathcal{E}}$.

Problem 3. Sensitivity analysis

Here we want to analyze a sensitivity of attitudes with respect to changes of environmental variable x. Through such an analysis we may find that attitudes are especially resistant to changes in certain directions or in certain positions. For example, in siting a waste processing facility, it is necessary to analyze a choice of its size, decide on a distance between facility and population centers, choose routes of the waste transportation, etc. Different population groups react differently on different options. Small changes in critical parameters may considerably affect the public attitudes, while substantial and possibly costly changes in non-critical parameters will not change the public response. The objective of the sensitivity analysis is to identify the critical parameters utilizing the knowledge of the direct dependencies and how these dependencies are affected by changes in environmental parameters.

In terms of our model the sensitivities of public responses is defined in terms of changes in response distribution $\mathcal{P}(\cdot)$ with respect to parameters x. This leads to the following formulation.

Given conditional distributions $H^i(z_i|z_{V_i};x)$ for $i = 1 : N$ and the values of x^t for $t = 0, 1, \dots, T-1$ estimate derivatives of distributions $\mathcal{P}(\zeta_i^T)$, $i = 1 : N$ with respect to $x^0, \dots, x^T$.

Note that distributions $\mathcal{P}(\cdot)$ depend on x indirectly through conditional distributions H, what leads to a challenging problems analyzed in next sections.

Problem 4. Social learning

As it is emphasized in social psychology people receive information from their social environment. A lack of connectivity between them develops clusters of people sharing similar views in a more heterogeneous population. Traditionally, it is assumed that a change in thought or behavior is only a reaction to some external factors or stimulus. The proposed model emphasizes the existence of intrinsic changes generated by interdependencies between individuals in the absence of any external factor. This dynamics may be perturbed by external factors activated at different times leading to more or less complex patterns of alternative dynamics. How can we learn the variety of alternative attitudinal developments and how we can characterize them? What are optimistic and pessimistic "scenarios" of such developments? Can attitudes reach critical levels? Answers to such questions depend not only on existing links between individuals but also on paths of activated external factors. The main problem is to use a model in order to learn possible alternative scenarios and their outcomes. For example, in the debates on siting a waste processing facility, there is a possibility to change sizes of facilities, their locations, premiums and other compensations in order to change public responses.

The power of a model is its ability to learn patterns of possible responses of the system without time consuming real observations and trial-and-error experiments. In our case the proposed model supports identification of paths of external factors leading to different outcomes, for example either decreasing a social tension, or achieving the worst case situation. In order to conduct such analysis we need to introduce a set of "performance" indicators or "score" functions distinguishing one trajectory of attitudes from another. For example, if c_{ij} is relative importance of attitude j by individuals of group i, then the cumulative score of a trajectory of attitudes can be expressed by the following score (performance) function:

$$F(x) = \sum_{t=0}^{T}\sum_{i=1}^{N}\sum_{j=1}^{M} c_{ij}\, p_j^{it}, \tag{6.27}$$

which implicitly depends on external factors x through conditional probabilities in (6.26).

The sensitivity analysis can indicate changes in x which lead to increasing value of an indicator $F(x)$. Using this information, it is possible to identify, for example, the worst or the best case sets of possible external factors.

The fundamental complexity of such type of problems is due to the probabilistic nature of $F(x)$ and implicit dependencies on variables x. Next sections discuss tools enabling to deal with the involved complexity.

6.7.3 Bayesian Networks and Markov Fields

The Bayesian net is a powerful tool which was developed primarily to deal with stochastic problems defined on acyclic graphs (see Fig. 6.8) We show that algorithms developed for the Bayesian nets constitute the building blocks for more general algorithms which can deal also with general graphs.

Bayesian nets are specifically designed for cases when the vector of random parameters ζ can have considerable dimension and/or it is difficult to come up with traditional parametric model of the joint distribution of random parameters.

The cause–effect structure is associated with the vector of random parameters $\zeta = (\zeta_1, \ldots, \zeta_i, \ldots, \zeta_N)$. That is, for any parameter ζ_i the set of indices V_i is selected such that the elements of subvector $\{\zeta_j\}_{j \in V_i}$ can be identified with "causes" of ζ_i. Vector ζ_i is changed in time leading to a random path or trajectory $\zeta^t, t = 0, 1, \ldots$. In the proposed model, the groups of a population, say in a given region, are represented as nodes of a graph. Direct dependencies between groups are represented by arrows indicating directions of communications. Node i of the graph has different random states ζ_i^t at time t reflecting different attitudes of the group. Thus the stochastic dynamics of attitudes is characterized by a random vector ζ^t with dependent components. The important feature is that changes in a component ζ_i^t are triggered only by its "neighbors" ζ_j^t, $j \in V_i$, apart from changes in external factors.

Such stochastic processes define Markov random fields. They can be regarded as a generalization of Markov processes to situations when the cause effect structure is not necessarily linearly ordered. Let us start the formal exposition by gathering the basic definitions which will be used in next sections. Consider a directed graph defined as a pair of two sets (V, U): nodes V and connecting them arrows (directed arcs) U. An example of such graph is given in Fig. 6.6 with $V = \{1, 2, \ldots, 9\}$. For each node $v \in V$ let us define the set of parents

$$c(v) = \{j \,|\, (j, v) \in E\}, \tag{6.28}$$

and the set of descendants $d(v)$ as the set of all nodes $\bar{v}$ from V for which there exists directed path which originates in $v$and terminates in $\bar{v}$.

Consider a set of random variables $\zeta_V = \{\zeta_v, v \in V\}$ indexed by nodes from V and defined on appropriate probability space $(\Omega, \mathbb{B}, \mathbb{P})$. Suppose that W is an arbitrary subset of V by $v \backslash \{v \cup d(v)\}$. A Markov random field ζ is characterized by the property

$$\mathbb{P}(\zeta_v | \zeta_{W \cup c(v)}) = \mathbb{P}(\zeta_v | \zeta_{c(v)}) \tag{6.29}$$

If G is a directed acyclic graph, then (ζ, G) which satisfies (6.29) is called *Bayesian network.*

Suppose now that $H(\zeta)$ is the joint distribution of ζ. Then it follows from the definition of the Bayesian network that

$$H(\zeta) = \prod_{v \in V} H^v(\zeta_v | \zeta_{c(v)}) \tag{6.30}$$

where $H^v(\zeta_v|\zeta_{c(v)})$ is a conditional distribution function of ζ_v given $\zeta_{c(v)}$.

Coming back to the model of Sect. 6.3 we see that it fits in the framework of Bayesian networks and Markov fields. The essential new feature is the dependence of the conditional distribution functions H^v on the vector of external variables x from a feasible set $X \subseteq R^n$:

$$H^v(\cdot) = H^v(\zeta_v|\zeta_{c(v)}, x) \tag{6.31}$$

The change of $x \in X$ affects the interactions between groups.

6.7.4 Sensitivity Analysis

The analysis aims at examining attitudes with respect to changes in external factors x. These changes are characterized by certain indicators or "score" functions such as (6.27) is a function of ζ and possibly x, e.g., $f(x, \zeta)$.

Several questions can now be considered. For example, how can we estimate the expected outcome

$$F(x) = \mathbb{E} f(x, \zeta) \tag{6.32}$$

at some point x? How sensitive is this value with respect to changes of parameters x? What are the most critical parameters? What is the value of the performance indicator at point $x + \delta x$ where δx is a small perturbation of x? These questions are not the trivial because each estimation of the value of $F(x)$ can be very time consuming taking into account indirect interdependencies of the network.

More precisely, in order to evaluate the sensitivity of $F(x)$ we need to develop algorithms for estimating the value of the gradient of this indicator. That is, for a given x we need to compute vector ξ such that

$$\mathbb{E}\xi = F_x(x) = \frac{d}{dx}\mathbb{E} f(x, \zeta)$$

One possibility is to use the finite differences:

$$\xi = \sum_{i=1}^{n} \frac{\hat{F}(x + \Delta e_i) - \hat{F}(x)}{\Delta} e_i,$$

where $\hat{F}(x)$ is an estimate of the value of function $F(x)$ at point x. In this case, however, it is necessary to compute at least $n + 1$ estimates of the performance measure which may be too demanding computationally.

Let us introduce simple but useful differentiation formulas for the gradient of function $F(x)$, which are used in stochastic optimization.

Theorem 6.1. *Suppose that*

1. Random variables ζ_v *have conditional densities* $h(\zeta_v|\zeta_{c(v)}, x)$

2. Functions $h(\zeta_v|\zeta_{c(v)}, x)$, $f(x,\zeta)$ *are differentiable with respect to* x *uniformly with respect to* ζ. *Then* $F(x)$ *defined in (6.32) is differentiable and*

$$F_x(x) = \mathbb{E}\left\{ f_x(x,\zeta) + f(x,\zeta)\sum_{v\in V}\frac{h_x(\zeta_v|\zeta_{c(v)}, x)}{h(\zeta_v|\zeta_{c(v)}, x)}\right\} \tag{6.33}$$

Proof.

From (6.30) we have the following expression for $F(x)$:

$$F(x) = \int f(x,\zeta)\prod_{v\in V} h(\zeta_v|\zeta_{c(v)}, x)\prod_{v\in V} d\zeta_v \tag{6.34}$$

Under assumptions of the theorem we can change the order of differentiation and integration, which yields:

$$\begin{aligned} F_x(x) &= \int \frac{d}{dx}\left(f(x,\zeta)\prod_{v\in V} h(\zeta_v|\zeta_{c(v)}, x)\right)\prod_{v\in V} d\zeta_v \\ &= \int f_x(x,\zeta)\prod_{v\in V} h(\zeta_v|\zeta_{c(v)}, x)\prod_{v\in V} d\zeta_v \\ &\quad + \sum_{w\in V}\int f(x,\zeta)h_x(\zeta_w|\zeta_{c(w)}, x))\prod_{v\in V,\ v\neq w} h(\zeta_v|\zeta_{c(v)}, x)\prod_{v\in V} d\zeta_v \\ &= \int\left(f_x(x,\zeta) + f(x,\zeta)\sum_{v\in V}\frac{h_x(\zeta_v|\zeta_{c(v)}, x)}{h(\zeta_v|\zeta_{c(v)}, x)}\right)\prod_{v\in V} h(\zeta_v|\zeta_{c(v)}, x)\prod_{v\in V} d\zeta_v \end{aligned}$$

The proof is completed. □

This theorem can also be proved by taking the logarithm from the expression under the integral in (6.35) and differentiating it. The theorem is similar to the Theorem 6.1 from [15].

Similar result holds when each of the random variables ζ_v takes finite number of values and instead of conditional densities we have conditional probabilities $P(\zeta_v|\zeta_{c(v)}, x)$. If these probabilities are differentiable with respect to x we obtain the following expression for $F_x(x)$:

$$F_x(x) = \mathbb{E}\left\{ f_x(x,\zeta) + f(x,\zeta)\sum_{v\in V}\frac{P_x(\zeta_v|\zeta_{c(v)}, x)}{P(\zeta_v|\zeta_{c(v)}, x)}\right\} \tag{6.35}$$

Note that the second terms in expressions (6.33) and (6.35) can be interpreted as sums of likelihood ratios [17, 27]. The calculation of exact values $F_x(x)$ is possible only in exceptional cases for simple networks.

Let us now present estimation algorithms which exploit the structure of Bayesian network and expressions (6.33) and (6.35) in order to obtain statistical estimates of $F(x)$ and $F_x(x)$. We shall make reference to discrete case and use (6.35); the continuous case is treated similarly. The simplest estimation scheme is the following:

$$\xi = \frac{1}{K}\sum_{k=1}^{K}\left(f_x(x,\zeta^k) + f(x,\zeta^k)\sum_{v\in V}\frac{P_x(\zeta_v^k|\zeta_{c(v)}^k, x)}{P(\zeta_v^k|\zeta_{c(v)}^k, x)}\right) \tag{6.36}$$

where $\zeta^k, \zeta_v^k, \zeta_{c(v)}^k, k = 1 : K, \ K \geq 1$ are independent observations of random vectors $\zeta, \zeta_v, \zeta_{c(v)}$ respectively. Stochastic vector ξ is termed as stochastic gradient. The estimation scheme (6.36) in the context of Bayesian networks is called also as Logic Sampling. The estimator (6.36) requires $K \geq 1$ observations ("scenarios") of attitudes ζ. This scenarios can be sequentially generated by using the Monte-Carlo simulation techniques. An arbitrary scenario generating cycle $k = 1 : K$ have the following simple steps:

1. *Initialization.* Select a node $v \in V$ such that $c(v)$ is the empty set, $c(v) = \emptyset$. Since the graph G is acyclic such nodes exist and we interpreted them as sources of opinions or leader nodes. Sample the component ζ_v from unconditional distribution $P(\zeta_v|\zeta_{c(v)}, x)$ and denote the result as ζ_v^k. Perform this step for all nodes with empty parent set $c(v)$.
2. *Selection.* Select a not yet sampled node $v \in V$ such that all random variables $\zeta_j, \ j \in c(v)$ have been sampled already during current scenario generating cycle. Suppose that the observation of ζ_j is $\zeta_j^k, j \in c(v)$.
3. *Sampling.* Sample random variable ζ_v and obtain an observation ζ_v^k. Steps 2 and 3 are repeated until all random components of ζ are sampled. In other words, until all groups reveal their opinions.

Note that by using observations $\zeta^k, k = 1 : K, \ K \geq 1$ along with the estimates of the gradient F_x we obtain the estimate of $F(x)$ itself. For example, $F(x)$ may correspond to the probability for ζ to belong to a certain desirable domain [for example, majority of attitudes 1 (yes) to the competing attitudes 0 (no)] we obtain the estimates of this probability and its sensitivity ξ.

There may be other sampling schemes known as Evidence Weighting [4] and Gibbs Sampling [16] which are particularly useful for the solution of the response interpretation problem (Problem 2 from Sect. 6.7.2). Such schemes use the structure of the network and a posteriori distributions instead of $P(\zeta_v|\zeta_{c(v)}, x)$. The estimates of stochastic gradient obtained above can be used for obtaining the optimal values of parameters using stochastic quasigradient methods like in [7–9].

6.7.5 General Interdependencies

Let us extend the analysis of the previous section to the case of cyclic graphs. The exposition follows closely the exposition of the previous section and therefore we concentrate on the new features only. The most important among these features is the dynamic aspect of attitudes change which in the case of conventional Bayesian network was reduced to analysis of their propagation from "leaders" to other groups.

Let us fix the time horizon $[0, l]$ during which we study the attitude dynamics and sensitivity estimates. Such estimates will be called l-links estimates. Generally, the variables x may change during this time period: $x = (x^0, \ldots, x^{l-1})$. For the sake of simplicity we shall derive our estimates for the case when $x^0 = x^1 = \ldots = x^{l-1}$ denoting this constant vector by x. The general case does not bring any conceptual difficulties and is treated similarly.

Consider again the case when conditional distributions have densities. Consider the following performance measure which is a generalization of the measure (6.34):

$$F(x) = \int f(x, \bar{\zeta^l}) h(x, \bar{\zeta^l}) d\bar{\zeta^l} \tag{6.37}$$

where for $t = 0, \ldots, l$ we denoted by $h(x, \bar{\zeta^t})$ the joint density of the random vector $\bar{\zeta^t} = (\zeta^0, \ldots, \zeta^t)$ with $\zeta^t = (\zeta_1^t, \ldots, \zeta_N^t)$ and

$$d\bar{\zeta^l} = \prod_{t=1}^{l} \prod_{v \in V} d\zeta_v^t$$

The density of $\bar{\zeta^t}$ is connected with the density of $\bar{\zeta}^{t-1}$ for $t = 1, \ldots, l$ with the following relation:

$$h(x, \bar{\zeta^t}) = h(x, \bar{\zeta}^{t-1}) \prod_{v \in V} h(\zeta_v^t | \zeta_{c(v)}^{t-1}, x) \tag{6.38}$$

with the initial distribution $h(x, \bar{\zeta^0})$ density being simply

$$h(x, \bar{\zeta^0}) = \prod_{v \in V} h(\zeta_v^0) \tag{6.39}$$

Combining expressions (6.37)–(6.39) we obtain the basic formula for the performance measure:

$$F(x) = \int f(x, \bar{\zeta^l}) \prod_{v \in V} h(\zeta_v^0) \prod_{t=1}^{l} \prod_{v \in V} h(\zeta_v^t | \zeta_{c(v)}^{t-1}, x) \prod_{t=1}^{l} \prod_{v \in V} d\zeta_v^t \tag{6.40}$$

Utilizing this expression similarly to the previous section we obtain the following result:

Theorem 6.2. *Suppose that*

1. Random variables ζ_v^t have conditional densities $h(\zeta_v^t | \zeta_{c(v)}^t, x)$ for $t = 1, \ldots, l$ and densities $h(\zeta^0)$ for $t = 0$.

2. *Functions* $h(x, \zeta_v^t|\zeta_{c(v)}^t)$, $f(x, \bar{\zeta^l})$ *are differentiable with respect to* x *uniformly with respect to* $\bar{\zeta^l}$. *Then* $F(x)$ *defined in (6.37) is differentiable and*

$$F_x(x) = \mathbb{E}\left\{ f_x(x, \bar{\zeta^l}) + f(x, \bar{\zeta^l}) \sum_{t=1}^{l} \sum_{v \in V} \frac{h_x(\zeta_v^t|\zeta_{c(v)}^{t-1}, x)}{h(\zeta_v^t|\zeta_{c(v)}^{t-1}, x)} \right\} \tag{6.41}$$

As in the previous section similar result holds when each of the random variables ζ_v^t takes a finite number of values, and instead of conditional densities we have conditional probabilities $P(x, \zeta_v^t|\zeta_{c(v)}^t, x)$. If these probabilities are differentiable with respect to x, then we obtain the following expression for $F_x(x)$:

$$F_x(x) = \mathbb{E}\left\{ f_x(x, \bar{\zeta^l}) + f(x, \bar{\zeta^l}) \sum_{t=1}^{l} \sum_{v \in V} \frac{P_x(\zeta_v^t|\zeta_{c(v)}^{t-1}, x)}{P(\zeta_v^t|\zeta_{c(v)}^{t-1}, x)} \right\} \tag{6.42}$$

The l-stage sensitivity estimate based on the logic sampling in the discrete case takes the form:

$$\xi = \frac{1}{K} \sum_{k=1}^{K} \left(f_x(x, \bar{\zeta}^{kl}) + f(x, \bar{\zeta}^{kl}) \sum_{t=1}^{l} \sum_{v \in V} \frac{P_x(\zeta_v^{kt}|\zeta_{c(v)}^{k,t-1}, x)}{P(x, \zeta_v^{kt}|\zeta_{c(v)}^{k,t-1}, x)} \right) \tag{6.43}$$

where $\bar{\zeta}^{kt}, \zeta_v^{kt}, \zeta_{c(v)}^{kt}$, $t = 0, 1, \ldots, l$, $k = 1 : K$, $K \geq 1$ are independent observations of random vectors $\bar{\zeta}^t, \zeta_v^t, \zeta_{c(v)}^t$ respectively. The vector ξ defines a stochastic gradient of the function $F(x)$ in (6.37).

Observations ζ^{kt}, $k = 1 : K$ of random vectors ζ^t, $t = 0, 1, \ldots, l$ can be simulated similar to acyclic graphs of the previous section. In fact, the study of l-stage interdependencies between groups (nodes) on general graphs can be reduced to the study of l-stage sensitivity estimates on acyclic graphs.

This reduction of l-stage sensitivity analysis to acyclic graphs provides a general approach to the study of the general nets with cyclic graphs. Keeping in mind this possibility let us describe the Monte-Carlo simulation procedure for the generation $k = 1 : K$ scenarios ζ, $t = 0, 1, \ldots, l$ describing l-stage propagation effects of public attitudes:

1. *Initialization.* Leader nodes of the equivalent acyclic graph (see Fig. 6.6) correspond to $t = 0$. Therefore, for all $v \in V$ sample the values ζ^{kt} of random variables ζ_v^t, $t = 0$.
2. *Selection.* Select all nodes $v \in V$ sequentially for $t = 1 : K$.
3. *Sampling.* For a sampled node v at step t sample random variables ζ_v^{kt+1} from distribution $P(\zeta_v^{t+1}|\zeta_t^v = \zeta_v^{kt}, x)$. Steps 2, 3 are repeated until all components ζ^{kt} of vectors ζ^t, $t = 0, 1, \ldots, l$ are sampled.

Scenarios ζ^{kt}, $t = 0, 1, \ldots, l$, $k = 1 : K$ of public attitudes allow to estimate vector $F_x(x)$ according to (6.43). As in the previous section this vector indicates directions of changes in x towards the increase of indicator $F(x)$.

6.8 Conclusion

In this paper we set the stage for further development of modeling and decision support tools for analysis of design and deployment of robust information infrastructure using as the case study one component of this infrastructure: advanced mobile data services. The methodologies utilized here, especially the portfolio theory and the Bayesian nets combined with the stochastic optimization, have a potential to be exploited also for solutions of similar problems arising in other types of infrastructure.

Acknowledgements Parts of this work has benefited from the support of the IST project IST-2005-027617 SPICE (partly funded by the European Union), and of the COST Action 293 *Graphs and algorithms in communication networks (GRAAL)*. The authors are grateful to Dr. Josip Zoric of Telenor for useful discussions and help in formulating Example 6.1.

Alexei Gaivoronski acknowledges the hospitality of the Integrated Modeling Environment Project, and the IIASA creative environment.

The authors are grateful to two anonymous referees who contributed with their comments to the improvement of the paper.

References

1. Audestad, J.-A., Gaivoronski, A.A., Werner, A.: Extending the stochastic programming framework for the modeling of several decision makers: pricing and competition in the telecommunication sector. Ann. Oper. Res. **142**, 19–39 (2006)
2. Birge, J.R., Louveaux, F.: Introduction to Stochastic Programming. Springer, New York (1997)
3. BIS: Amendment to the capital accord to incorporate market risks. Bank for International Settlements (1996)
4. Bishop, C.M.: Pattern Recognition and Machine Learning. Springer, New York (2006)
5. Colson, B., Marcotte, P., Savard, G.: Bilevel programming: A survey. 4OR **3**(2), 87–107 (2005)
6. Dempe, S.: Foundations of Bilevel Programming. Kluwer, Dordrecht (2002)
7. Ermoliev, Y.M.: Methods of Stochastic Programming. Nauka, Moscow (1976)
8. Ermoliev, Y.M.: Stochastic quasigradient methods: general theory. In: Encyclopedia of Optimization. Kluwer, Boston (2000)
9. Ermoliev, Y.M., Gaivoronski, A.A.: Stochastic quasigraient methods. SIAM/OPT Views-and-News (1995)
10. Ermoliev, Y.M., Wets, R.J.-B. (eds.): Numerical Techniques for Stochastic Optimization. Springer, Berlin (1988)
11. Faber, E., Ballon, P., Bouwman, H., Haaker, T., Rietkerk, O., Steen, M.: Designing business models for mobile ICT services. In: Proceedings of 16th Bled E-Commerce Conference, Bled, Slovenia, June 2003
12. Gaivoronski, A.A.: SQG: stochastic programming software environment. In: Wallace, S.W., Ziemba, W.T. (eds.) Applications of Stochastic Programming. SIAM & MPS, Philadelphia (2005)
13. Gaivoronski, A.: Stochastic optimization in telecommunications. In: Resende, M.G.C., Pardalos, P.M. (eds.) Handbook of Optimization in Telecommunications, chap. 27, pp. 761–799. Springer, Berlin (2006)
14. Gaivoronski, A.A., Pflug, G.: Value-at-risk in portfolio optimization: properties and computational approach. J. Risk **7**(2), 1–31 (2005)

15. Gaivoronski, A.A., Stella, F.: Stochastic optimization with structured distributions: the case of Bayesian Nets. Ann. Oper. Res. **81**, 189–211 (1998)
16. Geman, S., Geman, D.: Stochastic relaxation, gibbs distributions, and the Bayesian restoration of images. IEEE Trans. Pattern Anal. Mach. Intell. **6**, 721–741 (1984)
17. Glynn, P.W., Iglehart, D.L.: Importance sampling for stochastic simulations. Manage. Sci. **35**(11), 1367–1392 (1989)
18. Granat, J., Makowski, M.: Interactive specification and analysis of aspiration-based preferences. Eur. J. Oper. Res. **122**(2), 469–485 (2000)
19. Haaker, T., Kijl, B., Galli, L., Killström, U., Immonen, O., de Reuver, M.: Challenges in designing viable business models for context-aware mobile services. In: CICT Conference Papers, Copenaghen, 2006
20. Kall, P., Wallace, S.: Stochastic Programming. Wiley, New York (1994)
21. Luenberger, D.: Investment Science. Oxford University Press, Oxford (1998)
22. Makowski, M.: Management of attainable tradeoffs between conflicting goals. J. Comput. **4**, 1033–1042 (2009)
23. Makowski, M., Wierzbicki, A.: Modeling knowledge: model-based decision support and soft computations. In: Yu, X., Kacprzyk, J. (eds.) Applied Decision Support with Soft Computing. Springer, Berlin (2003)
24. Markowitz, H.: Portfolio Selection, 2nd edn. Blackwell, New York (1991)
25. Patriksson, M., Wynter, L.: Stochastic mathematical programs with equilibrium constraints. Oper. Res. Lett. **25**(4), 159–167 (1999)
26. Rockafellar, R.T., Ursayev, S.: Optimization of conditional value-at-risk. J. Risk **2**(3), 21–41 (2000)
27. Rubinstein, R.Y.: Sensitivity analysis and performance extrapolation for computer simulation models. Oper. Res. **37**, 72–81 (1989)
28. Scholtes, S., Stöhr, M.: Exact penalization of mathematical programs with equilibrium constraints. SIAM J. Control Optim. **37**, 617–652 (1999)
29. Trigeorgis, L.: Real Options: Managerial Flexibility and Strategy in Resource Allocation. MIT, Cambridge, MA (1996)
30. Wierzbicki, A., Makowski, M., Wessels, J. (eds.): Model-Based Decision Support Methodology with Environmental Applications. Kluwer, Dordrecht (2000)

Part III
Analysis and Optimization of Technical Systems and Structures Under Uncertainty

Chapter 7
Optimal Ellipsoidal Estimates of Uncertain Systems: An Overview and New Results

F.L. Chernousko

Abstract The method of ellipsoids for the guaranteed state estimation of uncertain dynamical systems is associated with optimal two-sided ellipsoidal bounds for reachable sets of the systems. Being based on the set-membership approach to uncertainties, the method can be regarded as a natural counterpart to well-known stochastic, or probabilistic, techniques. Basic concepts and results of the method are outlined, and certain results are presented. Various possible applications to problems in control, estimation, and observation are considered.

7.1 Introduction

Dynamical systems subjected to unknown but bounded perturbations appear in numerous applications. The set-membership approach that is a natural counterpart to the well-known stochastic, or probabilistic, one makes it possible to obtain guaranteed estimates on reachable sets and thus to evaluate the family of all possible trajectories of the perturbed system.

In the framework of the set-membership approach, the ellipsoidal estimation seems to be the most efficient technique. Among its advantages are the explicit form of approximations, invariance with respect to linear transformations, possibility of optimization, etc. The earlier results on the ellipsoidal estimation were presented in [33]. The concept of optimality for two-sided (inner and outer) approximating ellipsoids was first introduced in [1] and generalized, extended, and summarized in books [2, 3, 5].

In this paper, basic concepts and results of the method of optimal ellipsoids are outlined, and certain recent results are presented.

F.L. Chernousko
Institute for Problems in Mechanics, Russian Academy of Sciences,
pr. Vernadskogo, 101-1, 119526 Moscow, Russia,
e-mail: chern@ipmnet.ru

K. Marti et al. (eds.), *Coping with Uncertainty*, Lecture Notes in Economics and Mathematical Systems 633, DOI 10.1007/978-3-642-03735-1_7,

Dynamical systems subjected to bounded controls and/or perturbations are considered. For these systems, nonlinear differential equations are obtained that describe the evolution of the optimal ellipsoids representing two-sided (inner and outer) estimates for reachable sets. The approximating ellipsoids depend on the choice of the optimality criterion (e.g., volume of ellipsoids, sum of their squared axes), and on the notion of local/global optimality.

Various useful properties of the optimal approximating ellipsoids have been established. Asymptotic behavior of the ellipsoids near the initial point and at infinity have been studied. As a rule, the nonlinear equations for these ellipsoids are to be integrated numerically. However, certain explicit analytical solutions have been obtained.

Outer and inner ellipsoidal approximations can be used for various applications in control and estimation, including two-sided approximations for optimal control and differential games, analysis of practical stability and parameter excitation, state estimation in the presence of observation errors, control in the presence of uncertain perturbations, etc.

7.2 Reachable Sets

Consider a dynamical system subjected to control or disturbance and described by a system of ordinary differential equations

$$\dot{x} = f(x, u, t), \quad t \geq s \tag{7.1}$$

Here, $x = (x_1, \ldots, x_n)$ is the vector of state, t is time, s is the initial time instant, $u = (n_1, \ldots, u_m)$ is the vector of control or disturbance, and f is a given function. At each time instant, vector $u(t)$ should belong to the given set $U(t)$, so that

$$u(t) \in U(t) \subset R^m, \quad t \geq s \ . \tag{7.2}$$

The initial state belongs to a given initial set:

$$x(s) \in M \subset R^n \ . \tag{7.3}$$

An alternative description of the system (7.1)–(7.3) is given by the differential inclusion:

$$\dot{x} \in f(x, U(t), t), \quad x(s) \in M \ .$$

For systems under consideration, the notion of a reachable set is introduced.

The *reachable*, or *attainable*, set $D(t, s, M)$ of system (7.1) for $t \geq s$ is defined as the set of all end points $x(t)$ at the instant t of all state trajectories $x(\cdot)$ compatible

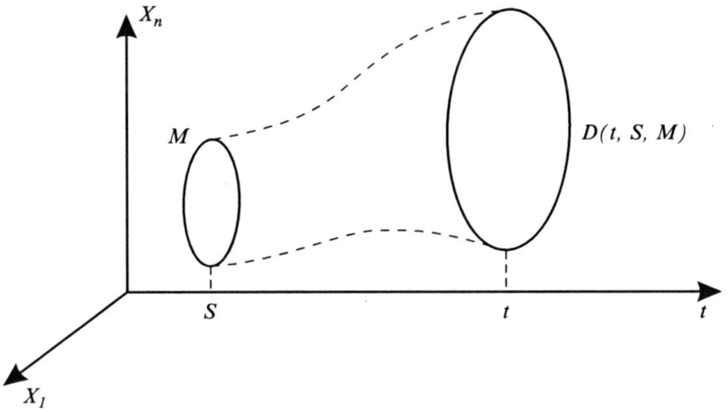

Fig. 7.1 Reachable set

with (7.1)–(7.3). The reachable set has the following evolutionary property

$$D(t, s, M) = D(t, \tau, D(\tau, s, M)) \tag{7.4}$$

that holds for all $\tau \in [s, t]$. Reachable sets are important for systems subjected to control or disturbance because they give a description of all possible states of the system at any given time t (Fig. 7.1).

In various applications, it is often required to verify whether a given state x^* is reachable at the given instant of time t^*, i.e., if the inclusion

$$x^* \in D(t^*, s, M) \tag{7.5}$$

holds. However, the practical determination of reachable sets in the n-dimensional space presents usually serious computational difficulties. In a number of cases, it is sufficient to obtain simple and efficient two-sided (inner and outer) bounds $D^-(t)$ and $D^+(t)$ on reachable sets such that the following inclusions

$$D^-(t) \subset D(t, s, M) \subset D^+(t) \tag{7.6}$$

are true for $t \geq s$. If the bounds D^- and D^+ are known, then the inclusions

$$x^* \in D^-(t^*), \quad x^* \in D^+(t^*) \tag{7.7}$$

can serve, respectively, as a sufficient and necessary conditions for the inclusion (7.5).

If we deal with a control system and u in (7.1) is a control to be chosen, then the inner bound $D^-(t)$ is important. The first inclusion (7.7) implies that the state x^* is reachable at the instant t^*, and there exists an admissible control $u(t)$ bringing the system to this state.

For an uncertain system, $u(t)$ is a disturbance, and the outer bound $D^+(t)$ provides a guaranteed estimate for all possible trajectories: the system cannot reach any state outside the set $D^+(t^*)$ at the instant t^*.

In the well-known stochastic (probabilistic) approach to uncertain systems, the Gaussian distribution plays a specific role and results in the simplest mathematical description. To some extent, ellipsoidal sets for the set-membership approach to uncertainties are analogous to the Gaussian distributions. In fact, the sets of constant probability for the n-dimensional Gaussian distribution are the surfaces of n-dimensional ellipsoids. Moreover, there is a certain similarity between stochastic systems subjected to the white noise and dynamical systems subjected to uncertain disturbances bounded by ellipsoids.

Denote by $E(a, Q)$ the following n-dimensional ellipsoid

$$E(a, Q) = \left\{x : (Q^{-1}(x - a), (x - a)) \leq 1\right\}, \tag{7.8}$$

where $a \in R^n$ is its center, Q is a symmetric positive definite $n \times n$ matrix, and $(.,.)$ denotes the scalar product of vectors.

Ellipsoids have a number of advantages as approximating sets. They provide a satisfactory approximation for a wide class of convex sets [1–3]. For any n-dimensional convex set D, there exists an ellipsoid $E(a, Q)$ such that

$$E(a, Q) \subset D \subset E(a, n^2 Q).$$

If the convex set is symmetric with respect to some point, then there exists an ellipsoid $E(a, Q)$ for which the inclusions.

$$E(a, Q) \subset D \subset E(a, nQ)$$

hold. These inclusions provide two-sided estimates for possible approximations of any convex set by means of ellipsoids.

The class of ellipsoids is invariant with respect to affine transformations. The class of parallelepipeds has the same invariance property but ellipsoids are defined by less number of parameters: $n + n(n+1)/2$ for ellipsoids, $n(n+1)$ for parallelepipeds. Note that rectangular parallelepipeds are defined by the same number of parameters as ellipsoids but they are not invariant with respect to affine transformations.

The last but not the least advantage of ellipsoids is the following one. Simple explicit formulas were obtained [1–3] for the basic algebraic operations for ellipsoidal sets, namely, for the multiplication by a constant, the sum, and intersection of ellipsoids. These operations are, in a certain sense, optimal and provide a basis for the ellipsoidal estimation of dynamical systems subjected to controls and/or uncertain disturbances.

The ellipsoidal state estimation occurs to be the most effective technique in the framework of the set-membership approach. In what follows, we will remind the basic concepts and results of this technique.

7.3 Ellipsoidal Bounds

Let us specify the general system (7.1) and consider a linear system of ordinary differential equations

$$\dot{x} = A(t)x + B(t)u + f(t), \quad t \geq s. \tag{7.9}$$

Here, $x \in R^n$ is the n-vector of state, $u \in R^m$ is the m-vector of control or unknown disturbances, A is an $n \times n$ matrix, B is an $n \times m$ matrix, and f is an n-vector. The matrices $A(t)$ and $B(t)$ as well as the vector $f(t)$ are given functions of time t for $t \geq s$.

Suppose the set $U(t)$ that bounds vector $u(t)$ in (7.2) is an ellipsoid. We have

$$u(t) \in E(0, G(t)), \quad t \geq s, \tag{7.10}$$

where $G(t)$ is a positive definite $m \times m$ matrix specified for $t \geq s$.

The initial set M in (7.3) is also supposed to be an ellipsoid. We have the following bound on the initial state:

$$x(s) \in M = E(a_0, Q_0), \tag{7.11}$$

where a_0 is a given n-vector, and Q_0 is a given positive definite $n \times n$ matrix.

Our goal is to obtain two-sided bounds (7.6) on reachable sets, where the bounding sets $D^-(t)$ and $D^+(t)$ are n-dimensional ellipsoids. In other words, we look for two families of n-dimensional ellipsoids:

$$E^-(t) = E(a^-(t), Q^-(t)), \quad E^+(t) = E(a^+(t), Q^+(t)) \tag{7.12}$$

defined, according to notation (7.8), by their centers $a^-(t), a^+(t)$ and positive definite $n \times n$ matrices $Q^-(t), Q^+(t)$ for $t \geq s$ and such that the following inclusions hold:

$$E^-(t) \subset D(t, s, M) \subset E^+(t), \quad t \geq s. \tag{7.13}$$

In addition, we require that the families of ellipsoids $E^-(t)$ and $E^+(t)$ possess the properties of subreachability and superreachability, respectively.

The families of ellipsoids $E^-(t)$ and $E^+(t)$ defined by (7.11) are called *subreachable* and *superreachable*, respectively, if

$$E^-(t) \subset D(t, \tau, E^-(\tau)), \quad E^+(t) \supset D(t, \tau, E^+(\tau))$$

for all $\tau \in [s, t]$. These properties are similar to the evolutionary property (7.4) of reachable sets. It occurs that subreachable and superreachable ellipsoids have certain advantages: they can be determined efficiently in a straightforward way and provide a sufficiently good approximations of reachable sets.

7.4 Optimality

To make the approximating ellipsoids (7.12) closer to reachable sets, it is quite natural to impose certain optimality conditions upon these ellipsoids. The optimality criteria should, to some extent, reflect the "size" of ellipsoids, and the inner ellipsoid E^- should be "larger", whereas the outer ellipsoid E^+"smaller", in the sense of the chosen criterion.

Let us characterize an ellipsoid $E(a, Q)$ by a scalar optimality criterion J which is a given function $L(Q)$ of the matrix Q, i.e., $J = L(Q)$. The function $L(Q)$ is defined for all symmetric positive definite matrices Q, is smooth and monotone. The monotonicity means that $L(Q_1) \geq L(Q_2)$, if $Q_1 - Q_2$ is a nonnegative definite matrix.

Consider some important particular cases of the general criterion $L(Q)$:

1. The volume of an ellipsoid [1–3] is given by

$$J = c_n(\det Q)^{1/2}, \tag{7.14}$$

   where c_n is a constant depending on n.
2. The sum of the squared semiaxes of an ellipsoid is equal to

$$J = \mathrm{Tr} Q.$$

3. A linear optimality criterion [3, 5]

$$J = \mathrm{Tr}(CQ), \tag{7.15}$$

   where C is a symmetric nonnegative definite $n \times n$ matrix, is a generalization of the previous case.
4. The following criterion

$$J = L(Q) = (Qv, v), \tag{7.16}$$

where v is a given nonzero n-vector, is a particular case of (7.15) with

$$C = v * v, \quad C_{ij} = v_i v_j, \quad i, j = 1, \ldots, n. \tag{7.17}$$

Here, the symbol $*$ denotes the dyadic product of vectors.

Criterion (7.16) has a clear geometric interpretation: it is related to the projection $\Pi_v(E)$ of the ellipsoid onto the direction of the vector v as follows:

$$\Pi_v(E) = 2(Qv, v)^{1/2}/|v|. \tag{7.18}$$

To prove (7.18), note that the projection $\Pi_v(E)$ is related to the support function $H_E(v)$ of the ellipsoid as follows:

$$\Pi_v(E) = [H_E(v) + H_E(-v)]/|v|. \tag{7.19}$$

Since the support function of the ellipsoid is given by

$$H_E(v) = (a, v) + (Qv, v)^{1/2}, \tag{7.20}$$

equation (7.18) follows immediately from (7.19) and (7.20).

By virtue of (7.18), the minimization of criterion (7.16) is equivalent to the minimization of the projection of the ellipsoid onto the direction of vector v. Other examples of optimality criteria are given in [3].

Below, we consider locally and globally optimal ellipsoids [3, 26].

A smooth family of ellipsoids $E^*(a(t), Q(t))$ is called *locally optimal*, if it is superreachable/subreachable and

$$dL(Q(\tau))/d\tau|_{\tau=t} \to \min/\max,$$

where the minimum/maximum is taken over all smooth families of superreachable/subreachable ellipsoids $E^{\pm}(t)$ such that $E^{\pm}(t) = E^*(t)$.

A smooth family of superreachable/subreachable ellipsoids is called *globally optimal* for a given $t = T$, if the minimum/maximum of $L(Q(T))$ over all superreachable/subreachable ellipsoids is attained on this family.

All definitions and results related to the optimal ellipsoids are true also for the case, where the criterion depends also on time t, so that $J = L(Q, t)$. For example, matrix C in (7.15) and vector v in (7.16) can depend on t: $C = C(t)$, $v = v(t)$.

Note that the volume of an ellipsoid as an optimality criterion (7.14) has a property that singles it out among all other criteria. The optimality of a given ellipsoid in the sense of volume remains intact under any affine transformation in R^n. Thus, the optimality in the sense of volume seems more "basic" property of an approximating ellipsoid that its optimality in the sense of other criteria.

7.5 Equations of Ellipsoids

We consider linear system (7.9) where $u(t)$ is subjected to constraint (7.10) and initial conditions are specified by (7.11).

The center $a^+(t)$ and the matrix $Q^+(t)$ of the outer locally optimal ellipsoid satisfy the following differential equations and initial conditions [3,5]:

$$\dot{a}^+ = A(t)a^+ + f(t), \quad a^+(s) = a_0, \tag{7.21}$$

$$\dot{Q}^+ = A(t)Q^+ + Q^+A^T(t) + hQ^+ + h^{-1}K, \quad Q^+(s) = Q_0. \tag{7.22}$$

Here, ^{T} denotes the transposed matrix, $K(t)$ is expressed via given matrices by the formula

$$K(t) = B(t)G(t)B^T(t), \tag{7.23}$$

and the following notation is used

$$h = \left[\mathrm{Tr}\left(\frac{\partial L}{\partial Q^+}K\right)/\mathrm{Tr}\left(\frac{\partial L}{\partial Q^+}Q^+\right)\right]^{1/2}. \tag{7.24}$$

Here, Tr is the trace of a matrix, and $\partial L/\partial Q^+$ is a symmetric matrix of partial derivatives $\partial L(Q)/\partial Q^+_{ij}$, $i, j = 1, \ldots n$.

Note that (7.21) for the vector $a(t)$ is linear and does not depend on the chosen optimality criterion $L(Q)$. By contrast, (7.22) for the matrix $Q^+(t)$ is nonlinear and depends on $L(Q)$ via (7.24) for h. For ellipsoids optimal in the sense of volume [see (7.14)], expression (7.24) becomes

$$h = \left\{n^{-1}\mathrm{Tr}\left[(Q^+)^{-1}K\right]\right\}^{1/2}. \tag{7.25}$$

For the linear criterion (7.15), we have

$$h = [\mathrm{Tr}(CK)/\mathrm{Tr}(CQ^+)]^{1/2}. \tag{7.26}$$

Further simplifications are possible for the criterion (7.16). Substituting C from (7.17) into (7.26), we obtain

$$h = [(Kv, v)/(Q^+v, v)]^{1/2}. \tag{7.27}$$

Equations for inner approximating ellipsoids locally optimal in the sense of volume [3,5] are as follows:

$$\begin{aligned} \dot{a}^- &= A(t)a^- + f(t), \quad a^-(s) = a_0, \\ \dot{Q}^- &= A(t)Q^- + Q^-A^T(t) + 2K^{1/2}(K^{-1/2}Q^-K^{-1/2})^{1/2}K^{1/2}, \\ Q^-(s) &= Q_0. \end{aligned} \tag{7.28}$$

Here, matrix $K(t)$ is again defined by (7.23). Note that equations for the centers of the inner and outer approximating ellipsoids (7.21) and (7.28) coincide, hence, $a^-(t) \equiv a^+(t)$ for $t \geq s$.

It occurs that (7.28) are true for inner approximating ellipsoids optimal in the sense of all criteria $L(Q)$ satisfying the conditions imposed in the beginning of Sect. 7.4. Thus, these equations have a universal nature.

Let us consider now globally optimal outer ellipsoids. The centers of these ellipsoids coincide with those of locally optimal ones and satisfy the initial value problem (7.21). The matrix $Q^+(t)$ of globally optimal ellipsoids satisfies equation and initial condition (7.22), where matrix K is defined by (7.23), whereas the scalar h is, instead of (7.24), given by the expression

$$h = [\mathrm{Tr}(PK)/\mathrm{Tr}(PQ^+)]^{1/2}. \tag{7.29}$$

Here, $P(t)$ is a symmetric positive definite matrix satisfying the following linear differential equation

$$\dot{P} = -PA(t) - A^T(t)P \tag{7.30}$$

and initial condition at $t = T$:

$$P(T) = [\partial L(Q^+)/\partial Q^+]_{t=T}. \tag{7.31}$$

Hence, we have a two-point boundary value problem for the pair of matrices Q^+ and P described by (7.22), (7.23), (7.29)–(7.31). For the linear criterion (7.15), the initial condition (7.31) is reduced to

$$P(T) = C(T). \tag{7.32}$$

Therefore, for outer ellipsoids globally optimal in the sense of criterion (7.15), the boundary value problem for matrices Q^+ and P becomes decoupled and reduces to two initial value problems: a linear one for $P(t)$ defined by (7.30) and (7.32), and a nonlinear one for $Q^+(t)$ defined by (7.22) and (7.29). First, the problem for $P(t)$ should be solved (from $t = T$ to $t = s$), and then the problem for $Q^+(t)$ (from $t = s$ to $t = T$).

Further simplifications are possible for criterion (7.16). For globally optimal outer ellipsoids, we have, on the strength of (7.17) and (7.32):

$$P(T) = v * v,$$

where v is a given constant n-vector. Let us introduce the adjoint n-vector $\psi(t)$ satisfying the initial value problem:

$$\dot{\psi} = -A^T(t)\psi, \quad \psi(T) = v. \tag{7.33}$$

Then we have [6, 7]

$$P(t) = \psi(t) * \psi(t).$$

Thus, in order to find the matrix $Q^+(t)$ of a globally optimal ellipsoid in the case of criterion (7.16), one is to solve first the linear n-dimensional initial value problem (7.33) for $\psi(t)$ (instead of $n(n+1)/2$-dimensional problem for $P(t)$ defined by (7.30) and (7.32)), then substitute

$$h = [(K\psi, \psi)/(Q\psi, \psi)]^{1/2}$$

into (7.22) and solve the resultant initial value problem for Q^+.

Thus, linear optimality criterion (7.15) and, especially, its particular case (7.16) lead to a considerable simplification of equations for optimal outer ellipsoids.

7.6 Transformation of the Equations

In what follows, we restrict ourselves mostly with outer approximating ellipsoids and omit the superscript $^+$, so that we denote below: $a^+(t) = a(t)$, $Q^+(t) = Q(t)$.

Equation (7.22) for locally optimal ellipsoids depends on two given matrices: $A(t)$ and $K(t)$, where the symmetric matrix $K(t)$ is given by (7.23). By the change of variables

$$Q = VQ_*V^T, \tag{7.34}$$

where $V(t)$ is an invertible $n \times n$ matrix to be specified below, and Q_* is a new matrix variable, it is possible to simplify (7.22) so that it will contain only one given matrix instead of two.

1. Let us define $\Phi(t)$ as the fundamental matrix of system (7.7):

$$\dot{\Phi} = A(t)\Phi, \quad \Phi(s) = I.$$

Here, I is the $n \times n$ identity matrix.
If we set

$$V(t) = \Phi(t), \tag{7.35}$$

then transformation (7.34) reduces (7.22) to the form

$$\dot{Q}_* = h_*Q_* + h_*^{-1}K_*(t), \quad Q_*(s) = Q_0, \tag{7.36}$$
$$K_*(t) = \Phi^{-1}(t)K(t)[\Phi^{-1}(t)]^T.$$

Here, h_* is given by (7.24) where Q should be replaced by $\Phi Q_* \Phi^T$. Explicit expressions [3, 5, 7] for h_* are presented below for criteria (7.14)–(7.16), respectively:

$$\begin{aligned} h_* &= [n^{-1}\mathrm{Tr}(Q_*^{-1}K_*)]^{1/2}, \\ h_* &= [\mathrm{Tr}(C_*K_*)/\mathrm{Tr}(C_*Q_*)]^{1/2}, \quad C_* = \Phi^T C \Phi, \\ h_* &= [(K_*v_*, v_*)/(Q_*v_*, v_*)]^{1/2}, \quad v_* = \Phi^T v. \end{aligned} \tag{7.37}$$

2. Let matrix $K(t)$ from (7.23) be positive definite. By taking

$$V(t) = [K(t)]^{1/2} \tag{7.38}$$

in (7.34), we convert (7.22) to the form

$$\begin{aligned} \dot{Q}_* &= A_*Q_* + Q_*A_*^T + h_*Q_* + h_*^{-1}I, \\ Q_*(s) &= K^{-1/2}(s)Q_0K^{-1/2}(s). \end{aligned} \tag{7.39}$$

Here, matrix $A_*(t)$ is given by

$$A_*(t) = K^{-1/2}(AK^{1/2} - dK^{1/2}/dt), \tag{7.40}$$

and explicit formulas [7] for h_* are given below for criteria (7.14)–(7.16), respectively:

$$\begin{aligned} h_* &= (n^{-1}\mathrm{Tr}Q_*^{-1})^{1/2}, \\ h_* &= [\mathrm{Tr}C_*/\mathrm{Tr}(C_*Q_*)]^{1/2}, \quad C_* = K^{1/2}CK^{1/2}, \\ h_* &= [(v_*, v_*)/\mathrm{Tr}(Q_*v_*, v_*)]^{1/2}, \quad v_* = K^{1/2}v. \end{aligned}$$

Thus, by choosing matrix V in (7.34) according to (7.35), we reduce (7.22) for matrix Q to the form (7.36) that corresponds to the case where $A = 0$. On the other hand, by taking matrix V in accordance with (7.38), we come to (7.39) where K is replaced by the unity matrix I. Therefore, we can restrict ourselves with considering (7.22) only in the cases where either $A = 0$ or $K = I$.

Similar simplifications take place also for (7.28) for inner ellipsoids.

Let us consider now equations for globally optimal outer ellipsoids [7] and use the change of variables (7.34) for Q and

$$P = (V^{-1})^T P_* V^{-1} \tag{7.41}$$

for P, where P_* is a new variable. Let us restrict ourselves with the case of linear criteria (7.15) and (7.16). If V is defined by (7.35), we obtain from (7.30) and (7.32) for criterion (7.15):

$$P_*(t) = \Phi^T(T)C(T)\Phi(T) = \mathrm{const}.$$

As a result, the equation for matrix $Q_*(t)$ coincides with (7.36), where

$$h_* = [\mathrm{Tr}(P_* K_*)/\mathrm{Tr}(P_* Q_*)]^{1/2}.$$

For criterion (7.16), h_* is determined by (7.37) with $v_* = \Phi^T(T)v$.

By defining V in (7.34) and (7.41) in accordance with (7.38), we again obtain for criterion (7.15), equations (7.39) for Q_*, equations

$$\dot{P}_* = -P_* A_* - A_*^T P_*, \quad P_*(T) = K^{1/2}(T)C(T)K^{1/2}(T)$$

for P_*, and the same expression (7.40) for A_*.

For criterion (7.16), (7.39) for Q_* and (7.40) for A_* still hold. Here, h_* is defined by

$$h_* = [(\psi_*, \psi_*)/(Q_* \psi_*, \psi_*)]^{1/2},$$

where the adjoint vector $\psi_*(t)$ satisfies the following initial value problem that replaces (7.33):

$$\dot{\psi}_* = -A_*^T \psi_*, \quad \psi_*(T) = K^{1/2}(T)v.$$

7.7 Properties of Optimal Ellipsoids

Outer approximating ellipsoids $E(a(t), Q(t))$ optimal in the sense of criterion (7.16) have the following properties:

1. Globally optimal ellipsoids touch reachable sets $D(t, s, M)$ for all $t \in [s, T]$ at points $x(t)$, where the normal to the boundary of these sets is parallel to the vector $\psi(t)$ defined by (7.33) [6]. In other words, these ellipsoids are tight in the sense of [16].
2. Globally optimal ellipsoids are also locally optimal for the vector $v(t) = \psi(t)$.
3. Locally optimal ellipsoids for the vector $v(t)$ defined by the initial value problem:

$$v(t) = \psi(t), \quad \dot{\psi} = -A^T(t)\psi, \quad \psi(s) = v^0, \tag{7.42}$$

where v^0 is an arbitrary vector, are also globally optimal for any terminal time instant $T \geq s$ and for the criterion $J = (Qv(T), v(T))$.

To construct these locally (and also globally) optimal ellipsoids, one is to solve the linear initial value problem (7.21) for $a^+(t)$ and also initial value problem consisting of (7.22) for Q^+ and (7.42) for $\psi(t)$. Here, the initial vector v^0 can be chosen arbitrarily, and different vectors v^0 correspond to different approximating ellipsoids touching reachable sets at different points.

Various properties of nonlinear equations (7.22) and (7.28) governing the evolution of locally optimal ellipsoids have been studied [1–3, 5–7, 26–28].

As a rule, the nonlinear differential equations for ellipsoids are to be integrated numerically. However, a number of explicit analytical solutions have been obtained both for locally [1–3, 5] and globally [6, 26–28] optimal ellipsoids.

Asymptotic behavior of the solution of (7.22) and (7.28) have been analyzed in the vicinity of the initial point $t = s$, if $Q_0 = 0$ [2, 3]. This case corresponds to the situation, where the initial set M in (7.11) is a given point $x(s) = a_0$. In this important case, (7.22) and (7.28) have a singularity [see also (7.24)–(7.27)], and the obtained asymptotic expansions of the solution are needed to start the numerical integration of equations near the initial point $t = s$ for the case where $Q_0 = 0$.

Also, asymptotic behavior for solutions of (7.22) and (7.28) at infinity ($t \to \infty$) have been analyzed [2, 3, 6, 26–28].

7.8 Generalizations

The method of ellipsoids has been extended to the case, where the parameters of the linear system (7.9) are uncertain and/or subjected to unknown but bounded perturbations. Consider the following system

$$\dot{x} = [A_0(t) + A_1(t)]x + f(t), \tag{7.43}$$

where $x \in R^n$ is the state, matrix $A_0(t)$ and n-vector $f(t)$ are given functions of time, whereas the matrix $A_1(t)$ is unknown, and its elements $a^1_{ij}(t)$ are bounded:

$$|a^1_{ij}(t)| \leq b_{ij}, \quad i, j = 1, \dots, n, \quad t \geq s. \tag{7.44}$$

Here, b_{ij} are given nonnegative numbers. The system described by (7.43) and (7.44) models the situation, where some parameters of the system are uncertain (fixed but unknown) or changeable, for example, in the case of parametric excitation.

Outer ellipsoidal estimates $E(a(t), Q(t))$ on the reachable sets of the system described by (7.43) and (7.44) have been obtained [4].

Suppose the initial data are given by (7.11). The equation for the center $a(t)$ of the approximating ellipsoid is still the same as (7.21), where $A(t)$ is replaced by $A_0(t)$, so that we have

$$\dot{a} = A_0(t)a + f(t), \quad a(s) = a_0.$$

Nonlinear matrix equation for $Q(t)$ differs from (7.22) and has the form

$$\begin{aligned}
\dot{Q} &= A_0 Q + Q A_0^T + hQ + h^{-1} R(a, Q), \qquad (7.45)\\
h &= [n^{-1}\mathrm{Tr}(Q^{-1}R)]^{1/2}, \quad R = \mathrm{diag}(R_1^2, \dots, R_n^2),\\
R_i &= \sum_{j=1}^{n} b_{ij}|a_j| + \left(\max_{\sigma} \sum Q_{jk} b_{ij} b_{ik} \sigma_{ij} \sigma_{ik}\right)^{1/2}.
\end{aligned}$$

Here, the maximum is taken over all $\sigma_{ij} = \pm 1,\ i, j = 1, \ldots, n$. Note that, in contrast to (7.22), the right-hand side of (7.45) for Q depends on vector a.

The method of ellipsoids can be extended also to nonlinear systems [2, 3]. The main idea is to construct a linear comparison system described by (7.9), (7.10), and (7.11), so that all possible motions of the original nonlinear system are within the reachable sets of the linear one. For example, consider a nonlinear system

$$\dot{x} = A(t)x + \varphi(u,t), \quad |\varphi(u,t)| \le \varphi_0(t), \quad t \ge s, \tag{7.46}$$

where u is the disturbance, and the absolute value of the nonlinearity $\varphi(u,t)$ is bounded by $\varphi_0(t)$ for all admissible u and $t \ge s : |\varphi(u,t)| \le \varphi_0(t)$.

Then, the following linear system

$$\dot{x} = A(t)x + w, \quad w \in E(0, \varphi_0^2(t)I) \tag{7.47}$$

can serve as a comparison system for (7.46). Here, w is a disturbance bounded by a ball.

An outer approximating ellipsoid $E(a(t), Q(t))$ for linear system (7.47) will provide an outer bound also for reachable sets of the original nonlinear system (7.46).

Similarly, inner ellipsoidal bounds for reachable sets of nonlinear systems can be obtained [2, 3].

The class of approximating sets can be extended: besides ellipsoids, also intersections and unions of several ellipsoids can be considered [2, 3]. Thus, the approximation of reachable sets can be improved significantly.

7.9 Applications

Two-sided ellipsoidal approximations of reachable sets can be used for the solution and approximation of various problems in control and state estimation. Here, we will only briefly mention these applications; see [3, 5] for details.

7.9.1 Two-Sided Estimates in Optimal Control

Consider the following optimal control problem for system (7.9) under the initial condition $x(s) = x^0$.

Find the control subject to constraint (7.10) that provides the minimum of the given terminal functional $J = F(x(T))$ at $t = T > s$.

Let us obtain two-sided approximating ellipsoids (locally or globally optimal) and evaluate the following minima of the function $F(x)$ over the ellipsoids

$$F^{\pm} = \min F(x), \quad x \in E(a^{\pm}(T), Q^{\pm}(T)).$$

Then the two-sided bounds $F^+ \le \min J \le F^-$ are true for the required minimum of the functional J. To obtain these bounds, one needs, first, to find the inner and outer ellipsoids, and, second, to solve the problems of nonlinear programming, namely, to minimize the function $F(x)$ over the ellipsoidal sets. Note that the first part of this procedure does not depend on the function $F(x)$; if we change this function, we are to change only the second part of the procedure.

7.9.2 Two-Sided Bounds on Time for the Time-Optimal Problem

Let the time-optimal problem ($T \to \min$) is set up for the system (7.9) under the initial condition $x(s) = x_0$. The terminal state is fixed: $x(T) = x^*$. Let us find the minimal time instants T^+ and T^- when the ellipsoids $E(a^\pm(t), Q^\pm(t))$ contain the point x^*:

$$T^\pm = \min t, \quad x^* \in E(a^\pm(t), Q^\pm(t)), \quad t \ge s.$$

Then the two-sided bounds $T^+ \le T \le T^-$ for T are true.

7.9.3 Suboptimal Control

Consider again a linear system described by (7.9) and (7.10) under the initial condition $x(s) = x^0$. Suppose that the given terminal state $x(T) = x^*$ at some instant T belongs to the inner ellipsoid $E(a^-(T), Q^-(T))$. Then there exists an admissible control $u(t)$ bringing the system to this terminal state at $t = T$. This control can be constructed efficiently [12, 13] and is defined by the following procedure. First, find the solution of the initial value problems (7.28) for functions $a^-(t)$ and $Q^-(t)$ that determine the inner approximating ellipsoid.

Denote

$$R(t) = K^{1/2}(K^{-1/2}Q^-K^{-1/2})^{1/2}K^{1/2}$$

and solve the auxiliary initial value problem

$$\begin{aligned} \dot{\psi} &= -A^T\psi - (Q^-)^{-1}R\psi \quad t \in [s, T], \\ \psi(T) &= [Q^-(T)]^{-1}[x^* - a^-(T)]. \end{aligned}$$

Then the admissible control $u(t)$ bringing system (7.9) from the initial state $x(s) = x^0$ to the prescribed terminal state $x(T) = x^*$ is given by the expression

$$u(t) = R(t)\psi(t), \quad t \in [s, T].$$

This control can be called suboptimal since, if the terminal time T is equal to the upper bound T^- for the time-optimal problem introduced in Sect. 7.9.2, this control brings our system to the prescribed terminal state at $t = T = T^-$.

7.9.4 Differential Games

Consider now a differential game of two players X and Y described by equations similar to (7.9) and (7.10):

$$\begin{aligned} X: \quad & \dot{x} = A_x(t)x + B_x(t)u + f_x(t), \\ & u \in E(0, G_x(t)), \quad x(s) = x^0; \\ Y: \quad & \dot{y} = A_y(t)y + B_y(t)v + f_y(t), \\ & v \in E(0, G_y(t)), \quad y(s) = y^0. \end{aligned}$$

Here, x and y are the state vectors of the players X and Y, respectively, u and v are their controls, and x^0 and y^0 are the respective initial states. The cost functional J is a given scalar function of the terminal states of the players at the prescribed time instant T:

$$J = \Phi(x(T), y(T)), \quad T > s.$$

Player X seeks to minimize J while player Y, which can be also identified with uncertain disturbances, opposes Y and tends to maximize J. The following inequalities

$$\max_{y \in D_y} \min_{x \in D_x} \Phi(x, y) \le J^* \le \min_{x \in D_x} \max_{y \in D_y} \Phi(x, y),$$

where D_x and D_y are the reachable sets of players X and Y, respectively, at the instant T, provide evident two-sided bounds on the optimal value J^* of the functional J that corresponds to optimal strategies of both players.

Using ellipsoidal bounds

$$E_x^\pm = E(a_x^\pm(t), Q_x^\pm(t)), \quad E_y^\pm = E(a_y^\pm(t), Q_y^\pm(t))$$

on reachable sets for players X and Y, we obtain the following two-sided estimates on J^*:

$$\Phi_1 \le J^* \le \Phi_2, \quad \Phi_1 = \max_{y \in E_y^-} \min_{x \in E_x^+} \Phi(x, y), \quad \Phi_2 = \min_{x \in E_x^-} \max_{y \in E_y^+} \Phi(x, y). \quad (7.48)$$

Suppose the pairs of points x_1^*, y_1^* and x_2^*, y_2^* are found where the maximin Φ_1 and minimax Φ_2 from (7.48) are attained, respectively. Using results of Sect. 7.9.3, we can find the control $u(t)$ bringing player X to the state $x_2^* \in E_x^-$ at $t = T$, and also the control $v(t)$ bringing player Y to the state $y_1^* \in E_y^-$ at $t = T$. If player X applies the open-loop control $u(t)$, then the value of functional J does not exceed Φ_2 under any admissible control of player Y. On the other hand, if player Y applies the open-loop control $v(t)$, than the value of functional J is not less than Φ_1 under any admissible control of player X. Thus, using approximating ellipsoids, we can obtain two-sided bounds on the value of the cost functional and determine open-loop controls of the players that ensure these bounds.

Approximating ellipsoids can be also used for obtaining two-sided bounds in games of pursuit-evasion. In this context, the rule of extremal aiming [14, 15] is used. Numerical example is presented in [11].

7.9.5 Control of Uncertain Systems

Consider a system subjected to both the control $u(t)$ and disturbance $v(t)$:

$$\dot{x} = A(t)x + B(t)u + C(t)v + f(t), \quad t \geq a. \tag{7.49}$$

Knowing the bound $v(t) \in E(0, G(t))$ on the disturbance, we can obtain the outer bound $x(t) \in E(a(t), Q(t))$ on all trajectories of system (7.49) subjected to the given control $u(t)$. Then, equations (7.21) for $a(t)$ and (7.22) for $Q(t)$ become

$$\dot{a} = Aa + Bu + f, \quad \dot{Q} = AQ + QA^T + hQ + h^{-1}K, \quad K = BGB^T$$

and can be considered as a control system for the whole set of possible trajectories. Various control problems can be set up for this system, and different control methods can be applied to these problems.

7.9.6 Other Applications

In case where $u(t)$ is a bounded disturbance in (7.9), the outer approximating ellipsoid provides an estimate on possible deviations of the trajectory caused by the disturbance. Such estimates are sometimes associated with the notion of "practical stability". For example, possible deviation of the trajectory of a moving body in the presence of wind disturbances can be evaluated.

Various applications of ellipsoidal bounds to parameter estimation are considered in [30, 31]. Aerospace applications of approximating ellipsoids are discussed in [25, 32].

7.9.7 State Estimation in the Presence of Observation Errors

Consider again the system subjected to uncertain disturbances and described by (7.9), (7.10), and (7.11). Suppose that, at the given time instants t_i, the results of observations

$$y_i = H_i x(t_i) + \xi_i, \quad t_i \geq s, \quad i = 0, 1, \ldots, \tag{7.50}$$

become available. Here, y_i are m-vectors of observation results, H_i are given $m \times n$ matrices, and ξ_i are m-dimensional observation errors subject to constraints $\xi_i \in E(0, L_i)$, where L_i are given symmetric positive definite $m \times m$ matrices. Thus, at the instants t_i, the state $x(t_i)$ of the system belongs to the intersection of two ellipsoidal sets:

$$\begin{aligned} & x(t) \in E(t_i) \cap \tilde{E}_i, \\ & \tilde{E}_i = \{x : (L_i^{-1}[H_i x(t_i) - y_i], [H_i x(t_i) - y_i]) \leq 1\}. \end{aligned} \tag{7.51}$$

Here, $E(t_i)$ is the outer ellipsoidal state estimate of the system based on all information available for $t < t_i$, and $\tilde{E}_i$ is the ellipsoid corresponding to the observation (7.50) at $t = t_i$.

To design the process of the ellipsoidal state estimation, we are to construct the ellipsoid $E(t_i + 0)$ that contains the intersection of ellipsoids (7.51). Then we can use $E(t_i + 0)$ as the initial ellipsoid for the next time interval (t_i, t_{i+1}), see Fig. 7.2. It is desirable to find $E(t_i + 0)$ that minimizes certain optimality criterion (see Sect. 7.4). Using optimal and suboptimal outer ellipsoidal bounds for the intersection of two ellipsoids [2, 3], the recursive procedure for the ellipsoidal state estimation in the presence of observation errors has been developed. The state estimation procedure

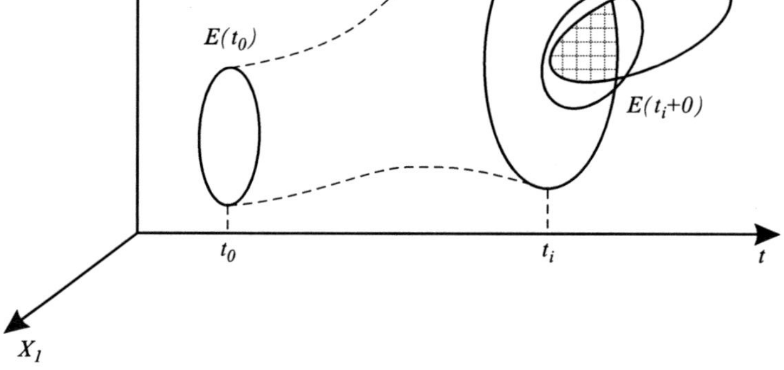

Fig. 7.2 Ellipsoidal state estimation

has been extended also to the case of continuous observations. Thus, the guaranteed analogue of the well-known Kalman filtering has been elaborated [2, 3, 5].

Applications of ellipsoidal technique to state estimation are considered in [17, 18, 25, 30, 31].

7.10 Ellipsoidal vs. Interval Analysis

There exists a vast literature on the interval analysis that is widely used in the computational mathematics [8,9,20,22]. In the interval analysis, to obtain the guaranteed error estimates, one operates with intervals $[x^-, x^+]$, instead of precise value x that is unknown but bounded, $x \in [x^-, x^+]$. When this approach is applied to vectors $x = (x_1, \dots, x_n)$, one is to deal with boxes

$$B = \{x : x_i^- \le x \le x_i^+, \quad i = 1, \dots, n\}, \tag{7.52}$$

i.e., rectangular parallelepipeds with sides parallel to coordinate axes.

The interval approach and its generalizations, where uncertainty domains are bounded by various polytopes, are also applied in the control and identification, see [10, 19, 21, 23, 24, 34].

Let us discuss and compare the method of ellipsoids with the interval methods.

One can consider several levels of generalization of interval methods. The sets of uncertainty can be bounded by: (a) boxes (7.52); (b) rectangular parallelepipeds; (c) arbitrary parallelepipeds; (d) polytopes. The class (d) of arbitrary polytopes seems to be too wide. As already mentioned in Sect. 7.2, the class (c), like the class of ellipsoids, is invariant with respect to linear transformations, but it requires almost two times more parameters for its description than the class of ellipsoids. The classes (a) and (b) are not invariant with respect to linear transformations.

In this context, the class of boxes (7.52) seems to be more attractive among classes (a)–(d) because it has the simplest description and requires only $2n$ parameters.

However, this class can lead to an undesirable but essential error increase.

Consider a simplest vectorial operation $y = Ax$, where x is defined by a box (7.52) and A is a given $n \times n$ matrix. As a result of this operation, box (7.52) is transformed to a parallelepiped P. To obtain a box B' for vector y, one needs to take a box that contains P, $B' \supset P$, and thus to expand the set of possible vectors y. In other words, the error estimate will become wider. In [29], the results of operation $y = Ax$ have been analyzed both for ellipsoidal and interval uncertainty bounds. It has been shown that the ellipsoidal bounds are frequently superior to interval ones and lead to tighter error estimates than those given by the interval analysis.

Thus, it seems that the method of ellipsoids can be of use for error estimation in numerical analysis.

7.11 Conclusions

The method of ellipsoids seems to be an efficient technique for the analysis of dynamical systems subjected to uncertain perturbations and observation errors. By means of this approach, exact and approximate solutions as well as reliable two-sided bounds for a number of basic problems in control and estimation can be obtained.

Acknowledgements This work was supported by the Russian Foundation for Basic Research (Project 08-01-00411) and by the Grant of Support for Leading Russian Scientific Schools (NSh-4315.2008.1).

References

1. Chernousko, F.L.: Optimal guaranteed estimates of indeterminacies using ellipsoids, Parts 1–3. Izv. Akad. Nauk SSSR, Tekh. Kibern. **3**, 3–11; **4**, 3–11; **5**, 5–11 (1980)
2. Chernousko, F.L.: Estimation of Phase State for Dynamical Systems. Nauka, Moscow (1988)
3. Chernousko, F.L.: State Estimation for Dynamic Systems. CRC, Boca Raton (1994)
4. Chernousko, F.L.: Ellipsoidal approximation of attainability sets of linear system with indeterminate matrix. J. Appl. Math. Mech. **60**(6), 921–931 (1996)
5. Chernousko, F.L.: What is ellipsoidal modelling and how to use it for control and state estimation? In: Elishakoff, I. (ed.) Whys and Hows in Uncertainty Modelling, pp. 127–188. Springer, Vienna (1999)
6. Chernousko, F.L., Ovseevich, A.I.: Some properties of optimal ellipsoids approximating reachable sets. Dokl. Math. **67**(1), 123–126 (2003)
7. Chernousko, F.L., Ovseevich, A.I.: Properties of the optimal ellipsoids approximating the reachable sets of uncertain systems. J. Optim. Theor. Appl. **120**(2), 223–246 (2004)
8. Goetz, A., Herzberger, J.: Introduction to Interval Analysis. Academic, New York (1983)
9. Jaulin, L., Kieffer, M., Didrit, O., Walter, E.: Applied Interval Analysis. Springer, London (2001)
10. Kieffer, M., Walter, E.: Interval analysis for guaranteed non-linear parameter and state estimation. Math. Comput. Model. Dyn. Syst. **11**(2), 171–181 (2005)
11. Klepfish, B.R.: Method of obtaining the two-side estimate on the time of pursuit. Izv. Akad. Nauk SSSR, Tekh. Kibern. **4**, 156–160 (1984)
12. Komarov, V.A.: Estimates on reachable sets and construction of admissible controls for linear systems. Dokl. Akad. Nauk SSSR **268**, 537–541 (1982)
13. Komarov, V.A.: Estimates on reachable sets for linear systems. Izv. Akad. Nauk SSSR, Ser. Mat. **48**, 865–879 (1984)
14. Krasovskii, N.N.: Game Problems of Meeting of Motions. Nauka, Moscow (1970)
15. Krasovskii, N.N., Subbotin, A.I.: Positional Differential Games. Springer, Berlin (1988)
16. Kurzhanski, A.B., Varaiya, P.: Ellipsoidal techniques for reachability analysis. Lecture Notes in Computer Science vol. 1790, pp. 202–214. Springer, Berlin (2000)
17. Maksarov, D.G., Norton, J.P.: State bounding with ellipsoidal set description of the uncertainty. Int. J. Control **65**, 847–866 (1996)
18. Maksarov, D.G., Norton, J.P.: Computationally efficient algorithms for state estimation with ellipsoidal approximations. Int. J. Adapt. Control Signal Process. **16**, 411–434 (2002)
19. Milanese, M., Norton, J., Piet-Lahanier, H., Walter, E. (eds.): Bounding Approaches to System Identification. Plenum, New York (1996)
20. Moore, R.E.: Methods and Applications of Interval Analysis. SIAM, Philadelphia (1979)

21. Nazin, S.A., Polyak, B.T.: Interval parameter estimation under model uncertainty. Math. Comput. Model. Dyn. Syst. **11**(2), 225–237 (2005)
22. Neumaier, A.: Interval Methods for Systems of Equations. Cambridge University Press, Cambridge (1990)
23. Norton, J.P. (ed.): Special issue on bounded-error estimation, 1. Int. J. Adapt. Control Signal Process. **8**(1) (1994)
24. Norton, J.P. (ed.): Special issue on bounded-error estimation, 2. Int. J. Adapt. Control Signal Process. **9**(2) (1995)
25. Norton, J.P.: Results to aid applications of ellipsoidal state bounds. Math. Comput. Model. Dyn. Syst. **11**(2), 209–224 (2005)
26. Ovseevich, A.I.: Limit behavior of attainable and superattainable sets. In: Modelling, Estimation, and Control of Systems with Uncertainty, pp. 324–333. Birkhäuser, Boston (1991)
27. Ovseevich, A.I.: On equations of ellipsoids approximating attainable sets. J. Optim. Theor. Appl. **95**(3), 659–676 (1997)
28. Ovseevich, A.I., Chernousko, F.L. : Methods of ellipsoidal estimation for linear control systems. In: Proc. 17th World Congress, The International Federation of Automatic Control, pp. 15345–15348. Seoul, Korea (2008)
29. Ovseevich, A.I., Taraban'ko, Yu.V., Chernousko, F.L.: A comparison of interval and ellipsoidal error bounds for vector operations. Dokl. Math. **71**(1), 127–130 (2005)
30. Polyak, B.T., Nazin, S.A., Durieu, C., Walter, E.: Ellipsoidal parameter or state estimation under model uncertainty. Automatica **40**, 1171–1179 (2004)
31. Pronzato, L., Walter, E.: Minimum-volume ellipsoids containing compact sets: application to parameter bounding. Automatica **30**, 1731–1739 (1994)
32. Rokityanskiy, D.Ya., Veres, S.M.: Application of ellipsoidal estimation to satellite control design. Math. Comput. Model. Dyn. Syst. **11**(2), 239–249 (2005)
33. Schweppe, F.C.: Uncertain Dynamic Systems. Prentice-Hall, Englewood Cliffs, NJ (1973)
34. Walter, E. (ed.): Special issue on parameter identification with error bound. Math. Comput. Simul. **32** (1990)

Chapter 8
Expected Total Cost Minimum Design of Plane Frames by Means of Stochastic Linear Programming Methods

Kurt Marti

Abstract Yield stresses, allowable stresses, moment capacities (plastic moments with respect to compression, tension and rotation), applied loadings, cost factors, manufacturing errors, etc., are not given fixed quantities in structural analysis and optimal design problems, but must be modeled as random variables with a certain joint probability distribution. Problems from plastic analysis and optimal plastic design are based on the convex yield (feasibility) criterion and the linear equilibrium equation for the stress (state) vector.

After the formulation of the basic mechanical conditions including the relevant material strength parameters and load components as well as the involved design variables (as, e.g., sizing variables) for plane frames, several approximations are considered: (1) approximation of the convex yield (feasible) domain by means of convex polyhedrons (piecewise linearization of the yield domain); (2) treatment of symmetric and non symmetric yield stresses with respect to compression and tension; (3) approximation of the yield condition by using given reference capacities.

As a result, for the survival of plane frames a certain system of necessary and/or sufficient linear equalities and inequalities is obtained. Evaluating the recourse costs, i.e., the costs for violations of the survival condition by means of piecewise linear convex cost functions, a linear program is derived for the minimization of the total costs including weight-, volume- or more general initial (construction) costs. Appropriate cost factors are given. Considering then the expected total costs from construction as well as from possible structural damages or failures, a stochastic linear optimization problem (SLP) is obtained. Finally, discretization of the probability distribution of the random parameter vector yields a (large scale) linear program (LP) having a special structure. For LP's of this type numerous, very efficient LP-solvers are available – also for the solution of very large scale problems.

K. Marti
Aero-Space Engineering and Technology, Federal Armed Forces University Munich, 85577 Neubiberg, Germany,
e-mail: kurt.marti@unibw-muenchen.de

K. Marti et al. (eds.), *Coping with Uncertainty*, Lecture Notes in Economics and Mathematical Systems 633, DOI 10.1007/978-3-642-03735-1_8,

8.1 Introduction

8.1.1 Plastic Analysis of Structures

Many materials, e.g., most of metals, have distinct, plastic properties, i.e., they are ductile, see, e.g., [10,23]. Even after the stress intensity attains the yield point stress, such materials can deform considerably without breaking. This implies that if the stress intensity at a certain point of a hyperstatic structure reaches the critical (yield) value, the structure does not necessarily fail or deform excessively. Instead, a certain amount of stress redistribution takes place and some further load increments can be supported. Structural failure does not occur before a kinematic mechanism of unconstrained plastic flow develops. Thus, the actual load-carrying capacity of a structure is higher (in some cases quite considerably) than that derived from classical elastic analysis. A crucial question for the engineer designing structures like buildings, bridges, etc., or structural components is to which extent a plastic deformation is permissible without leading to a failure of the structure, the component, resp., with respect to the expected load and material strength conditions. Applying standard methods, the load carrying capacity is determined using a certain code with general rules for safety evaluations. The use of such general rules may be very expensive in the safety evaluation and design of a structure. On the other hand, safety assessment and design based on stochastic optimization techniques, taking into account the available knowledge about random parameter variations, reduce the expected total project costs (primary costs, e.g., costs of construction, plus recourse cots, e.g., strengthening costs) considerably. Consequently, this way one obtains more robust (safe) information about the maximum load factors, hence, the carrying capacity, as well as about robust optimal designs. A further big advantage of stochastic optimization methods is the possibility of updates of the maximum load factors and robust optimal designs based on inspection, sampling and other posterior information about the probability distribution of the random parameters. For elastic-perfectly plastic materials, the ultimate load condition corresponding to complete collapse of the structure can be obtained through application of a pair of dual theorems [14,18,23]:

(ST) Static Theorem (lower bound or safe theorem) If any stress distribution throughout the structure can be found which is everywhere in equilibrium internally and balances the external loads, and at the same time does not violate the yield condition, those external loads will be carried safely by the structure.

(KT) Kinematic Theorem (upper bound or unsafe theorem)
Collapse occurs if a collapse mechanism, fulfilling the compatibility condition, exists such that the work done by the external loads is larger than the corresponding internal plastic work.
Limit analysis is concerned [5,8,10,14,16,18,19,22,24,33,34,36,37,40,41] with establishing the strength of a structure, i.e., its capacity for the supporting of loads. Hence, using the plastic ductility of structural materials in

improving the design of structures, limit analysis is not concerned with deformation: it can not therefore provide the load carrying capacity for a structure with elements that have a limited ductility or deformability, nor for a structure which becomes unstable because of the displacements induced by plastic deformation, see [10, 13, 18, 24].

8.1.2 Limit (Collapse) Load Analysis of Structures as a Linear Programming Problem

Assuming that the material behaves as an elastic-perfectly plastic material [17, 23, 32] a conservative estimate of the collapse load factor λ_T is based [5–8, 10, 13, 14, 22, 33, 40] on the following linear program:

$$\text{maximize } \lambda \tag{8.1a}$$

s.t.

$$F^L \leq F \leq F^U \tag{8.1b}$$

$$CF = \lambda R_0. \tag{8.1c}$$

Here, (8.1c) is the equilibrium equation of a statically indeterminate loaded structure involving an $m \times n$ matrix $C = (c_{ij}), m < n$, of given coefficients $c_{ij}, 1 \leq i \leq m, 1 \leq j \leq n$, depending on the undeformed geometry of the structure having n_0 members (elements). After taking into account the given support (boundary) conditions, we may suppose that rank $C = m$. Furthermore, R_0 is an external load m-vector, and F denotes the n-vector of internal forces and bending-moments in the relevant points (sections, nodes or elements) of lower and upper bounds F^L, F^U.

For a plane or spatial truss [25, 38] we have that $n = n_0$, the matrix C contains the direction cosines of the members, and F involves only the normal (axial) forces moreover,

$$F_j^L := \sigma_{yj}^L A_j, F_j^U := \sigma_{yj}^U A_j, j = 1, \ldots, n(= n_0), \tag{8.2}$$

where A_j is the (given) cross-sectional area, and $\sigma_{yj}^L, \sigma_{yj}^U$, respectively, denotes the yield stress in compression (negative values) and tension (positive values) of the j-th member of the truss. In case of a plane frame, F is composed of subvectors [38],

$$F^{(k)} = \begin{pmatrix} F_1^{(k)} \\ F_2^{(k)} \\ F_3^{(k)} \end{pmatrix} = \begin{pmatrix} t_k \\ m_k^+ \\ m_k^- \end{pmatrix}, \tag{8.3a}$$

where $F_1^{(k)} = t_k$ denotes the normal (axial) force, and $F_2^{(k)} = m_k^+, F_3^{(k)} = m_k^-$ are the bending-moments at the positive ("right"), negative ("left") end of the k-th member with respect to a certain chosen orientation of the members. In this case

F^L, F^U contain – for each member k – the subvectors

$$F^{(k)L} = \begin{pmatrix} \sigma_{yk}^L A_k \\ -M_{kpl} \\ -M_{kpl} \end{pmatrix}, F^{(k)U} = \begin{pmatrix} \sigma_{yk}^U A_k \\ M_{kpl} \\ M_{kpl} \end{pmatrix}, \tag{8.3b}$$

resp., where $M_{kpl}, k = 1, \ldots, n_0$, denotes the plastic moments (moment capacities) [17, 32] given by

$$M_{kpl} = \sigma_{yk}^U W_{kpl}, \tag{8.3c}$$

and $W_{kpl} = W_{kpl}(A_k)$ is the plastic section modulus of the cross-section of the k-th member (beam) with respect to the local z-axis. In order to omit instabilities, such as buckling, σ_{yk}^L can be selected by

$$\sigma_{yk}^L := -\kappa_k \sigma_{yk}^U \tag{8.3d}$$

with a certain reduction factor κ_k.

For a *space frame* [25, 38], corresponding to the k-th member (beam), F contains the subvector

$$F^{(k)} := (t_k, m_{kT}, m_{k\bar{y}}^+, m_{k\bar{z}}^+, m_{k\bar{y}}^-, m_{k\bar{z}}^-)^T, \tag{8.4a}$$

where t_k is the normal (axial) force, m_{kT} the twisting moment, and $m_{k\bar{y}}^+, m_{k\bar{z}}^+$, $m_{k\bar{y}}^-, m_{k\bar{z}}^-$ denote four bending moments with respect to the local $\bar{y}, z$-axis at the positive, negative end of the beam, respectively. Finally, the bounds F^L, F^U for F are given by

$$F^{(k)L} = \left(\sigma_{yk}^L A_k, -M_{kpl}^{\bar{p}}, -M_{kpl}^{\bar{y}}, -M_{kpl}^{\bar{z}}, -M_{kpl}^{\bar{y}}, -M_{kpl}^{z}\right)^T \tag{8.4b}$$

$$F^{(k)U} = \left(\sigma_{yk}^U A_k, M_{kpl}^{\bar{p}}, M_{kpl}^{\bar{y}}, M_{kpl}^{\bar{z}}, M_{kpl}^{\bar{y}}, M_{kpl}^{\bar{z}}\right)^T, \tag{8.4c}$$

where [17, 32]

$$M_{kpl}^{\bar{p}} := \tau_{yk} W_{kpl}^{\bar{p}}, M_{kpl}^{\bar{y}} := \sigma_{yk}^U W_{kpl}^{\bar{y}}, M_{kpl}^{\bar{z}} := \sigma_{yk}^U W_{kpl}^{\bar{z}}, \tag{8.4d}$$

are the plastic moments of the cross-section of the k-th element with respect to the local twisting axis, the local $\bar{y}-, \bar{z}$-axis, respectively. In (8.4d), $W_{kpl}^{\bar{p}} = W_{kpl}^{\bar{p}}(x)$ and $W_{kpl}^{\bar{y}} = W_{kpl}^{\bar{y}}(x), W_{kpl}^{\bar{z}} = W_{kpl}^{\bar{z}}(x)$, resp., denote the polar, axial modulus of the cross-sectional area of the k-th beam and τ_{yk} denotes the yield stress with respect to torsion; we suppose that $\tau_{yk} = \kappa_{\tau k} \sigma_{yk}^U$ with a reduction factor $\kappa_{\tau k}$. Moreover, x denotes the r-vector of design variables, see Sect. 8.1.1 for more details.

Remark 8.1. Possible plastic hinges [17, 24, 32] are taken into account by inserting appropriate eccentricities $e_{kl} > 0, e_{kr} > 0, k = 1, \ldots, n_0$, with $e_{kl}, e_{kr} << L_k$, where L_k is the length of the k-th beam.

Remark 8.2. Working with more general yield polygons [1, 40, 42], the stress condition (8.1b) is replaced by the more general system of inequalities

$$H(F_d^U)^{-1}F \leq h. \tag{8.5a}$$

Here, (H, h) is a given $\nu \times (n+1)$ matrix, and $F_d^U := (F_j^U \delta_{ij})$ denotes the $n \times n$ diagonal matrix of principal axial and bending plastic capacities

$$F_j^U := \sigma_{yk_j}^U A_{kj}, F_j^U := \sigma_{yk_j}^U W_{k_j pl}^{\kappa j}, \tag{8.5b}$$

where $kj, \kappa j$ are indices as arising in (8.3b)–(8.4d). The more general case (8.5a) can be treated by similar methods as the case (8.1b) which is considered here.

8.1.3 Plastic and Elastic Design of Structures

In the plastic design of trusses and frames [22,26,27,29,34,36] having n_0 members, the n-vectors F^L, F^U of lower and upper bounds

$$F^L = F^L(\sigma_y^L, \sigma_y^U, x), F^U = F^U(\sigma_y^L, \sigma_y^U, x), \tag{8.6}$$

for the n-vector F of internal member forces and bending moments $F_j, j = 1, \dots, n$, are determined [13, 22] by the yield stresses, i.e., compressive limiting stresses (negative values) $\sigma_y^L = \left(\sigma_{y1}^L, \dots, \sigma_{yn_0}^L\right)^T$, the tensile yield stresses $\sigma_y^U = \left(\sigma_{y1}^U, \dots, \sigma_{yn_0}^U\right)^T$, and the r-vector

$$x = (x_1, x_2, \dots, x_r)^T \tag{8.7}$$

of design variables of the structure. In case of trusses we have that, cf. (8.2),

$$\begin{aligned} F^L &= \sigma_{yd}^L A(x) = A(x)_d \sigma_y^L, \\ F^U &= \sigma_{yd}^U A(x) = A(x)_d \sigma_y^U, \end{aligned} \tag{8.8}$$

where $n = n_0$, and $\sigma_{yd}^L, \sigma_{yd}^U$ denote the $n \times n$ diagonal matrices having the diagonal elements $\sigma_{yj}^L, \sigma_{yj}^U$, respectively, $j = 1, \dots, n$, moreover,

$$A(x) = \left[A_1(x), \dots, A_n(x)\right]^T \tag{8.9}$$

is the n-vector of cross-sectional area $A_j = A_j(x), j = 1, \dots, n$, depending on the r-vector x of design variables $x_\kappa, \kappa = 1, \dots, r$, and $A(x)_d$ denotes the $n \times n$ diagonal matrix having the diagonal elements $A_j = A_j(x), 1 < j < n$.

Corresponding to (8.1c), here the equilibrium equation reads

$$CF = R_u, \tag{8.10}$$

where R_u describes [22] the ultimate load [representing constant external loads or self-weight expressed in linear terms of $A(x)$].

The *plastic design* of structures can be represented then [1,2] by the optimization problem

$$\min G(x), \tag{8.11a}$$

s.t.

$$F^L(\sigma_y^L, \sigma_y^U, x) \le F \le F^U(\sigma_y^L, \sigma_y^U, x) \tag{8.11b}$$

$$CF = R_u \tag{8.11c}$$

$$x \in D, \tag{8.11d}$$

where $G = G(x)$ is a certain objective function, e.g., the volume or weight of the structure, and $C \subset \mathbb{R}^+$ denotes the convex set of admissible design vecotrs x.

Remark 8.3. As mentioned in Remark 8.2, working with more general yield polygons, (8.11b) is replaced by the condition

$$H[F^U(\sigma_y^U, x)_d]^{-1} F \le h. \tag{8.11e}$$

For the *elastic design* we must replace the yield stresses σ_y^L, σ_y^U by the allowable stresses σ_a^L, σ_a^U and instead of ultimate loads we consider service loads R_s. Hence, instead of (8.11a–d) we have the related program

$$\min G(x), \tag{8.12a}$$

s.t.

$$F^L(\sigma_a^L, \sigma_a^U, x) \le F \le F^U(\sigma_a^L, \sigma_a^U, x), \tag{8.12b}$$

$$CF = R_s, \tag{8.12c}$$

$$x^L \le x \le x^U, \tag{8.12d}$$

where x^L, x^U still denote lower and upper bounds for x.

8.2 Plane Frames

For each bar $i = 1, \dots, B$ of a plane frame with member load vector $F_i = (t_i, m_i^+, m_i^-)^T$ we consider [37,41] the load at the negative, positive end

$$F_i^- := (t_i, m_i^-)^T, \; F_i^+ := (t_i, m_i^+)^T, \tag{8.13}$$

respectively.

Furthermore, for each bar/beam with rigid joints we have several plastic capacities: The plastic capacity N_{ipl}^{L} of the bar with respect to axial compression, hence, the maximum axial force under compression is given by

$$N_{ipl}^{L} = |\sigma_{yi}^{L}| \cdot A_i, \tag{8.14a}$$

where $\sigma_{yi}^{L} < 0$ denotes the (negative) yield stress with respect to compression and A_i is the cross sectional area of the ith element. Correspondingly, the plastic capacity with respect to (axial) tension reads:

$$N_{ipl}^{U} = \sigma_{yi}^{U} \cdot A_i, \tag{8.14b}$$

where $\sigma_{yi}^{U} > 0$ is the yield stress with respect to tension. Besides the plastic capacities with respect to the normal force, we have the moment capacity

$$M_{ipl} = \sigma_{yi}^{U} \cdot W_{ipl} \tag{8.14c}$$

with respect to the bending moments at the ends of the bar i.

Remark 8.4. Note that all plastic capacities have nonnegative values.

Using the plastic capacities (8.14a–c), the load vectors F_i^{+}, F_i^{+} given by (8.13) can be replaced by dimensionless quantities

$$F_i^{L-} := \left(\frac{t_i}{N_{ipl}^{L}}, \frac{m_i^{-}}{M_{ipl}}\right)^T, F_i^{U-} := \left(\frac{t_i}{N_{ipl}^{U}}, \frac{m_i^{-}}{M_{ipl}}\right)^T \tag{8.15a,b}$$

$$F_i^{L+} := \left(\frac{t_i}{N_{ipl}^{L}}, \frac{m_i^{+}}{M_{ipl}}\right)^T, F_i^{U+} := \left(\frac{t_i}{N_{ipl}^{U}}, \frac{m_i^{+}}{M_{ipl}}\right)^T \tag{8.15c,d}$$

for the negative, positive end, resp., of the ith bar.

Remark 8.5 (Symmetric yield stresses under compression and tension).
In the important special case that the absolute values of the yield stresses under compression (<0) and tension (>0) are equal, hence,

$$\sigma_{yi}^{L} = -\sigma_{yi}^{U} \tag{8.16a}$$

$$N_{ipl}^{L} = N_{ipl}^{U} =: N_{ipl}. \tag{8.16b}$$

The limit between the elastic and plastic state of the elements is described by the feasibility or yield condition: At the negative end we have the condition

$$F_i^{L-} \in K_i, F_i^{U-} \in K_i \tag{8.17a,b}$$

and at the positive end the condition reads

$$F_i^{L+} \in K_i, F_i^{U+} \in K_i. \tag{8.17c,d}$$

Here, K_i, $K_i \subset \mathbb{R}^2$, denotes the feasible domain of bar "i" having the following properties:

- K_i is a closed, convex subset of $\mathbb{R}^2$.
- The origin 0 of $\mathbb{R}^2$ is an interior point of K_i.
- The interior $\mathring{K}_i$ of K_i represents the elastic states.
- At the boundary ∂K_i yielding of the material starts.

Considering, e.g., bars with rectangular cross sectional areas and symmetric yield stresses, cf. Remark 8.5, K_i is given by $K_i = K_{0,sym}$, where [19,21]

$$K_{0,\mathrm{sym}} = \{(x, y)^T : x^2 + |y| \le 1\} \tag{8.18}$$

where $x = \frac{N}{N_{pl}}$ and $y = \frac{M}{M_{pl}}$, (see Fig. 8.1). Note that the symbols $x, y, (x, y)$, resp., denote in this Sect. 8.2 just real variables, a point in the real plane.

In case (8.18) and supposing symmetric yield stresses, the yield condition (8.17a–d) reads

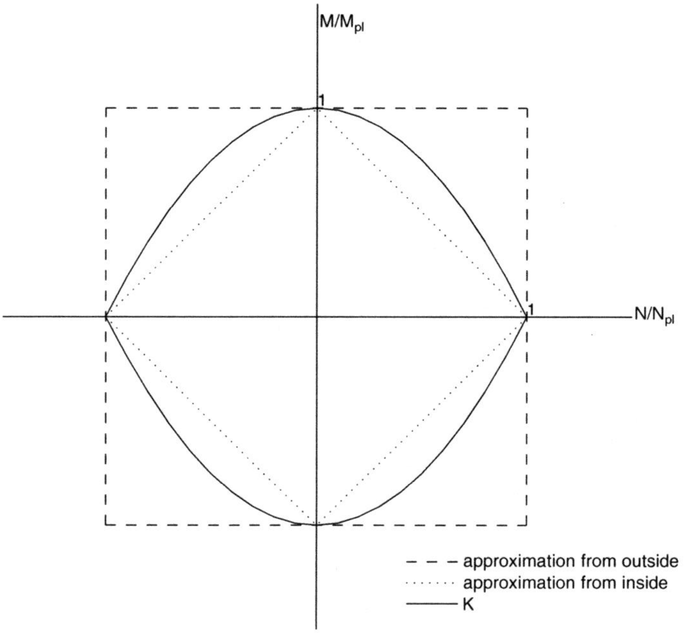

Fig. 8.1 Domain $K_{0,\mathrm{sym}}$ with possible approximations

$$\left(\frac{t_i}{N_{ipl}}\right)^2 + \left|\frac{m_i^-}{M_{ipl}}\right| \leq 1, \tag{8.19a}$$

$$\left(\frac{t_i}{N_{ipl}}\right)^2 + \left|\frac{m_i^+}{M_{ipl}}\right| \leq 1. \tag{8.19b}$$

Remark 8.6. Because of the connection between the normal force t_i and the bending moments, (8.19a,b) is also called "M–N-interaction".

If the M–N-interaction is not taken into account, $K_{0,\text{sym}}$ is approximated from outside, see Fig. 8.1, by

$$K^u_{0,\text{sym}} := \{(x, y)^T : |x|, |y| \leq 1\}. \tag{8.20}$$

Hence, (8.17a–d) are replaced, cf. (8.19a,b), by the simpler conditions

$$|t_i| \leq N_{ipl} \tag{8.21a}$$

$$|m_i^-| \leq M_{ipl} \tag{8.21b}$$

$$|m_i^+| \leq M_{ipl}. \tag{8.21c}$$

Since the symmetry condition (8.16a) does not hold in general, some modifications of the basic conditions (8.19a,b) are needed. In the non symmetric case $K_{0,xm}$ must be replaced by the intersection

$$K_0 = K_0^U \cap K_0^L \tag{8.22}$$

of two convex sets K_0^U and K_0^L. For a rectangular cross-sectional area we have

$$K_0^U = \left\{(x, y)^T : x \leq \sqrt{1 - |y|},\ |y| \leq 1\right\} \tag{8.23a}$$

for tension and

$$K_0^L = \left\{(x, y)^T : -x \leq \sqrt{1 - |y|},\ |y| \leq 1\right\} \tag{8.23b}$$

compression, where again $x = \frac{N}{N_{pl}}$ and $y = \frac{M}{M_{pl}}$ (see Fig. 8.2).

In case of tension, see (8.17b,d), from (8.23a) we obtain then the feasibility condition

$$\frac{t_i}{N^U_{ipl}} \leq \sqrt{1 - \left|\frac{m_i^-}{M_{ipl}}\right|} \tag{8.24a}$$

$$\frac{t_i}{N^U_{ipl}} \leq \sqrt{1 - \left|\frac{m_i^+}{M_{ipl}}\right|} \tag{8.24b}$$

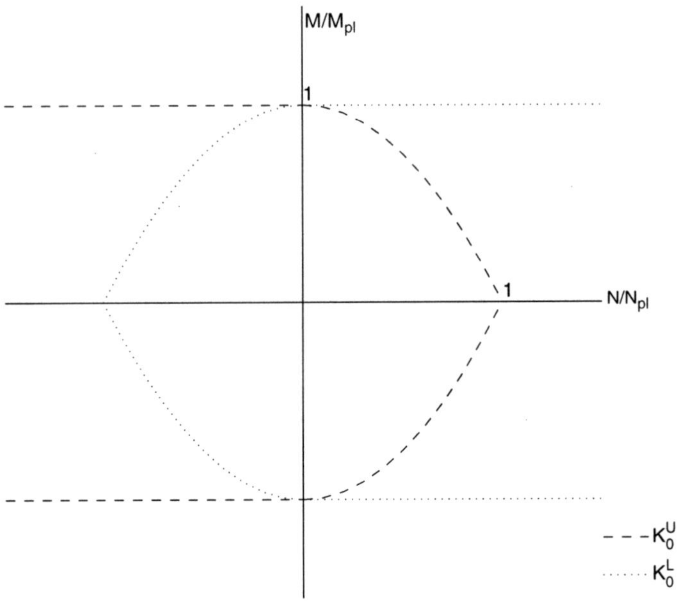

Fig. 8.2 Feasible domain as intersection of K_0^U and K_0^L

$$\left| \frac{m_i^-}{M_{ipl}} \right| \leq 1 \tag{8.24c}$$

$$\left| \frac{m_i^+}{M_{ipl}} \right| \leq 1. \tag{8.24d}$$

For compression, with (8.17a,c) and (8.23b) we get the feasibility condition

$$-\frac{t_i}{N_{ipl}^L} \leq \sqrt{1 - \left| \frac{m_i^-}{M_{ipl}} \right|} \tag{8.24e}$$

$$-\frac{t_i}{N_{ipl}^L} \leq \sqrt{1 - \left| \frac{m_i^+}{M_{ipl}} \right|} \tag{8.24f}$$

$$\left| \frac{m_i^-}{M_{ipl}} \right| \leq 1 \tag{8.24g}$$

$$\left| \frac{m_i^+}{M_{ipl}} \right| \leq 1. \tag{8.24h}$$

From (8.24a), (8.24e) we get

$$-N_{ipl}^{L}\sqrt{1-\left|\frac{m_i^-}{M_{ipl}}\right|}\leq t_i\leq N_{ipl}^{U}\sqrt{1-\left|\frac{m_i^-}{M_{ipl}}\right|}. \tag{8.25a}$$

and (8.24b), (8.24f) yield

$$-N_{ipl}^{L}\sqrt{1-\left|\frac{m_i^+}{M_{ipl}}\right|}\leq t_i\leq N_{ipl}^{U}\sqrt{1-\left|\frac{m_i^+}{M_{ipl}}\right|}. \tag{8.25b}$$

Furthermore, (8.24c), (8.24g) and (8.24d), (8.24h) yield

$$|m_i^-|\leq M_{ipl} \tag{8.25c}$$
$$|m_i^+|\leq M_{ipl}. \tag{8.25d}$$

For computational purposes, piecewise linearizations are applied [30] to the nonlinear conditions (8.25a,b), see also [3, 15]. A basic approximation of K_0^L and K_0^U is given by

$$K_0^{Uu}:=\{(x,y)^T: x\leq 1, |y|\leq 1\}=]\infty,1]\times[-1,1] \tag{8.26a}$$

and

$$K_0^{Lu}:=\{(x,y)^T: x\geq -1, |y|\leq 1\}=\ [-1,\infty]\times[-1,1] \tag{8.26b}$$

with $x=\frac{N}{N_{pl}}$ and $y=\frac{M}{M_{pl}}$ (see Fig. 8.3).

Since in this approximation the M–N-interaction is not taken into account, condition (8.17a–d) is reduced to

$$-N_{ipl}^{L}\leq t_i\leq N_{ipl}^{U} \tag{8.27a}$$
$$|m_i^-|\leq M_{ipl} \tag{8.27b}$$
$$|m_i^+|\leq M_{ipl}. \tag{8.27c}$$

8.2.1 Yield Condition in Case of $M-N$-Interaction

8.2.1.1 Symmetric Yield Stresses

Consider first the case

$$\sigma_{yi}^{U}=-\sigma_{yi}^{L}=:\sigma_{yi},\ i=1,\ldots,B. \tag{8.28a}$$

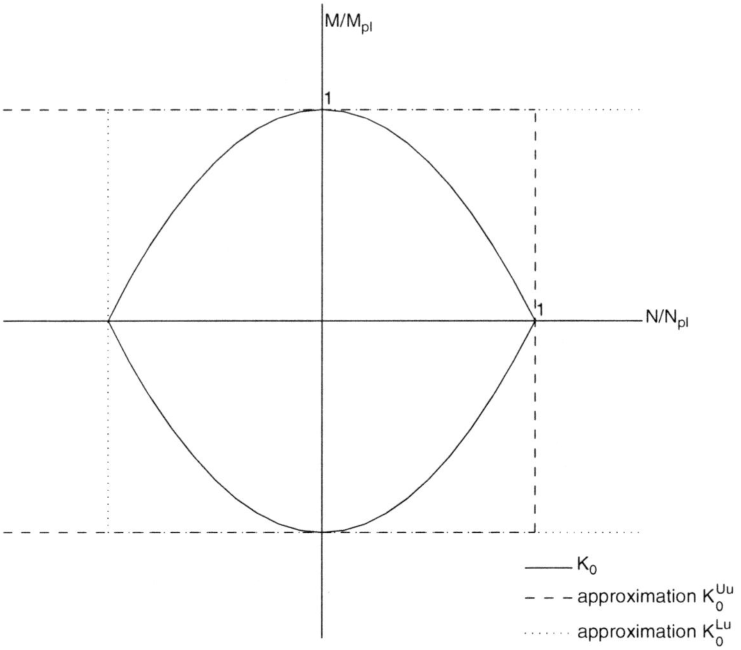

Fig. 8.3 Approximation of K_0 by K_0^{Lu} and K_0^{Uu}

Then,

$$N_{ipl}^L := |\sigma_{yi}^L| A_i = \sigma_{yi}^U A_i =: N_{ipl}^U, \tag{8.28b}$$

hence,

$$N_{ipl} := N_{ipl}^L = N_{ipl}^U = \sigma_{yi} A_i. \tag{8.28c}$$

Moreover,

$$M_{ipl} = \sigma_{yi}^U W_{ipl} = \sigma_{yi} W_{ipl} = \sigma_{yi} A_i \bar{y}_{ic}, \ i = 1, \dots, B, \tag{8.28d}$$

where $\overline{y}_{ic}$ denotes the arithmetic mean of the centroids of the two half areas of the cross-sectional area A_i of bar i.

Depending on the geometric from of the cross-sectional areal (rectangle, circle, etc.), for the element load vectors

$$F_i = \begin{pmatrix} t_i \\ m_i^+ \\ m_i^- \end{pmatrix}, i = 1, \dots, B, \tag{8.29}$$

of the bars we have the yield condition:

$$\left|\frac{t_i}{N_{ipl}}\right|^\alpha + \left|\frac{m_i^-}{M_{ipl}}\right| \leq 1 \quad \text{(negative end)} \tag{8.30a}$$

$$\left|\frac{t_i}{N_{ipl}}\right|^\alpha + \left|\frac{m_i^+}{M_{ipl}}\right| \leq 1 \quad \text{(positive end).} \tag{8.30b}$$

Here, $\alpha > 1$ is a constant depending on the type of the cross-sectional area of the ith bar. Defining the convex set

$$K_0^\alpha := \left\{ \begin{pmatrix} x \\ y \end{pmatrix} \in \mathbb{R}^2 : |x|^\alpha + |y| \leq 1 \right\}, \tag{8.31}$$

for (8.30a,b) we have also the representation

$$\begin{pmatrix} \frac{t_i}{N_{ipl}} \\ \frac{m_i^-}{M_{ipl}} \end{pmatrix} \in K_0^\alpha \quad \text{(negative end)} \tag{8.32a}$$

$$\begin{pmatrix} \frac{t_i}{N_{ipl}} \\ \frac{m_i^+}{M_{ipl}} \end{pmatrix} \in K_0^\alpha \quad \text{(positive end).} \tag{8.32b}$$

Piecewise Linearization of K_0^α

Due to the symmetry of K_0^α with respect to the transformation

$$x \to -x, y \to -y,$$

K_0^α is piecewise linearized as follows.

Starting from a boundary point of K_0^α, hence,

$$\begin{pmatrix} u_1 \\ u_2 \end{pmatrix} \in \partial K_0^\alpha \quad \text{with} \quad u_1 \geq 0,\ u_2 \geq 0, \tag{8.33a}$$

we consider the gradient of the boundary curve

$$f(x, y) := |x|^\alpha + |y| - 1 = 0$$

of K_0^α in the four points

$$\begin{pmatrix} u_1 \\ u_2 \end{pmatrix}, \begin{pmatrix} -u_1 \\ u_2 \end{pmatrix}, \begin{pmatrix} -u_1 \\ -u_2 \end{pmatrix}, \begin{pmatrix} u_1 \\ -u_2 \end{pmatrix}. \tag{8.33b}$$

We have

$$\nabla f \begin{pmatrix} u_1 \\ u_2 \end{pmatrix} = \begin{pmatrix} \alpha u_1^{\alpha-1} \\ 1 \end{pmatrix} \tag{8.34a}$$

$$\nabla f \begin{pmatrix} -u_1 \\ u_2 \end{pmatrix} = \begin{pmatrix} -\alpha(-(-u_1))^{\alpha-1} \\ 1 \end{pmatrix} = \begin{pmatrix} -\alpha u_1^{\alpha-1} \\ 1 \end{pmatrix} \tag{8.34b}$$

$$\nabla f \begin{pmatrix} -u_1 \\ -u_2 \end{pmatrix} = \begin{pmatrix} -\alpha(-(-u_1))^{\alpha-1} \\ -1 \end{pmatrix} = \begin{pmatrix} -\alpha u_1^{\alpha-1} \\ -1 \end{pmatrix} \tag{8.34c}$$

$$\nabla f \begin{pmatrix} u_1 \\ -u_2 \end{pmatrix} = \begin{pmatrix} \alpha u_1^{\alpha-1} \\ -1 \end{pmatrix}, \tag{8.34d}$$

where

$$\nabla f \begin{pmatrix} -u_1 \\ -u_2 \end{pmatrix} = -\nabla f \begin{pmatrix} u_1 \\ u_2 \end{pmatrix} \tag{8.35a}$$

$$\nabla f \begin{pmatrix} u_1 \\ -u_2 \end{pmatrix} = -\nabla f \begin{pmatrix} -u_1 \\ u_2 \end{pmatrix}. \tag{8.35b}$$

Furthermore, in the two special points

$$\begin{pmatrix} 0 \\ 1 \end{pmatrix} \quad \text{and} \quad \begin{pmatrix} 0 \\ -1 \end{pmatrix}$$

of ∂K_0^α we have, cf. (8.34a), (8.34d), resp., the gradients

$$\nabla f \begin{pmatrix} 0 \\ 1 \end{pmatrix} = \begin{pmatrix} 0 \\ 1 \end{pmatrix} \tag{8.36a}$$

$$\nabla f \begin{pmatrix} 0 \\ -1 \end{pmatrix} = \begin{pmatrix} 0 \\ -1 \end{pmatrix}. \tag{8.36b}$$

Though $f(x, y) = |x|^\alpha + |y| - 1$ is not differentiable at

$$\begin{pmatrix} 1 \\ 0 \end{pmatrix} \quad \text{and} \quad \begin{pmatrix} -1 \\ 0 \end{pmatrix},$$

we define

$$\nabla f \begin{pmatrix} 1 \\ 0 \end{pmatrix} := \begin{pmatrix} 1 \\ 0 \end{pmatrix} \tag{8.36c}$$

$$\nabla f \begin{pmatrix} -1 \\ 0 \end{pmatrix} := \begin{pmatrix} -1 \\ 0 \end{pmatrix}. \tag{8.36d}$$

Using a boundary point $\begin{pmatrix} u_1 \\ u_2 \end{pmatrix}$ of K_0^α with $u_1, u_2 > 0$, the feasible domain K_0^α can be approximated from outside by the convex polyhedron defined as follows.

From the gradients (8.36a–d) we obtain next to the already known conditions (no M–N-interaction):

$$\nabla f \begin{pmatrix} 0 \\ 1 \end{pmatrix}^T \left(\begin{pmatrix} x \\ y \end{pmatrix} - \begin{pmatrix} 0 \\ 1 \end{pmatrix} \right) = \begin{pmatrix} 0 \\ 1 \end{pmatrix}^T \begin{pmatrix} x \\ y-1 \end{pmatrix} \le 0$$

$$\nabla f \begin{pmatrix} 0 \\ -1 \end{pmatrix}^T \left(\begin{pmatrix} x \\ y \end{pmatrix} - \begin{pmatrix} 0 \\ -1 \end{pmatrix} \right) = \begin{pmatrix} 0 \\ -1 \end{pmatrix}^T \begin{pmatrix} x \\ y+1 \end{pmatrix} \le 0$$

$$\nabla f \begin{pmatrix} 1 \\ 0 \end{pmatrix}^T \left(\begin{pmatrix} x \\ y \end{pmatrix} - \begin{pmatrix} 1 \\ 0 \end{pmatrix} \right) = \begin{pmatrix} 1 \\ 0 \end{pmatrix}^T \begin{pmatrix} x-1 \\ y \end{pmatrix} \le 0$$

$$\nabla f \begin{pmatrix} -1 \\ 0 \end{pmatrix}^T \left(\begin{pmatrix} x \\ y \end{pmatrix} - \begin{pmatrix} -1 \\ 0 \end{pmatrix} \right) = \begin{pmatrix} -1 \\ 0 \end{pmatrix}^T \begin{pmatrix} x+1 \\ y \end{pmatrix} \le 0.$$

This yields

$$\begin{aligned} y - 1 &\le 0 \\ -1(y+1) &\le 0 \\ x - 1 &\le 0 \\ -1(x+1) &\le 0 \end{aligned}$$

or

$$|x| \le 1 \tag{8.37a}$$
$$|y| \le 1. \tag{8.37b}$$

Moreover, with the gradients (8.34a–d), cf. (8.35a,b), we get the additional conditions

$$\begin{aligned} &\nabla f \begin{pmatrix} u_1 \\ u_2 \end{pmatrix}^T \left(\begin{pmatrix} x \\ y \end{pmatrix} - \begin{pmatrix} u_1 \\ u_2 \end{pmatrix} \right) \\ &= \begin{pmatrix} \alpha u_1^{\alpha-1} \\ 1 \end{pmatrix}^T \begin{pmatrix} x - u_1 \\ y - u_2 \end{pmatrix} \le 0 \qquad \text{(1st quadrant)} \end{aligned} \tag{8.38a}$$

$$\begin{aligned} &\nabla f \begin{pmatrix} -u_1 \\ -u_2 \end{pmatrix}^T \left(\begin{pmatrix} x \\ y \end{pmatrix} - \begin{pmatrix} -u_1 \\ -u_2 \end{pmatrix} \right) \\ &= -\begin{pmatrix} \alpha u_1^{\alpha-1} \\ 1 \end{pmatrix}^T \begin{pmatrix} x + u_1 \\ y + u_2 \end{pmatrix} \le 0 \qquad \text{(3rd quadrant)} \end{aligned} \tag{8.38b}$$

$$\nabla f \begin{pmatrix} -u_1 \\ u_2 \end{pmatrix}^T \left(\begin{pmatrix} x \\ y \end{pmatrix} - \begin{pmatrix} -u_1 \\ u_2 \end{pmatrix} \right)$$
$$= \begin{pmatrix} -\alpha u_1^{\alpha-1} \\ 1 \end{pmatrix}^T \begin{pmatrix} x + u_1 \\ y - u_2 \end{pmatrix} \leq 0 \qquad \text{(2nd quadrant)} \tag{8.38c}$$
$$\nabla f \begin{pmatrix} u_1 \\ -u_2 \end{pmatrix}^T \left(\begin{pmatrix} x \\ y \end{pmatrix} - \begin{pmatrix} u_1 \\ -u_2 \end{pmatrix} \right)$$
$$= - \begin{pmatrix} -\alpha u_1^{\alpha-1} \\ 1 \end{pmatrix}^T \begin{pmatrix} x - u_1 \\ y + u_2 \end{pmatrix} \leq 0 \quad \text{(4th quadrant).} \tag{8.38d}$$

This means

$$\alpha u_1^{\alpha-1} x - \alpha u_1^{\alpha} + y - u_2 = \alpha u_1^{\alpha-1} x + y - \left(\alpha u_1^{\alpha} + u_2\right) \leq 0 \tag{8.39a}$$
$$-\left(\alpha u_1^{\alpha-1} x + \alpha u_1^{\alpha} + y + u_2\right) = -\left(\alpha u_1^{\alpha-1} x + y + \left(\alpha u_1^{\alpha} + u_2\right) -\right) \leq 0 \tag{8.39b}$$
$$-\alpha u_1^{\alpha-1} x - \alpha u_1^{\alpha} + y - u_2 = -\alpha - u_1^{\alpha-1} x + y - \left(\alpha u_1^{\alpha} + u_2\right) \leq 0 \tag{8.39c}$$
$$-\alpha u_1^{\alpha-1} x - \alpha u_1^{\alpha} - y - u_2 = \alpha - u_1^{\alpha-1} x - y - \left(\alpha u_1^{\alpha} + u_2\right) \leq 0. \tag{8.39d}$$

With

$$\alpha u_1^{\alpha} + u_2 = \nabla f \begin{pmatrix} u_1 \\ u_2 \end{pmatrix}^T \begin{pmatrix} u_1 \\ u_2 \end{pmatrix} =: \beta(u_1, u_2) \tag{8.40}$$

we get the equivalent constraints

$$\begin{aligned} \alpha u_1^{\alpha-1} x + y - \beta(u_1, u_2) &\leq 0 \\ \alpha u_1^{\alpha-1} x + y + \beta(u_1, u_2) &\geq 0 \\ -\alpha u_1^{\alpha-1} x + y - \beta(u_1, u_2) &\leq 0 \\ \alpha u_1^{\alpha-1} x - y - \beta(u_1, u_2) &\leq 0. \end{aligned}$$

This yields the double inequalities

$$|\alpha u_1^{\alpha-1} x + y| \leq \beta(u_1, u_2) \tag{8.41a}$$
$$|\alpha u_1^{\alpha-1} x - y| \leq \beta(u_1, u_2). \tag{8.41b}$$

Thus, a point $u = \begin{pmatrix} u_1 \\ u_2 \end{pmatrix} \in \partial K_0^{\alpha},\ u_1 > 0, u_2 > 0$, generates therefore the inequalities

$$-1 \leq x \leq 1 \tag{8.42a}$$
$$-1 \leq y \leq 1 \tag{8.42b}$$
$$-\beta(u_1, u_2) \leq \alpha u_1^{\alpha-1} x + y \leq \beta(u_1, u_2) \tag{8.42c}$$
$$-\beta(u_1, u_2) \leq \alpha u_1^{\alpha-1} x - y \leq \beta(u_1, u_2). \tag{8.42d}$$

Obviously, each further point $\hat{u} \in \partial K_0^\alpha$ with $\hat{u}_1 > 0, \hat{u}_2 > 0$ yields additional inequalities of the type (8.42c,d).

Condition (8.42a–d) can be represented in the following vectorial form:

$$-\begin{pmatrix} 1 \\ 1 \end{pmatrix} \leq I \begin{pmatrix} x \\ y \end{pmatrix} \leq \begin{pmatrix} 1 \\ 1 \end{pmatrix} \tag{8.43a}$$

$$-\beta(u_1, u_2) \begin{pmatrix} 1 \\ 1 \end{pmatrix} \leq H(u_1, u_2) \begin{pmatrix} x \\ y \end{pmatrix} \leq \beta(u_1, u_2) \begin{pmatrix} 1 \\ 1 \end{pmatrix}, \tag{8.43b}$$

with the matrices

$$I = \begin{pmatrix} 1 & 0 \\ 0 & 1 \end{pmatrix}, \quad H(u_1, u_2) = \begin{pmatrix} \alpha u_1^{\alpha-1} & 1 \\ \alpha u_1^{\alpha-1} & -1 \end{pmatrix}. \tag{8.44}$$

Choosing a further boundary point $\hat{u}$ of K_0^α with $\hat{u}_1 > 0, \hat{u}_2 > 0$, we get additional conditions of the type (8.43b).

Using (8.42a–d), for the original yield condition (8.32a,b) we get then the approximative feasibility condition:

1. Negative end of the bar

$$-N_{ipl} \leq t_i \leq N_{ipl} \tag{8.45a}$$

$$-M_{ipl} \leq m_i^- \leq M_{ipl} \tag{8.45b}$$

$$-\beta(u_1, u_2) \leq \alpha u_1^{\alpha-1} \frac{t_i}{N_{ipl}} + \frac{m_i^-}{M_{ipl}} \leq \beta(u_1, u_2) \tag{8.45c}$$

$$-\beta(u_1, u_2) \leq \alpha u_1^{\alpha-1} \frac{t_i}{N_{ipl}} - \frac{m_i^-}{M_{ipl}} \leq \beta(u_1, u_2). \tag{8.45d}$$

2. Positive end of the bar

$$-N_{ipl} \leq t_i \leq N_{ipl} \tag{8.45e}$$

$$-M_{ipl} \leq m_i^+ \leq M_{ipl} \tag{8.45f}$$

$$-\beta(u_1, u_2) \leq \alpha u_1^{\alpha-1} \frac{t_i}{N_{ipl}} + \frac{m_i^+}{M_{ipl}} \leq \beta(u_1, u_2) \tag{8.45g}$$

$$-\beta(u_1, u_2) \leq \alpha u_1^{\alpha-1} \frac{t_i}{N_{ipl}} - \frac{m_i^+}{M_{ipl}} \leq \beta(u_1, u_2). \tag{8.45h}$$

Defining

$$\Gamma^{(i)} := \begin{pmatrix} \frac{1}{N_{ipl}} & 0 & 0 \\ 0 & \frac{1}{M_{ipl}} & 0 \\ 0 & 0 & \frac{1}{M_{ipl}} \end{pmatrix}, \quad F_i = \begin{pmatrix} t_i \\ m_i^+ \\ m_i^- \end{pmatrix}, \tag{8.46}$$

conditions (8.45a–h) can be represented also by

$$-\begin{pmatrix}1\\1\\1\end{pmatrix} \le \Gamma^{(i)} F_i \le \begin{pmatrix}1\\1\\1\end{pmatrix} \tag{8.47a}$$

$$-\beta(u_1,u_2)\begin{pmatrix}1\\1\\1\\1\end{pmatrix} \le \begin{pmatrix}\alpha u_1^{\alpha-1} & 1 & 0\\ \alpha u_1^{\alpha-1} & -1 & 0\\ \alpha u_1^{\alpha-1} & 0 & 1\\ \alpha u_1^{\alpha-1} & 0 & -1\end{pmatrix}\Gamma_i F_i \le \beta(u_1,u_2)\begin{pmatrix}1\\1\\1\\1\end{pmatrix}. \tag{8.47b}$$

Multiplying (8.45a,c,d,g,h) with N_{ipl}, due to

$$\frac{N_{ipl}}{M_{ipl}} = \frac{\sigma_{yi} A_i}{\sigma_{yi} W_{ipl}} = \frac{\sigma_{yi} A_i}{\sigma_{yi} A_i \bar{y}_{ic}} = \frac{1}{\bar{y}_{ic}}, \tag{8.48}$$

for (8.45a,c,d,g,h) we also have

$$-\beta(u_1,u_2)N_{ipl} \le \alpha u_1^{\alpha-1} t_i + \frac{m_i^-}{\bar{y}_{ic}} \le \beta(u_1,u_2)N_{ipl} \tag{8.49a}$$

$$-\beta(u_1,u_2)N_{ipl} \le \alpha u_1^{\alpha-1} t_i - \frac{m_i^-}{\bar{y}_{ic}} \le \beta(u_1,u_2)N_{ipl} \tag{8.49b}$$

$$-\beta(u_1,u_2)N_{ipl} \le \alpha u_1^{\alpha-1} t_i + \frac{m_i^+}{\bar{y}_{ic}} \le \beta(u_1,u_2)N_{ipl} \tag{8.49c}$$

$$-\beta(u_1,u_2)N_{ipl} \le \alpha u_1^{\alpha-1} t_i - \frac{m_i^+}{\bar{y}_{ic}} \le \beta(u_1,u_2)N_{ipl}. \tag{8.49d}$$

8.2.2 Approximation of the Yield Condition by Using Reference Capacities

According to (8.31), (8.32a,b) for each bar $i = 1,\dots,B$ we have the condition

$$\begin{pmatrix}\frac{t}{N_{pl}}\\ \frac{m}{M_{pl}}\end{pmatrix} \in K_0^\alpha = \left\{\begin{pmatrix}x\\y\end{pmatrix} \in \mathbb{R}^2 : |x|^\alpha + |y| \le 1\right\}$$

with $(t,m) = (t_i, m_i^\pm)$, $(N_{pl}, M_{pl}) = (N_{ipl}, M_{ipl})$.

Selecting fixed reference capacities

$$N_{i0} > 0,\, M_{i0} > 0,\, i = 1,\dots,B,$$

related to the plastic capacities N_{ipl}, M_{ipl}, we get

$$\left|\frac{t_i}{N_{ipl}}\right|^{\alpha} + \left|\frac{m_i^{\pm}}{M_{ipl}}\right| = \left|\frac{t_i}{N_{i0}} \cdot \frac{1}{\frac{N_{ipl}}{N_{i0}}}\right|^{\alpha} + \left|\frac{m_i^{\pm}}{M_{i0}} \cdot \frac{1}{\frac{M_{ipl}}{M_{i0}}}\right|.$$

Putting

$$\rho_i = \rho_i(a(\omega), x) := \min\left\{\frac{N_{ipl}}{N_{i0}}, \frac{M_{ipl}}{M_{i0}}\right\}, \tag{8.50}$$

we have

$$\frac{\rho_i}{\frac{N_{ipl}}{N_{i0}}} \le 1, \; \frac{\rho_i}{\frac{M_{ipl}}{M_{i0}}} \le 1$$

and therefore

$$\left|\frac{t_i}{N_{ipl}}\right|^{\alpha} + \left|\frac{m_i^{\pm}}{M_{ipl}}\right| = \left|\frac{t_i}{\rho_i N_{i0}}\right|^{\alpha} \cdot \left|\frac{\rho_i}{\frac{N_{ipl}}{N_{i0}}}\right|^{\alpha} + \left|\frac{m_i^{\pm}}{\rho_i M_{i0}}\right| \cdot \left|\frac{\rho_i}{\frac{M_{ipl}}{M_{i0}}}\right| \le \left|\frac{t_i}{\rho_i N_{i0}}\right|^{\alpha} + \left|\frac{m_i^{\pm}}{\rho_i M_{i0}}\right|. \tag{8.51}$$

Thus, the yield condition (8.30a,b) or (8.32a,b) is guaranteed by

$$\left|\frac{t_i}{\rho_i N_{i0}}\right|^{\alpha} + \left|\frac{m_i^{\pm}}{\rho_i M_{i0}}\right| \le 1$$

or

$$\begin{pmatrix} \frac{t_i}{\rho_i N_{i0}} \\ \frac{m_i^{\pm}}{\rho_i M_{i0}} \end{pmatrix} \in K_0^{\alpha}. \tag{8.52}$$

Applying the piecewise linearization described in Sect. 8.2.1 to condition (8.52), we obtain, cf. (8.45a–h), the approximation stated below. Of course, conditions (8.45a,b,e,f) are not influenced by this procedure. Thus, we find

$$-N_{ipl} \le t_i \le N_{ipl} \tag{8.53a}$$

$$-M_{ipl} \le m_i^- \le M_{ipl} \tag{8.53b}$$

$$-M_{ipl} \le m_i^+ \le M_{ipl} \tag{8.53c}$$

$$-\beta(u_1, u_2) \le \alpha u_1^{\alpha-1} \frac{t_i}{\rho_i N_{i0}} + \frac{m_i^-}{\rho_i M_{i0}} \le \beta(u_1, u_2) \tag{8.53d}$$

$$-\beta(u_1, u_2) \le \alpha u_1^{\alpha-1} \frac{t_i}{\rho_i N_{i0}} - \frac{m_i^-}{\rho_i M_{i0}} \le \beta(u_1, u_2) \tag{8.53e}$$

$$-\beta(u_1, u_2) \le \alpha u_1^{\alpha-1} \frac{t_i}{\rho_i N_{i0}} + \frac{m_i^+}{\rho_i M_{i0}} \le \beta(u_1, u_2) \tag{8.53f}$$

$$-\beta(u_1, u_2) \le \alpha u_1^{\alpha-1} \frac{t_i}{\rho_i N_{i0}} - \frac{m_i^+}{\rho_i M_{i0}} \le \beta(u_1, u_2). \tag{8.53g}$$

Remark 8.7. Multiplying with ρ_i we get quotients $\frac{t_i}{N_{i0}}, \frac{m_i^{\pm}}{M_{i0}}$ with fixed denominators.

Hence, multiplying (8.53d–g) with ρ_i, we get the equivalent system

$$-N_{ipl} \le t_i \le N_{ipl} \tag{8.54a}$$
$$-M_{ipl} \le m_i^- \le M_{ipl} \tag{8.54b}$$
$$-M_{ipl} \le m_i^+ \le M_{ipl} \tag{8.54c}$$
$$-\beta(u_1,u_2)\rho_i \le \alpha u_1^{\alpha-1}\frac{t_i}{N_{i0}} + \frac{m_i^-}{M_{i0}} \le \beta(u_1,u_2)\rho_i \tag{8.54d}$$
$$-\beta(u_1,u_2)\rho_i \le \alpha u_1^{\alpha-1}\frac{t_i}{N_{i0}} - \frac{m_i^-}{M_{i0}} \le \beta(u_1,u_2)\rho_i \tag{8.54e}$$
$$-\beta(u_1,u_2)\rho_i \le \alpha u_1^{\alpha-1}\frac{t_i}{N_{i0}} + \frac{m_i^+}{M_{i0}} \le \beta(u_1,u_2)\rho_i \tag{8.54f}$$
$$-\beta(u_1,u_2)\rho_i \le \alpha u_1^{\alpha-1}\frac{t_i}{N_{i0}} - \frac{m_i^+}{M_{i0}} \le \beta(u_1,u_2)\rho_i . \tag{8.54g}$$

Obviously, (8.54a–g) can be represented also in the following form:

$$|t_i| \le N_{ipl} \tag{8.55a}$$
$$|m_i^-| \le M_{ipl} \tag{8.55b}$$
$$|m_i^+| \le M_{ipl} \tag{8.55c}$$
$$\left|\alpha u_1^{\alpha-1}\frac{t_i}{N_{i0}} + \frac{m_i^-}{M_{i0}}\right| \le \beta(u_1,u_2)\rho_i \tag{8.55d}$$
$$\left|\alpha u_1^{\alpha-1}\frac{t_i}{N_{i0}} - \frac{m_i^-}{M_{i0}}\right| \le \beta(u_1,u_2)\rho_i \tag{8.55e}$$
$$\left|\alpha u_1^{\alpha-1}\frac{t_i}{N_{i0}} + \frac{m_i^+}{M_{i0}}\right| \le \beta(u_1,u_2)\rho_i \tag{8.55f}$$
$$\left|\alpha u_1^{\alpha-1}\frac{t_i}{N_{i0}} - \frac{m_i^+}{M_{i0}}\right| \le \beta(u_1,u_2)\rho_i . \tag{8.55g}$$

By means of definition (8.50) of ρ_i, system (8.55a–g) reads

$$|t_i| \le N_{ipl} \tag{8.56a}$$
$$|m_i^-| \le M_{ipl} \tag{8.56b}$$
$$|m_i^+| \le M_{ipl} \tag{8.56c}$$
$$\left|\alpha u_1^{\alpha-1}\frac{t_i}{N_{i0}} + \frac{m_i^-}{M_{i0}}\right| \le \beta(u_1,u_2)\frac{N_{ipl}}{N_{i0}} \tag{8.56d}$$
$$\left|\alpha u_1^{\alpha-1}\frac{t_i}{N_{i0}} + \frac{m_i^-}{M_{i0}}\right| \le \beta(u_1,u_2)\frac{M_{ipl}}{M_{i0}} \tag{8.56e}$$

$$\left|\alpha u_1^{\alpha-1}\frac{t_i}{N_{i0}} - \frac{m_i^-}{M_{i0}}\right| \le \beta(u_1, u_2)\frac{N_{ipl}}{N_{i0}} \tag{8.56f}$$

$$\left|\alpha u_1^{\alpha-1}\frac{t_i}{N_{i0}} - \frac{m_i^-}{M_{i0}}\right| \le \beta(u_1, u_2)\frac{M_{ipl}}{M_{i0}} \tag{8.56g}$$

$$\left|\alpha u_1^{\alpha-1}\frac{t_i}{N_{i0}} + \frac{m_i^+}{M_{i0}}\right| \le \beta(u_1, u_2)\frac{N_{ipl}}{N_{i0}} \tag{8.56h}$$

$$\left|\alpha u_1^{\alpha-1}\frac{t_i}{N_{i0}} + \frac{m_i^+}{M_{i0}}\right| \le \beta(u_1, u_2)\frac{M_{ipl}}{M_{i0}} \tag{8.56i}$$

$$\left|\alpha u_1^{\alpha-1}\frac{t_i}{N_{i0}} - \frac{m_i^+}{M_{i0}}\right| \le \beta(u_1, u_2)\frac{N_{ipl}}{N_{i0}} \tag{8.56j}$$

$$\left|\alpha u_1^{\alpha-1}\frac{t_i}{N_{i0}} - \frac{m_i^+}{M_{i0}}\right| \le \beta(u_1, u_2)\frac{M_{ipl}}{M_{i0}} \tag{8.56k}$$

Corresponding to Remark 8.7, the variables

$$t_i, m_i^+, m_i^-, A_i \text{ or } x$$

enters linearly. Increasing the accuracy of approximation by taking a further point $\hat{u} = (\hat{u}_1, \hat{u}_2)$ with the related points $(\hat{u}_1, -\hat{u}_2), (-\hat{u}_1, \hat{u}_2), (-\hat{u}_1, -\hat{u}_2)$, we obtain further inequalities of the type (8.55d–g), (8.56d–g) respectively.

8.3 Stochastic Optimization

Due to (8.1c), (8.10), (8.11c), (8.12c), the $3B$ – vector

$$F = (F_1^T, \ldots, F_B^T)^T \tag{8.57a}$$

of all interior loads fulfills the equilibrium condition

$$CF = R \tag{8.57b}$$

with the external load vector R and the equilibrium matrix C.

In the following we collect all random model parameters [31, 35, 39], such as external load factors, material strength parameters, cost factors, etc., into the random ν-vector

$$a = a(\omega),\ \omega \in (\Omega, \mathcal{A}, \mathcal{P}), \tag{8.58a}$$

on a certain probability space $(\Omega, \mathcal{A}, P)$.

Thus, since in some cases the vector R of external loads depend also on the design r-vector x, we get

$$R = R\left(a(\omega), x\right). \tag{8.58b}$$

Of course, the plastic capacities depend also on the vectors x and $a(\omega)$, hence,

$$N_{ipl}^{\Gamma} = N_{ipl}^{\Gamma}\left(a(\omega), x\right), \ \Gamma = L, U \tag{8.58c}$$

$$M_{ipl} = M_{ipl}\left(a(\omega), x\right). \tag{8.58d}$$

We assume that the probability distribution and/or the needed moments of the random parameter vector $a = a(\omega)$ are known [2,28,31,42].

The remaining deterministic constraints for the design vector x are represented by

$$x \in \mathcal{D} \tag{8.59}$$

with a certain convex subset D of $\mathbb{R}^r$.

8.3.1 Violation of the Yield Condition

Consider in the following an interior load distribution F fulfilling the equilibrium condition (8.58b).

According to the analysis given in Sect. 8.2, after piecewise linearization, the yield condition for the ith bar can be represented by an inequality of the type

$$H\Gamma^{(i)}\Big(a(\omega), x\Big) \leq h^{(i)}\Big(a(\omega), x\Big), i = 1, \ldots, B, \tag{8.60}$$

with matrices $H, \Gamma^{(i)} = \Gamma^{(i)}\Big(a(\omega), x\Big)$ and a vector $h^{(i)} = h^{(i)}\Big(a(\omega), x\Big)$ as described in Sect. 8.2.

In order to take into account violations of condition (8.60), we consider the equalities

$$H\Gamma^{(i)}F_i + z_i = h^{(i)}, i = 1, \ldots, B. \tag{8.61}$$

If

$$z_i \geq 0, \text{ for all } i = 1, \ldots, B, \tag{8.62a}$$

then (8.60) holds, and the yield condition is then fulfilled too, or holds with a prescribed accuracy.

However, in case

$$z_i \not\geq 0 \text{ for some bars } i \in \{1, \ldots, B\}, \tag{8.62b}$$

the survival condition is violated at some points of the structure. Hence, structural failures may occur. The resulting costs Q of failure, damage and reconstructure of the frame is a function of the vectors z_i, $i = 1, \ldots, B$, defined by (8.61). Thus, we have

$$Q = Q(z) = Q(z_1, \ldots, z_B), \tag{8.63a}$$

where

$$z := (z_1^T, z_2^T, \ldots, z_B^T)^T, \tag{8.63b}$$

$$z_i := h^{(i)} - H\Gamma^{(i)} F_i, \; i = 1, \ldots, B. \tag{8.63c}$$

8.3.2 *Cost Function*

Due to the survival condition (8.62a), we may consider cost functions Q such that

$$Q(z) = 0, \text{ if } z \geq 0, \tag{8.64}$$

hence, no (recourse) costs arise if the yield condition (8.60) holds.

In many cases the recourse or failure costs of the structure are defined by the sum

$$Q(z) = \sum_{i=1}^{B} Q_i(z_i), \tag{8.65}$$

of the element failure costs $Q_i = Q_i(z_i), i = 1, \ldots, B$.

Using the representation

$$z_i = y_i^+ - y_i^-, \quad y_i^+, y_i^- \geq 0, \tag{8.66a}$$

the member cost functions $Q_i = Q_i(z_i)$ are often defined [20, 31] by considering the linear function

$$q_i^{-T} y_i^- + q_i^{+T} y_i^+ \tag{8.66b}$$

with certain vectors q_i^+, q_i^- of cost coefficients for the evaluation of the condition $z_i \geq 0, z_i \ngeq 0$, respectively.

The cost function $Q_i = Q_i(z_i)$ is then defined by the minimization problem

$$\min q_i^{-T} y_i^- + q_i^{+T} y_i^+ \tag{8.67a}$$

$$\text{s.t. } y_i^+ - y_i^- = z_i \tag{8.67b}$$

$$y_i^-, y_i^+ \geq 0. \tag{8.67c}$$

If $z_i := (z_{i1}, \ldots, z_{i\mu})^T, y_i^\pm := (y_{i1}^\pm, \ldots, y_{i\mu}^\pm)^T, i = 1, \ldots, B$ then (8.67a–c) can also be represented by

$$\min \sum_{l=1}^{\mu} (q_{il}^- y_{il}^- + q_{il}^+ y_{il}^+) \tag{8.68a}$$

$$\text{s.t. } y_{il}^+ - y_{il}^- = z_{il}, \quad l = 1, \ldots, \mu \tag{8.68b}$$

$$y_{il}^-, y_{il}^+ \geq 0, \quad l = 1, \ldots, \mu. \tag{8.68c}$$

Since the pairs of variables $(y_{il}^-, y_{il}^+), l = 1, \ldots, \mu$, are not connected with each other by the constraints, and the objective function is separabel with respect to these pairs of variables, (8.68a–c) can be decomposed into μ separated minimization problems

$$\min\ q_{il}^- y_{il}^- + q_{il}^+ y_{il}^+ \tag{8.69a}$$

$$\text{s.t. }\ y_{il}^+ - y_{il}^- = z_{il} \tag{8.69b}$$

$$y_{il}^-, y_{il}^+ \geq 0, \tag{8.69c}$$

for the pairs of variables $(y_{il}^-, y_{il}^+), l = 1, \ldots, \mu$.

Under the condition

$$q_{il}^- + q_{il}^+ \geq 0, l = 1, \ldots, \mu, \tag{8.70}$$

the following result holds:

Lemma 8.1. *Suppose that (8.70) holds. Then the minimum value function $Q_{il} = Q_{il}(z_{il})$ of (8.69a–c) is a piecewise linear, convex function given by*

$$Q_{il}(z_{il}) := \max\{q_{il}^+ z_{il}, -q_{il}^- z_{il}\}. \tag{8.71}$$

Hence, the member cost functions $Q_i = Q_i(z_i)$ reads

$$Q_i(z_i) = \sum_{l=1}^{\mu} Q_{il}(z_{il}) = \sum_{l=1}^{\mu} \max\{q_{il}^+ z_{il}, -q_{il}^- z_{il}\}, \tag{8.72a}$$

and the total cost function $Q = Q(z)$ is given by

$$Q(z) = \sum_{i=1}^{B} Q_i(z_i) = \sum_{i=1}^{B} \sum_{l=1}^{\mu} \max\{q_{il}^+ z_{il}, -q_{il}^- z_{il}\}. \tag{8.72b}$$

8.3.3 Choice of the Cost Factors

Under elastic conditions the change $\Delta\sigma$ of the total stress σ and the change ΔL of the element length L are related by

$$\Delta L = \frac{L}{E} \Delta\sigma, \tag{8.73a}$$

where E denotes the modulus of elasticity.

Assuming that the neutral axis is equal to the axis of symmetry of the element, for the total stress $\Delta\sigma$ in the upper ("+") lower ("−") fibre of the boundary we have

$$\Delta\sigma = \frac{\Delta t}{A} \pm \frac{\Delta m}{W}, \tag{8.73b}$$

where W denotes the axial modulus of the cross-sectional area of the element (beam).

Representing the change ΔV of volume of an element by

$$\Delta V = A \cdot \Delta L, \tag{8.74a}$$

then

$$\begin{aligned} \Delta V &= A \cdot \Delta L = A \cdot \frac{L}{E}\Delta\sigma = A \cdot \frac{L}{E}\left(\frac{\Delta t}{A} \pm \frac{\Delta m}{W}\right) \\ &= \frac{L}{E}\Delta t \pm \frac{L}{E}\frac{A}{W}\Delta m = \frac{L}{E}\Delta t \pm \frac{L}{E}\frac{1}{\frac{W}{A}}\Delta m = \frac{L}{E}\Delta t \pm \frac{L}{E} \cdot \frac{1}{\bar{y}_c}\Delta m, \end{aligned} \tag{8.74b}$$

where $\bar{y}_c$ is the cross-sectional parameter as defined in (8.28d).

Consequently, due to (8.73a,b), for the evaluation of violations Δt of the axial force constraint we may use a cost factor of the type

$$\Gamma_K := \frac{L}{E}, \tag{8.75a}$$

and an appropriate cost factor for the evaluation of violations of moment constraints reads

$$\Gamma_M = \frac{L}{E} \cdot \frac{1}{\bar{y}_c}. \tag{8.75b}$$

8.3.4 Total Costs

Denoting by

$$G_0 = G_0\Big(a(\omega), x\Big) \tag{8.76a}$$

the primary costs, such as weighted negative load factors, material costs, costs of construction, etc., the total costs including failure or recourse costs are given by

$$G = G_0\big(a(\omega), x\big) + Q\big(z\big(a(\omega), x, F(\omega)\big)\big). \tag{8.76b}$$

Hence, the total costs $G = G\big(a(\omega), x, F(\omega)\big)$ depend on the vector $x = (x_1, \ldots, x_r)^T$ of design variables, the random vector $a(\omega) = \big(a_1(\omega), \ldots, a_\nu(\omega)\big)^T$ of model parameters and the random vector $F = F(\omega)$ of all internal loadings.

Minimizing the expected total costs, we get the following stochastic optimization problem (SOP) of recourse type [20]

$$\min E\Big(G_0\big(a(\omega), x\big) + Q\big(z\,(a(\omega), x, F(\omega))\big)\Big) \tag{8.77a}$$

$$\text{s.t. } H\Gamma^{(i)}\Big(a(\omega), x\Big)F_i(\omega) + z_i(\omega) = h^{(i)}\Big(a(\omega), x\Big) \text{ a.s.,} \quad i = 1, \dots, B \tag{8.77b}$$

$$CF(\omega) = R\Big(a(\omega), x\Big) \text{ a.s.} \tag{8.77c}$$

$$x \in D. \tag{8.77d}$$

Using representation (8.65), (8.67a–c) of the recourse or failure cost function $Q = Q(z)$, problem (8.77a–d) takes also the following equivalent from

$$\min E\left(G_0\Big(a(\omega), x\Big) + \sum_{i=1}^{B}\left(q_i^-(\omega)^T y_i^-(\omega) + q_i^+(\omega)^T y_i^+(\omega)\right)\right) \tag{8.78a}$$

$$\text{s.t. } H\Gamma^{(i)}\Big(a(\omega), x\Big)F_i(\omega) + y_i^+(\omega) - y_i^-(\omega) = h^{(i)}\Big(a(\omega), x\Big) \text{ a.s.,} \quad i = 1, \dots, B \tag{8.78b}$$

$$CF(\omega) = R\,(a(\omega), x) \quad \text{a.s.} \tag{8.78c}$$

$$x \in D,\ y_i^+(\omega), y_i^-(\omega) \geq 0 \quad \text{a.s.,} \quad i = 1, \dots, B. \tag{8.78d}$$

Remark 8.8. Stochastic optimization problems of the type (8.78a–d) are called “two-stage stochastic programs” or “stochastic problems with recourse”.

In many cases the primary cost function $G_0 = G_0\Big(a(\omega), x\Big)$ represents the volume or weight of the structural, hence,

$$G_0\Big(a(\omega), x\Big) := \sum_{i=1}^{B} \Gamma_i(\omega)V_i(x) = \sum_{i=1}^{B} \Gamma_i(\omega)L_i A_i(x), \tag{8.79}$$

with certain (random) weight factors $\Gamma_i = \Gamma_i(\omega)$.

In case

$$A_i(x) := u_i x_i \tag{8.80}$$

with fixed sizing parameters $u_i, i = 1, \dots, B$, we get

$$G_0\,(a(\omega), x) = \sum_{i=1}^{B} \Gamma_i(\omega)L_i u_i(x) = \sum_{i=1}^{B} \Gamma_i(\omega)L_i h_i x_i = c\,(a(\omega))^T\, x, \tag{8.81a}$$

where

$$c\left(a(\omega)\right) := \left(\Gamma_1(\omega)L_1u_1, \ldots, \Gamma_B(\omega)L_Bu_B\right)^T. \tag{8.81b}$$

Thus, in case (8.80), $G_0 = G_0\Big(a(\omega), x\Big)$ is a linear function of x.

8.3.5 Discretization Methods

The expectation in the objective function of the stochastic optimization problem (8.78a–d) must be computed numerically. One of the main methods is based on the discretization of the probability distribution $P_{a(\cdot)}$ of the random parameter ν-vector $a = a(\omega)$, hence,

$$P_{a(\cdot)} \approx \mu := \sum_{k=1}^{s} \alpha_k \epsilon_{a^{(k)}} \tag{8.82a}$$

with

$$\alpha_k \geq 0,\ k = 1, \ldots, s,\ \sum_{k=1}^{s} \alpha_k = 1. \tag{8.82b}$$

Corresponding to the realizations $a^{(k)}, k = 1, \ldots, s$, of the discrete approximate (8.82a,b), we have the realizations $y_i^{-(k)}, y_i^{+(k)}$ and $F^{(k)}, F_i^{(k)}, k = 1, \ldots, s$, of the random vectors $y_i^-(\omega), y_i^+(\omega), i = 1, \ldots, B$, and $F(\omega)$. then

$$E\left(G_0\big(a(\omega), x\big) + \sum_{i=1}^{B}\Big(q_i^{-T} y_i^-(\omega) + q_i^{+T} y_i^+(\omega)\Big)\right)$$
$$\approx \sum_{k=1}^{s} \alpha_k \left(G_0(a^{(k)}, x) + \sum_{i=1}^{B}(q_i^{-T} y_i^{-(k)} + q_i^{+T} y_i^{+(k)}\right). \tag{8.83a}$$

Furthermore, the equilibrium equation (8.77c) is approximated by

$$CF^{(k)} = R(a^{(k)}, x),\ k = 1, \ldots, s, \tag{8.83b}$$

where $F^{(k)} := \left(F_1^{(k)^T}, \ldots, F_B^{(k)^T}\right)^T$, and we have, cf. (8.78d), the nonnegativity constraints

$$y_i^{+(k)},\ y_i^{-(k)} \geq 0,\ k = 1, \ldots, s,\ i = 1, \ldots, B. \tag{8.83c}$$

Thus, (SOP) (8.78a–d) is reduced to the parameter optimization problem

$$\min\ \overline{G}_0(a^{(k)}, x) + \sum_{i=1}^{B} \alpha_k \left(q_i^{-T} y_i^{-(k)} + q_i^{+T} y_i^{+(k)}\right) \tag{8.84a}$$

$$\text{s.t.}\quad H\Gamma^{(i)}(a^{(k)},x)F_i^{(k)} + y_i^{+(k)} - y_i^{-(k)} = h^{(i)}(a^{(k)},x),$$
$$i = 1,\dots,B,\ k = 1,\dots,s \tag{8.84b}$$
$$CF^{(k)} = R(a^{(k)},x),\ k = 1,\dots,s \tag{8.84c}$$
$$x \in D,\ y_i^{+(k)},\ y_i^{-(k)} \geq 0,\ k = 1,\dots,s,\ i = 1,\dots,B. \tag{8.84d}$$

A further important class of methods for computing expectations and probabilities, hence, multiple integrals, occurring in stochastic optimization problems, reliability analysis and reliability-based optimal design (RBO), cf. [9, 11, 30, 31], are simulation methods, such as Monte Carlo Simulation (MCS) procedures. Simulation techniques are used especially in cases with only few information about the analytical properties of the underlying technical device, e.g., in case of analytically almost unavailable limit state functions. In principle, MSC is a very simple technique which is widely applicable on the one hand, but may have a very low efficiency of estimation on the other hand. Hence, several improvements were considered in the last time, such as Advanced Monte Carlo Simulation techniques: Variance reduction methods reducing the sampling error, based, e.g., on importance sampling methods, direction sampling, subset simulation, etc., see, e.g., [4]. Further improvements can be obtained by combining these simulation/estimation techniques with Response Surface Methods (RSM) for estimating unknown functions using regression techniques and advanced nonlinear programming procedures (NLP), cf. [12].

8.3.6 Complete Recourse

According to Sect. 8.3.1, the evaluation of the violation of the yield condition (8.60) is based on (8.61), hence

$$H\Gamma^{(i)}F_i + z_i = h^{(i)}, i = 1,\dots,B.$$

In generalization of the so-called "simple recourse" case (8.67a–c), in the "complete recourse" case the deviation

$$z_i = h^{(i)} - H\Gamma^{(i)}F_i$$

is evaluated by means of the minimum value $Q_i = Q_i(z_i)$ of the linear program, cf. (8.68a–c)

$$\min\ q^{(i)T}y^{(i)} \tag{8.85a}$$
$$\text{s.t.}\quad M^{(i)}y^{(i)} = z_i \tag{8.85b}$$
$$y^{(i)} \geq 0. \tag{8.85c}$$

Here, $q^{(i)}$ is a given cost vector and $M^{(i)}$ denotes the so-called recourse matrix [20].

We assume that the linear equation (8.85b) has a solution $y^{(i)} \geq 0$ for each vector z_i. This property is called "complete recourse".

In the present case the stochastic optimization problem (8.77a–c) reads

$$\min E\left(G_0\Big(a(\omega), x\Big) + \sum_{i=1}^{B} q^{(i)T} y^{(i)}(\omega)\right) \tag{8.86a}$$

$$\text{s.t. } H\Gamma^{(i)}\Big(a(\omega), x\Big)F_i(\omega) + M^{(i)}y^{(i)}(\omega) = h^{(i)}\Big(a(\omega), x\Big) \text{ a.s.,} \quad i = 1, \ldots, B \tag{8.86b}$$

$$CF(\omega) = R\,(a(\omega), x) \quad \text{a.s.} \tag{8.86c}$$

$$x \in D,\ y^{(i)}(\omega) \geq 0 \quad \text{a.s.,} \quad i = 1, \ldots, B. \tag{8.86d}$$

As described in Sect. 8.3.5, problem (8.86a–d) can be solved numerically by means of discretization methods and application of linear/nonlinear programming techniques.

References

1. Arnbjerg-Nielsen, T.: Rigid-ideal plastic model as a reliability analysis tool for ductil structures. Ph.D. Dissertation. Technical University of Denmark, Lyngby (1991)
2. Augusti, G., Baratta, A., Casciati, F.: Probabilistic Methods in Structural Engineering. Chapman and Hall, London (1984)
3. Bullen, P.S.: Handbook of Means and Their Inequalities. Kluwer, Dordrecht (2003)
4. Charmpis, D.C., Schuëller, G.I.: Using Monte Carlo simulation to treat physical uncertainties in structural reliability. In: Marti, K., et al. (ed.) Coping with Uncertainties. LNEMS, vol. 581, pp. 67–83. Springer, Berlin, Heidelberg, New York (2006)
5. Charnes, A., Lemke, C.E., Zienkiewicz, O.C.: Virtual work, linear programming and plastic limit analysis. Proc. R. Soc. A **25**, 110–116 (1959)
6. Cocchetti, G., Maier, G.: Static shakedown theorems in piecewise linearized plasticity. Arch. Appl. Mech. **68**, 651–661 (1988)
7. Corradi, L., Zavelani, A.: A linear programming approach to shakedown analysis of structures. Comput. Methods Appl. Mech. Eng. **3**, 37–53 (1974)
8. Damkilde, L., Krenk, S.: Limits – a system for limit state analysis and optimal material layout. Comput. Struct. **64**(1–4), 709–718 (1997)
9. Ditlevsen, O., Madsen, H.O.: Structural Reliabilitx Methods. Wiley, New York (1996)
10. Doltsinis, I.: Elements of Plasticity. WIT, Southhampton (1999)
11. Frangopol, D.M.: Reliability-based optimum structural design. In: Sundarajan, C. (ed) Probabilistic Structural Mechanics Handbook, pp. 352–387. Chapman and Hall, New York (1995)
12. Gasser, M., Schuëller, G.I.: Some basic principles in reliability-based optimisation (RBO) of structures and mechanical components. In: Marti, K., Kall, P. (eds.) Stochastic Programming Methods and Technical Applications. Lecture Notes in Economics and Mathematical Systems, vol. 458, pp. 80–103. Springer, Berlin (1998)
13. Haftka, R.T., Gürdal, Z., Kamat, M.P.: Elements of structural optimization. Kluwer, Dordrecht (1990)
14. Han, H., Reddy, B.D.: Plasticity – Mathematical Theory and Numerical Analysis. Springer, New York (1999)
15. Hardy, G.H., Littlewood, J.E., Pólya, G.: Inequalities. Cambridge University Press, London (1973)

16. Heitzer, M.: Traglast- und Einspielanalyse zur Bewertung der Sicherheit passiver Komponenten. Dissertation RWTH Aachen (1999)
17. Hodge, P.G.: Plastic Analysis of Structures. McGraw-Hill, New York (1959)
18. Kaliszky, S.: Plastizitätslehre. VDI, Düsseldorf (1984)
19. Kaliszky, S.: Plasticity: Theory and Engineering Applications. Elsevier, Amsterdam (1989)
20. Kall, P.: Stochastic Linear Programming. Springer, Berlin, Heidelberg, New York (1976)
21. Kemenjarzh, J.A.: Limit Analysis of Solids and Structures. CRC, Boca Raton (1996)
22. Kirsch, U.: Structural Optimization. Springer, Berlin, Heidelberg, New York (1993)
23. König, J.A.: Shakedown of Elastic-Plastic Structures. Elsevier, Amsterdam (1987)
24. Lawo, M.: Optimierung im konstruktiven Ingenieurbau. Vieweg, Braunschweig (1987)
25. Lawo, M., Thierauf, G.: Stabtragwerke, Matrizenmethoden der Statik und Dynamik. Teil I: Statik. F. Vieweg, Braunschweig (1980)
26. Madsen, H.O., Krenk, S., Lind, N.C.: Methods of Structural Safety. Dover, Mineola, NY (1986, 2006)
27. Marti, K.: Stochastic optimization methods in structural mechanics. ZAMM **70**, 742–T745 (1990)
28. Marti, K.: Approximation and derivatives of probabilities of survival in structural analysis and design. Struct. Optim. **13**, 230–243 (1997)
29. Marti, K.: Stochastic optimization methods in optimal engineering design under stochastic uncertainty. ZAMM **83**(11), 1–18 (2003)
30. Marti, K.: Reliability Analysis of Technical Systems / Structures by means of Polyhedral Approximation of the Safe / Unsafe Domain. GAMM Mitteilungen **30**(2), 211–254 (2007)
31. Marti, K.: Stochastic Optimization Methods, 2nd edn. Springer, Berlin, Heidelberg, New York (2008)
32. Neal, B.G.: The Plastic Methods of Structural Analysis. Chapman and Hall, London (1965)
33. Nielsen, M.P.: Limit Analysis and Concrete Plasticity. Prentice Hall, Englewood Cliffs, NJ (1984)
34. Prager, W., Shield, R.T.: A general theory of optimal plastic design. J. Appl. Mech. **34**(1), 184–186 (1967)
35. Purek, L.: Probabilistische Beurteilung der Tragsicherheit bestehender Bauten. Dissertation ETH Nr. 11880 (1997)
36. Smith, D.L. (ed.): Mathematical Programming Methods in Structural Plasticity. CISM Courses and Lectures, vol. 299. Springer, Vienna (1990)
37. Smith, D.L., Munro, J.: Plastic analysis and synthesis of frames subject to multiple loadings. Eng. Optim. **2**, 145–157 (1976)
38. Spiller, W.R.: Automated Structural Analysis: An Introduction. Pergamon, New York (1972)
39. Thoma, K.: Stochastische Betrachtung von Modellen für vorgespannte Zugelemente. IBK Bericht Nr. 287, Hochschulverlag AG an der ETH Zürich (2004)
40. Tin-Loi, F.: On the optimal plastic synthesis of frames. Eng. Optim. **16**, 91–108 (1990)
41. Tin-Loi, F.: Plastic limit analysis of plane frames and grids using GAMS. Comput. Struct. **54**, 15–25 (1995)
42. Zimmermann, J.J., Corotis, R.B., Ellis, J.H.: Stochastic programs for identifying significant collaps models in structural systems. In: Der Kiureghian, A., Thoft-Christensen, P. (eds.) Reliability and optimization of structural systems 1990, pp. 359–365. Springer, Berlin, Heidelberg, New York (1991)

Part IV
Analysis and Optimization of Economic and Engineering Systems Under Uncertainty

Chapter 9
Uncertainty in the Future Nitrogen Load to the Baltic Sea Due to Uncertain Meteorological Conditions

Jerzy Bartnicki

Abstract The Norwegian Meteorological Institute has a long-term project with HELCOM Commission for regular calculation of annual atmospheric deposition of nitrogen to the Baltic Sea. In 2005, the institute received an additional project from HELCOM with the aim of estimating atmospheric nitrogen deposition to six sub-basins and catchments of the Baltic Sea for the year 2010, using nitrogen emission projections according to agreed emission ceilings under the EU National Emission Ceilings (NEC) Directive and the Gothenburg Protocol. Since, the meteorology for 2010 is unknown, model calculations were performed for four selected years with different meteorology: 1996, 1997, 1998 and 2000, which are available in the database. Final deposition values for the year 2010 were calculated as an average over the four selected years. In this way we were able to estimate the uncertainty restricted to meteorological variability. The ranges between minimum and maximum of calculated depositions to sub-basins and catchments are large indicating significant variation of the deposition depending on meteorological conditions.

9.1 Introduction

The Helsinki Commission, or HELCOM is the governing body of the "Convention on the Protection of the Marine Environment of the Baltic Sea Area" – known as the Helsinki Convention [7]. The important role of HELCOM is to protect the marine environment of the Baltic Sea from all sources of pollution through intergovernmental co-operation between Denmark, Estonia, the European Community, Finland, Germany, Latvia, Lithuania, Poland, Russia and Sweden, which are the Contracting Parties to HELCOM. HELCOM's vision for the future is a healthy Baltic Sea environment with diverse biological components functioning in balance, resulting in a good ecological status and supporting a wide range of sustainable economic and social activities (http://www.helcom.fi/).

J. Bartnicki
Norwegian Meteorological Institute, P.O. Box 43 Blindern, 0313 Oslo, Norway,
e-mail: jerzy.bartnicki@met.no

K. Marti et al. (eds.), *Coping with Uncertainty*, Lecture Notes in Economics and Mathematical Systems 633, DOI 10.1007/978-3-642-03735-1_9,

Eutrophication is one of the major problems for the Baltic Sea. Since the 1800s, the Baltic Sea has changed from an oligotrophic clear-water sea into a eutrophic marine environment. Nitrogen and phosphorus are among the main growth-limiting nutrients and as such do not pose any direct hazards to marine organisms. Eutrophication, however, is a condition in an aquatic ecosystem where high nutrient concentrations stimulate growth of algae which leads to imbalanced functioning of the system. It is mainly caused by a significant nutrient load to the Baltic Sea (http://www.helcom.fi/).

The nutrient inputs entering the Baltic Sea are either airborne or waterborne. The main pathways of nutrient input to the Baltic Sea are:

- Direct atmospheric deposition on the Baltic Sea water surface
- Riverine inputs of nutrients to the sea. Rivers transport nutrients that have been discharged or lost to inland surface waters within the Baltic Sea catchment area
- Point sources discharging directly to the sea.

Atmospheric deposition of nitrogen accounts for approximately 30% of the total nitrogen load to the Baltic Sea. Therefore, nitrogen has been regularly monitored by analysing the results of measurements and model calculations in co-operation with the Co-operative Programme for Monitoring and Evaluation of the Long-range Transmission of Air pollutants in Europe (EMEP). The main objective of the EMEP is to regularly provide Governments and subsidiary bodies under the Convention on Long-range Transboundary Air Pollution [8] with qualified scientific information to support the development and further evaluation of the international protocols on emission reductions negotiated under the Convention (http://www.emep.int/). A co-operation between EMEP and HELCOM was established already in 1996. Since then, three EMEP Centres have published joint reports estimating annual supply of nitrogen heavy metals and persistent organic pollutants to the Baltic Sea. Updated emissions, as well as results of measurements and their analysis are also included in the annual EMEP reports for HELCOM. Such an annual report has been recently prepared for HELCOM in 2007 [1].

In 2004 HELCOM established a project using models to assess the implications of different policy scenarios on nutrient inputs and the resulting eutrophication status in order to indicate the most cost-effective measures for the different sub-regions of the Baltic Sea. Most of the models considered under HELCOM have been ecological models related to the assessment of effects to the sea. The basis of the effect models ware the scenarios of activities on land. Therefore, the aim was to link management scenario models with ecological models in order to assess measures for reducing nutrient inputs. The aim of the project was to assess for HELCOM Contracting Parties the impact of different agricultural policy scenarios on nutrient inputs in the Baltic Sea catchment area and on the eutrophication status of the Baltic Sea. The final aim was to enable the identification of cost-effective measures in the different parts of the Baltic Sea catchment area which requires achieving a good ecological status throughout the Baltic Sea area [2].

In 2005 HELCOM considered airborne nitrogen pollution and decided to include nitrogen air depositions into the ongoing HELCOM Project. To evaluate the implications of different policy scenarios on nutrient inputs, HELCOM agreed that EMEP should be charged with the task of assessing the changes in atmospheric nitrogen deposition under the condition that nitrogen emission targets according to the Gothenburg protocol to the LRTAP Convention and the EU NEC Directive as projected for 2010 are fulfilled. Based on the above decision, EMEP has received a project from HELCOM.

As requested by HELCOM, all nitrogen depositions as well as source allocation budgets have been calculated for the six sub-basins and catchments of the Baltic Sea for the year 2010 using the EMEP Unified model [10] (http://www.emep.int/). Names and acronyms of the sub-basins and catchments used in the model calculations are given below:

1. Gulf of Bothnia (GUB)
2. Gulf of Finland (GUF)
3. Gulf of Riga (GUR)
4. Baltic Proper (BAP)
5. Belt Sea (BES)
6. The Kattegat (KAT)

Depositions and source allocation budgets have been also calculated for the entire basin and the entire catchment of the Baltic Sea. Geographical borders of sub-basins and catchments used in the computations are shown in Fig. 9.1.

Emissions and meteorological fields are the main input into the EMEP Unified model which calculates transport and deposition of air pollutants over Europe. For the historical calculations, meteorological fields are available from the Numerical Weather Prediction model HIRLAM [3] which is run operationally at the Norwegian Meteorological Institute in Oslo. Since the meteorological fields are not known for 2010, we have used the meteorological fields from the past years, which are

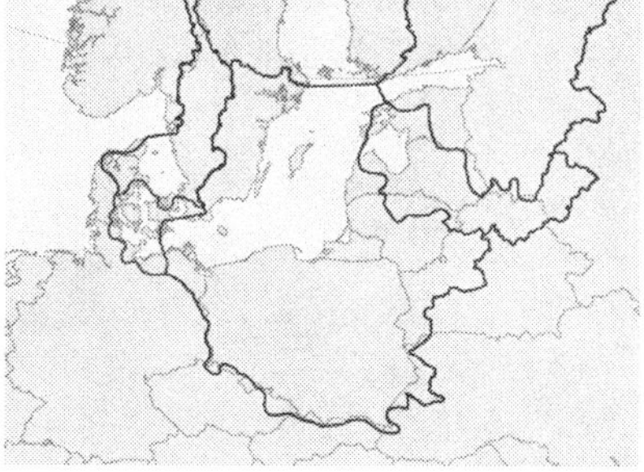

Fig. 9.1 Locations of six sub-basins and catchments of the Baltic Sea for which the nitrogen depositions for 2010 have been calculated

available in the EMEP database. Variability between different meteorological years in the past create significant uncertainty in the model results for 2010. Analysis of this specific uncertainty is the main subject of the current paper.

9.2 Nitrogen Emissions

Emission input to the EMEP Unified model requires information about annual nitrogen oxides and ammonia emissions from all sources located in the EMEP domain.

Projections for 2010 assume nitrogen oxides and ammonia national emissions as specified in the NEC Directive [9] for selected 15 EU countries. National nitrogen oxides and ammonia emissions as specified in the Gothenburg Protocol [4] are assumed for all countries listed therein except for those already mentioned in the NEC Directive. Emissions for the Russian Federation and Estonia are taken from the HELCOM publication [6]. Finally, 2010 projections of nitrogen emissions due to international ship traffic on the Europeans seas are taken from the EMEP database following Entec projections [5]. All remaining emission sources of nitrogen are taken from the EMEP database http://www.emep.int/.

9.2.1 National Emission Ceilings According to EU NEC Directive

2010 national emission ceilings for nitrogen oxides (units: kt of NO_2 per year) and ammonia (units: kt of NH_3 per year) according to the EU NEC Directive [9] are shown in Table 9.1.

The national emission ceilings presented in Table 9.1 were designed with the aim of broadly meeting the interim environmental objectives set out in Article 5 of the Directive. It is supposed that the Community area with depositions of nutrient nitrogen in excess of the critical loads will be reduced by about 30% compared to 1990.

Table 9.1 2010 national emission ceilings for nitrogen oxides (units: kt of NO_2 per year) and for ammonia (units: kt of NH_3 per year) according to the EU NEC Directive [9]

Country	NO_x	NH_3	Country	NO_x	NH_3
Austria	103	66	Belgium	176	74
Denmark	127	69	Finland	170	31
France	810	780	Germany	1,051	550
Greece	44	73	Ireland	65	116
Italy	90	419	Luxembourg	11	7
Netherlands	260	128	Portugal	250	90
Spain	847	353	Sweden	148	57
United Kingdom	1,167	297	**EC15**	**6,519**	**3,110**

Table 9.2 2010 national emission ceilings of nitrogen oxides (units: kt of NO_2 per year) and ammonia (units: kt of NH_3 per year) according to the 1999 Gothenburg Protocol [4]

Country	NO_x	NH_3	Country	NO_x	NH_3
Armenia	46	25	Austria	107	66
Belarus	255	158	Belgium	181	74
Bulgaria	266	108	Croatia	87	30
Czech Republic	286	101	Denmark	127	69
Finland	170	31	France	860	780
Germany	1,081	550	Greece	344	73
Hungary	198	90	Ireland	65	116
Italy	1,000	419	Latvia	84	44
Liechtenstein	0.37	0.15	Lithuania	110	84
Luxembourg	11	7	Netherlands	266	128
Norway	156	23	Poland	879	468
Portugal	260	108	Republic of Moldova	90	42
Romania	437	210	Russian Federation (PEMA)	265	49
Slovakia	130	39	Slovenia	45	20
Spain	847	353	Sweden	148	57
Switzerland	79	63	Ukraine	1,222	592
United Kingdom	1,181	297	**European Community**	**6,671**	**3,129**

Among the EU-15 countries, United Kingdom (1,167 kt NO_2) and Germany (1,051 kt NO_2) are the main NO_2 emitters in 2010. France (780 kt NH_3) and Germany (550 kt NH_3) are the main NH_3 sources.

During the EU accession process national emission ceilings were reduced for Latvia to 61 kt NO_2. This most recent value for Latvia [6] and the nitrogen emissions shown in Table 9.2 were used for the EMEP model run. The same emission sources are also considered in the Gothenburg Protocol, however NEC emissions of nitrogen oxides and ammonia projected for 2010 are lower or the same as those projected by the Gothenburg Protocol presented in the next section.

9.2.2 National Emission Ceilings According to Gothenburg Protocol

The national emission ceilings for nitrogen oxides and ammonia specified in the Protocol to Abate Acidification, Eutrophication and Ground-level Ozone done in Gothenburg, Sweden on 30 November 1999 [4] are shown in Table 9.2.

For the Russian Federation the Protocol specifies only the emission ceilings for the so-called Pollutant Emissions Management Area (PEMA). In the EMEP domain, 2010 nitrogen oxide and ammonia emissions for the entire Russian Federation are 2,653 kt NO_2 and 1,179 kt NH_3, respectively.

Compared to emission levels in 1990, main reductions of nitrogen oxides emissions should occur in the Czech Republic (61%), Germany (60%), Sweden

(56%) and United Kingdom (56%). Main reductions of ammonia emissions in 2010 compared to 1990 are expected in Denmark (43%), Netherlands (43%), Slovakia (37%) and Czech Republic (35%).

9.2.3 Nitrogen Emission Projections Used in the Model Runs

Projected national annual total emissions of nitrogen oxides and ammonia used in the model runs are shown in Table 9.3 for 2010. All anthropogenic sources in the EMEP area are taken into account in Table 9.3, including emissions from the HELCOM parties and from the international ship traffic on the Baltic Sea.

The numbers in Table 9.3 are based on total emission targets for each country, including areas outside the Baltic Sea Catchment, which explains high values for, e.g. Russia and Germany. Not all emissions in the Contracting Parties end up in the Baltic Sea. For 2010, the total nitrogen oxide emissions target in the EMEP area is approximately 19.7 million tones of NO_2, whereas the total ammonia emissions target in the EMEP area is about 7.3 million tones of NH_3.

Table 9.3 2010 national total emissions of nitrogen oxides (units: kt of NO_2 per year) and ammonia (units: kt of NH_3 per year) used in the model runs

Source	NO_x	NH_3	Source	NO_x	NH_3
Albania	27.9	25.9	Armenia	46.0	25.0
Austria	102.9	66.2	Azerbaijan	43.0	25.0
Baltic Sea	457.5	0.0	Belarus	254.9	157.8
Belgium	176.1	74.4	Black Sea	154.8	0.0
Bosnia and Herz.	54.0	17.3	Bulgaria	265.5	107.6
Croatia	97.3	30.1	Cyprus	21.1	6.3
Czech Republic	286.1	101.4	Denmark	126.7	69.3
Estonia	59.0	28.4	Finland	170.3	30.8
France	810.3	780.3	Georgia	30.0	97.0
Germany	1,051.2	549.8	Greece	344.5	72.5
Hungary	198.4	89.5	Iceland	30.0	3.0
Ireland	65.0	115.8	Italy	990.1	418.88
Kazakhstan	50.2	19.0	Latvia	60.8	43.8
Lithuania	110.4	84.4	Luxembourg	11.0	7.1
Macedonia	5.9	1.4	Mediterranean Sea	2,382	80.0
Moldova, Republic of	90.6	41.6	Netherlands	260.3	128.3
North Africa	96.0	235.0	North Sea	862.4	0.0
Norway	156.0	23.0	Norway	156.0	23.0
Poland	878.3	467.9	Remaining Asiatic ar.	9.0	178.0
Remaining NE Atlantic	740.5	0.0	Romania	436.5	209.7
Russian Federation	2,653.0	1,178.8	Serbia and Montenegro	167.8	69.3
Slovakia	130.2	19.9	Spain	148.4	57.3
Switzerland	79.3	62.6	Turkey	851.8	240.6
Ukraine	1,222.2	591.9	United Kingdom	1,167.4	296.6
EMEP	**19,686.3**	**7,345.2**			

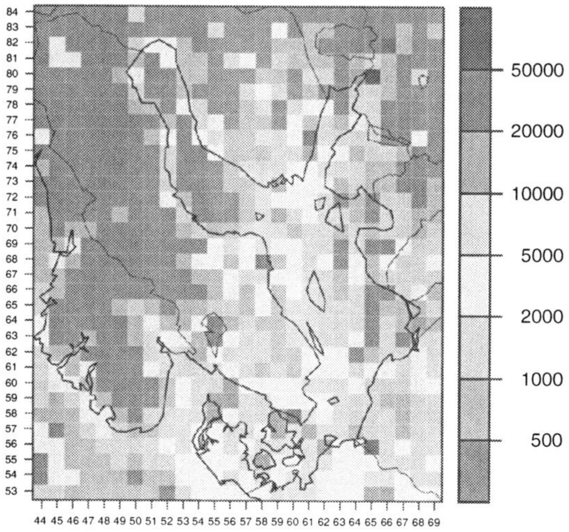

Fig. 9.2 Spatial distribution of nitrogen oxide emissions around the Baltic Sea for 2010, used in the EMEP model computations. Units: tonnes of NO_2 per year and per 50 km × 50 km grid cell

The total nitrogen oxides emissions and ammonia emissions targets in the HELCOM Parties are estimated to be 5.3 million tones of NO_2 and 2.5 million tones of NH_3, respectively.

Table 9.3 exhibits that the Russian Federation and ship traffic on the Mediterranean Sea are the largest emitters of nitrogen oxides in 2010, with 2,653 and 2,383 kt NO_2, respectively. These two sources are approximately twice as large than the next on the list: Ukraine (1,222 kt NO_2), United Kingdom (1,167 kt NO_2), Germany (1,051 kt NO_2) and Italy (990 kt NO_2).

The Russian Federation (1,179 kt NH_3), France (780 kt NH_3), Ukraine (592 kt NH_3) and Germany (550 kt NH_3) belong to the four highest emitters of ammonia in 2010. The emission targets for the rest of the countries are much lower with Poland (550 kt NH_3) and Italy (468 kt NH_3) taking the fifth and sixth places, respectively.

To run the EMEP model, the emissions should be distributed in the model grid system. Maps with the 2010 nitrogen oxides and ammonia emissions in 2010 spatially distributed over the Baltic Sea region are shown in Figs. 9.2 and 9.3 respectively.

9.3 Computed Nitrogen Depositions for 2010

Nitrogen emission inventories described in the previous chapter have been used in order to estimate nitrogen depositions in 2010. Since, the meteorology for 2010 is unknown, model calculations had to be performed with existing meteorological data from the past years. Meteorological data for the years 1996, 1997, 1998 and 2000 and emission projections for the year 2010 have been used for the EMEP Unified model runs. Final deposition values for the year 2010 were calculated as the average over the four selected years.

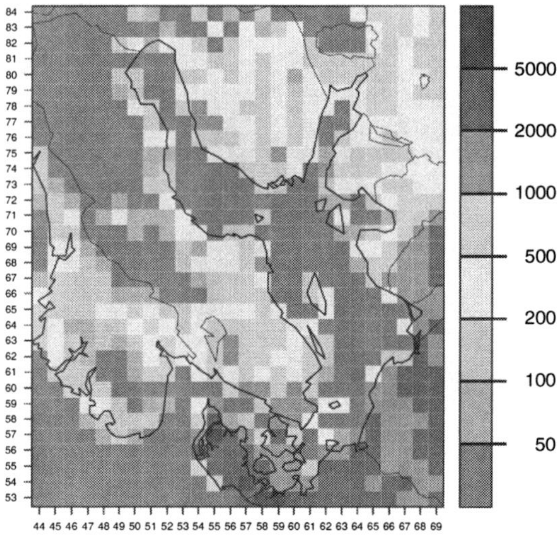

Fig. 9.3 Spatial distribution of ammonia emissions around the Baltic Sea for 2010, used in the EMEP model computations. Units: tonnes of NH_3 per year and per 50 km × 50 km grid cell

9.3.1 Unified EMEP Model

The Unified EMEP model is an Eulerian model that has been developed at the Meteorological Synthesizing Centre West of EMEP (EMEP/MSC-W) for simulating atmospheric transport and deposition of acidifying and eutrophying compounds as well as photo-oxidants in Europe. The model has been documented in the EMEP Status Report [10]. Here we only give a short description of the basic features of the model. Model details as well as recent changes and updates can be found on the EMEP web site http://www.emep.int/.

The model domain covers Europe and the Atlantic Ocean. The model grid (of the size 170 × 133) has a horizontal resolution of 50 km at 60°N, which is consistent with the resolution of emission data reported to EMEP. In the vertical, the model has 20 sigma layers reaching up to 100 hPa. Approximately 10 of these layers are placed below 2 km to obtain a high resolution for the boundary layer which is of special importance for the long range transport of air pollution.

The EMEP unified model uses 3-hourly resolved meteorological data from the PARLAM-PS model, a dedicated version of the HIRLAM (High Resolution Limited Area Model) Numerical Weather Prediction model operational at the Norwegian Meteorological Institute [3].

The emission input consists of gridded national annual emissions of sulphur dioxide, nitrogen oxides, ammonia, non-methane volatile organic compounds (VOC) and carbon monoxide. They are available in each of the 50 km × 50 km model grid cells. These emissions are distributed temporally according to monthly and daily factors derived from data provided by the University of Stuttgart (IER).

Concentrations of 71 species are computed in the Unified EMEP model (56 are advected, 15 are short-lived and not advected). The sulphur and nitrogen chemistry is coupled to the photo-chemistry, which allows a more sophisticated description of, e.g. the oxidation of sulphur dioxide to sulphate.

Dry deposition is calculated using the resistance analogy and is a function of the pollutant type, meteorological conditions and surface properties. Parameterization of wet deposition processes includes both in-cloud and sub-cloud scavenging of gases and particles using scavenging coefficients.

9.3.2 Calculated Depositions to Sub-basins and Catchments of the Baltic Sea

Calculated nitrogen depositions to sub-basins and catchments of the Baltic Sea are given in Tables 9.4 and 9.5, respectively. Calculated depositions are shown for each sub-basin and catchment, and for four meteorological years used in the computations. The deposition in 2010 was calculated as an average over four meteorological years.

Locations of minimum and maximum deposition are slightly different for different sub-basins and catchments and for different deposition types. However, for most sub-basins and catchments, the lowest deposition values can be observed for the meteorological year 1997 and the highest for the meteorological year 2000.

For all meteorological years and all sub-basins and catchments, wet nitrogen deposition is significantly higher, sometimes two times higher than dry deposition of nitrogen.

Depositions of oxidized and reduced nitrogen to all sub-basins are approximately on the same level, but deposition of oxidized nitrogen is slightly higher than deposition of reduced nitrogen in the north and in the middle (GUB, GUF, GUR and BAP sub-basins). In the south (BES and KAT sub-basins) deposition of reduced nitrogen is higher. Deposition of oxidized nitrogen is also higher than deposition of reduced nitrogen to the entire basin of the Baltic Sea. In general, there is more oxidized nitrogen then reduced nitrogen deposition into sub-basins of the Baltic Sea.

In the case of catchments, oxidized nitrogen deposition is higher than reduced nitrogen deposition for GUB, GUF and KAT catchments. For GUR, BAP, BES catchments and the for entire catchment of the Baltic Sea, the relation between oxidized and reduced nitrogen deposition is opposite. In general, there is more reduced nitrogen then oxidized nitrogen deposition into catchments of the Baltic Sea.

9.4 Uncertainty Due to Meteorological Variability

Inspection of Tables 9.4 and 9.5 gives the first impression of uncertainty restricted to meteorological variability. To asses the variability of computed depositions due to varying meteorological conditions the standard deviation (σ_{N-1}) was calculated for each of the deposition type and each of sub-basin and catchment of the Baltic Sea.

The ranges between minimum and maximum can also indicate the uncertainty of computed depositions due to variable meteorological conditions. This is an

Table 9.4 2010 nitrogen depositions to sub-basins of the Baltic Sea and to the entire basin of the Baltic Sea calculated with the help of the meteorology from four different years and as average over these years. Units: kt N

Sub-basin	Meteo	Oxidized dry	Oxidized wet	Reduced dry	Reduced wet	Total
GUB	1996	6.1	11.8	2.9	8.9	29.7
GUB	1997	5.2	10.2	2.4	6.8	24.6
GUB	1998	6.5	15.0	2.8	11.0	35.3
GUB	2000	6.9	18.7	3.9	15.7	45.1
GUB	Mean	6.2	13.9	3.0	10.6	33.7
GUF	1996	2.9	5.2	1.9	5.4	15.3
GUF	1997	2.5	4.1	1.6	3.7	11.9
GUF	1998	3.0	5.5	2.1	4.9	15.4
GUF	2000	3.1	6.4	2.3	5.5	17.3
GUF	Mean	2.9	5.3	2.0	4.9	15.0
GUR	1996	2.2	3.3	1.8	3.7	11.0
GUR	1997	1.9	3.3	1.4	3.2	9.8
GUR	1998	2.2	3.5	1.8	3.4	10.9
GUR	2000	2.4	3.9	2.0	3.9	12.2
GUR	Mean	2.2	3.5	1.7	3.6	11.0
BAP	1996	23.1	40.1	23.3	39.7	126.2
BAP	1997	19.6	33.3	20.2	31.4	104.4
BAP	1998	22.2	47.1	23.9	43.8	136.9
BAP	2000	22.9	47.1	26.3	44.1	140.4
BAP	Mean	21.9	41.9	23.4	39.7	127.0
BES	1996	2.6	4.9	5.1	6.3	18.9
BES	1997	2.4	3.9	5.6	5.4	17.3
BES	1998	2.5	5.8	v5.5	7.8	21.6
BES	2000	2.5	5.8	6.1	7.3	21.7
BES	Mean	2.5	5.1	5.6	6.7	19.9
KAT	1996	2.5	4.4	3.5	5.0	15.4
KAT	1997	2.3	4.1	3.4	4.6	14.4
KAT	1998	2.3	5.4	3.5	5.7	16.9
KAT	2000	2.6	6.5	4.0	6.7	19.8
KAT	Mean	2.4	5.1	3.6	5.5	16.6
BAS	1996	39.4	69.7	38.5	68.9	216.6
BAS	1997	33.9	58.9	34.6	55.0	182.4
BAS	1998	38.7	82.3	39.5	76.5	237.0
BAS	2000	40.3	88.4	44.5	83.2	256.5
BAS	Mean	38.1	74.8	39.3	70.9	223.1

important indicator of uncertainty from the practical point of view in the decision-making process. The relative ranges and relative standard deviations were expressed in per cent of the mean value of the calculated deposition. The results are presented in Tables 9.6 and 9.7 for sub-basins and catchments, respectively.

Table 9.5 2010 nitrogen depositions to six catchments of the Baltic Sea and to the entire catchment of the Baltic Sea calculated with the help of the meteorology from four different years and as average over these years. Units: kt N

Catchment	Meteo	Oxidized dry	Oxidized wet	Reduced dry	Reduced wet	Total
GUB	1996	36	48	14	39	137
GUB	1997	31	44	12	31	118
GUB	1998	34	58	13	44	150
GUB	2000	42	74	19	59	193
GUB	Mean	36	56	15	43	149
GUF	1996	47	70	22	69	207
GUF	1997	41	66	23	63	194
GUF	1998	51	76	27	72	227
GUF	2000	56	82	31	75	244
GUF	Mean	49	74	26	70	218
GUR	1996	23	34	25	46	128
GUR	1997	20	37	23	49	130
GUR	1998	23	37	26	51	136
GUR	2000	25	36	29	46	136
GUR	Mean	23	36	26	48	132
BAP	1996	128	187	198	273	786
BAP	1997	118	178	202	264	761
BAP	1998	132	209	206	293	840
BAP	2000	138	205	225	279	848
BAP	Mean	129	195	208	277	809
BES	1996	7	11	12	15	46
BES	1997	7	9	13	13	42
BES	1998	7	14	13	19	52
BES	2000	7	14	14	18	53
BES	Mean	7	12	13	16	48
KAT	1996	18	19	14	21	72
KAT	1997	16	18	14	20	68
KAT	1998	16	23	14	24	78
KAT	2000	18	30	16	31	96
KAT	Mean	17	23	15	24	78
BAS	1996	258	370	286	463	1,376
BAS	1997	233	353	288	441	1,314
BAS	1998	264	417	300	503	1,483
BAS	2000	286	442	334	507	1,569
BAS	Mean	260	395	302	478	1,436

The relative ranges are roughly twice as high as the standard deviations and these two measures of uncertainty are well correlated for all kinds of nitrogen depositions. The relative ranges can be interpreted as the worst case scenario when the meteorology in the reference year and in the year 2010 gives the most different results in

Table 9.6 Relative ranges and standard deviations (in brackets) of nitrogen depositions to sub-basins of the Baltic Sea and to the entire basin of the Baltic Sea, calculated for 2010 with meteorology from four different years. Units: per cent of mean nitrogen deposition over four years

Sub-basin	Oxidized dry	Oxidized wet	Reduced dry	Reduced wet	Total
GUB	28 (12)	61 (27)	48 (21)	84 (38)	61 (36)
GUF	22 (10)	42 (17)	35 (15)	38 (17)	36 (15)
GUR	24 (10)	18 (8)	38 (16)	19 (8)	22 (9)
BAP	16 (7)	33 (16)	26 (11)	32 (15)	28 (13)
BES	6 (2)	38 (18)	17 (7)	36 (16)	22 (11)
KAT	10 (5)	47 (21)	15 (7)	38 (17)	32 (14)
BAS	17 (8)	39 (18)	25 (10)	40 (17)	33 (14)

Table 9.7 Relative ranges and standard deviations (in brackets) of nitrogen depositions to the catchments of the Baltic Sea and to the entire catchment of the Baltic Sea, calculated for 2010 with meteorology from four different years. Units: per cent of mean nitrogen deposition over four years

Sub-basin	Oxidized dry	Oxidized wet	Reduced dry	Reduced wet	Total
GUB	30 (13)	53 (24)	46 (20)	64 (27)	50 (21)
GUF	31 (13)	22 (10)	36 (16)	17 (7)	23 (10)
GUR	20 (8)	10 (4)	21 (9)	10 (5)	6 (3)
BAP	16 (7)	16 (8)	13 (6)	10 (4)	11 (5)
BES	3 (1)	39 (18)	16 (6)	35 (16)	22 (11)
KAT	16 (8)	52 (24)	15 (7)	45 (20)	35 (15)
BAS	20 (8)	22 (10)	16 (7)	14 (7)	18 (8)

calculated nitrogen deposition. This is the maximum uncertainty estimated from the four years meteorological database. This can be the case if only one meteorological year is taken as a basis for verification of the emission reductions effects in 2010.

The relative standard deviation shows the deviation from the mean deposition over four years. Therefore, the averaging effect is already included in this measure of uncertainty.

For all sub-basins except GUR the uncertainty of computed wet deposition of both oxidized and reduced nitrogen is much higher than the uncertainty of computed dry deposition. In the extreme case of BES sub-basin, uncertainty of computed wet deposition is several times higher than the uncertainty of computed dry deposition for this sub-basin. The reason is a patchy and irregular pattern of precipitation fields, both in space and time.

The largest uncertainty due to meteorological variability can be noticed for GUB (Gulf of Bothnia) sub-basin and catchment with relatively low nitrogen deposition and significant contribution of nitrogen from the long-range transport to the deposition. On the other hand, for BES and KAT sub-basin uncertainty of computed dry oxidized deposition is relatively small, because these sub-basins are located close to the large emissions sources of nitrogen

The uncertainties of the depositions due to variable meteorology are rather large, both for sub-basins and catchments of the Baltic Sea. However, uncertainties of nitrogen deposition to the catchments are in general lower than the uncertainties of nitrogen deposition to sub-basins of the Baltic Sea. The reason is a larger area of the catchment compared to the area of corresponding sub-basin and therefore more likely compensation of different types of uncertainties for the catchments.

Taking into account that uncertainties in nitrogen deposition caused by changing meteorology are larger than the expected emission reductions, one cannot expect a proportional impact of emission reduction on the deposition for one specific year 2010. For example, 10% reduction of nitrogen emission in all EMEP sources can give 6% increase of nitrogen deposition in the entire basin of the Baltic Sea in 2010, if meteorological conditions are the same as in the year 2000. However, the uncertainty due to variable meteorology is decreasing when more years are included in the verification period for the effects of emission reductions.

9.5 Conclusions

The largest uncertainty due to meteorological variability in computed nitrogen deposition for 2010 can be noticed for GUB (Gulf of Bothnia) sub-basin and catchment with relatively low deposition fluxes.

The relative ranges and deviations of the depositions due to variable meteorology are rather large compared to expected emission reductions, both for sub-basins and catchments of the Baltic Sea. For the deposition of oxidized and reduced nitrogen to sub-basins and to the entire basin of the Baltic Sea there is more uncertainty in wet than in dry deposition. The uncertainty of nitrogen deposition to the catchments are in general lower than the uncertainty of nitrogen deposition to sub-basins of the Baltic Sea.

The main conclusion from the decision making perspective is that even significant efforts in reducing nitrogen emissions do not need to become immediately visible in modelled and measured nitrogen depositions to the Baltic Sea due to inter-annual variability of meteorological conditions. In order to notice significant deposition changes caused by the emission reduction it is necessary to wait at least several years. It is very important to take this type of uncertainty into account in future policy actions concerning the improvement of the Baltic Sea environment (http://www.helcom.fi/).

References

1. Bartnicki, J., Gusev, A., Fagerli, H.: Atmospheric supply of nitrogen, lead, cadmium, mercury and dioxins/furans to the Baltic Sea in 2005. EMEP MSC-W Technical Report 3/2007. Meteorological Synthesizing Centre West of EMEP (2007)
2. Bartnicki, J., van Loon M.: Estimation of atmospheric nitrogen deposition to the Baltic Sea in 2010 based on agreed emission ceilings under the EU NEC Directive and the Gothenburg Protocol. Met.no research note No. 26, ISSN-1503-8009. Norwegian Meteorological Institute, Oslo, Norway (2005)

3. Bjørge, D., Skalin, R.: PARLAM – the parallel HIRLAM version at DNMI. Research Report No. 27, ISSN 0332-9879. Norwegian Meteorological Institute, Oslo, Norway (1995)
4. ECE/EB/AIR/85: Handbook for the 1979 convention on long-range transboundary air pollution and its protocols. Economic Commision for Europe. United Nations, Geneva, Switzerland. 1999 Protocol to Abate Acidification, Eutrophication and Ground-level Ozone. UN Economic Commission for Europe. Palais des Nations, 1211 Geneva 10, Switzerland (2004)
5. ENTEC: Quantification of emissions from ships associated with ship movements between ports in the European Community. Study for the European Commission (2002)
6. HELCOM: Airborne nitrogen loads to the Baltic Sea. Helsinki Commission. Baltic Marine Environmrnt Protection Commission (2005)
7. Helsinki Convention: Convention on the protection of the marine environment of the Baltic Sea Area, 1992 (Helsinki Convention). Helsinki Commission. Katajanokanlaituri 6 B, 00160 Helsinki, Finland (2004)
8. LRTAP Convention: 1979 Convention on long-range transboundary air pollution. UN Economic Commission for Europe. Palais des Nations, 1211 Geneva 10, Switzerland (1979)
9. NEC: Directive 2001/81/EC of the European Parliament and of the Council of 23 October 2001 on national emission ceilings for certain atmospheric pollutants. Official Journal of the European Communities L 309/22 (2001)
10. Simpson, D., Fagerli, H., Jonson, J.E., Tsyro, S., Wind, P., Tuovinen, J.-P.: The EMEP unified Eulerian model. Model Description, EMEP MSC-W report 1/2003. The Norwegian Meteorological Institute, Oslo, Norway (2003)

Chapter 10
Planning Sustainable Agricultural Development Under Risks

G. Fischer, T. Ermolieva, and L. Sun

Abstract In this paper we show that explicit treatment of risks and uncertainties is an essential element in planning sustainable agricultural development. Introduction of risks and uncertainties in production models considerably alter strategies for achieving robust outcomes. We discuss stochastic models that may assist to derive optimal agricultural production allocation and expansion within environmental and health risk indicators. Approaches are illustrated with the example of spatially-explicit livestock production allocation in China to 2030.

10.1 Introduction

Global change, economic-demographic and urbanization growth, changing consumption preferences alter the structure of agricultural production systems. In particular, they promote industrial agriculture geared towards making use of economies of scale to produce the highest output at the lowest cost. Although intensification has shown many positive effects, there are significant disadvantages, risks, and costs involved. Undesirable impacts of intensification include environmental pollution, input-intensive mono-cropping, and the marginalization and decline of smallholder farms, causing abandonment of land and migration of rural population to cities. These are further exacerbated by various risks such as climate change and variability, natural catastrophes, market distortions and instabilities.

Alone environmental impacts and health hazards associated with intensive agricultural production have increased awareness and established the need to identify pathways towards sustainable agriculture.

This paper aims to show that adequate accounting and treatment of risks and uncertainties is a necessary condition for planning sustainable agricultural development. Naturally that considerations of risks may considerably alter production and consumption decisions. This fact is illustrated in Sect. 10.2 with a stylized model of

T. Ermolieva (✉), G. Fischer, and L. Sun
Institute for Applied Systems Analysis, Laxenburg, Austria
e-mail: ermol@iiasa.ac.at

K. Marti et al. (eds.), *Coping with Uncertainty*, Lecture Notes in Economics and Mathematical Systems 633, DOI 10.1007/978-3-642-03735-1_10,

two agricultural producers characterized by different levels of efficiency and exposure to risks. The example captures, in a nutshell, the features of a geographically detailed and dynamic model for agricultural production planning under risks and uncertainties, as adopted for the analysis of livestock production development in China to 2030 [12, 14].

Now, a growing share of livestock products in China is coming from industrial and specialized enterprises associated with hazardous pollution of the atmosphere, water and soil resources, which becomes a critical environmental issue [17,25]. Traditional livestock systems represented a natural farming cycle; livestock was kept on grass areas or in confined places close to farmland. Primary sources of feed were grass, feed from fodder crops and other crops, household wastes and crop residues. In these systems, livestock waste and manure were considered valuable sources of nutrients for crop production or for fuel. The manure was recycled efficiently, causing minimal environmental degradation and pollution. With the introduction of large-scale industrial livestock production, especially of pigs and poultry, this closed cycle is collapsing. Intensive livestock production enterprises are located close to meat markets, near urban areas, and in these locations there is much more livestock concentrated than land can support for proper manure recycling.

Geographical allocation of animals and the levels of intensification at which livestock is kept, differently affect the occurrences and spread of livestock diseases. In a sense, increasing specialization and intensification of livestock production is associated with newly emerging diseases (e.g., possibly SARS, avian flu) that can threaten human health.

Concentration of intensive livestock production is an important cause of environmental pollution and health hazards. When coinciding with intensive crop cultivation, the problem of pollution through excess nutrients from livestock operations is further exacerbated by imbalanced fertilizer application. Over-supply of nutrients may lead to toxic nitrate pollution in the water supply and may cause eutrophication of surface water. The trend is alarming and in some locations, without appropriate measures, it may turn irreversible. The analysis in [14] has shown that the development of China livestock production sector cannot just continue along past intensification trends. The goal of this paper is to discuss model-based approaches to guide decisions regarding the inevitable and significant future expansion of livestock production with respect to economic conditions at locations accounting for sustainability and risk indicators. Indicators of sustainability and risks are defined by various interdependent factors including the spatial distribution of people and incomes, the current levels of livestock production and intensification, and the conditions and current use of land resources. Combinations of these factors are used in proposed models to distinguish different locations by the degree of their risk exposure in order to achieve robust solutions.

In Sect. 10.3 we introduce a stochastic spatially explicit and dynamic simulation model used for planning livestock and crop production expansion coherently with projected demand increases to 2030. It allows for spatio-temporal and risk-adjusted analysis of production developments under alternative socio-economic, demographic, and technological scenarios. This allows to address not only environmental and social concerns, but also investigate innovative policies offering new viable opportunities to farmers, agricultural workers, consumers, and markets. The approach has also been discussed in [13] with alternative scenario settings.

The meaningful specification of risk indicators and constraints to define alternative allocation scenarios is often constrained by the paucity of data at required resolutions. In this case, specific downscaling (disaggregating) and upscaling (aggregating) procedures [11] provide a tool for estimation of dependencies between the geographical factors, constraints, and economic-environmental policy responses. In Sect. 10.3 we analyze the main features of these procedures for spatial production allocation with respect to risks and suitability constraints in locations. Section 10.4 introduces a new stochastic optimization approach for planning production allocation when some of the risks in the model of Sect. 10.3 are explicitly taken into account by stochastic scenarios. In fact, the Sect. 10.3 and Sect. 10.4 distinguish two types of uncertainties: endogenous uncertainties associated with behavioral principles regarding production expansion and exogenous uncertainties associated with parameters of models. Section 10.5 describes alternative allocation scenarios and presents selected numerical results. Section 10.6 concludes and indicates directions for future work.

10.2 Cooperation and Co-existence for Risk Sharing

Over the last 20 years, China's demand and production of livestock products has increased remarkably due to rapid development of the national economy, urbanization, rising living standards, and population growth [5]. Increasing incomes and changing consumption preferences have boosted production and have shifted the composition of producers towards specialized enterprises with a number of advantages: they are more feed efficient and profitable, flexible in terms of management, may better adjust and comply to legislation, and, in general benefit from economies of scale. In a sense, these trends follow the Ricardo's assertion [24] that trading nations gain from production specialization and intensification. Accordingly, we may expect that production should be undertaken by the most efficient agent, with intensified production on large farms. This is true only under idealized conditions when risks are not accounted for.

In reality, agricultural production facilities may be exposed to various risks, but also may cause different negative impacts. Depending on the location and intensity, values of the facilities are interdependent subject to contingencies, and are determined endogenously. For this reason, of particular interest are production chains with large and small units to stabilize the aggregate production. Contrary to Ricardo, the less efficient and intensive producer may provide the supply of production and enhance market stability, say, if the producer's risks are different and weakly or even negatively correlated with others. Such diversification of producers by scale and location hedges against economic and environmental risks, improves welfare and ensures continuous supply of agricultural products to markets. Explicit accounting of risks may considerably alter the composition of production units and their intensification levels in a chain.

Let us illustrate this with a stylized model of only two producers, $i = 1, 2$, which in Sect. 10.4 will be extended to a multi-producer case. Let x_i denote the production level of i-th producer and assume that only one good is produced, e.g., meat; c_i is the cost per unit of produce. The product can also be imported from an external source with price b per unit of produce. Assume $c_1 < c_2 < b$, i.e., the cheapest source is the first producer. The production has to satisfy the exogenous inelastic demand d of a given region.

In the *absence of risks*, the model is formulated as the minimization of the total cost function:

$$c_1 x_1 + c_2 x_2 \tag{10.1}$$

subject to

$$\begin{aligned} x_1 + x_2 = d, \\ x_1 \geq 0, x_2 \geq 0, \end{aligned} \tag{10.2}$$

where x_1, x_2 are production capacities. The optimal solution to the problem is $x_1^* = d, x_2^* = 0$, i.e., the production is undertaken by the more efficient producer, which accords with Ricardo's views.

In case of *risk exposure*, the endogenous supply (10.2) is expressed, for example, as a linear function

$$a_1 x_1 + a_2 x_2 = d, \tag{10.3}$$

where a_1, a_2 are contingencies or "supply" shocks to x_1, x_2, e.g., due to outbreaks of diseases, weather risks, or other hazardous events. We assume that a_1, a_2 are random variables $0 \leq a_i \leq 1$, which may reduce the supply from $i = 1, 2$. If endogenous supply $a_1 x_1 + a_2 x_2$ falls short of demand d, the residual amount $d - a_1 x_1 - a_2 x_2$ must be obtained from external sources at unit import cost b. The planning of production capacities x_1, x_2 can be evaluated from the minimization of total production costs and potential import cost, i.e., the minimization of the function

$$F(x) = c_1 x_1 + c_2 x_2 + bE \max\{0, d - a_1 x_1 - a_2 x_2\},$$

where $x_1 \geq 0$, $x_2 \geq 0$ and the expected import cost when the demand d exceeds the supply $a_1 x_1 + a_2 x_2$ is $bE \max\{0, d - a_1 x_1 - a_2 x_2\}$. In this case, the role of a less efficient producer for stabilizing supply is clearly visible.

Assume that only the efficient producer is at risk, that is $a_2 = 1$. Let function $F(x)$ have continuous derivatives, e.g., the cumulative distribution function of a_1 has a continuous density function. It is easy to see that the optimal positive decisions $x_1^* > 0$, $x_2^* > 0$ exist in the case when partial derivatives meet $F_{x_1}(0,0) < 0$, $F_{x_2}(0,0) < 0$. We have $F_{x_1}(0,0) = c_1 - bEa_1$, $F_{x_2}(0,0) = c_2 - b$ and, perhaps counter intuitively, the less efficient, but without risks producer 2 is active unconditionally (since $c_2 - b < 0$). The cost efficient producer 1 is inactive in the case $c_1 - bEa_1 \geq 0$, leaving production entirely to the higher-cost producer 2. Only in the case $c_1 - bEa_1 < 0$ both producers are active. Hence, in this example the less cost-efficient producer is able to stabilize the aggregate production in the presence of contingencies affecting the more cost-effective producer 1.

To derive the market share of the producer 2, take the derivative

$$F_{x_2}(x, x_2) = c_2 - bP[d > a_1 x + x_2]$$

according to optimality conditions of stochastic minimax problems [8]. This means that the optimal production level $x_2^* > 0$ of producer 2 is a quantile defined by the equation $P[d > a_1 x_1^* + x_2^*] = c_2/b$, assuming $x_1^* > 0$ (otherwise $x_2^* = d$). It also depends on x_1^* and all conditions ensuring a positive share x_1^* of producer 1. Although not at risk ($a_2 = 1$), the optimal production level of producer 2 is defined by (10.3) through interdependencies among producers participating in the same market with demand d. Let us now consider the case when both producers are at risks, i.e., $a_1 \neq 1$, $a_2 \neq 1$. The existence of optimal positive production of both producers follows from similar equations

$$F_{x_1}(0, 0) = c_1 - bEa_1 < 0,$$

$$F_{x_2}(0, 0) = c_2 - bEa_2 < 0.$$

The structure of optimal solution is similar to the case when only one producer is at risk. In particular, there may be a situation where $c_2 - bEa_2 \geq 0$, when producer 2 is inactive, but the cost effective producer 1 is active now with the insurance provided by the external source (import or borrowing).

Apart from exogenous risks, the production and the market are subject to endogenous risks dependent on the level of x_1, x_2. Negative impacts of production increase and intensification cause contamination of water, soil, air in the densely populated areas, which may incur uncertain, possibly highly non-linear costs, increasing with increasing x_1, x_2. In this case, the cooperation and market sharing may be unconditionally advantageous, as the following case illustrates.

Let us now consider a case when costs are increasing non-linear functions, for the sake of simplicity, *quadratic*, $c_1 x_1^2 + c_2 x_2^2$, $c_1 < c_2$, and there are no production distortions, i.e., $a_1 = a_2 = 1$. The problem is to minimize

$$c_1 x_1^2 + c_2 x_2^2$$

subject to the demand-supply constraints

$$x_1 + x_2 = d, x_1 \geq 0, x_2 \geq 0.$$

At least one producer must be active, say producer 1. It is easy to see from the standard optimality condition that the optimal level is $x_1^* = \frac{c_1}{c_1+c_2} d$. Therefore, the optimal level for producer 2 is $x_2^* = \frac{c_2}{c_1+c_2} d$. In other words, both $x_1^* > 0$, $x_2^* > 0$, i.e., unconditional on the cost effectiveness of the producer 1, the increasing non-linear production costs require co-existence and cooperation of both producers. These examples emphasize that market shares are to a larger extent determined by the production costs, the import price, and the contingencies of producers. In fact, the less efficient but with lower risk, producer will likely have a higher share than a more efficient, but with higher risk exposure, producer. For the sake of simplicity, in the above examples the contingencies are characterized by a probability distribution. In reality, the contingencies, e.g., livestock diseases, environmental pollution, demand fluctuations, economic instabilities, have complex geographical and temporal patterns of occurrences, are subject to spatial interactions. Their mutual probability distribution functions may not be analytically tractable and thus require stochastic simulation models as presented in Sect. 10.3 and downscaling procedures allowing for estimation of required values based on all available auxiliary statistics and model-derived results. Risk exposures are often characterized by certain standards commonly imposed as additional safety constraints on admissible values of some indicators, e.g., constraints on ambient standards in the pollution control.

10.3 Agricultural Planning Under Risks

10.3.1 A Simulation Model

The stochastic and dynamic livestock and crop production model developed by the Land Use Change and Agriculture (LUC) program at the International Institute for Applied Systems Analysis (IIASA) [10,12] integrates demographic, economic, agricultural and environmental modeling components. The IIASA model is essentially an accounting GIS-based model, which allows to incorporate inherent processes in an endogenized manner.

The model is developed with the aim to assist in planning sustainable agricultural developments combining various national, subnational and regional interacting agricultural activities, production, processing, consumers. Together with reasonable scales of biophysical modeling, this allows for production planning within limited resources and possibilities to improve or recover production potentials, against uncertainties of weather, climate change, market situation or other risks such as the contamination of land or pasture. Simplicity of model's structure enables to incorporate individual and collective risks combined with proper equity, fairness and safety constrains, which leads to welfare generating policies. Contrary to traditional linear programming [2] and general equilibrium approaches [19], the model allows to deal with economies of scales, time dynamics and increasing returns. This phenomenon is typical for practical problems of production and resource allocation, however the discussion of these topics is beyond the scope of the paper. In contrast to general equilibrium and standard growth theory, the proposed risk-adjusted approach permits to deal with issues involving externalities, inherent uncertainties, non-monetary values such as environmental degradation, non-marketable risks of high consequences, social heterogeneities regarding various representative agents.

Allocation of production facilities have to reasonably confirm to the distribution of current and future consumers including evaluation of the option make-vs.-buy typically addressed in spatial production planning models [15, 18, 20]. This implies the analysis of main production and demand driving forces such as population growth, urbanizations, energy provision, infrastructure, markets and market access. The discussed model can easily address regional "behavioral" aspects of production planning if these are determined by criteria other than pure cost-benefit or risks analysis. For example, rebalancing production allocation procedure in Sect. 10.3.3 allows to account for heterogeneous cultural traditions, complex interactions of behavioral, socio-economic, cultural and technological factors [7, 26], or specific fairness and equity considerations [23].

Within a project on "Policy Decision Support for Sustainable Adaptation of China's Agriculture to Globalization" (CHINAGRO [17]), the model included specifics of China agricultural developments and has been applied for the spatial analysis of future livestock sector expansions. Using alternative economic and demographic projections [4, 16, 17, 21, 22], the model estimates per capita demand increases and consumption of major agricultural products, e.g., cereals, meat, milk, etc. Demand patterns differ between urban and rural areas, between geographical regions, and vary with income. Thus, with increasing incomes higher quality low-fat meat, e.g., poultry is preferred. In fact, evolution of consumption is modeled as a function of group-specific per capita income increases by applying income elasticities and distinguishing urban and rural consumers.

Agricultural supply is represented at county level, i.e., for about 2,430 spatial units. Smallholders and specialized livestock farms adjust the livestock herd structure and production in response to the demand increase and the changes of consumption patterns. The model distinguishes the following livestock types: poultry, pigs, dairy, cattle, buffaloes, yaks, sheep and goats, and other large animals (combining horses, donkeys, and camels). To examine the current situation and the production intensification trends, modeling of livestock production considers three management systems: traditional, specialized/industrial, and grazing.

In the environmental module, the environmental loads caused by intensive crop and livestock production are evaluated against admissible environmental and health thresholds (which can be proposed by stakeholders and environmental experts). Indicators used for measuring environmental impacts and human health risk are: the density of livestock, nutrients from manure and chemical fertilizers in excess of a location's nutrient uptake by crop production, urbanization share, density of population, and others. Combinations of these and other factors (see Sect. 10.5) reflect different degrees of socio-economic and environmental risk exposures and can be used to guide sustainable production allocation. The model simulates different paths of demand increase, which induces respective location-specific production adjustments. In some locations, the environmental and health risk indicators may already exceed admissible thresholds, which signals that further production growth in these locations should not take place. This raises the question of how to adjust the composition and allocation of livestock production facilities in response to increasing demand but without exacerbating environmental and health problems. The detailed description of the model and the allocation procedure [12] is rather lengthy. Therefore, in the following we provide only rather aggregate representation of their main constraints.

10.3.2 A Simplified Production Model

When planning livestock development, the objective is to allocate the foreseeable increases of demand for livestock products among the locations and the main production systems in the best possible way while accounting for various risks. In the following model the risks are treated as constraints restricting production expansion. In Sect. 10.4 we introduce a stochastic model that allows to account for risks and uncertainties in a more explicit manner.

Denote the expected national demand increase (to be satisfied by supply increase) in livestock product i by d_i, $i = \overline{1,m}$. Let x_{ijl} be the unknown supply increase in product i at location j and by management system l. In its simplest form, the problem is to find x_{ijl} satisfying the following system of equations:

$$\sum_{l,j} x_{ijl} = d_i, \tag{10.4}$$

$$x_{ijl} \geq 0 \tag{10.5}$$

$$\sum_{i} x_{ijl} \leq b_{jl}, l = \overline{1,L}, j = \overline{1,n}, i = \overline{1,m}, \tag{10.6}$$

where b_{jl} are thresholds aggregating environmental and health risks and imposing limitations to expand production in system l and location j. Equation (10.6) restricts prevalence of specific production systems. For example, the dominance of industrial systems in a location inevitably leads to intensification of feeding operations, the need in recycling facilities, etc. For simplicity of presentation, the constraint (10.6) captures only production side. Apart from b_{jl}, there may be additional limits on x_{ijl}, $x_{ijl} \leq r_{ijl}$, which can be associated with legislation, for example, to restrict production within a production belt, or to exclude from urban or protected areas, etc. Thresholds b_{jl} and r_{ijl} may either indicate that livestock in excess of these values is strictly prohibited or it incurs penalties such as taxes or premiums, for mitigation of the risks, say, livestock disease outbreaks or environmental pollution. Equations (10.4)–(10.6) belong to the type of transportation problems. However, there may be more general constraints of type $\sum_{ij} a_{ijl} x_{ijl} \leq d_i$ as in Sect. 10.2, $0 \leq a_{ijl} \leq 1$, which require extensions of the proposed approach.

In general, there exist infinitely many solutions of (10.4)–(10.6). The aim is to derive a solution that ensures appropriate balance between the efficiency and the risks. We can distinguish two sources of uncertainties generating potential risks: behavioral or endogenous uncertainties associated with allocation of new production capacities and exogenous uncertainties related to parameters of the model. In this section we consider only the first type of uncertainties. Section 10.4 addresses the second type of constraints.

The information on the current production facilities, threshold values b_{jl}, r_{ijl}, and costs are used to derive a prior probability q_{ijl} reflecting our belief that a unit of demand d_i should be allocated to management system l in location j. The use of priors is consistent with spatial economic theory (see discussion, e.g., in [7, 20, 26]). The likelihood q_{ijl} can be inversely proportional to production costs and inherent risks r_{ijl} [6, 7]. In Sect. 10.3.3 we show how it is used in a rebalancing procedure to determine the solution of (10.4)–(10.6) relying on behavioral, in a sense, risk-averse and cost-minimizing principles defined by this prior as in (10.10).

10.3.3 A Rebalancing Production–Allocation Algorithm

For simplicity of exposition, let us renumerate all pairs (l, j), $l = \overline{1, L}$, $j = \overline{1, n}$ by $k = \overline{1, K}$. In this new notation, the problem is formulated as finding y_{ik} satisfying constraints:

$$\sum_k y_{ik} = d_i \tag{10.7}$$

$$y_{ik} \geq 0, \tag{10.8}$$

$$\sum_k y_{ik} = b_k, i = \overline{1, m}, k = \overline{1, K} \tag{10.9}$$

consistent with a prior q_{ik} belief that a unit of demand for product i should be supplied by activity k. For instance, it is reasonable to allocate more livestock to locations with a larger demand increase, higher productivity, or better feed access. Assume that this preference structure is expressed in prior q_{ik}, $\sum_k q_{ik} = 1$ for all i. In this case, the initial amount of production i allocated to k can be derived as $q_{ik}d_i$. But this may lead to violation of constraints (10.9). Sequential rebalancing [11] proceeds as follows. Relying on prior probability q_{ik}, the *expected* initial allocation of d_i to k is $y^0_{ik} = q_{ik}d_i$, $i = \overline{1, m}$. However, this allocation may not satisfy constraint $\sum_i y^0_{ik} \leq b_k$, $j = \overline{1, n}$. Derive the relative imbalances $\beta^0_k = b_k / \sum_i y^0_{ik}$ and update $z^0_{ik} = y^0_{ik}\beta^0_k$, $i = \overline{1, m}$. Now the constraint $\sum_i y_{ik} \leq b_k$ is satisfied, $k = 1, 2, \ldots$, but the estimate z^0_{ik} may cause imbalance for (10.7), i.e., $\sum_k z^0_{ik} \neq d_i$.

Continue with calculating $\alpha_i^0 = d_i / \sum_k z_{ik}^0$, $i = \overline{1,m}$, and updating $y_{ik}^1 = z_{ik}^0 \alpha_i^0$, an so on. The estimate y_{ik}^s can be represented as

$$q_{ik}^s = (q_{ik}\beta_k^{s-1})/(\sum_j q_{ik}\beta_k^{s-1}),$$

where $i = \overline{1,m}$, $k = 1,2,\ldots$.

Assume $y^s = y_{ik}^s$ has been calculated. Find $\beta_k^s = \overline{b}_k / \sum_i y_{ik}^s$ and $q_{ik}^{s+1} = (q_{ik}\beta_j^s / \sum_i q_{ik}\beta_j^s)$, $i = \overline{1,m}$, $k = 1,2,\ldots$, and so on.

In this form the procedure can be viewed as a redistribution of required supply d_i among producers $k = 1,2,\ldots$ by applying sequentially adjusted q_{ik}^{s+1}, e.g., by using a Bayesian type of rule for updating the prior distribution:

$$q_{ik}^{s+1} = q_{ik}\beta_k^s / \sum_i q_{ik}\beta_k^s,$$

were $q_{ik}^0 = q_{ik}$.

The update is done on an observation of imbalances of basic constraints rather than observations of random variables. A rebalancing procedure, similar to the one described above for Hitchcock–Koopmans transportation constraints (10.7)–(10.9), was proposed by G.V. Sheleikovskii (see a proof and references in [3]) for estimation of passenger flows between regions. A proof of its convergence to the optimal solution maximizing the cross-entropy function

$$\sum_{i,k} y_{ik} ln \frac{y_{ik}}{q_{ik}} \tag{10.10}$$

is given in [11] for rather general types of constraints. It should be noted that in our model we use equality constraints (10.9). The general inequality constraints are reduced to this model by introduction of a fictitious demand constraint.

10.4 Stochastic Production Allocation Model

The approach presented in Sect. 10.3 evaluates the increase of livestock production relying on individual behavioral principles set by priors. There, the risks are characterized in a simplified deterministic way by imposing certain standards as additional "safety" constraints. In general, these constraints may depend on some scenarios of potential future shocks. The behavioral uncertainty in Sect. 10.3 can also be treated in a stochastic manner as allocation of random demand $d_i(\omega)$ among points $k = \overline{1,K}$ with respect to the prior q_{ik}, which is a topic of a separate paper.

Let us consider now a more general multi-producer model in a stochastic environment analogous to the Example of Sect. 10.2. We may assume that there is a coordinating agency. The goal of this agency is to maximize the overall performance of the production chain with large and small units to stabilize the aggregate production and increase the facility values. Suppose that the agency has to determine levels of livestock product i in locations k in order to meet stochastic demand $d_i(\omega)$, where $\omega = (\omega_1, \omega_2, \ldots)$ is a vector of all contingencies affecting demand and production. Naturally to assume that the decision on production expansion has to be made before the information on contingencies arrives. In this case, the total ex-ante production may not exactly correspond to the real demand, i.e., we may face both oversupplies and shortfalls. In other words, the amount of production y_{ik}, $k = \overline{1, K}$, which is planned ex-ante to satisfy the demand $d_i(\omega)$, $y_i(\omega) = \sum a_{ik}(\omega) y_{ik}$ may underestimate $(y_i(\omega) < d_i(\omega))$ or overestimate $(y_i(\omega) > d_i(\omega))$ the real demand $d_i(\omega)$ under revealed contingencies and the safety constraints imposed by strict thresholds b_k in (10.9). The constraint (10.9) necessitates, in general, additional supply of ex-ante production $z_i \geq 0$ from external sources (say, through international trade). It may also require the ex-post redistribution of the production from internal producers, $k = \overline{1, K}$, to eliminate arising shortfalls and oversupplies in locations. For now, let us ignore these ex-post redistributional aspects assuming that the most significant impacts are associated with ex-ante decisions y_{ik} and z_i. In fact, the presented further model can be easily extended to represent the ex-post adjustments of decisions y_{ik}, z_i, as well as temporal aspects of production planning.

Let c_{ik} be the unit production cost. In more general model formulation, c_{ik} may also include the unit transportation cost for satisfying location-specific demand. Then the model of production planning among the facilities can be formulated as the minimization of the total cost function:

$$f(y, z) = \sum_{i,k} c_{ik} y_{ik} + \sum_{i=1}^{m} e_i z_i,$$

subject to constraints (10.8), (10.9), and the following additional safety constraints

$$P[\sum_{k=1}^{K} a_{ik}(\omega) y_{ik} + z_i \geq d_i(\omega)] \geq p_i, z_i \geq 0, i = \overline{1, m}, \tag{10.11}$$

where $e_i > 0$, $i = \overline{1, m}$, denotes the unit import cost. A safety level p_i, $0 < p_i < 1$, defines (ensures) the stability of the supply-demand relations for all possible scenarios (contingencies) ω. The introduction of constraints of type (10.11) is a standard approach for characterizing stability in case of the insurance business, operations of nuclear power plants and other risky activities especially when involving catastrophic risks [9]. Safety constraints of type (10.11) are usually used in cases where impacts of random interruptions can not be easily evaluated. In this case, the value p_i is selected such that an expected shortfall occurs only, say, once in 100 month, i.e., $p_i = 0.01$.

The main methodological challenge is concerned with the lack of convexity of constraints (10.11). Yet, the remarkable fact is that the model defined by

(10.8)–(10.11) can be effectively solved by linear programming methods due to the following equivalent convex form of this model. Let us consider the minimization of the expectation function

$$F(y,z) = f(y,z) + \sum_{i=1}^{m} \alpha_i E \max\{0, d_i(\omega) - \sum_{k=1}^{K} a_{ik}(\omega) y_{ik} - z_i\}, \quad (10.12)$$

subject to constraints (10.8), (10.9), and $z_i \geq 0, i = \overline{1,m}$. The minimization of function $F(y,z)$ is a rather specific case of stochastic minimax models analyzed (both optimality conditions and solution procedures) in [8]. In particular, if $F(y,z)$ has continuous derivatives with respect to z_i, e.g., the probability distribution function of ω has continuous density function, then

$$\frac{\partial F}{\partial z_i} = e_i - \alpha_i E I(d_i(\omega) - \sum_{k=1}^{K} a_{ik}(\omega) y_{ik} - z_i \geq 0)$$

where $I(\xi \geq 0)$ is the indicator function: $I(\xi \geq 0) = 1$, if $\xi \geq 0$, and $I(\xi \geq 0) = 0$ otherwise. Therefore, we can rewrite $\frac{\partial F}{\partial z_i}$ as

$$\frac{\partial F}{\partial z_i} = e_i - \alpha_i P[d_i(\omega) - \sum_{k=1}^{K} a_{ik}(\omega) y_{ik} - z_i \geq 0], \quad (10.13)$$

which allows to establish connections between the original model defined by (10.8)–(10.11) and the minimization of convex function $F(y,z)$ defined by (10.12).

Assume (y^*, z^*) minimizes $F(y,z)$ subject to constraints (10.8), (10.9), and $z_i \geq 0, i = \overline{1,m}$. Assume also that $e_i < \alpha_i, i = \overline{1,m}$. Then from (10.13) it follows that for all i with positive components $z_i^* > 0$, i.e., when $\frac{\partial F}{\partial z_i} = 0$, the optimal solution (y^*, z^*) satisfies the following safety constraints

$$P[d_i(\omega) - \sum_{k=1}^{K} a_{ik}(\omega) y_{ik} - z_i \geq 0] = e_i/\alpha_i. \quad (10.14)$$

Moreover, for all i with $z_i^* = 0$, i.e., when $\frac{\partial F}{\partial z_i} \geq 0$, the optimal (y^*, z^*) satisfies the following safety constraint

$$P[d_i(\omega) - \sum_{k=1}^{K} a_{ik}(\omega) y_{ik} \geq 0] \leq e_i/\alpha_i. \quad (10.15)$$

If we choose α_i as $e_i/\alpha_i = 1 - p_i$, i.e., $\alpha_i = e_i/(1 - p_i)$, then (10.14)–(10.15) become equivalent to the safety constraint (10.11) of the original model (10.8)–(10.11). In other words, the minimization of convex function $F(y,z)$ defined by

(10.12) subject to (10.8), (10.9), and $z_i \geq 0$, $i = \overline{1,m}$, yields the optimal solution of the original model (10.8)–(10.11). Efficient computational procedures for solving stochastic minimax problems with objective functions defined as in (10.12) can be found in [8, 25]. In particular, the paper [25] discussed the applicability of linear programming methods in cases where the original model defined by a general probability distributions of ω can be sufficiently approximated by models with discrete probability distributions. This paper establishes also important connections between the minimization of (10.12)-type functions and Conditional-Value-at-Risk risk measure.

The minimization of function (10.12) can also be solved by a stochastic quasi-gradient method [8]. In applying this method to minimization of (10.12), the differentiability of $F(y)$ and any assumption on probability distribution of ω is not required. Also, the probability distribution of ω may only be given implicitly. For instance, only observations of random $d_i(\omega)$ and $a_{ik}(\omega)$ may be available or only a Monte Carlo procedure (pseudo-sampling simulation model such as described in Sect. 10.3.1) is used to simulate supply and demand. In the following section we illustrate some applications by using only the rebalancing algorithm described in Sect. 10.3.3; elaboration of the outlined stochastic allocation algorithm is a topic for future implementation.

10.5 Numerical Experiments

The model in Sect. 10.3 is used in the analysis of current and plausible future livestock production allocation and intensification in China. Namely, in each time period the simulation model generates levels and geographic distribution of demand for livestock products coherent with urbanization processes [22], demographic change [4] and expected growth of incomes [16, 17]. Production allocation and intensification levels are projected from the base year data for the main livestock types (pigs, poultry, sheep, goat, cattle) and management systems (grazing, industrial/specialized, traditional) at the level of counties (about 2,500 administrative units). For production allocation, we used the sequential rebalancing procedure described in Sect. 10.3.3. Two scenarios of future production allocation corresponding to different priors q_{ik}, $i = \overline{1,m}$, $k = \overline{1,K}$, are compared: (1) an intensification scenario, when production is allocated proportionally to the geographical patterns of demand increases, and (2) a risk-adjusted scenario that combines the preference structure as defined by the geographical distribution of demand with indicators of environmental pressure.

Intensification scenario. Currently, common practice is to allocate intensive livestock production in areas with good access to consumers, close to high demand and high population density [1]. In many practical problems of large dimensionality, to describe the "profitability" of a location it has been standard practice to use an ad hoc but reasonable measure referred to as market access function. The typical market access function measures the potential of location k as a weighted sum of purchasing power of all other locations in some vicinity of the given k. The weights are defined either as a function of distance or as a function of other factors, say, costs or losses. In these studies, each county is characterized by its market access calculated as a weighted sum of demand for product i in nearby counties within some vicinity. Values $\tilde{\Delta}_{ik}$, determine a profit-based prior q_{ik}, $k = \overline{1, K}$ for allocation of demand increase among production units in locations as it is described in Sect. 10.3.3.

Risk-adjusted scenario. The objective of this scenario is to care for the balance between profitability of the agricultural production, rural welfare, and the respect of nature and the environment. Challenges of spatially-explicit planning for sustainability are related to the choice of adequate location-specific indicators to guide rural development within defined socio-economic and environmental objectives. While information on economic and livelihood conditions at location may be available from statistics and census data, estimation of agricultural pollution and health risks (for example, related to livestock diseases) is a more challenging task.

The agricultural pollution falls into the category of non-point source pollution, which is geographically disperse, and the likelihood of disease occurrences is determined by a combination of factors. Measurements of the pollution level, health risks, and related impacts or losses are hardly possible as they depend on multiple highly uncertain socio-economic and environmental factors: weather patterns, population density, level of development, agricultural inputs and intensification levels, etc. In many practical situations when the target variable is impossible or impractical to measure, it is possible to use context-specific proxies or even a set of proxies that can considerably well represent the state of the non-measurable variable (see, e.g., [13]).

For planning sustainable agricultural developments, the profit-driven prior of the "intensification scenario" is adjusted with such variables as nutrients in excess of crop uptakes, density of livestock biomass, etc., are used to characterize environmental risks. Health norms and associated health risks are introduced by a combination of urbanization share (share of urban population in total population) and availability of non-residential area suitable for further production expansion in each location. In general, allocation prior is defined by a compound probability distribution function of relevant variables.

The intensification scenario implicitly minimizes the transportation costs as the production concentrates in the proximity of large markets in urban areas with high demand. In the alternative scenario, which is a compromise between the demand driven production allocation and the considerations of health and environmental risks, the production is shifted to more distant locations characterized by availability of cultivated land, lower livestock and population density, which increases transportation. However, the measure of goodness for the scenarios accounts not only for the transportation cost but also includes environmental and health risk proxies. Thus, the two scenarios are compared with respect to number of people in China's regions exposed to different categories of environmental risks.

Environmental risks are measured in terms of environmental pressure in relation to the coincidence of three factors: density of confined livestock, human population density, and availability of cultivated land. For this purpose some 2,434 counties

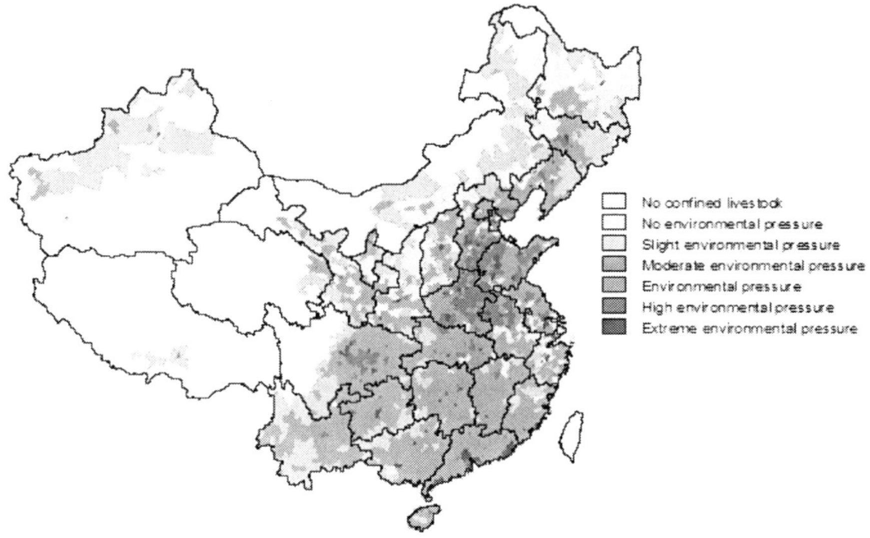

Fig. 10.1 Environmental pressure from confined livestock production, 2000

were classified as follows into seven categories, namely: (a) *No confined livestock*, i.e., counties in scarcely populated areas (desert or mountain/plateau) and with very little confined livestock; (b) *No environmental pressure*, i.e., counties with substantial crop production but with little confined livestock; (c) *Slight environmental pressure* counties with low environmental pressure from confined livestock production; (d) *Moderate environmental pressure*, i.e., counties with moderate environmental pressure from confined livestock production; (e) *Environmental pressure*, i.e., counties with substantial urbanization and environmental pressure from confined livestock production; (f) *High Environmental pressure*, i.e., counties with substantial urbanization and high environmental pressure from livestock production, and (g) *Extreme environmental pressure* i.e., counties with high degree of urbanization coinciding with high environmental pressure from confined livestock production. Figure 10.1 presents the above classification of environmental pressure for the year 2000. Figure 10.1 indicates that currently (i.e., year 2000) hot-spots of environmental pressure are located mainly in provinces covering the North China Plain, the Sichuan basin, and several locations along the coast of South China. Locations of livestock production concentrate around or in the vicinity of areas where the livestock demand grows fast, e.g., highly populated and urban areas.

Figure 10.2 presents diagrams of the distribution of current population against the mapped classes of severity of environmental pressures from livestock. The left diagram shows absolute numbers, i.e., million people per class and region. The diagram on the right gives shares of population within each region falling into respective classes. For year 2000, the estimates suggest that about 20% of China's population lives in counties characterized as having high or extreme severity of environmental pressure from intensive livestock production. In the "intensification" scenario, by

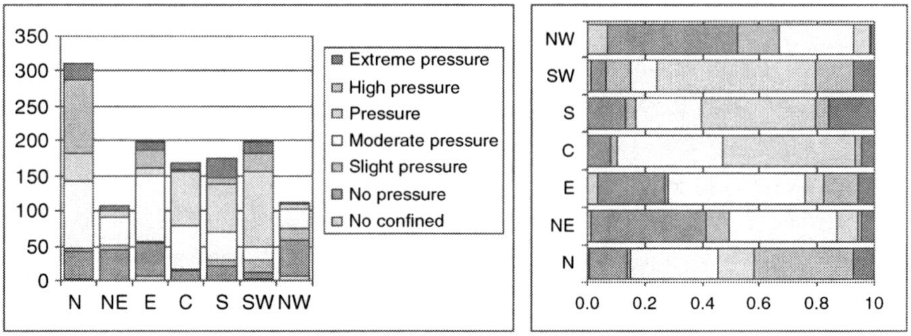

Fig. 10.2 Absolute (million people) and relative (share of total population) distribution of population according to classes of severity of environmental pressure from livestock, 2000. The *label on the horizontal axis* indicate China regions: *N, NE, E, C, S, SW, NW* stand for North, North-East, East, Center, South, South-West, North-West, respectively

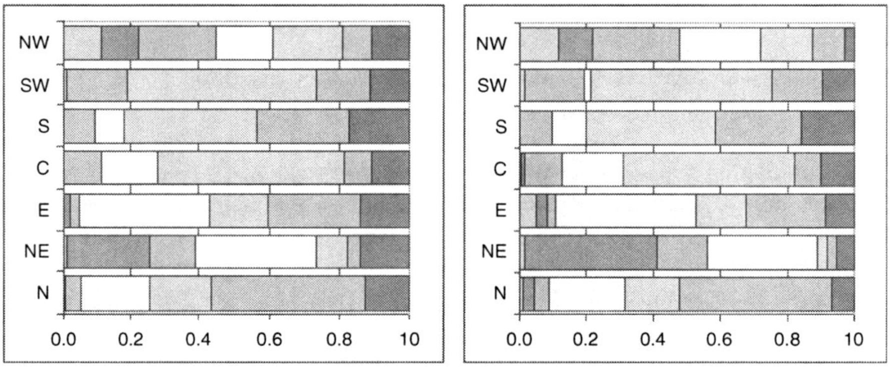

Fig. 10.3 Relative (share of total population) distribution of population according to classes of severity of environmental pressure from livestock, 2030: **a** "intensification" scenario, **b** risk-adjusted scenario

2030 this population share increases to 36% (Fig. 10.3a), i.e., from one-fifth in 2000 to about one-third in 2030. Looking only at the highest pressure class, the South region appears to have the largest number of people and the highest population share in such unfavorable environmental pressure, about 38 million or 22% of population in 2000 increasing to nearly 45 million or 17% in 2030. The region with the highest occurrence of people (both absolute and relative) in the two highest pressure classes is the North region, with more than 40% of the population. In 2030, the estimated share becomes 57%, followed by the South region with 27% population in two highest pressure classes in 2000 and with 44% in 2030. Looking at the second allocation scenario, the positive changes are quite visible (see Fig. 10.3b). The estimate of people living in highest pressure class for South region changes from 45 to 42 million. Percentage of population in the two highest pressure classes for the same region varies between scenarios as 43 for the bad and 41 for the good. For

North region, 56.5% of total population will live in two highest pressure classes in 2030 in bad scenario and only 52.3 – in good, In North East, the highest pressure classes change percentage from about 18 to less than 10. The intensification scenario (1) implicitly minimizes the transportation costs as production concentrates in the vicinity of urban areas with high demand. In the alternative scenario (2), the production is shifted to more distant locations characterized by availability of cultivated land, lower livestock and population density, but at the expense of additional transportation. Environmental sustainability aspects of the two scenarios were compared with respect to the share of people in China's regions exposed to different severity classes of environmental risks. Environmental risks are measured in terms of environmental pressure in relation to the coincidence of three factors: density of confined livestock, human population density, and availability of cultivated land. For year 2000, the estimates suggest that about 20% of China's population lives in counties characterized as having high or extreme severity of environmental pressure from intensive livestock production. In the "intensification" scenario, by 2030 this population share increases to 37%, while in the second, environmentally friendly scenario, it stays below 30%. To finally compare the two scenarios, it is necessary to "normalize" gains due to improved life conditions with expenses of additional transportation.

10.6 Conclusions

This paper addresses some important aspects of agricultural production planning under risks, uncertainties and incomplete information. When planning agricultural developments, the objective is to allocate the foreseeable increases of demand in the best possible way while accounting for various risks associated with production and suitability criteria for profitability, transport, health and environmental impacts. Models for production allocation under risks and uncertainties may have considerable implications. In particular, the allocation of livestock production away from urban peripheries where pressure is highest to regions where feed grains are in abundance could decrease the income gaps between the regions. Similarly, establishment of agricultural pollution regulations, e.g., taxation, at locations with high environmental loads may change the balance of agricultural market attracting imports from abroad.

In Sect. 10.3, the production allocation procedure is proposed for situations when the available information is given in the form of aggregate values without providing necessary local perspectives. Therefore, the main issue is to downscale these values to the local levels consistently with location specific behavioral principles based on some priors. Yet, many practical situations may require more rigorous probabilistic treatment of priors and safety constraints. Section 10.4 proposes an allocation mechanism with more general treatment of uncertainties and risks based on principles of stochastic optimization. This is a promising approach for a coordinating agency aiming to improve the overall performance of the production chain.

By diversifying large and small units the agency may stabilize the aggregate production and increase "utility" of individual facilities. The application of this allocation procedure is a topic for future research.

References

1. Anderson, J.E., van Wincoop, E.: Gravity with gravitas: A solution to the border puzzle. Am. Econ. Rev. **93**(1), 170–192 (2003)
2. Arrow, K., Hurwicz, L., Uzawa, H. with contribution by Chenery, H., Johnson, S., Karlin, S., Marschak, T., Solow, R.: Studies in Linear and Non-linear Programming. Stanford University Press, Stanford, CA (1958)
3. Bregman, L.M.: Proof of the convergence of Sheleikhovskii's method for a problem with transportation constraints. J. Comput. Math. Math. Phys. **7**(1), 191–204 (1967). (Zhournal Vychislitel'noi Matematiki, USSR, Leningrad, 1967)
4. Cao, G.Y.: The future population of China: prospects to 2045 by place of residence and by level of education. Interim Report IR-00-026. International Institute for Applied Systems Analysis, Laxenburg, Austria (2000)
5. CCAP: Estimates of province-level meat production and consumption, 1980–1999: Database prepared for CHINAGRO project. Center for Chinese Agricultural Policy, Beijing (2002)
6. Ermoliev, Y., Fischer, G.: Spatial modeling of resource allocation and agricultural production under environmental risk and uncertainty. Working Paper WP-93-11. International Institute for Applied Systems Analysis, Laxenburg, Austria (1993)
7. Ermoliev, Y., Leonardi, G.: Some proposals for stochastic facility location models. Math. Model. **3**, 407–420 (1982)
8. Ermoliev, Y., Wets, R. (eds.): Numerical Techniques for Stochastic Optimization. Computational Mathematics. Springer, Berlin (1988)
9. Ermolieva, T., Ermoliev, Y.: Catastrophic risk management: flood and seismic risks case studies. In: Wallace, S.W., Ziemba, W.T. (eds.) Applications of Stochastic Programming. MPS-SIAM Series on Optimization, Philadelphia (2005)
10. Fischer, G., Ermoliev, Y., Keyzer, M., Rosenzweig, C.: Simulating the socio-economic and biogeophysical driving forces of land-use and land-cover change: The IIASA land-use change model. WP-96-010 (1996)
11. Fischer, G., Ermolieva, T., Ermoliev, Y., van Velthuizen, H.: Sequential downscaling methods for estimation from aggregate data. In: Marti, K., Ermoliev, Y., Makowski, M., Pflug, G. (eds.) Coping with Uncertainty: Modeling and Policy Issue. Springer, Berlin (2006)
12. Fischer, G., Ermolieva, T., Ermoliev, Y., van Velthuizen, H.: Livestock production planning under environmental risks and uncertainties. J. Syst. Sci. Syst. Eng. **15**(4), 385–399 (2006)
13. Fischer, G., Ermolieva, T., Ermoliev, Y., Sun, L.: Risk-adjusted approaches for planning sustainable agricultural development. Stochastic Environmental Research and Risk Assessment. Stoch. Environ. Res. Risk Assess. **23**, 441–450 (2009). doi:10.1007/s00477-008-0231-9
14. Fischer, G., Ermolieva, T., van Velthuizen, H.: Livestock production and environmental risks in China: Scenarios to 2030. FAO/IIASA Research Report. International Institute for Applied Systems Analysis, Laxenburg, Austria (2006)
15. Fujita, M., Krugman, P., Venables, A.J.: The Spatial Economy: Cities, Regions, and International Trade. MIT, Cambridge, MA (1999)
16. Huang, J., Liu, H.: Income growth and lifestyle. CHINAGRO Internal Working Paper WP 1.7. Center for Chinese Agricultural Policy, Chinese Academy of Sciences, Beijing (2003)
17. Huang, J., Zhang, L., Li, Q., Qiu, H.: CHINAGRO project: National and regional economic development scenarios for China's food economy projections in the early 21st century. Report to Center for Chinese Agricultural Policy. Chinese Academy of Sciences. Draft (2003)
18. Hotelling, H.: The economics of exhaustible resources. J. Polit. Econ. **39**, 137–175 (1931)

19. Kaldor, N.: Causes of Growth and Stagnation in The World Economy. Published by the Press Syndicate of the University of Cambridge, The Pitt Building. The Estate of Nicholas Kaldor (1996)
20. Karlqvist, A., Lundqvist, L., Snickars, F., Weibull, J.W.: Studies in Regional Science and Urban Economics: Spatial Interaction Theory and Planning Models, vol. 3. North-Holland, Amsterdam (1978)
21. Keyzer, M.A., van Veen, W.: A summary description of the CHINAGRO-welfare model. CHINAGRO report. SOW-VU, Free University, Amsterdam (2005)
22. Liu, S., Li, X., Zhang, M.: Scenario analysis of urbanization and rural–urban migration in China. Interim Report IR-03-036. International Institute for Applied Systems Analysis, Laxenburg, Austria (2003)
23. Manne, A.S.: Investments for Capacity Expansions: Size, Location, and Time-Phasing. George Allen and Unwin, London (1967)
24. Ricardo, D.: On Protection in Agriculture. John Murray, London (1822)
25. Rockafellar, T., Uryasev, S.: Optimization of conditional value-at-risk. J. Risk **2**, 21–41 (2000)
26. Wilson, A.G.: Entropy in Urban and Regional Modelling. Pion, London (1970)

Chapter 11
Dealing with Uncertainty in GHG Inventories: How to Go About It?

Matthias Jonas, Thomas White, Gregg Marland, Daniel Lieberman, Zbigniew Nahorski, and Sten Nilsson

Abstract The assessment of greenhouse gases emitted to and removed from the atmosphere is high on both political and scientific agendas. Under the United Nations Framework Convention on Climate Change, Parties to the Convention publish annual or periodic national inventories of greenhouse gas emissions and removals. Policymakers use these inventories to develop strategies and policies for emission reductions and to track the progress of these policies. However, greenhouse gas inventories (whether at the global, national, corporate, or other level) contain uncertainty for a variety of reasons, and these uncertainties have important scientific and policy implications. For scientific, political, and economic reasons it is important to deal with the uncertainty of emissions estimates proactively. Proper treatment of uncertainty affects everything from our understanding of the physical system to the economics of mitigation strategies and the politics of mitigation agreements. A comprehensive and consistent understanding of, and a framework for dealing with, the uncertainty of emissions estimates should have a large impact on the functioning and effectiveness of the Kyoto Protocol and its successor. This chapter attempts to

M. Jonas (✉), and S. Nilsson
International Institute for Applied Systems Analysis (IIASA), Schlossplatz 1, 2361 Laxenburg, Austria,
e-mail: jonas@iiasa.ac.at, marland@iiasa.ac.at, nilsson@iiasa.ac.at

T. White
Canadian Forest Service, Pacific Forestry Centre, 506 West Burnside Road, Victoria, BC, Canada V8Z 1M5,
e-mail: thos_white@hotmail.com

G. Marland
Carbon Dioxide Information Analysis Center, Oak Ridge National Laboratory, Bethel Valley Road, P.O. Box 2008, Building 1509, Oak Ridge, TN 37831-6335, USA
e-mail: marlandgh@ornl.gov

D. Lieberman
ICF International, 1725 Eye St. NW., Suite 1000, Washington, DC 20006, USA (currently at Chevron Corporation, 6001 Bollinger Canyon Rd, K2339, San Ramon, CA 94583, USA),
e-mail: dlieberman@chevron.com

Z. Nahorski
Systems Research Institute of the Polish Academy of Sciences, ul. Newelska 6, 01-447 Warsaw, Poland,
e-mail: zbigniew.nahorski@ibspan.waw.pl

K. Marti et al. (eds.), *Coping with Uncertainty*, Lecture Notes in Economics and Mathematical Systems 633, DOI 10.1007/978-3-642-03735-1_11,

pull together relevant fragments of knowledge, allowing us to get a better picture of how to go about dealing with the uncertainty in greenhouse gas inventories.

11.1 Introduction

The assessment of greenhouse gases (GHGs) emitted to and removed from the atmosphere is high on both political and scientific agendas. Under the United Nations Framework Convention on Climate Change (UNFCCC), Parties to the Convention (so-called Annex I countries) have published annual or periodic national inventories of GHG emissions and removals since the mid 1990s. Policymakers use these inventories to develop strategies and policies for emission reductions and to track the progress of these policies. Where formal commitments to limit emissions exist, regulatory agencies and corporations rely on inventories to establish compliance records. Scientists, businesses, the public, and other interest groups use inventories to better understand the sources and trends in emissions. Table 11.1 provides general background information on the six GHGs, or groups of gases, considered under the Kyoto Protocol and their global emissions as reported by the Intergovernmental Panel on Climate Change (IPCC) in its assessment reports for the late 1990s and beyond.

GHG inventories contain uncertainty for a variety of reasons – for example, the lack of availability of sufficient and appropriate data and the techniques to process them. Uncertainty has important scientific and policy implications. However, until recently, relatively little attention has been devoted to how uncertainty in emissions estimates is dealt with and how it might be reduced. Now this situation is changing, with uncertainty analysis increasingly being recognized as an important tool for improving national, sectoral, and corporate inventories of GHG emissions and removals [5] (see also [6] and [7]).

At present, Parties to the UNFCCC are encouraged, but not obliged, to include with their periodic reports of in-country GHG emissions and removals, estimates of the uncertainty associated with these emissions and removals; consistent with the Intergovernmental Panel on Climate Change's (IPCC) good practice guidance reports [8,9]. Inventory uncertainty is monitored, but not regulated, under the Kyoto Protocol [5].

We argue that it makes a big difference in the framing of policies whether or not uncertainty is considered: reactively, because there is a need to do so; or proactively, because difficulties are anticipated. We follow [7, p. 2–3] (see also [5]) who clearly state that uncertainty estimates are not intended to dispute the validity of national GHG inventory figures. Although the uncertainty of emissions estimates underscores the lack of accuracy that characterizes many source and sink categories, its consideration can help to establish a more robust foundation on which to base policy.

According to the IPCC good practice guidance reports (notably, [8, p. 6.5]), uncertainty analysis is intended to help "improve the accuracy of inventories in the future and guide decisions on methodological choice." Uncertainty analyses function as indicators of opportunities for improvement in data measurement, data

Table 11.1 The six GHGs, or groups of gases, considered under the Kyoto Protocol to the UNFCCC [1, Annex A] and their global emissions as reported by the IPCC in its Third and Fourth Assessment Reports for the late 1990s and beyond. The GWP (last column) describes the global warming potential for a given GHG. It allows expressing the emissions of a non-CO_2 GHG in terms of CO_2-equivalent, which is the amount of CO_2 that causes the same global warming when measured over a specified timescale (generally, 100 years). The relative uncertainty ranges within which Annex I countries generally report the emissions of these GHGs are specified in Table 11.3

Kyoto gas	Global emissions [2, Table 4.1], [4, Tables 7.1, 7.6, 7.7]		GWP[a] [3, Table TS.2]
	Anthropogenic late 1990s[b]	Natural late 1990s[b]	
	2000–2005:	*2000–2005:*	
Carbon dioxide (CO_2)	7.2 ± 0.3 Pg C 26.4 Pg CO_2-eq.	-3.1 ± 0.8 Pg C[c] 11.4 Pg CO_2-eq.	1
	1996–2001:	*1996–2001:*	
Methane (CH_4)	428 Tg[d] 9.0 Pg CO_2-eq.	168 Tg[e] 3.5 Pg CO_2-eq.	21 (25)
	1990s:	*1990s:*	
Nitrous oxide (N_2O)	6.7 Tg N[f] 6.5 Pg CO_2-eq.	11.0 Tg N[g] 10.7 Pg CO_2-eq.	310 (298)
Hydroflurocarbons (HFCs)	See below	None	See below
Perfluorocarbons (PFCs)	See below	Negligible	See below
Sulfur hexafluoride (SF_6)	~6 Gg 0.14 Pg CO_2-eq.	Negligible	23,900 (22,800)
Important HFCs and PFCs:			
HFC-23 (CHF_3)	~7 Gg 0.08 Pg CO_2-eq.	None	11,700 (14,800)
HFC-134a(CH_2FCF_3)	~25 Gg 0.03 Pg CO_2-eq.	None	1,300 (1,430)
HFC-152a (CH_3CHF_2)	~4 Gg 0.56 Tg CO_2-eq.	None	140 (124)
PFC-14 (CF_4)	~15 Gg 0.10 Pg CO_2-eq.	Negligible	6,500 (7,390)
PFC-116 (C_2F_6)	~2 Gg 0.02 Pg CO_2-eq.	None	9,200 (12,200)

[a] The net global warming potential (GWP) refers to a time horizon of 100 years. The GWPs stem from the IPCC Second Assessment Report (Climate Change 1995) as these are used for reporting under the UNFCCC. The most recent GWP updates for the IPCC Fourth Assessment Report (Climate Change 2007) are reported in addition (in parentheses)
[b] If not indicated otherwise
[c] Uptake: net atmosphere-to-land flux; and atmosphere-to-ocean flux
[d] Emissions: coal mining; gas, oil and industry; ruminants; rice agriculture; and biomass burning
[e] Emissions: wetlands and termites
[f] Emissions: fossil fuel combustion and industrial processes; agriculture; biomass and biofuel burning; human excreta; rivers, estuaries and coastal zones; and atmospheric deposition
[g] Emissions: soils under natural vegetation; oceans; and atmospheric chemistry

collection, and calculation methodology. Only by identifying elements of high uncertainty can methodological changes be introduced to address them. Currently, most countries that perform uncertainty analyses do so for the express purpose of improving their future estimates; and the rationale is generally the same at the corporate and other levels. Estimating uncertainty helps to prioritize resources and to take precautions against undesirable consequences. Depending on the intended purpose of an inventory, however, this may not be the extent of the utility of uncertainty analysis. Another rationale for performing uncertainty analysis is to provide a policy tool, a means to adjust inventories or analyze and compare emission changes in order to determine compliance or the value of a transaction. While some experts find the quality of uncertainty data associated with national inventories insufficient to use for these purposes, others offer justification for conducting uncertainty analyses to inform and enforce policy decisions. Some experts suggest revising the system of accounting on which current reduction schemes are based, while others seek to incorporate uncertainty measurements into emission and emission change analysis procedures. The latter could offer policy makers enhanced knowledge and additional insight on which to base GHG emission reduction measures.

We follow the proactive track in dealing with uncertainty. In Sect. 11.2 we look into the question of why uncertainty matters in general. Sections 11.3 and 11.4 elaborate on Sect. 11.2. In Sect. 11.3 we provide an overview of the state-of-the-art of analyzing emission changes in consideration of uncertainty. We envision this analysis taking place in accordance with, not independent of, a dual-constrained (bottom-up/top-down) verification framework in Sect. 11.4. We summarize our findings in Sect. 11.5.

11.2 Does Uncertainty Matter?

Reference [5] (see also [7]) offers a number of reasons why the consideration of uncertainty in GHG inventories is important:

- Understanding the basic science of GHG sources and sinks requires an understanding of the uncertainty in their estimates.
- Schemes to reduce human-induced global climate change rely on confidence that inventories of GHG emissions allow the accurate and transparent assessment of emissions and emission changes.
- Uncertainty is higher for some aspects of a GHG inventory than for others. For example, past experience shows that, in general, methods used to estimate nitrous oxide (N_2O) emissions are more uncertain than those for methane (CH_4) and much more uncertain than those for carbon dioxide (CO_2). Whether in multi-gas, cross-sectoral, international comparisons, trading systems, or in compliance mechanisms, approaches to uncertainty analysis need to be robust and standardized across sectors and gases, as well as among countries.

Uncertainty analysis helps to understand uncertainties: better science helps to reduce them. Better science needs support, encouragement and investment. Full carbon accounting (FCA) – or full accounting of emissions and removals, including all

GHGs – in national GHG inventories is important for advancing the science. FCA is a prerequisite for reducing uncertainties in our understanding of the global climate system. From a policy viewpoint, FCA could be encouraged by including it in reporting emissions, but it might be separated from targets for reducing emissions. Future climate agreements will become more robust if there is explicit accounting for the uncertainties associated with emission estimates. Hence, understanding uncertainty matters in many ways.

11.3 State of the Art of Analyzing Uncertain Emission Changes

In this section we elaborate on Sect. 11.2 by looking into the state-of-the-art of analyzing changes in emissions and removals of GHGs in consideration of uncertainty. From a physical (measurability) point of view, the uncertainty surrounding emission changes becomes more important in relative terms the smaller are the changes in the emissions, that is, the smaller the dynamics that they exhibit. Two options exist to avoid situations of great uncertainty vs. small change: (1) allowing more time so that greater emission changes can materialize; and (2) increasing measurability, e.g., by focusing on GHG emissions that can be grasped with "sufficient" certainty so that their changes are still "significant" in spite of the uncertainty. (Alternatively, emissions that do not possess these characteristics should be treated differently, e.g., separately from single-point emission targets (see above), or only in connection with targets that are defined as emission intervals or corridors.) Given that renegotiating the commitment times under the Kyoto Protocol cannot happen, Option 1 is not considered further. Option 2 requires the application of techniques that allow analyzing emission changes quantitatively (i.e., on an intra-technique basis) and qualitatively (i.e., on an inter-technique basis). Any of these techniques can be applied to GHG emissions individually, that is, they allow a detailed and thorough comparison of agreed or realized changes in emissions.

While pursuing the analysis of uncertain emission changes (also termed emission signals), we typically refer to the country scale, the principal reporting unit for reporting GHG emissions and removals under the Kyoto Protocol, but we could also refer to any other spatio-thematic scale. Our main motivation for studying the uncertainty of country-scale emissions estimates is the still unresolved issue of compliance (see also [10]). For most countries the emission changes agreed on under the Kyoto Protocol are of the same order of magnitude as, or smaller than, the uncertainty that underlies their combined CO_2 equivalent emissions estimates (compare the right column of Table 11.2 with the second column of Table 11.3).[1] Techniques are not in place to analyze uncertain emission signals from various points

[1] The issue of great uncertainty vs. small change also arises for small, intermediate, reduction targets. For instance, the EU discusses annual reduction steps in the context of an overall (EU-wide) GHG emission reduction of 20% by 2020 compared to 2005 [11, p. 7]. These steps follow a linear reduction path and are small (<2% per year; not compounded).

Table 11.2 Countries included in Annex B to the Kyoto Protocol (KP) and their emission limitation and reduction commitments (commitment period for all countries: 2008–2012; for the ISO Country Code for country abbreviations see below). The individual commitments have to be seen in context, i.e., vis-à-vis the uncertainty that underlies the reporting of emissions at the country scale (see Table 11.3). *Sources:* [14, Article 3.8, Annex B], [1, Decision 11/CP.4], [15], [16], [17, National Inventory Submissions], [18, Sect. 2.b]

Country group	Annex B country	Base year(s) for CO_2, CH_4, N_2O (for HFCs, PFCs, SF_6)	KP commitment %
1a	See below[a]	1990 (1995)	
	See below[b]	1990 (1990)	
1b	RO	1989 (1989)	92
1c	BG	1988 (1995)	
1d	SI	1986 (1995)	
2	US[c]	1990 (1990)	93
3a	JP	1990 (1995)	
	CA	1990 (1990)	94
3b	PL	1988 (1995)	
3c	HU	1985–1987 (1995)	
4	HR	1990 (1995)	95
5a	RU	1990 (1995)	100
5b	NZ, UA	1990 (1990)	
6	NO	1990 (1990)	101
7	AU	1990 (1990)	108
8	IS	1990 (1990)	110

[a]Country Group 1a: BE, CZ, DE, DK, EC (= EU-15; the EU-27 does not have a common Kyoto target), EE, ES, FI, GR, IE, LT, LU, LV, MC, NL, PT, SE, UK. Member States of the EU-27 but without individual Kyoto targets: CY, ML. Listed in the Convention's Annex I but not included in the Protocol's Annex B: BY and TR (BY and TR were not Parties to the Convention when the Protocol was adopted). BY requested becoming an Annex B country by amendment to the Kyoto Protocol at CMP 2 in 2006. (CMP = Conference of the Parties serving as the Meeting of the Parties to the Kyoto Protocol.) BY's base years and emission reduction commitments are 1990 (1995) and 92%, respectively

[b]Country Group 1a: AT, CH, FR, IT, LI, SK

[c]Country Group 2: The US has indicated its intention not to ratify the Kyoto Protocol. The US reports all its emissions with reference to 1990. However, information on 1990 in its national inventory submissions does not reflect or prejudge any decision that may be taken in relation to the use of 1995 as base year for HFCs, PFCs and SF_6 in accordance with Article 3.8 of the Kyoto Protocol

Abbreviations: AT Austria, *AU* Australia, *BE* Belgium, *BG* Bulgaria, *BY* Belarus, *CA* Canada, *CH* Switzerland, *CY* Cyprus, *CZ* Czech Republic, *DE* Germany, *DK* Denmark, *EC* European Community, *EE* Estonia, *ES* Spain, *FI* Finland, *FR* France, *GR* Greece, *HR* Croatia, *HU* Hungary, *IE* Ireland, *IS* Iceland, *IT* Italy, *JP* Japan, *LI* Liechtenstein, *LT* Lithuania, *LU* Luxembourg, *LV* Latvia, *MT* Malta, *MC* Monaco, *NL* Netherlands, *NO* Norway, *NZ* New Zealand, *PL* Poland, *PT* Portugal, *RO* Romania, *RU* Russian Federation, *SE* Sweden, *SI* Slovenia, *SK* Slovak Republic, *TR* Turkey, *UA* Ukraine, *UK* United Kingdom, *US* United States

Table 11.3 Emissions and/or removals of GHGs, or combinations of GHGs, classified according to their relative uncertainty ranges (reference: country scale). The bars of the arrows indicate the dominant uncertainty range for these emissions and removals, while the tops of the arrows point at the neighboring uncertainty ranges, which cannot be excluded but appear less frequently. LULUCF stands for the direct human-induced land-use, land-use change, and forestry activities stipulated by Articles 3.3 and 3.4 under the Kyoto Protocol [14]. The arrows are based on the total uncertainties that are reported for the Member States of the EU-25 [19] and the expertise available at IIASA's Forestry Program (cf. http://www.iiasa.ac.at/Research/FOR/unc_bottomup.html) and elsewhere (e.g., [20, Sects. 2.3.7, 2.4.1], [9, Sect. 5.2]). *Source:* [21, Table 1], modified

Class	Relative uncertainty (%) for 95% confidence interval	Classification of emissions and/or removals
1	0–5	⇓ CO_2 from fossil fuel (plus cement)
2	5–10	⇕ All Kyoto GHGs
3	10–20	⇕ Plus LULUCF
4	20–40	⇓
5	>40	⇓ CO_2 net terrestrial (>80%)

of view, ranging from signal quality (defined adjustments, statistical significance, detectability, etc.) to the way uncertainty is addressed (trend uncertainty or total uncertainty). Any such technique, if implemented, could "make or break" compliance, especially in cases where countries claim fulfillment of their reduction commitments. As already mentioned above, inventory uncertainty is monitored, but not regulated, under the Protocol. It remains to be seen whether the current status of ignoring uncertainty and abstaining from specifying clear rules on how to go about it will survive in the long term (see Compliance under the Kyoto Protocol in [12]) .

The analysis of emission signals in consideration of uncertainty can take three forms involving a multitude of techniques: (1) preparatory, (2) midway, and (3) retrospective signal analysis. Preparatory signal analysis allows generating useful knowledge that one would ideally wish to have at hand before negotiating international environmental agreements such as the Kyoto Protocol or its successors. For instance, it is important to know beforehand how great the uncertainties could be, depending on the desired level of confidence in the emission signal. What is the signal one wishes to detect and what is the risk one is willing to tolerate in meeting an agreed emission limitation or reduction commitment? To this end, it is generally assumed that (1) the emissions path between the base year and commitment year/period is a straight line (this is only a boundary condition, not a restriction); and (2) our knowledge of uncertainty in the commitment year/period will be as good as today's, in relative (qualitative) terms. Preparatory signal analysis allows factoring in the change in uncertainty, which can be due to learning and/or can result from structural changes in the emitters. However, researchers only begin to grasp these two determinants and to discriminate between them [22]. Handling the change in uncertainty is within reach but more data and research are needed. Being able to estimate the change in uncertainty is important in setting appropriate emission reduction targets, but one must not forget that preparatory signal analysis has not yet been applied in its simplest form to Kyoto commitments.

The state-of-the-art of preparatory signal analysis is well summarized by [23] (see also [21,24–26]), who compare six of the most widely discussed techniques.[2] In addition, preparatory signal analysis also allows monitoring the success of a country in reducing its emissions along a prescribed emissions target path between its base year and commitment year/period. This positive feature opens up a range of policy-relevant applications.[3]

Midway and retrospective signal analysis are less advanced than preparatory signal analysis. So far, midway signal analysis still focuses on emissions rather than on emissions changes. Midway signal analysis is an attempt to assess information on an emissions path at some point in time between base year and commitment. It considers a signal's path realized so far vis-à-vis a possible path toward the agreed emission limitation or reduction commitment. In this process, the dynamical moments (velocity, acceleration, etc.) of the historical and envisioned paths are compared, and this indicates (first-order control) whether or not it is likely to achieve the emission commitment. Midway signal analysis generally incorporates information from emissions prior to the base year to determine the signal's dynamical moments more accurately. The techniques explored so far to grasp the dynamics of, mostly, fossil-fuel CO_2 emissions encompass: polynomial regressions [29]; integral transforms [30]; and smoothing splines, parametric modeling and geometric Brownian motion modeling [31]. A related technique based on the analysis of short-term vs. long-term attainability and controllability has been followed by [32] and [33].

Retrospective signal analysis of emission changes becomes important when countries seek to assess their actual achievements in the commitment year/period. We distinguish between two fundamentally different approaches: static and dynamic. The static approach is identical to the one taken under preparatory signal analysis except that the agreed emission limitation or reduction commitment is replaced by the actual emission achievement. The emission signal is evaluated in terms of uncertainty, detectability or statistical significance, risk, etc. In contrast, the dynamic approach additionally considers how the emission signal has actually evolved between the base year and the commitment year/period, taking its dynamics into account. Here, expertise gained under midway analysis can serve as a platform as it also aims at evaluating full emission paths.

In their commensurability exercise, with its focus on six preparatory signal analysis techniques, Jonas and colleagues [23] concluded that a single best technique does not, and most likely will not, exist.[4] This is because the available techniques

[2] It is noted that attempts exist to put one of these six techniques to analyze uncertain emission changes, the verification time concept, on a stochastic basis (see Ermolieva et al., herein; and also [27] and [28]). It is correct to say that this technique still undergoes scientific scrutiny and awaits adjustment in order to operate in a preparatory mode.

[3] See http://www.iiasa.ac.at/Research/FOR/unc_overview.html for an overview on IIASA's monitoring reports and the countries that are monitored.

[4] For the authors' study and numerical results see http://www.iiasa.ac.at/Research/FOR/unc_prep.html. Referring readers to this website facilitates easy replication for follow-up studies or, as in this case, avoiding duplication.

suffer from inconsistencies in dealing with uncertainty that are not scientific but that are related to the way the Kyoto Protocol was designed. One technique, e.g., allows a country with a smaller emission reduction commitment to gain an advantage over a country with a greater emission reduction commitment;[5] while another technique forces countries a priori to realize detectable signals before they are permitted to make economic use of their excess emission reductions.[6] Jonas and colleagues [23] stress that these "inconsistencies" are the consequence of the Kyoto policy process running ahead of science, leaving us with the awkward problem of choosing between bad or undesirable alternatives in applying preparatory signal analysis. We can simply ignore uncertainty knowing that, e.g., emission markets will then lack scientific credibility or we can give preference to one preparatory technique over another knowing that none is ideal in satisfying the Protocol's political cornerstones. As two of the most important shortfalls on the side of policymaking the authors [23] identify (1) the overall neglect of uncertainty confronting experts with the finding that for most countries the agreed emission changes are of the same order of magnitude as the uncertainty that underlies their combined CO_2 equivalent emissions (compare the right column of Table 11.2 with the second column of Table 11.3); and (2) the existence of nonuniform emission limitation or reduction commitments that were determined "off the cuff" (i.e., derived via horse-trading) and did not result from rigorous scientific considerations.[7] From a purely quantitative point of view, the first shortfall is of greater relevance than the second one.

---

[5] See, for instance, the so-called undershooting (Und) concept: Excel file available via *numerical results* to [23] at http://www.iiasa.ac.at/Research/FOR/unc_prep.html: Worksheet *Undershooting 4*:
column C = Kyoto commitments δ_{KP} for country groups 1–8 (see also Table 11.2);
column E = the accepted risk α that a country's true emissions in the commitment year/period are equal to, or greater than, the country's true Kyoto target (risk α can be grasped although true emissions and targets derived from them are unknown by nature); and
columns F–N or U–AC (restricted to rows 14–16) = presumed relative uncertainty ρ of the country's reported emissions.
The Und concept requires undershooting of the countries' Kyoto targets in the commitment year in order to handle and decrease risk α (see columns F–N, rows $\geq$ 17, for the required undershooting). Varying δ_{KP} while keeping the relative uncertainty ρ and the risk α constant (e.g., at $\rho = 15\%$ and $\alpha = 0.3$) exhibits that countries complying with a smaller δ_{KP} are better off than countries that must comply with a greater δ_{KP} (see columns U–AC, rows $\geq$ 17, for the modified emission limitation or reduction target, which is the sum of the agreed target under the Kyoto Protocol plus the required undershooting). Such a situation is not in line with the spirit of the Kyoto Protocol!

[6] See, for instance, the so-called combined undershoot and verification time (Und&VT) concept: Excel file available via *numerical results* to [23] at http://www.iiasa.ac.at/Research/FOR/unc_prep.html: Worksheet *Und&VT 1:* Fig. 1 therein. The Und&VT concept requires a priori detectable emission reductions, not limitations. That is, it requires the Protocol's emission limitation or reduction targets to be corrected for nondetectability through the introduction of an initial or obligatory undershooting so that the countries' emission signals become detectable before the countries are permitted to make economic use of their excess emission reductions. This nullifies, de facto, the politically agreed targets under the Kyoto Protocol!

[7] The situation would be different if the nonuniformity of the emission limitation or reduction commitments would be the outcome of a rigorously based process resulting in a straightforward

However, it appears that the first shortfall will vanish soon as mankind is increasingly under pressure to adopt and realize greater emission reductions in the mid to long-term [36], [37, Decision 1/CP.13]. Notwithstanding, we would still be left with the problem of which analysis technique to give preference to. This discussion has not even started.

11.4 How to Deal with Uncertainty?

In this section we elaborate on Sect. 11.2 from a holistic point of view. Our starting point is FCA (or more generally, full accounting of all emissions and removals of all GHGs). We consider FCA a prerequisite for constraining and reducing uncertainties in our understanding of the global climate system. A dual-constrained (verified) full carbon analysis can compare the sum of Earth-based measurements of flows to and from the atmosphere with atmosphere-based evaluation of exchanges with the Earth. As specified by [5], a verified FCA, including all sources and sinks of both the technosphere and terrestrial biosphere considered continuously over time, would allow the research and inventory communities to:

- Present a real picture of emissions and removals at continental and smaller scales.
- Avoid ambiguities generated by such terms as "managed biosphere," "base-line activities," and "additionality." Elimination of splitting the terrestrial biosphere into directly human-impacted (managed) and not directly human-impacted (natural) parts would be advantageous from a verification point of view as there is no atmospheric measurement that can discriminate between the two [21, Sect. 3].
- Make available reliable and comprehensive estimates of uncertainties that cannot be achieved using the current approach under the UNFCCC and Kyoto Protocol [8, 9]. It is impossible to estimate the reliability of any system output if only part of the system is considered. The tacit assumptions underlying the Protocol are that man's impact on nature, the not-accounted remainder under the Protocol, is irrelevant and inventory uncertainty only matters from a relative point of view over space and time, not an absolute one. However, this approach is highly problematic because biases (i.e., discrepancies between true and reported emissions), typically resulting from partial accounting, are not uniform across space and time. In addition, man's impact on nature need not be constant or negligible.[8]

rule that applies equally to all countries as would be the case, for instance, under the so-called contraction and convergence approach (e.g., [34, Sect. 2.3.2], [35]).

[8] In their recent study [38, Table 1] show that, making use of global carbon budget data between 1959 and 2006, the efficiency of natural carbon sinks to remove atmospheric CO_2 has declined by about 2.5% per decade. Although this decline may look modest, it represents a mean net "source" to the atmosphere of 0.13 PgC y^{-1} during 2000–2006. In comparison, a 5% reduction in the mean global fossil emissions during the same time period yields a net "sink" of 0.38 PgC y^{-1}. Thus, deteriorating natural carbon sinks as a result of climate change or man's direct impact exhibit the potential to offset efforts to reduce fossil fuel emissions. This shows that man's impact on nature

FCA is essential for good science. However, it would be for policymakers to decide how FCA is used. FCA could be used for "crediting" in the sense of the Kyoto Protocol (i.e., for compliance) or only for "accounting" and scientific understanding as required under the UNFCCC. Given that the treatment of the land use, land-use change, and forestry (LULUCF) sector in general poses a number of characteristic challenges (see box), we prefer FCA accounting under which, however, the Kyoto compliance accounting as required under the Protocol would form a logical and consistent subset.

Uncertainty in the LULUCF sector

Expressing uncertainties in the LULUCF sector can be challenging because of:

- The complexities and scales of the systems being modeled.
- The fact that human activity in a given year can impact emissions and removals in these systems over several years, not just the year in which the activity took place.
- These systems being strongly affected by inter-annual, decadal, and long-term variability in climate.

Knowledge of the temporal dynamics of systems – what has happened in the past, and how actions in the present will affect emissions/removals in the future – is important; gaps in this knowledge add to uncertainties about the immediate impacts of human activities.

Approaches to estimating emissions and removals in the LULUCF sector frequently involve the use of detailed data and computer models to simulate the complex functional relationships that exist in natural systems. But a consequence of using more detailed methods is that the estimation of uncertainty also comes more into play. However, despite conceptual and technical challenges, powerful tools for combining different kinds of information from multiple sources are becoming available and are increasingly being used by modelers to reduce uncertainties in the LULUCF sector.

These tools allow modelers to increase their focus on model validation and on reconciling results from alternative approaches. However, one key barrier remains. Reporting under the UNFCCC and Kyoto Protocol provides only a partial account of what is happening in the LULUCF sector. To close the verification loop would require the adoption of FCA.

Despite improvements in approaches to estimating uncertainty in emissions and removals in the LULUCF sector, some challenges remain. The treatment of this sector in future policy regimes requires special consideration.

---

is indeed not negligible and stresses the need to look at the entire system, that is, to develop a FCA system where emissions and removals and their trends are monitored in toto.

FCA is expected to facilitate the reconciliation of two broad accounting approaches: top-down and bottom-up accounting. Top-down accounting takes the point of view of the atmosphere. It relies on observations of atmospheric CO_2 concentrations, changes in concentrations, and atmospheric modeling to infer fluxes from land and ocean sources. Bottom-up accounting takes the opposite perspective. It relies on observations of stock changes or fluxes at the Earth's surface and infers the change in the atmosphere. FCA – estimating all land and ocean-based fluxes, whether human-induced or not – is necessary to reconcile the top-down and bottom-up approaches.

While methods of both top-down and bottom-up accounting have improved in recent years, both approaches still have areas of weakness. Investment in research is needed to tackle these limitations, improve the FCA approach, and hence reduce uncertainties (see also [39]).

Last, but not least, it must be kept in mind that verification of emission estimates does not necessarily imply detection of emission signals (e.g., decreased emissions) over time. It is the detection of emission signals that is needed to complement the bottom-up/top-down accounting of GHGs and the prime goal of this research community to close the existing gaps in the accounting.

Thus, from a policy perspective, there are pressing issues regarding how uncertainty can be dealt with through uncertainty analysis techniques and improvements to science. The implications for policymakers working to reduce human impacts on the global climate include [5]:

- Uncertainty analysis helps to understand uncertainties: better science helps to reduce uncertainties. Better science needs support, encouragement, and investment. FCA in national GHG inventories is important for advancing the science. It could be included in reporting but separated from targets for reducing emissions.
- Uncertainty is inherently higher for some aspects of an inventory than for others. For example, the LULUCF sector has higher uncertainties than the fossil fuel sector and estimates of N_2O emissions tend to be more uncertain than those of CH_4 and CO_2. This raises the possibility that in designing future policy agreements some components of a GHG inventory could be treated differently than others.[9]
- Improving inventories requires one approach; improving emissions trading mechanisms another. Inventories will be improved by increasing their scope to include FCA. In contrast, one option for improving emissions trading mechanisms would be to reduce their scope. Currently, emissions trading mechanisms may include estimation methodologies with varying degrees of uncertainty but they do not explicitly consider uncertainty or treat it in a standardized fashion. There are two options for improving this situation. The first option, as mentioned, is to reduce

[9] This view of treating subsystems individually and differently runs counter to the approach typically taken. The tendency has been to treat subsystems collectively and equally and to dispose over a wide range of options in order to minimize costs or maximize benefits resulting from the joint emissions reduction of GHGs and air pollutants (e.g., [40], [41, (77)], [42]).

the scope of emissions trading mechanisms – by excluding uncertain methodologies or more uncertain GHGs – to make them more manageable (see also [43]). The second option is to retain the scope of emissions trading mechanisms but to adopt a standardized approach to estimating uncertainty. But we could not guarantee that the latter approach would eventually withstand large biases resulting, e.g., from a mismatch in the bottom-up vs. top-down accounting.

In the context of pricing uncertainty, it needs to be mentioned that uncertainty is an inherent part of any emissions accounting and that it will play an important role in both the scientific understanding of emissions and in their political treatment. At present, however, uncertainty does not play a role in trading of emissions credits. Ultimately, uncertainty can be borne by either the buyer or seller of any asset, and it should be agreed in advance of any exchange how this is to be dealt with. Risky or uncertain assets will be traded at a discount to the extent that the risk and uncertainty are to be assumed by the buyer.

Literature on treating scientific uncertainty upfront in financial markets is already emerging (e.g., [26, 44, 45]), but this has not yet been applied widely to GHG emissions credits. For now it appears that buyers of emissions credits generally accept credits without uncertainty and the seller is obligated to ensure that the credits are fulfilled.

With the current system of trading in allowances and credits, neither buyers nor sellers have much incentive to reduce the uncertainty associated with emissions inventory or reduction estimates; to do so might impact the single-point emission (or reduction) estimate, thus directly affecting the value of allowances or credits. For example, a highly uncertain emission reduction estimate that is biased high will tend to be worth more (claiming greater reductions), presupposing the market's willingness-to-pay, than the same reduction figured more accurately and with greater uncertainty. This suggests the possibility that other, more complex, pricing mechanisms than the current cap-and-trade system might exist and would be better able to deal with uncertainty by, for example, monetizing (i.e., rewarding) increased confidence. Such a system might also differentiate between different types of emissions and/or reductions and their uncertainties.

11.5 Conclusions

We see a clear rationale for conducting and improving uncertainty analysis:

- Uncertainty analysis improves the monitoring of GHG emissions. Uncertainty analysis helps to understand uncertainties and encourages better science that will help to reduce uncertainty.
- Better science requires the adoption of FCA. More investment in research is needed to reconcile the bottom-up and top-down accounting approaches that are fundamental to FCA and that dual-constrain uncertainty. FCA may only be used for "accounting" but with the Kyoto compliance accounting as a logical

and consistent subset used for "crediting." We anticipate that within a few years scientists will overcome still existing bottom-up/top-down accounting gaps for the Kyoto GHGs at the scale of continents. Scientists may even be able to downscale validated, and verified (dual-constrained), emissions estimates to the scale of countries or groups of countries. That is, scientists will be able to verify (correct) politically driven (mis-)accounting reported annually bottom-up under the Kyoto Protocol and its successor.

- Some GHG emissions and removals estimates are more uncertain than others. Options exist to address this issue, and these could be incorporated in the design of future policy regimes. These options also include (1) the option of not splitting the terrestrial biosphere into a directly human-impacted (managed) and a not-directly human-impacted (natural) part to avoid sacrificing bottom-up/top-down verification; and (2) the option of not pooling subsystems, including sources and sinks, with different relative uncertainties but treating them individually and differently.
- We expect the treatment of scientific uncertainty in emission trading markets to gain relevance. Neither buyers nor sellers of GHG emission credits have a strong interest to let this issue go unresolved.
- The issue of compliance also goes unresolved and requires directing attention to the appropriate treatment of emission changes in consideration of uncertainty. Currently, signal analysis is still treated independently of bottom-up/top-down verification, but scientists will eventually be able to make the two consistent and to go hand-in-hand.

References

1. FCCC: Report of the conference of the parties on its fourth session, Buenos Aires, 2–14 November 1998. Addendum. Part two: action taken by the conference of the parties at its fourth session. Tech. Rep. FCCC/CP/1998/16/Add.1, UN Framework Convention on Climate Change (FCCC), Bonn, Germany (1999). http://unfccc.int/resource/docs/cop4/16a01.pdf
2. Prather, M., Ehhalt, F., Dentener, F., Derwent, R., Dlugokencky, E., Holland, E., Isaksen, I., Katima, J., Kirchhoff, V., Matson, P., Midgley, P., Wang, M.: Atmospheric chemistry and greenhouse gases. In: Houghton, J.T., et al. (eds.) Climate Change 2001: The Scientific Basis. Contribution of Working Group I to the Third Assessment Report of the Intergovernmental Panel on Climate Change, pp. 239–287. Cambridge University Press, Cambridge (2001). http://www.ipcc.ch/ipccreports/tar/wg1/index.htm
3. Solomon, S., Manning, D.Q.M., Alley, R., Berntsen, T., Bindoff, N., Chen, Z., Chidthaisong, A., Gregory, J., Hegerl, G., Heimann, M., Hewitson, B., Hoskins, B., Joos, F., Jouzel, J., Kattsov, V., Lohmann, U., Matsuno, T., Molina, M., Nicholls, N., Overpeck, J., Raga, G., Ramaswamy, V., Ren, J., Rusticucci, M., Somerville, R., Stocker, T., Whetton, P., Wood, R., Wratt, D.: Technical summary. In: Solomon, S., et al. (eds.) Climate Change 2007: The Physical Science Basis. Contribution of Working Group I to the Fourth Assessment Report of the Intergovernmental Panel on Climate Change, pp. 19–91. Cambridge University Press, Cambridge (2007)
4. Denman, K., Brasseur, G., Chidthaisong, A., Ciais, P., Cox, P., Dickinson, R., Hauglustaine, D., Heinze, C., Holland, E., Jacob, D., Lohmann, U., Ramachandran, S., da Silva Dias, P., Wofsy, S., Zhang, X.: Couplings between changes in the climate system and biogeochemistry,

In: Solomon, S., et al. (eds.) Climate Change 2007: The Physical Science Basis. Contribution of Working Group I to the Fourth Assessment Report of the Intergovernmental Panel on Climate Change, pp. 499–587. Cambridge University Press, Cambridge (2007)

5. IIASA: Uncertainty in greenhouse gas inventories. IIASA Policy Brief 01, International Institute for Applied Systems Analysis, Laxenburg, Austria (2007). http://www.iiasa.ac.at/Publications/policy-briefs/pb01-web.pdf
6. International Petroleum Industry Environmental Conservation Association: Greenhouse gas emissions estimations and inventories. Summary Report, Addressing uncertainty and accuracy (2007). http://www.ipieca.org/activities/climate_change/downloads/publications/Uncertainty.pdf
7. Lieberman, D., Jonas, M., Winiwarter, W., Nahorski, Z., Nilsson, S.: Accounting for climate change: introduction. In: Lieberman, D., et al. (eds.) Accounting for Climate Change. Uncertainty in Greenhouse Gas Inventories – Verification, Compliance, and Trading. Springer, Dordrecht (2007). doi:10.1007/s11267-006-9120-8
8. Penman, J., Kruger, D., Galbally, I., Hiraishi, T., Nyenzi, B., Emmanuel, S., Buendia, L., Hoppaus, R., Martinsen, T., Meijer, J., Miwa, K., Tanabe, K., (eds.): Good practice guidance and uncertainty management in national greenhouse gas inventories. Institute for Global Environmental Strategies, Hayama, Kanagawa, Japan (2000). http://www.ipcc-nggip.iges.or.jp/public/gp/english/
9. Penman, J., Gytarsky, M., Hiraishi, T., Krug, T., Kruger, D., Pipatti, R., Buendia, L., Miwa, K., Ngara, T., Tanabe, K., Wagner, F. (eds.): Good Practice Guidance for Land Use, Land-Use Change and Forestry. Institute for Global Environmental Strategies, Hayama, Kanagawa, Japan (2003). http://www.ipcc-nggip.iges.or.jp/public/gpglulucf/gpglulucf.htm
10. Gupta, J., Olsthoorn, X., Rotenberg, E.: The role of scientific uncertainty in compliance with the Kyoto Protocol to the climate change convention. Environ. Sci. Policy **6**(6), 475–486 (2003). doi:10.1016/j.envsci.2003.09.001
11. COM: Proposal for a directive of the European Parliament and of the council amending directive 2003/87/ec so as to improve and extend the greenhouse gas emission allowance trading system of the community. Tech. Rep. 16 final, Commission of the European Communities, Brussels, Belgium (2008). http://eur-lex.europa.eu/LexUriServ/LexUriServ.do?uri=COM:2008:0016:FIN:EN:PDF
12. FCCC: Compliance under the Kyoto Protocol. Secretariat to the UN Framework Convention on Climate Change (FCCC), Bonn, Germany (2008). http://unfccc.int/kyoto_protocol/compliance/items/2875.php
13. FCCC: Report of the conference of the parties on its second session, Geneva, 8–19 July 1996. Addendum. Part two: action taken by the conference of the parties at its second session. Tech. Rep. FCCC/CP/1996/15/Add.1, UN Framework Convention on Climate Change (FCCC), Bonn, Germany (1996). http://unfccc.int/resource/docs/cop2/15a01.pdf
14. FCCC: Report of the conference of the parties on its third session, Kyoto, 1–11 December 1997. Addendum. Part two: action taken by the conference of the parties at its third session. Tech. Rep. FCCC/CP/1997/7/Add.1, UN Framework Convention on Climate Change (FCCC), Bonn, Germany (1998). http://unfccc.int/resource/docs/cop3/07a01.pdf
15. FCCC: Kyoto Protocol: countries included in Annex B to the Kyoto Protocol and their emission targets. Secretariat to the UN Framework Convention on Climate Change (FCCC), Bonn, Germany (2008). http://unfccc.int/kyoto_protocol/items/3145.php
16. FCCC: Kyoto Protocol base year data. Secretariat to the UN Framework Convention on Climate Change (FCCC), Bonn, Germany (2008). http://unfccc.int/ghg_data/kp_data_unfccc/base_year_data/items/4354.php
17. FCCC: National inventory submissions 2008. Secretariat to the UN Framework Convention on Climate Change (FCCC), Bonn, Germany (2008). http://unfccc.int/national_reports/annex_i_ghg_inventories/national_inventories_submissions/items/4303.php
18. COM: Report from the commission. Assigned amount report of the European Union. Tech. Rep. 799 final, Commission of the European Communities (COM), Brussels, Belgium (2006). http://eur-lex.europa.eu/LexUriServ/LexUriServ.do?uri=COM:2006:0799:FIN:EN:PDF

19. EEA: Annual European Community greenhouse gas inventory 1990–2005 and inventory report 2007. Tech. Rep. 7, European Environment Agency (EEA), Copenhagen (2007). http://reports.eea.europa.eu/technical_report_2007_7/en/
20. Watson, R., Noble, I., Bolin, B., Ravindranath, N., Verardo, D., Dokken, D., (eds.): Land use, land-use change, and forestry. Cambridge University Press, Cambridge (2000). http://www.ipcc.ch/ipccreports/sres/land_use/index.htm
21. Jonas, M., Nilsson, S.: Prior to an economic treatment of emissions and their uncertainties under the Kyoto Protocol: scientific uncertainties that must be kept in mind. In: Lieberman, D., et al. (eds.) Accounting for Climate Change. Uncertainty in Greenhouse Gas Inventories – Verification, Compliance, and Trading, p. 159. Springer, Dordrecht (2007). doi:10.1007/s11267-006-9113-7
22. Hamal, K.: Reporting GHG emissions: change in uncertainty and its relevance for the detection of emission changes. Interim Report, International Institute for Applied Systems Analysis, Laxenburg, Austria (2008). http://www.iiasa.ac.at/Admin/YSP/2007abstracts.pdf
23. Jonas, M., Gusti, M., Jeda, W., Nahorski, Z., Nilsson, S.: Comparison of preparatory signal detection techniques for consideration in the (post-) Kyoto policy process. In: 2nd International Workshop on Uncertainty in Greenhouse Gas Inventories, pp. 107–134, Laxenburg, Austria (2007). http://www.ibspan.waw.pl/ghg2007/par.htm
24. Jonas, M., Nilsson, S., Bun, R., Dachuk, V., Gusti, M., Horabik, J., Jeda, W., Nahorski, Z.: Preparatory signal detection for Annex I countries under the Kyoto Protocol – a lesson for the post-Kyoto policy process. Interim Report IR-04-024, International Institute for Applied Systems Analysis, Laxenburg, Austria (2004). http://www.iiasa.ac.at/Publications/Documents/IR-04-024.pdf
25. Gillenwater, M., Sussman, F., Cohen, J.: Practical policy applications of uncertainty analysis for national greenhouse gas inventories. In: Lieberman, D., et al. (eds.) Accounting for Climate Change. Uncertainty in Greenhouse Gas Inventories – Verification, Compliance, and Trading. Springer, Dordrecht (2007). doi:10.1007/s11267-006-9118-2
26. Nahorski, Z., Horabik, J., Jonas, M.: Compliance and emissions trading under the Kyoto Protocol: rules for uncertain inventories. In: Lieberman, D., et al. (eds.) Accounting for Climate Change. Uncertainty in Greenhouse Gas Inventories – Verification, Compliance, and Trading. Springer, Dordrecht (2007). doi:10.1007/s11267-006-9112-8
27. Hudz, H.: Verification times underlying the Kyoto Protocol: consideration of risk. Interim Report IR-02-066, International Institute for Applied Systems Analysis, Laxenburg, Austria (2003). http://www.iiasa.ac.at/Publications/Documents/IR-02-066.pdf
28. Hudz, H., Jonas, M., Ermolieva, T., Bun, R., Ermoliev, Y., Nilsson, S.: Verification times underlying the Kyoto Protocol: consideration of risk. Background data for IR-02-066. International Institute for Applied Systems Analysis, Laxenburg, Austria (2003). http://www.iiasa.ac.at/Research/FOR/vt_concept.html
29. Gusti, M., Jeda, W.: Carbon management: a new dimension of future carbon research. Interim Report IR-02-006, International Institute for Applied Systems Analysis, Laxenburg, Austria (2002). http://www.iiasa.ac.at/Publications/Documents/IR-02-006.pdf
30. Dachuk, V.: Looking behind the Kyoto Protocol: can integral transforms provide help in dealing with the verification issue? Interim Report IR-02-046, International Institute for Applied Systems Analysis, Laxenburg, Austria (2003). http://www.iiasa.ac.at/Publications/Documents/IR-02-046.pdf
31. Nahorski, Z., Jeda, W.: Processing national CO_2 inventory emissions data and their total uncertainty estimates. In: Lieberman, D., et al. (eds.) Accounting for Climate Change: Uncertainty in Greenhouse Gas Inventories – Verification, Compliance, and Trading. Springer, Dordrecht (2007). doi:10.1007/s11267-006-9114-6
32. Smirnov, A.: Attainability analysis of the dice model. Interim Report IR-05-049, International Institute for Applied Systems Analysis, Laxenburg, Austria (2005). http://www.iiasa.ac.at/Publications/Documents/IR-05-049.pdf
33. Pivovarchuk, D.: Consistency between long-term climate targets and short-term abatement policies: attainability analysis technique. Interim Report IR-08-017, International Institute for Applied Systems Analysis, Laxenburg, Austria (2008)

34. WBGU: Climate protection strategies for the 21st century: Kyoto and beyond. Special report, German Advisory Council on Global Change (WBGU), Bremerhaven, Germany (2003). http://www.wbgu.de/wbgu_sn2003_engl.html
35. Pearce, F.: Saving the World, Plan B. New Sci. **180**(2425), 6–7 (2003). http://www.newscientist.com/article/dn4467-greenhouse-gas-plan-b-gaining-support.html
36. CANA: Scorecard of the Bali climate talks. Climate Action Network Australia, Ultimo, NSW, Australia (2007). http://www.cana.net.au/international/Bali_scorecard_back_page_final.pdf
37. FCCC: Report of the conference of the parties on its thirteenth session, Bali, 3–15 December 2007. Addendum. Part two: action taken by the conference of the parties at its thirteenth session. Tech. Rep. FCCC/CP/2007/6/Add.1, UN Framework Convention on Climate Change (FCCC), Bonn, Germany (2007). http://www.unisdr.org/eng/risk-reduction/climate-change/docs/Bali-Action-plan.pdf
38. Canadell, J., Le Quere, C., Raupach, M., Field, C., Ciais, P., Conway, T., Gillett, N., Houghton, R., Marland, G.: Contributions to accelerating atmospheric CO_2 growth from economic activity, carbon intensity, and efficiency of natural sinks. PNAS **104**(47), 18866–18870 (2007). doi:10.1073/pnas.0702737104
39. Pearce, F.: Kyoto promises are nothing but hot air. New Sci. **190**(2557), 10–11 (2006). http://www.newscientist.com/article/mg19025574.000-kyoto-promises-are-nothing-but-hot-air.html
40. McCarl, B., Schneider, U.: Climate change: Greenhouse gas mitigation in US agriculture and forestry. Science **294**(5551), 2481–2482 (2001). doi:10.1126/science.1064193
41. Hansen, J.: Defusing the global warming time bomb. Sci. Am. **290**(3), 68–77 (2004). http://www.columbia.edu/~jeh1/hansen_timebomb.pdf
42. APD: Air pollution and greenhouse gases. Atmospheric Pollution and Economic Development (APD) Program, International Institute for Applied Systems Analysis, Laxenburg, Austria (2008). http://www.iiasa.ac.at/rains/ghg.html
43. Monni, S., Syri, S., Pipatti, R., Savolainen, I.: Extension of EU emissions trading scheme to other sectors and gases: consequences for uncertainty of total tradable amount. In: Lieberman, D., et al. (eds.) Accounting for Climate Change. Uncertainty in Greenhouse Gas Inventories – Verification, Compliance, and Trading. Springer, Dordrecht (2007). doi:10.1007/s11267-006-9111-9
44. Ermoliev, Y., Michalevich, M., Nentjes, A.: Markets for tradeable emission and ambient permits: a dynamic approach. Environ. Res. Econ. **15**(1), 39–56 (2000). http://www.springerlink.com/content/x7385ul217736h24/fulltext.pdf
45. Godal, O., Ermoliev, Y., Klassen, G., Obersteiner, M.: Carbon trading with imperfectly observable emissions. Environ. Res. Econ. **25**(2), 151–169 (2003). http://www.springerlink.com/content/r7wl2242j0725354/fulltext.pdf

Chapter 12
Uncertainty Analysis of Weather Controlled Systems

K.J. Keesman and T. Doeswijk

Abstract The indoor climate of many storage facilities for agricultural produce is controlled by mixing ambient air with the air flow through the store room. Hence, the indoor climate is affected by the ambient weather conditions. Given hourly fluctuating energy tariffs, weather forecasts over some days are required to effectively anticipate. Hence, typically a real-time optimal control strategy results. As weather forecasts are uncertain, predicted model outputs and related costs of the control strategy become uncertain. Usually, a medium-range weather forecast for a period of some days consists of an ensemble of forecasts. Hence, the uncertainty in the weather forecast is known a priori. In addition to this, in past-performance studies where weather forecasts and observed weather variables are given, an a posteriori evaluation of the forecast errors can be made as well. The objective of this study is to evaluate the uncertainty in the costs related to weather forecast errors and uncertainty, given the control inputs. In a simulation case-study with real weather forecasts and observed weather, it appeared that only slight cost increases can be expected due to errors and uncertainties in weather forecasts if the optimal control problem is calculated every 6–12 h in a receding horizon context.

12.1 Introduction

Indoor climate in greenhouses, office buildings and storage facilities for agricultural produce are generally affected by outdoor weather conditions [2, 9, 13], for instance by heat transfer through the boundaries, solar radiation and ventilation with outdoor air. In addition to this, the indoor climate is also affected by the respiration and evapotranspiration of living creatures and biological material, or by active heating or cooling. Typically, in practice, indoor temperatures and relative humidities are

K.J. Keesman (✉) and T. Doeswijk
Systems and Control Group, Wageningen University, P.O. Box 17, 6700 AA Wageningen, The Netherlands,
e-mail: karel.keesman@wur.nl, timo.doeswijk@wur.nl

K. Marti et al. (eds.), *Coping with Uncertainty*, Lecture Notes in Economics and Mathematical Systems 633, DOI 10.1007/978-3-642-03735-1_12,

controlled by feedback controllers with P or PI structure (see [14] for an overview) using direct measurements of temperature and relative humidity. In this study, we focus on indoor climate control that uses ambient air in the air flow through the (store) room. Hence, optimal control strategies that are able to anticipate on future changes of the weather conditions or on hourly-daily changes in energy tariffs allow, in principle, a better cost-effective performance. Clearly, in such control strategies weather forecasts become important and uncertainties in weather forecasts then lead to uncertain predictions of the indoor climate. Moreover, the cost function related to any kind of optimal control becomes subject to uncertainties and errors in the weather forecasts. The sensitivity of the model outputs and the related costs therefore needs to be investigated. In what follows, we restrict ourselves to a so-called receding horizon optimal control (RHOC) strategy in discrete-time. The basic idea behind RHOC is that a finite-time optimization problem over the horizon H, with or without inequality constraints, is solved, i.e.

$$u(k, k+H) = \arg\min \phi(x(k+H)) + \sum_{\kappa=k}^{k+H} \mathcal{L}(x(\kappa), u(\kappa), d(\kappa)) \qquad (12.1)$$

$$s.t. \quad x(k+1) = f(x(k), u(k), d(k)), \quad g(x(k), u(k)) \leq 0$$

$$x_0 = x(k)$$

Subsequently, only the first $l < H$ inputs are implemented and at $k + l$, given an update of x_0, the optimization problem is solved again, but now on the time interval $[k+l, k+l+H]$. For both air conditioned buildings [7] and potato storage facilities [8, 10] it was shown that a good weather forecast reduces the cost function almost as much as a perfect weather forecast. In these studies short term weather forecasts (1–2 days ahead) were used. However, because of the slow dynamics of products in storage facilities, knowledge of medium-range predictions of the indoor climate may be profitable. Hence, this study focuses on the effect of errors and uncertainties in medium-range (up to 10 days ahead) weather forecasts on the indoor climate predictions, in real-time, and associated product temperature and costs.

Nowadays, the medium-range weather forecast [11] consists of an ensemble of 50 different weather forecasts. All 50 ensemble members have an equal probability of occurring. Hence, the uncertainty, or variation, in the weather forecasts is known a priori. This knowledge can then be used to evaluate a calculated optimal control solution by calculating the differences in costs related to each of the ensemble members. Next to this, with observed weather given, in a post-performance analysis the optimal control solution can also be evaluated a posteriori.

The objective of the study is to evaluate the effect of errors and uncertainties in the weather forecasts on the costs of the calculated optimal control problem of a (potato) bulk storage facility.

This chapter is structured as follows: first a brief description of the bulk storage model and the cost function is given. Next, some information about the weather forecasts is provided. Then, open loop and closed loop evaluations using weather forecasts and observations are given in subsequent sections. Finally, the results are discussed and conclusions are drawn.

12.2 Preliminaries

12.2.1 *Bulk Storage Model*

Models that describes the dynamics of the product in a bulk storage facility can be found in, e.g. [6, 8, 9, 15–19]. The model developed in [9] was simplified to make it suitable for use in an RHOC algorithm [10]. A brief description of this discrete-time dynamic model is presented in this subsection. The model is given by

$$\begin{aligned} x_T(k+1) &= x_T(k) + p_1 + [d_1(k) - x_T(k)] \\ &\quad \Big(p_3 + p_2 p_5 u_1(k)[p_7 + (1-p_7)u_2(k)]\Big) \end{aligned} \tag{12.2}$$

$$\begin{aligned} x_{CO_2}(k+1) &= x_{CO_2}(k) + p_1 p_4 + [d_2(k) - x_{CO_2}(k)] \\ &\quad \Big(p_2 u_1(k)[p_7 + (1-p_7)u_2(k)] + p_6\Big) \end{aligned} \tag{12.3}$$

where x_T represents the product temperature, x_{CO_2} the CO_2 concentration in the bulk, $d_1 = T_{wb,ext}$ the ambient wet bulb temperature, d_2 the ambient CO_2 concentration, $u_1 = u_{mix}$ the fraction of ambient air in the air flow, $u_2 = u_{vent}$ the fraction of maximum possible internal ventilation, and p a vector containing physical and design parameters. Hence, the control input is defined by: $u = [u_1, u_2]^T$. The sample time used for this model is 1 h.

The temperature of the bulk can be measured and thus (12.2) is regularly updated. The CO_2-concentration, however, is difficult to measure in practice. The store room is, therefore, basically controlled on the product temperature. In the model-based RHOC strategy, the CO_2-concentration is calculated using (12.3) without any measurement correction. The CO_2-concentration is taken into account via constraints to avoid too high concentrations that lead to damage of the product. This approach suffices, because of the fast dynamics of the CO_2-concentration, when the room is ventilated. That is, during ventilation the CO_2-concentration quickly reaches the known external CO_2-concentration, so that large modeling errors are avoided. Consequently, right after a period of ventilation, $x_{CO_2}(k)$ is equal to $d_2(k)$.

12.2.2 *Weather Forecasts*

The medium-range weather forecasts have been provided by Weathernews Benelux. These forecasts contain hourly forecasts of weather variables up to 10 days ahead. The ensemble weather forecasts used in this paper consist of 50 ensemble members with equal chance of occurring. In what follows, the ensemble mean is used as the nominal weather forecast.

New weather forecasts and actual weather observations become available every 24 and 6 h, respectively. It has been shown in [3] that short-term weather forecasts (up to 36 h ahead) can be improved by using more frequent, local observations in combination with Kalman filtering techniques. A similar method is used here to correct the medium-range weather forecasts every 6 h.

12.2.3 Cost Function

To give some physical insight into the control objectives of the potato storage facility, the elements of the cost function are mentioned below. How the weighting factors in the cost function are chosen is beyond the scope of this paper. In our application, the following control objectives are to be fulfilled:

- The temperature of the bulk must be kept as close as possible to a pre-specified reference temperature T_{ref} (i.e. minimize $||x_T(k) - T_{ref}||$).
- The temperature must always be kept above a specified minimum temperature T_{min} (i.e. inequality constraint: $x_T(k) > T_{min}$)
- The temperature may not decrease faster than a specified limit T_Δ within 24 h (i.e. inequality constraint: $x_T(k-24) - x_T(k) < T_\Delta$).
- The weight loss of the product due to evaporation must be as small as possible (i.e. minimize $\sum_{k=0}^{H} f_1(x_T(k), u(k), d(k))$ with H the prediction horizon.
- The CO_2-concentration must always be kept below a specified maximum $CO_{2,max}$ (i.e. inequality constraint: $x_{CO_2} < CO_{2,max}$)
- The energy costs related to ventilation must be as small as possible (i.e. minimize $\sum_{k=0}^{H} f_2(u_2(k))$).

12.2.4 Receding Horizon Optimal Control

Given the model (12.2)–(12.3) with corresponding initial states, a weather forecast containing $d_1(k)$ for $k = 1, \ldots, H$ and assuming $d_2(k) = 0.0314$ to be constant, control trajectories of u_1 (fraction of ambient air in the air flow) and u_2 (fraction of maximum possible internal ventilation) can be calculated such that a cost function on the time interval $[0, H]$ is minimized according to:

$$\min_u J = \min_u \phi(x(H)) + \sum_{k=0}^{H} \mathcal{L}(x(k), u(k), E[d(k)]) \qquad (12.4)$$

where the expected value (denoted by $E[\cdot]$) is taken because the weather forecast is a stochastic variable. In our application of bulk storage in a store room there are no final costs at the end of the prediction horizon, i.e. $\phi(x(H)) = 0$. From the preceding subsection we notice that $\mathcal{L}(\cdot)$ contains a weighted combination of $||x_T(k) - T_{ref}||$, $f_1(\cdot)$ and $f_2(\cdot)$. If this open-loop control problem is solved repeatedly every l hours (with $l < H$) given the updated (or measured) states and weather forecasts the control loop is closed. As mentioned before, this type of control strategy is called receding horizon optimal control (RHOC).

In the following, the RHOC solution with the nominal weather forecast is taken as the reference point for the uncertainty evaluation. For the evaluation a period from 13 April to 2 June 2005 (almost 51 days) had be chosen. The medium-range weather forecasts have been obtained for location "De Bilt, The Netherlands". The forecast horizon of the weather forecast also determines the maximum horizon of the RHOC strategy ($H = 217$ h). In fact, one would expect a maximum horizon of $(10 \times 24 + 1)$ h. However, the medium-range forecasts are released with a delay of 1 day and thus $H = 241 - 24 = 217$ h. During the evaluation period every 6 h a new optimal control trajectory was calculated (i.e. $l = 6$) with an updated weather forecast, according to the procedure suggested in [3]. For each optimal control run the initial conditions of the states were set to the corresponding measured values. In an RHOC framework the calculated optimal inputs for the pre-specified control interval (l) of 6 h are implemented and recalculated when after 6 h new information becomes available. Consequently, in our application the optimal control input trajectories are calculated over an interval of 217 h and updated every 6 h. In total 203 optimal control trajectories, given the nominal weather forecasts, are calculated (four times a day, almost 51 days). An example of a calculated control trajectory with the accompanying predicted state and cost evolution is presented in Fig. 12.1. The reference temperature (T_{ref}) in this case was 7°C. Notice that the calculated open loop optimal control inputs are in antiphase with the external wet bulb temperature. As expected, the product temperature (T_p) follows the ventilation pattern. Notice, furthermore, that with increasing temperature differences between the product temperature and the external wet bulb temperature also the costs increase accordingly.

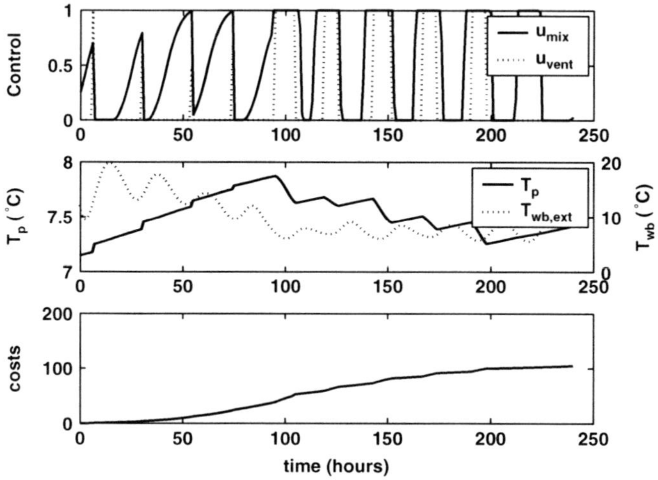

Fig. 12.1 Optimal control output starting from 1 May 2005. The sub-figures show respectively: calculated optimal controls; predicted product temperature and external wet bulb temperature; predicted cumulative costs

12.3 Weather Forecast Uncertainty and Error Analysis

In this section, we investigate several possibilities to evaluate the weather forecasts errors and uncertainties on the costs. First, the change in costs is investigated when observations of the weather are used instead of the nominal forecast. Hence, this analysis, based on 203 control trajectories with associated costs, evaluates the effect of errors in the weather forecasts on the costs. Second, the effect of individual ensemble members on the costs is investigated. Thus, this type of analysis evaluates the maximum effect of variation in the ensemble on the absolute costs.

12.3.1 Open Loop Evaluation

Let us start by evaluating the effect on the absolute costs when observed instead of nominal forecasted weather data is used. Recall that the prediction horizon is 217 h, which defines the maximum length of the evaluation. However, in what follows the summation variable r is introduced to account for intermediate changes in the costs. If now the model with fixed optimal control trajectories is ran again but with observed weather (d_1 and d_2) instead of nominal forecasted weather data, a change in costs is observed. In Fig. 12.2, 203 differences between calculated (using forecasts) and realized (using observations) running costs, as a function of the summation variable r, are presented. Herein, the running costs are defined by:

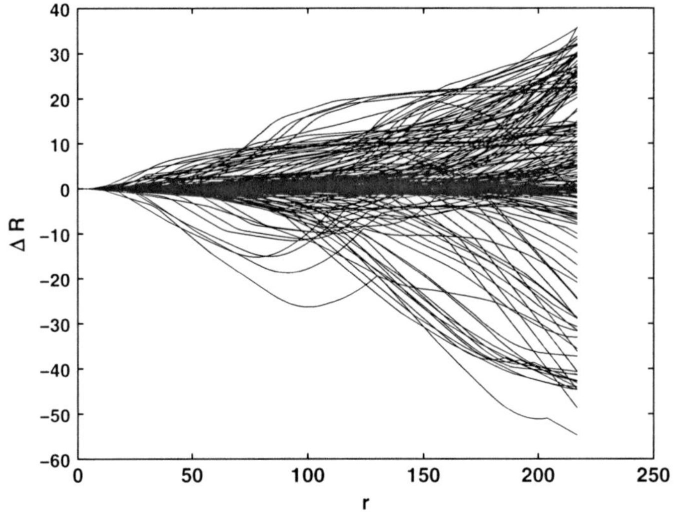

Fig. 12.2 Difference in calculated absolute costs and realized costs for each of the 203 optimal control runs

$$R(r) = \sum_{k=0}^{r} \mathcal{L}\left(x(k), u(k), d(k)\right) \quad \text{for } r = 1, \ldots, H \tag{12.5}$$

For what follows, we define $J \triangleq R(H)$.

From Fig. 12.2, it can be seen that the total cost difference, as expected, can be both positive and negative. This implies that, given the optimal control trajectory based on forecasts, the realized weather can reduce the total costs more than was expected from the forecasted weather. However, it does not mean that, if the observed weather was used in the RHOC calculations (which is only possible afterwards), the calculated optimal control trajectory generates minimal costs. Furthermore, histograms derived from Fig. 12.2 for different r show that the mode is around $\Delta R = 0$, and that the frequency distributions are rather skewed with a "thin" tail for ΔR negative.

Since the total absolute costs J change for every optimal control run, because of changing initial states and changing weather forecasts after each 6 h, the relative change in the costs, i.e. $\Delta R^{rel} = \frac{R_{obs} - R_{fct}}{J_{obs}}$, is calculated as well and presented in Fig. 12.3. In addition to the absolute or relative differences, we noticed that the absolute costs R can change dramatically (not shown here) over the evaluation period from 13 April to 2 June, because the outdoor temperature significantly increases. Especially, the cost criterion term $||x_{k,T} - T_{ref}||$ then increases. Relatively to the costs at the beginning or the end of the prediction horizon H, however, the change in costs does not seem to change that dramatically.

Using the observed weather to calculate the costs is a useful a posteriori tool, since it evaluates the positive and negative costs when controlling a system under expected disturbances. Nevertheless, it should be emphasized that the conclusions

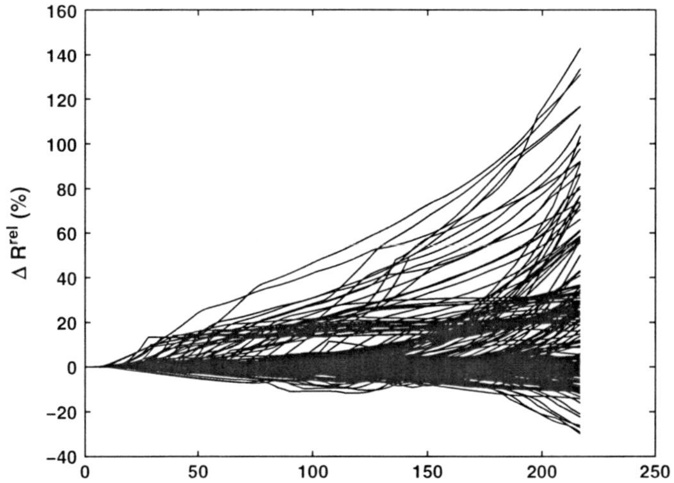

Fig. 12.3 Difference in calculated relative costs and realized costs, for $r = 1, \ldots, H$, for each of the 203 optimal control runs

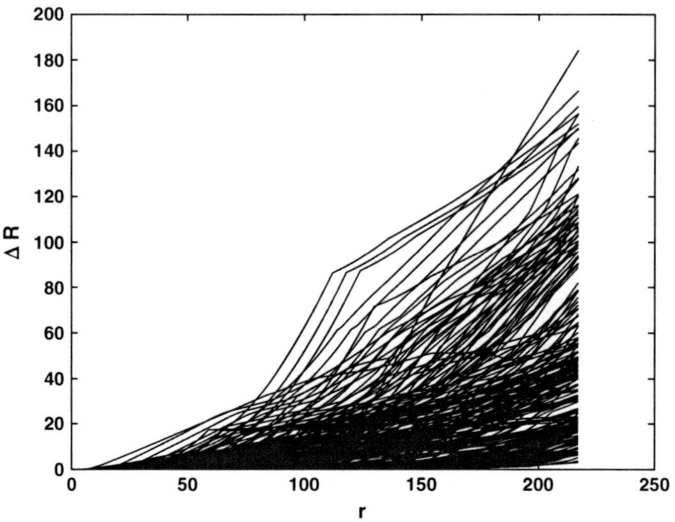

Fig. 12.4 Difference in calculated absolute costs and maximum costs based on the ensemble related to each of the 203 optimal control runs

certainly depend on the weather data and thus the type of weather. For a full analysis of the posterior behavior, other periods over a couple of years should have been evaluated, as well. However, such analysis was out of the scope of this study. To evaluate the uncertainty of the costs a priori, however, other information about the uncertainty of the weather forecast is needed. As mentioned before, the uncertainty of the weather forecast is embedded within the ensemble (see Sect. 12.2.2). By calculating the costs related to each of the ensemble members the worst case scenario can be evaluated, i.e. the ensemble member for which the costs are highest of all for $r = H$. In Fig. 12.4 the worst-case cost differences are presented, again as a function of r with $r = 1, \ldots, H$. Since only the worst case differences, related to the largest difference at $r = H = 217$, are presented, only 203 out of 50×203 runs are shown. It can be seen here that the worst-case scenarios always lead to increased costs, as expected, and these costs are considerably larger than the realized costs (Fig. 12.2). The question remains what the effect of a receding horizon control strategy on the uncertainty in costs is. This question will be answered in the next subsection.

12.3.2 Closed Loop Evaluation

In the specific RHOC implementation every 6 h, when actual weather data and updated forecasts become available, the optimal control trajectories are recalculated from (12.4) and implemented. Hence, only the control inputs for the first 6 h are

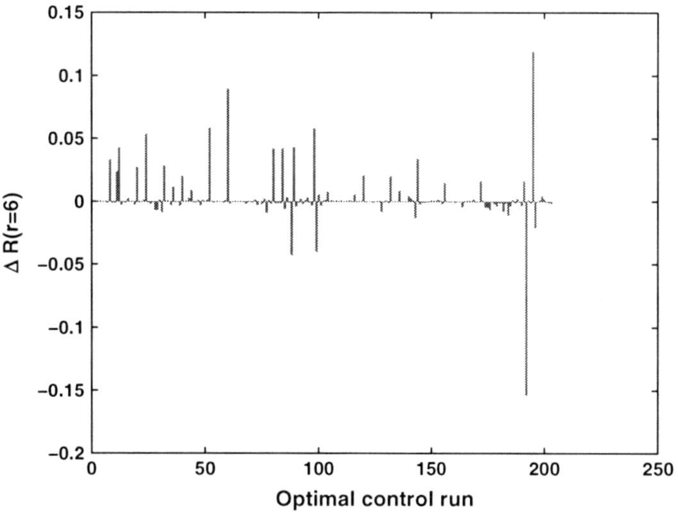

Fig. 12.5 Difference in calculated costs and realized costs after 6 h for each of the 203 optimal control runs

effectively used in the control strategy. This also implies that the uncertainties and errors in the cost function after 6 h needs to be evaluated. In Fig. 12.5 the differences between calculated costs, after 6 h because of the updates and using nominal weather, and actual costs based on observed weather are presented for each of the 203 control runs. Hence, this allows an a posteriori evaluation of the forecast errors. The small differences in costs after 6 h can be clearly seen from this figure. Notice that the small magnitude of the differences could also have been seen from Fig. 12.4. If all the costs shown in Fig. 12.5 are summed the additional costs due to errors in the nominal weather forecasts are known for the given period from 13 April to 2 June 2005.

From the a priori known weather forecast uncertainties, which can be directly obtained from the ensemble predictions, the worst-case scenario can be calculated as before. It appeared that the differences in calculated costs and maximum costs after 6 h (not shown here) based on the ensemble related to each of the 203 optimal control runs are very small, $\Delta R(r = 6) < 0.2$. However, the question remains how ΔR will evolve when the control horizon ℓ is chosen larger. In Table 12.1 the maximum (relative) deviation in the cost function, obtained from the weather forecast ensemble, is given for different control intervals (l).

Hence, for this case study it can be concluded that applying feedback every 6 h will reduce the uncertainty in the calculated costs, due to weather forecast errors and uncertainty, tremendously. It can also be seen from Table 12.1 that the difference in costs up to 24 h remains relatively small. In other words, in case of a communication failure between the optimization algorithm and the control computer of less than, e.g. 1 day no manual intervenience is required. However, it should be realized that these statements are only valid for our specific application in the pre-specified

Table 12.1 Additional costs of the realized costs and the worst-case scenarios for weather forecast uncertainty

ℓ	$\Sigma(\Delta J)/J$ (%)
6	0.37
12	1.15
18	2.05
24	3.17
48	8.0
72	14.2

evaluation period. Nevertheless, as can be seen from Figs. 12.2–12.4, the weather forecasts do not show much variation in the first 12 h. This phenomenon has been observed for other periods, as well. Thus, for $l \leq 12$, we conclude that in general the effect of the weather forecast uncertainty on the costs, as defined in (12.4), is marginal. Hence, under the assumption that CO_2 is never limiting, a control horizon of $l = 12$ instead of 6 would suffice in practice. Clearly, the weather forecast ensemble can be used for an a priori uncertainty evaluation on the costs. But we would be more flexible if we could anticipate on the forecast uncertainty. This idea will be further exploited in the next section.

12.4 Discussion

If in open loop the calculated costs based on the weather forecast and the *actual* costs, given observed weather and realized controls, are compared, it is evident that the costs can both increase and decrease. Not only the total costs are of importance but also the relative costs. During a long warm period the calculated costs will increase significantly. The absolute increase or decrease due to uncertain weather forecasts can then be quite large whereas the relative change may be reasonable. The opposite may also occur, i.e. large relative change versus a low absolute change.

Prior knowledge about the uncertainty of the forthcoming weather is very useful to study uncertainties in the near-future costs. From Figs. 12.2 and 12.4 it can be seen that, for this specific case, the (expected) cost increase is in general much larger for the a priori forecast uncertainty than for the a posteriori forecast error. Similar conclusions can be drawn in the closed loop case, see Fig. 12.5. From these figures it can be seen that the ensemble, indeed, is a useful tool to evaluate the uncertainty of the cost function a priori.

Recall that in optimal control theory a cost function, like (12.4), is minimized by adjusting the control inputs u, see, e.g. [12]. In case of the storage facility, genuine optimal control trajectories could be calculated if the (near) future weather would be exactly known in advance. However, future weather forecasts are never exactly known. A common approach, as we introduced in the previous subsections, is to define a cost function on the basis of the nominal disturbance trajectory (d^0, i.e. the nominal weather forecast), solve the optimization problem and evaluate the uncertainties in the costs. Alternatively, an approach based on implicit formulation of uncertainties in the cost function could have been chosen, such that an effective uncertainty reduction can be obtained. A classical example of this approach is minimum variance control [1].

For instance, in the case of our storage facility, minimum variance control would lead to $u = 0$, that is no ventilation of the store room, because the uncertainty in the product temperature is directly related to the weather forecast. Clearly, the product temperature uncertainty increases during ventilation. Hence, minimum variance control is not an option in our case. Thus, future work should focus on appropriate extensions of the cost function (12.4), for instance in line with [5] or [4].

12.5 Concluding Remarks

In post-harvest storage of agricultural produce, optimal control strategies can be used to anticipate on future weather conditions. In the simulation case-study with real weather forecasts and observed weather it appeared that there are only slight cost increases due to uncertainty and errors in weather forecasts if the optimal control problem is calculated every 6 h within a RHOC framework. In our application, even an increase of the control horizon to 24 h leads to a maximum increase of less then 5%.

Acknowledgements We are indebted to Weathernews Benelux for providing the weather forecasts and Agrotechnology and Food Innovations for providing the storage model and the accompanying RHOC solutions. This research is part of the EETK01120 project "Weather in Control".

References

1. Åström, K.: Introduction to Stochastic Control Theory. Academic, New York (1970)
2. Crawley, D.B., Lawrie, L.K., Winkelmann, F.C., Buhl, W.F., Huang, Y.J., Pedersen, C.O., Strand, R.K., Liesen, R.J., Fisher, D.E., Witte, M.J., Glazer, J.: Energyplus: creating a new-generation building energy simulation program. Energy Buildings **33**(4), 319–331 (2001)
3. Doeswijk, T.G., Keesman, K.J.: Adaptive weather forecasting using local meteorological information. Biosyst. Eng. **91**(4), 421–431 (2005)
4. Doeswijk, T.G., Keesman, K.J., vanStraten, G.: Impact of weather forecast uncertainty in optimal climate control of storehouses. In: K. Gottschalk (ed.) 4th IFAC Workshop on Control Applications in Post-harvest and Processing Technology, Bornimer Agrartechnische Berichte, vol. 55, pp. 46–57. Potsdam, Germany (2006)
5. Ermoliev, Y., Wets, R. (eds.): Numerical Techniques for Stochastic Optimization. Springer, New York (1988)
6. Gottschalk, K.: Mathematical modelling of the thermal behaviour of stored potatoes and developing of fuzzy control algorithms to optimise the climate in storehouses. Acta Hortic. **10**, 331–340 (1996)
7. Henze, G.P., Kalz, D.E., Felsmann, C., Knabe, G.: Impact of forecasting accuracy on predictive optimal control of active and passive building thermal storage inventory. HVAC&R Res. **10**(2), 153–178 (2004)
8. Keesman, K.J., Peters, D., Lukasse, L.J.S.: Optimal climate control of a storage facility using local weather forecasts. Control Eng. Pract. **11**(5), 505–516 (2003)

9. Lukasse, L.J.S., de Kramer-Cuppen, J., van der Voort, A.: A physical model to predict climate dynamics in ventilated bulk-storage of agricultural produce. Int. J. Refrig. **30**(1), 195–204 (2007)
10. Lukasse, L.J.S., van der Voort, A., de Kramer-Cuppen, J.: Optimal climate control to anticipate future weather and energy tariffs. In: K. Gottschalk (ed.) 4th IFAC Workshop on Control Applications in Post-harvest and Processing Technology, Bornimer Agrartechnische Berichte, vol. 55, pp. 109–122. Potsdam, Germany (2006)
11. Palmer, T.N., Barkmeijer, J., Buizza, R., Petroliagis, T.: The ECMWF ensemble prediction system. Meteor. Appl. **4**(4), 301–304 (1997)
12. Stengel, R.F.: Optimal Control and Estimation. Dover, New York (1994)
13. Tap, F.: Economics-based optimal control of greenhouse tomato crop production. Ph.D. thesis, Wageningen University, The Netherlands (2000)
14. van Mourik, S.: Modelling and control of systems with flow. Ph.D. thesis, University of Twente, The Netherlands (2008)
15. van Mourik, S., Ploegaert, J., Zwart, H., Keesman, K.: Performance analysis of a temperature controlled bulk storage room. In: Proc. 8th International IFAC Symposium on Dynamics and Control of Process Systems, vol. 3, pp. 223–228. Cancun, Mexico (2007)
16. van Mourik, S., Zwart, H., Keesman, K.: Switching control for post-harvest food storage. In: Proc. AgriControl 2007, pp. 189–194. Osijek, Croatia (2007)
17. Verdijck, G.: Product quality control. Ph.D. thesis, University of Eindhoven (2003)
18. Xu, Y., Burfoot, D.: Predicting condensation in bulks of foodstuffs. J. Food Eng. **40**, 121–127 (1999)
19. Xu, Y., Burfoot, D.: Simulating the bulk storage of foodstuffs. J. Food Eng. **39**, 23–29 (1999)

Chapter 13
Estimation of the Error in Carbon Dioxide Column Abundances Retrieved from GOSAT Data

Mitsuhiro Tomosada, Koji Kanefuji, Yukio Matsumoto, Hiroe Tsubaki, and Tatsuya Yokota

Abstract In this chapter, the estimation of error in the method used to retrieve carbon dioxide (CO_2) column abundances obtained from the Greenhouse Gases Observing Satellite (GOSAT) is presented. GOSAT will be the first satellite dedicated to primarily observe CO_2 and methane (CH_4), which are considered to be major greenhouse gases, from space. The column abundances of CO_2 and CH_4 can also be retrieved using IMG data and SCIAMACY data. The retrieved CO_2 and CH_4 column abundances can be evaluated using spectral fitting residuals. On the other hand, there are uncertainties in the retrieved column abundances due to the error in the input parameters, such as the temperature profile and the water vapor profile during spectral observation. The estimated error of the retrieved column abundances using IMG data and SCYAMACY data is not affected by the error in input parameters. In this study, the retrieval error due to input parameters using factorial experiments is described. Finally, some application results are given.

13.1 Introduction

Global environmental problems, such as global warming, desertification and ozone depletion, have recently become major concerns. The Intergovernmental Panel on Climate Change (IPCC) recently said that "it is *extremely unlikely* (<5%) that the global pattern of warming during the past half century can be explained without

M. Tomosada (✉), K. Kanefuji, and H. Tsubaki
The Institute of Statistical Mathematics, Tokyo, Japan,
e-mail: tomosada@ism.ac.jp, kanefuji@ism.ac.jp, tsubaki@ism.ac.jp

Y. Matsumoto
Association of International Research Initiatives for Environmental Studies, Tokyo, Japan,
e-mail: y-matsu@airies.or.jp

T. Yokota
National Institute for Environmental Studies, Ibaraki, Japan,
e-mail: yoko@nies.go.jp

K. Marti et al. (eds.), *Coping with Uncertainty*, Lecture Notes in Economics and Mathematical Systems 633, DOI 10.1007/978-3-642-03735-1_13,

external forcing, and *very unlikely* that it is due to known natural external causes alone. Greenhouse gas forcing has very *likely* caused most of the observed global warming over the last 50 years" [13]. Water vapor is the most important greenhouse gas, followed by carbon dioxide (CO_2). Methane (CH_4), nitrous oxide (N_2O), ozone (O_3), and several other gases present in the atmosphere in small amounts also contribute to the greenhouse effect [13]. Gases such as CO_2, CH_4, N_2O, and O_3, which are present in low concentrations in the atmosphere ($<0.03\%$ [21]) are referred to as "trace gases". Therefore, an understanding of the characteristics of CO_2 on a global scale is crucial to solving global environmental problems.

Observing trace gases that are distributed very sparsely across the Earth's surface is difficult. Despite the fact that oceans account for roughly two-thirds of the Earth's surface, there are very few observation stations located on the oceans. Thus, it is important to develop methods that allow the observation of trace gases over the Earth. Satellite remote sensing provides one of the most useful methods for global atmospheric observations despite several limitations. Firstly, satellites cannot record data if clouds obscure the sensor's view. Secondly, observation time is determined based on the local time when a satellite's orbit is sun-synchronous polar orbit. Thirdly, solar radiance and Earth irradiance at specific wavelengths change due to the effects of trace gases. Since the trace gas contribution to the spectrum is so small, it is extremely difficult to precisely detect the physical quantity of a trace gas using satellite remote sensing from the measured spectrum. Furthermore, the spectral difference occurs at very narrow wavelengths of the specific solar radiance bands. Therefore, more advanced techniques are required to obtain spectral data with high resolution and high sensitivity.

The Interferometric Monitor for Greenhouse gases (IMG) aboard the Advanced Earth Observing Satellite (ADEOS), launched in August 1996, was the first high resolution nadir infrared instrument allowing the simultaneous measurement from space of a series of tropospheric trace gases: H_2O, CO_2, N_2O, CH_4, CFCs, O_3, and CO_3 [1, 7]. IMG was designed to measure the terrestrial thermal radiation in an infrared spectrum at a high resolution. The Scanning Imaging Absorption Spectrometer for Atmosphere Cartography (SCIAMACY) onboard the European Environmental Satellite (ENVISAT) observes the atmosphere with moderate spectral resolution. An algorithm is currently being developed and refined primarily for the retrieval of CH_4, CO, CO_2, H_2O, N_2O, and O_2 columns using SCIAMACY data [5]. In Japan, the Greenhouse Gases Observing Satellite (GOSAT) will be launched in early 2009. In the US, the Orbiting Carbon Observatory (OCO) satellite will also be launched in late 2008. GOSAT and OCO will also perform passive nadir observations in the near-infrared spectral region, but they will be optimized for CO_2 observations. It is expected that GOSAT and OCO will improve knowledge of CO_2 surface fluxes. In the GOSAT project, a global map of CO_2 and CH_4 column abundances will be distributed to users. Given the scope of these aims, it is important to clarify the precision of the GOSAT's CO_2 column data. I. Aben quantifies the effect of aerosols and thin cirrus clouds in the atmosphere, which is the largest error source for CO_2 measurements from space. This study presents a whole evaluation method. Two types of error for the retrieved column abundance data are considered. Errors derived from the spectral fitting residuals (internal errors) are obtained analytically. For the evaluation of the retrieved column abundance derived from the IMG data and the SCIAMACY data, the error is primarily internal error. In this study, errors derived from input parameters (external errors), which are used in the retrieval stage, are analyzed by examining factorial experiments, which are mainly used for quality control.

The structure of this paper is as follows. In Sect. 13.2, trace gas observations from satellite sensors are introduced, and the role of GOSAT in trace gas measurement is presented. The GOSAT mission is then described, and the previous error analysis by satellite remote sensing is presented. In Sect. 13.3, an evaluation method of GOSAT's CO_2 column abundance data is described in detail. Certain measurement geometry and atmospheric conditions are set, and the retrieval precision results are presented as one example. The results presented in this paper are not officially used but are referred to as the basic concepts in the GOSAT project, since the retrieval algorithm of the operational data processing is currently in the coding and tuning stage. The practical retrieval precision shall be made public at a future date. Section 13.4 concludes this paper.

13.2 Trace Gas Measurement by Satellite Remote Sensing

Trace gas measurement using satellite remote sensing is presented in Sect. 13.2.1, and the role of GOSAT in trace gas measurement from space is explained in Sect. 13.2.2. Lastly, the current evaluation methods of error in retrieved trace gas amounts are presented in Sect. 13.2.3.

13.2.1 Observations of Trace Gases with Various Sensors

Passive sensors measure solar radiance directly from the sun or reflected from the Earth. Both methods detect physical quantities of trace gases based on the differences in the spectrum observed from the trace gases in the atmosphere. Solar occultation is a technique that measures direct sunlight using limb observation of the Earth from the satellite. Observing the spectrum reflected from the Earth's surface is called "down-looking observation". It is necessary to use sensors with a high signal-to-noise ratio (*S*/*N*) and high spectral resolution. Advances in sensor technology combined with diffraction grating have allowed a Fourier Transform Spectrometer (FTS) system to detect the physical quantities of trace gases. For example, solar occultation sensors include the Improved Limb Atmospheric Spectrometer (ILAS) onboard the ADEOS satellite and the Stratospheric Aerosol and Gas Experiment (SAGE) and Halogen Occultation Experiment (HALOE) onboard the Upper Atmosphere Research Satellite (UARS). Solar occultation observations detect trace gases in the stratosphere at or above a height of 10 km. Below an altitude of 10 km, trace gas values are not observed precisely, and the horizontal resolution is quite low. Moreover, since concentrations of CO_2 in the stratosphere are lower than in the troposphere, it is difficult to ascertain global CO_2 distributions using solar occultation technology.

Down-looking observations of trace gases are retrieved by both FTS and grating sensors. The Infrared Interferometer Spectrometer (IRIS) onboard the NIMBUS-3 satellite launched at 1969 was the first FTS type sensor. The IRIS instrumentation was a modified version of a classical Michelson Interferometer, and measures radiance in narrow ($5\,cm^{-1}$) spectral intervals from 400 to $2{,}000\,cm^{-1}$ (5–25 μm) [9]. Two portions of the spectrum are used: 11 intervals each within the CO_2 band at $667\,cm^{-1}$ and the atmospheric window at $900\,cm^{-1}$. However, when NIMBUS-3 was launched, IRIS spectral resolution was lower than the width of trace gas absorption. Thus, it was considered that nadir observation of the tropospheric constituents was difficult, and that observation of greenhouse gases was not a significant objective.

ENVISATwith SCIAMACHYwas launched in March 2002 in a sun-synchronous polar orbit. SCIAMACHY is outfitted with a grating spectrometer for measuring solar radiation reflected off the surface of the Earth, backscattered from the atmosphere, transmitted through the atmosphere, or emitted from the atmosphere in the ultraviolet, visible, and NIR spectral regions (240–1,750 nm, 1,940–2,040 nm, 2,265–2,380 nm) at moderate spectral resolutions (0.2–1.4 nm) [6]. CH_4, CO, CO_2, N_2O, and O_2 are retrieved from the SCIAMACHY near-infrared and visible spectra [5]. The standard deviation of the dry air column averaged mixing ratio XCO_2 within 1% latitude bands is ± 10 ppmv or $\pm 2.7\%$ for measurements over land. The spatial (horizontal) resolution of SCIAMACHY nadir measurements depends on the orbital position and the spectral interval. The ground pixel (footprint) size is approximately identical at each orbital position. For example, the footprint is $60 \times 30\,km^2$ for an integration time of 0.25 s, and $120 \times 30\,km^2$ for 0.5 s.

The IMG sensor is a Fourier Transform type infrared spectrometer (FTIR) that was developed to measure greenhouse gases. The ADEOS satellite was in a near-polar sun-synchronous orbit, a 41-day recurrent polar orbit. IMG was developed for measuring greenhouse gases in the atmosphere, particularly in the troposphere, and was designed to measure the terrestrial thermal radiation in the infrared spectral region of $640–3{,}000\,cm^{-1}$ at a high resolution of $0.1\,cm^{-1}$ after reaching apodosis [12]. The footprint size is 8 km × 8 km, and the main atmospheric constituents, H_2O, CO_2, O_3, N_2O, CO, CH_4, and HNO_3 were retrieved. Error bars for the retrieved CO_2 at 1,000, 500, and 200 hPa, which are recognized as the 1-standard deviation (1-sigma) for analysis, are 6.03%, 3.93%, and 3.09%, respectively [14].

13.2.2 GOSAT Mission

The GOSAT project is a joint project of the Japan Aerospace Exploration Agency (JAXA), the Ministry of Environment (MOE), and the National Institute for Environmental Studies (NIES) [8,22]. JAXA is responsible for the satellite, its instrument

Table 13.1 Overview of GOSAT and orbit

Parameter	Value
(a) Body	
Size	Main body: 3.7 m (hight) × 1.8 m (width) × 2.0 m (depth)
@@	Include solar paddle: 13.7 m
Weight	1,750 kg
Electric power	3.5 kW
Life span	5 years
(b) Orbit	
Orbit	Sun-synchronous polar orbit
Altitude	666 km
Inclination	98.05°
Period of revolution	14.66 orbits/day
Equatorial crossing time	Nominally 1 p.m. (±15 min) local time (descending node)
Repear coverage	3 days

development, and its operation, while MOE is involved in instrument development, and NIES is responsible for satellite data retrieval and data distribution. GOSAT will be launched in early 2009. An overview of the GOSAT satellite and the orbit is presented in Table 13.1.

Higher retrieval precisions are expected when the gas amount information is retrieved from the optimal wavelength band for the gas. The GOSAT mission will make the first global, space-based measurements of CO_2 with the precision, resolution, and coverage needed to characterize CO_2 sources and sinks at the regional scale. GOSAT carries two sensors: TANSO-FTS (Thermal And Near infrared Sensor for carbon Observation – Fourier Transform Spectrometer) that obtains solar radiance reflected from the ground surface; and TANSO-CAI (Thermal And Near infrared Sensor for carbon Observation – Cloud and Aerosol Imager) that obtains cloud and aerosol information of observed areas of TANSO-FTS. TANSO-FTS, which detects the gas absorption spectra of the solar short wavelength infrared (SWIR) reflected on the Earth's surface, as well as of the thermal infrared (TIR) radiated from the ground and the atmosphere. TANSO-FTS-SWIR data is obtained for three spectral bands, 0.76, 1.6, and 2 μm band with a 0.2 cm^{-1} resolution. Each band spectrum is shown in Fig. 13.1. Polarization data for the two axes are also obtained for each spectral band. The CO_2 absorption bands near 1.6 and 2.0 μm are quite important because they provide a significant amount of information near the Earth's surface where changes in CO_2 concentrations are most apparent. The 0.76 μm band is also important to detect ground surface pressure and cirrus cloud information. If satellite observations are to improve over the existing ground network, monthly averaged column data at a precision of 1% or better, for an $8° \times 10°$ footprint are needed [2]. The GOSAT mission was designed to observe CO_2 density with 1% relative precision averaged in a certain period and with 10 km × 10 km footprint size during the first commitment period of the Kyoto Protocol (2008–2012).

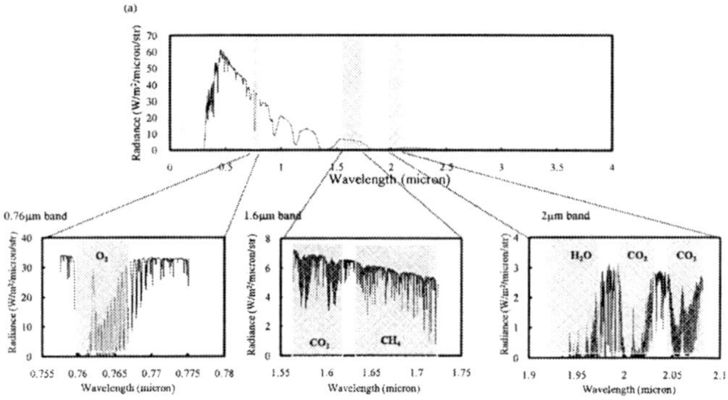

Fig. 13.1 Three-spectral bands of the GOSAT TANSO-FTS-SWIR (**b**) and solar spectral radiance (**a**)

NIES developed a GOSAT Data Handling Facility (DHF) for processing GOSAT data in routine operation. After data reception and Level 1 processing by JAXA, GOSAT data will be transferred to the DHF. TANSO-FTS data will provide information for spectral analysis, while TANSO-CAI data will be used to generate cloud and aerosol information. Later, these data will be combined to calculate CO_2 and CH_4 column abundances at observation points with no or only thin clouds and aerosol layers. Furthermore, an atmospheric transport model will be used with the obtained distribution of CO_2 column abundances to estimate global distributions of CO_2 fluxes, as well as to generate three-dimensional distributions of CO_2 concentrations. Retrieved CO_2 column abundances will be distributed to users. It is important for the users to know the precision of the retrieved column abundances. Therefore, it is necessary to clarify the precision of the retrieved CO_2 column abundance.

13.2.3 Previous Error Analysis

In this section, error analysis of the column abundances derived from SCIAMACHY data and IMG data is introduced.

SCIAMACHY [4, 5]

An algorithm to retrieve trace gas vertical columns is first introduced. The Weighting Function Modified Differential Optical Absorption Spectroscopy (WFM-DOAS) is based on fitting the logarithm of a linearized radiative transfer model I_i^{mod} plus a low-order polynomial P_i to the logarithm of the observed sun-normalized radiance I_i^{obs}, which is the ratio of the observed nadir radiance and the solar irradiance spectrum. Index i refers to the detector pixel number i. Provided there exists an appropriate spectra fitting window, I_i^{obs}, depends on the true but unknown vertical columns of the trace gases of interest (components of vector $\mathbf{V}^T$). The WFM-DOAS equation can be written as follows:

$$\ln I_i^{obs}(\mathbf{V}^t) - \ln I_i^{mod}(\bar{\mathbf{V}}) + \sum_{j=1}^{J} \frac{\partial \ln I_i^{mod}}{\partial V_j} |_{\mathbf{V}_j} \times (\hat{\mathbf{V}}_j - \bar{\mathbf{V}}_j) + P_i(a_m)$$
$$\equiv \| RES_i \|^2 \rightarrow min. \qquad (13.1)$$

The fit parameters are the desired trace gas vertical columns $\hat{V}^j$ and the polynomial coefficients a_m. The errors of the retrieved columns have been calculated as follows

$$\sigma_{\hat{V}_j} = \sqrt{(C_x)_{jj} \times \sum_i RES_i^2/(nv - m)}, \qquad (13.2)$$

where $(C_x)_{jj}$ is the j-th diagonal element of the covariance matrix, nv is the number of spectral points in the fitting window, and m is the number of linear fit parameters.

The errors derived from the error of input parameters are examined. A preliminary version of the WFM-DOAS algorithm was implemented based on a fast look-up table approach. The fast look-up table approach introduces quantifiable errors [5], which are related to the (rather sparse) grid selected for the reference spectra look-up table: solar zenith angle interpolation, scan angle correction, and surface elevation (pressure). In addition, factors affecting the calculated model spectrum are surface albedo, surface pressure, aerosols, vertical CO_2 profiles, water vapor, and temperature. When varied alone, each of these parameters creates an error in the column of typically less than 2% [2]. Although the effect of each parameter on the retrieved column abundance is evaluated, the total error on the retrieved column abundance, which includes the error source written above, is not evaluated. Furthermore, a modified retrieval algorithm called the (FSI)-WFM-DOAS was developed. The (FSI)-WFM-DOAS algorithm generates a reference spectrum for every single SCIAMACHY measurement to obtain the best linearization point for retrieval. Analysis of the (FSI)-WFM-DOAS retrievals with respect to the ground-based, FTIR instrumentation (located at Egbert, Canada) reveals that the overall bias of the CO_2 columns retrieved by the FSI algorithm is <4.0% with a monthly precision of 1 × 1 gridded data close to 1.0% [3].

IMG [14]

The retrieval algorithm uses the maximum likelihood method to take into account the statistical prior knowledge into the least squares method. An iterative calculation is used, so that the estimator $\hat{\mathbf{X}}$ will converge to the global minimum point of the error curve.

$$\mathbf{e}_y^T \mathbf{S}_\epsilon^{-1} \mathbf{e}_y + \mathbf{e}_x^T \mathbf{S}_a^{-1} \mathbf{e}_x, \qquad (13.3)$$

$$\mathbf{e}_y = (\mathbf{y} - \mathbf{y}_0) - \mathbf{K}(\mathbf{X} - \mathbf{X}_0), \qquad (13.4)$$

$$\mathbf{e}_x = (\mathbf{X} - \mathbf{X}_0), \qquad (13.5)$$

where $\mathbf{y}$, which has as many rows as the number nv channel, is a vector of observations (radiance); $\mathbf{X}_0$ obtained from the analysis of climatological data is used as an initial guess for $\hat{\mathbf{X}}$; $\mathbf{y}_0$ is a vector of radiance associated with $\mathbf{X}_0$; $\mathbf{K}$ is the Jacobian matrix for the nv channels and m estimation parameters; $\mathbf{S}_\epsilon$ is the error covariance matrix and includes the observed noise and forward model error; and $\mathbf{S}_a$ is the covariance matrix of $\mathbf{X}_0$. In an iterative method, the k-th estimator $\hat{\mathbf{X}}_k$ is used as the linearization point for the next iteration. $\hat{\mathbf{X}}_{k+1}$ is calculated as follows:

$$\hat{\mathbf{X}}_{k+1} - \hat{\mathbf{X}}_k = \mathbf{C}_k(\mathbf{y} - \mathbf{y}_k) + (\mathbf{I} - \mathbf{C}_k\mathbf{K}_k)(\mathbf{X}_0 - \hat{\mathbf{X}}_k) \tag{13.6}$$

$$\mathbf{C}_k = (\mathbf{K}_k^T\mathbf{S}_{\epsilon,k}^{-1}\mathbf{K}_k + \mathbf{S}_a^{-1})^{-1}\mathbf{K}_k^T\mathbf{S}_{\epsilon,k}^{-1}, \tag{13.7}$$

where $\mathbf{C}_k$, $\mathbf{K}_k$, and $\mathbf{S}_{\epsilon,k}$ are calculated using $\hat{\mathbf{X}}_k$. Iteration can reduce the nonlinearity effects, and the final estimation is obtained when sufficient convergence is achieved. The covariance of estimation errors $\mathbf{S}_x$ is written as

$$\mathbf{S}_x = (\mathbf{K}^T\mathbf{S}_\epsilon^{-1}\mathbf{K} + \mathbf{S}_a^{-1})^{-1}. \tag{13.8}$$

It should be noted that the evaluation of the retrieved column abundances using SCIAMACHY data and IMG data are performed above. However, input parameters may suffer when column abundances are retrieved. For example, the temperature profile and water vapor amount profile in the atmosphere at the time of spectrum observation affect the intensity of irradiance that is noted by the sensor. Therefore, the temperature profile and water vapor amount should be known precisely in order to retrieve CO_2 column abundances. However, it is difficult to know these input parameters. Therefore, it is better to consider that the retrieved column abundances have an error due to the error of the input parameters. Thus, the precision of the retrieved column abundances is included in this study.

13.3 Error Evaluation and Results

Section 13.3.1 introduces the retrieval algorithm used for this study, which is based on the Rodgers method [20], while Sect. 13.3.2 provides a discussion of the error evaluation method for this retrieval algorithm. Finally, Sect. 13.3.3 provides an example of the observation scenario, such as measurement geometry and the results of error evaluation. It should be noted that this retrieval algorithm may be modified in the future.

13.3.1 Retrieval Method

CO_2 and CH_4 column abundances are retrieved separately. The retrieval method of CH_4 column abundances will be different from that of CO_2 column abundances in terms of the unknown and input parameters. This section focuses on the retrieval

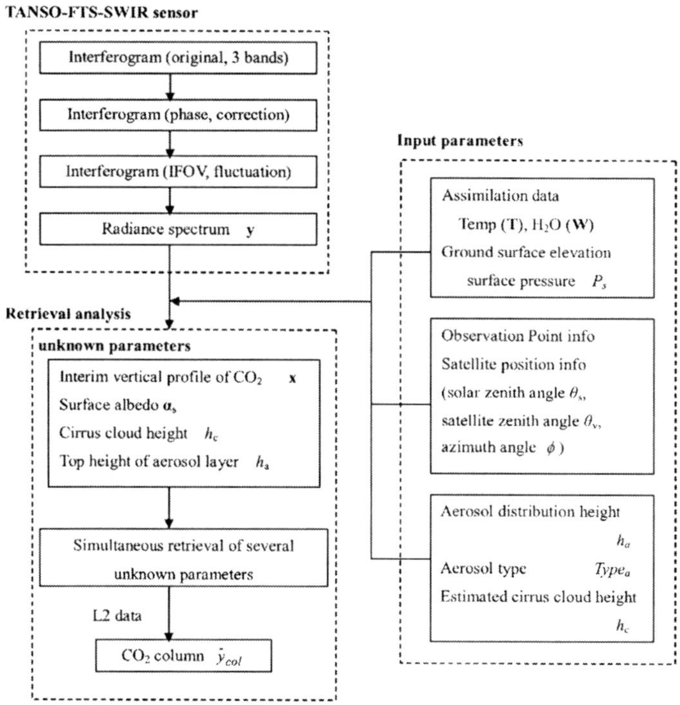

Fig. 13.2 Flow until a CO_2 column abundances is retrieved

of CO_2 column abundance. Figure 13.2 depicts the retrieval test flow until the CO_2 column abundance is retrieved. In this study, the unknown parameters are CO_2 density for each layer **x**, albedo of the observation region α_s, cirrus cloud optical depth at 0.55 μm τ_c, and aerosol optical depth at 0.55 μm τ_a. Clouds were assumed to be thin cirrus. Since albedo varies according to wave number, albedos were estimated at 5 wave number points in the band at even intervals used in the retrieval.

Input parameters are as follows. The radiance spectrum obtained by TANSO-FTS is denoted as **y**. Input parameters related to measurement geometry are the solar zenith angle θ_s, the satellite's zenith angle θ_v, and azimuth angle between the sun and satellite ϕ. Cloud height h_c is a parameter that provides information about cloud conditions. Parameters related to atmospheric conditions are surface pressure P_s, temperature profile **T**, and water vapor amount **W**. Aerosol parameters are aerosol height distribution (h_a) and aerosol type ($Type_a$). In practice, representing aerosols is difficult due to the different optical characteristics derived from varying aerosol compositions. Therefore, aerosols are expressed as typical rural and urban aerosols. The relationship between unknown and input parameters is represented by introducing the following function f with error $\mathbf{e}_X$

$$(\mathbf{x}\alpha_s\tau_c\tau_a)^T = f(\mathbf{y}, \theta_s, \theta_v, \phi, h_c, P_s, \mathbf{T}, h_a, \mathbf{W}, Type_a) + \mathbf{e}_X. \quad (13.9)$$

The quantities to be retrieved can be represented by the state vector $\mathbf{X}(=(\mathbf{x}\ \alpha_s\ \tau_c\ \tau_a)^T)$. $\mathbf{y}$ is approximated by some forward model $F(\mathbf{X})$. The relationship between the radiance spectrum $\mathbf{y}$ and the state vector $\mathbf{X}$ is written with error covariance $\mathbf{e}_y$

$$\mathbf{y} = F(\mathbf{X}) + \mathbf{e}_y. \tag{13.10}$$

A linear problem is one for which the forward model is linear: $\mathbf{y}=\mathbf{K}\mathbf{X}$ where $\mathbf{K}=\partial F(\mathbf{X})/\partial \mathbf{X}$. Gaussian statistics usually provide a good approximation for the errors in real measurement. The maximum posteriori approach is considered for this nonlinear problem. The Bayesian solution for the linear problem can be easily modified for an inverse problem in which the forward model is a general function of the state vector $\mathbf{X}$. The measurement error $\mathbf{e}_y$ in (13.10) is assumed to be Gaussian, and there is a prior estimate with a Gaussian error

$$-2\ln P(\mathbf{X}|\mathbf{y}) = [\mathbf{y} - F(\mathbf{X})]^T \mathbf{S}_\epsilon^{-1} [\mathbf{y} - F(\mathbf{X})] + [\mathbf{X} - \mathbf{X}_0]^T \mathbf{S}_a^{-1} [\mathbf{X} - \mathbf{X}_0] + c, \tag{13.11}$$

where $P(q)$ is expressed as a probability distribution function *pdf* of q, and c is constant. As in the linear case, the task in the nonlinear case is to find a best estimate $\hat{\mathbf{X}}$ and an error characteristic that describes *pdf* well enough for practical purposes. The Gauss–Newton method gives

$$\mathbf{X}_{k+1} = \mathbf{X}_k + (\mathbf{S}_a^{-1} \mathbf{K}_k^T \mathbf{S}_\epsilon^{-1} \mathbf{K}_k)^{-1} \{\mathbf{K}_k^T \mathbf{S}_\epsilon^{-1} [\mathbf{y} - F(\mathbf{X}_k)] - \mathbf{S}_a^{-1} [\mathbf{X}_k - \mathbf{X}_0]\}, \tag{13.12}$$

where k is the iteration number, $\mathbf{X}_0$ is a prior of the state vector, $\mathbf{S}_a$ is the associated covariance matrix of $\mathbf{X}_0$, and $\mathbf{S}_\epsilon$ is the error covariance.

nl refers to the number of layers and vector $\hat{\mathbf{x}}$ $(=\hat{\mathrm{x}}_1\hat{\mathrm{x}}_2 \ldots \hat{\mathrm{x}}_{nl})^T)$ refers to the retrieved CO_2 density for all layers. CO_2 column abundances $\hat{y}_{col}$ is obtained from $\hat{\mathbf{x}}$ by [10].

$$\hat{y}_{col} = \int_0^{z_{top}} \hat{x} n_a dz, \tag{13.13}$$

where z is referred to the altitude, z_{top} is the altitude at top layer, and n_a is the number of density of air.

13.3.2 Error Evaluation

Since the CO_2 column abundance evaluation method is the same as that for CH_4 column abundance, the term "column abundance" in this section indicates both CO_2 and CH_4 column abundance. The variance of the retrieved column abundances, σ^2, is estimated, and the error bar is assumed to be provided by 1-sigma analysis, $\sigma(=\sqrt{\sigma^2})$. Internal and external errors are also considered. Internal error is calculated from the residuals for spectral fitting using the Gauss–Newton approach in (13.12). σ_{int}^2 denotes the variance of retrieved column abundances due to internal error. External error, which is the error of the retrieved column abundances due to the errors in the input parameters, is written as σ_{ext}^2. The variance of total error σ^2 is defined as the sum of σ_{int}^2 and σ_{ext}^2. σ^2 is given by

$$\sigma^2 = \sigma_{int}^2 + \sigma_{ext}^2. \quad (13.14)$$

Internal and external errors are calculated in the following section.

Internal Error

Column abundances $\hat{y}_{col}$ are obtained numerically using the trapezoid rule to integrate (13.13) as

$$\hat{y}_{col}(\hat{\mathbf{x}}) = \mathbf{h}^T\hat{\mathbf{x}}, \quad (13.15)$$

where

$$\mathbf{h} = \begin{pmatrix} 1/2n_{a,1}\Delta z_1 \\ 1/2n_{a,2}(\Delta z_1 + \Delta z_2) \\ \vdots \\ 1/2n_{a,nl-1}\Delta z_{nl-1} \end{pmatrix},$$

Δz_l expresses $z(l+1)-z(l)$ for a certain layer l. The variance of the random variable q is written as $Var[q]$, and the variance of $\hat{y}_{col}(\hat{\mathbf{x}})$ is given by (13.15) as

$$Var[\hat{y}_{col}(\hat{\mathbf{x}})] = Var[\mathbf{h}^T\hat{\mathbf{x}}] = \mathbf{h}^T Var[\hat{\mathbf{x}}]\mathbf{h} = \sigma_{int}^2, \quad (13.16)$$

where $Var[\hat{\mathbf{x}}]$ is given as

$$Var(\hat{\mathbf{x}}) = (\mathbf{K}^T\Sigma^{-1}\mathbf{K} + \mathbf{S}_a^{-1})^{-1}. \quad (13.17)$$

It is assumed that the elements of $\mathbf{y}$ are independent of each other, and Σ is given by $\Sigma = \psi\mathbf{S}_\epsilon$. ψ is written as

$$\psi = \frac{(\mathbf{y} - F(\hat{\mathbf{X}}))\mathbf{S}_\epsilon^{-1}(\mathbf{y} - F(\hat{\mathbf{X}}))}{nv - m}, \quad (13.18)$$

where nv is the number of channels in $\mathbf{y}$ when $\mathbf{X}$ is retrieved, and m is the number of unknown parameters. Consequently, σ_{int}^2 can be obtained by

$$\sigma_{int}^2 = \mathbf{h}^T(\frac{\mathbf{K}^T\mathbf{S}_\epsilon^{-1}\mathbf{K}}{\psi} + \mathbf{S}_a^{-1})^{-1}\mathbf{h}. \quad (13.19)$$

External Error

Next, the effects on retrieved CO_2 column abundances derived from the errors of input parameters are considered. External error was not reflected in the error evaluations of the retrieved CO_2 column abundances derived from SCIAMACHY and IMG. The levels of all factors in the experiment represent only a random subset of the possible factor levels of interest [15]. Randomization is a procedure whereby factor-level combinations are assigned to a test sequence in such a way that every factor-level combination has an equal chance of being assigned to any experiment unit or position in the test sequence. In a completely randomized design, all the factor-level combinations in the experiment are randomly assigned to experimental units. However, complete factorial experiments cannot always be conducted due to economic, time, and other constraints. This paper makes use of an orthogonal array to obtain fractional factorial experiments, since using orthogonal arrays is a widely-accepted method used in industry, with σ_{ext}^2 obtained from factorial experiments using an orthogonal array. The model terms that represent random factor levels are themselves random variables. When error factors are denoted by $A, B, \ldots,$ the model can be expressed as follows:

$$\hat{y}_{col} = \mu(A, B, \ldots) + e_{col} = \mu + \alpha + \beta + \cdots + e_{col}, \tag{13.20}$$

where e_{col} ($\sim N(0, \sigma_e^2)$) represents random error effects, α, β, ... represent the contribution to $\hat{y}_{col}$ derived from factors A, B, ..., respectively, and "$\sim N(0,\sigma_e^2)$" indicates that the function has a normal distribution with mean 0 and variance σ_e^2. All of the random variables for A, B, ... are assumed to be statistically independent. When the factor effects are all random, the response $\hat{y}_{col}$ has a common normal distribution with mean $\mu(A,B,\ldots)$ and variance σ_{ext}^2

$$\hat{y}_{col} \sim N(\mu(A, B, \ldots), \sigma_{ext}^2). \tag{13.21}$$

σ_{ext}^2 is obtained from the experimental results based on an orthogonal array. The number of tests is represented by N, and σ_{ext}^2 is obtained by

$$\sigma_{ext}^2 = \frac{1}{N} \sum_{n=1}^{N} (\hat{y}_{col,n} - \bar{y}_{col})^2, \tag{13.22}$$

where $\bar{y}_{col}$ is the mean of retrieved column abundances from N tests. That is,

$$\bar{y}_{col} = \frac{1}{N} \sum_{n=1}^{N} \hat{y}_{col,n}. \tag{13.23}$$

13.3.3 Error Evaluation Results

Setting Parameters

Table 13.2 shows the measurement geometry and atmospheric conditions. The atmosphere is represented using the US Standard model. It was assumed that there are

Table 13.2 Measurement geometry and atmospheric condition during observation

	Parameter	Value
(a) Observation geometry	Solar zenith angle	30.0°
	Satellite zenith angle	0° (nadir observation)
	Atmospheric model	US Standard model
	CO_2 density	380 ppmv for all layers
(b) Atmospheric condition	Cirrus height	10 km
	Aerosol type	Rural type aerosol
	Aerosol optical depth	0.1

thin cirrus clouds with 0.2 optical depth at 0.55 μm and at a middle height of 10 km, and that aerosols are distributed uniformly from the Earth's surface to an altitude of 3 km with 0.1 optical depth at 0.55 μm. CO_2 density is low at the higher altitude, and it is difficult to retrieve CO_2 density at such altitudes precisely. Therefore, the number of layers nl is set to 15 from the Earth's surface to 15 km at 1-km intervals. It is assumed that the observation area on the Earth's surface is covered with conifer forests. These parameters are given as true values of the estimation parameters when observation spectra are calculated by a radiative transfer code.

CO_2 column abundances are retrieved from the 1.6 μm band only to reduce the amount of time required. The 1.6 μm GOSAT spectrum $\mathbf{y}$ is calculated in the following manner. Solar incident radiance every 0.01 cm^{-1} from 6,180 to 6,380 cm^{-1} is calculated using the HSTAR (High resolution RSTAR [17, 18]) code, which calculates the spectral radiance of a spectral band using the line-by-line method according to cloud and aerosol conditions. Polarization bands are not used here, and solar radiance multispectral scattering in the atmosphere is taken into consideration. An optical filter function of the 1.6 μm band of the sensor is applied to the incident radiance data. Interfered radiance, called an interferogram, is obtained from the FTS by calculating an inverse Fourier transformation. The spectrum is obtained every 0.2 cm^{-1} using a Fourier transformation. Then, sensor noise corresponding to $S/N = 300$ is added to the obtained spectrum, since the sensor efficiency is $S/N = 300$. It is assumed that the sensor characteristic is known based on an examination before the satellite is launched. Essentially, this yields a maximum spectrum y_{max} and standard deviation of noise σ_ϵ obtained by $y_{max}/300$. Noise is assumed to follow a normal distribution with mean value 0 and variance σ_ϵ^2, and is calculated for every 0.2 cm^{-1} of the spectrum.

$\mathbf{S}_a$ and $\mathbf{S}_\epsilon$ in (13.12) is set as follows. Knowing $\mathbf{S}_a$ correctly is difficult in practice. In the case where the elements of $\mathbf{S}_a$ are small, retrieval precision greatly depends on $\mathbf{S}_a$. Therefore, this retrieval algorithm uses a larger $\mathbf{S}_a$ than that which is normally practical. $\mathbf{S}_a$ variance terms are shown in Table 13.3. Covariance terms of $\mathbf{S}_a$ are set to 0, and it is assumed that $\mathbf{S}_\epsilon$ is derived only from sensor efficiency. Then, all $\mathbf{S}_\epsilon$ variance terms are set to σ_ϵ^2, and $\mathbf{S}_\epsilon$ covariance terms are set to 0.

The number of factors in the experiment to obtain σ_{ext}^2 is set as follows. The nine input parameters of (13.9) are candidate factors in the experiment. The direction of the sun is calculated using either an "Astronomical Almanac [19]" or "Japanese

Table 13.3 $\mathbf{S}_a$ variance terms

				$\mathbf{x}$	(ppmv^2)			
Layer	1	2	3	4	5	6	7	8
Value	97.31^2	86.79^2	77.41^2	69.04^2	61.57^2	54.92^2	48.98^2	43.68^2
Layer	9	10	11	12	13	14	15	
Value	38.96^2	34.75^2	30.99^2	27.64^2	24.65^2	21.99^2	19.61^2	

			α_s			τ_c	τ_a
No.	1	2	3	4	5	1	1
Value	1^2	1^2	1^2	1^2	1^2	0.2^2	0.2^2

Ephemeris [11]" with high precision. Satellite positions are also obtained quite accurately from orbit information acquired by a global positioning system. Consequently, errors of θ_s, θ_v, and ϕ are not considered in this study, leaving six input parameters, which, except for θ_s, θ_v, and ϕ, are used as error factors. Table 13.4 shows the levels for each set error factor. Levels for each error factor are set as follows. Levels for each factor are set to the accuracies of each input parameter. Input parameters of temperature profile $\mathbf{T}$, water vapor profile $\mathbf{W}$, and surface pressure P_s are used as provided by six hourly (1.25° × 1.25°) Global Analysis data (GANAL) from the Japan Meteorological Agency. It is assumed that the errors of GANAL data for $\mathbf{T}$ and $\mathbf{W}$ are fewer than 2 K and 30% of $\mathbf{W}$, respectively. P_s precision is assumed to be 3 hPa, and P_s error is also assumed to be the error of the estimated surface elevation of the observation area. "Temperature profile shift" levels are set to 0 K (level 1), −2 K (level 2), and +2 K (level 3). CO_2 column abundance retrievals for each level are conducted by adding levels to the temperature profile of the GANAL data for all layers. "Water vapor amount shift" levels are set to 0% (level 1), −30% (level 2), and +30% (level 3) of the GANAL data. Retrievals for each level are carried out by adding levels to the water vapor amount of the GANAL data similar to the method for the "temperature profile shift". Atmospheric pressures for each layer from the surface to the top of the atmosphere are changed according to hydrostatic equilibrium using information on temperature shift and water vapor shift for each level. Furthermore, the shifts of atmospheric pressure at the surface are set to 0 hPa (level 1), −3 hPa (level 2), and +3 hPa (level 3). Cases where aerosol height distributions differ from the practical distribution when the spectrum was obtained were also considered with "aerosol height distribution" levels uniformly set to 3 km from the surface (level 1), 2 km from the surface (level 2), and 4 km from the surface (level 3). GOSAT is designed to estimate cirrus cloud height from the *A*-band of oxygen and saturated absorption region of water vapor in the 2.0 μm band before CO_2 column abundances are retrieved. Cloud top pressure changes based on cloud height, and it is assumed the precision of the estimated cirrus height is ± 1 km. "Shift of cirrus cloud height" levels are set to 0 km (level 0), −1 km (level 2), and +1 km (level 3). Experiments for each level are conducted by adding the level to the cloud height that is regarded as the estimated cirrus height. For the factors given above, there are

Table 13.4 Trace gas observation sensors onboard satellite

Label	Factor		Level		
			1	2	3
A	Cirrus height (km)	h_c	0	−1	+1
B	Surface pressure (hPa)	P_S	+ 0	−3	+ 3
C	Temperature profile shift (K)	**T**	0	−2	+2
D	Aerosol height distribution (km)	h_a	0–3	0–2	0–4
E	Water vapor amount shift (%)	**W**	0	−30	+30
F	Aerosol type	$Type_a$	Rural	Urban	

Table 13.5 Orthgonal array L18

Text no.	Level						Text no.	Level					
	A	*B*	*C*	*D*	*E*	*F*		*A*	*B*	*C*	*D*	*E*	*F*
1	1	1	1	1	1	1	10	1	1	3	3	2	2
2	1	2	2	2	2	1	11	1	2	1	1	3	2
3	1	3	3	3	3	1	12	1	3	2	2	1	2
4	2	1	1	2	2	1	13	2	1	2	3	1	2
5	2	2	2	3	3	1	14	2	2	3	1	2	2
6	2	3	3	1	1	1	15	2	3	1	2	3	2
7	3	1	2	1	3	1	16	3	1	3	2	3	2
8	3	2	3	2	1	1	17	3	2	1	3	1	2
9	3	3	1	3	2	1	18	3	3	2	1	2	2

Table 13.6 Initial values of unknown parameters

Unknown parameter	Value
CO_2 density profile	370 ppmv for all layers
Surface albedo	0.3 for all spectral points
Cirrus optical depth	0.05
Aerosol optical depth	0.1

three levels. Since the surface of the observation area is covered by conifer forest, "aerosol type" levels are set to rural type (level 0) and urban type (level 1), giving a total number of two possible levels. A major difference between "rural" and "urban" aerosols is the amount of black carbon particles. Table 13.5, "L18 (Latin square 18) orthogonal array" presents the orthogonal array used to obtain σ_{ext}^2. The number of tests N is 18. Initial values of estimation parameters are presented in Table 13.6. CO_2 column abundances are retrieved using (13.12). Since the solar Fraunhofer line mask and the water vapor mask are applied to 0.2 cm^{-1} resolution spectrum, the number of channels in $\mathbf{y}$ is 232 when CO_2 column abundances are retrieved. The iteration is conducted until the change of ψ is less than 10^{-5}.

Results

The evaluation results based on the input parameters are shown below. It is assumed that **T**, **W**, and P_s of the GANAL data are obtained to the same accuracy as the true values (US Standard values): the aerosols are distributed from the surface to 3 km; the estimated cloud height is 10 km; and the true CO_2 column abundance, which is not known in practical observation, is 7.109×10^{21} mol · cm^{-2}. Retrieved CO_2 column abundance is 7.118×10^{21} mol · cm^{-2}, and the variance of the internal error, σ^2_{int}, using (13.16) is 4.451×10^{38} mol · cm^{-2}. The variance of the external error is obtained by experiments following the orthogonal design table in Table 13.5. Retrieved CO_2 column abundances $\hat{y}_{col}$ for each test are shown in Fig. 13.3. A mean value of $\hat{y}_{col}$ from 18 tests is 7.135×10^{21} mol · cm^{-2}, and σ^2_{ext} is 1.133×10^{38} mol · cm^{-2}. The variance of total error is 5.584×10^{38} mol · cm^{-2}, with an error bar of 2.363×10^{19} mol · cm^{-2}. The CO_2 column abundance is $7.118 \times 10^{21} \pm 0.024 \times 10^{21}$ mol · cm^{-2}.

True CO_2 column abundances are covered by the error bars. The error ratio, which is calculated by $\sigma \times 100/\hat{y}_{col}$, is 0.33%. It is conceivable that a precision of less than 1% error was achieved for the retrieved CO_2 column abundance in this observation scenario. Since σ^2_{ext}, which is obtained by experiments using orthogonal arrays, takes a long time to calculate, σ^2_{ext} will be obtained in advance. Therefore, it is necessary to obtain σ^2_{ext} under several scenarios. GOSAT is designed to retrieve CO_2 column abundance data by both nadir observation and observations in the cross-track direction, so error evaluation needs to be conducted by altering solar zenith angle, satellite zenith angle, and azimuth angle between the sun and satellite using numerical simulation. Furthermore, in addition to rural and urban type aerosols, there are other types of aerosols, such as yellow sand and sea spray type aerosols. Therefore, it is necessary that retrieval precision is evaluated for different aerosol conditions. GOSAT will also take spectrum measurements using sun glint from large bodies of water.

Since the retrieval algorithm to be used in the DHF is currently being developed and coded, it is important to know the contribution of input parameter precision

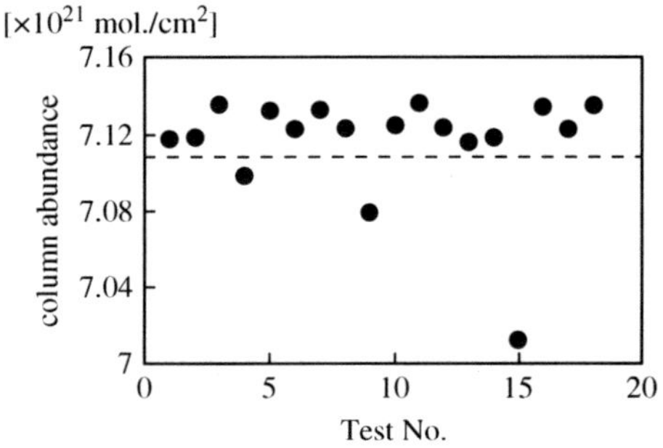

Fig. 13.3 Retrieved CO_2 column abundances for each test following orthogonal array

Table 13.7 Initial values of unknown parameters

Level	Factor	$S\times10^{39}$	η	$V\times10^{39}$	F-value	ρ (%)
A	Cirrus height	0.019	2	0.010	0.11	0.952
B	Surface pressure	0.119	2	0.059	0.70	5.816
C	Temperature profile shift	0.559	2	0.279	3.30	27.393
D	Aerosol height distribution	0.213	2	0.107	1.26	10.449
E	Water vapor amount shift	0.397	2	0.198	2.34	19.448
F	Aerosol type	0.224	1	0.224	2.65	11.003
	Error	0.509	6	0.085		24.938
	Total	2.039	17	0.120		

to the retrieved CO_2 column abundances to improve the retrieval precision, that is, it was assumed that the retrieval precision is improved by reducing the largest factor's contribution to the retrieved CO_2 column abundances. Next, the magnitude of contribution to the retrieved CO_2 column abundance for each factor is calculated derived based on an analysis of variance (ANOVA) using retrieved CO_2 column abundances shown in Fig. 13.3, and the results are presented in Table 13.7. S, η, V, F-value, and ρ in Table 13.7 stand for sums of squares, degrees of freedom, mean squares, F-value, and the rate of contribution (%), respectively. The sum of squares for factor "Total" S_{Total} is calculated by (13.22) $\times N$. When Q represents factors A, B, C, D, and E, which have three levels, the sums of squares S_Q are

$$S_Q = 6\sum_{q=1}^{3}(\hat{y}_{col,q} - \bar{y}_{col})^2, \tag{13.24}$$

where q refers to the level, and $\bar{y}_{col,q}$ is the mean value of $\hat{y}_{col}$ when the level of factor "Q" is q. When Q represents the factor F, which has two levels,

$$S_Q = 9\sum_{q=1}^{2}(\bar{y}_{col,q} - \bar{y}_{col})^2. \tag{13.25}$$

Furthermore, the "Error" factor of sum of squares S_{Error} is obtained by

$$S_{Error} = S_{Total} - S_A - S_B - S_C - S_D - S_E - S_F. \tag{13.26}$$

and V, F-value, and ρ of factor q are obtained by S_q/η, V_q/V_{Error}, and S_q/S_{Total}, respectively.

Based on Table 13.7, "Temperature profile shift" is the largest contributor, followed by "water vapor amount shift". On the other hand, "cirrus height" and "surface pressure" are low contributors when the estimated cirrus height error and the surface pressure shift are given within ±1 km and ±3 hPa. Therefore, these two factors are not included in the error, and the number of factors can be reduced. Thus, it can be concluded that ANOVA effectively examines the contribution of each error term to the retrieved CO_2 column abundances. Prior to launch of the satellite, it is necessary to conclusively determine the contribution of each of the error factors on the retrieved CO_2 column.

13.4 Conclusions

The Greenhouse Gases Observing Satellite (GOSAT), which has been slated to be launched in early 2009, will be the world's first satellite to collect data on the concentrations of carbon dioxide (CO_2) and methane (CH_4), two major greenhouse gases, from space. GOSAT features an onboard down-looking FTS sensor. The data collected will be distributed to users in the form of digital data and global maps of CO_2 and CH_4 column abundances.

Clarifying the precision of the retrieved CO_2 and CH_4 column abundances for GOSAT users is an important component in developing an understanding of global warming mechanisms. This paper describes in detail an evaluation method to clarify GOSAT data precision by setting a measurement scenario and performing a numerical simulation to judge the precision of retrieved CO_2 column abundance data. The evaluation method presented considers two types of error: internal error derived from spectral fitting residuals, and external error derived from the error of input parameters when the column abundances are retrieved. Internal errors were calculated analytically, while the external errors, which are not considered in the error estimation using IMG data and SCIAMACHY data, were obtained using factorial experiments.

The following observation scenario provided the basis for this analysis. Cirrus clouds and aerosols were present in the atmosphere and CO_2 column abundance data was retrieved using nadir observation. The spectra calculated were assumed to be GOSAT-measured spectra. Eighteen experiments were conducted by changing combinations of levels according to an orthogonal array, yielding 18 CO_2 column abundances obtained using numerical simulation. The precision of the CO_2 column abundances was evaluated, and the results were shown. Based on the above scenario, less than 1% precision was demonstrated for the retrieved CO_2 column abundances. Analysis of variance was applied to determine the magnitude of the contribution to the retrieved CO_2 column abundances. In the future, the magnitude of contribution will be examined for several measurement scenarios.

References

1. Aben, I., Hasekamp, O., Hartmann, W.: Uncertainties in the space-based measurements of CO_2 columns due to scattering in the Earth's atmosphere. J. Quant. Spectrosc. Radiat. Transf. **104**, 450–459 (2007)
2. Barkley, M.P., Frieß, U., Monks, P.S.: Measuring atmospheric CO_2 from space full spectral initiation (FSI) WFM-DOAS. Atmos. Chem. Phys. **6**, 3517–3534 (2006)
3. Barkley, M.P., Monks, P.S., Frieß, U., Mittermeier, R.I., Fast, H., Körner, S., Heimann, M.: Comparisons between SCIAMACHY atmospheric CO_2 retrieved using (FSI) WFM-DOAS

to ground based FTIR data and the TM3 chemistry transport model. Atmos. Chem. Phys. **6**, 4483–4498 (2006)
4. Bovensmann, B., Ahlers, B., Buchwitz, M., Frerick, J., Gottwald, M., Hoogeveen, R., Kaiser, J.W., Kleipool, Q., Krieg, E., Lichtenberg, G., Mager, R., Meyer, J., Schlesier, A., Sioris, C., Skupin, J., Savigny, C.v., Wuttke, M.W., Burrows, J.P.: SCIAMACHY in-flight instrument performance. In: Proceedings of the Envisat Calibration Review (SP-520). ESA Publications Division, Noordwijk, The Netherlands (2002)
5. Buchwitz, M., Burrows, J.P.: Retrieval of CH_4, CO, and CO_2 total column abundancess from SCIAMACHY near-infrared nadir spectra: Retrieval algorithm and first results. In: SPIE 10th International Symposium Remote Sensing, 8–12 September, Barcelona, Spain (2003)
6. Buchwitz, M., Schneising, O., Burrows, J.P., Bovensmann, H., Reuter, M., Notholt, J.: First direct observation of the atmospheric CO_2 year-to-year increase from space. Atmos. Chem. Phys. **7**, 4249–4256 (2007)
7. Clerbaux, C., Hadji-Lazaro, J., Turquety, S., Megie, G., Coheur, P.-F.: Trace gas measurement from infrared satellite for chemistry and climate application. Atmos. Chem. Phys. **3**, 1495–1508 (2003)
8. Hamazaki T., Kuze, A., Kondo, K.: Sensor System for Greenhouse Gas Observing Satellite (GOSAT). In: Int. Symp. on Optical Science and Technol., 49th SPIE Annual Meeting, Conf. 5543, Denver, CO, USA (2004)
9. Hanel, R.A., Schlachman, B., Clark, F.D., Prokesh, C.H., Taylor, J.B., Wilson, W.M., Chaney, L.: The Nimbus V Micheson Interferometer. Appl. Opt. **9**(8), 1767–1774 (1970)
10. Hobbs, P.V.: Introduction to Atmospheric Chemistry, pp. 4–8. Cambridge University Press, Cambridge (2000)
11. Hydrographic and Oceanographic Department (2008) Japanease ephemeris
12. Imasu, R.: Global distribution feature of methane and carbon monoxide as observed by DEOS/IMG sensor. In: ASSFTS8, 16–18 November, Toulouse, France (1998)
13. IPCC: Forth Assessment Report. Working Group I Report The Physical Science Basis of Climate Change, sect. 9. Intergovernmental Panel on Climate Change (2007)
14. Kobayashi, H.: Interferometric monitor for greenhouse gases (IMG) project technical report. IMG mission operation & Verification committee (1999)
15. Mason, R.L., Gunst, R.F., Hess, J.L.: Statistical Design and Analysis of Experiments, p. 354. Wiley, New York (1995)
16. McCormick, M.P., Winker, D.M., Browell, E.V., Coakley, J.A., Gardner, C.S., Hoff, R.M., Kent, G.S., Melfi, S.H., Menzies, R.T., Platt, C.M.R., Randall, D.A., Reagan, J.A.: Scientific investigations planned for the Lidar In-Space Technology Experiment (LITE). Bull. Am. Meteorol. Soc. **74**(2), 205–214 (1993)
17. Nakajima, T., Tanaka, M.: Matrix formulations for the transfer of solar radiation in a plane-parallel scattering atmosphere. J. Quant. Spectrosc. Radiat. Transf. **35**, 13–21 (1986)
18. Nakajima, T., Tanaka, M.: Algorithm for radiative intensity calculations in moderately thick atmospheres using a truncation approximation. J. Quant. Spectrosc. Radiat. Transf. **40**, 51–69 (1988)
19. Nautical Almanc Office (US): Astronomical almanac for the year 2008 and its companion. Naval Observatory, Nautical Almanac Office US Defense Department (2008)
20. Rodgers, C.D.: Inverse Method for Atmospheric Sounding: Theory and Practice. World Science, Singapore (2000)
21. Schneider, S.H.: Encyclopedia of Climate and Weather. Oxford University Press, Oxford (1996)
22. Yokota, T., Oguma, H., Morino, I., Inoue, G.: A nadir looking SWIR FTS to monitor CO_2 column density for Japanese GOSAT project. In: Proc. 24th ISTS (selected papers), pp. 887–889. JSASS (2004)

Breinigsville, PA USA
12 December 2009
229098BV00003B/13/P

Nonlinear and Complex Dynamics

José António Tenreiro Machado · Dumitru Baleanu
Albert C.J. Luo

Nonlinear and Complex Dynamics

Applications in Physical, Biological, and Financial Systems

Springer

José António Tenreiro Machado
Department of Electrical Engineering
Institute of Engineering of Polytechnic of Porto
Porto, Portugal
jtm@isep.ipp.pt

Dumitru Baleanu
Faculty of Art and Sciences
Department of Mathematics and Computer Sciences
Cankaya University
Ankara, Turkey
dumitru@cankaya.edu.tr

Albert C.J. Luo
Department of Mechanical and Industrial Engineering
Southern Illinois University
Edwardsville, IL, USA
aluo@siue.edu

ISBN 978-1-4614-0230-5 e-ISBN 978-1-4614-0231-2
DOI 10.1007/978-1-4614-0231-2
Springer New York Dordrecht Heidelberg London

Library of Congress Control Number: 2011934799

Printed on acid-free paper

Springer is part of Springer Science+Business Media (www.springer.com)

Preface

Nonlinear Science and Complexity (NSC) presently plays a dynamic and active role in nonlinear physics and mathematics. In recent years, the significant achievement in such fields including nonlinear dynamics, chaos, discontinuous systems, fractional dynamics, economical, social and biological systems, and many other apparently unrelated fields, proved to be manifestations of the NSC paradigm.

Bearing these facts in mind, it was decided to organize a scientific event dedicated to these emerging concepts. This book presents recent developments, discoveries, and progresses in NSC. The aims of the selected papers are to present the fundamental and frontier theories and techniques for modern science and technology, and to stimulate more research interest for exploration of nonlinear science and complexity. The studies focus on fundamental theories and principles, analytical and symbolic approaches, and computational techniques in nonlinear physical science and nonlinear mathematics. This book, *Nonlinear Dynamics of Complex Systems: Applications in Physical, Biological and Financial Systems*, addresses specifically these areas.

The editors hope that this collection of chapters may be useful and fruitful for scholars, researchers, and advanced technical members of the industrial laboratory facilities for developing new tools and products.

The editors thank the Rector and the President of the Board of Trustee of Cankaya University as well as the Scientific and Technological Research Council of Turkey for the support needed to hold the discussions and debates and all colleagues for sharing their expertise andknowledge.

Porto, Portugal — José António Tenreiro Machado
Ankara, Turkey — Dumitru Baleanu
Edwardsville, IL, USA — Albert C.J. Luo

Contents

Part I
Nonlinear and Complex Dynamics: Applications in Physical Systems

On the Mechanisms of Natural Transport in the Solar System

Yuan Ren, Josep J. Masdemont, Gerard Gómez, and Elena Fantino

1 Introduction

Natural transport is a phenomenon in which particles of natural material transfer from their original orbit to very distant locations in the solar system. It is considered a crucial issue in the material exchange between terrestrial planets. The PCR3BP is a chaotic system, as the motion of the third body is highly sensitive to initial conditions and parameter values. In this system, the natural transport shows very different behaviors from those observed in the two-body model due to the long-term perturbations caused by the second primary and the effect of close passages. The natural transport has been investigated by many authors, in the general framework as well as relative to specific cases: for the Earth–Moon transport the reader is referred to [1, 2, 8], for the Mars–Earth transport to [4–6], and in general to [3]. In this paper, two types of natural transport in the PCR3BP are investigated: the short-time natural transport within two coplanar coupled PCR3BPs, i.e. direct connections between the manifolds of a pair of PCR3BPs, and the long-time natural transport in the PCR3BP, based on the analysis of the chaotic motion in this model.

For a definition of the adopted dynamical models (i.e. the PCR3BP, and the coupled PCR3BPs), the reader is referred to [7,9, 10]. The analysis of the short-time natural transport between planets in solar system is presented in Sect. 2), whereas in Sect. 3 the process to determine the long-time natural transport is described. The natural transport from Mars to the Earth is used as an example throughout the paper. However, the method can be extended to other pairs of planets or other three-body systems.

Y. Ren (✉)
IEEC & Departament de Matemàtica Aplicada I, ETSEIB, Universitat Politècnica de Catalunya, Diagonal 647, 08028 Barcelona, Spain
e-mail: yuan.ren@upc.edu

J.A.T. Machado et al., *Nonlinear and Complex Dynamics: Applications in Physical, Biological, and Financial Systems*, DOI 10.1007/978-1-4614-0231-2_1,

2 Short-Time Natural Transport

The short-time natural transport is based on the existence of heteroclinic connections between libration point orbits (LPOs), and in particular Lyapunov orbits, of a pair of consecutive Sun–planet PC3BPs, i.e. such that the two involved planets are on consecutive orbits in the solar system. Initially moving on the unstable manifold of an LPO of the departure PC3BP, the third body passes onto the stable manifold of an LPO of the arrival PC3BP and eventually reaches the vicinity of the corresponding planet. The dynamical model switch happens in the region where the two given invariant manifolds intersect in phase space. The existence of such intersection is the necessary condition for the specific short-time transport to take place. Figure 1 illustrates the outer branches of the stable and unstable manifolds of a Sun–Earth L_2 Lyapunov orbit and the inner branches of the stable and unstable manifolds of a Sun–Mars L_1 Lyapunov orbit. For convenience, each invariant manifold has been drawn in the respective synodic reference frame, a simplification that in the present case does not affect the values of the distances of the two objects from the Sun. Figure 1 shows that the two sets of objects stay well separated from each other. Computations extended over much longer times indicate that the intersection between a Sun–Mars and a Sun–Earth invariant manifold never occurs.

The manifolds of other types of LPOs look similar. In this respect, they can be represented and replaced by the invariant manifolds of the corresponding libration points. Figure 2 shows the invariant manifolds of the L_1 and L_2 points of a generic PCR3BP. Their minimum and maximum distances from the center of mass of the

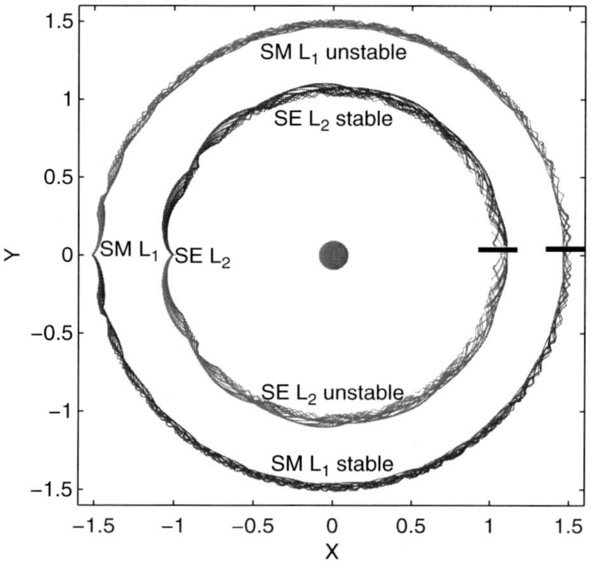

Fig. 1 Branches of stable and unstable invariant manifolds, respectively of periodic orbits around L_1 in the Sun–Mars PCR3BP and L_2 in the Sun–Earth PCR3BP. This plot has been obtained by superimposing the respective synodic reference frames

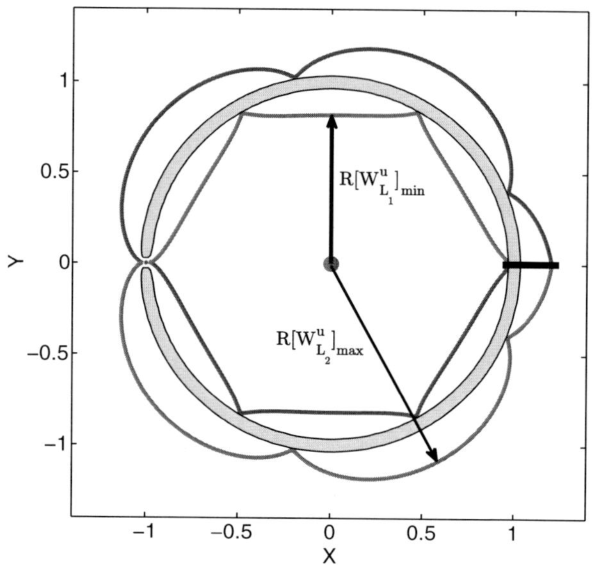

Fig. 2 Branches of the stable and unstable invariant manifolds of the L_1 and L_2 libration points in a PCR3BP. The minimum and maximum distances from the center of mass of the system are labelled $R[W^u_{L_1}]_{\min}$ and $R[W^u_{L_2}]_{\max}$, respectively

Table 1 The minimum and maximum heliocentric distances of the invariant manifolds of the L_1 and L_2 points of the Sun–planet PC3BPs. Quantities expressed in astronomical units. The superscript 'a' indicate the consecutive Sun–planet systems which satisfy the necessary condition for the short-time transfer

	$R\left[W^{u(s)}_{L_1}\right]$		$R\left[W^{u(s)}_{L_2}\right]$	
PCR3BP	Min	Max	Min	Max
Sun–Mercury	0.37547	0.38562	0.38858	0.39918
Sun–Venus	0.67152	0.71658	0.73011	0.78024
Sun–Earth	0.92328	0.98998	1.01008	1.08488
Sun–Mars	1.46684	1.51644	1.53094	1.58331
Sun–Jupiter	3.02493	4.85550	5.56589	9.46402[a]
Sun–Saturn	6.63467[a]	9.12494	9.99818	14.04464
Sun–Uranus	15.87518	18.75222	19.69233	23.49574[a]
Sun–Neptune	23.35734[a]	29.33805	30.89615	37.25573

system are labelled $R[W^u_{L_1}]_{\min}$ and $R[W^u_{L_2}]_{\max}$, respectively. When referred to a Sun–planet PCR3BP, such quantities constitute a good approximation of the minimum and maximum distances from the Sun, given that the mass ratio of any Sun–planet system is always a very small number. Table 1 reports the values of these quantities for all the Sun–planet PCr3BPs (except Sun–Pluto). The necessary condition for the short-time transfer can then be translated into the following inequality:

$$R_{\max}\left[W^{u(s)}_{L_2}\right]_{\text{inner}} \geq R_{\min}\left[W^{u(s)}_{L_1}\right]_{\text{outer}}. \tag{1}$$

It states that the short-time transport between consecutive Sun–planet systems is possible only if the maximum heliocentric distance of the inner problem is larger or equal to the minimum heliocentric distance of the outer problem, thus expressing in a quantitative way the intersection condition. According to Table 1, only two short-time connections are possible in the solar system, i.e. Jupiter–Saturn and Uranus–Neptune (denoted by asterisks). We conclude that not only is the short-time transport between Mars and the Earth not possible but also that this is the most common situation in our planetary system.

3 Long-Time Natural Transport

The short-time transport concept cannot comprehensively explain the exchange of natural material throughout our solar system, and it completely fails in the case of the terrestrial planets. This imposes to evaluate alternative mechanisms. The scenario proposed in this contribution is the long-time transport based on the study of the chaoticity within a PCR3BP, and in particular the analysis of transport between fixed points on the Poincaré maps or lobe dynamics.

The Mars-to-Earth transport is used as an example. Apart from the Sun, the gravity of the Earth, Mars and Jupiter are the dominant forces in the region between the orbits of Mars and the Earth (Fig. 3). For this reason, the long-time natural transport here discussed is based on the Sun–Jupiter PCR3BP. The chaoticity of the system is analysed by considering Poincaré maps drawn at several energy levels.

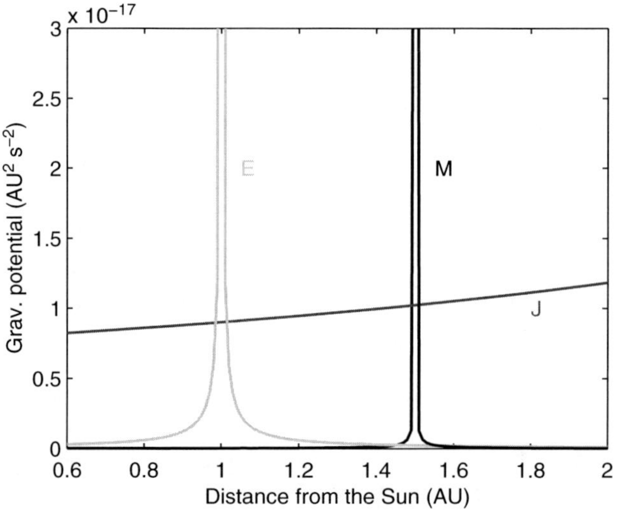

Fig. 3 The relative importance among the gravity of the Earth, Mars and Jupiter in the heliocentric region including the orbits of the Earth and Mars

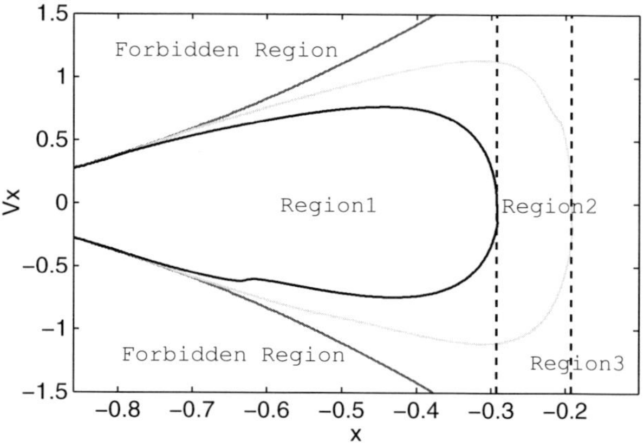

Fig. 4 Subdivision of the initial conditions distributed on the Poincaré section for $C = 3.03$ into regions describing the extension (perihelion, aphelion) of the corresponding heliocentric inertial orbits relative to the orbits of the Earth and Mars

The adopted Poincaré section is the $x\dot{x}$ plane in the Sun–Jupiter PC3BP synodic barycentric reference frame, supplemented by the relations $\tan^{-1}(y/x) = 180^\circ$ and $\dot{y} < 0$. In this way, a point on the Poincaré section corresponds to a well defined state, given that $y = 0$, and $\dot{y}$ is straightforwardly determined by the previous conditions, once the Jacobi constant C is given. By forward propagating a given initial state over a short time (i.e. $<2\pi$ in Sun–Jupiter adimensional time units), a short trajectory segment in synodic coordinates is obtained, corresponding to an almost complete elliptical orbit in inertial space. By propagating a dense set of initial conditions for $C = 3.03$, we have determined the perihelion and aphelion distances of the corresponding inertial orbits, based on which we have divided the Poincaré section into regions (Fig. 4): the two "forbidden" regions correspond to impossible motion at the given energy level; "region 1" contains initial states of inertial orbits that do not intersect the orbits of the Earth and Mars; the initial states in "region 2" produce orbits which only intersect the orbit of Mars; finally, the initial states in "region 3" correspond to motions which have intersections with the orbit of the Earth. Therefore, finding transport from "region 2" to "region 3" on the Poincaré section is equivalent to finding transport from Mars to the Earth in configuration space.

3.1 *Poincaré Maps*

Figure 5 shows four Poincaré maps for the Sun–Jupiter PCR3BP, characterized by decreasing value of C. Each of them has been obtained by setting a mesh grid on the Poincaré section, forward propagating the corresponding initial states and marking their successive intersections with the section. The four plots have been obtained from an initial mesh grid of 35 by 25 points such that $-0.86048 \leq x \leq -0.15$ and $-0.2 \leq \dot{x} \leq 0.2$, with $C = 3.14$, 3.09, 3.06 and 3.03, respectively. They illustrate the first 10^3 returns. As such, they give an insight into the structure of the phase

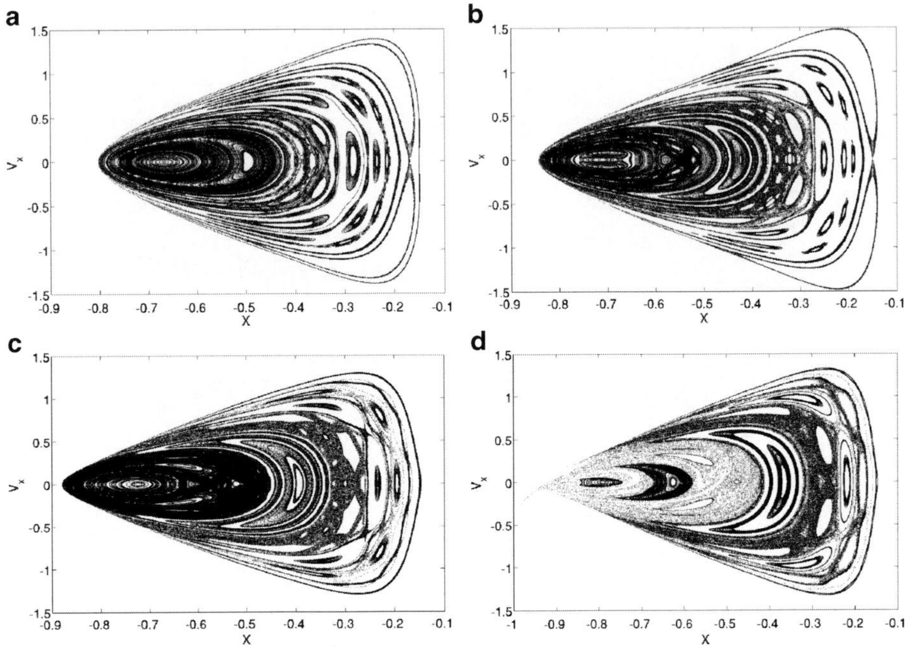

Fig. 5 Poincaré maps of the Sun–Jupiter PCR3BP corresponding to $C = 3.14$ (**a**), 3.09 (**b**), 3.06 (**c**), 3.03 (**d**)

space at varying energy. In particular, it is possible to detect islands (periodic orbits) and fixed points at their centers. When transformed into configuration space, the periodic orbits correspond to invariant tori and the fixed points to periodic orbits. The only possible transfer between different regions of the Poincaré map can only happen through the gaps in the chain of islands, i.e. through the chaotic sea. A close look at the four plots of Fig. 5 shows that the size of the chaotic sea increases as the Jacobi constant decreases (i.e. the energy increases), and in particular when $C = 3.03$, "region 2" and a part of "region 3" are in the chaotic sea, thus suggesting that at this energy level the long-time natural transport from Mars to the Earth may be possible.

3.2 *Fixed Points and Their Manifolds*

Fixed points can either be found at the center of the islands or in the chaotic sea, the latter case being associated to the transport. The most efficient way to identify the fixed points on the Poincaré map is by either the shooting method or the parallel shooting method, respectively when the order (periodicity, or number of distinct successive returns) of the fixed point is lower or greater than five, the reason being the inability of the shooting method to converge to a high-order fixed point, with the tendency of detecting the nearest fixed point of order less than five.

The shooting method aims at zeroing the difference

$$\mathbf{F} = \begin{bmatrix} x \\ \dot{x} \end{bmatrix}_{t_f} - \begin{bmatrix} x \\ \dot{x} \end{bmatrix}_{t_0} \tag{2}$$

between the final state (i.e. after a chosen number of returns) and the initial state on the Poincaré section. By setting $\mathbf{x} = \begin{bmatrix} x \\ \dot{x} \end{bmatrix}_{t_0}$, the variation of $\mathbf{F}$ can be expressed as

$$\frac{\mathrm{d}\mathbf{F}}{\mathrm{d}\mathbf{x}} = (\Phi - \mathbf{I})\,\Delta\mathbf{x}, \tag{3}$$

where $\mathbf{I}$ is the 2×2 identity matrix. Φ is a 2×2 matrix obtained as

$$\Phi_{2\times2} = \begin{bmatrix} \dfrac{\partial x_f}{\partial x_0} & \dfrac{\partial x_f}{\partial \dot{x}_0} & \dfrac{\partial x_f}{\partial \dot{y}_0} \\ \dfrac{\partial \dot{x}_f}{\partial x_0} & \dfrac{\partial \dot{x}_f}{\partial \dot{x}_0} & \dfrac{\partial \dot{x}_f}{\partial \dot{y}_0} \end{bmatrix}_{2\times3} \begin{bmatrix} 1 & 0 \\ 0 & 1 \\ \dfrac{\partial \dot{y}_0}{\partial x_0} & \dfrac{\partial \dot{y}_0}{\partial \dot{x}_0} \end{bmatrix}_{3\times2}. \tag{4}$$

The 2×3 matrix in the above product represents the variation from the initial to the target Poincaré section. $\partial \dot{y}_0/\partial x_0$ and $\partial \dot{y}_0/\partial \dot{x}_0$ in the 3×2 matrix on the right account for the dependence of $\dot{y}_0$ on x_0 and $\dot{x}_0$ (through C). $\mathbf{F}$ is driven to zero by Newton–Raphson. In practice, the process is divided into two steps: a first search is conducted with a large tolerance because different initial guesses tend to converge to the same fixed point, and the use of a strict tolerance would only increase the computation time. Then, the distance between the final points is checked: whenever two points are closer than twice the tolerance, one of them is simply the duplicate of the other and is eliminated. A successive application of the shooting method with a stricter tolerance allows to refine the estimation of the fixed points that have survived the first search.

The parallel shooting method introduces additional, intermediate Poincaré sections, as illustrated in Fig. 6: here the order of the fixed point is 7, and two intermediate sections are set (although for this order one section would be sufficient). Sections 2 and 3 represent respectively the second and the fourth crossing of the trajectory through the Poincaré section. The mesh grids are set on all the sections and the nodes on different sections are combined together to form a set of initial guesses. The side effect of this new method is a considerable increase in the computation time, but the strategy allows to detect fixed points to any order.

The stability of a fixed point depends on the modulus of the eigenvalues of its monodromy matrix. Since the monodromy matrix is a 2×2 matrix, it has two eigenvalues. A saddle point is a fixed point for which the two eigenvalues have modulus greater and smaller than unity, respectively. There are stable and unstable

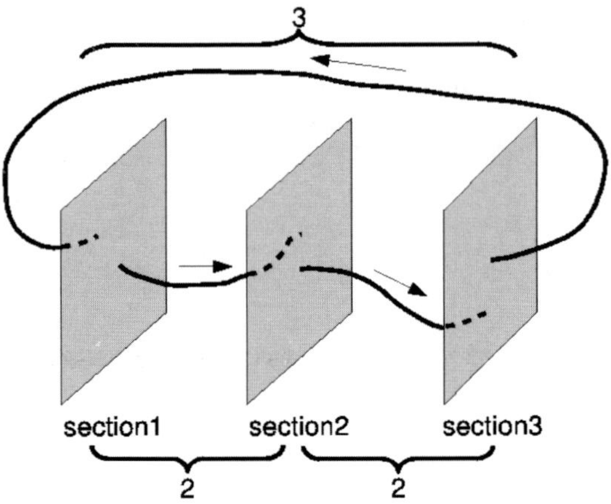

Fig. 6 Sketch of the parallel shooting method

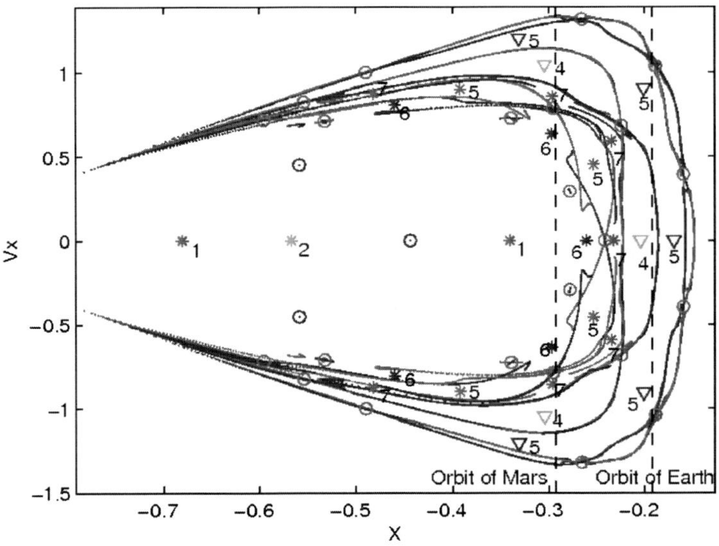

Fig. 7 Fixed points and their invariant manifolds in the Sun–Jupiter PCR3BP at $C = 3.0$: saddle points (∘), stable fixed points (⋆ and ▽) and their invariant manifolds

manifolds associated to a saddle point and the eigenvectors corresponding to the stable and unstable eigenvalues constitute the linear approximation of the stable and unstable manifolds, respectively.

Figure 7 shows all the fixed points identified in the Sun–Jupiter PCR3BP for $C = 3.0$. All the fixed points in the center of the islands are stable, whereas the saddle points appear at the gaps of the island chain.

Also the invariant manifolds of the fixed points are represented. There are five sets of saddle points. Their position on the Poincaré section, their eigenvalues and

Table 2 Position $(x, \dot{x})$, stable and unstable eigenvalues (E^s and E^u) and period (in adimensional time units, 1 TU $\approx 12/2\pi$ years = 1.9 years) for the detected saddle points in the Sun–Jupiter PCR3BP

Order	4	4	5	7	8
	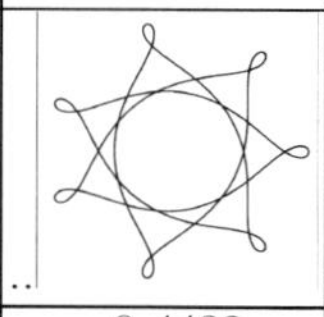	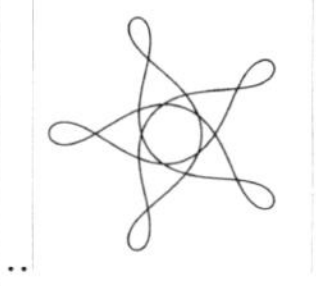	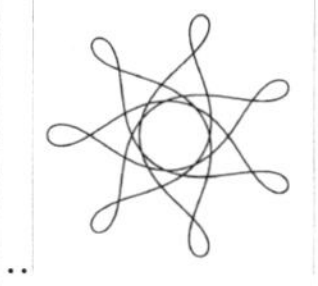	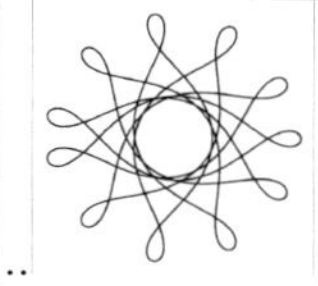	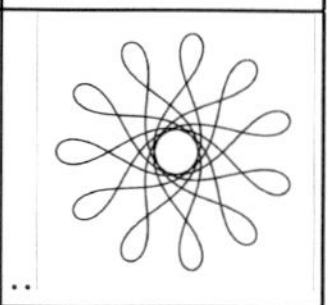
x	− 0.4432	− 0.2240	− 0.2412	− 0.3388	− 0.1591
$\dot{x}$	+0.0000	− 0.6839	+0.0000	+0.7269	+0.3938
E^u	− 1.3040	− 6.2870	+11.730	− 2.7930	+4.8820
E^s	− 0.7666	− 0.1590	+0.0852	− 0.3579	+0.2049
Period	25.1	12.6	18.9	31.5	25.2

the period of the corresponding periodic orbit in configuration space are reported in Table 2: an n-periodic orbit is identified by n fixed points, but for the sake of brevity, the table provides data only for one of them, the others being the result of iterating $n-1$ times.

3.3 Lobe Dynamics

Figure 7 shows that the manifolds of one of the order-4 saddle points have intersections with the orbit of the Earth, the manifolds of the order-5 saddle point have intersections with the orbit of Mars, and these manifolds have intersections with each other. Thus, the natural transport can be realized with the aid of these manifolds: the particle leaves Mars following the unstable manifold of the order-5 orbit, then switches to the stable manifold of the order-4 orbit at the intersection of these two manifolds, passes through the gap in the chain of islands, simultaneously switches to the unstable manifold of the order-4 orbit and intersects the orbit of the Earth. The region enclosed by the stable and unstable manifolds is called "lobe." The transport between regions of the phase space can be completely described by the dynamical evolution of the lobe. Hence, by choosing a set of initial states in the lobe, backward and forward propagation are performed until the transport from "region 2" to "region 3" (i.e. from the orbit of Mars to that of the Earth) is achieved. Figure 8 illustrates an example of lobe evolution: T represents the initial states near the intersection of the manifolds of the order-4 and order-5 saddle points. After 21 iterations in the forward direction, some of these states move from "region 2" to "region 3." However, since such states were initially already very close to "region 3" (i.e. the perihelions of their initial orbits were very close to the orbit of the Earth), it means that the effect of the gravity of Jupiter is minor in this case.

By an additional backward propagation, an initial orbit with the perihelion far from the orbit of the Earth is sought, thus obtaining an overall transfer with a

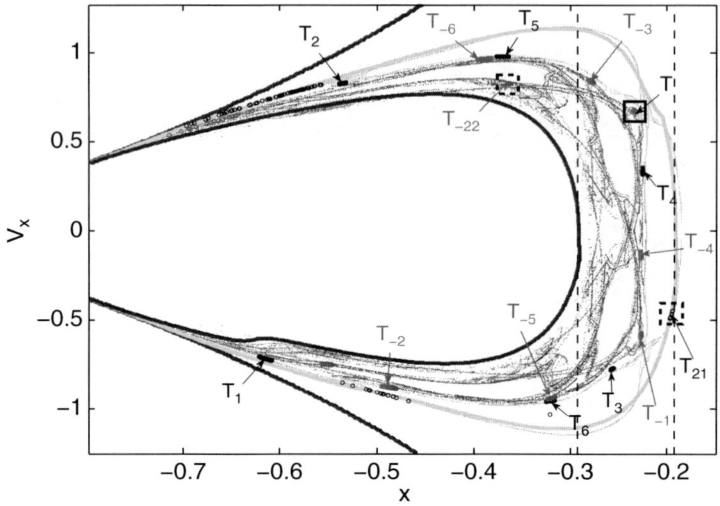

Fig. 8 Lobe evolution

Fig. 9 The Mars-to-Earth transfer orbit corresponding to the long-time natural transport described in the text for $C = 3.0$. In the synodic barycentric reference frame

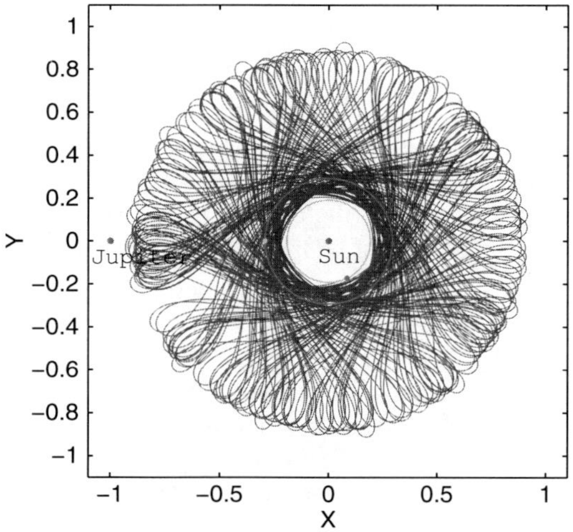

significant effect of the perturbations of Jupiter. Figures 9 and 10 shows the orbit of a long-time natural transport from Mars to the Earth found by this procedure. The inertial frame view shows that the initial trajectory does not have any intersection with the orbit of the Earth. Then, as a result of the gravity of Jupiter, the orbit gradually evolves, eventually reaching the vicinity of the Earth's orbit. This specific transport involves 161 iterations corresponding to 671.32 time units in the Sun–Jupiter system, or 1282.6 years. We note, however, that the transfer orbit found by this method only guarantees a connection between the orbits of the Earth and

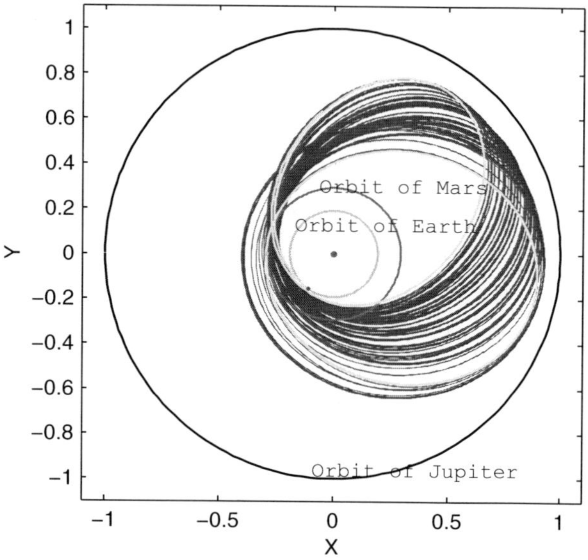

Fig. 10 The Mars-to-Earth transfer orbit corresponding to the long-time natural transport described in the text for $C = 3.0$. In the inertial heliocentric reference frame

Mars, not between the two planets. It certainly proves the possibility of the natural transport at this specific energy level, but the real transport from the surface of Mars to that of the Earth may require much longer times.

4 Conclusions

Two techniques for the analysis of the natural transport in the solar system have been presented. Each of them has a preferential framework of application. The first method achieves the short-time transport, with application to the design of spacecraft trajectories between Jupiter and Saturn and tours of Jupiter's icy moons. The second method is the long-time alternative for the cases in which the manifolds of the coupled PCR3BPs do not present direct intersection, such as the Mars-to-Earth case. Our numerical simulations prove that the second method is capable of describing the transfer of the martian ejecta with aphelion near the orbit of Jupiter to the region of the Earth's orbit. Nevertheless, the most appropriate framework for the study of the low-energy, natural transport between, e.g. Mars and the Earth is the n-body model (see [4, 6]), since the close passages at the Earth and Mars (not accounted for in the Sun–Jupiter PCR3BP) may be relevant. An attempt at describing the natural transport according to dynamical models of growing complexity (three-body, quasi-bicircular, n-body models) will be the subject of a subsequent publication.

Acknowledgements E. Fantino and Y. Ren have been supported by the Marie Curie Actions Research and Training Network AstroNet MCRTN-CT-2006-035151. G. Gómez and J.J. Masdemont have been partially supported by the grants MTM2006-05849/Consolider, MTM2009-06973 and 2009SGR859. The authors also acknowledge the use of EIXAM, the UPC Applied Math cluster system for research computing (see http://www.ma1.upc.edu/eixam/), and in particular Pau Roldan for providing technical support in the use of the cluster.

References

1. Bollt, E.M., Meiss, J.D.: Controlling chaotic transport through recurrence. Physica D: Nonlinear Phenomena. **81**(3), 280–294 (1995)
2. Bollt, E.M., Meiss, J.D.: Targeting chaotic orbits to the Moon through recurrence. Physics Letters A. **204**(5/6), 373–378 (1995)
3. Dellnitz, M., Junge, O., Koon, W.S., Lekien, F., Lo, M.L., Marsden, J.E., Padberg, K., Preis, R., Ross, S.D., Thiere, B.: Transport in dynamical astronomy and multibody problem. International Journal of Bifurcation and Chaos. **15**(3), 699–727 (2005)
4. Gladman, B.: Destination Earth: Martian meteorite delivery. Icarus. **130**, 228–246 (1997)
5. Gladman, B.: The exchange of impact ejecta between terrestrial planets. Science. **217**, 1387–1392 (1996)
6. Gladman, B.: Delivery of planetary ejecta to Earth. Cornell University. (1996)
7. Gómez, G., Jorba, A., Llibre, J., Martinez, R., Masdemont, J., Simó, C.: Dynamics and Mission Design near Libration Points, Vol. I-IV. World Scientific Publishing Co., Singapore (2001)
8. Meiss, J.D., Ott, E.: Markov tree model of transport in area-preserving maps. Physica D: Nonlinear Phenomena. **20**(2/3), 387–402 (1986)
9. Parker, T.S., Chua, L.O.: Practical Numerical Algorithms for Chaotic Systems. Springer-Verlag, Berlin (1989)
10. Szebehely, V.: Theory of Orbits in the Restricted Problem of Three Bodies. Academic Press, New York (1967)

A Method to Design Efficient Low-Energy, Low-Thrust Transfers to the Moon

Giorgio Mingotti, Francesco Topputo, and Franco Bernelli-Zazzera

1 Introduction

Low-energy transfers to the Moon are being studied since the rescue of the Japanese spacecraft Hiten in 1991 [3]. In essence, a low-energy lunar transfer reduces the hyperbolic excess velocity upon Moon arrival, typical of a patched-conics approach. This process is called ballistic capture, and relies on a better exploitation of the gravitational nature ruling the transfer problem instead of the classic Keplerian decomposition of the solar system. The reduced speed relative to the Moon sets the trajectory to low energy levels, which in turn imply a reduced propellant mass needed to stabilize the spacecraft around the Moon.

It is known that the dynamical mechanism governing a class of exterior low energy transfers to the Moon is related to the structure of the invariant manifolds associated with the Lyapunov orbits about the collinear libration points [2]. In particular, a systematic method for the construction of low energy transfers using the knowledge of the phase space of the Sun–Earth and Earth–Moon systems is given in the works of the *Caltech and the Barcelona groups* [8, 9, 14].

It is worth mentioning that previous works have faced the combination of n-body dynamics with low-thrust propulsion. An interior ballistic capture state using low-thrust propulsion was found in [1]. This approach paved the way for the design of the trajectory for ESA's SMART-1 mission [16]. Earth–Venus transfers have been obtained in [6] by combining invariant manifold dynamics and low-thrust, with set oriented methods. Moreover, there are also some examples of the integration of dynamical system theory and optimal control problems for the design of efficient low-energy, low-thrust transfers to the halo orbits [12, 13].

G. Mingotti (✉)
Institut für Industriemathematik, Universität Paderborn, Warburger Str. 100, 33098 Paderborn, Germany
e-mail: mingotti@math.uni-paderborn.de

J.A.T. Machado et al., *Nonlinear and Complex Dynamics: Applications in Physical, Biological, and Financial Systems*, DOI 10.1007/978-1-4614-0231-2_2,

In this paper, both efficient two-impulse transfers and the low-thrust version of the transfers described in [9] are presented, all of them starting from the same LEO with an impulsive maneuver given by the launcher. It is in fact possible to further reduce the propellant necessary to send a spacecraft to the Moon by exploiting both the simultaneous gravitational attractions of the Sun, the Earth, and the Moon, and the high specific impulse provided by the low-thrust engines (above 1,000 s). Nevertheless, including the low-thrust is not trivial and asks for a number of issues to be faced. It is of great importance to, for instance, overcome the loss of Jacobi integral, finding subsets of the phase space that lead to low-thrust ballistic capture (playing the separatrix-like role of the stable manifold associated with $L2$ Lyapunov orbit of the Earth–Moon system), and summarize, using as few parameters as possible, all the reachable orbits that it is possible to target with the finite thrust magnitude available, like low lunar orbits (LLOs).

The purpose of this work is therefore to formulate a systematic approach for the design of efficient pure low energy as well as mixed invariant-manifold low-thrust transfers to low orbits around the Moon. Then, a comparison between the trajectories computed and some solutions found in literature is presented.

The remainder of the paper is organized as follows: first, a brief recall of the dynamics involved in the problem is given. Then, a description of the trajectory design strategy to low Moon orbits and the introduction of special *attainable sets* are given. Finally, the optimization problem is formulated and later the optimal solutions are discussed.

2 Design Strategy

With the *coupled restricted three-body problems approximation*, the four-body dynamics, characterizing the low-energy lunar transfers, is decomposed into two RTBPs, and the invariant manifolds of the Lyapunov orbits are computed. It is possible to show that, with a suitably chosen Poincaré section, the trajectory design is restricted to the selection of a single point on this section [9]. Flown backward, this initial condition gives rise to a trajectory close to the stable and unstable manifolds of the Lj, for $j = 1, 2$, Lyapunov orbits of the Sun–Earth system; integrated forward, a transit, lunar ballistic capture orbit (i.e. an orbit contained inside the stable manifold tube-like structure of the $L2$ Lyapunov orbit of the Earth–Moon system) is achieved. A small Δv maneuver is eventually needed at this patching point, in order to match the energies of the two stages. With this approach, it is possible to find efficient Earth–Moon transfers like the ones described in [3].

The transfers studied in this work are defined as follows. The spacecraft is assumed to be initially on a circular parking orbit around the Earth at a height $h_\mathrm{E} =$ 167 km; then an impulsive maneuver, Δv_E, carried out by the launch vehicle, places the spacecraft on a translunar trajectory, performing a translunar insertion TLI. Two different typologies of mission are investigated, with respect to the propulsion adopted: (a) low-energy two-impulse transfers to LLOs: after the insertion, the spacecraft flies ballistically under the dynamics of the problem until the Moon neighborhood, where a second impulsive maneuver inserts it on a stable low altitude orbit; (b) low-energy low-thrust transfers to LLOs: after the launch, the spacecraft can only rely on its low-thrust propulsion to reach a stable low-altitude orbit around the Moon.

As far as it concerns these transfers, the final orbit has moderate eccentricity, e, and periapsis/apoapsis, r_p/r_a, prescribed by the mission requirements. The transfer terminates when the spacecraft is at the periapsis of the final orbit around the Moon. While both e and r_p/r_a are assumed to be given, the orientation, i.e. the argument of periapsis, ω, of the final orbit around the Moon is not fixed.

In general, the two transfer families are achieved by optimizing, in a four-body scenario, a first guess derived in the coupled three-body problems approximation. From the prospective of this model, the transfer trajectory is conceived as made up of two distinct portions: the first, called Earth escape stage, is built in the Sun–Earth model, SE, whereas the second, called Moon capture stage, is defined in the Earth–Moon model, EM.

2.1 *The Planar Circular Restricted Three-Body Problem*

The motion of the spacecraft, m_3, is studied in the gravitational field generated by the mutual circular motion of two primaries of masses m_1, m_2, respectively, about their common center of mass (see Fig. 3a). It is assumed that m_3 moves in the same plane of m_1, m_2 under the following equations [19]:

$$\ddot{x} - 2\dot{y} = \frac{\partial \Omega}{\partial x}, \qquad \ddot{y} + 2\dot{x} = \frac{\partial \Omega}{\partial y}, \tag{1}$$

where the auxiliary function is

$$\Omega(x,y,\mu) = \frac{1}{2}(x^2+y^2) + \frac{1-\mu}{r_1} + \frac{\mu}{r_2} + \frac{1}{2}\mu(1-\mu), \tag{2}$$

and $\mu = m_2/(m_1+m_2)$ is the mass parameter of the three-body problem. Equation (1) is written in a barycentric rotating frame with nondimensional units: the angular velocity of m_1, m_2, their distance, and the sum of their masses are all set to the unit value. It is easy to verify that the primary of mass $1-\mu$, is located at $(-\mu,0)$, whereas the smaller primary μ, is located at $(1-\mu,0)$; thus, the distances between m_3 and the primaries are:

$$r_1^2 = (x+\mu)^2 + y^2, \qquad r_2^2 = (x+\mu-1)^2 + y^2. \tag{3}$$

For fixed μ, the Jacobi integral reads

$$J(x,y,\dot{x},\dot{y}) = 2\Omega(x,y,\mu) - (\dot{x}^2+\dot{y}^2), \tag{4}$$

and, for a given energy C, it defines a three-dimensional manifold

$$F(C) = \{(x,y,\dot{x},\dot{y}) \in \mathscr{R}^4 | J(x,y,\dot{x},\dot{y}) - C = 0\}, \tag{5}$$

foliating the four-dimensional phase space. The projection of $F(C)$ on the configuration space (x,y) defines the Hill's curves bounding the allowed and forbidden regions associated with prescribed values of C. The vector field defined by (1) has five well-known equilibrium points, known as the Lagrange points, labeled Lj, $j = 1,\ldots,5$. This study deals with the portion of the phase space surrounding the two collinear points $L1$ and $L2$. In a linear analysis, these two points behave like the product *saddle* $\times$ *center*. Thus, there exists a family of retrograde Lyapunov orbits and two-dimensional stable and unstable manifolds emanating from them [5, 10].

The system governed by (1) is used alternatively to describe the motion of the spacecraft either in the Sun–Earth (SE) or in the Earth–Moon (EM) system. The mass-parameter value assumed for these models are $\mu_{\text{SE}} = 3.0034 \times 10^{-6}$ and $\mu_{\text{EM}} = 1.2150 \times 10^{-2}$, respectively.

As for the SE model, the generic periodic orbit about Lj, $j = 1,2$, is referred to as γ_j, whereas its stable and unstable manifolds are labeled $W^s(\gamma_j)$, $W^u(\gamma_j)$. Dealing with the EM model, the generic periodic orbit about Lj, $j = 1,2$, is called λ_j, while its stable and unstable manifolds are named $W^s(\lambda_j)$, $W^u(\lambda_j)$.

3 Earth Escape Stage

If a value of Jacobi constant in the SE model, C_{SE}, is suitably chosen, there exists a unique Lyapunov orbit about both $L1$ and $L2$, labeled γ_1 and γ_2, respectively. Assuming the energy values $C_{\text{SE}} \ll C_2$ such that both γ_1 and γ_2 exist, the Hill's regions are opened at both $L1$ and $L2$. Without any loss of generality, the Earth escape stage is constructed considering the dynamics around $L2$; using $L1$ instead of $L2$ is straightforward. The stable and unstable manifolds associated with γ_2, $W^s(\gamma_2)$ and $W^u(\gamma_2)$, are computed starting from the Lyapunov orbit until a certain surface of section is reached.

Aiming at exploiting the structure of both $W^s(\gamma_2)$ and $W^u(\gamma_2)$, two surfaces of section are introduced to study their cuts at different stages. Section S_A, making an angle φ_A (clockwise) with the x-axis and passing through the Earth, is considered to cut $W^s(\gamma_2)$, whereas section S_B, inclined by φ_B (counterclockwise) on the x-axis and passing through the Earth, is assumed for $W^u(\gamma_2)$ (see Fig. 1a for the two manifolds topology, and see Fig. 2a for the two angles). The corresponding section curves, $\partial\Gamma_2^s$, $\partial\Gamma_2^u$, represented on the $(r_2,\dot{r}_2)$-plane, are diffeomorphic to circles (see Fig. 1b, $\partial\Gamma_2^s$, $\partial\Gamma_2^u$ are plotted on the $(r_2,\dot{r}_2)$-plane as $r_2 = y$, $\dot{r}_2 = \dot{y}$ for $x = 1-\mu$, $\varphi_A = \varphi_B = \pi/2$).

Both Poincaré sections represent two-dimensional maps for the flow of the RTBP. Indeed, any point on these sections uniquely defines an orbit. This property

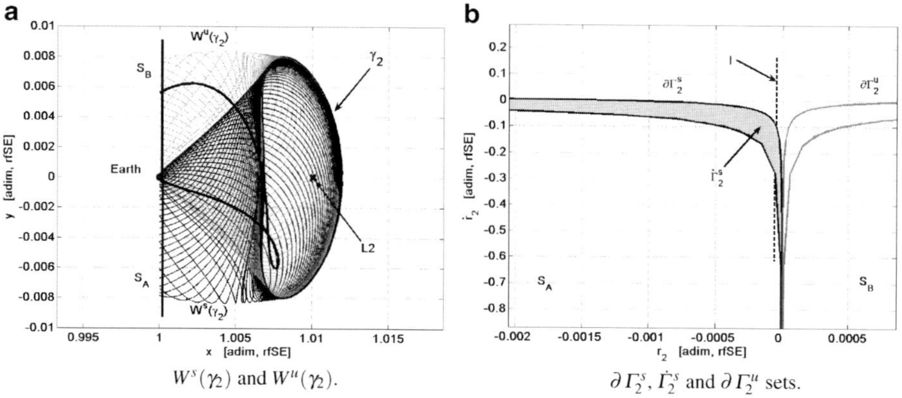

Fig. 1 Stable and unstable manifolds $W^s(\gamma_2)$, $W^u(\gamma_2)$ associated with the $L2$ Lyapunov orbit γ_2, and their section curves $\partial\gamma_2^s$, $\partial\gamma_2^u$, respectively. In (**a**), the *bold line* stands for a sample Earth escape trajectory. In (**b**), the set $\dot{\Gamma}_2^s$ (*gray*) is made up by the points of S_A that lie inside $\partial\gamma_2^s$, whereas the line l (*dashed*) is the locus of points being at $h_E = 167$ km altitude above the Earth surface

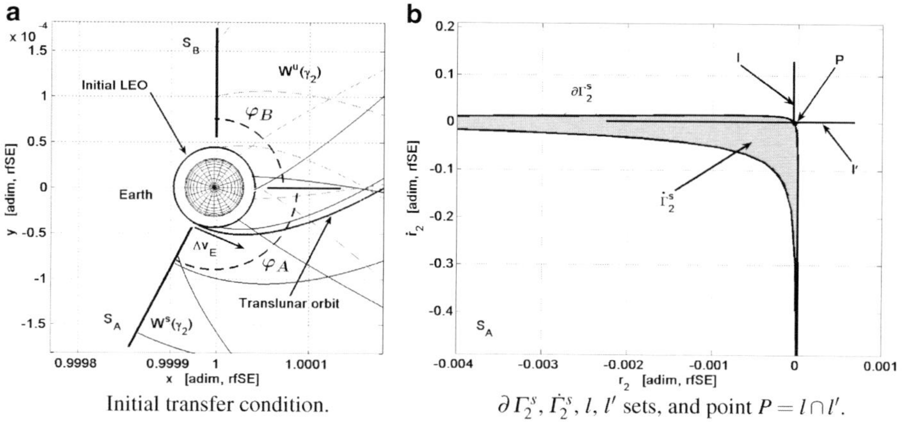

Fig. 2 Earth escape trajectory performed with a tangential Δv_E maneuver and its associated section point P

holds as both $F(C_{SE})$ and $S_{A,B}$ lower the dimension of the phase space to two. By definition, points on $\partial\Gamma_2^s$ generate orbits that asymptotically approach γ_2 in forward time. Points inside Γ_2^s give rise to transit orbits that pass from the Earth region to the exterior region, whereas points outside Γ_2^s correspond to nontransit orbits (the manifolds act as separatrices for the states of motion [5, 10]).

Candidate trajectories for Earth–Moon transfers are nontransit orbits close to both $W^s(\gamma_2)$ and $W^u(\gamma_2)$. This property is wanted, since the existence of $W^s(\gamma_2)$ and $W^u(\gamma_2)$ has to be exploited, although the transfer orbit does not exactly lie on any invariant subset. Let $\dot{\Gamma}_2^s$ be the set of points in the $(r_2, \dot{r}_2)$-plane that are enclosed by $\partial\Gamma_2^s$, and $\bar{\Gamma}_2^s$ the closed set made up of $\partial\Gamma_2^s \cup \dot{\Gamma}_2^s$ (see Fig. 1b). Points on $\bar{\Gamma}_2^s$ have to be avoided as they lead to either transit or asymptotic orbits. On the contrary, all the points that lie on

$$l = \{(r_2, \dot{r}_2) \in S_A,\ (r_2, \dot{r}_2) \notin \bar{\Gamma}_2^s | r_2 = R_{\mathrm{E}} + h_{\mathrm{E}}\} \tag{6}$$

are translunar candidate orbits as they intersect the initial parking orbit (R_{E} is the radius of the Earth). This intersection occurs in the configuration space only, as the initial parking orbit and the translunar trajectory have two different energy levels.

The pair $\{C_{\mathrm{SE}},\ \varphi_A\}$ uniquely defines the curve $\partial\gamma_2^s$ on S_A: C_{SE} stands for the orbit γ_2, whereas φ_A defines the surface of section S_A to cut the first intersection of $W^s(\gamma_2)$. Thus, $\{C_{\mathrm{SE}},\ \varphi_A\}$ are used to define the first guess Earth escape stage. In order to obtain efficient transfer trajectories, the lowest possible initial instantaneous maneuver, Δv_{E}, is searched. It is necessary to define its components: a first contribution to the Δv_{E} amount is related to the radial term Δv_r, while the second tangential contribution Δv_t is needed to fill the gap ΔC between the energy of the initial parking orbit, C_E, and C_{SE} (i.e. $\Delta C = C_E - C_{\mathrm{SE}}$). It is possible to show that $\Delta v(\Delta C, \varphi_A) = \Delta v_t(\Delta C) + \Delta v_r(\varphi_A)$, and it is even possible to lower Δv_r to zero by properly tuning φ_A. This approach leads to initial tangential maneuvers, i.e. the initial Δv_{E} is aligned with the velocity of the circular parking orbit around the Earth. The search is therefore restricted to the point $P \in S_A$ defined by $P = l \cap l'$, where l' is the set of points having zero radial velocity with respect to the Earth

$$l' = \{(r_2, \dot{r}_2) \in S_A,\ (r_2, \dot{r}_2) \notin \bar{\Gamma}_2^s | \dot{r}_2 = 0\}. \tag{7}$$

Point P does not exactly lie on the stable manifold (but outside), and can be found sufficiently close to $W^s(\gamma_2)$ by suitably tuning φ_A (see Fig. 2b). In practice, since at this stage a first guess solution is constructed to be later optimized, orbits sufficiently close to P can also be considered. In particular, points $P' \in S_A$ such that $\|P' - P\| \leq \varepsilon$ are also taken into account, where ε is a certain prescribed distance.

A number of P' points can be generated by tuning the angle φ_A. These points, flown forward, generate orbits that are close to $W^s(\gamma_2)$ until the region about γ_2 is reached. From this point on, the orbits get close to $W^u(\gamma_2)$, and their intersection with S_B is studied. The set labeled $\mathscr{E}_{\mathrm{SE}}$, $\mathscr{E}_{\mathrm{SE}} \in S_B$, stands for the set of orbits close to $W^u(\gamma_2)$ whose preimage $\mathscr{E}_{\mathrm{SE}}^{-1}$, $\mathscr{E}_{\mathrm{SE}}^{-1} \in S_A$, is made up by P' points. Earth escape trajectories defined on $\mathscr{E}_{\mathrm{SE}}^{-1}$, $\mathscr{E}_{\mathrm{SE}}$ are taken into account, as the latter intersects a special subset leading to orbits at the Moon (see Fig. 4b, where $\mathscr{E}_{\mathrm{SE}}$ is reported).

It is worth noticing that the parking orbit is defined at a lower energy level than orbits on $\mathscr{E}_{\mathrm{SE}}^{-1}$, therefore the instantaneous velocity change Δv_{E} is required to place the translunar trajectory on $F(C_{\mathrm{SE}}) \cap \mathscr{E}_{\mathrm{SE}}^{-1}$. In practice, this maneuver is provided by the launch vehicle once the spacecraft is on the Earth parking orbit. A mission profile including this parking orbit is in fact consistent with major architecture requirements, as summarized in [15].

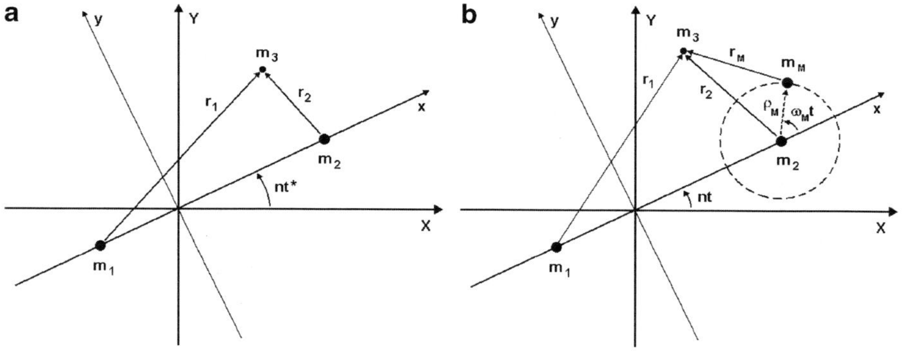

Classic circular planar restricted three-body problem.

Moon-perturbed Sun–Earth RTBP.

Fig. 3 Mathematical models to described the physics of the problem

The design of the SE phase is reduced in this way to the determination of just the two sets $\mathcal{E}_{\mathrm{SE}}^{-1}$, $\mathcal{E}_{\mathrm{SE}}$, and to the computation of Δv_{E}. This first phase of the transfer is constructed using (1). The SE trajectory does not make use of low-thrust propulsion, and the phase space structure of the ballistic PCRTBP is exploited.

3.1 The Moon-Perturbed Sun–Earth Restricted Three-Body Problem

When the gravitational attraction of the Moon is taken into account, (1) are augmented aiming at introducing the dynamics of the Moon in an autonomous fashion, leading to the formulation of the bicircular restricted four-body problem, BRFBP (see Fig. 3b). The dynamical system moves from the fourth order to the fifth one. Some assumptions are taken into account, recalling that the orbits of the primaries show low eccentricity values (≈ 0.01, ≈ 0.04), and the Moon inclination with respect to the ecliptic plane is small ($\approx 5\,\mathrm{deg}$): (a) the Sun and the Earth are revolving in circular orbits around their center of mass; (b) the Earth–Moon barycenter is moving in a circular orbit around the center of mass of the Sun–Earth–Moon system.

Assuming all the hypotheses written above, the planar equations of motion are:

$$\ddot{x} - 2\dot{y} = \frac{\partial \Omega_{\mathrm{M}}}{\partial x}, \qquad \ddot{y} + 2\dot{x} = \frac{\partial \Omega_{\mathrm{M}}}{\partial y}, \qquad \dot{\theta} = \omega_{\mathrm{M}} \tag{8}$$

where the subscripts denote the partial derivative of the auxiliary function

$$\Omega_{\mathrm{M}}(x,y,\theta) = \Omega(x,y,\mu_{\mathrm{SE}}) + \frac{m_{\mathrm{M}}}{r_M} - \frac{m_{\mathrm{M}}}{\rho_{\mathrm{M}}^2}\,(x\cos\theta + y\sin\theta). \tag{9}$$

The quantity $\Omega(x,y,\mu_{SE})$ stands for the classic CRTBP potential expressed by (2), while the remaining part represents the gravitational perturbation of the Moon.

The dimensionless physical constants introduced to describe the Moon influence are in agreement with those of the SE model. Thus, the distance between the Moon and the Earth is $\rho_M = 2.5721 \times 10^{-3}$, the mass of the Moon is $m_M = 3.6942 \times 10^{-8}$, and its angular velocity with respect to the SE rotating frame is $\omega_M = 1.2367 \times 10^1$. The location of the Moon is therefore at $(1-\mu_{SE}+\rho_M\cos\theta, \rho_M\sin\theta)$, such that:

$$r_M^2 = (x - 1 + \mu_{SE} - \rho_M \cos\theta)^2 + (y - \rho_M \sin\theta)^2. \tag{10}$$

According to the differential equation (8), the system does not admit the existence of any libration point or integral of motion. Anyway, as the Moon can be considered as a small perturbation of the Sun–Earth model, a qualitative global analysis about the motion of the spacecraft is proposed, assuming the restricted four-body model as a perturbation of the invariant objects of the classic RTBP. If the points belonging to the escape set $\mathscr{E}_{SE}$ are backwards integrated under the dynamics associated with (8), the topology of their trajectories in the configuration space is only slightly and negligible different. The main variations appear associated with the preimage set $\mathscr{E}_{SE}^{-1}$. If the trajectories pass near the Moon, the points $P' \in S_A$ show almost the same phase–space coordinates $(r_2, \dot{r}_2)$ as before, while the tangential velocity reduces significantly, with respect to the classic Sun–Earth PCRTBP computation. In details, when a lunar flyby is explicitly taken into account while designing the Earth escape stage, the difference between the energy level of the parking orbit at a height $h_E = 167$ km and the one of the orbits on $\mathscr{E}_{SE}^{-1}$ is lowered. This means that a reduced instantaneous velocity change Δv_E is now required to place the translunar trajectory on $F(C_{SE}) \cap \mathscr{E}_{SE}^{-1}$.

4 Low-Thrust Propulsion and Attainable Sets

Once the initial transfer stage is defined, the final one that leads to the Moon is required. A general approach useful to mathematically describe impulsive maneuvers, low-thrust arcs. The core of the formulation is based on a perturbed version of the classic PCRTBP. The perturbation is the low-thrust propulsion, and when its driving parameters are wisely tuned, different types of transfer arcs as well as several final conditions at the Moon can be investigated.

To model the *controlled* motion of m_3 under both the gravitational attractions of m_1, m_2, and the low-thrust propulsion, the following differential equations are considered:

$$\ddot{x} - 2\dot{y} = \frac{\partial \Omega}{\partial x} + \frac{T_x}{m}, \quad \ddot{y} + 2\dot{x} = \frac{\partial \Omega}{\partial y} + \frac{T_y}{m}, \quad \dot{m} = -\frac{T}{I_{sp}\, g_0}, \tag{11}$$

where $T = \sqrt{T_x^2 + T_y^2}$ is the thrust magnitude, I_{sp} the specific impulse of the engine and g_0 the gravitational acceleration at sea level. The ballistic motion (1) is represented by a fourth-order system, while the controlled motion (11) is described by a fifth-order system of differential equations. Continuous variations of the spacecraft mass, m, are taken into account when low-thrust propulsion is considered. This causes a singularity arising when $m \to 0$, beside the well-known singularities given by impacts of m_3 with m_1 or m_2.

The thrust law $\mathbf{T}(t) = \{T_x(t), T_y(t)\}^\top$, $t \in [t_i, t_f]$, in (11) is not given, but rather in this approach it represents an unknown that is found when the optimal control problem is solved (t_i and t_f are the initial and final times, respectively). It is determined such that a certain state is targeted and, at the same time, a certain objective function is minimized. However, in order to build first guess solutions, at this stage the profile of $\mathbf{T}$ over time is described. In particular, using tangential thrust, attainable sets can be defined in the same fashion as reachable sets are defined in [6].

Let $\mathbf{y}_i$ be a vector representing a generic initial state, i.e. $\mathbf{y}_i = \{x_i, y_i, \dot{x}_i, \dot{y}_i, m_i\}^\top$, and let $\phi_{\mathbf{T}(\tau)}(\mathbf{y}_i, t_i; t)$ be the flow of system of (11) at time t, starting from $(\mathbf{y}_i, t_i)$ and considering the thrust profile $\mathbf{T}(\tau)$, $\tau \in [t_i, t]$. The latter has to be taken within proper bounds that are typically given by technological constraints. This condition usually reads $T(t) \leq T_{\max}$, where $T_{\max}$ is the maximum available thrust magnitude. With this notation, it is possible to define the generic point of a tangential low-thrust trajectory through

$$\mathbf{y}(t) = \phi_{\overline{\mathbf{T}}}(\mathbf{y}_i, t_i; t), \tag{12}$$

where $\overline{\mathbf{T}} = \overline{T}\,(\mathbf{v}/v)$, $v = \sqrt{\dot{x}^2 + \dot{y}^2}$, $\mathbf{v} = \{\dot{x}, \dot{y}\}^\top$. Equation (12) represents the flow of the differential system governed by (11), when constant tangential thrust of magnitude $\overline{T}$ is considered. With given $\overline{T}$, tangential thrust maximizes the variation of Jacobi energy, which is the only property that has to be dealt with when designing trajectories in the RTBP. (In [11], a comparison between tangential thrust in either rotating or inertial frame is proposed). The low-thrust orbit, at time t, can be expressed as

$$\gamma_{\overline{\mathbf{T}}}(\mathbf{y}_i, t) = \{\phi_{\overline{\mathbf{T}}}(\mathbf{y}_i, t_i; \tau) | \tau < t\}, \tag{13}$$

where the dependence on the initial state $\mathbf{y}_i$ is kept. The attainable set, at time t, can be defined as

$$\mathscr{A}_{\overline{\mathbf{T}}}(t) = \bigcup_{\mathbf{y}_i \in \mathscr{Y}} \gamma_{\overline{\mathbf{T}}}(\mathbf{y}_i, t), \tag{14}$$

where $\mathscr{Y}$ is a domain of admissible initial conditions. Attainable set in (14) is associated with a generic $\mathscr{Y}$; this set can be suitably defined for the three different types of transfers at hand. Thanks to the definition of $\mathscr{A}_{\overline{\mathbf{T}}}(t)$, low-thrust propulsion can be incorporated in a three-body frame, using the same methodology developed for the invariant manifolds [9]. More specifically, invariant manifolds are replaced by attainable sets which are manipulated (i.e. intersected) to find a transfer point on a suitable surface of section. The idea is to mimic the role that invariant

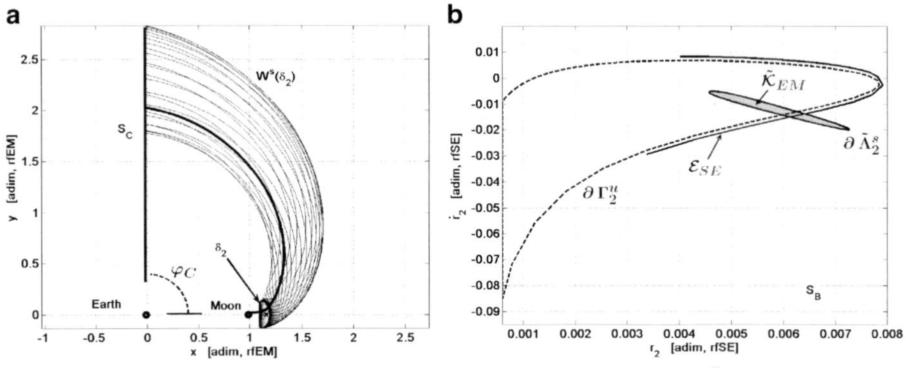

$W^s(\lambda_2)$. $\mathscr{E}_{SE}$ and $\tilde{\mathscr{K}}_{EM}$ sets.

Fig. 4 Stable manifold $W^s(\lambda_2)$ and its section curve $\partial\tilde{\Lambda}_2^s$. The latter is used to define the set of orbits that lead to Moon capture $\tilde{\mathscr{K}}_{\mathrm{EM}}$. In (**a**), the bold line stands for a sample Moon capture trajectory

manifolds have in trajectory design with the use of attainable sets. This approach can be adapted to design either two-impulse transfers as well as low-thrust transfer to the Moon.

4.1 Moon Ballistic Capture Stage

In analogy with what was described for the Earth escape stage in Sect. 3, by fixing a suitable value of the Jacobi constant in the EM model, C_{EM}, a unique Lyapunov orbit about both $L1$ and $L2$, named λ_1 and λ_2, respectively, can be defined. Restricting the energy to $C_{\mathrm{EM}} \ll C_2$, both λ_1 and λ_2 exist, and the Hill's regions are opened at both $L1$ and $L2$. In order to reach the final orbit about the Moon from the exterior, a capture via $L2$ is considered (see Fig. 4a). This means that the Moon ballistic capture stage is constructed by exploiting the dynamics around $L2$. The stable manifold associated with λ_2, $W^s(\lambda_2)$, is computed starting from λ_2 and integrating backward until a certain surface of section is reached. Section S_C, making an angle φ_C (counterclockwise) with the x-axis and passing through the Earth, is considered to cut $W^s(\lambda_2)$ ($\varphi_C = \pi/2$ in Fig. 4a). The corresponding section curve, $\partial\Lambda_2^s$ (computed on the $(r_1, \dot{r}_1)$-plane in the EM model), is diffeomorphic to a circle. The set $\mathscr{K}_{\mathrm{EM}} = \dot{\Lambda}_2^s$ is defined, where $\dot{\Lambda}_2^s \in S_C$ is the set of points inside $\partial\Lambda_2^s$, set that leads to the Moon capture.

The set $\mathscr{K}_{\mathrm{EM}}$ is defined on section S_C in the EM model. However, it is possible to represent $\mathscr{K}_{\mathrm{EM}}$ on S_B defined in the SE model through the transformation $\tilde{\mathscr{K}}_{\mathrm{EM}} = \mathscr{M}(\mathscr{K}_{\mathrm{EM}})$. The operator $\mathscr{M}$ maps states on S_C (EM model) to states on S_B (SE model), provided the two angles φ_C, φ_B. The same conversion is also applied to $\partial\Lambda_2^s$, in order to obtain $\partial\tilde{\Lambda}_2^s = \mathscr{M}(\partial\Lambda_2^s)$ on section S_B from section S_C.

This transformation is basically made up of a rotation and a scaling of the variables in proper units (see Fig. 4b where $\tilde{\mathscr{K}}_{\rm EM}$ and $\partial \tilde{\Lambda}_2^s$ are reported).

Considering only section S_B, it is possible to define the ballistic low-energy Earth–Moon transfers as the orbits belonging to the set $\mathscr{E}_{\rm SE} \cap \tilde{\mathscr{K}}_{\rm EM}$. The sets $\mathscr{E}_{\rm SE}$ and $\tilde{\mathscr{K}}_{\rm EM}$ are characterized by different values of the Jacobi constant, $C_{\rm SE}$ and $C_{\rm EM}$, respectively. In addition, any point in $\tilde{\mathscr{K}}_{\rm EM}$ has a different value of Jacobi constant as $\mathscr{M}$ is not energy preserving. Thus, in order to join together orbits on $\mathscr{E}_{\rm SE}$ and orbits on $\tilde{\mathscr{K}}_{\rm EM}$, an impulsive maneuver at the patching point may be required.

This approach is the one known in literature to design Earth–Moon low-energy transfers with impulsive maneuvers [9]. Indeed, another impulsive maneuver, beside the one performed at the patching point, is necessary upon Moon arrival to place the spacecraft into a stable orbit around the Moon. This approach is based on either (1) or (11) with $T = 0$, and allows to define low energy transfers like those described in [2, 3, 8].

Two-impulse transfers to the Moon are defined as follows. The spacecraft is assumed to be initially on a parking orbit about the Earth with given eccentricity, e, and perigee/apogee altitude, h_p/h_a. The argument of perigee, ω_E, of this orbit is not fixed. The transfer begins when the spacecraft is at the perigee of this orbit. An impulsive maneuver, carried out by the launch vehicle, places the spacecraft onto a translunar trajectory; from this point on, the spacecraft flies ballistically under the RTBP dynamics until it reaches the Moon neighborhood. Then, a second impulsive maneuver is required to insert the spacecraft into a stable orbit around the Moon. This orbit has moderate eccentricity, e, and periapsis/apoapsis altitude h_p/h_a, prescribed by the mission requirements. The transfer terminates when the spacecraft is at the periapsis of this orbit. While both e and r_p/r_a are given, the orientation, i.e. the argument of periapsis, $\omega_{\rm M}$, of the final orbit around the Moon is not fixed.

The global transfer is designed recalling the coupled RTBPs approximation. The initial orbit is a trajectory belonging to the set $\mathscr{E}_{\rm SE}$, whereas the second part is defined using a suitable attainable set in the EM model, according to the formalism introduced in Sect. 4 through (14). As this set constitutes the second part of the trajectory, it is made up of ballistic orbits that are integrated backward, i.e. considering the thrust magnitude $T = 0$. More specifically, as both eccentricity and apsidal altitude are prescribed, the final state of the transfers (i.e., the periapsis point of the orbit about the Moon) is a function of the argument of periapsis and of the final tangential $\Delta v_{\rm M}$ impulsive maneuver required to place the spacecraft into a stable lunar orbit, i.e. $\mathbf{y}_f = \mathbf{y}_f(\omega_{\rm M}, \Delta v_{\rm M})$, see Fig. 5a.

According to this approach, the domain of admissible final states becomes

$$\mathscr{Y}^M = \{\mathbf{y}_f(\omega_{\rm M}, \Delta v_{\rm M}) | \omega_{\rm M} \in [0, 2\pi], \Delta v_{\rm M} \in [0, +\infty]\}, \qquad (15)$$

and the attainable set, for some $t \geq 0$ (i.e. $-t$ is a backward integration), containing ballistic capture trajectories with impulsive stabilization is

$$\mathscr{A}_{\mathbf{B}}^M(-t) = \bigcup_{\mathbf{y}_f \in \mathscr{Y}^M} \gamma_{\mathbf{B}}(\mathbf{y}_f(\omega_{\rm M}, \Delta v_{\rm M}), -t). \qquad (16)$$

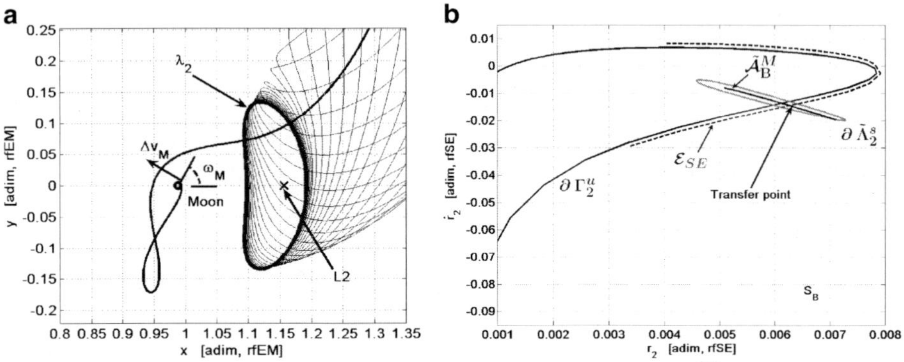

Sample impulsive capture trajectory. $\mathscr{E}_{SE}$ and $\tilde{\mathscr{A}}^M_{\mathbf{B}}(-t)$ sets.

Fig. 5 The first guess impulsive capture solution as the transfer point $\mathscr{B}^M_{-t} = \mathscr{E}_{\mathrm{SE}} \cap \tilde{\mathscr{A}}^M_{\mathbf{B}}(-t)$, the latter reported on section S_B in (**a**)

Each generic Moon capture orbit written in (16), at time $-t$, can be expressed as

$$\gamma_{\mathbf{B}}(\mathbf{y}_i, -t) = \{\phi_{\mathbf{B}}(\mathbf{y}_i, t_i; -\tau) | -\tau > -t\}, \tag{17}$$

where $\mathbf{B} = T\,(\mathbf{v}/v)$, $v = \sqrt{\dot{x}^2 + \dot{y}^2}$, $\mathbf{v} = \{\dot{x}, \dot{y}\}^\top$, and assuming $T = 0$. Since the first part of the transfer is defined on $\mathscr{E}_{\mathrm{SE}}$, the transfer points, if any, that generate two-impulse transfers are contained in the set

$$\mathscr{B}^M_{-t} = \mathscr{E}_{\mathrm{SE}} \cap \tilde{\mathscr{A}}^M_{\mathbf{B}}(-t). \tag{18}$$

Once again, the transformation $\mathscr{M}$ is required to map $\mathscr{A}^M_{\mathbf{B}}(-t)$, computed in the EM model, into $\tilde{\mathscr{A}}^M_{\mathbf{B}}(-t)$, defined in the SE model, see Fig. 5b. At this stage just first guesses are defined, states with small, tolerable mismatches can be admitted in $\mathscr{B}^M_{-t}$ as the discontinuities are spread in the subsequent optimization step.

4.2 Moon Low-Thrust Capture Stage

Dealing with the design of low-energy, low-thrust transfers to the Moon, the initial and final conditions at the Earth and the Moon, respectively, are assumed in the same way as described in Sect. 4.1 for the two-impulse trajectories. Actually, this case differs from the previous one in the fact that low-thrust contribution is taken into account. As part of the coupled RTBPs approximation, in connection with the Earth escape stage, the continuously propelled capture stage is used with the low-thrust term preliminary acting only in the EM model. Thus, the initial orbit is a trajectory belonging to the set $\mathscr{E}_{\mathrm{SE}}$, whereas the second part is defined using

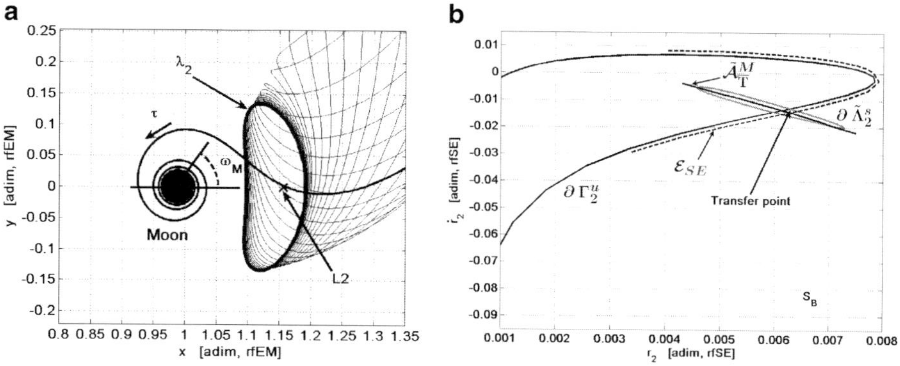

Sample low-thrust capture trajectory. $\mathscr{E}_{SE}$ and $\mathscr{A}^{\tilde{M}}_{\mathbf{T}}(-t)$ sets.

Fig. 6 The first guess low-thrust capture solution as the transfer point $\mathscr{T}^{M}_{-t} = \mathscr{E}_{\text{SE}} \cap \mathscr{A}^{\tilde{M}}_{\mathbf{T}}(-t)$, the latter reported on section S_B in (**b**)

a suitable attainable set. As this set constitutes the second part of the trajectory, it is made up of tangential low-thrust orbits that are integrated backward. More specifically, as both eccentricity and apsidal altitude are prescribed, the final state of the transfers (i.e. the periapsis point of the orbit about the Moon) is a function of the argument of periapsis, i.e. $\mathbf{y}_f = \mathbf{y}_f(\omega_{\text{M}})$, as shown in Fig. 6a. The domain of admissible final states therefore is

$$\mathscr{Y}^{M} = \{\mathbf{y}_f(\omega_{\text{M}}) | \omega_{\text{M}} \in [0, 2\pi]\}, \tag{19}$$

and the attainable set, for some $t \geq 0$ (i.e. $-t$ is a backward integration), containing low-thrust, ballistic capture trajectories is

$$\mathscr{A}^{M}_{\mathbf{T}}(-t) = \bigcup_{\mathbf{y}_f \in \mathscr{Y}^{M}} \gamma_{\overline{\mathbf{T}}}(\mathbf{y}_f(\omega_{\text{M}}), -t). \tag{20}$$

Since the first part of the transfer is defined on $\mathscr{E}_{\text{SE}}$, the transfer points, if any, that generate low-energy, low-thrust transfers are contained in the set

$$\mathscr{T}^{M}_{-t} = \mathscr{E}_{\text{SE}} \cap \mathscr{A}^{\tilde{M}}_{\mathbf{T}}(-t). \tag{21}$$

The transformation $\mathscr{M}$ is required to map $\mathscr{A}^{M}_{\mathbf{T}}(-t)$, computed in the EM model, into $\mathscr{A}^{\tilde{M}}_{\mathbf{T}}(-t)$, defined in the SE model, as shown in Fig. 6b. It is worth mentioning that first guess solutions are being generated in this step. These preliminary solutions have to be later optimized in a four-body context. Thus, small discontinuities can be tolerated when looking for the transfer point. This means that it is possible to intersect two states such that $\|\mathbf{y}_A - \mathbf{y}_E\| \leq \varepsilon$, where $\mathbf{y}_E \in \mathscr{E}_{\text{SE}}$, $\mathbf{y}_A \in \mathscr{A}^{\tilde{M}}_{\mathbf{T}}(-t)$, and ε is a prescribed tolerance. The greater ε is, the higher number of first guess solutions is found; however, ε should be kept sufficiently small to permit the convergence of the subsequent optimization step.

5 Trajectory Optimization

Once feasible and efficient first guess solutions are achieved, combining attainable sets with Earth-escape sets, an optimal control problem is stated in the BRFBP framework. The model used to take into account low-thrust propulsion and the gravitational attractions of all the celestial bodies involved in the design process (i.e. the Sun, the Earth, and the Moon) is

$$\ddot{x} - 2\dot{y} = \frac{\partial \Omega_S}{\partial x} + \frac{T_x}{m}, \quad \ddot{y} + 2\dot{x} = \frac{\partial \Omega_S}{\partial y} + \frac{T_y}{m}, \quad \dot{\theta} = \omega_S, \quad \dot{m} = -\frac{T}{I_{\mathrm{sp}}\, g_0}. \tag{22}$$

This is a modified version of the classic bicircular four-body problem [17] (see Fig. 7a) and, in principle, incorporates the perturbation of the Sun into the Earth–Moon PCRTBP described by (1). The four-body potential Ω_S reads

$$\Omega_S(x, y, \theta) = \Omega(x, y, \mu_{\mathrm{EM}}) + \frac{m_S}{r_S} - \frac{m_S}{\rho_S^2}(x\cos\theta + y\sin\theta). \tag{23}$$

The dimensionless physical constants introduced to describe the Sun perturbation are in agreement with those of the EM model. Thus, the distance between the Sun and the Earth–Moon barycenter is $\rho_S = 3.8878 \times 10^2$, the mass of the Sun is $m_S = 3.2890 \times 10^5$, and its angular velocity with respect to the EM rotating frame is $\omega_S = -9.2518 \times 10^{-1}$. The Sun is located at $(\rho_S \cos\theta, \rho_S \sin\theta)$, and therefore the Sun-spacecraft distance is calculated as

$$r_S^2 = (x - \rho_S \cos\theta)^2 + (y - \rho_S \sin\theta)^2. \tag{24}$$

This low-thrust version of the BRFBP is represented by the sixth-order system of differential equation (22).

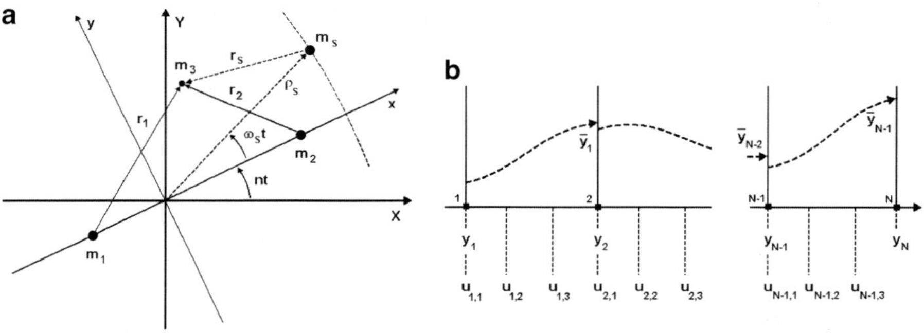

Sun-perturbed Earth–Moon RTBP. Direct multiple shooting scheme.

Fig. 7 Optimization process, dynamical model and integration scheme

According to the formalism proposed by Betts [4], the controlled BRFBP described by (22) is written in the first-order form

$$\begin{aligned}\dot{x} &= v_x \\ \dot{y} &= v_y \\ \dot{v}_x &= 2v_y + \Omega_{Sx} + T_x/m \\ \dot{v}_y &= -2v_x + \Omega_{Sy} + T_y/m \\ \dot{\theta} &= \omega_S \\ \dot{m} &= -T/(I_{\text{sp}}\, g_0),\end{aligned} \tag{25}$$

with $v_x = \dot{x}$ and $v_y = \dot{y}$. In a compact explicit form, system (25) reads

$$\dot{\mathbf{y}} = \mathbf{f}[\mathbf{y}(t), \mathbf{T}(t), \mathbf{p}, t], \tag{26}$$

where $\mathbf{f}$ stands for the vector field, $\mathbf{T} = \{T_x, T_y\}^\top$ is the thrust vector, and $\mathbf{y} = \{x, y, \dot{x}, \dot{y}, \theta, m\}^\top$ is the state vector. In addition, the dynamics is also allowed to incorporate some constant parameters $\mathbf{p}$ useful for the definition of the optimal control problem. The aim is finding, according to the standard optimal control theory, the guidance law $\mathbf{T} = \mathbf{T}(t)$, $t \in [t_i, t_f]$, that minimizes a prescribed scalar performance index or objective function

$$J = J(\mathbf{y}, \mathbf{T}, \mathbf{p}, t), \tag{27}$$

while satisfying certain mission constraints. These constraints are represented by the two boundary conditions, defined at the end points of the optimal control problem, and by the inequality conditions, defined along the whole arc. These last quantities are derived specifically for each of the three types of mission to the Moon that are investigated.

The optimal control problem (OCP) is then transcribed into a nonlinear programming, NLP, problem using a direct approach. This method, although suboptimal, generally shows robustness and versatility, and does not require explicit derivation of the necessary conditions of optimality. Moreover, direct approaches offer higher computational efficiency and are less sensitive to variation of the first guess solutions [4]. Furthermore, a multiple shooting scheme is implemented. With this strategy the BRFBP dynamics presented by (25) is forward integrated within $N-1$ intervals (in which $[t_i, t_f]$ is uniformly split), i.e. the time domain is divided in the form $t_i = t_1 < \cdots < t_N = t_f$, and the solution is discretized over the N grid nodes (see Fig. 7b). The continuity of position, velocity, and mass is imposed at their ends [7], in the form of defects $\eta_j = \bar{\mathbf{v}}_j - \mathbf{v}_{j+1} = 0$, for $j = 1, \ldots, N-1$. The quantity $\bar{\mathbf{v}}_j$ stands for the result of the integration, i.e. $\bar{\mathbf{v}}_j = \phi(\mathbf{v}_j, \mathbf{p}, t)$, $t_j \leq t_{j+1}$, and is made up of state variables and control variables (i.e. $\mathbf{v}_j = \{\mathbf{y}_j, \mathbf{T}_{j,k}\}^\top$, for $k = 1, \ldots, M-1$). The control law $\mathbf{T}(t)$ is described within each interval by means of cubic spline functions. The algorithm computes the value of the control at mesh points, satisfying both boundary and path constraints, and minimizing the performance index.

Dynamics described by (22) are highly nonlinear and, in general, lead to chaotic orbits. In order to find accurate optimal solutions without excessively increasing the computational burden, an adaptive nonuniform time grid has been implemented. Thus, when the trajectory is close to either the Earth or the Moon the grid is automatically refined, whereas in the intermediate phase, where a weak vector field governs the motion of the spacecraft, a coarse grid is used. The optimal solution found is assessed a posteriori by forward integrating the optimal initial condition (with RK scheme of the eighth order) and by cubic interpolation of the discrete optimal control solution. Not only the low-thrust portion, but rather the whole transfer trajectory is discretized and optimized, so allowing the low-thrust to act also in regions preliminary made up by coast arcs.

5.1 Two-Impulse Problem Statement

As far as it concerns low energy transfers to the Moon, designed by means of two impulsive maneuvers at both ends of the trajectory, the dynamical system described by (25) is characterized by the thrust magnitude always forced to be equal to zero, in order to describe the coast arcs which compose the whole Earth–Moon transfer.

According to the NLP problem recalled in Sect. 5, the variable vector is

$$\mathbf{x} = \{(\mathbf{y},\mathbf{B})_1,\ldots,(\mathbf{y},\mathbf{B})_N,t_1,t_N\}^\top, \tag{28}$$

where $\mathbf{B} = \{0,0\}^\top$ is forced, i.e. the thrust magnitude is $T = 0$ along the whole time domain of the transfer.

The initial conditions read:

$$\psi_i(\mathbf{y}_1,t_1) := \begin{cases} (x_1+\mu)^2+y_1^2-r_i^2 = 0 \\ (x_1+\mu)(\dot{x}_1-y_1)+y_1(\dot{y}_1+x_1+\mu) = 0, \end{cases} \tag{29}$$

which force the first $\mathbf{y}_1$ state of the transfer to belong to a circular orbit of radius $r_i = R_\text{E}+h_\text{E}$, where R_E and h_E stand for the Earth radius and the orbit altitude with respect to the Earth, respectively. In analogy, the final conditions at the Moon are:

$$\psi_f(\mathbf{y}_N,t_N) := \begin{cases} (x_N-1+\mu)^2+y_N^2-r_f^2 = 0 \\ (x_N-1+\mu)(\dot{x}_N-y_N)+y_N(\dot{y}_N+x_N-1+\mu) = 0, \end{cases} \tag{30}$$

according to which, the final $\mathbf{y}_N$ state of the transfer is on a circular orbit of radius $r_f = R_\text{M}+h_\text{M}$, where R_M and h_M stand for the Earth radius and the orbit altitude with respect to the Moon, respectively. Conditions written in (29) and (30) mean that in the reference frame centered in the Earth and the Moon, respectively, the position and velocity vectors are orthogonal [21]. The nonlinear equality constraint vector, made up of the boundary conditions and the ones representing the dynamics, is therefore written as follows:

$$\mathbf{c}(\mathbf{x}) = \{\psi_i, \eta_1, \dots, \eta_{N-1}, \psi_f\}^\top. \tag{31}$$

Moreover, aiming at avoiding the collision with the two primaries, the following inequality constraints are imposed:

$$\Psi_j^c(\mathbf{y}_j) := \begin{cases} R_\mathrm{E}^2 - (x_j+\mu)^2 - y_j^2 \le 0 \\ R_\mathrm{M}^2 - (x_j - 1 + \mu)^2 - y_j^2 \le 0, \end{cases} \qquad j = 2, \dots, N-1. \tag{32}$$

Finally, the flight time is searched to be positive, i.e.

$$\Psi^t = t_1 - t_N \le 0. \tag{33}$$

The complete inequality constraint vector therefore reads:

$$\mathbf{g}(\mathbf{x}) = \{\Psi_2^c, \dots, \Psi_{N-1}^c, \Psi^t\}^\top. \tag{34}$$

As for the performance index to minimize, this is a scalar that represents the two velocity variations at the beginning and at the arrival of the transfer, i.e. $J(\mathbf{x}) = \Delta v_1 + \Delta v_N$, where

$$\Delta v_1 = \sqrt{(\dot{x}_1 - y_1)^2 + (\dot{y}_1 + x_1 + \mu)^2} - v_i, \tag{35}$$

and where

$$\Delta v_N = \sqrt{(\dot{x}_N - y_N)^2 + (\dot{y}_N + x_N - 1 + \mu)^2} - v_f. \tag{36}$$

In summary, the NLP problem for the two-impulse transfers is formulated as follows:

$$\min_{\mathbf{x}} J(\mathbf{x}) \text{ subject to } \begin{array}{l} \mathbf{c}(\mathbf{x}) = 0, \\ \mathbf{g}(\mathbf{x}) \le 0. \end{array} \tag{37}$$

5.2 Low-Thrust Problem Statement

Dealing with low-energy low-thrust transfers to the Moon, their design is performed by means of an initial impulsive TLI maneuver, followed by a ballistic arc closed by a low-thrust capture around the Moon.

According to the NLP problem recalled in Sect. 5, the variable vector is

$$\mathbf{x} = \{(\mathbf{y}, \mathbf{T})_1, \dots, (\mathbf{y}, \mathbf{T})_N, t_1, t_N\}^\top, \tag{38}$$

where, in this case, the thrust vector is properly $\mathbf{T} = (T_x, T_y)^\top$. Actually, even if the first guess control law is assigned aligned with the synodic velocity of the spacecraft (tangential thrust of magnitude $\overline{T}$), the optimization process acts on all the variable of (38), and it is free to modify the control history, within the technological limits of the engine.

Then, the equality constraint vector, made up of initial and final conditions, together with the dynamics defect η satisfaction, is proposed in the same way as (31) of Sect. 5.1. Dealing with the inequality constraints, beside the ones preventing the collision of the spacecraft with the two primaries and the one that represents the positive time evolution of the trajectory, a path constraint is added. This is introduced to model the saturation of the low-thrust engine. Thus, the inequality $T(t) \leq T_{\max} = 0.5\,\mathrm{N}$, $t \in [t_i, t_f]$, is imposed along the whole transfer.

The optimal control step aims at finding the guidance law, $\mathbf{T}(t)$, $t \in [t_i, t_f]$, that minimizes the hybrid performance index

$$J = \rho \Delta v_1 + \int_{t_i}^{t_f} \frac{T(t)}{I_{\mathrm{sp}}\, g_0}\, \mathrm{d}t; \tag{39}$$

the first part stands for the initial impulsive translunar insertion maneuver (provided by the launcher and recalling (35)), while the second contribution corresponds to the propellant mass, $m_p = m_i - m(t_f)$, needed to perform the transfer. The tuning parameter ρ is a weight quantity introduced to wisely balance the two contributions in the objective index computing. It is worth noticing that the optimization process is performed on the NLP variables described by (38), which are already dimensionless and easy to deal with as they show the same order of magnitude. In order to eliminate possible discontinuities, arising when the intersection of the orbits is not exact, each first guess is first preprocessed to minimize

$$J = \frac{1}{2} \int_{t_i}^{t_f} \mathbf{u}^\top \mathbf{u}\, \mathrm{d}t, \tag{40}$$

with $\mathbf{u}(t) = \mathbf{T}(t)/m(t)$. The resulting solution is later used to minimize the objective function described by (39). Numerical experiments have shown that performing an intermediate step with objective function of (40) better enforces the satisfaction of boundary conditions and path constraints [11].

Finally, the NLP problem for the low-energy low-thrust transfers is formulated as proposed by (37) at the end of Sect. 5.1.

6 Optimized Transfer Solutions

In this section the transfer solutions arising from the optimization process are presented. In Sect. 1, two families of trajectories are discussed, according to different types of propulsion system. In the following, the optimized transfers to LLOs are proposed in terms of some relevant performance parameters.

Table 1 Two-impulse transfers and low-energy low-thrust transfers to LLOs. A set of impulsive reference solutions found in literature is also reported

Type	Δv_i [m/s]	Δv_f [m/s]	f_f [adim.]	f_t [adim.]	Δt [days]
Sol.1	3211	–	0.061	0.683	271
Sol.2	3203	–	0.050	0.681	145
Sol.3	3169	–	0.046	0.675	103
Sol.4	3143	650	0.198	0.724	88
Yag	3137	718	0.216	0.730	44
WSB	3161	677	0.206	0.729	90–120
BP	3232	721	0.217	0.739	∞
BE	3161	987	0.284	0.756	55–90
L1	3265	629	0.192	0.734	255
Hoh	3143	848	0.250	0.742	5
Min	3099	622	0.191	0.718	–

Yag two-impulse transfers [21], *WSB* weak stability boundary, *BP* bi-parabolic, *Hoh* Hohmann, *BE* bi-elliptic [3], *L1* via *L*1 transit orbits [20], *Min* theoretical minimum [18]

6.1 Trajectories to Low Orbits Around the Moon

Optimal two-impulse and low-energy low-thrust solutions are presented. These transfers start from a circular parking orbit at an altitude of $h_E = 167$ km around the Earth, and end at circular orbit around the Moon, at an altitude of $h_M = 100$ km. The results are shown in Table 1 as follows: the first three solutions (i.e. sol.1, sol.2, sol.3) correspond to the low-energy low-thrust transfers, while solution sol.4 represents a two-impulse low-energy transfer. Then, solutions below the horizontal line are some reference impulsive transfers found in literature [3, 18, 20, 21]; all of them begin from the same LEO and arrive at the same LLO.

Table 1 is organized as follows: the second column Δv_i stands for the initial impulsive maneuver that inserts the spacecraft onto the translunar trajectory. For the solutions computed in this thesis, they are a direct output of the optimization process, described in Sect. 5. The third column Δv_f represents the final impulsive maneuver that permits a stable permanent capture into a circular parking orbit around the Moon. This comes out from the optimization step for sol.4; it is not present for the low-thrust transfers, whereas for the reference solutions this term takes into account all the impulsive maneuvers necessary to carry out the transfers except for Δv_i (i.e. for an Hohmann transfer, only the second burn necessary to place the spacecraft into the final orbit about the Moon is considered; for a WSB transfer, this term has to take into account the possible mid-course maneuver as well as the final maneuver needed to place the spacecraft into the final orbit about the Moon; similar arguments apply also to the bi-elliptic and bi-parabolic transfers).

Then, the fourth column f_f is a direct output (i.e. it is part of the performance index to minimize, divided by $m_{TLI} = 1{,}000$ kg) only for the first three solutions, and stands for the propellant mass fraction required to perform the transfers, after the

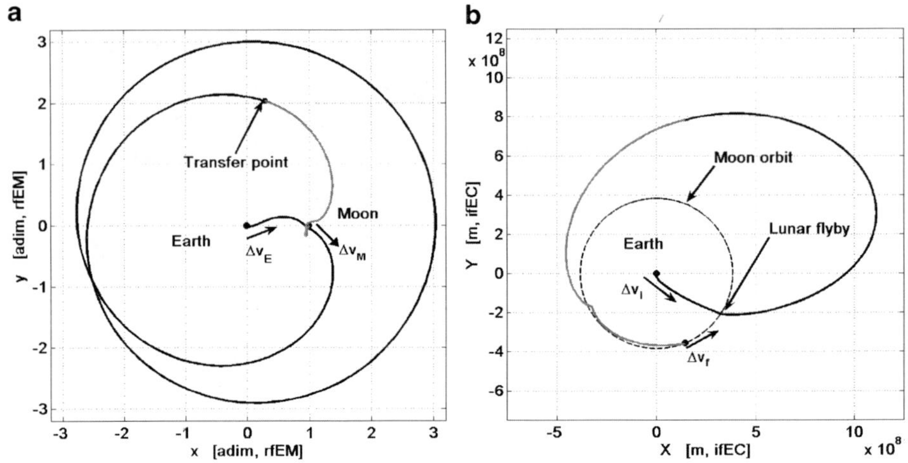

Optimized two-impulse trajectory, inertial geocentric reference frame.

First guess two-impulse trajectory, rotating EM reference frame.

Fig. 8 Optimized two-impulse transfer to a low orbit around the Moon, corresponding to solution 4 in Table 1

translunar insertion. As for the others, the value written in Table 1 is the propellant mass fraction associated with the impulsive Δv_f maneuver. This, through the rocket equation reads

$$f_f = \frac{m_p}{m_{\mathrm{TLI}}} = 1 - \exp\left(-\frac{\Delta v_f}{I_{\mathrm{sp}}^{\mathrm{ht}} g_0}\right), \tag{41}$$

where $I_{\mathrm{sp}}^{\mathrm{ht}} = 300\,\mathrm{s}$ is assumed as the typical specific impulse related to high-thrust chemical engines. The fifth column f_t represents the overall mass fraction necessary to complete the Earth–Moon transfers. Even if for low-energy low-thrust solutions, according to the design of the Earth escape stage described in Sect. 3, the initial Δv_i is given by the launch vehicle. For a sake of a fair comparison, the cost of this maneuver is considered, as written below:

$$f_t = \frac{m_p}{m_i} = \left[1 - \exp\left(-\frac{\Delta v_i}{I_{\mathrm{sp}}^{\mathrm{ht}} g_0}\right)\right] + \frac{1}{m_i}\int_{t_i}^{t_f} \frac{T(t)}{I_{\mathrm{sp}}^{\mathrm{lt}} g_0}\,\mathrm{d}t, \tag{42}$$

where $I_{\mathrm{sp}}^{\mathrm{lt}} = 3{,}000\,\mathrm{s}$ is assumed as the specific impulse related to low-thrust electrical engines. Finally, the last column on the right stands for the transfer time. Dealing with the last line, the minimum theoretical cost is computed via energetic considerations, and no solutions corresponding to such solution exist.

First of all, few considerations about the first three low-energy low-thrust transfers: moving from sol.1 to sol.3, the performances of the solutions become better, both in terms of Δv_i and Δt. This is due because the first guess trajectory related to

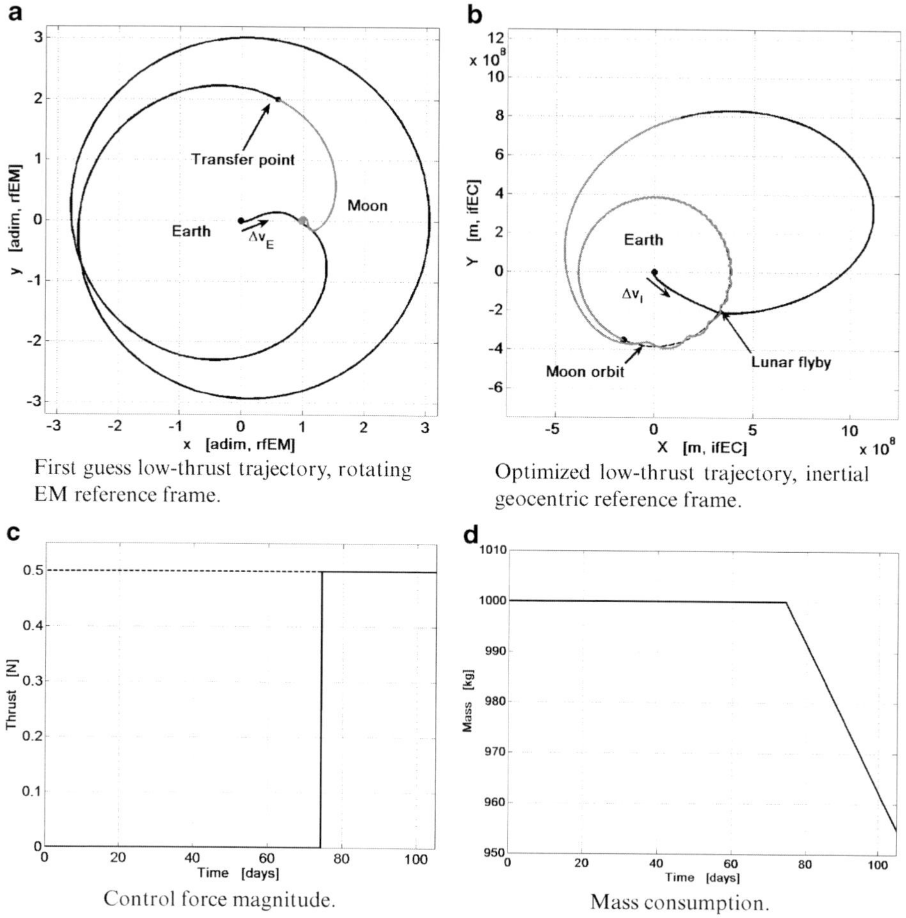

First guess low-thrust trajectory, rotating EM reference frame.

Optimized low-thrust trajectory, inertial geocentric reference frame.

Control force magnitude.

Mass consumption.

Fig. 9 Optimized low-energy low-thrust transfer to a low orbit around the Moon corresponding to solution 3 in Table 1

sol.3 (in particular the Earth escape stage) is designed taking advantage explicitly of a lunar flyby at the beginning of the transfer, thanks to the Moon-perturbed Sun–Earth restricted three-body model. Moreover, the plane where suitable Poincaré sections are considered and where the set $\mathscr{T}^M_{-t} = \mathscr{E}_{\mathrm{SE}} \cap \tilde{\mathscr{A}}^M_{\mathbf{T}}(-t)$ is defined, is chosen tuning shrewdly the φ_B and φ_C angles, in order to reduce the flight time.

In any case, sol.3, shown in Fig. 9, offers the lowest value of the overall mass consumption (see f_t). This happens for two reasons: first the fact that the low-thrust, $I_{\mathrm{sp}}^{\mathrm{lt}}$, is one order of magnitude greater than $I_{\mathrm{sp}}^{\mathrm{ht}}$. Second, the first guess solutions exploit deeply the dynamics of the RTBPs where they are designed, and later of the Earth–Moon BRFBP where they are optimized. Moreover, these trajectories take explicitly advantage of the initial lunar flyby. The latter can be seen as a kind of aid in the translunar orbit insertion, as it reduces the Δv_i required for that maneuver.

The two-impulse trajectory corresponding to sol.4, shown in Fig. 8, acknowledges these remarks, as it shows the lowest global $\Delta v = 3,793\,\mathrm{m/s}$ (with travel time $\Delta t = 88\,\mathrm{days}$). These considerations are in total accordance with what can be inferred from the rocket equation (see (42)). In order to reduce the total mass consumption, Δv has to decrease and I_{sp} has to increase. Finally, as far as it concerns the transfer times, the computed solutions presented turn out to be comparable with standard low energy transfers in terms of time of flight, although the latter are outperformed in terms of propellant mass.

Moreover, as far as it concerns sol.3, in Fig. 9c and in Fig. 9d are reported the control history profile and the mass consumption, respectively. It is evident that the thrust magnitude is always below the saturation limit (dashed line in Fig. 9c). Furthermore, when the engine is on duty (at the end of the trajectory in order to perform the low-thrust capture around the Moon), it works at the maximum level: the control profile is optimal as it recalls an *on–off* structure.

7 Conclusions

In this paper, two different technique to design Earth-to-Moon transfers have been investigated. The optimized solutions are revealed to be efficient, both in terms of propellant mass consumption and flight time.

References

1. Belbruno, E.: Lunar Capture Orbits, a Method of Constructing Earth Moon Trajectories and the Lunar Gas Mission. In: AIAA, DGLR, and JSASS, International Electric Propulsion Conference, 19th, Colorado Springs, CO, May 11-13, 1987. 10 p. (1987)
2. Belbruno, E.: The Dynamical Mechanism of Ballistic Lunar Capture Transfers in the Four-Body Problem from the Perspective of Invariant Manifolds and Hill'S Regions. Barcelona, Spain (1994)
3. Belbruno, E., Miller, J.: Sun-Perturbed Earth-to-Moon Transfers with Ballistic Capture. Journal of Guidance, Control and Dynamics **16**, 770–775 (1993)
4. Betts, J.: Survey of Numerical Methods for Trajectory Optimization. Journal of Guidance, Control and Dynamics **21**(2), 193–207 (1998)
5. Conley, C.: Low Energy Transit Orbits in the Restricted Three-Body Problem. SIAM Journal of Applied Mathematics **16**, 732–746 (1968)
6. Dellnitz, M., Junge, O., Post, M., Thiere, B.: On Target for Venus: Set Oriented Computation of Energy Efficient Low Thrust Trajectories. Celestial Mechanics and Dynamical Astronomy **95**, 357–370 (2006)
7. Enright, P., Conway, B.: Discrete Approximations to Optimal Trajectories Using Direct Transcription and Nonlinear Programming. Journal of Guidance, Control and Dynamics **15**, 994–1002 (1992)
8. Gómez, G., Koon, W., Lo, M., Marsden, J., Masdemont, J., Ross, S.: Invariant Manifolds, the Spatial Three-Body Problem, and Space Mission Design. Advances of the Astronautical Sciences **109**, 3–22 (2001)

9. Koon, W., Lo, M., Marsden, J., Ross, S.: Low Energy Transfer to the Moon. Celestial Mechanics and Dynamical Astronomy **81**, 63–73 (2001)
10. Llibre, J., Martinez, R., Simó, C.: Transversality of the Invariant Manifolds associated to the Lyapunov Family of Periodic Orbits near L_2 in the Restricted Three-Body Problem. Journal of Differential Equations **58**, 104–156 (1985)
11. Mingotti, G., Topputo, F., Bernelli-Zazzera, F.: Numerical Methods to Design Low-Energy, Low-Thrust Sun-Perturbed Transfers to the Moon. In: Proceedings of 49th Israel Annual Conference on Aerospace Sciences, Tel Aviv–Haifa, Israel, pp. 1–14 (2009)
12. Mingotti, G., Topputo F. Bernelli-Zazzera, F.: Combined Optimal Low-Thrust and Stable-Manifold Trajectories to the Earth–Moon Halo Orbits. In: American Institute of Physics Conference Proceedings, vol. 886, pp. 100–110 (2007)
13. Ozimek, M., Grebow, D., Howell, K.: Design of Solar Sail Trajectories with Applications to Lunar South Pole Coverage. Journal of Guidance, Control and Dynamics **32**(6) (2009)
14. Parker, J., Lo, M.: Shoot The Moon 3D. Advances in the Astronautical Sciences **123** (2006)
15. Perozzi, E., Di Salvo, A.: Novel Spaceways for Reaching the Moon: an Assessment for Exploration. Celestial Mechanics and Dynamical Astronomy **102**, 201–218 (2008)
16. Schoenmaekers, J., Horas, D., Pulido, J.: SMART-1: With Solar Electric Propulsion to the Moon. In: Proceedings of the 16th International Symposium on Space Flight Dynamics (2007)
17. Simó, C., Gómez, G., Jorba, A., Masdemont, J.: The Bicircular Model near the Triangular Libration Points of the RTBP. In: From Newton to Chaos, pp. 343–370 (1995)
18. Sweetser, T.: An Estimate of the Global Minimum DV Needed for Earth-Moon Transfer. In: Proceedings of the 1st AAS/AIAA Spaceflight Mechanics Meeting, pp. 111–120 (1991)
19. Szebehely, V.: Theory of Orbits: the Restricted Problem of Three Bodies. Academic Press New York (1967)
20. Topputo, F., Vasile, M., Bernelli-Zazzera, F.: Earth-to-Moon Low Energy Transfers Targeting L1 Hyperbolic Transit Orbits. Annals-New York Academy of Sciences **1065**, 55 (2005)
21. Yagasaki, K.: Sun-Perturbed Earth-to-Moon Transfers with Low Energy and Moderate Flight Time. Celestial Mechanics and Dynamical Astronomy **90**(3), 197–212 (2004)

Low-Energy Earth-to-Halo Transfers in the Earth–Moon Scenario with Sun-Perturbation

Anna Zanzottera, Giorgio Mingotti, Roberto Castelli, and Michael Dellnitz

1 Introduction

This paper deals with the design of trajectories connecting LEOs (low Earth orbits) with halo orbits around libration points of the Earth–Moon CRTBP [1] using impulsive maneuvers. The interest for such transfers comes mainly from the importance of halo orbits as possible locations for lunar far-side data relay satellites; in particular, a satellite evolving on a halo orbit will always maintain line of sight contact with the Earth and Moon's far side [6].

Indeed, it is taken into account a $h_{\mathrm{E}} = 167\,\mathrm{km}$ altitude parking orbit and a family of halo orbits associated to $L2$ Lagrangian point. As widely used in literature, suitable first guess trajectories are derived exploiting the *coupled circular restricted three-body problem approximation* [2, 7, 9]. This method consists in the superposition of two different CRTBPs, namely the SE (Sun–Earth) and EM (Earth–Moon) restricted problems. The transfer trajectories are obtained patching together on convenient Poincaré sections the invariant manifolds related to periodic orbits of the two systems. Such transfers are achieved if the Poincaré maps of the two problems intersect almost exactly in the configuration space.

In the current case, intersections between the stable manifolds – related to EM halo orbits – and Earth escape trajectories – integrated in the planar SE CRTBP – have to be detected. This is done using the software package GAIO [4]: n-dimensional Poincaré maps are replaced by a collection of n-dimensional *boxes*, each one identified by a vector containing its center and the radii in each dimension. The implemented box covering structure allows to deal with flows of different dimensions in an efficient way. Moreover, the accuracy of theintersections in the

A. Zanzottera (✉)
Dipartimento di Matematica F.Brioschi, Piazza Leonardo da Vinci, 33, 20133 Milano, Italy
e-mail: anna.zanzottera@mail.polimi.it

J.A.T. Machado et al., *Nonlinear and Complex Dynamics: Applications in Physical, Biological, and Financial Systems*, DOI 10.1007/978-1-4614-0231-2_3,

configuration space, as well as the discontinuities in terms of Δv, are controlled through the parameters of the box covering. This exploration is systematically performed combining various halo orbits and Earth-escaping trajectories, with a choice of different Poincaré sections.

Then, first guess solutions are optimized through a direct-method approach and multiple-shooting technique [3], in the framework of the Sun-perturbed Earth–Moon bicircular four-body problem [11]. Finally, trajectories with single-impulse and two-impulse maneuvers are presented and compared with results already known in literature [10].

2 Dynamical Models

In this section, we present the dynamical system that we will use to study the motion of a spacecraft in a field of three massive bodies.

2.1 Circular Restricted Three-Body Problem

The circular restricted three-body problem (CRTBP) [1] studies the motion of a massless particle P moving in the gravitational field of two main primaries, with masses $m_1 > m_2$. The primaries are supposed to move under their mutual gravity in circular orbits around the center of mass and their motion is not affected by the third particle. In this paper, P stands for the spacecraft while the role of primaries is played by the Sun and the Earth (SE CRTBP) or by the Earth and the Moon (EM CRTBP).

In a rotating reference frame where the units of measure are normalized so that the distance between the primaries, the modulus of their angular velocity and the total mass are equal to 1, the motion for the third body is governed by the equation

$$\begin{cases} \ddot{x} - 2\dot{y} = \Omega_x \\ \ddot{y} + 2\dot{x} = \Omega_y \\ \ddot{z} = \Omega_z \end{cases} \tag{1}$$

where $\Omega(x,y,z) = \frac{1}{2}(x^2+y^2) + \frac{1-\mu}{r_1} + \frac{\mu}{r_2} + \frac{1}{2}\mu(1-\mu)$ is the effective potential of the system and the subscripts denote partial derivatives. Here $\mu = m_2/(m_1+m_2)$ is the mass ratio, while $r_1^2 = (x+\mu)^2 + y^2 + z^2$ and $r_2^2 = (x-1+\mu)^2 + y^2$ are the distances from the spacecraft respectively to the larger and the smaller primary. The system (1) admits a first integral of motion, the Jacobi integral, defined as:

$$C(x,y,z,\dot{x},\dot{y},\dot{z}) = 2\Omega(x,y,z) - (\dot{x}^2 + \dot{y}^2 + \dot{z}^2) \tag{2}$$

As C^* varies, the family of five-dimensional energy manifolds

$$\mathcal{M}(C^*) = \{(x,y,z,\dot{x},\dot{y},\dot{z}) : C(x,y,z,\dot{x},\dot{y},\dot{z}) = C^*\}$$

is a foliation of the six-dimensional phase space. For every C^* the solution in the configuration space of the equation $C^* = 2\Omega(x,y,z)$ detects the zero-velocity surface, which bounds the Hill's region where the motion is possible.

The system (1) admits five equilibrium points, referred to as Lagrange points and denoted with Li, $i : 1\ldots 5$: three of them $L1, L2, L3$ lie on the x-axis and represent collinear configurations, while $L4, L5$ correspond to equilateral configurations.

The topology of the Hill's region changes in correspondence to the values C_i of the Jacobi constant relative to the libration points, allowing to open necks between different regions on the configuration space [8].

The collinear libration points, as well as the continuous families of planar and spatial periodic orbits surrounding them, have a saddle-center type stability character: the invariant manifold related to these orbits act as separatrices in the energy manifold and provide dynamical channels in the phase space useful for design low cost spacecraft trajectories.

2.2 *Bicircular Restricted Four-Body Problem*

The bicircular four-body model (BCRFBP) is a restricted four-body problem where two of the primaries (the Earth and the Moon) are moving in circular orbit around their center of mass B that is at the same time orbiting, together with the last mass (the Sun), around the barycenter of all the system. The motion of the primaries is supposed to be coplanar and with constant angular velocity. The equations of motion of the BCRFBP are written in the EM synodical reference frame and the physical quantities are normalized as in the EM CRTBP. Let m_s be the mass of the Sun, R_s the distance between the Sun and the origin of the frame B and ω_s the angular velocity of the Sun. Moreover let $\theta(t)$ be the angle between the Earth–Moon line and the Sun, and r_s the Sun-spacecraft distance (see Fig. 1):

$$r_s^2 = (x - R_s\cos\theta)^2 + (y - R_s\sin\theta)^2 + z^2. \tag{3}$$

The motion of the massless particle solves the system of differential equations

$$\begin{cases} \ddot{x} - 2\dot{y} = \Omega_{\mathrm{Bx}} \\ \ddot{y} + 2\dot{x} = \Omega_{\mathrm{By}} \\ \ddot{z} = \Omega_{\mathrm{Bz}}, \end{cases} \tag{4}$$

where $\Omega_B = \Omega + \frac{m_s}{r_s} - \frac{m_s}{R_s^2}(x\cos\theta + y\sin\theta)$. In literature it is common to refer to the dynamical system (4) as the Sun–Perturbed Earth–Moon Restricted Three-Body problem [11].

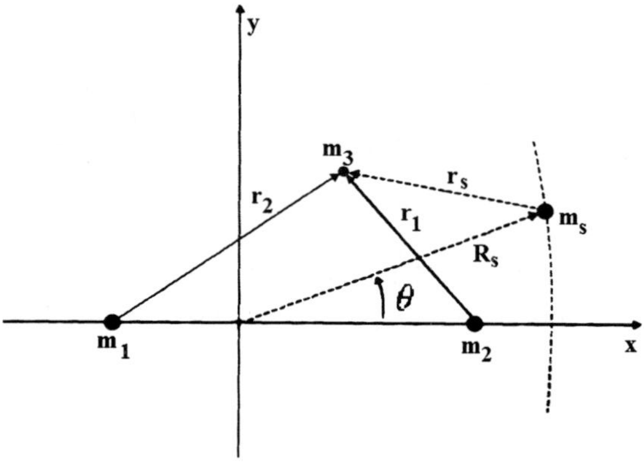

Fig. 1 Bicircular four body model

3 Trajectory Design

The aim of this paper is to find trajectories in the bicircular model, starting from a LEO and targeting a halo orbit around $L2$ in the EM-restricted problem. Since the phase space of the four-body system is poor of useful dynamical properties, the design is first performed in the CRTBP and then the initial guess trajectories are optimized to be solution of the bicircular model.

The model adopted as an approximation of the Sun–Earth–Moon-spacecraft restricted four-body problem is the so-called Patched Restricted Three-Body Problem. It consists the superposition of two CRTBPs with a common primary, namely the Sun–Earth CRTBP and the Earth–Moon CRTBP. The structure of the invariant manifolds associated with periodic orbits around the collinear libration points provides natural transfers from and to the smaller primaries. Then, by means of a Poincaré section the two legs of the trajectory are joined together, eventually with an impulsive maneuver, yielding a low-energy ballistic transfer.

According to this procedure, the design of LEO-to-halo trajectories is conceived in two different stages: the *Earth escape stage* and the *halo capture stage*. In the first part the planar SE CRTBP is exploited to leave a LEO orbit of $h_E = 167\,\text{km}$ of altitude, then the invariant manifold structure in the three-dimensional EM CRTBP is followed to obtain a natural transfer to halo orbits. The Poincaré section is chosen as a hyperplane in the phase space passing through the Earth and perpendicular to the (x, y) plane. The inclinations with respect to the positive x-semiaxis in the SE and EM synodical reference frame are denoted respectively with φ_{SE} and φ_{EM} and represent two of the design parameters.

3.1 Earth Escape Stage

As above mentioned, the departure leg of the transfer consists in a trajectory leaving a LEO orbit and integrated in planar SE CRTBP until the Poincaré section is

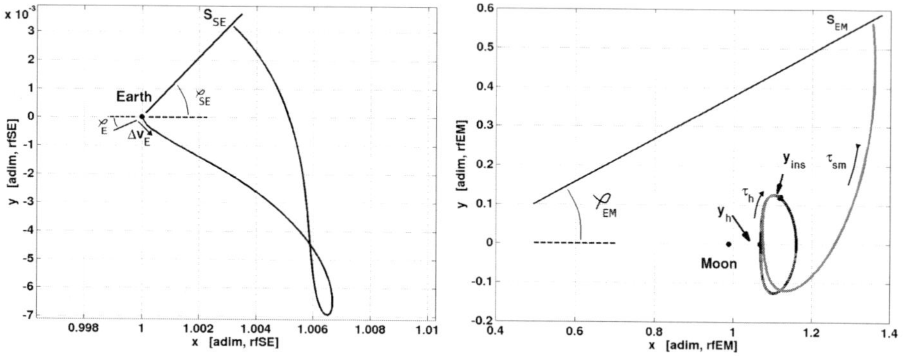

Fig. 2 The two stages of the Earth-to-halo trajectory design. (*left*) Earth escape trajectory. (*right*) Halo arrival trajectory

reached. The launch point $\mathbf{y}_{\mathrm{E}}(\varphi_{\mathrm{E}}, \Delta v_{\mathrm{E}})$ is identified by the angle φ_{E} that represents the position on LEO and by the Δv_{E} maneuver applied to insert the spacecraft into the translunar trajectory (see Fig. 2(left)). The maneuver is chosen tangential to the parking orbit with magnitude in the range $I = [3200, 3285]\,\mathrm{m/s}$. The choice of the interval I follows from the fact that the Jacobi constant, associated to the spacecraft, in the SE restricted model, needs to be decreased from the value related to a LEO orbit, $C_{\mathrm{SE}} \approx 3.07053$, to a value just below C_2, that has been shown to reveal good opportunities for the design.

For a large set of value of $\varphi_{\mathrm{E}} \in [0, 2\pi]$ and every integer value of $\Delta v_{\mathrm{E}} \in I$, the initial state $\mathbf{y}_{\mathrm{E}}(\varphi_{\mathrm{E}}, \Delta v_{\mathrm{E}})$ is forward integrated in the SE CRTBP at most for one unit of time. For a value of the angle φ_{SE}, let P_{SE} be the set of all the intersections that such paths have with the selected Poincaré section.

3.2 *Halo Capture Stage*

The second phase of the design exploits the (exterior) stable branch of the EM manifold related to a target halo orbit λ_2 around $L2$. The stable manifold $W^s(\lambda_2)$ is integrated backwards starting from λ_2 until the surface of the section is reached: let P_{EM} the associated Poincaré map.

The halo orbit is identified by the nominal point $\mathbf{y}_h$ where the plane $\{y = 0\}$ is intersected with positive $\dot{y}$, while the state $\mathbf{y}(t)$ of a generic point on the stable manifold is determined by two time parameters $(\tau_h, \tau_{\mathrm{sm}})$. Indeed, denoting with ϕ the flow, let $\mathbf{y}_{\mathrm{ins}} = \phi(\mathbf{y}_h, 0; \tau_h)$ be a point along the Halo orbit $(\tau_h \geq 0)$ and $\tilde{\mathbf{y}}_{\mathrm{ins}}$ a slight shift of $\mathbf{y}_{\mathrm{ins}}$ along the stable direction of the linearized system. The time $\tau_{\mathrm{sm}} < 0$ is a parameter of the trajectory $\mathbf{y}_{\mathrm{sm}} = \phi(\tilde{\mathbf{y}}_{\mathrm{ins}}, 0; \tau_{\mathrm{sm}})$ on the stable manifold (see Fig. 2(right)).

3.3 Box Covering Technique

As the parameters $\varphi_{\rm SE}$ and $\varphi_{\rm EM}$ vary, the design of the trajectory is restricted to the detection on the Poincaré section of two points, $\mathbf{y}_{\rm SE} \in P_{\rm SE}$ and $\mathbf{y}_{\rm EM} \in P_{\rm EM}$. First the map $P_{\rm SE}$ is transformed in EM synodical coordinate system, being $\beta = \varphi_{\rm SE} - \varphi_{\rm EM}$ the angle between the Moon and the Sun-Earth line. Then, in order to obtain a feasible transfer, the points $\mathbf{y}_{\rm SE}$ and $\mathbf{y}_{\rm EM}$ has to match almost exactly in configuration space. Moreover, they should minimize the distance in velocity space. Note that while it is possible to achieve intersections in configuration space between $P_{\rm SE}$ and the subset $P_{\rm EM} \cap \{|z| = 0\}$, a Δv maneuver it's always necessary. Indeed on the $\{z = 0\}$ plane the out of plane component of the velocity of the point $\mathbf{y}_{\rm EM}$ it's always different from zero.

The search of candidate transfer points is performed using a box covering structure implemented in the software package GAIO (Global Analysis of Invariant Objects), see [4] for a detailed description. An n-dimensional box $\mathscr{B}(C,R)$, identified by a center $C = (C_1, \cdots, C_N) \in \mathbb{R}^N$ and a vector of radii $R = (r_1, \cdots, r_n) \in \mathbb{R}^N$, is defined as

$$\mathscr{B}(C,R) = \cap_{i=1}^{N} \{(x_1, x_2, \ldots .x_N) \in \mathbb{R}^N : |x_i - C_i| < r_i\}.$$

Starting from an initial box $\mathscr{B}_0$ containing the projection of $P_{\rm EM}$ on the configuration space, a multiple subdivision process is carried out to create families $\mathscr{F}_k$ of smaller boxes $\{\mathscr{B}_k\}$ with the property to cover $\mathscr{B}_0$, i.e. $\bigcup \mathscr{B}_k = \mathscr{B}_0$. In the kth subdivision step each rectangle $\mathscr{B}(C,R)$ of the existing collection is subdivided along the j-th coordinate, where j can vary cyclically or can be fixed by the user. Once the radii of the boxes in $\mathscr{F}_k$ reach a prescribed size $\bar{\sigma}$, the Poincaré map is inserted: only those boxes $\mathscr{B}_k$ with nonempty intersection with $P_{\rm EM}$ are stored, otherwise removed. Denoting with $\mathscr{F}$ the collection of the remaining boxes, the feasibility condition previously discussed is fulfilled choosing the possible transfer points in the intersection $\mathscr{F} \cap P_{\rm SE}$.

In the numerical simulation here presented, the Poincaré maps are at first transformed in EM cylindrical coordinates $(r_1, \theta, z, \dot{r}_1, r_1 \dot{\theta}, \dot{z})$ centered on the Earth, then inserted into a collection of boxes with radii $\bar{\sigma}$, at most equal to 10^{-4} in r_1 and z coordinates. In Fig. 3(left), the intersection – in the space of configurations – of the Earth escape trajectories ($P_{\rm SE}$ black line) and the stable manifold related to the final halo target (grey curve $W^s(\lambda_2)$) is shown. The black dots ($W^s_{z=0}(\lambda_2)$) stand for the intersection of $P_{\rm SE}$ with the $\{z = 0\}$ plane. According to Fig. 3(right), it is possible to detect the candidate transfer points as the ones corresponding to $P_{\rm SE} \cap W^s(\lambda_2)$.

4 Trajectory Optimization

This section gives firstly a brief introduction of the trajectory optimization approach used in this work, then formulates in details the minimization problems later solved.

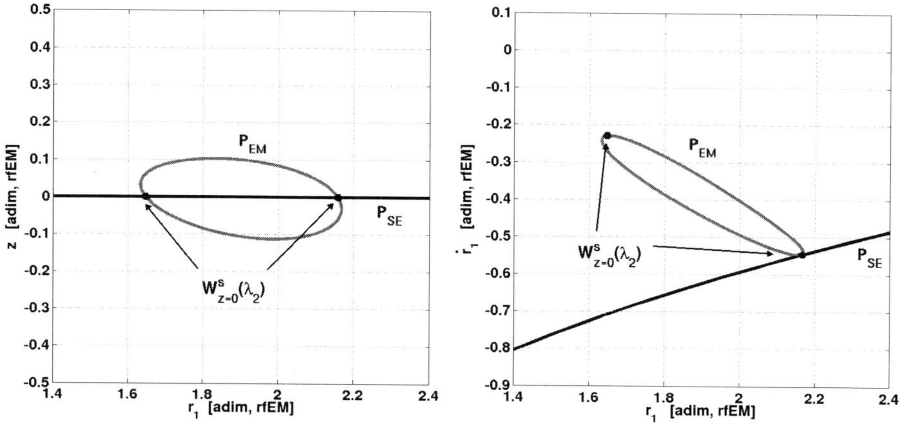

Fig. 3 Transfer point definition on suitable Poincaré maps. (*left*) $P_{\mathrm{SE}} \cap W^s(\lambda_2)$ in the r_1, z coordinates. (*right*) $P_{\mathrm{SE}} \cap W^s(\lambda_2)$ in the $r_1, \dot{r}_1$ coordinates

Once feasible and efficient first guess solutions are achieved, combining the two legs of the transfer, an optimization problem is stated. A given objective function is minimized taking into account the dynamic of the process. The dynamical model used to consider the gravitational attractions of all the celestial bodies involved in the design process (i.e. the Sun, the Earth, and the Moon) is the spatial BCRFBP described by (4) and here written in an autonomous fashion:

$$\begin{cases} \ddot{x} - 2\dot{y} = \Omega_{\mathrm{Bx}} \\ \ddot{y} + 2\dot{x} = \Omega_{\mathrm{By}} \\ \ddot{z} = \Omega_{\mathrm{Bz}} \\ \dot{\theta} = \omega_s. \end{cases} \tag{5}$$

According to the formalism proposed by [3], the BCRFBP described by (5) is written in the first-order form

$$\begin{aligned} \dot{x} &= v_x \\ \dot{y} &= v_y \\ \dot{z} &= v_z \\ \dot{v}_x &= 2v_y + \Omega_{\mathrm{Bx}} \\ \dot{v}_y &= -2v_x + \Omega_{\mathrm{By}} \\ \dot{v}_z &= \Omega_{\mathrm{Bz}} \\ \dot{\theta} &= \omega_s, \end{aligned} \tag{6}$$

with $v_x = \dot{x}$, $v_y = \dot{y}$ and $v_z = \dot{z}$. In a compact explicit form, system (6) read as

$$\dot{\mathbf{y}} = \mathbf{f}[\mathbf{y}(t), \mathbf{p}, t], \tag{7}$$

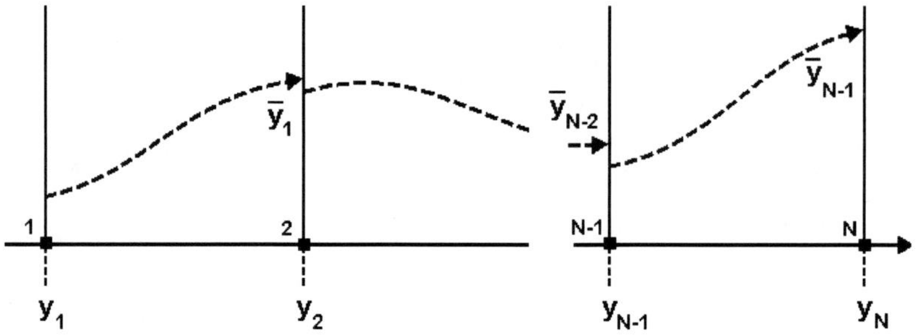

Fig. 4 Direct multiple shooting scheme

where **f** stands for the vector field and the state vector is $\mathbf{y} = \{x, y, z, \dot{x}, \dot{y}, \dot{z}, \theta\}^\top$. The aim is finding $\mathbf{y} = \mathbf{y}(t)$, $t \in [t_i, t_f]$, that minimizes a prescribed scalar performance index or objective function

$$J = J(\mathbf{y}, \mathbf{p}, t), \tag{8}$$

while satisfying certain mission constraints. These constraints are represented by the two boundary conditions, defined at the end points of the optimization problem, and by the inequality conditions, defined along the whole arc. These last quantities are derived specifically for the mission investigated. Moreover, **p** stands for a vector which brings together some free parameters useful for the optimization process.

The optimization problem, OP, is then transcribed into a nonlinear programming, NLP, problem using a direct approach. This method, although suboptimal, generally shows robustness and versatility, and does not require explicit derivation of the necessary conditions of optimality. Moreover, direct approaches offer higher computational efficiency and are less sensitive to variation of the first guess solutions [3]. Furthermore, a multiple shooting scheme is implemented. With this strategy the BCRFBP dynamics presented by (5) is forward integrated within $N-1$ intervals (in which $[t_i, t_f]$ is uniformly split), i.e. the time domain is divided in the form $t_i = t_1 < \cdots < t_N = t_f$, and the solution is discretized over the N grid nodes (see Fig. 4). The continuity of position and velocity is imposed at their ends [5], in the form of defects $\eta_j = \bar{\mathbf{y}}_j - \mathbf{y}_{j+1} = 0$, for $j = 1, \ldots, N-1$. The quantity $\bar{\mathbf{y}}_j$ stands for the result of the integration, i.e. $\bar{\mathbf{y}}_j = \phi(\mathbf{y}_j, \mathbf{p}, t)$, $t_j \leq t_{j+1}$. The algorithm computes the value of the states at mesh points, satisfying both boundary and path constraints, and minimizing the performance index.

Dynamics described by (5) are highly nonlinear and, in general, lead to chaotic orbits. In order to find accurate optimal solutions without excessively increasing the computational burden, an adaptive nonuniform time grid has been implemented. Thus, when the trajectory is close to either the Earth or the Moon the grid is automatically refined, whereas in the intermediate phase, where a weak vector field governs the motion of the spacecraft, a coarse grid is used. The optimal solution found is assessed a posteriori by forward integrating the optimal initial condition (with a Runge–Kutta 8th order scheme).

4.1 Two-Impulse Problem Statement

In this section, the approach previously described is exploited to obtain optimal transfers with two-impulsive maneuvers.

According to the NLP formalism recalled, the variable vector is

$$\mathbf{x} = \{\mathbf{y}_1, \ldots, \mathbf{y}_N, \mathbf{p}, t_1, t_N\}^\top, \tag{9}$$

where $\mathbf{p} = \{\tau_h, \tau_{\text{sm}}\}$, is made up of two free optimization parameters useful to describe the final condition of the transfer (see Fig. 2).

The initial conditions read:

$$\psi_i(\mathbf{y}_1, t_1) := \begin{cases} (x_1+\mu)^2 + y_1^2 + z_1^2 - r_i^2 = 0 \\ (x_1+\mu)(\dot{x}_1 - y_1) + y_1(\dot{y}_1 + x_1 + \mu) + z_1\dot{z}_1 = 0, \end{cases} \tag{10}$$

which force the first $\mathbf{y}_1$ state of the transfer to belong to a circular orbit of radius $r_i = R_{\text{E}} + h_{\text{E}}$, where R_{E} and h_{E} stand for the Earth radius and the orbit altitude with respect to the Earth, respectively. The transfer ends when the spacecraft flies on the stable manifold related to the final halo. In details, only the continuity in terms of position is imposed, so that the final condition reads

$$\psi_f = \bar{\mathbf{y}}_N - \bar{\mathbf{y}}_{\text{sm}} = 0, \tag{11}$$

where it is worth noting that $\bar{\mathbf{y}}_N = \{x_N, y_N, z_N\}^\top$ and $\bar{\mathbf{y}}_{\text{sm}} = \{x_{\text{sm}}, y_{\text{sm}}, z_{\text{sm}}\}^\top$. This means that, after the initial impulsive maneuver, a second one is required to inject the spacecraft onto the stable manifold that takes it ballistically to the final halo orbit associated.

The nonlinear equality constraint vector, made up of the boundary conditions and the ones representing the dynamics, is therefore written as follows:

$$\mathbf{c}(\mathbf{x}) = \{\psi_i, \eta_1, \ldots, \eta_{N-1}, \psi_f\}^\top. \tag{12}$$

Moreover, aiming at avoiding the collision with the two primaries, the following inequality constraints are imposed:

$$\Psi_j^c(\mathbf{y}_j) := \begin{cases} R_{\text{E}}^2 - (x_j+\mu)^2 - y_j^2 - z_j^2 \le 0 \\ R_{\text{M}}^2 - (x_j - 1 + \mu)^2 - y_j^2 - z_j^2 \le 0, \end{cases} \qquad j = 2, \ldots, N-1. \tag{13}$$

Finally, the flight time is searched to be positive, i.e.

$$\Psi^t = t_1 - t_N \le 0. \tag{14}$$

The complete inequality constraint vector therefore reads:

$$\mathbf{g}(\mathbf{x}) = \{\Psi_2^c, \dots, \Psi_{N-1}^c, \Psi^t\}^\top. \tag{15}$$

The performance index to minimize is a scalar that represents the two velocity variations at the beginning and at the final node of the transfer, i.e. $J(\mathbf{x}) = \Delta v_1 + \Delta v_N$. In details,

$$\Delta v_1 = \sqrt{(\dot{x}_1 - y_1)^2 + (\dot{y}_1 + x_1 + \mu)^2 + (z_1)^2} - v_i, \tag{16}$$

assuming $v_i = \sqrt{(1-\mu)/r_i}$ as the velocity along the initial circular parking orbit, and

$$\Delta v_N = \sqrt{(\dot{x}_N - \dot{x}_{\mathrm{sm}})^2 + (\dot{y}_N - \dot{y}_{\mathrm{sm}})^2 + (\dot{z}_N - \dot{z}_{\mathrm{sm}})^2}, \tag{17}$$

which represents the discontinuity in terms of velocity between the translunar trajectory and the stable manifold related to the final halo.

In summary, the NLP problem for the two-impulse transfers is formulated as follows:

$$\min_{\mathbf{x}} J(\mathbf{x}) \text{ subject to } \begin{aligned} \mathbf{c}(\mathbf{x}) &= 0, \\ \mathbf{g}(\mathbf{x}) &\le 0. \end{aligned} \tag{18}$$

4.2 Single-Impulse Problem Statement

Aiming to obtain single-impulse trajectories, the equality constraint vector is defined, as in the previous paragraph, by (12) with the exception of the final condition. In this case, at the final point of the transfer, the whole dynamical state is forced to be equal to the one associated to the stable manifold:

$$\psi_f = \mathbf{y}_N - \mathbf{y}_{\mathrm{sm}} = 0, \tag{19}$$

where $\mathbf{y}_N = \{x_N, y_N, z_N, \dot{x}_N, \dot{y}_N, \dot{z}_N\}^\top$ and $\mathbf{y}_{\mathrm{sm}} = \{x_{\mathrm{sm}}, y_{\mathrm{sm}}, z_{\mathrm{sm}}, \dot{x}_{\mathrm{sm}}, \dot{y}_{\mathrm{sm}}, \dot{z}_{\mathrm{sm}}\}^\top$.

Instead, the inequality constraint vector is defined in the same way as in the two-impulse scenario.

Dealing with the objective index to minimize, this is made up of only the initial velocity maneuver, i.e. $J(\mathbf{x}) = \Delta v_1$, where

$$\Delta v_1 = \sqrt{(\dot{x}_1 - y_1)^2 + (\dot{y}_1 + x_1 + \mu)^2 + (z_1)^2} - v_i. \tag{20}$$

Finally, the NLP problem for the low-energy low-thrust transfers is formulated as proposed by (18) at the end of the previous section.

5 Optimized Transfer Solutions

In this section, the transfers to halos obtained solving the optimization process are presented. In the previous two sections, two families of trajectories are discussed according to the number of impulsive maneuvers that are allowed. In the following, the optimized solutions are proposed in terms of some relevant performance parameters.

5.1 Trajectories to Halos

Optimal two-impulse and single-impulse solutions are presented. These transfers start from a circular parking orbit at an altitude of $h_{\mathrm{E}} = 167\,\mathrm{km}$ around the Earth, and reach a halo orbit around $L2$, with an out-of-plane amplitude of $A_z = 8{,}000\,\mathrm{km}$. Referring to Table 1: the first sol.1 corresponds to the two-impulse low energy transfer, while solution sol.2 represents a single-impulse low energy transfer. The last two solutions presented are some reference impulsive transfers found in literature.

More in details, Table 1 is so structured: the second column Δv_i stands for the initial impulsive maneuver that inserts the spacecraft onto the translunar trajectory. The third column Δv_f represents the final impulsive maneuver that permits the insertion of the spacecraft onto the stable manifold related to the target halo. The fourth column Δv_t represents the overall amount of impulsive maneuvers necessary to complete the Earth-to-halo transfers. Finally, the last column on the right stands for the transfer time.

An analysis of the table shows that the single-impulse sol.2 offers the lowest value of the overall impulsive maneuvers (see Δv_t). This happens because the first guess solution takes explicitly advantage of the initial lunar flyby (see Fig. 5(left)). The latter can be seen as a kind of aid in the translunar orbit insertion, as it reduces the Δv_i required for that maneuver. The lunar flyby performs a change of plane of the translunar trajectory that allows the insertion of the spacecraft onto the three-dimensional halo stable manifold without any other maneuver. Summarizing, the single-impulse trajectory corresponding to sol.2 acknowledges these remarks, as it shows the lowest global $\Delta v_t = 3161\,\mathrm{m/s}$ (with travel time $\Delta t = 98\,\mathrm{days}$).

Table 1 Two-impulse and single-impulse low energy transfers to halos around $L2$. A set of impulsive reference solutions found in literature is also reported [10]

Type	Δv_i (m/s)	Δv_f (m/s)	Δv_t (m/s)	Δt (days)
Sol.1	3110	214	3,324	97
Sol.2	3161	0	3,161	98
Parker.1	3132	618	3,750	–
Parker.2	3235	0	3,235	–

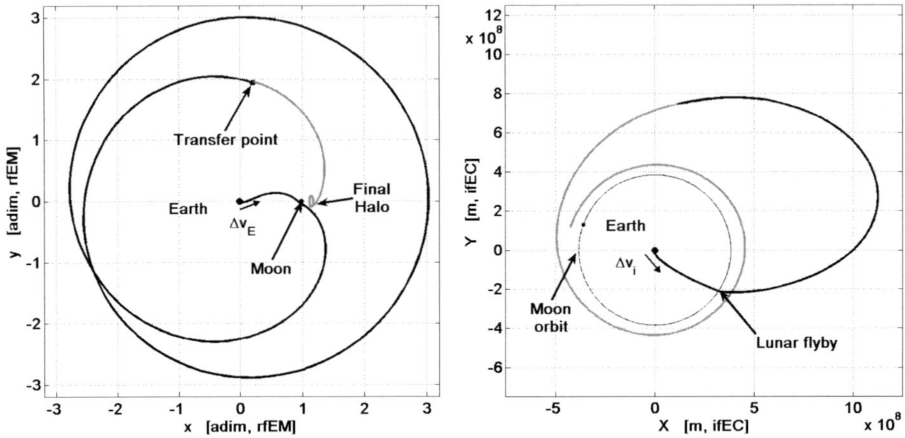

Fig. 5 First guess solution and optimized trajectory corresponding to sol.2 of Table 1. (*left*) First guess solution in Earth–Moon rotating frame. (*right*) Optimized trajectory in Earth centered inertial frame

6 Conclusion and Future Works

In this paper, a technique to design low-energy transfers to halo orbits has been investigated. Through the box-covering technique, an immediate definition of the transfer points in the phase space was possible, formalizing in a systematic way the intersection of three-dimensional manifolds. The optimization approach resulted robust and versatile, allowing to obtain single and two impulse transfer efficiently both in terms of flight time and Δv.

A more detailed analysis of the design process and further results will be presented and discussed in a future paper.

Acknowledgments This work was supported by the Marie Curie Actions Research and Training Network AstroNet, Contract Grant No. MCRTN-CT-2006-035151.

References

1. Szebehely, V.: Theory of Orbits: the Restricted Problem of Three Bodies, Academic Press New York (1967)
2. Belbruno, E.A. and Miller, J.K.: Sun-Perturbed Earth-to-Moon Transfers with Ballistic Capture. Journal of Guidance Control and Dynamics **16**, 770–775 (1993)
3. Betts, J.T.: Survey of Numerical Methods for Trajectory Optimization. Journal of Guidance control and dynamics **21**, 193–207 (1998)
4. Dellnitz, M. and Junge, O.: Set Oriented Numerical Methods for Dynamical Systems. Handbook of dynamical systems **2**, North-Holland (2002)

5. Enright, P.J. and Conway, B.A.: Discrete Approximations to Optimal Trajectories Using Direct Transcription and Nonlinear Programming. Journal of Guidance Control and Dynamics **15**,994–1002 (1992)
6. Farquhar, R.: Future Missions for Libration-Point Satellites. Astronautics and Aeronautics **7**, 52–56 (1969)
7. Gómez, G. and Koon, WS and Lo, MW and Marsden, JE and Masdemont, J. and Ross, SD: Connecting orbits and invariant manifolds in the spatial restricted three-body problem. Nonlinearity **17**, IOP Publishing (2004)
8. Koon, W.S. and Lo, M.W. and Marsden, J.E. and Ross, S.D.: Heteroclinic Connections Between Periodic Orbits and Resonance Transitions in Celestial Mechanics. Chaos: An Interdisciplinary Journal of Nonlinear Science **10**,(2000)
9. Koon, W. and Lo, M. and Marsden, J. and Ross, S.: Low Energy Transfer to the Moon. Celestial Mechanics and Dynamical Astronomy **81**, 63–73 (2001)
10. Parker, J.S.: Families of Low-Energy Lunar Halo Transfers. AAS/AIAA Spaceflight Dynamics Conference **90**, 1–20 (2006)
11. Simó, C. and Gómez, G. and Jorba, A. and Masdemont, J.: The Bicircular Model near the Triangular Libration Points of the RTBP. From Newton to Chaos pp. 343–370 (1995)

On the Relation Between the Bicircular Model and the Coupled Circular Restricted Three-Body Problem Approximation

Roberto Castelli

1 Introduction

Starting from the work of Belbruno and Miller [1], where the perturbation of the Sun has been shown to decrease the amount of fuel necessary for a Earth-to-Moon transfer, the Sun–Earth–Moon-spacecraft restricted 4-body model has been commonly adopted to describe the spacecraft motion.

One of the techniques used to approximate the 4-body dynamics, or in general the n-body problem, is the Coupled restricted three-body problem approximation: partial orbits from different restricted problems are connected into a single trajectory, yielding energy efficient transfers to the Moon [12], interplanetary transfers [6] or very complicated itineraries [10].

The procedure requires the choice of a Poincaré section where the phase spaces of the two different models have to intersect: the analysis of the Poincaré maps of the invariant manifolds reduces the design of the trajectory to the selection of a point on the section. The Poincaré section plays also a role in the accuracy of the approximation of the undertaken dynamical system: indeed the encounter with the Poincaré section is the criteria for switching from the first to the second restricted three-body problem. In the mentioned works, the Poincaré section is chosen *a priori* in order to accomplish certain design constraints or to simplify the selection of the connection point and it usually consists in a hyperplane passing through one of the primaries or lying along one of the coordinated axis.

Although it has been shown that for design purpose the solutions in a simplified model like the Circular restricted three-body problem (CR3BP) are very good approximations to real trajectories in the complicated and full system [15], this

R. Castelli (✉)
University of Paderborn, WarburgerStr. 100, 33098 Paderborn, Germany
e-mail: robertoc@math.upb.de

J.A.T. Machado et al., *Nonlinear and Complex Dynamics: Applications in Physical, Biological, and Financial Systems*, DOI 10.1007/978-1-4614-0231-2_4,

work deepens from a more theoretical point of view on the role played by the two restricted three-body problems in the approximation of the 4-body system.

The undertaken model considered here for the 4-body dynamics is the Bicircular model (BCP), [16], while the two restricted problems are the Earth–Moon CR3BP and the Sun–(Earth+Moon) CR3BP, where, in the last case, the Sun and the Earth–Moon barycenter act as primaries. The comparison of the mentioned systems leads to the definition of *Regions of Prevalence* in the space where one of the restricted problems performs, at least locally, the best approximation of the Bicircular model and therefore it should be preferred in designing the trajectory.

Then, setting the Poincaré section according to this prevalence, the coupled CR3BP approximation is implemented to design low energy transfers leaving Lyapunov orbits in the Sun–Earth system and leading to the Moon's region.

The plan of the paper is the following. In the next section, the CR3BP is briefly recalled and the equations of motion for the BCP in the inertial reference frame are written. Then, in Sect. 3, the comparison between the BCP and each one of the restricted problems is performed: this analysis enables to define, in Sect. 4, the regions of prevalence of the two restricted systems in the approximation of the 4-body model. Section 5 concerns the design of the transfer trajectory, while Sect. 6 deepens on the numeral scheme used to analyze the Poincaré maps for the selection of the connection points. Finally in the last section, some of the results are discussed.

2 Dynamical Models

2.1 Circular Restricted Three-Body Problem

The CR3BP is a simplified case of the general Three-Body Problem and models the motion of a massless particle under the gravitational influence of two bodies, with masses $M_1 < M_2$, that are revolving with constant angular velocity in circular orbit around their center of mass, see [17]. In the following, only the planar motion is considered.

In a rotating reference frame centered in the center of mass and with the units of measure normalized so that the total mass, the distance between the primaries and their angular velocity are equal to 1, the motion $z(t) = x(t) + \mathrm{i}y(t)$ of the massless particle evolves following the differential equation

$$\frac{\mathrm{d}^2 z}{\mathrm{d}t^2} + 2\mathrm{i}\frac{\mathrm{d}z}{\mathrm{d}t} - z = -\left[\frac{(1-\mu)(z+\mu)}{\| z+\mu \|^3} + \frac{\mu(z-(1-\mu))}{\| z-(1-\mu) \|^3}\right] \tag{1}$$

where $\mu = M_2/(M_1 + M_2)$ is the mass ratio. Note that the primaries are fixed on the x-axis and the bigger and the smaller mass are placed in positions $(-\mu, 0)$ and $(1-\mu, 0)$ respectively.

In (x,y) components, the equation of motion assumes the form

$$\ddot{x} - 2\dot{y} = \Omega_x, \qquad \ddot{y} + 2\dot{x} = \Omega_y$$

where $\Omega(x,y) = (x^2+y^2)/2 + (1-\mu)/r_1 + \mu/r_2 + \mu(1-\mu)/2$ is the potential function. The subscripts of Ω denote the partial derivatives, while $r_{1,2}$ are the distances between the moving particle and the primaries. The advantage to study the dynamics in a rotating frame is that the system (1) is Hamiltonian and autonomous and admits a first integral called Jacobi constant

$$J(x,y,\dot{x},\dot{y}) = -(\dot{x}^2+\dot{y}^2) + 2\Omega(x,y).$$

Therefore, the phase space is foliated in 3-dimensional energy manifolds

$$E(h) = \{(x,y,\dot{x},\dot{y}) \in R^4 : J(x,y,\dot{x},\dot{y}) = h\}$$

whose projections onto the configuration space are known as Hill's regions. For any fixed value of h, the Hill's region prescribes the region where the particle is allowed to move.

The potential Ω admits five critical points, the Lagrangian points $L_i, i = 1\dots5$, that represent equilibrium points for the vector field. The points $L_{4,5}$ correspond to equilateral triangle configurations, while the remaining are placed on the x-axis and correspond to collinear configurations of the masses. Of particular interest for mission design are L_1 and L_2 and the periodic orbits surrounding them that play the role of gates in the Hill's region, see for instance [11].

2.2 *Bicircular Model*

The Bicircular model (BCP), see [4], consists in a restricted four-body problem where two of the primaries are rotating around their center of mass, which is meanwhile revolving together with the third mass around the barycenter of the system. The massless particle is moving under the gravitational influence of the primaries and does not affect their motion. It is assumed that the motion of the primaries, like the motion of the test particle, is coplanar. The low eccentricity of the Earth's and Moon's orbit and the small inclination of the Moon's orbital plane, make the Bicircular a quite accurate model to describe the dynamics of a spacecraft in the Sun–Earth–Moon system; see for instance [18] and [14].

Referring to Fig. 1, let S, E, M be the positions of the three primaries, namely the Sun, the Earth, and the Moon while B and O indicate respectively the Earth–Moon barycenter and the total center of mass of the system. For a given time–space system of measure, let us define w_1 and w_2 the angular velocities of the couple S and B around O and of the couple E and M around B; L_S and L_M the distances

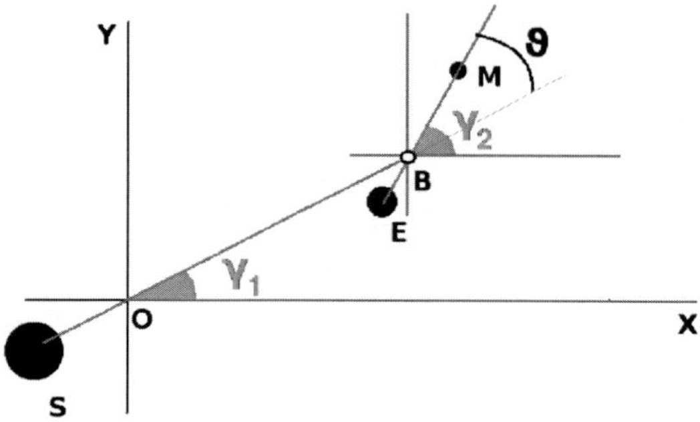

Fig. 1 Positions of the primaries in the inertial reference frame

$\overline{SB}$ and $\overline{EM}$; M_m, M_e, M_s the masses of the Moon, the Earth and the Sun and G the gravitational constant. Moreover let μ_s and μ_m be the mass ratios

$$\mu_m = \frac{M_m}{M_e + M_m}, \qquad \mu_s = \frac{M_e + M_m}{M_e + M_m + M_s} \,. \tag{2}$$

With respect to an inertial reference frame (X,Y) whose origin is fixed in the barycenter O and where τ denotes the time coordinate, the positions of the primaries are given by

$$\begin{aligned} S &= -\mu_s L_S \mathrm{e}^{\mathrm{i}(\varphi_0 + w_1 \tau)} \\ E &= (1-\mu_s) L_S \mathrm{e}^{\mathrm{i}(\varphi_0 + w_1 \tau)} - \mu_m L_M \mathrm{e}^{\mathrm{i}(\phi_0 + w_2 \tau)} \\ M &= (1-\mu_s) L_S \mathrm{e}^{\mathrm{i}(\varphi_0 + w_1 \tau)} + (1-\mu_m) L_M \mathrm{e}^{\mathrm{i}(\phi_0 + w_2 \tau)}. \end{aligned}$$

In order to lighten the notation, we define

$$\gamma_1(\tau) = \varphi_0 + w_1 \tau, \quad \gamma_2(\tau) = \phi_0 + w_2 \tau.$$

The motion $Z(\tau) = X(\tau) + \mathrm{i}Y(\tau)$ of the spacecraft is governed by the second order differential equation

$$\frac{\mathrm{d}^2 Z}{\mathrm{d}\tau^2} = -G \left[\frac{M_s(Z-S)}{\| Z-S \|^3} + \frac{M_e(Z-E)}{\| Z-E \|^3} + \frac{M_m(Z-M)}{\| Z-M \|^3} \right]. \tag{3}$$

Later on the BCP will be compared with two different restricted three-body problems: the $\mathrm{CR3BP_{EM}}$ and the $\mathrm{CR3BP_{SE}}$. The role of primaries is played by the Earth and the Moon in the first system and by the Sun and the barycenter B with mass $M_b = M_e + M_m$ in the latter. Three different reference frames and different units of measure are involved in the analysis: the inertial reference frame and the SE-synodical reference frame, whose origin is set in the center of mass O, and the EM-synodical reference frame centered on the point B.

2.3 Change of Coordinates

Following the notation previously adopted, let (X,Y,τ) be the space–time coordinates in the inertial reference frame and the small letters (x,y,t) the coordinates in the rotating systems. When necessary, in order to avoid any ambiguity, the subscript (x_s,y_s,t_s) and (x_m,y_m,t_m) are used to distinguish the set of coordinates in the CR3BP$_{\mathrm{SE}}$ and in the CR3BP$_{\mathrm{EM}}$ respectively. In complex notation

$$Z := X + \mathrm{i}Y, \qquad z_m := x_m + \mathrm{i}y_m, \qquad z_s := x_s + \mathrm{i}y_s$$

the relations between the inertial and the synodical coordinates are given by

$$Z = L_{\mathrm{S}} z_s \mathrm{e}^{\mathrm{i}\gamma_1}, \qquad \tau = \frac{t_s}{w_1}$$

$$Z = (1-\mu_s)L_{\mathrm{S}}\mathrm{e}^{\mathrm{i}\gamma_1} + L_{\mathrm{M}} z_m \mathrm{e}^{\mathrm{i}\gamma_2}, \qquad \tau = \frac{t_m}{w_2}.$$

Concerning the two synodical systems, the relation between the time coordinates t_s and t_m is easily derived

$$t_s = \frac{w_1}{w_2} t_m$$

while the transformation between the space coordinates (x_s,y_s) and (x_m,y_m) depends on the mutual position of the primaries. Let θ be the angle between the positive x_s-semiaxis and the positive x_m-semiaxis, see Fig. 1:

$$\theta(\tau) := \gamma_2 - \gamma_1 = \theta_0 + (w_2 - w_1)\tau.$$

For any value of θ, the position and the velocity of a particle in the two different synodical systems satisfy the relations

$$\begin{aligned} z_m &= \frac{L_{\mathrm{S}}}{L_{\mathrm{M}}} \mathrm{e}^{-\mathrm{i}\theta}(z_s - (1-\mu_s)) \\ \frac{\mathrm{d}z_m}{\mathrm{d}t_m} &= \frac{L_{\mathrm{S}}}{L_{\mathrm{M}}}\frac{w_1}{w_2} \mathrm{e}^{-\mathrm{i}\theta}\left[i\left(1-\frac{w_2}{w_1}\right)(z_s-(1-\mu_s)) + \frac{\mathrm{d}z_s}{\mathrm{d}t_s}\right] \end{aligned} \tag{4}$$

and

$$\begin{aligned} z_s &= \frac{L_{\mathrm{M}}}{L_{\mathrm{S}}} \mathrm{e}^{\mathrm{i}\theta} z_m + (1-\mu_s) \\ \frac{\mathrm{d}z_s}{\mathrm{d}t_s} &= \frac{L_{\mathrm{M}}}{L_{\mathrm{S}}}\frac{w_2}{w_1} \mathrm{e}^{\mathrm{i}\theta}\left[i\left(1-\frac{w_1}{w_2}\right)(z_m) + \frac{\mathrm{d}z_m}{\mathrm{d}t_m}\right]. \end{aligned}$$

A second integration provides the relations between the accelerations in the two systems:

$$\frac{d^2 z_s}{dt_s^2} = \frac{L_M}{L_S}\left(\frac{w_2}{w_1}\right)^2 e^{i\theta} \times \left[-\left(1-\frac{w_1}{w_2}\right)^2 z_m + 2i\left(1-\frac{w_1}{w_2}\right)\frac{dz_m}{dt_m} + \frac{d^2 z_m}{dt_m^2}\right]. \quad (5)$$

Moreover, as a consequence of the third Kepler's law,

$$w_1^2 L_S^3 = G(M_s + M_e + M_m), \; w_2^2 L_M^3 = G(M_e + M_m). \quad (6)$$

In this work, the Sun–Earth–Moon scenario is considered and the physical parameters adopted in the numerical simulations are set according with the Jet Propulsion Laboratory ephemeris (available on-line at http://ssd.jpl.nasa.gov/?constants). In particular the mass ratios are

$$\mu_s = 3.040423402066 \times 10^{-6}, \quad \mu_m = 0.012150581$$

being the masses of the bodies

$$M_s = 1.988924 \times 10^{30}\,\text{kg} \quad M_e = 5.973712 \times 10^{24}\,\text{kg}$$
$$M_m = 7.347686 \times 10^{22}\,\text{kg}.$$

In the inertial reference frame (X, Y, τ) the time is measured in seconds, the distances L_S and L_M are equal to

$$L_S = 149{,}597{,}870\,\text{km}, \; L_M = 384{,}400\,\text{km},$$

while the values of the angular velocities w_1 and w_2 are

$$w_1 = 1.99098898 \times 10^{-7}\,\frac{\text{rad}}{\text{s}}, \; w_2 = 2.6653174179 \times 10^{-6}\,\frac{\text{rad}}{\text{s}}.$$

3 The Comparison of the BCP with the CR3BPs

For applications in celestial mechanics and mission design, the dynamical model should be chosen to reproduce as better as possible the real force system. Moreover, it could be an advantage if the model presents useful mathematical properties so that dynamical system theory could be applied to heighten understanding and perspectives. These features are usually in conflict and, for mission design purposes, the Circular restricted three-body problem reveals to be a good compromise. Indeed, this system admits a rich phase portrait that could be used to construct energy-efficient spacecraft trajectories. In particular, the structure of certain invariant sets and associated invariant manifolds have been exploited to accomplish a wide class of mission requirements, [6, 7, 9, 10, 13, 18] and enables to heuristically explain other theories like the Weak Stability Boundary theory, formalized in the Coupled CR3BP model, [1–3, 8]. On the other side, in different situations like the Sun–Earth–Moon scenario, solutions in the restricted model have been shown to be good approximations to trajectory in the real system.

Following the Coupled CR3BP approach, a complete trajectory in a multibody system is built patching together selected arcs computed in different restricted three-body problems. The sequence of the restricted models and the criteria where to connect the different solution arcs are usually chosen a priori to accomplish the mission requirements and to facilitate the design. Indeed the Poincaré section where the intersection of the manifolds is analyzed is usually set on straight lines in the configuration space or on the boundary of the three-body sphere of influence as in [15].

The incoming analysis concerns the relation between the Bicircular 4-body system and a couple of circular restricted three-body problem and aims to define the coupled CR3BP approximation so that each restricted model is considered where it provides the best approximation of the more complicated system. The distance between the Bicircular model and each one of the CR3BP is estimated as the norm of the difference of the differential equations governing their dynamics, once they are written in the same reference frame and in the same units of measure. In particular, the comparison is performed in the synodical frame proper of the considered restricted problem, while the units of measure in both the cases will be the dimensional one.

3.1 Comparison with CR3BP$_{\mathrm{SE}}$

Denote with $\bar{z} = z_s L_{\mathrm{S}}$ the spatial coordinate in the Sun–Earth rotating reference frame, measured in km. Since $Z = \bar{z}\mathrm{e}^{\mathrm{i}\gamma_1}$, in this setting the positions of the primaries are given by

$$\begin{aligned}
\bar{S} &= -\mu_s L_{\mathrm{S}} \\
\bar{E} &= (1-\mu_s)L_{\mathrm{S}} - \mu_m L_{\mathrm{M}}\mathrm{e}^{\mathrm{i}(\gamma_2-\gamma_1)} \\
\bar{M} &= (1-\mu_s)L_{\mathrm{S}} + (1-\mu_m)L_{\mathrm{M}}\mathrm{e}^{\mathrm{i}(\gamma_2-\gamma_1)}.
\end{aligned}$$

The double derivative of $Z(\tau)$ turns into

$$\frac{\mathrm{d}^2 Z}{\mathrm{d}\tau^2} = \left(\frac{\mathrm{d}^2\bar{z}}{\mathrm{d}\tau^2} + 2\mathrm{i}w_1\frac{\mathrm{d}\bar{z}}{\mathrm{d}\tau} - w_1^2\bar{z}\right)\mathrm{e}^{\mathrm{i}\gamma_1}$$

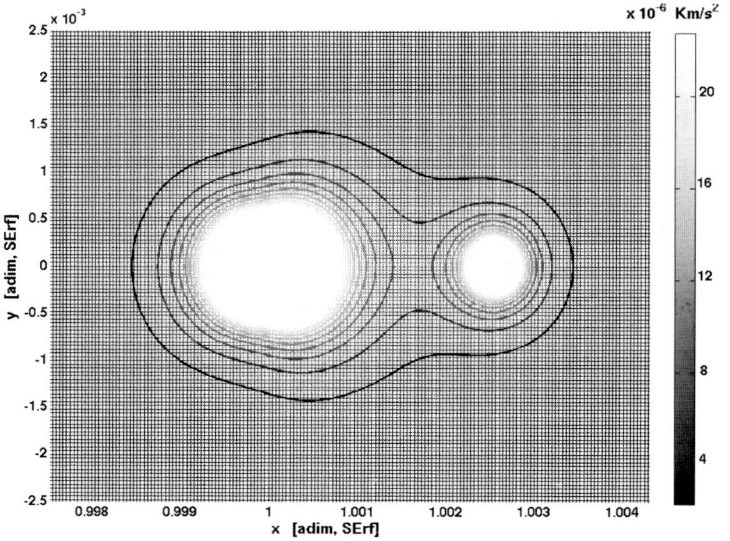

Fig. 2 Level curves of Δ_{SE} for $\theta = 0$

then, substituting the new variables into (3), the equation of motion associated to the BCP assumes the form

$$\frac{\mathrm{d}^2\bar{z}}{\mathrm{d}\tau^2} + 2\mathrm{i}w_1\frac{\mathrm{d}\bar{z}}{\mathrm{d}\tau} - w_1^2\bar{z} = -G\left[\frac{M_s(\bar{z}-\bar{S})}{\| \bar{z}-\bar{S} \|^3} + \frac{M_e(\bar{z}-\bar{E})}{\| \bar{z}-\bar{E} \|^3} + \frac{M_m(\bar{z}-\bar{M})}{\| \bar{z}-\bar{M} \|^3}\right]. \tag{7}$$

In the same time–space reference frame, the $\mathrm{CR3BP_{SE}}$ is described by

$$\frac{\mathrm{d}^2\bar{z}}{\mathrm{d}\tau^2} + 2\mathrm{i}w_1\frac{\mathrm{d}\bar{z}}{\mathrm{d}\tau} - w_1^2\bar{z} = -G\left[\frac{M_s(\bar{z}-\bar{S})}{\| \bar{z}-\bar{S} \|^3} + \frac{M_b(\bar{z}-\bar{B})}{\| \bar{z}-\bar{B} \|^3}\right].$$

It follows the difference between the two models

$$\begin{aligned}\Delta_{\mathrm{SE}}(\bar{z}) &= \| BCP - CR3BP_{\mathrm{SE}} \| \\ &= G\left\| -\frac{M_e(\bar{z}-\bar{E})}{\| \bar{z}-\bar{E} \|^3} - \frac{M_m(\bar{z}-\bar{M})}{\| \bar{z}-\bar{M} \|^3} + \frac{M_b(\bar{z}-\bar{B})}{\| \bar{z}-\bar{B} \|^3}\right\|.\end{aligned}$$

The gap Δ_{SE} arises because in the restricted three-body problem the Earth–Moon system is considered as a unique body concentrated in its center of mass rather than a binary system.

The distance between the two systems rapidly decreases to zero as the evaluation point is out of two disks around the primaries. For any different mutual position of the three primaries, the picture of Δ_{SE} is different but self-similar up to rotation around the point B; in Fig. 2, the value of Δ_{SE} is plotted for $\theta = 0$.

3.2 Comparison with CR3BP$_{\rm EM}$

Following the same procedure as before, the distance between the CR3BP$_{\rm EM}$ and the BCP is achieved. Again, let $\bar{z} = z_m L_{\rm M}$ be used to denote the complex coordinates in a rotating reference frame and dimensional units of measure. Reminding that the origin of the Earth–Moon rotating frame is placed in the barycenter B that is revolving around the center of mass O, the inertial coordinate Z and $\bar{z}$ are linked by the formula

$$Z = B + \bar{z}e^{i\gamma_2}, \qquad B = (1-\mu_s)L_{\rm S}e^{i\gamma_1}.$$

The positions of the primaries

$$\begin{aligned}\bar{S} &= (S-B)e^{-i\gamma_2} = -L_{\rm S}e^{i(\gamma_1-\gamma_2)}\\ \bar{E} &= (E-B)e^{-i\gamma_2} = -\mu_m L_{\rm M}\\ \bar{M} &= (M-B)e^{-i\gamma_2} = (1-\mu_m)L_{\rm M}\end{aligned}$$

and the acceleration of the particle

$$\frac{d^2Z}{d\tau^2} = \left(\frac{d^2\bar{z}}{d\tau^2} + 2iw_2\frac{d\bar{z}}{d\tau} - w_2^2\bar{z} - w_1^2(1-\mu_s)L_{\rm S}e^{i(\gamma_1-\gamma_2)}\right)e^{i\gamma_2}$$

yield the differential equation for the BCP in dimensional EM-synodical coordinates

$$\begin{aligned}&\frac{d^2\bar{z}}{d\tau^2} + 2iw_2\frac{d\bar{z}}{d\tau} - w_2^2\bar{z} - w_1^2(1-\mu_s)L_{\rm S}e^{i(\gamma_1-\gamma_2)}\\ &\quad = -G\left[\frac{M_s(\bar{z}-\bar{S})}{\|\bar{z}-\bar{S}\|^3} + \frac{M_e(\bar{z}-\bar{E})}{\|\bar{z}-\bar{E}\|^3} + \frac{M_m(\bar{z}-\bar{M})}{\|\bar{z}-\bar{M}\|^3}\right]. \end{aligned} \tag{8}$$

The term $-w_1^2(1-\mu_s)L_{\rm S}e^{i(\gamma_1-\gamma_2)}$ represents the centrifugal acceleration of B or, equivalently, the gravitational influence of the Sun on the Earth–Moon barycenter, indeed (2) and (6) imply $(1-\mu_s)w_1^2 = \frac{GM_s}{L_{\rm S}^3}$. The difference between (8) and

$$\frac{d^2\bar{z}}{d\tau^2} + 2iw_2\frac{d\bar{z}}{d\tau} - w_2^2\bar{z} = -G\left[\frac{M_e(\bar{z}-\bar{E})}{\|\bar{z}-\bar{E}\|^3} + \frac{M_m(\bar{z}-\bar{M})}{\|\bar{z}-\bar{M}\|^3}\right]$$

that describes the motion in the Earth–Moon restricted problem, gives the distance between the two models

$$\Delta_{\rm EM}(\bar{z}) = \| BCP - CR3BP_{\rm EM} \| = GM_s\left\|\frac{(\bar{S}-\bar{z})}{\|\bar{z}-\bar{S}\|^3} - \frac{(\bar{S}-\bar{B})}{\|\bar{S}-\bar{B}\|^3}\right\|.$$

The error originates because in the CR3BP$_{\rm EM}$ the influence of the Sun on the spacecraft is considered as the same influence that the Sun produces on the center B

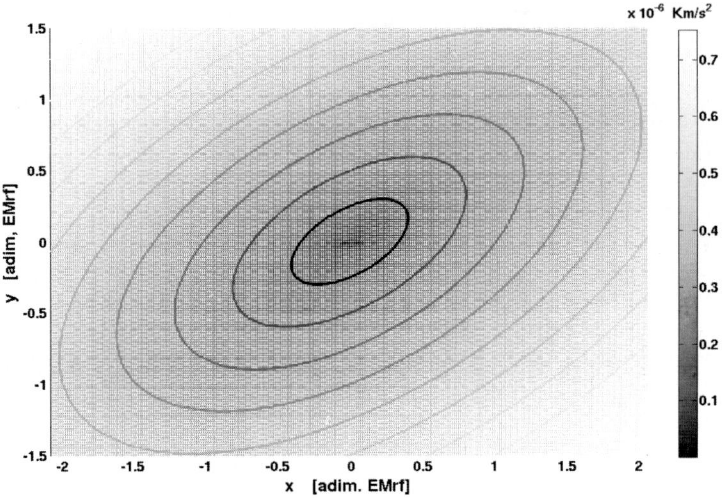

Fig. 3 Level curves of Δ_{EM} for $\theta = \pi/3$

of the rotating frame. Indeed the error vanishes whenever the spacecraft is placed in the origin of the reference frame and grows when it moves away, see Fig. 3.

4 Regions of Prevalence

The prevalence of each CR3BP is investigated according to which one produces the lowest error if it is considered in place of the BCP. To this aim, once a system of coordinates is chosen, let be defined the function

$$\begin{aligned}\Delta E(z) &= (\Delta_{\mathrm{SE}} - \Delta_{\mathrm{EM}})(z)\\ &= G\left\| -\frac{M_e(z-E)}{\| z-E \|^3} - \frac{M_m(z-M)}{\| z-M \|^3} + \frac{M_b(z-B)}{\| z-B \|^3} \right\|\\ &\quad -GM_s\left\| \frac{(S-z)}{\| z-S \|^3} - \frac{(S-B)}{\| S-B \|^3} \right\|.\end{aligned}$$

In any point z, one of the restricted models has to be preferred according with the sign of ΔE: where $\Delta E < 0$ the CR3BP$_{\mathrm{SE}}$ provides a better approximation of the BCP, otherwise the CR3BP$_{\mathrm{EM}}$.

For a given relative phase θ of the primaries, denote with $\Gamma(\theta)$ the zero level set of the function ΔE

$$\Gamma(\theta) := \{z : \Delta E(z) = 0\}.$$

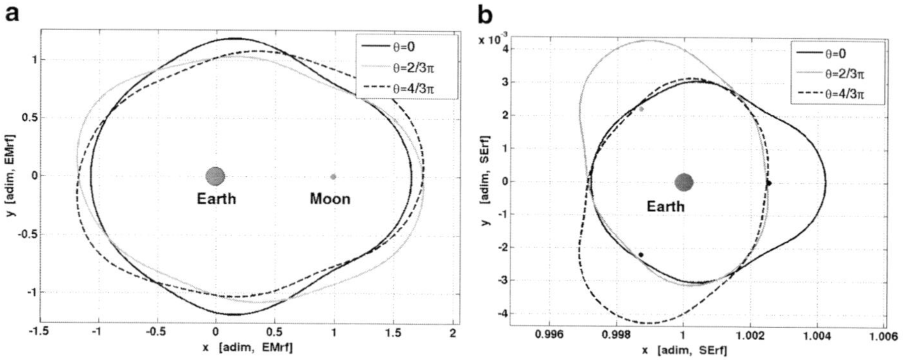

Fig. 4 $\Gamma(\theta)$ with $\theta = 0, 2/3\pi, 4/3\pi$ in EM and SE reference frame

Numerical simulations show that $\Gamma(\theta)$ is a closed simple curve: we refer to the two regions bounded by $\Gamma(\theta)$ as the *Regions of Prevalence* of the two restricted problems. In the bounded region $\Delta_{\mathrm{EM}} < \Delta_{\mathrm{SE}}$, while in the exterior region the opposite holds.

In Fig. 4, the zero level set of ΔE is drawn for different choices of the angle θ and in different system of coordinates. For any angle θ the Earth, the Moon as like as the L_1 and L_2 Lagrangian points related to the $\mathrm{CR3BP_{EM}}$ belong to the EM region of prevalence, while the $\mathrm{CR3BP_{SE}}$ Lagrangian points are placed in the exterior region.

5 The Coupled CR3BP Approximation

The Coupled CR3BP concerns the approximation of the four-body problem with the superposition of two Circular restricted three-body problems, [12]. Here the design of trajectories leaving a Lyapunov orbit around L_1 and L_2 in the $\mathrm{CR3BP_{SE}}$ and directed to the vicinity of the Moon is considered: therefore, denoting with $W^{(u),s}_{(SE),EM,i}(\gamma)$ any (un)-stable manifold related to Lyapunov orbits γ around L_i in the (SE) or EM restricted problem, the intersections of $W^s_{\mathrm{EM},2}(\gamma_1)$ with $W^u_{\mathrm{SE},1}(\gamma_2)$ and $W^u_{\mathrm{SE},2}(\gamma_2)$ have been exploited. According with the prevalence regions previously defined, the Poincaré section is set on the curve $\Gamma(\theta)$. The procedure to design the transfer trajectories is the following: first the angle θ is chosen and the curve $\Gamma(\theta)$ in both the synodical systems is set. After, for a couple of Lyapunov orbits γ_1, γ_2, the manifolds $W^s_{\mathrm{EM},2}(\gamma_1)$ and $W^u_{\mathrm{SE},1,2}(\gamma_2)$ are computed, each in their own coordinate frame, until the corresponding curve $\Gamma(\theta)$ is encountered. The resulting Poincaré map are then transformed into the same coordinates system and the analysis of their intersection leads to the selection on the point *Int* where the two solution arcs need to be connected.

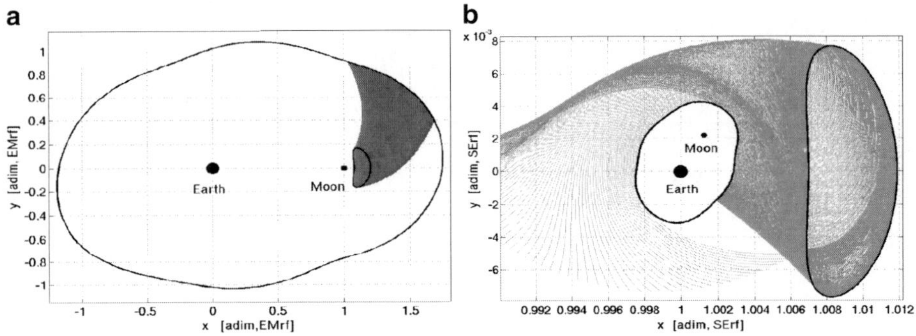

Fig. 5 Intersection of $W^u_{\mathrm{SE},2}(\gamma_2)$ and $W^s_{\mathrm{EM},2}(\gamma_1)$ with $\Gamma(\pi/3)$

As shown in Fig. 5a, for almost every Lyapunov orbits around L_2 in the EM system and every θ, the external stable manifold $W^s_{\mathrm{EM},2}(\gamma_1)$ invests completely the curve $\Gamma(\theta)$: the resulting Poincaré map, topologically equivalent to a circle, bounds the region $\mathscr{B}$ of those initial data leading to the Moon's region. On the other side, Fig. 5b, depending on the angle θ and on the selected Jacoby constant, the unstable manifolds $W^u_{\mathrm{SE},1,2}(\gamma_2)$ may intersect only partially the curve $\Gamma(\theta)$.

For our purpose, the point *Int* has to be selected in the set $\mathscr{B} \cap W^u_{\mathrm{SE},1,2}$. Indeed, patching together the trajectories obtained integrating the point *Int* backward in time in the CR3BP$_{\mathrm{SE}}$ and forward in the CR3BP$_{\mathrm{EM}}$, it follows an orbit that, starting from the SE-Lyapunov, after have passed through the EM Lyapunov gateway, will travel close to the Moon.

6 The Box-Covering Approach

In this section, the numerical technique used to detect the connecting point *Int* is discussed.

Once the intersection of the stable manifold relative to a Lyapunov orbit in the EM system with the curve $\Gamma(\theta)$ is computed, the four-dimensional Poincaré map is covered with box structures implemented in the software package GAIO (Global Analysis of Invariant Objects), see [5].

An N-dimensional box $B(C,R)$ is defined as a multidimensional rectangle and it is identified by a center $C = (C_1,\dots,C_N)$ and a vector of radii $R = (r_1,\dots,r_N)$:

$$B(C,R) = \bigcap_{i=1}^{N} \{(x_1,x_2,\dots,x_N) \in \mathscr{R}^N \ :\ |x_i - C_i| \le r_i\}.$$

Following a multiple subdivision process, starting from a box B_0, a larger family of smaller boxes is created with the property to cover B_0: the depth d of a family

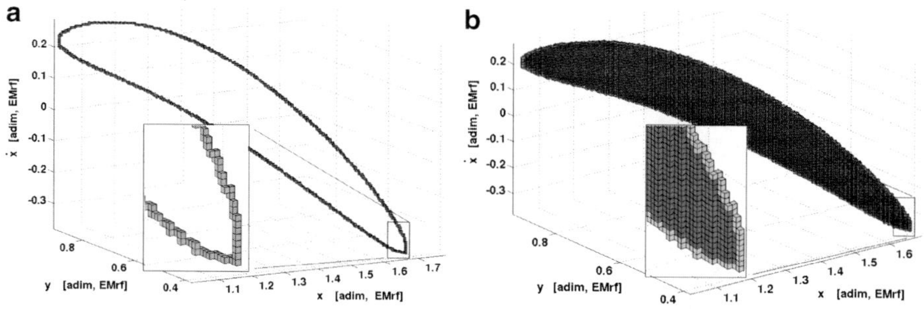

Fig. 6 Box covering of the Poincaré map of $W^s_{\mathrm{EM},2}$ and of the transfer region $\mathscr{B}$

denotes the number of such subdivision iterations. Once the family $\mathscr{F}(d)$ of boxes is created, the Poincaré map is therein inserted: only those boxes of $\mathscr{F}(d)$ containing at least one point of the Poincaré map are considered, the others are neglected, Fig. 6a. Denote with $\mathscr{P}$ the family of boxes used for the covering of the Poincaré map. In order to detect the points Int, the interior region $\mathscr{B}$ needs to be covered as well, see Fig. 6b. The definition of the centres of the boxes used to cover $\mathscr{B}$ is made "by columns": from the set of boxes in $\mathscr{P}$ whose centres have the same (x,y)-coordinates, let be selected the two boxes with the maximal $\dot{x}_{\max}$ and minimal $\dot{x}_{\min}$ value of the $\dot{x}$-coordinate. Then a new set of centres $\{C_k = (x,y,\dot{x}_k,\dot{y}_k)\}_{k=1}^K$ are defined, where $\dot{x}_k = \dot{x}_{\min} + k\Delta v$ and $\dot{y}_k$ is obtained from the Jacobi constant. Here Δv is twice the radius in the $\dot{x}$-direction of the boxes in $\mathscr{F}(d)$ and $K = (v_{\max} - v_{\min})/\Delta v$.

In the presented simulations, the covering is performed at $d = 32$: depending on the size of the Poincaré map the radii of the covering boxes result in the range $[4 \times 10^{-4}, 2 \times 10^{-3}]$ EM units.

Then, for a value of the Jacobi integral in the SE system, the Poincaré map of $W^u_{\mathrm{SE},1}(\gamma_2)$ or $W^u_{\mathrm{SE},2}(\gamma_2)$ is computed and, using (4), it is transformed in EM synodical coordinates, being θ the angle between the primaries. Finally, all those points of the SE Poincaré map lying in one of the boxes covering $\mathscr{B}$ are considered as transfer points.

7 Some Results

The existence of connection points is tested starting from a database of 60 Lyapunov orbits in the $\mathrm{CR3BP_{SE}}$ both around L_1 and L_2 and 60 Lyapunov orbits around L_2 in the $\mathrm{CR3BP_{EM}}$. The Jacobi constant varies in the range $[3.0004, 3.00084]$ for the SE system and in the interval $[3.053, 3.177]$ for the EM system and 32 equispaced values of θ have been considered.

Figure 7 concerns transfers leaving a Lyapunov orbit around L_1 in the Sun–Earth system: every dark sign marks a point in the intersection $\mathscr{B} \cap W^u_{\mathrm{SE},1}$, i.e. $\Delta V = 0$ connections. The coordinates represent the Jacobi constant of the connection point

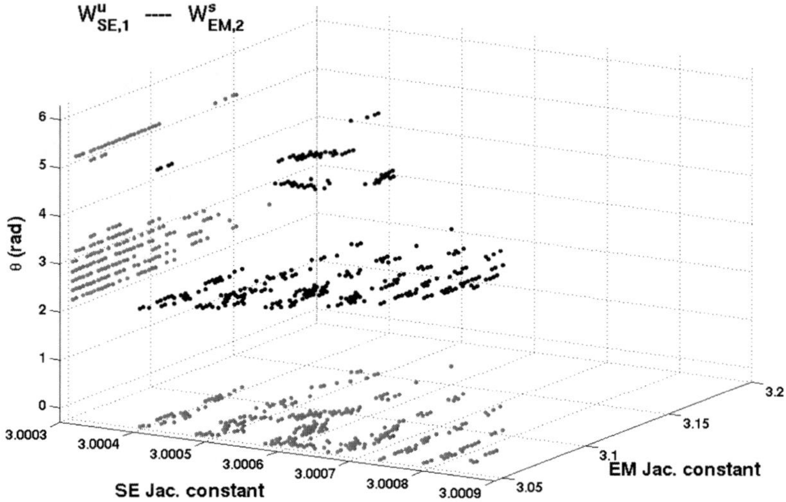

Fig. 7 Zero ΔV connections between $W^u_{\mathrm{SE},1}$ and $W^s_{\mathrm{EM},2}$

respectively in the SE and EM system and the angle θ of the Poincaré section $\Gamma(\theta)$ where the connection is detected. The lighter points are the projections of the previous ones onto the coordinates planes. Starting from one intersection, backward and forward integration in the two CR3BPs produce the complete transfer. If no differently specified, all the evaluations are done in the SE-synodical reference frame and in SE-units of measure. In the following figures, the darker and the lighter lines concern the pieces of trajectory integrated in the $\mathrm{CR3BP_{SE}}$ and in the $\mathrm{CR3BP_{EM}}$ respectively. In Fig. 8, the bigger picture depicts the orbit from a Lyapunov orbit around L_2 to the Moon region, while the smaller ones show the values of $\Delta_{\mathrm{SE}}(t)$ and $\Delta_{\mathrm{EM}}(t)$ evaluated along the trajectory. The integral $Total\ \Delta V = \int_{t_0}^{t_c} \Delta_{\mathrm{SE}}(t)\mathrm{d}t + \int_{t_c}^{t_{\mathrm{fin}}} \Delta_{\mathrm{EM}}(t)\mathrm{d}t$ is used as a measure of the overall distance between the Coupled CR3BP approximation and the Bicircular model along a selected trajectory. Here t_0 is the last time when the spacecraft is far from the Earth more than 2.5 times the Earth–Moon distance and t_{fin} is the first moment the spacecraft is 10,000 km close to the Moon, while t_c denotes the instant when the Poincaré section is crossed. Referring to Fig. 9, in the bigger figure the dotted line remarks the circle inside which the above integration starts, the small black circles show the position of the Moon when the spacecraft is on the section and at the end of the travel, the black line denotes the Poincaré section at the crossing time. In the upper of the smaller boxes the values of Δ_{EM} and Δ_{SE} are plotted together, while the last graph shows the value of the integral $Total\ \Delta V$. Starting from t_0 the error Δ_{SE} is integrated up to the crossing moment t_c, then two different integrations are done: in the first case the error Δ_{EM} is considered till the final time t_{fin} (lighter line), in the second case again Δ_{SE} is integrated for a short interval of time (darker line). The last figure depicts

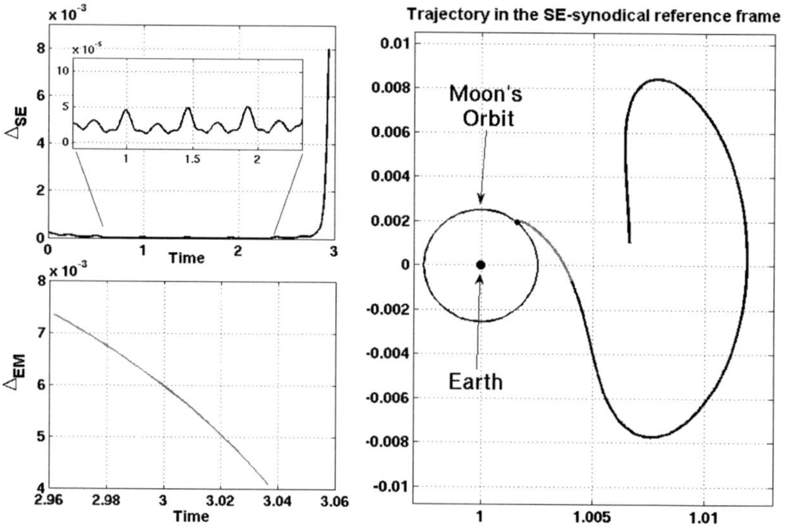

Fig. 8 Example of transfer trajectory and related errors Δ_{SE}, Δ_{EM}

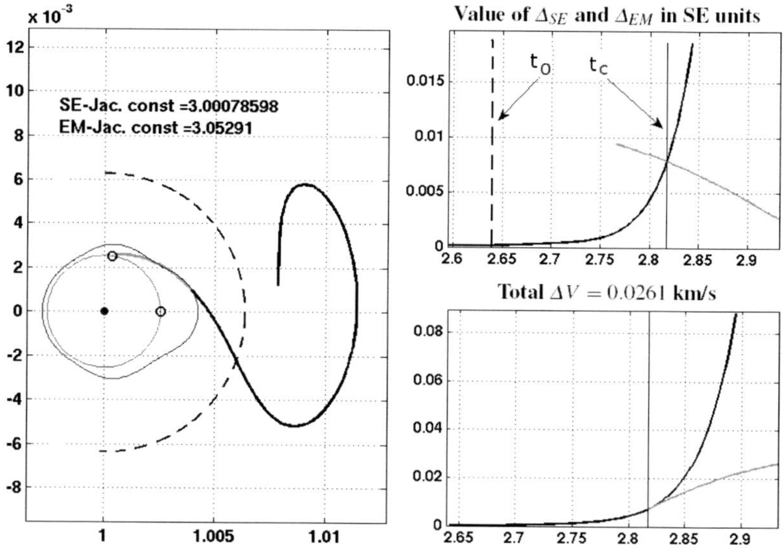

Fig. 9 Example of trajectory. Analysis of the error

the optimality of the choice of the Poincaré section $\Gamma(\theta)$ in minimizing the local approximation error.

A more detailed analysis and a comparison with trajectories designed adopting different Poincaré sections will be discussed in a future work.

References

1. Belbruno, E., Miller J.: Sun-perturbed Earth-to-Moon transfer with ballistic capture. Journal of Guidance Control and Dynamics, **16**, 1993.
2. Belbruno, E.A.:The dynamical mechanism of ballistic lunar capture transfers in the four body problem from the perspective of invariant manifolds and Hills regions. Institut dEstudis Catalans, CRM,Preprint No. 270, 1994.
3. Belbruno, E.: Capture Dynamics and Chaotic Motions in Celestial Mechanics: With Applications to the Construction of Low Energy Transfers. Princeton Univ. Press, 2004.
4. Cronin, J., Richards, P.B., Russell, L.H.: Some periodic solutions of a four-body problem. Icarus, **3**(5-6) 423–428 1964.
5. Dellnitz, M., Froyland, G., Junge, O.: The algorithms behind GAIO - Set oriented numerical methods for dynamical systems. In: Ergodic theory, analysis, and efficient simulation of dynamical systems, pages 145–174. Springer, 2000.
6. Dellnitz, M., Junge, O., Post, M., Thiere, B.: On target for Venus – Set oriented computation of energy efficient low thrust trajectories. Celestial Mech. Dynam. Astronom., **95**(1-4), 357–370, 2006.
7. Demeyer, J.R.C, and Gurfil, P.: Transfer from Earth to Distant Retrograde Orbits using Manifold Theory, AIAA Journal of Guidance, Control and Dynamics, **30**(5), 1261–1267, 2007.
8. F. Garca, F., Gómez, G.: A note on weak stability boundaries. Celestial Mech. Dynam. Astronom, **97**(2), 87–100, 2007.
9. Gómez, G., Jorba, A., Masdemont, J., Simo, C.: Study of the transfer from the Earth to a halo orbit around the equilibrium point L1. Celestial Mech. Dynam. Astronom., **56**(4), 541–562, 1993.
10. Gómez, G., Koon, W.S., Lo, M.W., Marsden, J.E., Masdemont, J., Ross, S.D.: Connecting orbits and invariant manifolds in the spatial restricted three-body problem. Nonlinearity, **17** 1571–1606, 2004.
11. Koon, W. S., Lo, M. W., Marsden, J. E., Jerrold, E., Ross, S. D.: Heteroclinic connections between periodic orbits and resonance transitions in celestial mechanics. Chaos, **10**(2), 427–469, 2000.
12. Koon, W. S., Lo, M. W., Marsden, J. E., Ross, S. D.: Low energy transfer to the Moon. Celestial Mech. Dynam. Astronom., **81**(1-2), 63–73, 2001.
13. Lo, M. W., Williams, B.G., Bollman, W.E., Han, D., Hahn, Y., Bell, J.L., Hirst, E.A., RCorwin, R.A., Hong,P. E., Howell, K. C., Barden, B., Wilson R.: Genesis Mission Design, The Journal of the Astronautical Sciences 49, 169–184, 2001.
14. Mingotti G., Topputo F., Bernelli-Zazzera F.: Low-energy, low-thrust transfers to the Moon. Celestial Mech. Dynam. Astronom., **105**(1-3) 61–74, 2009.
15. Parker, J.S.: Families of low-energy Lunar halo trasfer. Proceedings of the AAS/AIAA Space Flight Mechanics Meeting, 483–502, 2006.
16. Simó, C., Gómez, G., Jorba, À., Masdemont, J.: The Bicircular model near the triangular libration points of the RTBP. In: From Newton to chaos, volume 336 of NATO Adv. Sci. Inst. Ser. B Phys., pages 343–370. 1995.
17. Szebehely, V.: Theory of orbits, the restricted problem of three bodies. Academic Press, New York and London, 1967.
18. Yagasaki, K.: Sun-perturbed Earth-to-Moon transfers with low energy and moderate flight time. Celestial Mech. Dynam. Astronom., **90**(3-4): 197–212, 2004.

Adaptive Remeshing Applied to Reconfiguration of Spacecraft Formations

Laura Garcia-Taberner and Josep J. Masdemont

1 Introduction

Formation flying of spacecraft is a concept that has an important role in technology applications and continues growing in importance, mainly for science and astrophysical missions. The reason is that formation flying enables a set of some small (and cheaper) spacecraft to act as a virtual larger satellite, obtaining better information than a bigger one, with flexibility about the space observations that a formation can perform in the future.

Some formation flying missions are planed to be in orbits about the Earth, but also relevant science missions have big interest on the vicinity of libration points. The L_1 libration Sun–Earth+Moon point is nowadays the best place for observations of the Sun. The L_2 libration point is also an interesting place for deep space observations, where large telescopes or interferometry baselines could be located. This is the reason because projects like the Terrestrial Planet Finder [11] or Darwin [10] were planned to be in a libration point orbit about L_2.

The key technology that must be implemented in spacecraft formation flying is the control of a formation when doing observations. Mutual distances between spacecraft must be kept with high precision. A study of some of these control methodologies can be found in Farrar, Thein and Folta [1] and references therein.

But also there are many other technologies that have to be taken into account for formation flying. The one we consider in this paper is the reconfiguration of the formation. This technique is important in the lapses between observations. It can be necessary to change the orientation, the target point of the formation, or maybe also the shape or diameter of the cluster to give flexibility to the mission.

L. Garcia-Taberner (✉)
Departament d'Informàtica i Matemàtica Aplicada, Universitat de Girona,
Escola Politècnica Superior, 17071 Girona, Spain
e-mail: laura.garcia@ima.udg.edu

J.A.T. Machado et al., *Nonlinear and Complex Dynamics: Applications in Physical, Biological, and Financial Systems*, DOI 10.1007/978-1-4614-0231-2_5,

Some representative techniques of reconfigurations are the proximity maneuvering using artificial potential functions studied by McInnes [8], or the technique used by Hadaegh, Beard, Wang and McLain [6].

In this paper, we consider the reconfiguration of a formation by means of an optimization problem which uses the finite element methodology to obtain the controls that must be applied to each spacecraft. Additionally, our methodology can solve the problem of the transfer or deployment of the formation, which is another key problem in formation flying.

In this paper, we mainly focus on the obtention of a suitable optimal mesh using an adaptive remeshing strategy, that assures us that the error produced by the finite element methodology is small enough. Once the methodology gives us the trajectory and an optimal mesh, we check it considering a vector field of full JPL-ephemeris.

2 The FEFF Methodology

The main purpose of the FEFF methodology is to compute reconfigurations of a formation of spacecraft. In this paper, we consider that the spacecraft are in the vicinity of a halo orbit of 120,000 km of z-amplitude about L_2 in the Sun–Earth system, but the methodology can be applied to any other libration point orbit.

Here we present just a brief description of the basic FEFF methodology. More details about this point can be found in [2,4].

We consider a formation of N spacecraft which must perform a reconfiguration in a fixed time T. The spacecraft are in a small formation (i.e., the distance between them is only of a few hundreds of meters, both in the initial and the final configurations). Our objective is to find a trajectory for each of the spacecraft which guides it to the goal position, with the minimum fuel consumption and avoiding collisions with other spacecraft.

As the formations are small with respect to the amplitude of the halo orbit, we consider the linearized equations of motion about the nonlinear orbit. We compute the trajectories using these equations and then we will deal with the nonlinear part. For each of the spacecraft, we have the equation

$$\dot{\mathbf{X}}(t) = \mathbf{A}(t)\mathbf{X}(t), \tag{1}$$

where $\mathbf{A}(t)$ is a 6×6 matrix and $\mathbf{X}$ refers to the state of the spacecraft. The origin of the reference frame for the $\mathbf{X}$ coordinates is the nominal point on the base halo orbit at time t being the orientation of the coordinate axis parallel to the one of the RTBP.

The goal of the methodology is to find a set of optimal controls for each of the spacecraft. Including the initial and final states of the spacecraft in the reconfiguration problem and the controls, the equations that we deal with are

$$\begin{cases} \dot{\mathbf{X}}_i(t) = \mathbf{A}(t)\mathbf{X}_i(t) + \bar{\mathbf{U}}_i(t), \\ \mathbf{X}_i(0) = \mathbf{X}_i^0, \\ \mathbf{X}_i(T) = \mathbf{X}_i^T, \end{cases} \tag{2}$$

where $\mathbf{X}_i^0$ and $\mathbf{X}_i^T$ stand for the initial and final state of the i-th spacecraft of the formation, and $\bar{\mathbf{U}}_1,\ldots,\bar{\mathbf{U}}_N$, are the controls we are searching.

The key of procedure FEFF is that it uses the finite element methodology to obtain these controls (see [9] for references about the finite element method and [4] for a more detailed exposition about the FEFF methodology). Essentially, the time interval $[0,T]$ which we consider for the reconfiguration is split in M subintervals of the domain that we call elements. This mesh can have elements of different length and can be different for each of the spacecraft, depending on the nature of the trajectories of reconfiguration. We impose that controls are some maneuvers (in form of delta-v) that we apply in the points where two elements join (the nodes). The finite element methodology gives us a relation between the positions of the spacecraft in the nodes and the maneuvers, Δv, that we must apply.

Procedure FEFF reduces the reconfiguration problem to an optimization problem with constrains. The functional that must be minimized has to be related to fuel consumption, and since it is related to the sum of the norm of the delta-v, the functional we want to minimize is

$$J_1 = \sum_{i=1}^{N} \sum_{k=0}^{M_i} \rho_{i,k} ||\Delta \mathbf{v}_{i,k}||, \tag{3}$$

where $|| * ||$ denote the Euclidean norm and $\rho_{i,k}$ are weight parameters that can be used, for instance, to penalize the fuel consumption of selected spacecraft with the purpose of balancing fuel resources (here for clarity we consider that $\rho_{i,k}$ multiplies the modulus of the delta-v, but in a similar way we can impose a weight on each component).

The most important constraint in our problem is the avoidance of collision between spacecraft. This enters in the optimization problem as a constraint, imposing that each spacecraft is surrounded by a security sphere and the spheres of all the spacecraft cannot collide. We note that the partition of the time interval made by the finite element methodology gives us an efficient implementation to check this constraint.

3 Adaptive Remeshing Applied to Reconfigurations

When searching the optimal controls for the spacecraft, we must take into account two facts. One of them is related to ill conditioning problems and the other is related to the error due to the finite element approximation. Both of them lead us to think about a remeshing strategy.

When we consider the functional of equation (3), we see that it is ill conditioned to compute derivatives when delta-v values are small. But our objective is to find

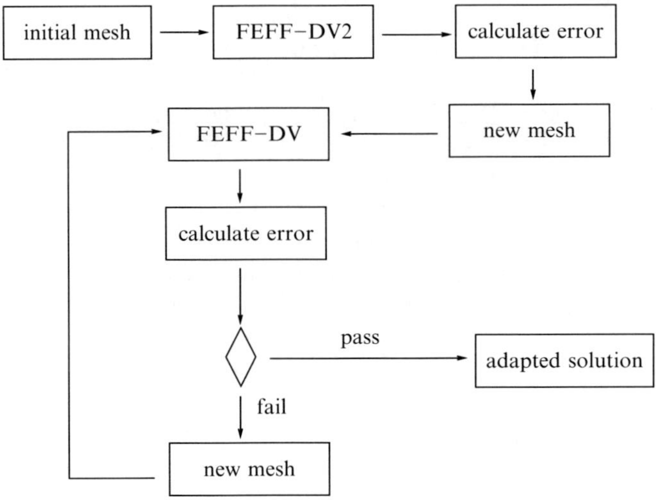

Fig. 1 Schema of the procedure of adaptive remeshing

small delta-v! In order to avoid this problem, we have two strategies: the first one consists on considering an alternative functional to minimize,

$$J_2 = \sum_{i=1}^{N} \sum_{k=0}^{M_i} \rho_{i,k} ||\Delta \mathbf{v}_{i,k}||^2, \tag{4}$$

which is also related to fuel consumption and it is not ill conditioned. The second one is based on considering a remeshing strategy to suppress the nodes with a small delta-v.

On the other hand, the approximation of the solution via the finite element method gives us some errors associated with the approximation we make on each one of the elements of the mesh.

To solve these two facts, we will consider an adaptive remeshing strategy applied to our reconfiguration problem. The general idea of adaptive remeshing is that, given a threshold value e, to find a mesh that provides an approximate solution with error (understood as the difference between the solution of the problem and its approximation inside of an element) less than e in some norm.

In Fig. 1, we give a schema of the general idea of our procedure. It has two different phases. In the first one, it starts computing the trajectories for the spacecraft using the finite element method with a given mesh. In this first phase the objective is to find a rough approximation of the solution, so we start with a small number of elements (a maximum of 10). All of these elements have the same length and all the spacecraft start with the same mesh. Once we have obtained the trajectories, it computes an estimation of the error. This error is essentially obtained by comparison between the gradient obtained using the finite element model and the one obtained by integration of the equations of motion. The second step of the procedure is an iterative process which performs adaptive remeshing, recomputes the approximate solution with the new mesh, and ends when the error is below a given tolerance. When remeshing is necessary, the new mesh is adapted using the estimation of errors of the previous mesh.

Adaptive remeshing methods penalize the elements where the error is considered big, dividing them in smaller elements. On the other hand, if the estimation of the error is small in an element, then this element is made bigger in the next iteration. Since, essentially our estimation of the error is related to the value of the delta-v maneuvers to be implemented, this method tends to increase the length of the elements which have associated small delta-v and tends to decrease the length of the elements which have associated big delta-v's.

Essentially to decide whether the current mesh is good enough or not we use a criterion which compares the modulus of the estimated error ($||e||$) with the total gradient of the solution,

$$||\bar{u}|| = \int_0^T \mathbf{v}_2 \mathrm{d}t.$$

We accept the mesh when

$$||e|| \leq \nu ||\bar{u}||,$$

where ν is the acceptability criteria. We discuss the value of ν taken on Sect. 4.

In order to compute the new mesh, we use the Li and Bettess remeshing strategy (see [7]). This strategy is based on the idea that the error distribution on an optimal mesh is uniform,

$$||\hat{e}_k|| = \nu ||\bar{u}|| / \sqrt{\hat{M}},$$

where ν is again the acceptability criteria, e_k is the computed error on element k, M is the number of elements of the mesh and the hat distinguishes the parameters of the new mesh. The strategy consists of finding the new length of the elements using the number of elements of the new mesh, $\hat{M}$. Let us denote d, the dimension of the problem and m, the maximum degree of the polynomials used in the interpolation. Then, according to Li and Bettess, the number of elements needed by the new mesh is,

$$\hat{M} = (\nu ||\bar{u}||)^{-d/m} \left(\sum_{k=1}^{M} ||e_k||^{d/(m+d/2)} \right)^{(m+d/2)/m}.$$

Since we work with linear elements in dimension one, we have $m = 1$ and $d = 1$. The recommended number of elements of the new mesh is

$$\hat{M} = (\nu ||\bar{u}||)^{-1} \left(\sum_{k=1}^{M} ||e_k||^{2/3} \right)^{3/2}.$$

Once we have the estimation of the number of elements, we can find the length of the new elements:

$$\hat{h}_k = \left(\frac{\nu||\bar{u}||}{\sqrt{\hat{M}}||e_k||} \right)^{1/m+d/2} h_k,$$

that in our case, turns out to be

$$\hat{h}_k = \left(\frac{\nu||\bar{u}||}{\sqrt{\hat{M}}||e_k||} \right)^{3/2} h_k.$$

4 Simulations with Adaptive Remeshing

When computing reconfigurations of spacecraft, we have two limiting cases: if there is no collision risk when the spacecraft follow a linear trajectory, we know that the optimal trajectory is a bang–bang control for each one of the spacecraft. This is the most critical case for our procedure, since the optimal maneuver consists in two delta-v: one at departure and another one at the arrival position. The remaining nodal delta-v must be zero, and so this is a case where the computation of derivatives for J_1 is very ill conditioned. On the other hand, for cases with collision risk, our methodology must tend to low thrust when the diameter of the mesh tends to zero. In this section, we present an example of each of these two cases.

4.1 A Bang–Bang Example

In this kind of problem, since there are no collision hazards, collision avoidance does not affect the trajectories of the spacecraft and the optimal trajectory for each spacecraft is independent from the others. For this reason, we can reduce the computations to obtain the optimal trajectory for a single spacecraft.

In order to exemplify the procedure, we consider a shift of a single spacecraft. We consider the reference frame for (2) aligned with respect to the RTBP reference frame, but with origin on the nominal point of the base halo orbit (when $t = 0$ this point corresponds to the "upper" position of the Halo orbit, this is when it crosses the RTBP plane $Y = 0$ with $Z > 0$). The initial condition for this example is taken 100 m far from the base nominal halo orbit in the X direction, and the goal is to transfer it to a symmetrical position with respect to the halo orbit in 8 h. This is to 100 m in the opposite X direction doing a shift of 200 m for the spacecraft transfer maneuver.

For this particular case, we know that the optimal solution is a bang–bang control with maneuvers of 0.69 cm/s at departure and arrival.

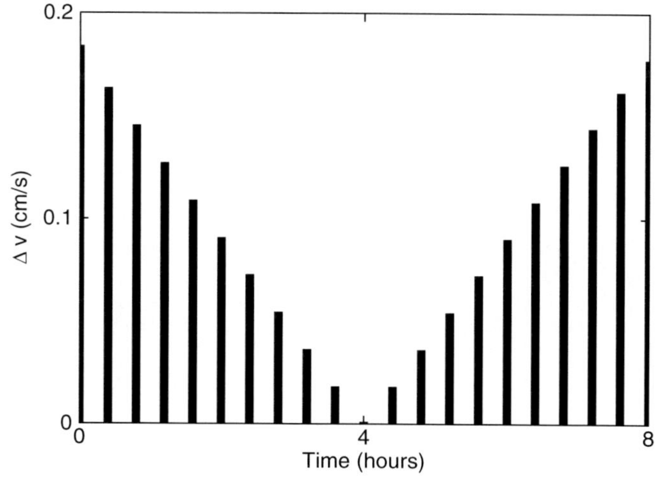

Fig. 2 Delta-v obtained with the minimization of functional (4) in the case of no collision risk. The methodology FEFF converges to a bang–bang solution with this initial seed

Our procedure starts by minimizing the functional (4) and obtains a trajectory with the delta-v profile of Fig. 2. This is a typical profile result for the reconfigurations without collision risk.

We compute now the reconfiguration considering different values for the parameter ν. We note that this parameter does not only appear in the acceptability criteria but it is also used to obtain the new mesh. If we take a small value of ν, we can end up with a mesh with a big number of nodes, which results in an optimization problem with a very large number of variables, that could be unsolvable in practice. Note that a mesh with 100 elements and a single spacecraft ends up with an optimization problem with 594 variables. In the other way around, if we use a big ν, we could end up accepting some meshes with big errors. In Table 1, we have a summary of the results obtained for different values of the parameter ν, the number of iterations needed to reach the bang–bang solution and the number of elements after the first iteration of the methodology.

We note that when ν is very small, there is no convergence. The case with ν equal to 0.0001 makes the optimal procedure awkward. When ν is 0.001, the final number of elements is greater than 1 (that we know is our final target number) although there are some very small delta-v. When ν is big, moreover, there is no convergence; the final mesh contains more elements than expected, because it passes the acceptability criteria before converging to the bang–bang control. We can conclude that, in this bang–bang case, the best values for ν are inside the range $[0.04, 0.06]$. With values larger than 0.06, the algorithm does not converge.

Table 1 Number of iterations necessary to obtain the bang–bang solution depending on ν. We have indicated by "fail" the cases where the procedure does not converge

ν	Elements iteration 1	Iterations
0.0001	3,008	Fail
0.001	301	Fail
0.002	149	25
0.005	61	16
0.01	31	14
0.02	15	10
0.03	11	6
0.04	7	4
0.05	6	4
0.06	4	2
0.07	4	Fail

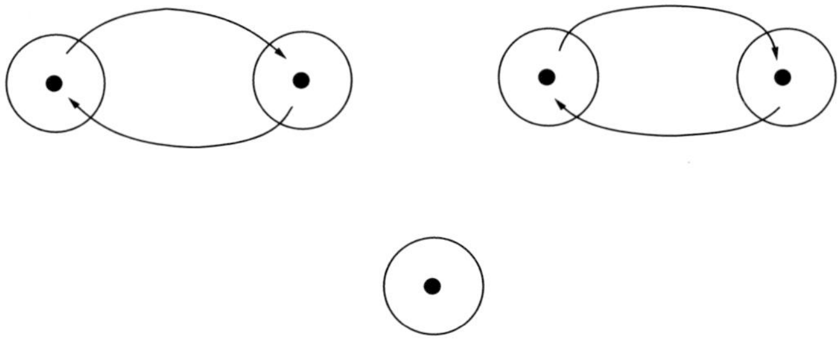

Fig. 3 Example of reconfiguration with collision risk: the switch of two pairs of spacecraft of the TPF formation

4.2 A Low-Thrust Example

In reconfigurations where the bang–bang trajectories end up with collisions with some of the spacecraft, FEFF obtains trajectories which of course are different from bang–bang. Our objective is to study whether these trajectories could tend to low thrust arcs.

For this case we consider a configuration based on the Terrestrial Planet Finder (TPF) model (see [11]). We assume that the satellites are initially contained in the local plane $Z = 0$, with the interferometry baseline aligned on the X axis. We simulate the swap between two pairs of satellites in the baseline: each inner satellite changes its location with the outer satellite which is closest to it in position (this is, inner satellites are maneuvered to attain outer positions and vice-versa as shown in Fig. 3). Again we consider 8 h for the reconfiguration. The process of switching positions has a collision risk and simple bang–bang controls are no longer valid.

As in the bang–bang case, we consider different values for the parameter ν. A discussion like in the previous section is also valid here: using a small ν, we can end up with a mesh with many elements. For example, taking $\nu = 0.0005$, in the first iteration we have around 1,000 elements. We do not only have the problem

Table 2 Number of iterations and elements obtained with the swapping example of TPF depending on ν

ν	N_1	Iteration	N_F
0.0001	3,504	Fail	
0.001	350	10	232
0.002	175	8	202
0.005	70	8	171
0.01	34	7	89
0.02	18	6	45
0.03	12	4	33
0.04	9	3	27
0.05	6	3	15
0.06	6	3	9
0.07	5	2	7

of having very small elements: the optimization problem that we end up with which has 29,970 variables and it is not desirable. Again, if we take a big ν, we can end up with a mesh with big errors or with a mesh with only a few elements.

In Table 2, we display a summary of the results obtained for different values of the parameter ν, the number of iterations until the methodology converges (Iter), the number of elements in the first iteration (N_1) and the number of elements in the last iterate (N_F).

As in the previous case, when ν is small, the number of elements is big, and the computation of the optimum is very expensive. Also, taking ν big, the number of elements may not be enough.

We note that now the best values are inside the range $[0.005, 0.05]$. With values larger than 0.05, the number of elements is very small and with values smaller than 0.005 the number of elements makes the computation much more expensive.

Since the value $\nu = 0.05$ is appropriated for the two cases, we consider it for our computations.

4.3 Considerations About the Value of ν

We have seen in the previous sections that a desirable value of ν must be in the range $[0.005, 0.05]$. This range gives us an idea of the value of ν we must choose.

We have applied the reconfiguration procedure to a test bench of 25 reconfigurations, which include switches between spacecraft located at opposite vertices of polygons (6), switches in the TPF formation (9) and parallel shifts (10) of different size with a number of spacecraft from 3 to 10. Ten of the reconfigurations are converging to a bang–bang solution, the other 15 reconfigurations are converging to low thrust. We have applied the methodology using different values of ν and we have computed the mean of the number of iterations of the adaptive process necessary to converge. The results can be seen in Table 3.

Again we conclude that the best value of ν is 0.05.

Table 3 Mean of the number of iterations as a function of ν for the 25 test bench reconfigurations

ν	0.005	0.01	0.02	0.03	0.04	0.05	0.055	0.06
It.	10.2	8.4	7.1	4.2	3.7	3.2	4.3	5.2

5 Dealing with Nonlinearities

We have pointed out that the size of the formations is very small (a few hundreds of meters) when comparing it to the size of the halo orbit. This fact gives us the possibility to work with linearized equations about the halo orbit. In this section, we present the results of some simulations to show that the use of linearized equations is really a good model.

Our objective in this section is to give a way to measure how the truncated nonlinear terms, as well as other perturbations, affect the nominal reconfiguration trajectory, and the corrections that should be applied to the nominal maneuvers (corrective maneuvers) in order to reach the mission goal. For this purpose, we consider the trajectories given by the FEFF methodology as the nominal path for the spacecraft and compute the corrective maneuvers that guide the spacecraft through the nominal nodal states.

The corrective maneuvers are computed using a similar strategy to [5]. The main idea is that on each element we have the nodal states given by the FEFF methodology, and the difference between these states and the true states is corrected with some small maneuvers. For more details, see [3] or [4].

In the same line as in previous computations, we consider two different cases: the ones which end up in low thrust trajectories (there are many elements on the mesh, and these elements are small) and the ones that end up in bang–bang (the mesh is formed by a single element which is the same as the reconfiguration time).

5.1 The Bang–Bang Case

Like we have pointed out in the previous section, in examples where there is no collision risk, we can focus in the results of a single spacecraft, because the results are independent.

In order to test the suitability of the linear approximation, we consider the shift of a spacecraft in the x direction (200 or 400 m in 8 or 24 h). In Table 4, we give the value of the delta-v given by FEFF methodology (Δv_L), the number of corrections that are performed on each element (n), the maximum of corrective maneuvers ($\Delta \hat{v}_{\mathrm{LJmax}}$), the total amount of corrective maneuvers ($\Delta \hat{v}_{\mathrm{LJ}}$), and the percentage of the corrective maneuvers with respect to the maneuvers given by FEFF.

We note that all these corrective maneuvers are very small, both in absolute values and in percentage.

Table 4 Corrective maneuvers for some cases of bang–bang. Model equations considered correspond to JPL ephemeris and all delta-v are given in cm/s

Case	Δv_L	n	$\Delta \hat{v}_{\mathrm{LJmax}}$	$\Delta \hat{v}_{\mathrm{LJ}}$	%
200 m	0.69	3	3.3×10^{-3}	6.7×10^{-3}	0.96
8 h	0.69	4	2.5×10^{-3}	6.1×10^{-3}	0.88
	0.69	5	2.2×10^{-3}	5.8×10^{-3}	0.84
	0.69	6	2.2×10^{-3}	5.6×10^{-3}	0.80
200 m	0.23	3	3.6×10^{-3}	7.5×10^{-3}	3.25
24 h	0.23	4	2.7×10^{-3}	6.7×10^{-3}	2.89
	0.23	5	2.2×10^{-3}	6.4×10^{-3}	2.77
	0.23	6	2.2×10^{-3}	6.1×10^{-3}	2.65
400 m	2.8	3	5.3×10^{-3}	9.7×10^{-3}	0.35
8 h	2.8	4	3.9×10^{-3}	9.2×10^{-3}	0.33
	2.8	5	3.1×10^{-3}	8.1×10^{-3}	0.29
	2.8	6	2.7×10^{-3}	7.5×10^{-3}	0.27

Table 5 Corrective maneuvers for some cases of low thrust. Model equations considered correspond to JPL ephemeris and all delta-v are given in cm/s

Case	Δv_L	n	$(\Delta v/l)_{\mathrm{Jmax}}$	Δv_J	%
Swap	0.63	3	9.1×10^{-3}	7.5×10^{-3}	1.19
2 sats	0.63	4	8.5×10^{-3}	6.9×10^{-3}	1.10
100 m	0.63	5	7.7×10^{-3}	6.7×10^{-3}	1.06
24 h	0.63	6	5.9×10^{-3}	6.4×10^{-3}	1.01
TPF	2.34	3	9.9×10^{-3}	1.3×10^{-2}	0.51
Swap	2.34	4	8.2×10^{-3}	1.1×10^{-2}	0.44
	2.34	5	6.4×10^{-3}	1.0×10^{-2}	0.42
	2.34	6	2.6×10^{-3}	9.2×10^{-3}	0.37
TPF	1.26	3	8.3×10^{-3}	1.0×10^{-2}	0.77
Symmetry	1.26	4	8.0×10^{-3}	9.4×10^{-3}	0.73
Baseline	1.26	5	5.9×10^{-3}	8.9×10^{-3}	0.66
	1.26	6	5.5×10^{-3}	8.1×10^{-3}	0.60

5.2 *The Low Thrust Case*

When we compute the corrective maneuvers with cases of low thrust we must take into account that the elements are very small, and we will do a lot of corrective maneuvers during the reconfiguration time. So, each one of these maneuvers is expected to be smaller than the ones in the case of bang–bang.

We have considered three cases of low-thrust examples, presented in Table 5. The first one is the swap of two spacecraft which are at a distance of 100 m. The reconfiguration time is 24 h. The second example is the TPF swap presented in the previous section. And the third example deals with another case based on the TPF formation: to change the position of the collector toward a point symmetric to the departure position with respect to the interferometry baseline. The parameters of the table are the same ones of the previous example, except for the maximum of the delta-v, $(\Delta v/l)_{\text{Jmax}}$. In this case, we give the maximum of delta-v divided by the length of the element instead, this is thrust acceleration as is usual for low-thrust trajectories.

We remark that delta-vs are again small, both in absolute value and in percentage. So, we can conclude that for small formations (of a few hundreds of meters) and small reconfiguration times (of a few hours), the corrective maneuvers are small and the linear approximation about the nonlinear orbit is good enough.

6 Conclusions

This paper presents a methodology to find trajectories for reconfigurations of spacecraft, using the finite element method.

We have shown that the strategy of adaptive remeshing is suitable for the problem. It converges toward a bang–bang solution when there is no collision risk (that is known in this case to be the optimal control), and when there exists collision risk in the reconfiguration process, the procedure tends to low-thrust arcs.

The computations have been done for formations of a few hundreds of meters, with reconfiguration times of a few hours. We have shown, through some general examples, that the linearized model about the nonlinear orbit is suitable for the nominal computations.

Acknowledgments This research has been supported by the MICINN-FEDER grant, MTM2009-06973, the CUR-DIUE grant 2009SGR859, and the European Marie Curie grant Astronet, MCRTN-CT-2006-035151.

References

1. Farrar, M., Thein, M. and Folta, D. C. (2008). A Comparative Analysis of Control Techniques for Formation Flying Spacecraft in an Earth/Moon-Sun L2-Centered Lissajous Orbit. *AIAA Paper* 2008–7358, Aug. 2008.
2. Garcia-Taberner, L. and Masdemont, J. J. (2009). Maneuvering Spacecraft Formations using a Dynamically Adapted Finite Element Methodology. *Journal of Guidance, Control, and Dynamics* Vol. 32, No. 5, pp. 1585–1597.
3. Garcia-Taberner, L. and Masdemont, J. J. (2010). FEFF methodology for spacecraft formations reconfiguration in the vicinity of libration points. *Acta Astronautica*, doi:10.1016.
4. Garcia-Taberner, L. (2010). Proximity Maneuvering of Libration Point Orbit Formations Using Adapted Finite Element Methods. *PhD Dissertation* Universitat Politècnica de Catalunya. Barcelona.
5. Gómez, G., Lo, M., Masdemont, J.J. and Museth, K. (2002). Simulation of Formation Flight near Lagrange Points for the TPF Mission. *Advances in the Astronautical Sciences* Vol. 109, 61–75.
6. Wang, P.K.C. and Hadaegh, F. Y. (1999). Minimum-fuel Formation Reconfiguration of Multiple Free-flying Spacecraft. *Journal of the Astronautical Sciences* Vol. 47, No. 1,2, pp. 77–102.

7. Li, L. Y., Bettess, P., Bull, J.W., Bond, T. and Applegarth, I. (1995). Theoretical formulation for adaptive finite element computations. *Communications in Numerical Methods in Engineering* 11, 858-868.
8. Badawy, A. and McInnes, C. R. (2008). On-Orbit Assembly Using Superquadric Potential Fields. *Journal of Guidance, Control, and Dynamics* Vol. 31, No. 1, pp. 30–43.
9. Reddy, J. N. (1993). An Introduction to the Finite Element Method. McGraw-Hill, New York.
10. The Science and Technology Team of Darwin and Alcatel Study Team. (2000) Darwin. The Infrared Space Interferometer. Concept and Feasibility Study Report. *European Space Agency, ESA-SCI* 12, iii+218.
11. The TPF Science Working Group. (1999) The Terrestrial Planet Finder: A NASA Origins Program to Search for Habitable Planets. *JPL Publication* 99-3, 1999.

A Cartographic Study of the Phase Space of the Elliptic Restricted Three Body Problem: Application to the Sun–Jupiter–Asteroid System

Cătălin Galeş

1 Introduction

In the last two decades, various numerical methods have been applied to celestial mechanics to distinguish between regular and chaotic motions. Depending on the mathematical model, the information offered and the amount of computation, each method has its own advantages and disadvantages. We shall briefly recall some widely used tools in investigating the dynamics of nonintegrable systems. Thus, the frequency map analysis [22, 23] has been used for studying small body dynamics by Nesvorný and Ferraz-Mello [31] and Celletti et al. [4]; the method of twist angles introduced by Contopoulos and Voglis [6] has been tested on the standard map and compared with other methods of analysis in [9]; the method of short time average of the stretching numbers [11] has been applied to the circular and elliptic restricted three-body problem in [35]; the fast Lyapunov indicator [12, 13] has been investigated in detail and applied to various dynamical problems; the MEGNO technique [5] has been utilized to perform a stability analysis of extra solar planets in [18, 19]; the time series given by the intervals between successive crossing of a given plane of section have been used to reconstruct the phase space in the restricted three-body problem [16].

In this chapter, we use the fast Lyapunov indicator (FLI) in order to perform a cartographic study of a given portion of the phase space of the restricted three-body problem. Introduced by Froeschlé et al. [12,13], this tool is easy to implement, cheap in computational time, and very sensitive for the detection of weak chaos and for distinguishing between regular resonant orbits and regular nonresonant ones. These features have been unveiled by testing it on symplectic mappings [10, 24] as well as on continuous flows [8]. Moreover, it was shown that FLI is a very useful numerical

C. Galeş (✉)
Faculty of Mathematics, Al. I. Cuza University of Iaşi, Romania
e-mail: cgales@uaic.ro

J.A.T. Machado et al., *Nonlinear and Complex Dynamics: Applications in Physical, Biological, and Financial Systems*, DOI 10.1007/978-1-4614-0231-2_6,

tool for revealing the geography of resonances (the so-called Arnold's web), for detecting the transition between the stable Nekhoroshev's regime and the diffusive Chirikov's one [14, 20, 36] and even for detection of diffusion along resonances (Arnold's diffusion) [15, 25, 26, 36]. These topics have been investigated in the framework of quasi-integrable Hamiltonian systems. In this sense, some model systems have been carefully chosen.

The FLI tool has been directly utilized to study the stability of extrasolar planets [33], to solve spacecraft preliminary trajectory design problems [38], to investigate the dynamics associated to nearly integrable dissipative systems [1].

In this chapter, we use the FLI tool to investigate the phase space of the restricted three body problem and we focus on the Sun–Jupiter–Asteroid system. We recall that FLI has been first applied to asteroidal motion. Froeschlé et al. [13] integrated a model consisting of the four giant planets and the Sun and studied the dynamics of the asteroids orbiting between the 3/1 and 5/2 Kirkwood gaps. Here, we implement the FLI tool for the elliptic restricted three-body model (the case Sun–Jupiter–infinitesimal mass) and prove that FLI reveals after a very short computational time, the entire structure of the phase space (the regular and chaotic regions, the geography of resonances, the libration regions, and other details closely connected with the degeneracy property of the restricted three-body problem).

This analysis is relevant in the study of the global dynamics in the asteroid belt. This topic, with particular attention to the mechanisms of chaotic diffusion, is extensively studied in literature (see for example the book of Morbidelli [27] and the review article by Tsiganis [37]). The theories of chaotic diffusion aim to predict the long-term behavior of ensembles of asteroids, rather than individual orbits. In this sense, the Lyapunov time, the size, and the shape of the chaotic regions and the diffusion coefficients are the main parameters needed to understand the long-term effect of chaotic diffusion in the asteroid belt. To compute them, analytical models and numerical studies of two-body mean motion resonances of different order have been accomplished (see [21,27,30,37] and references therein).

Here, a numerical study is given in the framework of the planar elliptic restricted three body problem in order to give an estimate of the size and the shape of the chaotic regions. We integrate the variational equations along with the equations of motion for a set of 500×500 test particles placed on a regular grid in the plane (a, e), where the semimajor axis a range from 1.5 to 6 AU, while the eccentricity $e \in [0, 0.5]$. The total time span covered by our integration has a maximum length of 8,400 y (700 periods of Jupiter). On the obtained dynamical maps, the structure of the above portion of the phase space is clearly displayed.

We recall that the phase space of the circular and elliptic restricted three-body problem has been investigated previously by Sándor et al. [35] by using the method of short time average of the stretching numbers. In the semimajor axis-eccentricity plane of the test particle, Sándor et al. [35] studied some features of the region $a \in$[3 AU, 4.5 AU], $e \in [0, 0.4]$ for three cases of the eccentricity of the primaries: $e_J = 0$, $e_J = 0.048$ and $e_J = 0.1$.

Another study [37] is based on the computation of the Lyapunov time in the framework of 2D and 3D elliptic three-body problem for 1 Gy in order to give an estimate of the removal time from different mean motion resonances.

Here, we consider the same problem, but we use a different method. The FLI is computationally cheap, robust, and easy to implement. In contrast with the above mentioned numerical methods, FLI is also very sensitive to detect weak chaos. The dynamical maps show not only the regular regions, the chaotic ones, and the geography of mean motion resonances of low order, but there are also displayed very thin layers associated with mean motion resonances of moderate order. Moreover, inside the mean motion resonances there are revealed other small chaotic lines which are due to the resonance splitting. We discuss our numerical results in the light of the theory of Hamiltonian systems.

2 Basic Formulation

In this section, we recall the definition of the FLI and describe the techniques utilized to integrate the equations of motion and the variational equations.

2.1 *The Fast Lyapunov Indicator (FLI)*

Let us consider the dynamical system

$$\frac{\mathrm{d}X}{\mathrm{d}t} = F(X(t)), \qquad X \in \mathbb{R}^n, \quad t \in \mathbb{R}, \tag{1}$$

where F is a continuously differentiable function. Given an initial condition $X(0) \in \mathbb{R}^n$, let us consider the evolution $V(t) \in \mathbb{R}^n$ of an initial vector $V(0) \in \mathbb{R}^n$ of norm 1, obtained by integrating the variational equations

$$\frac{\mathrm{d}V}{\mathrm{d}t} = \frac{\partial F}{\partial X}(X(t))V, \tag{2}$$

where $X(t)$ is the evolution of $X(0)$. Then, the maximum Lyapunov exponent is defined by

$$MLE(X(0)) = \lim_{t\to\infty} \frac{\ln \|V(t)\|}{t}. \tag{3}$$

Numerically, one works on finite times. Thus, one estimates MLE by computing the Lyapunov characteristic indicator defined by

$$LCI(t;X(0)) = \frac{\ln \|V(t)\|}{t}. \tag{4}$$

at a large time t. The fast Lyapunov indicator is defined by (see [10, 24])

$$FLI(T;X(0)) = \sup_{0<t<T} \ln \|V(t)\|. \quad (5)$$

The computation of FLI on a relatively short time is enough to discriminate between chaotic and regular orbits. The FLI of a regular orbit increases linearly, while for a chaotic orbit, the FLI increases exponentially. The FLI is very sensitive to detect weak chaos. Moreover, FLI may be used to discriminate among regular motion between nonresonant and resonant orbits. Because of the differential rotation, the norm of the vector V, asymptotically grows as $\|V(t)\| \cong \alpha t$, the coefficient α depending on the nature of invariant curve (torus or libration island). As against the MLE, there is a disadvantage. Namely, FLI depends on the initial conditions and if the system is Hamiltonian, on the choice of the canonical variables. But, once some reference orbits have been computed for which the chaotic (or regular) nature has been determined, the FLI allows the investigation of a large number of orbits and the computations of dynamical maps like the ones presented in this chapter.

2.2 The Restricted Three Body Problem

The three-body problem simplified by setting one mass to zero is widely known as the restricted three-body problem. In this case, the equations for the two primaries S (Sun) and J (Jupiter) decouple, so that their motion is Keplerian. According to the value of the eccentricity of their orbits we speak of the circular or elliptic problem. Here, we consider the elliptic problem. We define our system of units such that $G(m_\mathrm{S} + m_\mathrm{J}) = 1$, where G denotes the constant of gravitation, m_S the mass of the Sun, and m_J the mass of Jupiter. Let us introduce the notation

$$\mu = Gm_\mathrm{J} = \frac{m_\mathrm{J}}{m_\mathrm{S} + m_\mathrm{J}} = 9.54 \cdot 10^{-4}. \quad (6)$$

Then, the heliocentric equations of motion of Jupiter and the infinitesimal body are [29]

$$\begin{aligned} \frac{\mathrm{d}^2\mathbf{r}_\mathrm{J}}{\mathrm{d}t^2} &= -\frac{\mathbf{r}_\mathrm{J}}{\|\mathbf{r}_\mathrm{J}\|^3}, \\ \frac{\mathrm{d}^2\mathbf{r}}{\mathrm{d}t^2} &= -(1-\mu)\frac{\mathbf{r}}{\|\mathbf{r}\|^3} + \mu\left[\frac{\mathbf{r}_\mathrm{J} - \mathbf{r}}{\|\mathbf{r}_\mathrm{J} - \mathbf{r}\|^3} - \frac{\mathbf{r}_\mathrm{J}}{\|\mathbf{r}_\mathrm{J}\|^3}\right], \end{aligned} \quad (7)$$

where $\mathbf{r}_\mathrm{J}$ and $\mathbf{r}$ are their position vectors relative to the Sun. We have numerically integrated the above equations using as starter a single step method (Runge–Kutta), while a symmetric multistep method of 12th order (proposed by Quinlan and Tremaine [34]) performs most of the propagation. We recall that the numerical integration is subject to several sources of integration error, namely truncation

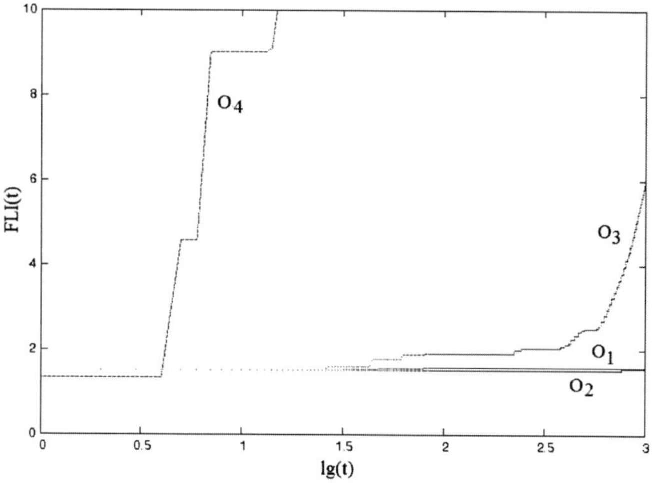

Fig. 1 Evolution of FLI as a function of $lg(t)$ for the orbits: O_1 (regular orbit), O_2 (regular resonant orbit), O_3 (weak chaotic orbit) and O_4 (strong chaotic orbit). The unit of time is the period of Jupiter

error, round-off error, error in the initial conditions and masses of the bodies, error in the physical model etc. Quinlan and Tremaine [34] showed that symmetric multistep methods reduce the truncation error and illustrated by numerical examples the superiority of the symmetric multistep methods over other methods. Based on those discussed in [34] concerning the optimum choice of the order and the coefficients of the integrator, we chose a 12th order method in our computations. For the corresponding variational equations, we have utilized again the Runge–Kutta method. For each initial condition, the total time span covered by the integration has a length of maximum 8,400 y (700 periods of Jupiter). The computations for a longer time (over 1,000 periods of Jupiter) did not change significantly the result. Therefore, we used a shorter computation time for the whole analysis. Moreover, since FLI increase exponentially with time for chaotic orbits, we stopped the integration if $ln(FLI)$ exceeded the value 1.8.

The initial conditions have been chosen such that the second primary (Jupiter) describes an elliptic motion with eccentricity $e_J = 0.048$ and at time $t = 0$, Jupiter is at perihelion. We have utilized a Cartesian coordinate system centered on the Sun with the x–axis pointing towards the perihelion of Jupiter.

We suppose the massless body moves in a coplanar heliocentric elliptic orbit perturbed by Jupiter. Its initial conditions are given once the semimajor axis a (in AU), the eccentricity e, the argument of perihelion ω (the angle between the perihelion line and the x line), and the mean anomaly M are prescribed.

In order to exemplify the behavior of FLI for different kind of orbits, we plotted the evolution of the values of FLI for four orbits: O_1 (regular orbit), O_2 (regular resonant orbit), O_3 (weak chaotic orbit), and O_4 (strong chaotic orbit) during 1,000 periods of Jupiter (see Fig. 1). The orbits correspond to the following initial conditions: O_1: $a = 2.43$ AU, $e = 0.15$, $\omega = 0^o$, $M = 0^o$; O_2: $a = 2.49$ AU, $e = 0.15$,

$\omega = 0^o, M = 0^o$; O_3: $a = 2.53AU, e = 0.15, \omega = 0^o, M = 0^o$; and O_4: $a = 4.7$ AU, $e = 0.15$, $\omega = 0^o$, $M = 0^o$. The regular orbit O_1 is close to the 3 : 1 resonance, O_2 is in the libration region of the 3 : 1 resonance, O_3 is on the separatrix of the 3 : 1 resonance, while O_4 is located in the unstable region (see Fig. 3). For the regular orbits O_1 and O_2, FLI increases linearly. The FLI of the chaotic orbits O_3 and O_4 increases exponentially, but with different rates.

3 Results

In this Section, we describe numerically the structure of a selected region of the phase space of the planar elliptic restricted three-body problem (PERTBP) by computing dynamical maps, such as the ones presented in Figs. 3–7. In order to give a theoretically based interpretation to our numerical analysis, we recall first some important definitions and analytical results.

The phase space of the PERTBP is four dimensional. Thus, from a theoretical point of view, to describe a portion of it, a dense network of points covering a subset of $\mathbb{R}^4$ should be investigated. However, since our model is a quasi-integrable Hamiltonian system, we resort to the space of the actions.

In the case of quasi-integrable Hamiltonian systems having a nondegenerate integrable part, the KAM theorem (see for example Celletti and Chierchia [2] for a recent description of the state of the art of the theory) assures the persistence of invariant tori carrying motion with diophantine frequencies, provided the perturbations are small enough. In other words, the nondegenerate integrable approximation H_0 gives a foliation of the phase space in invariant tori, the actions being constants and the angles circulating linearly with time. When a small perturbation εH_1 is added, the KAM theorem ensures that some invariant tori with diophantine frequencies continue to be invariant for the complete Hamiltonian $H_0 + \varepsilon H_1$. The size ε determines which tori continue to be invariant among all the unperturbed ones with diophantine frequencies.

Clearly, the integrable part of the PRTBP, namely the two-body problem is highly degenerate, i.e. the Hamiltonian does not depend on some action variables and so the hypotheses of the KAM theorem and also of the stability theorem of Nekhoroshev are not satisfied in their standard form. For this reason, the PRTBP exhibit very complicated dynamics. In the phase space, we identify: regular regions, chaotic areas, and resonant regions with its libration and chaotic zones. We discuss their size as function of the resonance order and the parameters entering into the perturbing function, in particular the argument of perihelion and the mean anomaly.

Let us recall now the Hamiltonian and the canonical variables of the planar restricted three-body problem. With our choice of dimensions and the Cartesian coordinate system, the modified Delaunay variables are [7, 27]

$$\begin{aligned} \Lambda &= \sqrt{(1-\mu)a}, & \lambda &= M + \omega, \\ \Omega &= \Lambda(1 - \sqrt{1-e^2}), & & -\omega. \end{aligned} \tag{8}$$

The autonomous Hamiltonian has the form

$$H = -\frac{(1-\mu)^2}{2\Lambda^2} + n_J\Lambda_J - \mu f(\Lambda,\Omega,\lambda,\omega,\lambda_J), \tag{9}$$

where $n_J = \dot{\lambda}_J$ is the mean motion of Jupiter, λ_J is the mean longitude, and Λ_J is the conjugate action corresponding to Jupiter. The disturbing function f, also depends on the constant elements of Jupiter's orbit (for a detailed description of the perturbation theories the reader is referred to the books of Murray and Dermott [29] and Ferraz-Mello [7]). The integrable part of H is strongly degenerate. The fast angle λ is nondegenerate, while the slow angle ω is degenerate.

Now, we recall that the resonances

$$(p+q)\dot{\lambda} - p\dot{\lambda}_J = 0, \qquad p,q \in \mathbb{Z} \tag{10}$$

are called mean motion resonances. The integer q is the order of resonance and determines its strength. It is known that asteroids can develop chaotic motion as a result of resonant perturbations, exerted by the major planets. The asteroid–Jupiter mean motion resonances of low order ($1 \leq q \leq 4$) force the resonant asteroids to become planet crossers on a short timescale (of order of a few million years), while in mean motion resonances of moderate order ($5 \leq q \leq 7-9$) the times required to become a planet crosser are much longer (from tens of millions of years to of order $1Gyr$) (see [27]).

In order to investigate numerically the topology of the phase space, we proceeded as follows: in the domain $0 \leq e \leq 0.5$, 1.5 AU $\leq a \leq$ 6 AU, of the action plane (Λ,Ω) we considered a grid of 500×500 equidistant initial conditions. We considered various choices of the initial angles ω and M. We report here only the figures obtained for: $\omega = 0^o, M = 0^o$ (Fig. 3); $\omega = 90^o, M = 0^o$ (Fig. 4); $\omega = 180^o, M = 0^o$ (Fig. 5); $\omega = 60^o, M = 0^o$ (Fig. 6); $\omega = -60^o, M = 0^o$ (Fig. 7); For each point on the grid, we calculated the final value of the logarithm of FLI for maximum 700 periods of Jupiter. The results are reported in the (Figs. 3–7), where the gray scale is used in such a way that white color corresponds to chaotic orbits, whereas the darker the color is, the more stable the orbit is.

Moreover, we represented the distribution of the main belt and Trojans asteroids (Fig. 2) using the osculating semimajor axis and eccentricity, in order to display the location of the mean motion resonances with Jupiter. We recall the location of some low order resonances: $4:1$ ($a = 2.06$ AU); $3:1$ ($a = 2.5$ AU); $5:2$ ($a = 2.82$ AU); $7:3$ ($a = 2.95$ AU); $2:1$ ($a = 3.277$ AU); $7:4$ ($a = 3.58$ AU); $5:3$ ($a = 3.7$ AU); $3:2$ ($a = 3.9$ AU); $1:1$ ($a = 5.2$ AU).

Figures 3–7 are in a good agreement with the analytical and numerical studies on the global dynamics of the asteroid belt (see [27, 29, 37] and references therein). Analyzing these maps, we may conclude the following: Depending on the initial proper elements a, e, ω, M, the infinitesimal mass can undergo chaotic or regular motions, and the phase space can be partitioned into the following zones:

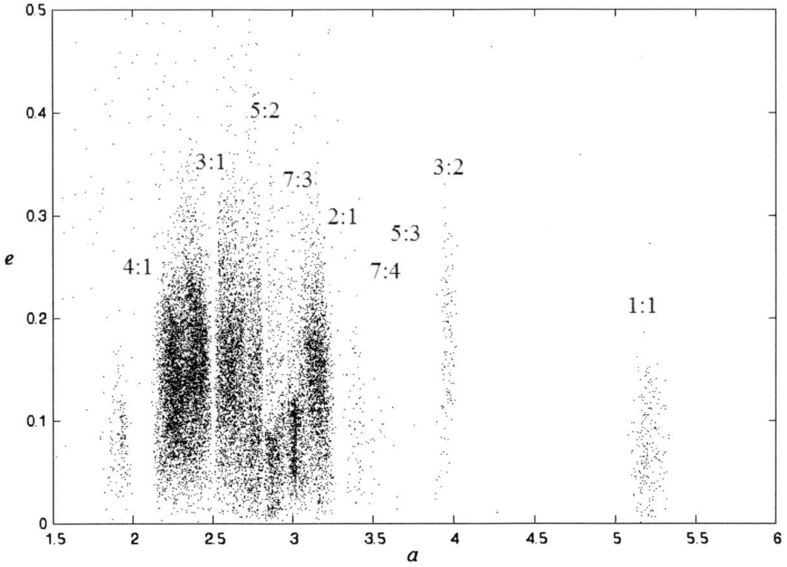

Fig. 2 Distribution of the main-belt and Trojans asteroids on the (a,e)-plane

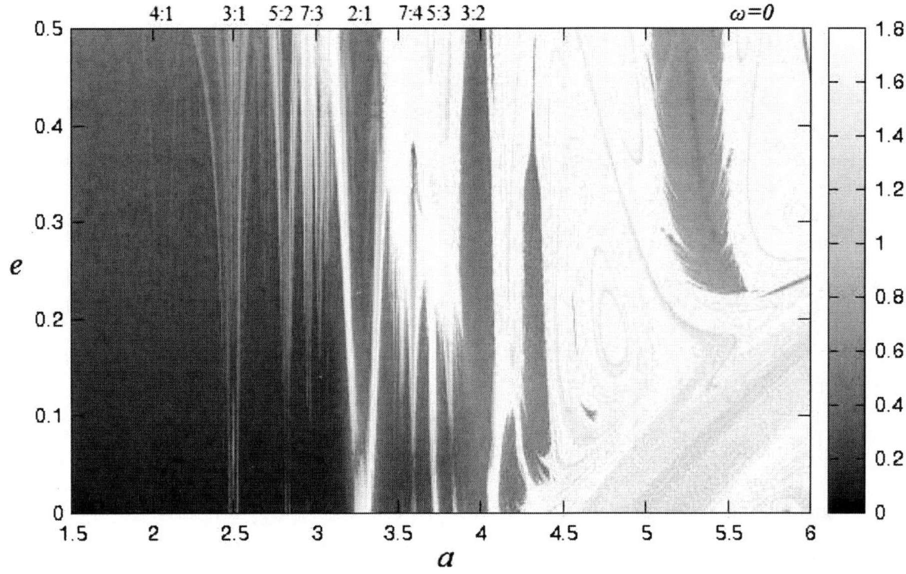

Fig. 3 Maps of the logarithm of the final value of FLI after 700 periods of Jupiter. The test particles were placed on a regular 500×500 grid in (a,e) plane and the initial angles was $\omega = 0^o$, $M = 0^o$ The values of $lg(FLI)$ are color-coded, according to the scale shown on the right (*black* = regular, *white* = chaotic)

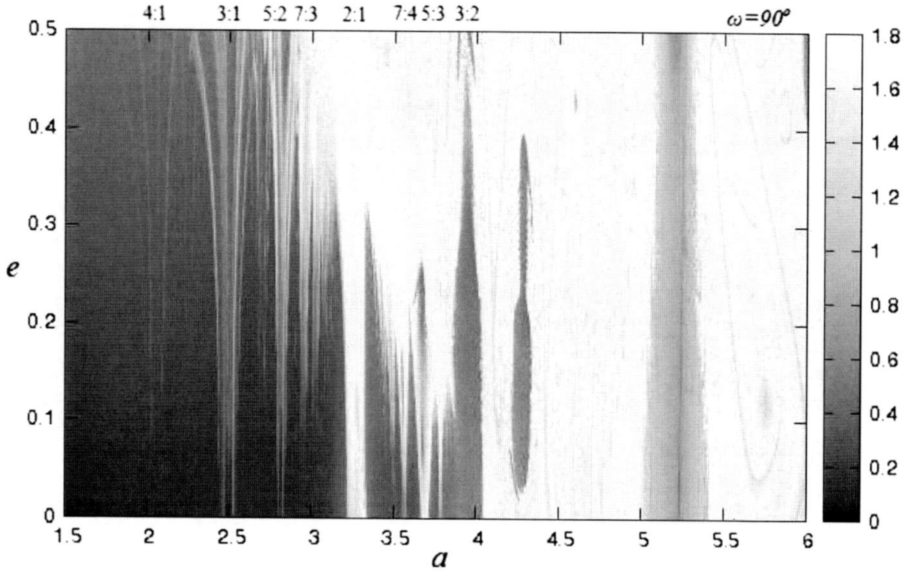

Fig. 4 Same as Fig. 3, but for $\omega = 90^o$, $M = 0^o$

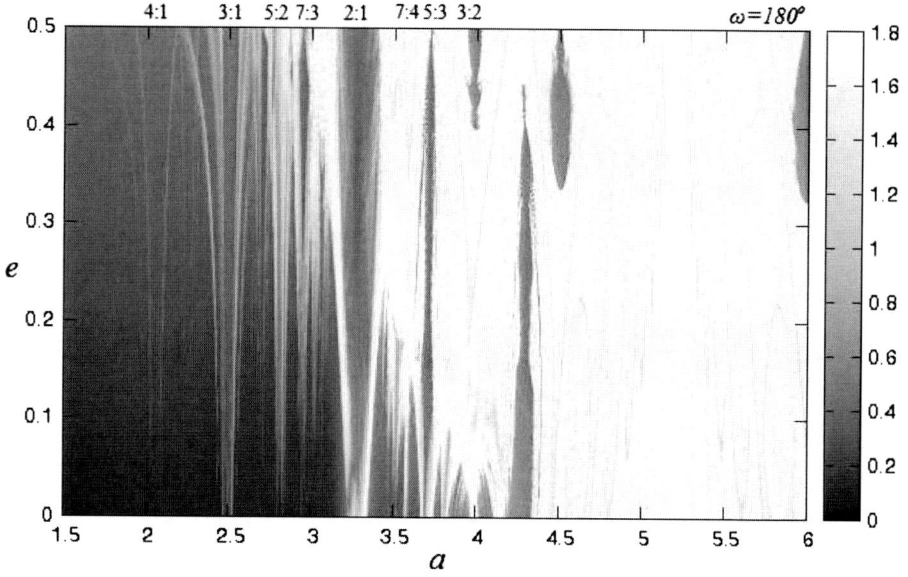

Fig. 5 Same as Fig. 3, but for $\omega = 180^o$, $M = 0^o$

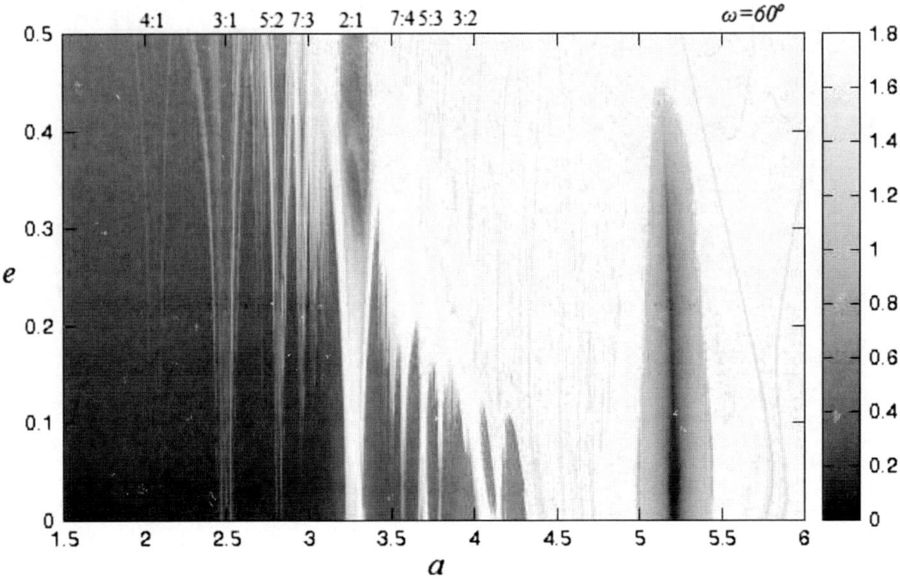

Fig. 6 Same as Fig. 3, but for $\omega = 60^o$, $M = 0^o$

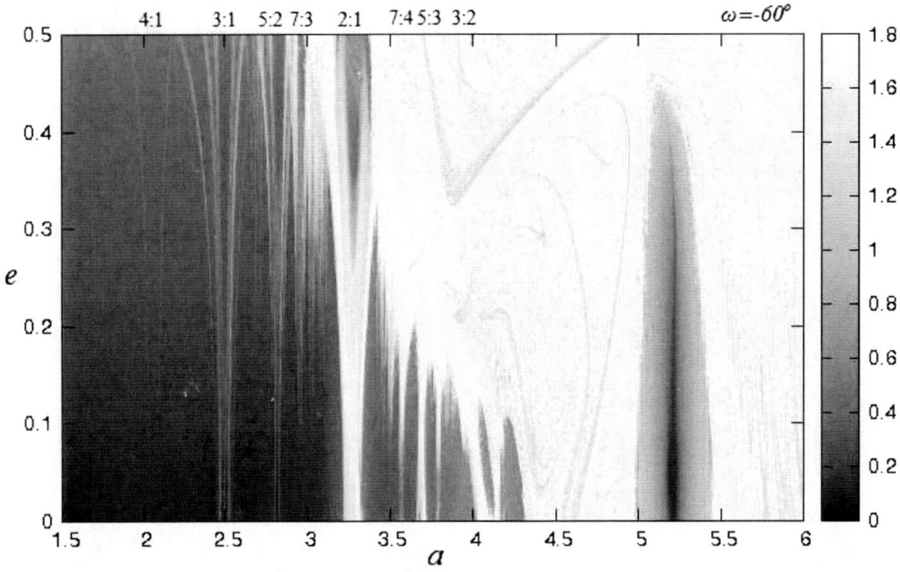

Fig. 7 Same as Fig. 3, but for $\omega = -60^o$, $M = 0^o$

The *stability region* located between 1.5 and 3.2 AU is distinguished on each map. Moreover, the maps computed for $\omega = 90^o$, $\omega = 60^o$ and $\omega^o = -60$ (Figs. 4, 6, and 7) show the regular zones associated with 1 : 1 jovian resonance, while on the Figs. 3 and 4 the regular region corresponding to 3 : 2 resonance is exhibited. Clearly, since we did not consider the perturbation exerted by Mars and Saturn, the stability regions displayed on our charts have a sensible bigger area than they should normally have when additional planets are taken into account. The results are in concordance with the analytical studies. Although many quasi-integrable systems, including the gravitational ones, are degenerate, some stability results have been proved by adapting the perturbation techniques to the considered model. For example, Morbidelli and Guzzo [28] studied the stability of the asteroid belt dynamical system and showed that a Nekhoroshev-like result might be constructed on domains that exclude all the mean motion resonances of order smaller than the logarithm of the mass of Jupiter. Moreover, they proved that Nekhoroshev result is strictly connected with the existence of a specific structure of the phase space even for a degenerate system. It is clear that the dark regions on the Figs. 3–7 match with these analytical studies.

Inside this domain, we expect to be many KAM tori. We recall that in the framework of the restricted circular planar three-body problem, Celletti and Chierchia [3] obtained a stability result for the asteroid Victoria ($a = 2.33$ AU, $e = 0.22$) by showing the existence of two nearby trapping KAM tori at the same energy level.

The *strong chaotic motions* arise as a consequence of the resonance overlap criterion. This criterion was discussed by Wisdom [39]. One could consider the phase space as being made of a succession of resonances, each independent of the others and having its own librational and chaotic regions, as in the case of the perturbed pendulum. An obvious example is the sequence of the first order interior resonances of the form $p+1 : p$. Since each resonance has a well-defined width in semimajor axis, and since the separation of adiacent resonances becomes smaller as the perturber is approached, there will come a point at which these overlap (for example, the resonances: 4 : 3 ($a = 4.29$ AU); 5 : 4 ($a = 4.48$ AU); 6 : 5 ($a = 4.6$ AU) etc.) As a consequence, we would expect a cleared zone in the asteroid belt beyond 4.3 AU. This is in good agreement with the observations (see Fig. 2). The chaotic zone predicted by the above described overlap criterion is obtained in each picture. For $a \geq 4.3$ AU a large white zone, whose shape depends on ω, is obtained by numerical integration.

The stability region is crossed by a series of "V"-shaped layers of various sizes. These layers correspond to the *mean motion resonances of low order* (4 : 1 ($a = 2.06$ AU); 3 : 1 ($a = 2.5$ AU); 5 : 2 ($a = 2.82$ AU); 7 : 3 ($a = 2.95$ AU); 2 : 1 ($a = 3.277$ AU); 7 : 4 ($a = 3.58$ AU); 5 : 3 ($a = 3.7$ AU); see also the Fig. 2). Moreover, between 2.5 and 3.27 AU, we can distinguish very thin white lines which for high eccentricity overlap with the mean motion resonance of low order. These lines correspond to moderate-order mean motion resonances. From a mathematical point of view, the dynamics inside mean motion resonances can be explained by analogy to a perturbed pendulum. In the pendulum case the phase space has two regions, the libration and the circulation zones, separated by separatrix. When the perturbing function is taken into account, the separatrix disappears. It place is taken by a chaotic region, whose size depends on the perturbation.

For each mean motion resonance of low order (see for example the resonance 3 : 1 ($a = 2.5$ AU)), we recognize its chaotic border (separatrix) and the libration region. Moreover, inside mean motion resonances we have other small chaotic lines, whose size and location depend again on ω. These regions are due by the contribution to the disturbing functions of the possible resonant arguments. Due to degeneracy of the problem, each resonance splits into a multiplet of resonances. For example, the angles associated with the 3 : 1 resonance are $\varphi_1 = 3\lambda_J - \lambda - 2\omega$, $\varphi_2 = 3\lambda_J - \lambda - \omega - \omega_J$ and $\varphi_3 = 3\lambda_J - \lambda - 2\omega_J$. Since ω have a small but nonzero frequency, φ_1, φ_2, and φ_3 have zero derivatives at different locations. Therefore, the 3 : 1 mean motion resonance splits in a natural way into a threeplet of resonances. The exact location of each component is given by $\dot{\varphi}_k = 0$, $k = 1,2,3$. The image of a given mean motion resonance varies from map to map since $\dot{\omega}$ depends on ω and M via Lagrange's equations.

Finally, we note that the libration region around mean motion resonance decreases with the order q and increases with the eccentricity e. In other words the layer around a mean motion resonance has a "V"-shape and occupies smaller area once the order q increases. These results are in agreement with the analytical perturbation theory (see, for example, the book of Murray and Dermott [29]), which guaranties higher terms in disturbing function once the order q is low and the eccentricity e is high. In fact, the coefficients of the resonant terms are proportional to e^q.

4 Conclusions

The two-body problem is highly degenerate. For this reason, even small perturbations of the two-body problem, like the restricted three-body problem, may exhibit very complicated dynamics. In this work, we considered the planar elliptic restricted three-body problem (the Sun–Jupiter–Asteroid system) and by using the fast Lyapunov indicator we studied numerically the global topology of the phase space. On each dynamical map, regular regions, chaotic zones, and "V"-shaped layers around the mean motion resonances of low order, predicted by analytical theories (see e.g. [27, 29, 37] and references therein) are revealed by the FLI in a very short computational time. The degeneracy of the problem, pointed out by the resonance splitting, is clearly illustrated in the Figs. 3–7. Secular and mean-motion resonances of low order are known to lead to fast chaotic transport of asteroid orbits on million-year time scales [17, 27].

On our dynamical maps, some thin layers associated with mean motion resonances of moderate order are displayed. These resonances together with the three-body mean motion resonances (asteroid–Jupiter–Saturn) (see [32]) form a dense network of thin chaotic layers throughout the asteroid belt, where small-amplitude variations of proper elements of asteroids accumulate slowly over time. This effect is known as chaotic diffusion [37]. The results obtained here for the PERTBP encourage the application of the FLI to study this fine chaotic structure of the asteroid belt.

Acknowledgment This work was supported from the POSDRU/89/1.5/S/49944 Project.

References

1. Celletti, A.: Weakly dissipative systems in Celestial Mechancs. Lect. Notes Phys. **729**, 67–90 (2008)
2. Celletti, A., Chierchia, L.: KAM tori for N-body problems (a brief history). Celest. Mech. Dynamical Astron. **95**, 117–139 (2006)
3. Celletti, A., Chierchia, L.: KAM stability and celestial mechanics. Memoirs of the American Mathematical Society vol. 187, n. 878 (2007)
4. Celletti, A., Froeschlé, C., Lega, E.: Frequency analysis of the stability of asteroids in the framework of the restricted three-body problem. Celest. Mech. Dynamical Astron. **90**, 245–266 (2004)
5. Cincotta, P., Simó, C.: Simple tools to study global dynamics in non-axisymetric galactic potentials-I. Astron. Astrophys. Suppl. Ser. **147**, 205–228 (2000)
6. Contopoulos, G., Voglis, N.: A fast method for distinguishing between order and chaotic orbits. Astron. Astrophys. **317**, 73–81 (1997)
7. Ferraz-Mello, S.: Canonical perturbations theories. Degenerate systems and resonance. Springer, New York (2007)
8. Fouchard, M., Lega, E., Froeschlé, Ch., Froeschlé, C.: On the relationship between fast Lyapunov indicator and periodic orbits for continuous flows. Celest. Mech. Dynamical Astron. **83**, 205–222, (2002)
9. Froeschlé, C., Lega, E.: Twist angles: a method for distinguishing islands, tori and weak chaotic orbits. Comparison with other methods of analysis. Astron. Astrophys. **334**, 355–362, (1998)
10. Froeschlé, C., Lega, E.: On the structure of symplectic mappings. The fast Lyapunov indicator: a very sensitive tool. Celest. Mech. Dynamical Astron. **78**, 167–195 (2000)
11. Froeschlé, C., Froeschlé, Ch., Lohinger, E.: Generalized Lyapunov characteristic indicators and corresponding Kolmogorov like entropy of the standard mapping. Celest. Mech. Dynamical Astron. **83**, 205–222 (1993)
12. Froeschlé, C., Lega, E., Gonczi, R.: Fast Lyapunov indicators. Application to asteroidal motion. Celest. Mech. Dynamical Astron. **67**, 41–62, (1997)
13. Froeschlé, C., Gonczi, R., Lega, E.: The fast Lyapunov indicator: a simple tool to detect weak chaos. Application to the structure of the main asteroidal belt. Planet. Space Sci. **45**, 881–886 (1997)
14. Froeschlé, C., Guzzo, M., Lega, E.: Graphical evolution of the Arnold web: from order to chaos. Science **289**, 2108–2110 (2000)
15. Froeschlé, C., Guzzo, M., Lega, E.: Analysis of the chaotic behaviour of orbits diffusing along the Arnold web. Celest. Mech. Dynamical Astron. **95**, 141–153 (2006)
16. Gidea, M., Deppe, F., Anderson, G.: Phase space reconstruction in the restricted three-body problem. In: New Trends in Astrodynamics and Applications III (Eds. E. Belbruno), AIP Conference Proceedings Volume 886, (2007)
17. Gladman, B. et al.: Dynamical lifetime of objects injected into asteroid belt resonances. Science **277**, 197–201 (1997)
18. Gozdziewski, K.: A dynamical analysis of the HD 37124 planetary system. Astron. Astrophys. **398**, 315–325 (2003)
19. Gozdziewski, K., Bois, E., Maciejewski, A.J., Kiseleva-Eggleton, L.: Global dynamics of planetary systems with the MEGNO criterion. Astron. Astrophys. **378**, 569–586 (2001)
20. Guzzo, M., Lega, E., Froeschlé, C.: On the numerical detection of the effective stability of chaotic motions in quasi-integrable systems. Physica D **163**, 1–25 (2002)
21. Holman, M., Murray, N.: Chaos in high order mean motion resonances in the outer asteroid belt. Astron. J. **112**, 1278–1293 (1996)
22. Laskar, J.: The chaotic motion of the Solar System. A numerical estimate of the size of the chaotic zones. Icarus **88**, 266–291, (1990)
23. Laskar, J., Froeschlé, C., Celletti, A.: The measure of chaos by the numerical analysis of the fundamental frequencies. Applications to the standard mapping. Physica D **56**, 253–269 (1992)

24. Lega, E., Froeschlé, C.: On the relationship between fast Lyapunov indicator and periodic orbits for symplectic mappings. Celest. Mech. Dynamical Astron. **81**, 129–147 (2001)
25. Lega, E., Guzzo, M., Froeschlé, C.: Detection of Arnold diffusion in Hamiltonian systems. Physica D **182**, 179–187 (2003)
26. Lega, E., Guzzo, M., Froeschlé, C.: Diffusion in Hamiltonian qusi-integrable systems. Lect. Notes Phys. **729**, 29–65 (2008)
27. Morbidelli, A.: Modern Celestial Mechanics. Aspects of the Solar System Dynamics. Taylor and Francis, London (2002)
28. Morbidelli, A., Guzzo, M.: The Nekhoroshev theorem and the asteroid belt dynamical system. Cel. Mech Dynamical Astron. **65**, 107–136 (1997)
29. Murray, C.D., Dermott, S.F.: Solar System Dynamics. Cambridge University Press, UK (1999)
30. Murray, N., Holman, M.: Diffusive chaos in the outer asteroid belt. Astron. J. **114**, 1246–1259 (1997)
31. Nesvorný, D., Ferraz-Mello, S.: On the asteroidal population of the first-order Jovian resonances. Icarus **130**, 247–258 (1997)
32. Nesvorný, D., Morbidelli, A.: Three-body mean motion resonances and the chaotic structure of the asteroid belt. Astron. J. **116**, 3029–3037 (1998)
33. Pilat-Lohinger, E.: Eccentric orbits in double stars. Proceedings of the 3rd Austrian-Hungarian Workshop on Trojan and related Topics (Eds. F. Freistetter, R. Dvorak and B. Érdi), 35–45 (2003)
34. Quinlan, G.D., Tremaine, S.: Symmetric multistep method for the numerical integration of planetary orbits. Astron. J. **100**, 1694–1700 (1990)
35. Sándor, Z., Balla, R., Téger, F., Érdi, B.: Short time Lyapunov indicators in the restricted three-body problem. Celest. Mech. Dynamical Astron. **79**, 29–40 (2001)
36. Todorovic, N., Lega, E., Froeschlé, C.: Local and global diffution in the Arnold web of a priori unstable systems. Celest. Mech. Dynamical Astron. **102**, 13–27 (2008)
37. Tsiganis, K.: Chaotic diffusion of asteroids. Lect. Notes Phys. **729**, 111–150 (2008)
38. Villac, B.F.: Using FLI maps for preliminary spacecraft trajectory design in multi-body environments. Celest. Mech. Dynamical Astron. **102**, 29–48 (2008)
39. Wisdom, J.: The resonance overlap criterion and the onset of stochastic behavior in the restricted three-body problem. Astron. J. **85**, 1122–1133 (1980)

Parameter Identification of the Langmuir Model for Adsorption and Desorption Kinetic Data

Dumitru Baleanu, Yeliz Yolcu Okur, Salih Okur, and Kasim Ocakoglu

1 Introduction

The humidity adsorption and desorption kinetic data of spin-coated 50 nm Ruthenium polypridyl complex (Ru-PC K314) film has been measured under relative humidity between 11% and 97% using Quartz Crystal Microbalance (QCM) technique. QCM has been extensively used for the determination and investigation of the kinetics of adsorption/desorption of adsorbate molecules for monolayer films [3, 4, 8]. QCM technique denotes a powerful approach for determining the sensing properties of materials before a sensor device design during development stages.

Langmuir model has been used successfully for monolayer films to analyze adsorption kinetics. For multilayer films, the Langmuir model cannot be used due to the diffusion of adsorbed molecules between layers. Therefore, it should be modified to determine the adsorption and desorption rates for the diffusion effect.

The chapter is organized as follows. In Sect. 2, we will give more information about the experiment. In Sect. 3, we briefly discuss about the celebrated Langmuir model, which was first introduced in [9]. In Sect. 4, we briefly mention about nonlinear regression analysis. In Sect. 5, we introduce the model that we propose for humidity adsorption and desorption kinetic data of spin-coated 50 nm Ruthenium polypridyl complex (Ru-PC K314) film and estimate the parameters of the model for the measured data. The last section constitutes a brief conclusion of the paper.

D. Baleanu (✉)
Department of Mathematics and Computer Science, Çankaya University, Ankara Turkey

First author is on leave from The National Institute for Laser, Plasma and Radiation, Physics, Institute of Space Sciences, Magurele Bucuresti, P.O. Box MG-23, R 76911, Romania
e-mail: dumitru@cankaya.edu.tr

J.A.T. Machado et al., *Nonlinear and Complex Dynamics: Applications in Physical, Biological, and Financial Systems*, DOI 10.1007/978-1-4614-0231-2_7,

2 Experimental

For the preparation of QCM electrodes, gold-coated quartz crystal electrodes were placed into ethanol and ultrasonically cleaned, then rinsed by deionized water. 1 mg/ml Ru-PC was dissolved in deionized water. 5 μl of solution was spin-coated on to quartz crystal with 2,000 rpm. After drying at room conditions, it was kept in dessicator at room temperature for 3 hours. Then the quartz crystal coated with Ru-PC film was used to record both the reference frequency at 11% and the frequency changes up to 97% relative humidity. The thicknesses of films were measured using a Dektak profilometer from Veeco and found to be 50 nm.

A Time-Resolved Electrochemical Quartz Crystal Microbalance (EQCM) with the model of CHI400A Series from CH Instruments (Austin, USA) has been used to measure the change in the resonance frequency of quartz crystals between gold electrodes via both serial and usb interface connected to a computer. The QCM works with oscillation frequencies between 7.995 and 7.950 MHz. The density (ρ) of the crystal is 2.684 g/cm^3, and the shear modulus (μ) of quartz is 2.947×10^{11} g/cm.s^2. Around oscillation frequency of 7.995-MHz, a net change of 1 Hz corresponds to 1.34 ng of materials adsorbed or desorbed onto the crystal surface of an area of 0.196 cm^2. For further information about QCM technique and its applications, see [2, 13, 14].

The signals coming from a QCM electrode and a commercial RH humidity sensor were simultaneously measured during the adsorption and desorption process. Both the relative humidity and temperature were also recorded during measurements while maintaining the temperature around 23°C. For this purpose, a EI-1050 selectable digital relative humidity and temperature probe with a response time of 4 s and a resolution of 0.03% RH was used with a USB controlled LabJack U12 ADC system combined with a single chip sensor module (SHT11) manufactured by Sensirion (Staefa, Switzerland).

[Ru^{II} (bis(4,7-diphenyl-1,10-phenanthroline-disulfonic acid disodium salt) (4,4′-dicarboxy-2,2′-bipyridine)], [Ru-PC K314] (Fig. 1), was synthesized according to procedure given in the literature [12]. [$RuCl_2$(p-cymene)]$_2$, 1,10-phenanthroline-disulfonic acid disodium salt and 4, 4′-dicarboxy-2, 2′-bipyridine were obtained from Aldrich. All organic solvents were purchased from Merck and Fluka, and used without further purification. The final metal complex was characterized as following: ^{1}H NMR (400 MHz, D_2O) δ ppm: 9.56 (t, $J = 1.7$ Hz, 1H, NCHCHC, L2), 9.43 (t, $J = 1.3$ Hz, 1H, NCHCHC, L2), 8.73 (s, 1H, NCHCH, L1), 8.62 (s, 1H, NCHCH, L1), 8.10 (t, $J = 1.1$ Hz, 1H, CCHC, L2), 8.02 (t, $J = 1.3$ Hz, 1H, CCHC, L2), 7.78 (d, $J = 3.6$ Hz, 1H, NCHCHC, L2), 7.73 (d, $J = 3.8$ Hz, 1H, NCHCHC, L2), 7.62 (m, $J = 1.4$ Hz, 4H, CCHCHC, L1), 7.44 (q, $J = 1.1$ Hz, 2H, NCHCHC, L1), 7.40 (q, $J = 1.2$ Hz, 2H, NCHCHC, L1), 7.24 (m, $J = 2.4$ Hz, 4H, $CCHCHCSO_3Na$, L1), 7.18 (m, $J = 2.6$ Hz, 4H, $CCHCHCSO_3Na$, L1), 6.95 (m, $J = 2.4$ Hz, 4H, $CCHCHCSO_3Na$, L1), 6.89 (m, $J = 2.4$ Hz, 4H, $CCHCHCSO_3Na$, L1). UV (MeOH); λ_{max}: 483 (1.38), 314 (2.66), 278 (4.21), 221 (4.28). MS (MALDI): m/z = 1419.1 [M+H]+. Anal. Calc. For $C_{60}H_{36}N_6Na_4O_{16}S_4Ru$ (1418.25): C, 50.81; H, 2.56; N, 5.93. Found: C, 50.79; H, 2.39; N, 5.88% [12].

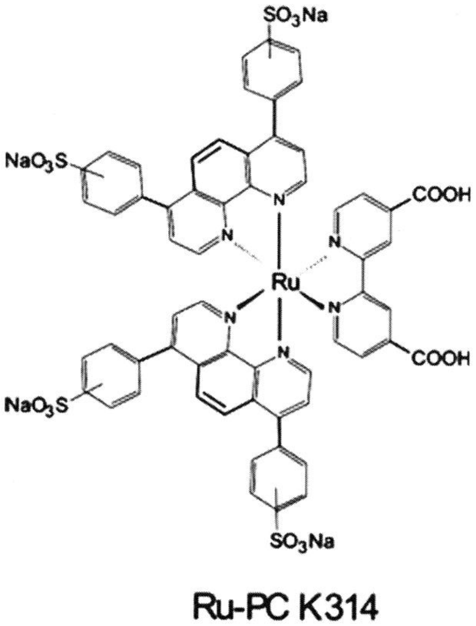

Fig. 1 Chemical structure of ruthenium polypridyl complex

3 Langmuir Model

In this work, we deal with Langmuir model, which was first developed by Irving Langmuir in 1916 [5,6,9]. In order to analyze the adsorption and desorption kinetics of gas vapor molecules onto organic or inorganic films, Langmuir adsorption isotherm model is applied [1, 7, 15, 17, 18]. As a result, the model describes the rate of surface reaction for forming a monolayer on the surface by using the below equation,

$$\frac{\mathrm{d}\theta}{\mathrm{d}t} = k_{\mathrm{a}}(1-\theta) - k_{\mathrm{d}}\theta. \tag{1}$$

Here, θ is a unitless quantity, which means the fraction of surface coverage, k_{a} and k_{d} denote the rate constants for the adsorption and desorption processes.

In this study, QCM has been used to measure the fractional coverage θ a function of time during the adsorption and desorption of water vapor molecules.

Hence, the difference between the oscillation frequency shift Δf of coated and uncoated QCM is directly proportional to the adsorbed mass of moisture molecules. The relationship between the surface adsorption kinetics and frequency shift (Δf) of QCM can be expressed as below

$$\frac{\Delta f}{\mathrm{d}t} = (k_{\mathrm{a}}C + k_{\mathrm{d}})\Delta f + k_{\mathrm{a}}C\Delta f_{\mathrm{max}}. \tag{2}$$

where C is the water vapor concentration in the air. Moreover, note that during the adsorption process, (Δf) is equal to max Δf_{max} for a very long time period.

4 Nonlinear Regression

The classical regression model assumes that the population regression function is the linear function of n independent variables, say x_i for $i = 1,\dots,n$ and the linear regression model can be written as in the following general matrix form:

$$y = \beta X + u,$$

where β is the vector of parameters, X is the vector of independent variables $x_1,\dots,x_n$, and u is the stochastic error term. The reason that we add a stochastic term is there are many unexpected and unobservable cases that will affect us to do some errors in modeling. The measurement error in calculations, missing variables in the model are the examples for such kind of unobservable errors. In order to have a realistic model, we must add a stochastic term to the model, the so-called disturbance term.

When we fit a model to our data, we obtain best-fit values that we can interpret in the context of the model. In many cases, the conditional expectation of the dependent variable is not a linear function of independent variables. Because of this reason, it is more practical to model with nonlinear systems to have more realistic models. For further information about nonlinear regression analysis, see [10] and [16].

5 The Main Result

The aim of this study is to fit the measured data of spin-coated 50 nm Ruthenium polypridyl complex (Ru-PC K314) film to a curve by an appropriate methodology. Polynomial fitting, cubic spline, and linear regression are some fundamental techniques for this procedure. However, most of them ignore the theoretical part of the study and just focus on curve fitting. We use nonlinear regression methodology to fit a curve regarding fundamentals of the theory. Moreover, it is showed that many type of the data are best analyzed by using nonlinear least squares [11].

We assume that the difference between the oscillation frequency shift $y := \triangle f$ of coated and uncoated QCM can be modeled by a nonlinear regression function as follows

$$y_t = f(t, y, k_\mathrm{a}, k_\mathrm{d}) + \varepsilon_t,$$

for all $t \in [0, T]$. Here, f is a nonlinear function of time and the process itself with parameters k_a and k_d. Indeed,

$$f(t, y, k_\mathrm{a}, k_\mathrm{d}) = \frac{k_\mathrm{a} C y_{\max}}{k_\mathrm{a} C + k_\mathrm{d}} + \left(y_0 - \frac{k_\mathrm{a} C y_{\max}}{k_\mathrm{a} C + k_\mathrm{d}} \mathrm{e}^{-(k_\mathrm{a} C + k_\mathrm{d})(t - t_0)} \right), \tag{3}$$

where t_0 and y_0 are the initial values for t and y, respectively. Therefore, for the adsorption data, we use the following model:

$$y_t = \frac{k_a C y_{max}}{k_a C + k_d} + \left(y_0 - \frac{k_a C y_{max}}{k_a C + k_d} e^{-(k_a C + k_d)(t - t_0)} \right) + \varepsilon_t, \tag{4}$$

where the expectation and the variance of ε is zero and a constant number, respectively, i.e.,

$$E[\varepsilon_t] = 0$$

and

$$Var(\varepsilon_t) = \sigma^2$$

for a positive constant σ.

Note that for modeling adsorption data Langmuir is convenient since

$$\lim_{t \to t_0} y_t = A_0,$$

$$\lim_{t \to \infty} y_t = \frac{k_a C_1 \triangle f_{max}}{k_a C_1 + k_d}.$$

However, for desorption data Langmuir model is not a good candidate because we have to guarantee that

$$\lim_{t \to t_0} y_t = y_{max},$$

$$\lim_{t \to \infty} y_t = M,$$

for a constant M which sufficiently small number and close to the minimum of the data series. Regarding all of these properties, we suggest an homogeneous exponential model for the desorption data instead of Langmuir model. We proposed the following one:

$$y_t = y_0 \exp(-(k_a C_2 + k_d)(t - t_0)) + \text{noise}. \tag{5}$$

5.1 Data Analysis

We have 3 adsorption and desorption cycles of data of Ruthenium polypridyl complex film, which is measured by QCM technique (Figs. 2–4). The starting nodes, the maximum values, and the length of the time series are summarized by the following tables (Tables 1–3). Here are the details of the data of Ruthenium polypridyl complex film:

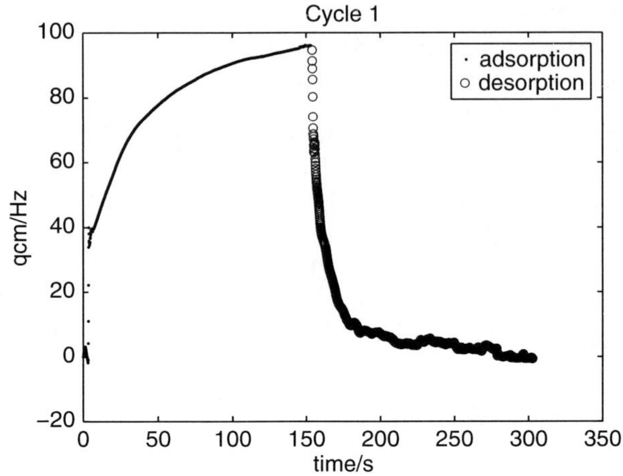

Fig. 2 Plot of observed data for adsorption and desorption for cycle 1

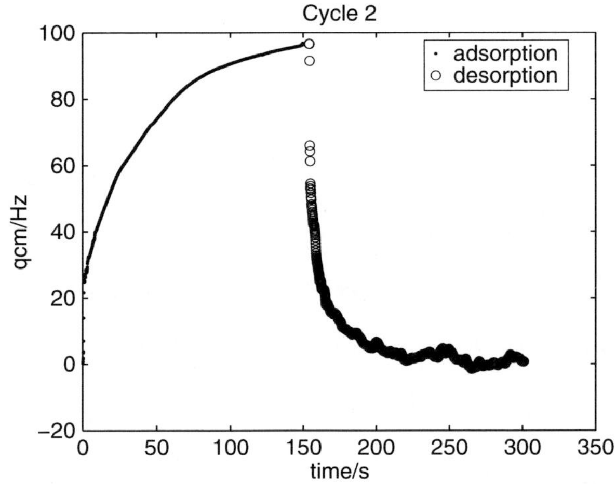

Fig. 3 Plot of observed data for adsorption and desorption for cycle 2

5.2 *Parameter Estimation*

Parameter estimation is the crucial part of mathematical modeling. An estimator attempts to approximate the unknown parameters using the measurements. In this paper, we use nonlinear least squares methodology to find the parameters of adsorption and desorption concentration rate. The parameters describe the physical setting such that it represents the distribution of the data.

However, we have realized two facts. First, the initial values of the original data does not satisfy the usual conditions of a function. Because of this reason we cut some initial points from the measured data in each cycle before estimating the parameters of the nonlinear least squares. Moreover, we realized that the desorption data series are noisy. So, first we denoise the original desorption series by wavelet

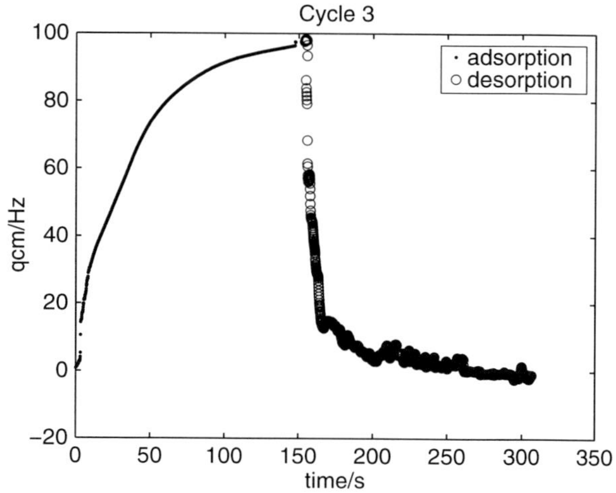

Fig. 4 Plot of observed data for adsorption and desorption for cycle 3

Table 1 Cycle 1

Properties	Adsorption	Desorption
Start	-3.81×10^{-7}	94.65
End	96.07	−0.63
Max	96.14	94.65
Length	1,532	1,482

Table 2 Cycle 2

Properties	Adsorption	Desorption
Start	0.18	96.57
End	96.67	0.75
Max	96.74	96.60
Length	1,499	1,469

Table 3 Cycle 3

Properties	Adsorption	Desorption
Start	0.81	97.56
End	97.43	−1.01
Max	97.53	98.33
Length	1,479	1,528

methodology. For each cycle, we have found the most appropriate wavelet family to denoise. It is explicitly specified at each table below. Every calculation is done by MATLAB® program (Tables 4– 6).

For each cycle, we calculate the error terms (which are exact values minus estimated values) and show that they have zero expectation with a constant variance. We also check for the heteroscedasticity of the error terms in order to guarantee that the error terms have constant variance. We found that no ARCH effects exist.

Table 4 Cycle 1 estimation

Properties	Adsorption	Desorption
k_a	32.499	32.499
k_d	$1.02e-004$	0.103
# Cut points	50	–
Wavelet family	–	"Haar"

Table 5 Cycle 2 estimation

Properties	Adsorption	Desorption
k_a	32.331	32.331
k_d	$1.074e-004$	0.155
# Cut points	9	–
Wavelet family	–	"db3"

Table 6 Cycle 3 estimation

Properties	Adsorption	Desorption
k_a	34.554	34.554
k_d	$8.441e-005$	0.131
# Cut points	34	–
Wavelet family	–	"rbio1.5"

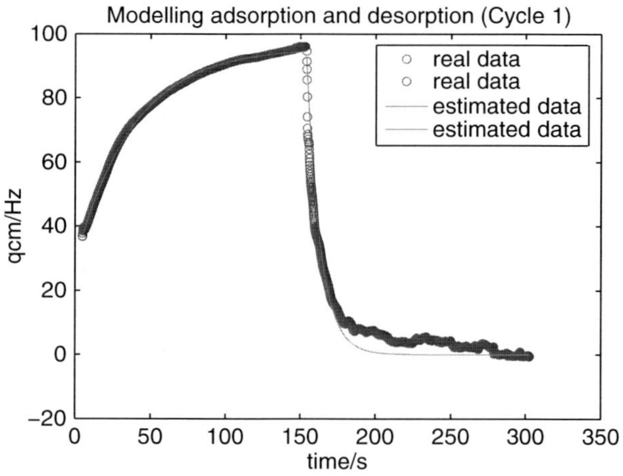

Fig. 5 Plot of estimated and observed data for cycle 1

Indeed, in each cycle of data, we observe that in the first cycle, the error terms has a mean of -1.1×10^{-15} and variance 0.0051. In the second cycle of data, the error terms has a mean of -0.9×10^{-15} and variance 0.0271. In the last cycle of data, their mean and variance are -0.9×10^{-15} and 0.05, respectively.

Let us show our results by plotting the estimated and observed data over time period. Note that for each cycle at the following graph, the blue circles show the original data of spin coated 50 nm Ruthenium polypridyl complex (Ru-PC K314) film, which was measured by QCM technique and the red line shows the estimated data by using the mathematical model introduced in the previous section (Figs. 5–7).

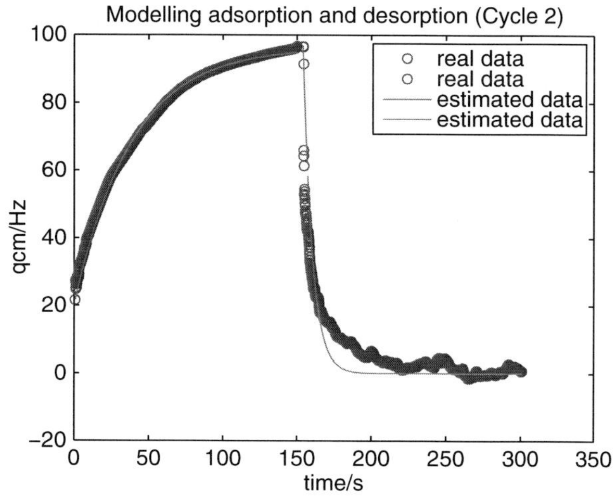

Fig. 6 Plot of estimated and observed data for cycle 2

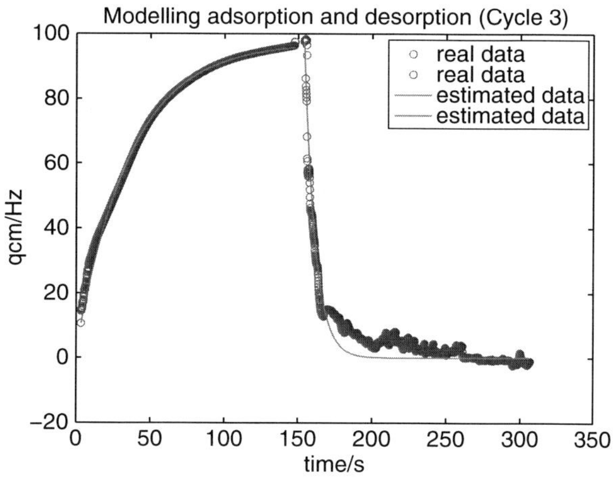

Fig. 7 Plot of estimated and observed data for cycle 3

6 Conclusion

In this paper, our aim is to find the parameters k_a and k_d of our models for given 3 cycles of data of Ruthenium polypridyl complex film, which is measured by QCM technique. We realized that the humidity desorption data of Ruthenium polyprid complex film is noisy so we denoise the original data by appropriate wavelet families by using MATLAB® program. For each cycle 1, 2, and 3, we use haar, db3, and rbio1.5 wavelet families, respectively. We realized that adsorption concentration rate is in between 32.5 and 34.5. Moreover, the desorption concentration rate is found in between 0.1 and 0.2. We also conclude that the humidity adsorption data of spin coated 50 nm Ruthenium polypridyl complex film is smooth and the Langmuir model with disturbance term is appropriate to model.

References

1. Gregg, S.J., Sing, K.S.W.: Adsorption, Surface Area and Porosity. Academic Press (1982)
2. Erol, A., Okur S., Comba, B., Mermer, O., Arikan, M.C.: Humidity Sensing Properties of ZnO Nanodots Synthesized by Sol-Gel Process. Sensors and Actuators B: Chemical **145/1**, 174–180 (2010)
3. Hartmann, J., Auge, J., Lucklum, R., Rsler, S., Hauptmann, P., Adler, B., Dalcanale, E.: Supramolecular interactions on mass sensitive sensors in gas phases and liquids. Sensors and Actuators B: Chemical **34**, 305–311 (1996)
4. Kalchenko, V.I., Koshets, I.A., Matsas, E.P., Kopylov, O.N., Solovyov, A.V., Kazantseva, Z.I., Shirshov, Y.M.: Calixarene-based QCM sensors array and its response to 265 volatile organic vapours. Journal of Materials Science **20**, 73–88 (2002)
5. Kankare, J., Vinokurov I.A.: Kinetics of Langmuirian adsorption onto planar, spherical, and cylindrical surfaces. Langmuir **15**, 5591–5599 (1999)
6. Kapoor A., Ritter, J.A., Yang, R.T.: An Extended Langmuir model for adsorption of gas mixtures on heterogeneous surfaces. Langmuir **6**, 660–664 (1990)
7. Karpovich, D.S., Blanchard, G.J.: Direct measurement of the adsorption kinetics of Alkanethiolate Self-Assembled Monolayers on a microcrystalline gold surface. Langmuir **10**, 3315-3322 (1994)
8. Koshets, I.A., Kazantseva Z.I., Shirshov, Y.M., Cherenok, S.A., Kalchenko, V.I.: Calixarene films as sensitive coatings for QCM-based gas sensors. Sensors and Actuators B: Chemical **106**, 177–181 (2005)
9. Langmuir, I.: The constitution and fundamental properties of solids and liquids. Journal of the American Chemical Society **38**, 2221–2295 (1916)
10. Motulsky J.H., Christopoulos A.: Fitting Models to Biological Data Using Linear and Nonlinear Regression: a Practical Guide to Curve Fitting. Oxford University Press US (2004)
11. Motulsky J.H., Ransnas L.A.: Fitting curves to data using nonlinear regression: a practical and nonmathematical review. The FASEB Journal **1**, 365–374 (1987)
12. Ocakoglu, K., Okur, S.: Humidity sensing properties of Novel Ruthenium Polypridyl Complex. Sensors and Actuators B: Chemical **151/1**, 223–228 (2010)
13. Okur, S., Kus, M., Ozel, F., Aybek, V., Yilmaz, M.: Humidity Adsorption Kinetics of Calix [4] arene derivatives measured using QCM. Talanta **81/1-2**, 248–251 (2010)
14. Okur, S., Kus, M., Ozel, F., Aybek, V., Yilmaz, M.: Humidity Adsorption Kinetics of water soluble Calix [4] arene derivatives measured using QCM technique. Sensors and Actuators B: Chemical **145/1**, 93–97 (2010)
15. Sauerbrey, G.: Verwendung von schwingquarzen zur w*ä*gung d*ü*nner schichten und zur mikrow*ä*gung. Z. Phys. **155**, 206–222 (1959)
16. Seber G.A.F., Wild C.J.: Nonlinear Regression. Wiley (2003)
17. Su, P.G., Chang, Y.P.: Low Humidity sensor based on a quartz-crystal microbalance coated with polyprrole/Ag/TiO_2 nanoparticles composite thin films. Sensors and Actuators B: Chemical **129**, 915–920 (2008)
18. Sun, Y.L., Wu, R.J., Huang, Y. C., Su, P.G., Chavali, M., Chen, Y.Z., Lin, C.C.: In situ prepared polypyrrole for low humidity QCM sensor and related theoretical calculation. Talanta **73**, 857–861 (2007)

Effects of Suspended Sediment on the Structure of Turbulent Boundary Layer

H.P. Mazumdar, S. Bhattacharya, and B.C. Mandal

1 Introduction

River flow is a class of turbulent boundary layer flow. It carries sediment. Suspended sediment is a portion of total sediment load carried by the rivers, and it plays a big role in morphological changes that occurs in rivers. Suspended sediment load is considered important in estimating the effects of land use changes and engineering practices in watercourses. Many areas of hydraulics and sediment transport require essentially knowledge of vertical velocity profile. Some insight into this problems have been gained from study of experimental results and evidence already presented by the profession [1,2,4,7,10,11,15]. In most of the above-mentioned papers, effects of the distribution of sediment concentration on the structure of turbulent boundary layer, specially the modification of the constants of the laws in the overlapping region and value of the wake parameter of the law of the wake are concerned.

Let us first analyze the structure of a single phase turbulent boundary layer as the results that to be drawn from such effort may subsequently be employed to find whether they could describe boundary layer structure with seeded particles. Further Coleman's [2] experimental data would be recalled for this purpose.

In the present analysis, it is admitted that the turbulent boundary layer may be split up [12] into two regions, namely: (I) the inner region where the law of the wall governs the flow and which is unaffected by outside manipulation (pressure changes, history, etc.) and (II) the outer region where the flow may be described by a law of the wake and velocity profiles are as such can take care of the outside manipulation of the turbulent boundary layer. Following Persen [12] a formal description on the

H.P. Mazumdar (✉)
Physics and Applied Mathematics Unit, Indian Statistical Institute,
203 B. T. Road, Kolkata–700108, India
e-mail: hpm@isical.ac.in

J.A.T. Machado et al., *Nonlinear and Complex Dynamics: Applications in Physical, Biological, and Financial Systems*, DOI 10.1007/978-1-4614-0231-2_8,

turbulent boundary layer will now be presented. Let us introduce the inner variables u^+, y^+ as

$$u^+ = \frac{u}{v_*}, \quad y^+ = \frac{y v_*}{\nu} \tag{1}$$

where u^+ is dimensionless velocity and y^+ is the dimensionless distance from the wall; u is the streamwise mean velocity; y is the distance from the wall; v_* is the friction velocity, defined by ${v_*}^2 = \tau_w/\rho$, τ_w being the wall shear stress; ρ and ν are the density and kinematic viscosity of the fluid.

One basic idea follows from dimensional consideration is that the law of the wall may be expressed in the form:

$$u^+ = f(y^+) \tag{2}$$

An objection to the law (2) is that it is too simple for describing the wake region (cf. [12]). Spalding [14] forwarded an analytical expression for $f(y^+)$ which led the law of the wall in the form:

$$y^+ = u^+ + A\left[\exp(\kappa u^+) - 1 - \kappa u^+ - (\kappa u^+)^2/2 - (\kappa u^+)^3/6 - (\kappa u^+)^4/24\right] \tag{3}$$

The formulation (3) has the advantage that it can be applied right from the wall. But Spalding's attempt to make the expression valid for the whole boundary layer with the choice of the constants κ and A, respectively, as $\kappa = 0.4$ and $A = 0.1108$ was too ambitious.

Persen [12] examined thoroughly the applicability of Spalding's formulation (3) against the voluminous data placed in the Stanford Conference (1968) on turbulent boundary layer and found the appropriate values of κ and A, respectively, as

$$A = 0.015, \quad \kappa = 0.53227, \tag{4}$$

which makes the Spalding formulation valid near to wall region only.

As discussed earlier, the turbulent boundary layer may assume to contain two regions, namely, (I). A 'near to wall region', which includes viscous sub-layer and wherein Spalding's formulation (3) is universally valid with the values of the constants κ and A, given in (4) and (II). A 'wake region' in which Persen proposed a formulation which allows a smooth matching of the velocity profile at the upper edge of wake region with the free stream velocity. In this concept, the joining point of the 'near to wall region' (inner region) with the wake region is mathematically well defined. Also, this has an advantage over the old concept of three regions (viscous sub-layer being one) where there are no well-defined conditions for their range of validity.

The establishment of Persen's theory depends on an experimentally supported relation between non-dimensional velocity $u^+ = \xi$ at the end point of the boundary layer and the corresponding non-dimensional distance from the wall $y^+ = {y_0}^+$. The point $(\xi, {y_0}^+)$ at the edge of the boundary layer has been shown by Persen [12] to lie on a curve called the locus of ξ. The quantity is important on the point of view that it may take care of manipulation from outside the boundary layer as well as the history of the boundary layer. It is worth mentioning that Persen's [12] theory has the special importance due to the fact that it is compatible with first principle of fluid mechanics. All these considerations lead us now to accept Persen's [12] theory and examine whether it is applicable to the open channel flow seeded with particles. In the present analysis, experimental data as measured by Coleman [2] for flow seeded with particles are brought into picture.

2 Coles' Wake Function (Coleman's Data)

Equation (3) being too simple for describing the wake region of the boundary layer and accordingly based on the idea of Coles [5] (3) should be replaced by

$$u^+ = f(y^+) + A(x)w(\eta) \tag{5}$$

where $A(x)$ is the amplitude function, $w(\eta)$ is the wake function and $\eta = y/\delta$, δ being the boundary layer thickness.

The wake function is generally defined as the difference between the measured data in the outer region of the boundary layer and values obtained from extension of logarithmic law in this region.The formulation proposed by Coles and Hirst [6] for the function $f(y^+)$ is

$$f(y^+) = \frac{1}{\kappa}\ln(y^+) + B \tag{6}$$

where

$$\kappa = 0.41, \quad B = 5 \tag{7}$$

This is the formulation valid for $y^+ > 50$. The method to be followed here will be somewhat different. Values of wake are determined as difference between the measured data in the outer region of the boundary layer and values obtained from (3) and (4) in that region. As $y^+ \to {y_0}^+$, $\eta \to 1$ one then obtains from (5) the relation

$$\xi = f({y_0}^+) + A(x)w(1) \tag{8}$$

The maximum value $w_{\max}$ and its position η_m at which it occurs are found by fitting a parabola through three points around the maximum value. Form now the wake function $w(\eta)$ is defined such that

$$w(\eta) = 1, \quad A(x) = \mathrm{w}_{\max} \tag{9}$$

Effect of concentration on $w_{\max}$ and η_m are shown in Figs. 1 and 2, respectively. It reveals that $w_{\max}$ increases as sediment concentration increases while sediment concentration has more or less no effect on η_m.

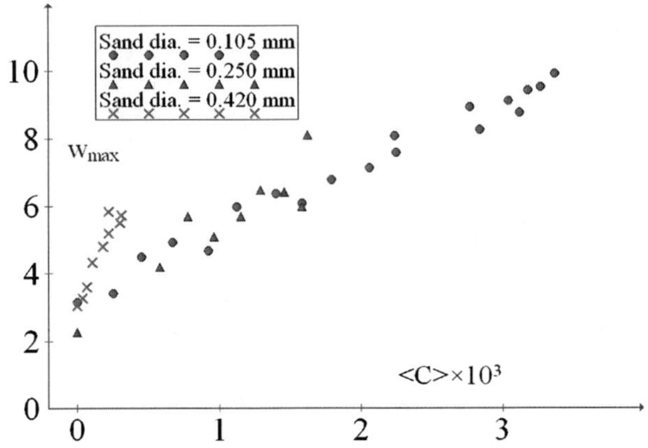

Fig. 1 Variation of w_{max} plotted as function of average sediment concentration

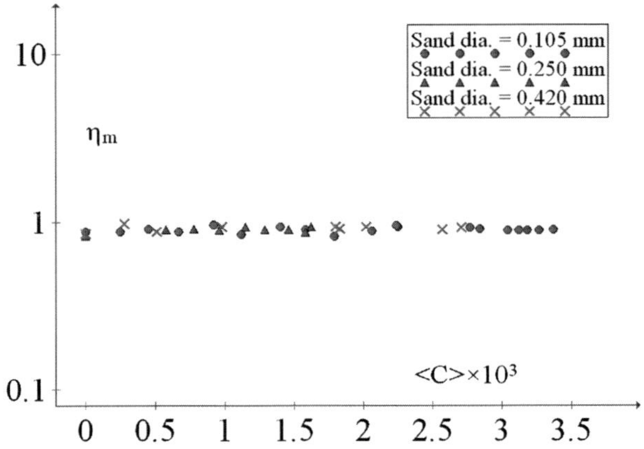

Fig. 2 Variation in the position η_m with average sediment concentration

3 Persen's Wake Law

Persen [12] replaced the Coles wake function with the expression

$$(u^+ - u_\infty{}^+)/(\xi - u_\infty{}^+) = \exp\left[-(y_0{}^+ - y^+)^2/\alpha^2\right] \tag{10}$$

where

$$\left.\begin{aligned} & u \to U_0 \text{ as } y \to \delta \\ & u^+ \to \frac{U_0}{v_*} = \xi \text{ as } y^+ \to \frac{\delta v_*}{v} = y_0^+ \end{aligned}\right\} \tag{11}$$

$u_\infty{}^+ =$ constant, U_0 is the outer velocity and

$$\frac{1}{\alpha^2} = [\ln(\xi - u_\infty{}^+) - \ln(u_1{}^+ - u_\infty{}^+)]/(y_0{}^+ - y_1{}^+)^2$$

Here, $(u_1{}^+, y_1{}^+)$ is the point where the law of the wall meets with the law of the wake. The boundary layer ends up at $y^+ = y_0{}^+$ and for which $u^+ = \xi$. Persen's law of the wake (10) has superiority over the Coles wake law as it exhibits a horizontal tangent at the outer edge $(u^+ = \xi,\ y^+ = y_0{}^+)$ of the boundary layer and that applies also for the manipulated boundary layer (adverse pressure gradient etc.).

4 Colman Data: Open Channel Flow Seeded With Particles

The experiments of Coleman have been dully exposed to the profession through published papers (cf. [2, 4, 11]) and discussion (cf. [9] and Reply of Coleman [3]).

The experiments were performed in a re-circulating flume with a plexiglass channel 15 m long and 356 mm wide. The flume was supported on jacks so that the channel slope could be adjusted to maintain uniform flow. A pitot static tube on a vertical traverse mechanism was located at a position on the flume channel centerline 12 m from the channel entrance. The maximum outside diameter of the tube was 16 mm and the diameter of the impact leg opening was 3.2 mm. The impact leg could be isolated and could be used for taking suspended sediment samples. Uniform flow was maintained with constant discharge $0.064\,\mathrm{m}^3/\mathrm{s}$ and constant depth 169 mm with a standard deviation of 1.69 mm but with a systematic increase in sediment suspension. Sand particles of average diameters, namely 0.105 mm, 0.210 mm and 0.420 mm were used for seeding in the three series of experiments. No roughness elements were installed and no sand was allowed to deposit on the bed. Thus, any changes observed in the velocity profiles could be attributed to increase in suspended sediment only.

Water temperature varied within a narrow range around 23°C with a standard deviation of 1.2°C. So, value of viscosity and density of water at 23°C are considered throughout the analysis. The variation of kinematic viscosity is important in the vertical extent as it is dependent on the concentration distribution of the suspended sediments. The kinemetic viscosity is modified for scaling purposes. We accept the relation between kinemetic viscosity and concentration as given below [2, 8]:

$$\nu = \frac{\mu_w(1 + 2.5C + 6.25C^2 + 15.62C^3)}{\rho_w + (\rho_s - \rho_w)C} \tag{12}$$

where ν the kinemetic viscosity of the sand water mixture, μ_w and ρ_w are the molecular viscosity and density of clear water, ρ_s is the density of sand and C is the local volumetric concentration which was expressed by Coleman in volume of sediment per unit volume of sediment water mixture.

In Coleman's [2] experiment, the aspect ratio (width: height) of the channel was about 2:1. Consequently, the influence of the side wall may be present in the centre plane and, accordingly velocity deep phenomenon was found to occur in this experiment.

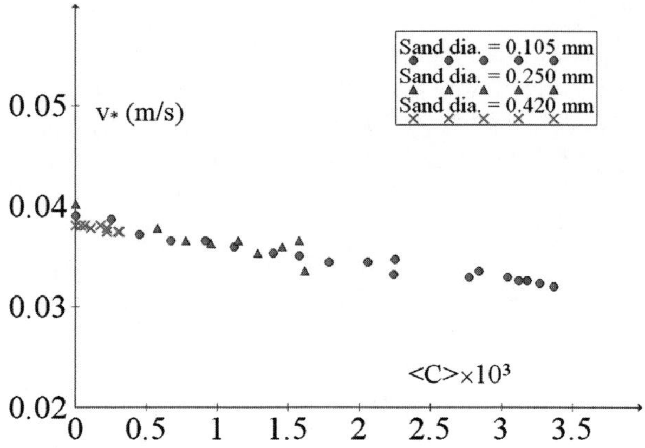

Fig. 3 Wall shear stress as a function of average sediment concentration

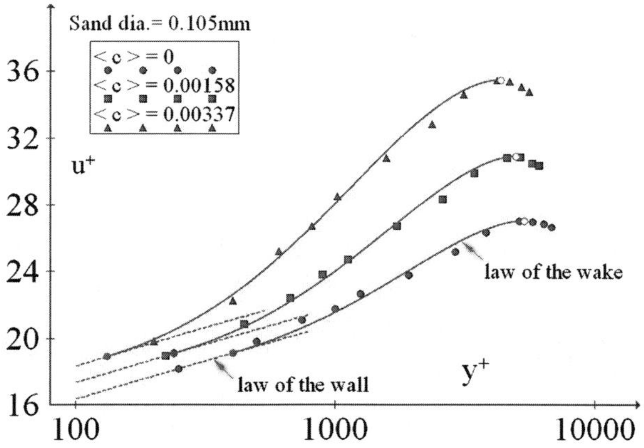

Fig. 4 Example on how the measured data and the theoretical curves correlate. The points $(u_1{}^+, y_1{}^+)$ and $(\xi, y_0{}^+)$ are shown as closed bold circle and open circle, respectively

The determination of wall shear stress or friction velocity v_* (m/s) is important as it is used for scaling purposes. The values of friction velocity v_* are taken from table published in Persen and Coleman [13]. Figure 3 shows the variation of wall shear stress with average sediment concentration $< C >$. It is observed that wall shear stress is a decreasing function of concentration.

Once the outer edge condition ξ and $y_0{}^+$ are known $u_1{}^+$ and $y_1{}^+$ can be determined equating slope of (3) and (10) at the point $u_1{}^+$, $y_1{}^+$. The values of $u_\infty{}^+$ are allowed to float in the exercise of obtaining a best fit of Coleman's [2] experimental data with the theoretical curve. The exercise leads to an appropriate value 40 for $u_\infty{}^+$.

In this case, ξ is the non-dimensional maximum velocity and $y_0{}^+$ is the position of ξ. The results of three profiles are shown in Fig. 4 where a comparison between the theoretical profile and measured data is shown for the series of experiments with 0.105 mm sand. The profiles have been selected for the case of no seeding to the case of maximum seeding. The comparison between the theoretical curve

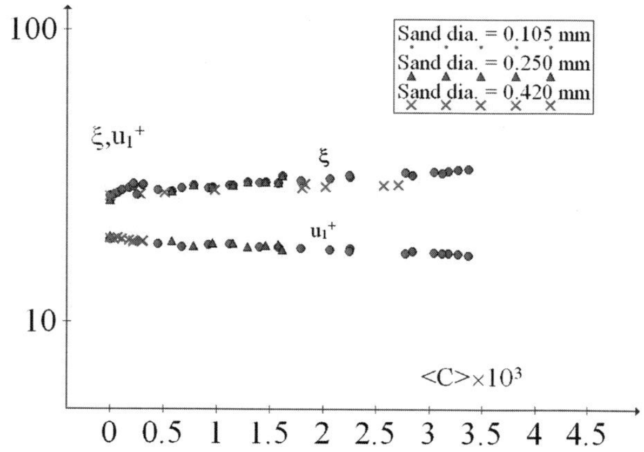

Fig. 5 ξ and $u_1{}^+$ as function of average concentration

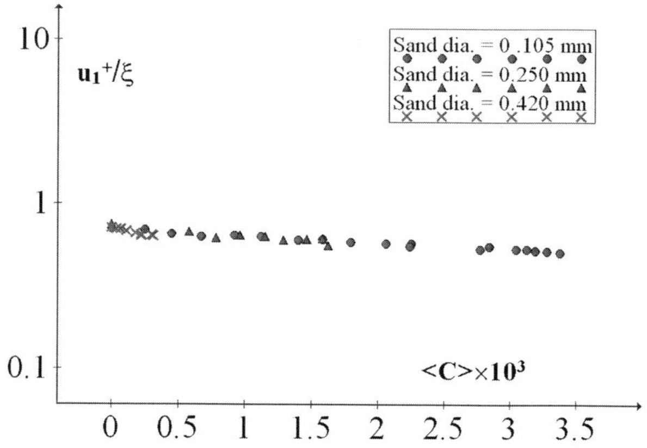

Fig. 6 The ratio between $u_1{}^+$ and ξ as a function of average concentration

and experimental data (Fig. 4) shows that most of the data lie in the wake region. Inspection of Fig. 4 indicates clearly that the portion of the near to wall region decreases with the increase of concentration.

For all cases correlation coefficient '*r*' is seen to be greater than 0.99, which justifies the use of the theoretical velocity profile derived for the single phase flow to the case of turbulent channel flow seeded with particles.

The values of ξ and $u_1{}^+$ are plotted against average concentration, $< C >$ in Fig. 5. It reveals that ξ is an increasing function of concentration while $u_1{}^+$ is a decreasing function of concentration.

The values of the ratio $u_1{}^+/\xi$ are plotted against average concentration in Fig. 6. It is a decreasing function of concentration. This indicates clearly that if one follows the path of velocity profile he will notice the shifting over of the flow within the boundary layer from the law of the wall to the law of the wake occurs early as average concentration increases i.e. contribution of wake region to total mass flux increases as average concentration increases.

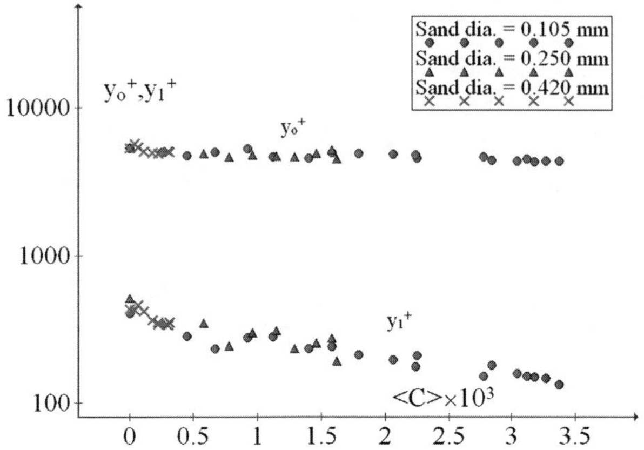

Fig. 7 The values of $y_1{}^+$ and $y_0{}^+$ as function of average concentration

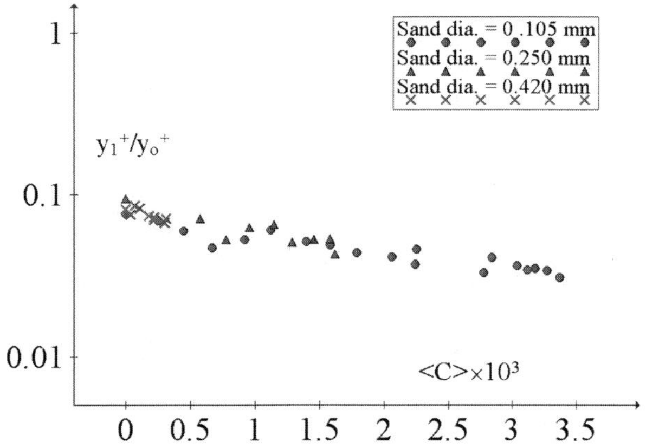

Fig. 8 The ratio between $y_1{}^+$ and $y_0{}^+$ as function of average concentration

The values of $y_1{}^+$ and $y_0{}^+$ are plotted against $< C >$ in Fig. 7. Both $y_1{}^+$ and $y_0{}^+$ decreases as concentration increases but at different rates. Figure 8 shows the ratio $y_1{}^+/y_0{}^+$ decreases as concentration increases. In fact, the ratio $y_1{}^+/y_0{}^+$ for each station is a measure of the non-dimensional quantity y/δ from the wall below which law of the wall (3), (4) governs the flow and above which law of the wake (10) takes over. The result may physically be interpreted that the near to wall region represents a decreasing portion of the total boundary layer as concentration increases.

The values of ξ are plotted against $y_0{}^+$ in Fig. 9. It is the locus of ξ under certain conditions. In other conditions, the locus of ξ will be different.

The values of the constant $u_\infty{}^+$ turned out to be independent of the concentration distribution of the suspended sediment. The present examination shows that the influence of particle diameter on velocity profile appears to be insignificant.

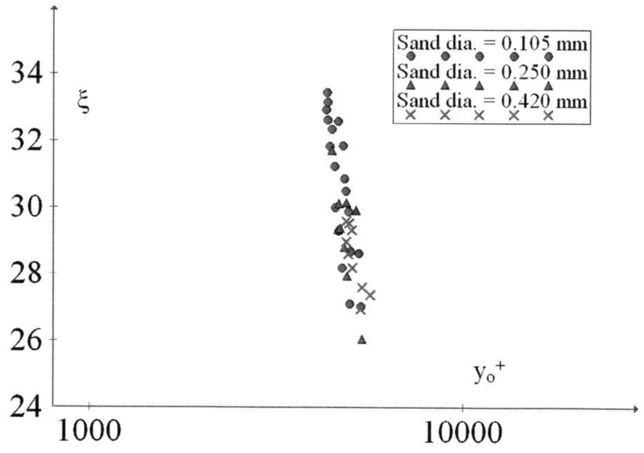

Fig. 9 Semi-log plot of locus of ξ

5 Conclusion

We now discuss the results as follows:

a. Flow field with sediment suspension are also describable through the inner variables.
b. A model applicable for turbulent boundary layer, i.e. law of the wall (3), (4) and law of the wake (10) is found equally applicable for describing the channel flow seeded with particles. It is worth mentioning that in the present analysis, value of the viscosity as related to sediment concentration, i.e. relation (10) has been used.
c. Shear velocity is a slowly decaying function of concentration.
d. Maximum velocity of the flow, expressed in its non-dimensional form increases as sediment concentration increases, while, the non-dimensional wall distance at maximum velocity of the flow decreases as sediment concentration increases.
e. With the increase of sediment concentration, the near to wall region is found to be a decreasing portion of the total boundary layer.
f. With the increase of sediment concentration, the ratio between the mass flux in the wake region and the same occurs in the inner region increases.
g. Maximum value of the wake ($w_{\max}$) increases with sediment concentration.
h. Present investigation shows that sediment diameter has no significant influence on the velocity profile, but it has influence on sediment-carrying capacity of channel.

References

1. Cioffi, F., Gallerano, F.: Velocity and concentration profiles of particles in a channel with movable and erodible bed. J. Hydr. Res. **29**, 387–401 (1991)
2. Coleman, N. L.: (1981). Velocity profiles with suspended sediment. J. Hydr. Res. **19(3)**, 211–229 (1981)
3. Coleman, N. L.: Reply, J. Hydr. Res. **22:4**, 275–289 (1984)

4. Coleman, N. L.: (1986). Effects of suspended sediment on the open channel-channel velocity distribution. Water Resources Research, AGU. **22(10)**, 1377–1384 (1986)
5. Coles, D.E.: The law of the wake in the turbulent boundary layer. J. Fluid Mech. **1**, 191–226 (1956)
6. Coles, D. E., Hirst, E. A.: Proc. Computation of Turbulent Boundary layers. Stanford Univ. (1968)
7. Elata, C., Ippen, A. T.: (1961). The dynamics of open channel flow with suspensions of neutrally buoyant particles. Technical Report. Hydrodynamics Laboratory. MIT. **45**, (1961)
8. Graf, E. H.: Hydraulics of sediment transport. McGraw-Hill Book Co., Inc., New York (1971)
9. Gust, G.: Discussion: Velocity profiles with suspended sediment. J. Hydr. Res. **22:4**, 263–275 (1984)
10. Muste, M., Patel, V. C.: (1991). Velocity profiles for particles and liquid in open- channel flow with suspended sediment. J. Hydr. Engr. ASCE. **123(9)**, 742–751 (1991)
11. Parker, G., Coleman, N. L.: Simple model of sediment laden flows. J. Hydr. Engr. ASCE. **112(5)**, 356–374 (1986)
12. Persen, L. N.: The turbulent boundary layer and the closer problem. Proc. AGARD Conf. on Turbulent Boundary Layer –Experiments, theory and modelling. 17–1 to 17–14 (1974)
13. Persen, L. N., Coleman, N. L.: The influence of suspended particles on turbulent shear flow. Turbulence Modification in Dispersed Multiphase Flow. FED. **80**, 81–86 (1990)
14. Spalding, D. B.: A single formulae for the law of the wall. J. Appl. Mech., **28**, Ser. E. 455–458 (1961)
15. Vanoni, V. A.: Transportation of suspended sediment by running water. Trans. ASCE. **111**, 67–133 (1946)

A Renormalization-Group Study of the Potts Model with Competing Ternary and Binary Interactions

Nasir Ganikhodjaev, Seyit Temir, Selman Uğuz, and Hasan Akin

1 Introduction

Consideration of spin models with multispin interactions has proved to be fruitful in many fields of physics, ranging from the determination of phase diagrams in metallic alloys and exhibition of new types of phase transition to site percolation. Systems exhibiting spatially modulated structures, commensurate or incommensurate with the underlying lattice, are of current interest in condensed matter physics [1]. Among the idealized systems for modulated ordering, the axial next-nearest–neighbor Ising (ANNNI) model, originally introduced by Elliot [2] to describe the sinusoidal magnetic structure of Erbium, and the chiral Potts model, introduced by Ostlund [3] and Huse [4] in connection with monolayers adsorbed on rectangular substrates, have been studied extensively by a variety of techniques. A particularly interesting and powerful method is the study of modulated phases through the measure-preserving map generated by the mean-field equations, as applied by Bak [5] and Jensen and Bak [6] to the ANNNI model. The main drawback of the method lies in the fact that thermodynamic solutions correspond to stationary but unstable orbits. However, when these models are defined on Cayley trees, as in the case of the Ising model with competing interactions examined by Vannimenus [10], it turns out that physically interesting solutions correspond to the attractors of the mapping. This simplifies the numerical work considerably, and detailed study of the whole phase diagram becomes feasible. Apart from the intrinsic interest attached to the study of models on trees, it is possible to argue that the results obtained on trees provide a useful guide to the more involved study of their counterparts on crystal lattices. The ANNNI model, which consists of an Ising model with nearest-neighbor interactions

N. Ganikhodjaev (✉)
Kulliyyah of Science, International Islamic University Malaysia,
25200 Kuantan, Malaysia
e-mail: gnasir@iiu.edu.my

J.A.T. Machado et al., *Nonlinear and Complex Dynamics: Applications in Physical, Biological, and Financial Systems*, DOI 10.1007/978-1-4614-0231-2_9,

augmented by competing next-nearest–neighbor couplings acting parallel to a single axis direction, is perhaps the simplest nontrivial model displaying a rich phase diagram with a Lifshitz point and many spatially modulated phases. There has been a considerable theoretical effort to obtain the structure of the global phase diagram of the ANNNI model in the $T-p$ space, where T is temperature and $p = -J_p/J_1$ is the ratio between the competing exchange interactions. On the basis of numerical mean-field calculations, Bak and von Boehm [7] suggested the existence of an infinite succession of commensurate phases, the so-called devil's staircase, at low temperatures. This mean-field picture has been supported by low-temperature series expansions performed by Fisher and Selke [8]. At the paramagnetic-modulated boundary analytic mean-field calculations show that the critical wave number varies continuously and vanishes at the Lifshitz point. A phase diagram of a model describes a morphology of phases, stability of phases, transitions from one phase to another and corresponding transitions line. A Potts model just as an Ising model on a Cayley tree with competing interactions has recently been studied extensively because of the appearance of nontrivial magnetic orderings (see [9–11,14,15,19–22] and references therein). The Cayley tree is not a realistic lattice; however, its amazing topology makes the exact calculation of various quantities possible. For many problems, the solution on a tree is much simpler than on a regular lattice and is equivalent to the standard Bethe–Peierls theory [16]. On the Cayley tree one can consider two type of triple neighbors: prolonged and two-level (definitions see below). In the case of the Ising model with competing nearest–neighbor interactions J and prolonged next-nearest–neighbor interactions J_p Vannimenus [10] was able to find new modulated phases, in addition to the expected paramagnetic and ferromagnetic ones. From this result follows that Ising model with competing interactions on a Cayley tree is real interest since it has many similarities with models on periodic lattices. In fact, it has many common features with them, in particular the existence of a modulated phase, and shows no sign of patological behavior – at least no more than mean-field theories of similar systems [10]. Moreover, detailed study of its properties was carried out with essentially exact results, using rather simple numerical methods. This suggest that more complicated models should be studied on trees, with the hope to discover new phases or unusual types of behavior. The important point is that statistical mechanics on trees involve nonlinear recursion equations and are naturally connected to the rich world of dynamical systems, a world presently under intense investigation [10]. The model (1) with $J_p = 0$ was considered in [20, 21] and proved that phase diagram of this model contains ferromagnetic and antiferromagnetic phases.

In this paper, we consider the Potts model with competing triple nearest–neighbor interactions. The Potts model [17] was introduced as a generalization of the Ising model to more than two components and encompasses a number of problems in statistical physics (see, e.g. [18]) recently. In [19], the phase diagram of the three states Potts model with nearest–neighbor interactions J and prolonged next-nearest-neighbors interactions J_p was described.

2 The Main Results

2.1 The Model

A Cayley tree Γ^k of order $k \geq 1$ is an infinite tree, i.e., a graph without cycles with exactly $k+1$ edges issuing from each vertex. Let denote the Cayley tree as $\Gamma^k = (V, \Lambda)$, where V is the set of vertices of Γ^k, Λ is the set of edges of Γ^k. Two vertices x and y, $x,y \in V$ are called *nearest–neighbors* if there exists an edge $l \in \Lambda$ connecting them, which is denoted by $l = <x,y>$. The distance $d(x,y), x,y \in V$, on the Cayley tree Γ^k, is the number of edges in the shortest path from x to y. For a fixed $x^0 \in V$, we set

$$W_n = \{x \in V | d(x,x^0) = n\}, V_n = \{x \in V | d(x,x^0) \leq n\}$$

and L_n denotes the set of edges in V_n. The fixed vertex x^0 is called the 0-th level and the vertices in W_n are called the n-th level. For the sake of simplicity, we put $|x| = d(x,x^0)$, $x \in V$. Two vertices $x,y \in V$ are called *the next-nearest–neighbors* if $d(x,y) = 2$. Three vertices x,y, and z are called a triple of neighbors and they are denoted by $<x,y,z>$, if $<x,y>, <y,z>$ are nearest neighbors. The triple of vertices x,y, and z is called *prolonged* if $x \in W_n, y \in W_{n+1}$ and $z \in W_{n+2}$ for some nonnegative integer n and is denoted by $\widetilde{<x,y,z>}$. The triple of vertices $x,y,z \in V$ that are not prolonged is called *two-level* since $|x| = |z|$ and are denoted by $<x,\bar{y},z>$.

Below we will consider a semi-infinite Cayley tree Γ_+^2 of second order, i.e. an infinite graph without cycles with 3 edges issuing from each vertex except for x^0 which has only 2 edges.

For the three-state Potts model with spin values in $\Phi = \{1,2,3\}$, the relevant Hamiltonian with competing binary nearest–neighbor and ternary interactions has the form

$$H(\sigma) = -J_p \sum_{\widetilde{<x,y,z>}} \delta_{\sigma(x)\sigma(y)\sigma(z)} - J_1 \sum_{<x,y>} \delta_{\sigma(x)\sigma(y)}, \qquad (1)$$

where $J_p, J_1 \in R$ are coupling constants and δ is the Kronecker symbol. Here, the generalized Kronecer's symbol $\delta_{\sigma(x)\sigma(y)\sigma(z)}$ is

$$\delta_{\sigma(x)\sigma(y)\sigma(z)} = \begin{cases} 1 \text{ if } \sigma(x) = \sigma(y) = \sigma(z) \\ 0 \text{ otherwise.} \end{cases}$$

Let $a = \exp(J_1/T); b = \exp(J_p/T)$, where T is the temperature.

The model (1) with $J_p = 0$ was considered in [12, 13, 20, 21] and proved that phase diagram of this model contains ferromagnetic and antiferromagnetic phases. Below we consider model (1) with $J_p \neq 0$, and describe its phase diagram.

2.2 Basic Equations

In order to produce the recurrent equations, we consider the relation of the partition function on V_n to the partition function on subsets of V_{n-1}. Given the initial conditions on V_1, the recurrence equations indicate how their influence propagates down the tree. Let $Z^{(n)}(i,j)$ be a partition function on V_n with the configuration (i,j) on an edge $< x^0, x >$, where $x \in W_1$ with $i,j = 1,2,3$ and $Z^{(n)}(i_1,i_0,i_2)$ be the partition function on V_n where the spin in the root x^0 is i_0 and the two spins in the proceeding ones are i_1 and i_2, respectively. There are 27 different partition functions $Z^{(n)}(i_1,i_0,i_2)$ and the partition function $Z^{(n)}$ in volume V_n can be written as follows

$$Z^{(n)} = \sum_{i_1,i_0,i_2=1}^{3} Z^{(n)}(i_1,i_0,i_2).$$

Then through a direct calculation one gets the following equalities:

$$\begin{aligned}
Z^{(n)}(1,1,1) &= a^2 Z^{(n)}(1,1)Z^{(n)}(1,1),\\
Z^{(n)}(1,1,2) &= aZ^{(n)}(1,1)Z^{(n)}(1,2),\\
Z^{(n)}(1,1,3) &= aZ^{(n)}(1,1)Z^{(n)}(1,3),\\
Z^{(n)}(2,1,1) &= aZ^{(n)}(1,2)Z^{(n)}(1,1),\\
Z^{(n)}(2,1,2) &= Z^{(n)}(1,2)Z^{(n)}(1,2),\\
Z^{(n)}(2,1,3) &= Z^{(n)}(1,2)Z^{(n)}(1,3),\\
Z^{(n)}(3,1,1) &= aZ^{(n)}(1,3)Z^{(n)}(1,1),\\
Z^{(n)}(3,1,3) &= Z^{(n)}(1,3)Z^{(n)}(1,3),\\
Z^{(n)}(3,1,2) &= Z^{(n)}(1,3)Z^{(n)}(1,2),\\
Z^{(n)}(1,2,1) &= Z^{(n)}(2,1)Z^{(n)}(2,1),\\
Z^{(n)}(1,2,2) &= aZ^{(n)}(2,1)Z^{(n)}(2,2),\\
Z^{(n)}(1,2,3) &= Z^{(n)}(2,1)Z^{(n)}(2,3),\\
Z^{(n)}(2,2,1) &= aZ^{(n)}(2,2)Z^{(n)}(2,1),\\
Z^{(n)}(2,2,2) &= a^2 Z^{(n)}(2,2)Z^{(n)}(2,2),\\
Z^{(n)}(2,2,3) &= aZ^{(n)}(2,2)Z^{(n)}(2,3),\\
Z^{(n)}(3,2,1) &= Z^{(n)}(2,3)Z^{(n)}(2,1),\\
Z^{(n)}(3,2,2) &= aZ^{(n)}(2,3)Z^{(n)}(2,2),\\
Z^{(n)}(3,2,3) &= Z^{(n)}(2,3)Z^{(n)}(2,3),\\
Z^{(n)}(1,3,1) &= Z^{(n)}(3,1)Z^{(n)}(3,1),\\
Z^{(n)}(1,3,2) &= Z^{(n)}(3,1)Z^{(n)}(3,2),\\
Z^{(n)}(1,3,3) &= aZ^{(n)}(3,1)Z^{(n)}(3,3),\\
Z^{(n)}(2,3,2) &= Z^{(n)}(3,2)Z^{(n)}(3,2),\\
Z^{(n)}(2,3,1) &= Z^{(n)}(3,2)Z^{(n)}(3,1),\\
Z^{(n)}(2,3,3) &= aZ^{(n)}(3,2)Z^{(n)}(3,3),\\
Z^{(n)}(3,3,1) &= aZ^{(n)}(3,3)Z^{(n)}(3,1),\\
Z^{(n)}(3,3,2) &= aZ^{(n)}(3,3)Z^{(n)}(3,2),\\
Z^{(n)}(3,3,3) &= a^2 Z^{(n)}(3,3)Z^{(n)}(3,3).
\end{aligned}$$

We can select only six variables $Z^{(n)}(1,1,1)$, $Z^{(n)}(2,1,2)$, $Z^{(n)}(3,1,3)$, $Z^{(n)}(1,2,1)$, $Z^{(n)}(2,2,2)$, $Z^{(n)}(3,3,3)$, and with the introduction of new variables

$$u_1^{(n)} = \sqrt{Z^{(n)}(1,1,1)},\ u_2^{(n)} = \sqrt{Z^{(n)}(2,1,2)},$$
$$u_3^{(n)} = \sqrt{Z^{(n)}(3,1,3)},\ u_4^{(n)} = \sqrt{Z^{(n)}(1,2,1)},$$
$$u_5^{(n)} = \sqrt{Z^{(n)}(2,2,2)},\ u_6^{(n)} = \sqrt{Z^{(n)}(3,3,3)},$$

straightforward calculations show that

$$\begin{aligned}
u_1^{(n+1)} &= a\left(bu_1^{(n)} + u_2^{(n)} + u_3^{(n)}\right)^2,\\
u_2^{(n+1)} &= \left(u_3^{(n)} + u_4^{(n)} + u_5^{(n)}\right)^2,\\
u_3^{(n+1)} &= \left(u_2^{(n)} + u_4^{(n)} + u_6^{(n)}\right)^2,\\
u_4^{(n+1)} &= \left(u_1^{(n)} + u_2^{(n)} + u_3^{(n)}\right)^2,\\
u_5^{(n+1)} &= a\left(u_3^{(n)} + u_4^{(n)} + bu_5^{(n)}\right)^2,\\
u_6^{(n+1)} &= a\left(u_2^{(n)} + u_4^{(n)} + bu_6^{(n)}\right)^2.
\end{aligned} \tag{2}$$

The total partition function is given in terms of (u_i) by

$$Z^{(n)} = \left(u_1^{(n)} + u_2^{(n)} + u_3^{(n)}\right)^2 + \left(u_3^{(n)} + u_4^{(n)} + u_5^{(n)}\right)^2 + \left(u_2^{(n)} + u_4^{(n)} + u_6^{(n)}\right)^2. \tag{3}$$

Note that, for boundary condition $\bar{\sigma}_n(V \setminus V_n) \equiv 1$ we have $Z^{(n)}(2,1,2) = Z^{(n)}(3,1,3)$, and $Z^{(n)}(2,2,2) = Z^{(n)}(3,3,3)$, i.e., $u_2 = u_3$ and $u_5 = u_6$.

For discussing the phase diagram, the following choice of reduced variables is convenient:

$$x = \frac{u_2 + u_4}{u_1 + u_6},$$
$$y_1 = \frac{u_1 - u_6}{u_1 + u_6},$$
$$y_2 = \frac{u_2 - u_4}{u_1 + u_6}.$$

The variable x is just a measure of the frustration of the nearest–neighbor bonds and is not an order parameter like y_1, and y_2. Equations (2) yield:

$$x' = \frac{1}{aD}\left[(2x+y_2+1)^2+(y_1+y_2)^2\right];$$
$$y_1' = \frac{2}{D}(2x+y_2+b)(by_1+y_2);$$
$$y_2' = -\frac{2}{aD}(2x+y_2+1)(y_1+y_2); \tag{4}$$

where

$$D = (b+2x+y_2)^2+(by_1+y_2)^2.$$

Below we use numerical methods to study its detailed behavior.

2.3 Morphology of the Phase Diagram

It is convenient to know the broad features of the phase diagram before discussing the different transitions in more detail. This can be achieved numerically in a straightforward fashion. The recursion relations (4) provide us the numerically exact phase diagram in $(T/J_1, -J_p/J_1)$ space. Let $T/J_1 = \alpha, -J_p/J_1 = \beta$ and, respectively, $a = \exp(\alpha^{-1}), b = \exp(-\alpha^{-1}\beta)$. Starting from initial conditions

$$x^{(1)} = \frac{1+a^2}{(a^3b^2+a)},$$
$$y_1^{(1)} = \frac{a^2b^2-1}{a^2b^2+1},$$
$$y_2^{(1)} = \frac{1-a^2}{(a^3b^2+a)},$$

that corresponds to boundary condition $\bar{\sigma}^{(n)} \equiv 1$, one iterates the recurrence relations (4) and observes their behavior after a large number of iterations. In the simplest situation, a fixed point (x^*, y_1^*, y_2^*) is reached. It corresponds to a paramagnetic phase if $y_1^* = 0, y_2^* = 0$ or to a ferromagnetic phase if $y_1^*, y_2^* \neq 0$.

Secondary, the system may be periodic with period p, where case $p = 2$ corresponds to antiferromagnetic phase and case $p = 4$ corresponds to so-called antiphase, that denoted $< 2 >$ for compactness. Finally, the system may remain aperiodic. The distinction between a truly aperiodic case and one with a very long period is difficult to make numerically. Below we consider periodic phases with period p where $p \leq 12$. All periodic phases with period $p > 12$ and aperiodic phase we will consider as modulated phase. The resultant phase diagram is shown that in Fig. 1.

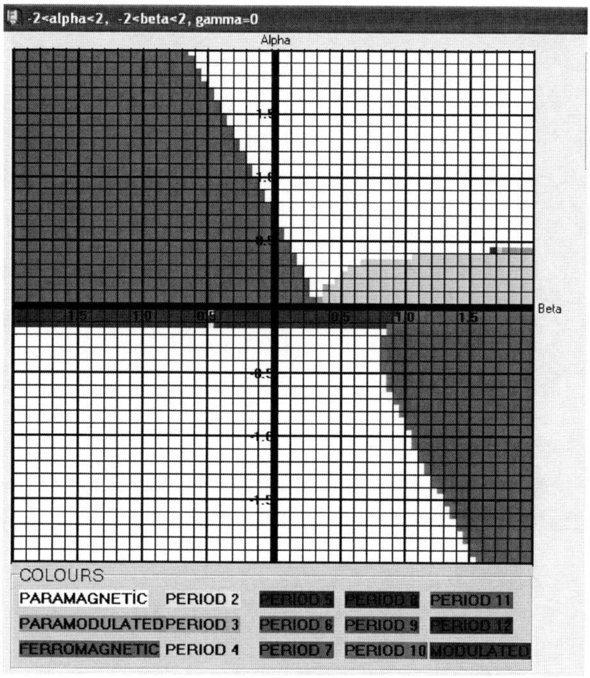

Fig. 1 Phase diagram of the Potts model (1) (color online)

2.4 *Average Magnetization*

Since the form of spins of the Potts model is unessential one can replace the set of spin values $\{1,2,3\}$ by the centered set $\{-1,0,1\}$. Then the average magnetization $\tilde{m}$ for the nth generation is given by

$$\tilde{m} = -\frac{4(y_1+y_2)(2x+y_2+1)}{(2x+y_1+2y_2+1)^2+2(2x+1-y_1)^2}. \tag{5}$$

A definition of the wavevector that is convenient for numerical purposes is

$$q = \lim_{N\to\infty}\left(\frac{1}{2}\frac{n}{N}\right),$$

where n is the number of times the magnetization (5) changes sign during N successive iterations [10].

A typical graphs of q versus T are drawn in Figs. 2 and 3 for $\beta = 0.2$ below the multicritical point and $\beta = 0.4$ above the multicritical point respectively. These figures show how the wavevector vary a function of T between its value at the paramodulated transition and its value in the $< 2 >$ phase is equal to $\frac{1}{4}$.

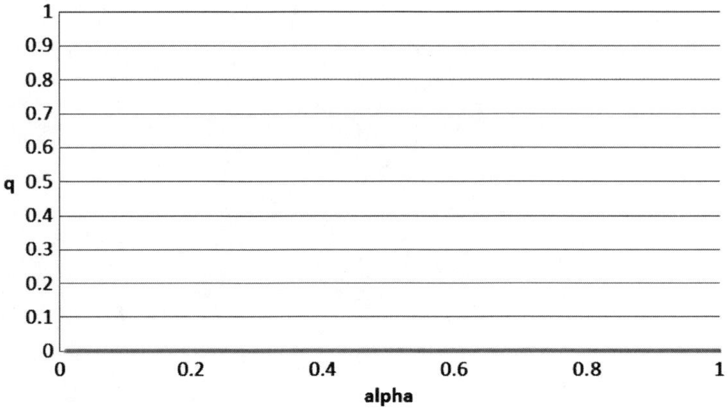

Fig. 2 Variation of the wavevector q versus, for $\beta = 0.2$

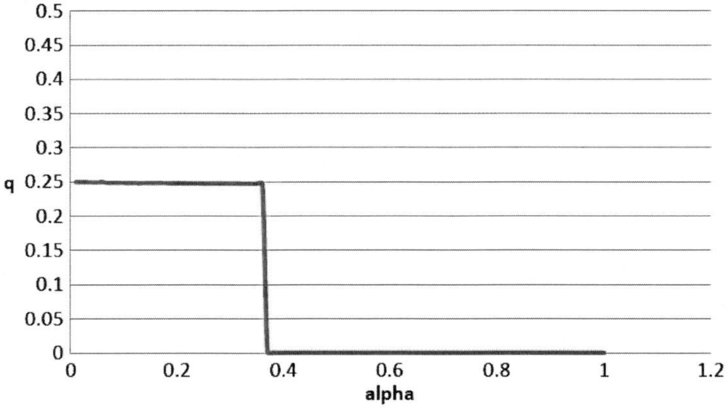

Fig. 3 Variation of the wavevector q versus, for $\beta = 0.8$

3 Conclusion

In [19] proved that the phase diagram of Potts model with competing nearest–neighbor and prolonged next-nearest–neighbor interactions consists of five phases: ferromagnetic, paramagnetic, modulated, antiphase, and paramodulated phases. We have found the phase diagram of the Potts model with competing prolonged ternary and binary nearest–neighbor interactions on the Cayley tree of second order and show that it consists of three phases only: ferromagnetic, paramagnetic, and antiphase with period 4. Thus for considered model, one can reach periodic phase with period 4 only, i.e., the set of modulated phases has simplest structure.

Acknowledgements The work is supported by The Scientific and Technological Research Council of Turkey-TUBITAK (Project No: 109T678).

References

1. P. Bak., Rep. Prog. Phys., 45, 587-629 (1982).
2. R.J. Elliott, Phys. Rev, 124, 346-353 (1961).
3. S. Ostlund, Phys. Rev. B, 24, 398-405 (1981).
4. D.A. Huse, Phys. Rev. B, 24, 5180-5194, 1981.
5. P. Bak, Phys. Rev. Lett., 46, 791-794 (1981).
6. M.H. Jensen and P. Bak, Phys. Rev. B, 27, 6853-6868 (1983).
7. P. Bak and J. von Boehm, Phys. Rev. B, 21, 5297-5308 (1980).
8. M.E. Fisher and W. Selke, Phys. Rev. Lett., 44, 1502-1505 (1980).
9. S. Coutinho, W. A. M. Morgado, E. M. F. Curado, L. da Silva, Physical Review B, 74(9): 094432-1-7 (2006).
10. J. Vannimenus, Z. Phys. B, 43, 141-148 (1981).
11. Mariz A.M., Tsallis C. and Albuquerque E.L., Journal of Statistical Physics, 40, pp. 577-592. (1985).
12. Monroe J.L., Journal of Statistical Physics, V. 67, pp. 1185-2000 (1992).
13. Monroe J.L., Physics Letters A, V. 88, pp. 80-84 (1994).
14. C.R. da Silva and S. Coutinho, Physical Review B, 34, pp. 7975-7985 (1986).
15. S. Inawashiro, C.J. Thompson, Physics Letters , 97A, 245-248 (1983).
16. Katsura, S., Takizawa, M., Prog. Theor. Phys., 51, 82-98 (1974).
17. R.B. Potts, Proc. Cambridge Philos. Soc., 48, 106 (1952).
18. F. Y. Wu, Rev. Mod. Phys., 54, 235-268 (1982).
19. N.N. Ganikhodjaev, F.M. Mukhamedov, C.H. Pah, Physics Letters A, 373, 33-38 (2008).
20. Ganikhodjaev N.N., Temir S. and Akin H., Cubo. A Mathematical Journal., 7, pp. 37-48 (2005).
21. Ganikhodjaev N.N., Akin H. and Temir S., Turkish Journal of Mathematics, 31, pp. 229-238 (2007).
22. Ganikhodjaev N.N., Temir S. and Akin H., Journal of Statistical Physics, 137, pp. 701-715. (2009).

The Mechanical Properties of CaX_6 ($X = B$ and C)

Sezgin Aydin and Mehmet Şimşek

1 Introduction

Graphite intercalation compounds (GICs) are produced by placing foreign atoms between two-dimensional sheets of graphite [1], and due to these sheets they possess interesting two-dimensional properties [2]. Therefore, these compounds have attracted a considerable interest and their physical properties were studied in a wide framework [3–5]. In alkali metal GICs, two-dimensional high conductivity is appeared through charge transfer from s-electron of intercalated atoms to carbon 2p π band of the graphene layers [2]. However, some of the alkali-metal GICs are superconductors (CaC_6) and generally their superconducting temperature is low (11.5 K) [6].

In this study, structural, electronic and mechanical properties (such as bulk modulus, shear modulus, Young modulus and hardness) of CaC_6 compound were investigated by using first-principle calculations within density functional theory. Replacing carbon atom to boron atom in the structure, it was studied how change the physical properties of the structure.

Crystal structure of CaX_6 compound is shown in Fig. 1 [7]. There is one Ca atom and six carbon atoms in the unit cell. Space group is R-3m (No:166).

S. Aydin (✉)
Department of Physics, Gazi University, Teknikokullar, 06500, Turkey
e-mail: sezginaydin@gazi.edu.tr

J.A.T. Machado et al., *Nonlinear and Complex Dynamics: Applications in Physical, Biological, and Financial Systems*, DOI 10.1007/978-1-4614-0231-2_10,

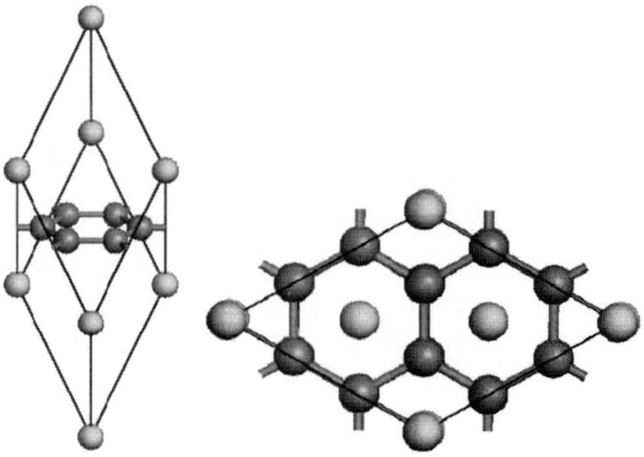

Fig. 1 Crystal structure of rhombohedral CaC_6. *Light grey and dark grey spheres* stand for calcium and carbon atoms, respectively

2 Computational Method

Our calculations are based on the first-principle plane-wave calculations. All calculations were performed by using CASTEP simulation package [8] within the density functional theory. In calculations, all the atomic coordinates and unit cell parameters were relaxed for minimization of Broyden, Fletcher, Goldfarb, Shanno (BFGS) scheme. The Vanderbilt ultrasoft pseudopotential [9] was used to model the ion–electron interactions, and exchange-correlation effects were treated within the generalized gradient approximation (GGA) by the Perdew–Burke—Ernzerhof (PBE) [10, 11] exchange correlation functional. The plane wave cut-off energy of 500 eV was employed, and the special k-points were generated by Monkhorst–Pack scheme, and they were chosen as $10 \times 10 \times 10$. For the convergence, the ultra-fine setup of software package was chosen, i.e. all calculations were converged in following qualities together (a) when the maximum ionic Hellman–Feynman force was below 0.01 eV/Å, (b) maximum displacement was below 5.0×10^{-4} Å, (c) maximum energy change was below 5.0×10^{-6} eV/atom and (d) maximum stress was below 0.02 GPa.

After optimization of the unit cell, cohesive energy is calculated, $E_{\text{coh}} = (E_{\text{Total}} - E_{\text{ica}} - 6E_{\text{iX}})/n$, which is a key parameter to discuss energetic stability of the structure, where E_{Total} is total energy of the unit cell, E_{ica} and E_{iX} are energies of an isolated Ca atom and X atom, respectively. n is the total number of atoms in the unit cell.

Elastic constants are determined by using stress-strain method [12], and mechanical properties such as Bulk modulus, shear modulus and Young modulus are calculated as functions of elastic constants within Reuss–Voigt–Hill approximation [13]. For hexagonal/rhombohedral crystals, there are five independent elastic constants; c_{11}, c_{33}, c_{44}, c_{12} and c_{13}. Reuss (R) and Voigt (V) bulk modulus and shear modulus are given as,

$$\begin{aligned}
B_V &= \frac{1}{9}\left[2(c_{11}+c_{12})+4c_{13}+c_{33}\right],\\
G_V &= \frac{1}{30}\left[M+12c_{44}+12c_{66}\right],\\
B_R &= C^2/M,\\
G_R &= \frac{5}{2}(C^2 c_{44}c_{66})/\left[3B_V c_{44} c_{66}+C^2(c_{44}+c_{66})\right],
\end{aligned}$$

where the abbreviations are $C^2 = (c_{11}+c_{12})c_{33} - 2c_{13}^2$, $M = c_{11}+c_{12}+2c_{33} - 4c_{13}$, and $c_{66} = (c_{11}-c_{12})/2$. While Reuss value is minimum limit, Voigt value is maximum limit for quantity. And, mechanical stability criteria are

$$c_{44} > 0,\ c_{11} > |c_{12}|,\ (c_{11}+2c_{12})c_{33} > 2c_{13}^2.$$

Other hand, hardness is one of the important mechanical properties of a material, and can be calculated by Simunek's method [14]. In this method, hardness of a material is defined as

$$\begin{aligned}
H &= \frac{c}{\Omega}n\left[\prod_{i,j=1}^{n} N_{ij}S_{ij}\right]^{1/n} \mathrm{e}^{-\sigma f_e},\\
f_e &= 1-\left[k\left(\prod_{i=1}^{k} e_i\right)^{1/k} \Big/ \Sigma_{i=1}^{k} e_i\right]^2,
\end{aligned}$$

where C and σ are constants, Ω is volume of unit cell, n is denoted the number of different bond type, k corresponds to the number of atoms in the system, and the number N_{ij} counts inter-atomic bonds in unit cell. S_{ij} is bond strength of individual bonds (d_{ij}) between atoms i and j defined as, $S_{ij} = \sqrt{e_i e_j}/(n_i n_j d_{ij})$; and $e_i = Z_i/R_i$, where Z_i is the valance electron number and R_i is atomic radius of the atom i. In calculation, $C = 1{,}450$ and $\sigma = 2.8$ values are used and the atomic radii for different elements are taken from Pearson text book [15].

3 Results and Discussion

The calculated structural parameters, elastic constants (c_{ij}), bulk modulus (B), shear modulus (G) and Young modulus (E) are listed in Table 1. Cohesive energies of both compounds are negative; thus, they are energetically stable. Cohesive energy of CaC_6 is smaller than that of CaB_6. Therefore, we can expect that CaC_6 is more stable than CaB_6. The calculated structural parameters for CaC_6 agree well with the literature, but calculated bulk modulus in this study is higher than the value of 103 GPa in [16].

Table 1 Calculated structural parameters, elastic constants (c_{ij}), bulk modulus (B), shear modulus (G) and Young modulus

	CaC_6	CaB_6
a (Å)	5.142	5.333
	5.170 [7]	
α (°)	49.40	56.00
	49.55 [7]	
V (Å^3)	71.939	97.354
E_{coh} (eV/atom)	−8.512	−5.763
c_{11}	645	184
c_{33}	84	88
c_{44}	78	42
c_{12}	88	132
c_{13}	87	49
c_{66}	278	26
B	147	90
	103 [16]	
G	122	35
E	287	93

However, all mechanical properties of CaC_6 listed in Table 1 are higher than those of CaB_6. This is an expected result, because CaC_6 is more stable than CaB_6. From calculated elastic constants, it was shown that CaC_6 and CaB_6 are mechanically stable. Other words, CaB_6 can be crystallized in CaC_6-type structure with lower mechanical properties.

Calculated band structure and density of states for CaX_6 compounds are shown in Figs. 2 and 3, respectively. It is shown from Figs. 2 and 3 that CaC_6 and CaB_6 have metallic character. From Fig. 3, X atoms are dominant on DOS. B p-orbitals possess more density of states than C p-orbitals. The hybridization between Ca s-orbitals and C p-orbitals is stronger than that of B-p orbitals.

Finally, hardness which is one of the important mechanical properties, is investigated. Micro-hardness for CaC_6 and CaB_6 is calculated as 13.4 GPa and 8.8 GPa, respectively. CaC_6 is harder than CaB_6. For CaX_6 compounds, it was shown that X–X bonds are stronger than Ca–X bonds. It is concluded that graphene layer in the structures plays important role on the mechanical properties.

In conclusion, the structural, electronic and mechanical properties of CaX_6 ($X = B$ and C) were investigated by using first-principle calculations. From calculated cohesive energies and elastic constants, both compounds are energetically and mechanically stable. Our calculations showed that CaB_6 can be crystallized in CaC_6-type structure. However, graphene layer in the structures plays important role on the electronic and mechanical properties. From hardness analysis, CaX_6 compounds are hard material (not superhard).

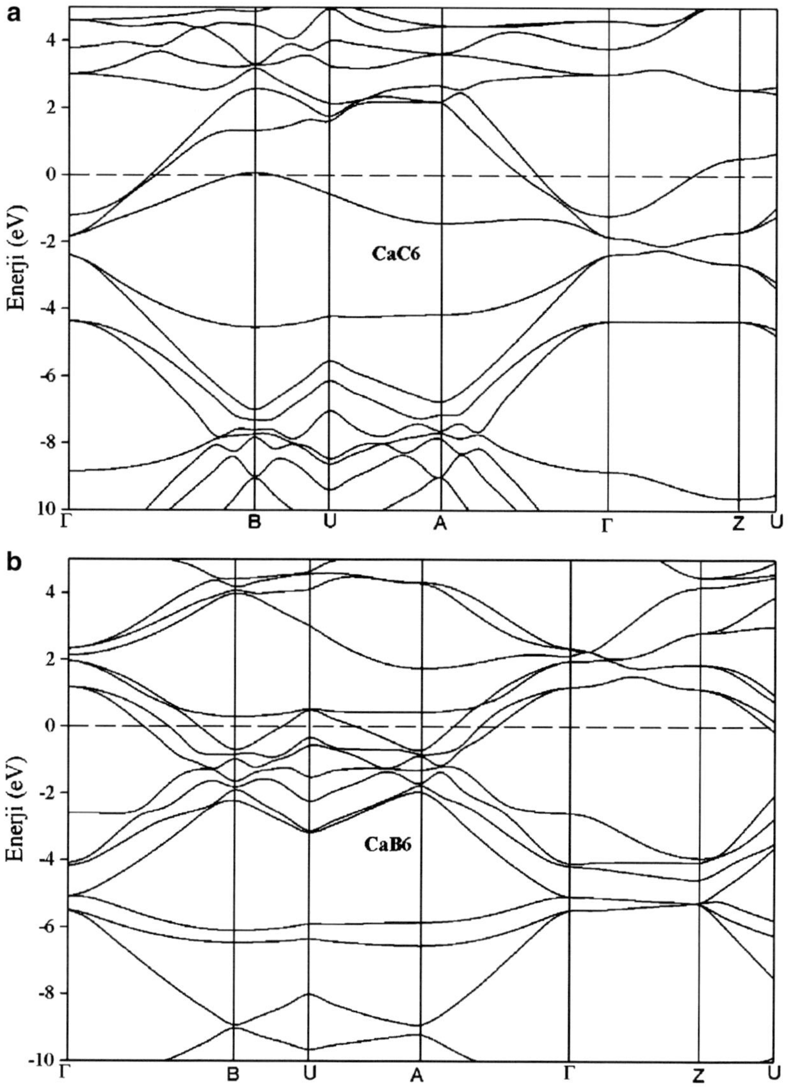

Fig. 2 Calculated band structure for (**a**) CaC_6 and (**b**) CaB_6

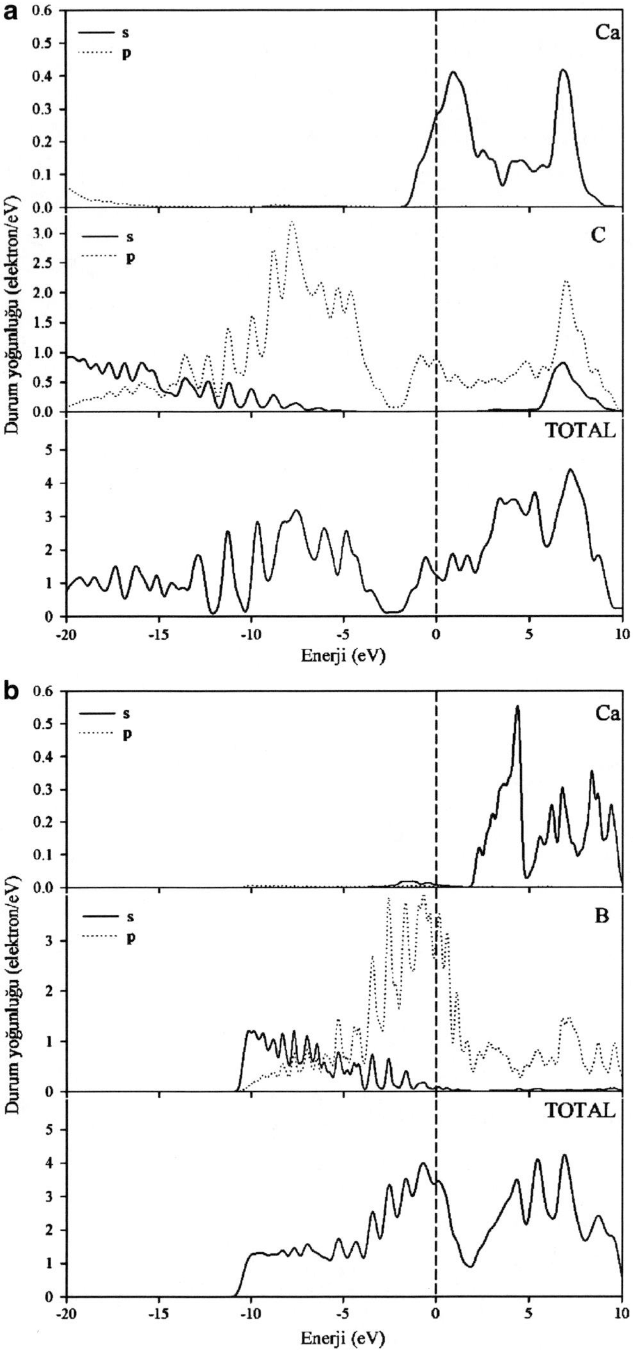

Fig. 3 Calculated partial density of states for (**a**) CaC_6 and (**b**) CaB_6

References

1. R. Singh, S. Prakash, Carbon 48, 1341–1344 (2010)
2. H. Okazaki et al., Physica C 469, 1041–1044 (2009)
3. M.S. Dresselhaus, D. Dresselhaus, Adv. Phys. 51, 1 (2002)
4. U. Mizutani et al., Phys. Rev. B 17, 3165 (1978)
5. S.L. Molodtsov et al., Phys. Rev. B 53, 16621 (1996)
6. T.E. Weller et al., Nat. Phys. 1, 39 (2005)
7. N. Emery et al., Solid State Sci 10, 466–470 (2008)
8. M. D. Segall, P. J. D. Lindan, M. J. Probert, C. J. Pickard, P. J. Hasnip, S. J. Clark, M. C. Payne, J. Phys.: Cond. Matt., 14, 2717 (2002)
9. D. Wanderbilt, Phys. Rev. B, 41, 7892 (1990)
10. J. P. Perdew, J. A. Chevary, S. H. Vosko, K. A. Jackson, M. R. Pederson, D. J. Singh, and C. Fiolhais, Phys. Rev. B 46, 6671 (1992)
11. J. P. Perdew, K. Burke, M. Ernzerhof, Phys. Rev. Lett., 77, 3865 (1996)
12. S.Q. Wu, Z.F. Hou, Z.Z. Zhu, Solid State Comm. 143, 425–428 (2007)
13. Zhi-Jian Wu, Er-Jun Zhao, Hong-Ping Xiang, Xian-Feng Hao, Xiao-Juan Liu, and J. Meng, Phys. Rev. B 76, 054115 (2007)
14. A. Simunek, Phys. Rev. B 75, 172108 (2007)
15. W. B. Pearson, The Crystal Chemistry and Physics of Metals and Alloys (Wiley, New York, p.151, Table 4-4 (1972)
16. J. S. Kim et al., Phys. Rev. B 74, 214513 (2006)

The System Design of an Autonomous Mobile Waste Sorter Robot

Ahmet Mavus, Sinem Gozde Defterli, and Erman Cagan Ozdemir

1 Introduction

Recycling is receiving increased attention in the environmental debate. This makes the first step of recycling, which is classification of waste materials, much more important. Manual sorting is obviously slow, tedious and unhealthy. Therefore, development of automatic techniques is inevitable. In their development, multidisciplinary approaches are applied because of great variety of parameters influencing the operation such as size, chemical properties, physical properties, etc.

First, while doing research for the general information about automatic waste sorting, some patents and commercial products are found in which more or less, the same techniques are applied. These existing automatic waste sorters are big plants including giant machines contrary to our case, which is a really little sorter. The sorting techniques used in these plants in a small-size sorter can be listed as follows: (for non-ferrous metals) eddy-current method, inductive sensors, electromagnetic and dual energy X-ray transmission sensor; (for ferrous metals) magnets; (for plastics) air classifiers, hydro cyclones, and float/sink baths, IR spectroscopy, selective dissolution, triboelectric charging; (for glasses) pneumatic separator. In addition to separation techniques how waste materials will be handled in operation, how they place in containers are crucial topics to search on. Moreover, energy management and automatic control systems are also important subjects to be investigated.

In this study, our design project is about recyclable waste sorting problem which is the subject of 2010 ASME Student Design Contest [1]. The aim is to design, build, and test a system capable of rapidly and accurately sorting the four waste

A. Mavus (✉)
Middle East Technical University, Department of Mechanical Engineering,
Ankara, 06531 Turkey
e-mail: ahmetmavus@hotmail.com

J.A.T. Machado et al., *Nonlinear and Complex Dynamics: Applications in Physical, Biological, and Financial Systems*, DOI 10.1007/978-1-4614-0231-2_11,

materials into distinct waste containers. This system must operate autonomously and be capable of both material identification and waste handling. A semi-rigid waste container (containing twelve waste products) is provided, specifically [1]:

1. Three empty plastic bottles $D = 75$ mm (± 20 mm), $L = 220$ mm (± 20 mm)
2. Three empty aluminum cans $D = 65$ mm(± 20 mm), $L = 120$ mm (± 20 mm)
3. Three empty steel containers $D = 75$ mm(± 20 mm), $L = 110$ mm (± 20 mm)
4. Three empty glass containers $D = 60$ mm (± 20 mm), $L = 95$ mm (± 20 mm)

1.1 Sorting and Working Principles

In the structure of the design, it is supposed to identify, handle and sort some kinds of wastes such as ferrous–non ferrous wastes, glass, and plastics. In order to do that, we use a kind of robot which has only one arm, one long hopper, one inductive sensor, one capacitive sensor, one magnetic switch, one encoder and Programmable Logic Controller (PLC) for control mechanism. These parts are connected to the main machine, and there are four bins for each type of waste. First, the capacitive sensor defines whether a waste exists or not. Then the inductive sensor and magnetic switch are used for detecting non-ferrous and ferrous wastes. The encoder is used for sorting glass and plastic bottles, since the capacitive gives sign "on", but the inductive sensor and the magnetic switch give "off" signal, at this time encoder is used for counting steps while translational motion of robot passing the waste. The lengths of the plastic bottles are much larger than the lengths of glass bottles, so this remarkable difference can be used for sorting these bottles by counting step signals from encoder and separating them into different containers. Therefore, sorting logic is defined as follows:

Case 1: Capacitive sensor is "on", inductive sensor is "on" but magnetic switch is "off" concludes that the waste is *non-ferrous metal.*

Case 2: Capacitive sensor is "on", inductive sensor is "on" but magnetic switch is "on" concludes that the waste is *ferrous metal.*

Case 3: Capacitive sensor is "on", inductive sensor is "off" but magnetic switch is "off" and encoder gives larger value than the limit concludes that the waste is *plastic material.*

Case 4: Capacitive sensor is "on", inductive sensor is "off" but magnetic switch is "off" and encoder gives smaller value than the limit concludes that the waste is *glass material.*

The hopper is used for arranging wastes, the arm is for picking the wastes (cf. Fig. 1), and these four bins are lying through the translational motion of robot. At the beginning of the process, the wastes are poured onto hopper by hand. After finishing the pouring of the waste process, the robot starts to work.

In the beginning, the capacitive sensor gives the information that there is a waste here by giving signal "on". At the same time, the inductive sensor and magnetic

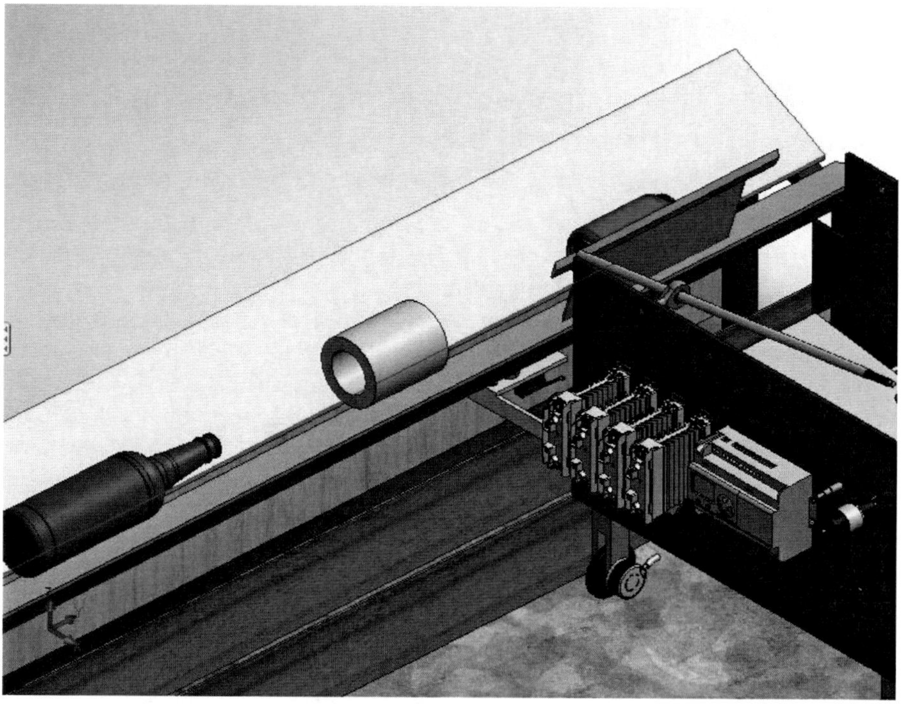

Fig. 1 The wastes on the V-shaped hopper and the robot

switch detect non-ferrous and ferrous metals. If inductive sensor gives "on" but magnetic switch gives "off", this means that a non-ferrous waste exists here; on the other hand, both the inductive sensor and magnetic switch give "on" signals, this means that there is a ferrous waste here. After all wastes are detected, the robot arm begins to collect wastes, and put them into correct bin by the help of the sloped surface driven by linear actuator and geometric design of the robot. All wastes are collected by the robot arm. The robot arm picks these wastes into the machine one by one. At the end, the wastes are put into their containers, and the operation of the machine is completed.

2 Properties of the Designed System

First of all, this machine is considered as a robot and mainly, the system has 3 degrees of freedom. One of them is the translational motion of the robot along the hopper. The second one is related to the arm mechanism to be used for the process of taking the wastes from hopper to the inside of the robot to separate. The third one is the translational motion of the turning block mechanism for the sloped surface driven by the linear actuators.

2.1 The Degrees of Freedom

If it is needed to be more precise, a detailed explanation can be given about the degrees of freedom of the mechanism listed above. First, the robot moves to the right or to the left by the aid of a strong direct current (DC) motor. The necessary torque is calculated afterwards. Specifically, this torque is calculated by putting the friction coefficient of the ground and weight of the robot (all parts inside the robot plus the part on the robot and the skeleton) into the perspective. According to this torque value, a suitable DC motor is selected.

The other important point of the arm mechanism of the robot is that the selection of the dimensions of the links of the inverted slider-crank mechanism as an arm. These links of the mentioned mechanism are selected for mainly carrying the waste with the maximum weight. However, in addition to that, selected links should not have contact with the sloped surface. Since the wastes are taken one by one, the heaviest parts are the ferrous wastes (steel parts), which are dominant in this process considering the durability of the mechanism. The third degree of freedom is about the motion of turning block mechanism of the sloped surface. This motion is used as sorting glass and plastic wastes. When the capacitive sensor gives signal as "on", but the inductive sensor and the magnetic switch give signal as "off", this means that the type of waste recognized is either plastic or glass. Since it is known that the length of the plastic wastes are much larger than the length of glass wastes, this remarkable difference is used for sorting by counting the step signals given from the encoder which is placed on the shaft of the driven wheel of the robot. So, the process of sorting the plastic and glass bottles are done by this principle.

2.2 The Hopper Design

The hopper design is also crucial for the system. Since it carries all the wastes, it should have the necessary strength. Moreover, its length is determined by adding the maximum lengths of the wastes plus a clearance value so that when they are poured to the hopper, they are arranged along the length of the hopper and they do not overlap. This is very important when identification process is put into the perspective. However, it should also be noted that its length may be 2 m maximum. Hence, overlap actually is inevitable.

For the hopper, the following conditions have to be considered:

(a) It should be wide enough not to let waste to fall down during pouring.
(b) It should be in a geometry not disturbing the arm.
(c) It should have enough alignment to guide waste in appropriate manner.

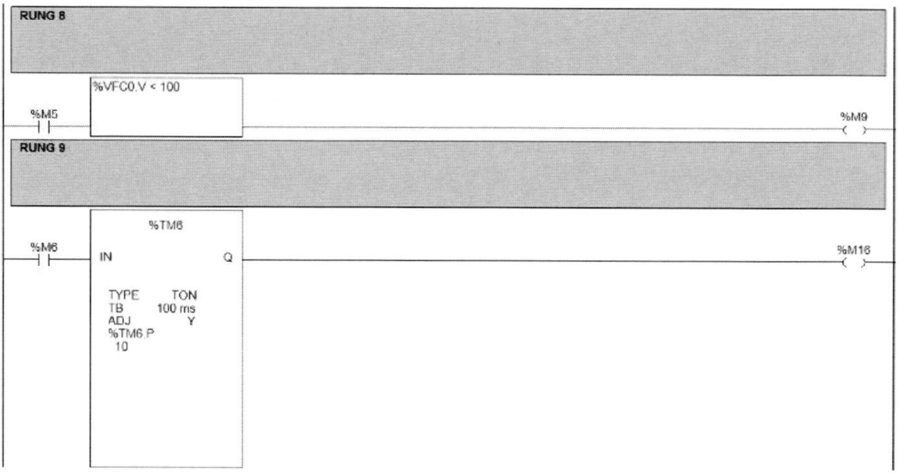

Fig. 2 A part of case definitions in the ladder diagram of PLC algorithm

2.3 *The Control Mechanism*

The control of the slider crank arm mechanism, driven wheel DC motor, arm motor, encoders, linear actuator is achieved by PLC. PLCs provide ease and flexibility of control based on programming and executing simple logic instructions (often in ladder diagram form) [2].

For the inductive sensor and capacitive sensor which are used in our mechanism, the range of sensor is not significant since its head is on the underneath of the long hopper part. Therefore, accuracy may be increased. The key point is actually when using sensors, motors, encoder and linear actuator, computer interaction should be achieved. To be more specific, our system includes a main computer and all the motors, sensors, encoder, magnetic switch and linear actuator are controlled by the main controller that is PLC. In order to connect motors and sensors to the computer, PLC is easy to implement and allows us to do any changes in the algorithm. For this reason, these kinds of control systems or derivatives can be used to obtain accurate results.

PLC algorithm is written by first considering the cases like if capacitive sensor is "on", the inductive sensor is "on" but the magnetic switch is "off", the case is that the waste is non-ferrous material such as called "M2". After defining the cases as mentioned above, the actions are taken into consideration in the second part of the algorithm. All these cases are defined in the ladder diagram where a part of it can be seen in Fig. 2.

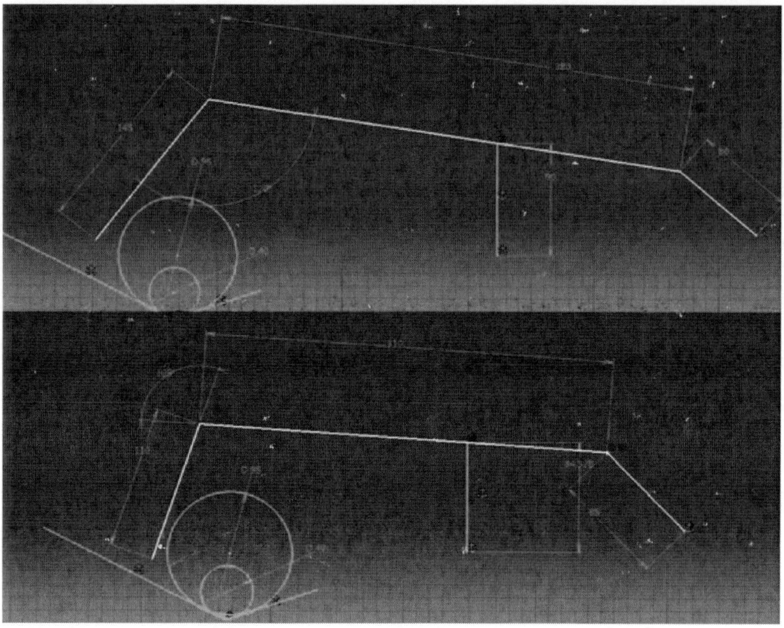

Fig. 3 Geometric design of the inverted slider crank arm mechanism

2.4 The Motions of the Inverted Slider Crank Arm Mechanism

The slider crank arm mechanism has two different motions such as inward and outward motion. These types of motions are done according to the detection of the waste material. If the waste material is plastic, the arm mechanism does outward motion and if the waste material is non-ferrous and glass, the arm does the inward motion to the robot. When the material of the waste is detected as ferrous, then arm does not move and the ferrous waste stays on the hopper since the weight of ferrous wastes can damage the arm mechanism, which is undesirable.

For the arm, the items given below have to be satisfied:

(a) It should, in a position in its turn, lay within the given geometric restrictions (see Fig. 3).
(b) It should cover whole waste range, not skipping some small wastes.
(c) It should not, in a position in its turn, intersect with other components.

2.5 Motor Selection

The choice of the suitable motors is also vital. The most critical motor is the one located under the base plate to drive the robot. It is chosen by considering the total torque acting on the wheels. The strongest motor is the driving motor as stated at the beginning of the mechanism. There is also other motor namely, the DC motor actuating the arm mechanism, this motor does not have to apply huge torques as the main driving motor since the weight of the wastes is very small when compared to the total weight of the robot.

3 Engineering Calculations of the Geometric Design

3.1 The Inverted Slider Crank Arm Mechanism

The arm mechanism (as seen in Fig. 3) is the most crucial part of the robot since it is used for taking the detected wastes and sending them to the correct container. For the arm of the robot, four-bar and inverted slider crank are preferred mechanisms and the inverted slider crank is the first candidate since its motion curve seems more appropriate [3].

In Excel, as seen in Table 1 and Fig. 4, parametric model of inverted slider crank is created. By changing the values of parameters, a suitable motion curve is generated. Since there is no limitation about the design of the long hopper, it is achieved according to the design requirements of the arm mechanism.

3.2 Turning Block Mechanism for the Sloped Surface Driven by the Linear Actuator

This mechanism is used for separating glass bottles into their specified container and driven by a linear actuator which is controlled by PLC according to taken data from sensors and encoders. If the control logic, which is mentioned before for the glass bottles is satisfied, then linear actuator becomes active and turning block mechanism for sloped surface moves toward to the long V-shaped hopper. The arm mechanism takes the glass on the sloped surface and the glass bottle moves along the surface and falls into the corresponding container. The geometric design

Table 1 Position analysis of arm mechanism in Excel

		Links	x (mm)	y (mm)	Prismatic joint	
x_{fixed}	345.4	–	0	0	−354.995	97.007
y_{fixed}	70.4	A_0	40.358	−110.884	−318.793	79.995
Crank	118	A	−345.4	70.4	−335.805	43.793
Slider	495	B_0	−407.639	99.648	−372.007	60.806
Coupler arm	105	B	−488.274	32.396	−354.995	97.007
Force angle α_3	65	C	−457.556	58.016	–	–

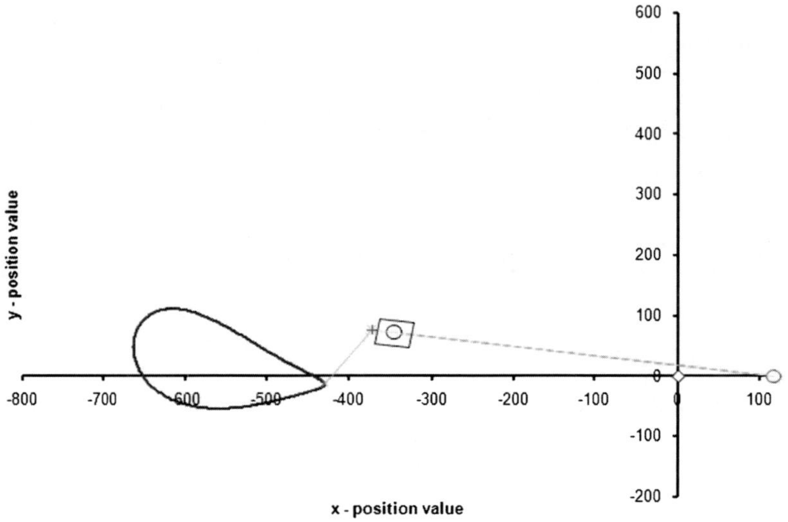

Fig. 4 Kinematic modelling of the arm mechanism

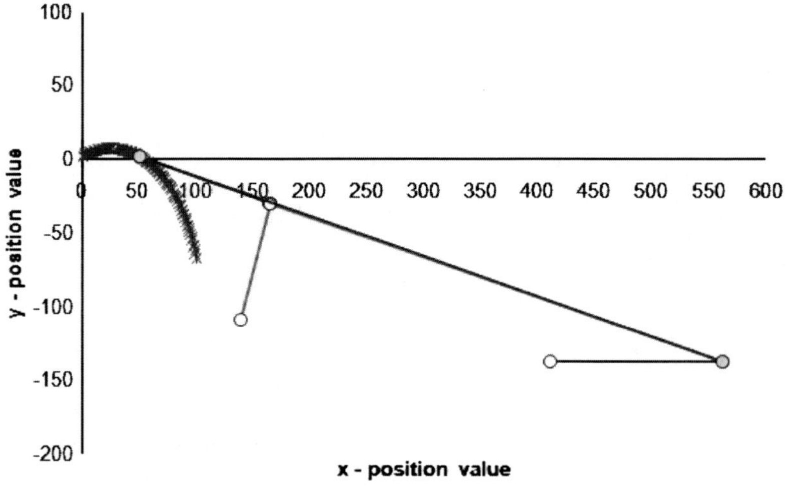

Fig. 5 Visualization of the turning block mechanism after kinematic synthesis

of turning block mechanism for the sloped surface is achieved according to hopper and arm mechanism's design in order to get a high accuracy. Thus, the motion analysis of turning block mechanism for the sloped surface is done in Excel Macro and presented in Fig. 5. Moreover, Fig. 6 shows the side view of the turning block mechanism in the robot.

Fig. 6 Side view of the turning block mechanism

3.3 *Shaft Design of Wheels*

Considering the construction materials and the control components of the robot, the total weight is calculated a 300 N. Focusing on all five wheels of the robot, the static analysis gives that the most critical force exerts on the wheel shaft is 90 N. Also, the shaft design is performed according to this critical force value.

$$F_w = 90\,\text{N}, \quad d_s = 5.25\,\text{mm}, \quad L = 19.56\,\text{mm}, \quad f = 0.5, \quad d_w = 62\,\text{mm}.$$

The material properties are taken as: AISI 1080

$$S_{\text{ut}} = 615.4\,\text{MPa}, \quad S_y = 375.8\,\text{MPa}, \quad E = 200\,\text{GPa}.$$

(S_{ut} : Ultimate Tensile Stress, S_y : Yield Stress, E: Modulus of Elasticity)

After determining the torque, moment and direct force on each shaft on the wheels, bending and shear stresses are evaluated. Von Misses Stresses are evaluated to find alternating and mean components of the shear and bending stresses. Moreover after determining the design factors (size, temperature, reliability, and so on), the process for finding safety factor begins. The corresponding formulas are given in the following equations [4].

$$\sigma_{a1} := \sqrt{\sigma_{\text{abending}}^2 + 3\tau_a^2} \tag{1}$$

$$\tau_m := 16\frac{T_w}{\pi d_s^3} \tag{2}$$

$$\sigma_m := 0 \tag{3}$$

$$\sigma_{m1} := \sqrt{\sigma_m^2 + 3\tau_m^2} \tag{4}$$

$$S_{e1} := 0.5S_{\text{ut}} \tag{5}$$

The endurance strength limit is calculated as:

$$S_e = k_a.k_c.k_d.k_f.S_{e1} \tag{6}$$

$$S_e = 1.853 \times 10^2\,\text{MPa} \tag{7}$$

$$n := \frac{1}{\left(\frac{\sigma_{a1}}{S_e} + \frac{\sigma_{m1}}{S_y}\right)} \tag{8}$$

At the end, it is defined that the safety factor, which is denoted by n, should be at least 2 and after calculations are done we obtained that $n = 2.542$.

Hence, this design for shafts of wheels is safe enough to satisfy the needs of the design.

3.4 The Design of Structural Members

The structural members carry sweeper motor for arm mechanism, spherical bearing connection, PLC controller box and relays. Since the exact location for the arm connection point to the structural members should be determined in detail, more durable body frame is satisfied and space minimization is provided, it is better to use special designed laser cut 5 mm thick sheet metal.

4 Conclusion

Before deciding to use this design, it is thought about lots of different designs workability. At the beginning of the concept selection, there are lots of alternative concept designs. First, it is tried to use some mechanical ways to separate the wastes. But in order to use these mechanical ways, huge amount of space for the machine is needed and large amount of energy is required for running these mechanical systems. According to these design specifications, the space is strictly limited, and it is only permitted to use dry cells as energy sources in the system so it should be better to give up thinking of usage of mechanical ways to separate the wastes. Second, it is tried to use some piston arrangement to separate the wastes but in this design, it is needed lots of small pistons, and compressors to run these pistons, also the system needs lots of energy to run these pistons. It is known that dry cells are not powerful enough to run these compressors, also compressors occupy large amount of space. Another disadvantage of that system is that these pistons are hard to find, and it is too expensive. Finally, we tried to find a system which occupies small amount of space and also economic. In order to fulfill these conditions, we decided to use a kind of robot.

In a nut shell, the robot has some specialties. These specifications are weight, time, energy, user friendliness, accuracy and space occupation [5]. It is chosen that the materials of the system with considering these specifications. There are a few motors in this system. These motors should not be heavy, should occupy small amount of space, and should consume small amount of energy and should satisfy torque and revolution requirements. Also, the process has to be completed in five minutes so that the robot should make the process quick. Our system has to be user friendly. In our design, the user will only pour the waste, push the button, and wait until the wastes separated. The system has to occupy small amount of space. A robot can easily fulfill all of these specifications. Further effort is going to be made in order to apply for patent of this design project and to provide automation of the robot for industrial needs. Since this autonomous mobile waste sorter robot design provides an innovation to the sorting problem in recycling industry, we hope that this new and original project is going to be patented.

Acknowledgements This work is partially supported by the Scientific and Technical Research Council of Turkey.

References

1. ASME (2010) Problem Description. In: The 2010 ASME Student Design Competition - EARTH SAVER: Autonomous Material Sorter. http://files.asme.org/asmeorg/Events/Contests/DesignContest/18197.pdf.
2. Warnock IG (1988) Programmable controllers: Operation and Application. Prentice Hall, London, UK
3. Soylemez E (1999) Mechanisms. Middle East Tecnical University Press Cooperation, Ankara, Turkey
4. Budynas RG, Nisbett JK (2006) Shigley's Mechanical Engineering Design, 8th edition. McGraw-Hill, Boston MA
5. Dieter GE (2000) Engineering Design, 3rd edition. McGraw Hill, Boston MA

Nomenclature

d: Diameter (mm)
d_s: Diameter of the Shaft (mm)
d_w: Diameter of the Wheel (mm)
L: Length (mm)
A: Cross Sectional Area (mm^2)
E: Modulus of Elasticity (GPa)
S_y: Yield Strength (MPa)
S_t: Tensile Strength (MPa)
S_{ut}: Ultimate Strength (MPa)
S_e: Endurance Limit (MPa)
n: Safety Factor
k: Yield Strength Modification Factor
k_a: Surface Condition Modification Factor
k_c: Load Modification Factor
k_d: Temperature Modification Factor
k_f: Miscellaneous Effect Modification Factor
F: Force (N)
F_w: Force on the Wheel (N)
f: Friction Coefficient
M: Moment ($N.m$)
T: Torque ($N.m$)
σ: Tensile Stress (MPa)
σ_m: Mean Tensile Stress (MPa)
σ_a: Alternating Tensile Stress (MPa)
τ: Shear Stress (MPa)
τ_m: Mean Shear Stress (MPa)
τ_a: Alternating Shear Stress (MPa)

Evidence of the Wave Phase Coherence for Freak Wave Events

Alexey Slunyaev

1 Introduction

Irregular waves arise in many important physical problems and are the subject of investigation. The study of irregular wave dynamics and statistics is relevant for correct physical understanding and for practical applications as well. Sea waves are an example of inherently stochastic waves. They are often understood as a combination of quasi-sinusoidal waves with independent random uniformly distributed phases (the Gaussian sea). If waves were linear and random, they would possess the Gaussian probability distribution function due to the central limit theorem.

The difference between the Gaussian sea approximation and the real sea results in recognizing the problem of *freak wave* or *rogue wave* phenomenon, see reviews [12, 19, 20]. The attempts to integrate the wave nonlinearity effect into the statistical models have been undertaken many times with a certain success. However, each time these endeavors employ the condition of weak nonlinearity, which is questionable when applied to freak wave events. Meanwhile, obtaining the statistical description of freak waves for the case of a given wave energy spectrum is the cherished aim of the researchers.

Kinetic approach is conventional and well established for the study of random wave spectrum evolution. The kinetic theory is weakly nonlinear, uses closure assumptions, thus, is eventually capable of describing only near-Gaussian processes, and it disregards wave phases. The stochastic approach employs dynamical models, which resolve the wave phases. Then the relations between the spectral and

A. Slunyaev (✉)
Institute of Applied Physics, Nizhny Novgorod, 603950, Russia

Keele University, Keele, ST5 5BG, UK
e-mail: Slunyaev@hydro.appl.sci-nnov.ru

J.A.T. Machado et al., *Nonlinear and Complex Dynamics: Applications in Physical, Biological, and Financial Systems*, DOI 10.1007/978-1-4614-0231-2_12,

statistical wave characteristics may be established. Instead of computing the kinetic equations, the stochastic approach has become very popular due to computer power progress and building large and well-equipped experimental facilities.

The stochastic modeling requires definition of the initial wave field. Typically, the initial condition is defined in the form of wave fields obeying some spectrum with random uniformly distributed phases. A great number of numerical and laboratory experiments prove that this initial condition undergoes strong evolution at the initial stage, what changes the average spectrum of the waves, see [5–7, 14, 16, 21–25, 27–29, 32] among others. Thus, the wave fields at this stage cannot be considered statistically stationary. Limited fetches (due to the limitation of the laboratory facility or short numerical simulations) may prevent the achievement of the stationary state at all.

The most striking nonlinear effect, which is now believed able to cause rogue waves is the Benjamin-Feir instability (otherwise, the side-band or the modulational instability, see for instance [17, 36]), which leads to the generation of intense wave groups from uniform wave trains in deep-water conditions. This effect for unidirectional waves has been confirmed many times by means of numerical simulations, and also by laboratory measurements. Meanwhile, it is known to be weakened and even cancelled, when broad-band waves or random waves are concerned.

The Benjamin-Feir index (BFI) was introduced by Onorato et al. [22] and Janssen [16] to measure the strength of the nonlinear self-modulation effects for a given wave energy spectrum. This compound spectral parameter is in agreement with its dynamic counterpart, which follows from the weakly nonlinear weakly modulated theory for surface waves (the framework of the nonlinear Schrödinger equation, NLS). Then the BFI corresponds to the similarity parameter of the NLS equation, otherwise, the soliton number.

The NLS equation is a unique mathematical model due to the property of integrability. Its solutions have been suggested to describe real freak waves in the ocean [2, 13, 15, 26]. The so-called *breather* solutions of the NLS equation are actually solitary waves interacting with other background waves. The similarity of the breather solutions and the large-wave events observed in numerical simulations has been pointed out many times [8, 9, 15]. The existence of long-living strongly nonlinear wave groups similar to the envelope solitons has been reported recently for numerical simulations of fully nonlinear equations for hydrodynamics [10, 30].

The framework of the NLS equation for unidirectional waves requires the conditions of weak nonlinearity and narrow spectrum. In the general case, the solitary-like patterns are supposed to show themselves as coherent wave structures, what implies nonzero correlation between the Fourier modes. The dynamics of a single four-wave resonance quartet was studied within the framework of the Zakharov equation by means of the analytic solution in [18, 33]; in [33] ensembles of the wave quartets were considered as well and compared versus the results of the kinetic approach. It was shown, in particular, that initially random phases develop a significant coherence in the course of evolution. Specific spectrum shape was suggested by [1] as a possible early-warning criterion for the rogue wave danger.

In this paper, we deal with the deep-water limit, and the unidirectional case of surface sea waves, which is the most investigated. The wave trains are generally supposed narrow-band; the physics is governed by the free wave components, while the bound waves (Stokes corrections) under the narrow-band assumption may be trivially obtained on the basis of the free waves.

The coherence might be revealed in the dynamics of the strongly interacting components (when they may be singled out) such as resonance quartets or soliton-like wave groups. We show that the wave coherence can be revealed globally in a stochastic wave field. A more detailed description of the background of the present study is given in [31].

2 Stochastic Numerical Simulations of Modulationally Unstable Wave Fields

In this section, we summarize the results of numerical simulations, performed within the frameworks of the nonlinear Schrödinger equation, its high-order generalization (the Dysthe equation with the exact deep-water linear dispersion law taken into account [11, 34]), and the fully nonlinear simulations of the Euler equations in conformal variables [35]. The algorithms are briefly described in [30]. Nonbreaking waves are considered; 100 wave ensembles are used for the averaging.

In all cases, the initial wave realizations are defined in the form of a linear superposition of Stokes waves with random phases, similar to those described in [28], which obey the Gaussian spectrum. The carrier wavenumber is chosen $k_0 \approx 1.78\ \mathrm{m}^{-1}$. Due to the deep-water conditions, the mean wave period is estimated through the linear dispersion law as $T_0 = 1.5$ s. The NLS model is used to solve waves with a moderate initial steepness, $k_0\eta_{\mathrm{rms}} \approx 0.042$ (where η_{rms} is the root-mean-square surface wave displacement defined on the basis of the free wave component) for different initial spectrum widths. The fully nonlinear simulations are performed for one initial spectrum width, $\nu/k_0 \approx 0.076$, where ν is defined as the second moment of the average spectrum. In terms of the introduced parameters, the BFI may be defined as [16]

$$\mathrm{BFI} = 2\sqrt{2}k_0^2\frac{\eta_{\mathrm{rms}}}{\nu}, \tag{1}$$

and the characteristic nonlinear time we define as

$$T_{\mathrm{nl}} = \frac{1}{\omega_0 k_0^2 \eta_{\mathrm{rms}}^2}. \tag{2}$$

In the course of evolution the spectrum width, ν and the BFI evolve. The variable BFI for the simulations is shown in Fig. 1 as the function of the scaled time. The results of the NLS simulations are represented by solid lines, and the results of the fully nonlinear simulations are represented by dashed lines.

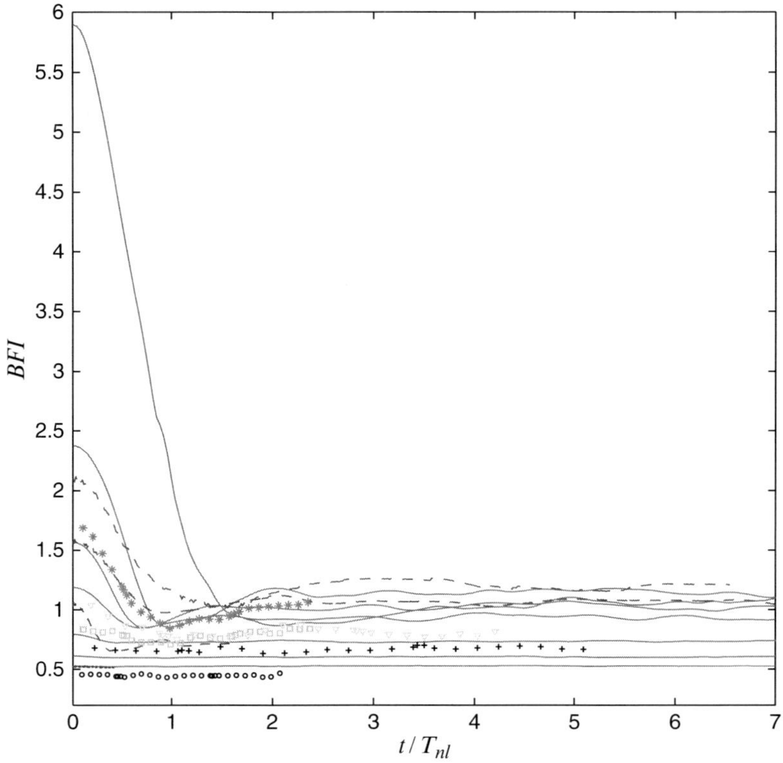

Fig. 1 The temporal dependences of the BFI versus the scaled time for numerical simulations (*lines*) with different initial spectrum widths and wave intensities, and for the laboratory measurements [29] (*symbols*) with different spectral widths and shapes

The results of laboratory measurements [29] are given in Fig. 1 with symbols. In the laboratory experiments, the spatial wave evolution was considered and surface elevation time series were retrieved at different distances. The distances are recomputed to the corresponding times in Fig. 1 supposing that the waves propagate with the group velocity of the carrier wave.

It is evident that the curves corresponding to the two sets of numerical simulations and the results of laboratory measurements well agree when represented in the scaled variables. For sufficiently long times kinds of steady states are achieved in all the cases.

The Alber theory for narrow-band weakly nonlinear random waves [3, 16] predicts the cancellation of the Benjamin-Feir instability effect for BFI < 1. It may be seen in Fig. 1 that this threshold describes well the qualitative difference in the evolution of the BFI. The wave fields with initially large values of the BFI tend to the state, which seems to be marginally stable. The wave fields with small values of the BFI remain practically unchanged during the evolution.

3 The Evidence of Phase Coherence

The spectral phases are implied by the concept of the Gaussian sea to be uniformly distributed. Indeed, the phase distribution observed in our simulations may be considered uniform (see details in [31]). However, the nonlinear wave phases are obviously not fully independent as it is in the linear approximation.

The phase coherence supports existence of coherent wave structures. Some exact solutions of the NLS equations are discussed in this context in [1, 31]. When many nonlinear coherent structures are present in the wave field, the wave dynamics is supposed to be quite complicated.

We apply the following correlation function to make the coherence between the Fourier phases evident:

$$
\begin{aligned}
R(\delta,t) &= \frac{R_1}{R_2}, \\
R_1 &= \left| \sum_{n=1}^{N} S_n(k_0+\delta)S_n(k_0-\delta)S_n^*(k_0+\delta')S_n^*(k_0-\delta') \right|, \\
R_2 &= \sum_{n=1}^{N} \left| S_n(k_0+\delta)S_n(k_0-\delta)S_n^*(k_0+\delta')S_n^*(k_0-\delta') \right|, \\
\delta &= \frac{2\pi}{L}m, \quad \delta' = \frac{2\pi}{L}(m+1), \quad m \geq 0.
\end{aligned}
\tag{3}
$$

In (3) k_0 is the dominant wavenumber, δ and δ' specify the wavenumber offsets according to the wavenumber discretization (integer m counts the spectral nodes), L is the computational domain length. The summation is performed over all N realizations.

The results of the computation of the autobicoherence function (3) for the conditions $k_0\eta_{\mathrm{rms}} \approx 0.042$, $\nu/k_0 \approx 0.076$ (and BFI ≈ 1.56) are presented in Fig. 2 by the colour intensity. Different models are simulated: the NLS equation (Fig. 2a), the Dysthe equation (Fig. 2b), and the Euler equations in conformal variables (Fig. 2c). The horizontal axis shows the normalized time, the vertical axis represents the wavenumber offset. The temporal dependence of the average spectrum width ν is given over the diagram (the solid line), and the spectrum shape at $t = 0$ for $k \geq k_0$ is shown to the right of the diagram (the line with circles); both are given for the reference.

The case of the NLS equation simulation is shown in Fig. 2a. It is obvious that the initial condition corresponds to an insignificant correlation (dark area near $t = 0$). But with time the correlation for a sufficiently large offset quickly grows up to the unity. While the Fourier modes which are sufficiently far from the spectral peak are thus shown to be absolutely correlated, the most energetic area of the spectrum close to the spectral peak is uncorrelated (at least with respect to the autobicorrelation function (3)).

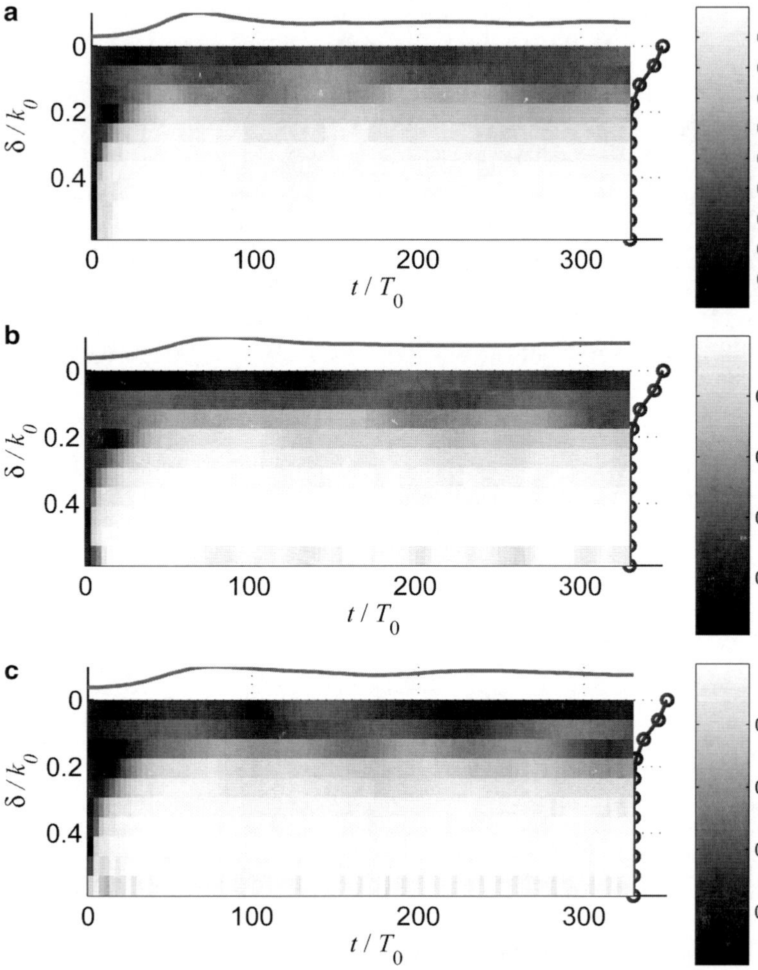

Fig. 2 Diagrams of the correlation estimator $R(\delta,t)$ shown by the color intensity. The temporal dependence of the average spectrum width is given over the diagram (*the solid line*) for the reference, and the spectrum shape at $t = 0$ is shown to the right of the diagram (*the line with circles*) for the reference. The initial condition for simulation $\mathrm{BFI}(t = 0) = 1.56$ is computed in different frameworks: the NLS equation (**a**), the Dysthe model (**b**), and the fully nonlinear simulations (**c**)

The obtained result is not an artifact of the NLS approximation, but is confirmed in the simulations of the Dysthe model (Fig. 2b) as well as the fully nonlinear equations (Fig. 2c). Some difference between the diagrams Fig. 2a–c may be noticed, but are not significant.

Steeper wave conditions ($k_0\eta_{\mathrm{rms}} = 0.056$) are concerned in Fig. 3 for the same initial spectrum width $\nu/k_0 \simeq 0.076$ (BFI = 2.08). The waves are simulated by means of the fully nonlinear model. Although the correlation picture in Fig. 3 is less sharp than in Fig. 2, the level of the phase coherence is again very high. It is

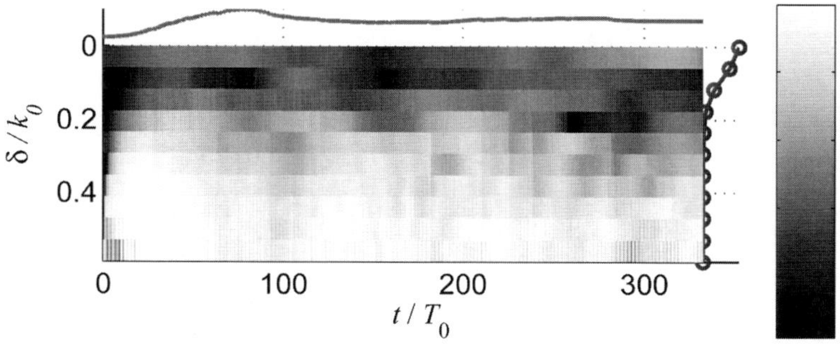

Fig. 3 The diagram of the correlation estimator $R(\delta,t)$ shown by the color intensity. The temporal dependence of the average spectrum width is given over the diagram (*the solid line*) for the reference, and the spectrum shape at $t = 0$ is shown to the right of the diagram (*the line with circles*) for the reference. The fully nonlinear simulation with BFI$(t = 0) = 2.08$ is reported

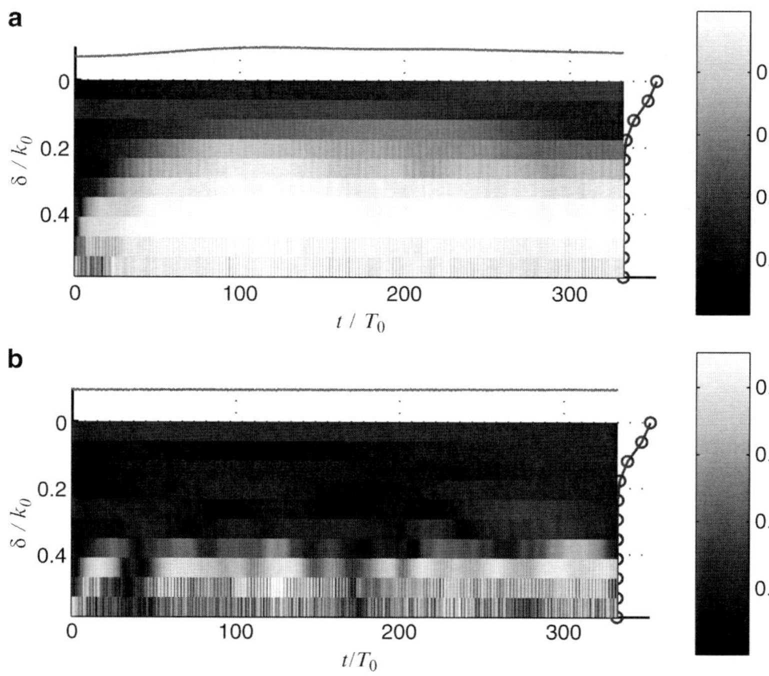

Fig. 4 Diagrams of the correlation estimator $R(\delta,t)$ shown by the color intensity. The temporal dependence of the average spectrum width is given over the diagram (*the solid line*) for the reference, and the spectrum shape at $t = 0$ is shown to the right of the diagram (*the line with circles*) for the reference. The fully nonlinear simulations with BFI$(t = 0) = 1.04$ (**a**), and BFI$(t = 0) = 0.52$, (**b**) are reported (different wave intensities)

also significant that in the course of evolution about one half of the realizations resulted in very steep waves; these simulations were not taken into account after the steep event occurrence. Thus, in a certain sense, less coherent wave fields compose the statistical data at longer times.

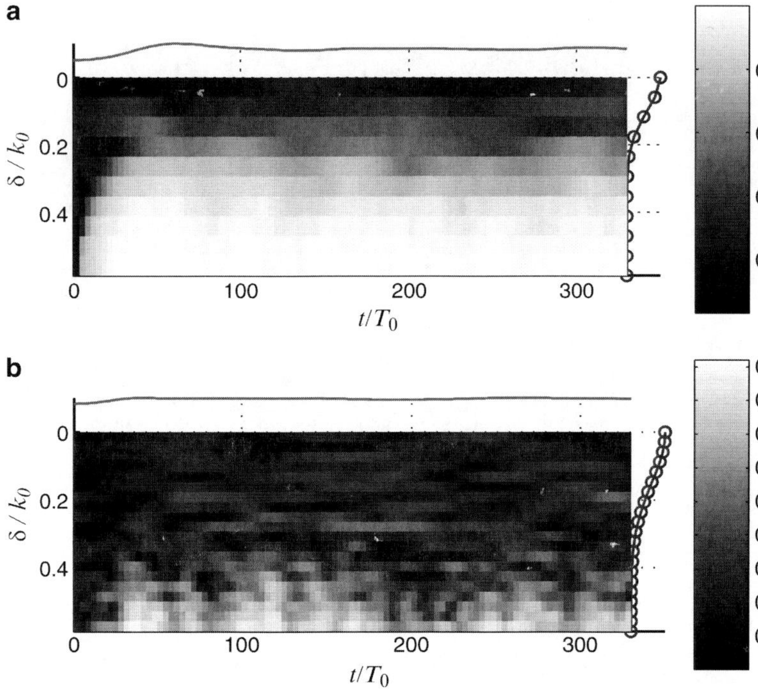

Fig. 5 Diagrams of the correlation estimator $R(\delta,t)$ shown by the colour intensity. The temporal dependence of the average spectrum width is given over the diagram (*the solid line*) for the reference, and the spectrum shape at $t = 0$ is shown to the right of the diagram (*the line with circles*) for the reference. The NLS equation simulations with $\mathrm{BFI}(t=0) = 1.19$ (**a**), and $\mathrm{BFI}(t=0) = 0.79$ (**b**) are reported (different spectrum widths)

Different wave amplitudes with the same initial spectrum width are considered in Fig. 4 within the fully nonlinear framework with the initial values of the BFI equal to 1.04 and 0.52 (Fig. 4a,b, respectively). The level of the correlation and the interval of wave numbers, where the coherence is revealed are noticeably smaller for the case displayed in Fig. 4b in comparison with Fig. 4a.

In Fig. 5 wave fields with the same root-mean-square surface displacement but different initial spectrum widths are considered. The typical wave amplitude is modest, and the NLS equation is solved to obtain the results. Similar to Fig. 4, the initial BFIs were chosen equal to 1.19 and 0.79 in Fig. 5a,b correspondingly. Again, the coherence is much less evident, if $\mathrm{BFI} < 1$ (Fig. 5b), than if $\mathrm{BFI} > 1$ (Fig. 5a).

A single example based on a laboratory experiment data is given in Fig. 6. The intense irregular wave groups were generated by the wavemanker at one wall of the tank, and then they propagate, see [29]. The initial spectrum had a rectangular shape, and the initial BFI was greater than 1, thus occurrence of large waves associated with rogue events was observed for the distance about 70 m. It is very well seen in the figure that there is a bright spot between $50\ \mathrm{m} < x < 100\ \mathrm{m}$, what manifests a

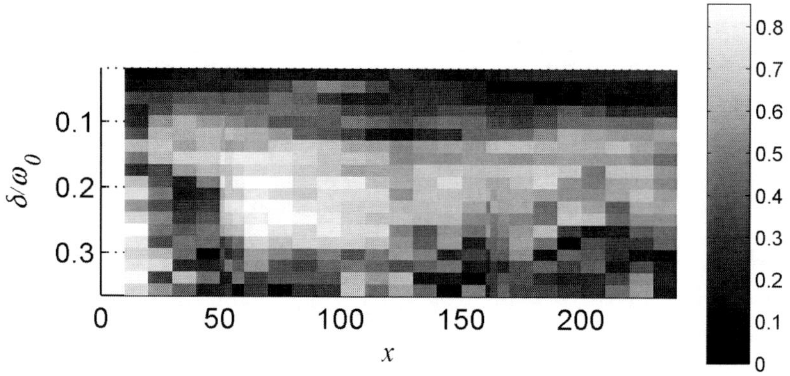

Fig. 6 The correlation estimator R shown by the colour intensity for laboratory wave modeling. The initial condition ($x = 0$) is characterized by $\mathrm{BFI}(t = 0) = 1.71$ ($\eta_{\mathrm{rms}} = 0.03$ m, the wave period is about 1.5 s, and the typical wave length is about 3.5 m; the frequency spectrum width is $\nu_\omega/\omega_0 = 0.045$). Coordinate x shows the distance from the wave maker in meters

strong coherence between spectral harmonics. It goes without saying that laboratory data is more difficult for analysis due to the instrumental imperfectness and natural presence of noise in the data. However, the estimator (3) turns out able to capture the coherence.

4 Conclusion

The stochastic approach to the study of intense sea waves, and in particular the freak or rogue waves, has become quite popular during the recent years. The irregular waves are usually defined as a linear superposition of Stokes waves with random phases following the conventional concept of a Gaussian or near-Gaussian sea. Although the random phase assumption is acknowledged to be violated due to nonlinearity, the obtaining of the phase correlation functions for realistic waves from general principles is a hard task. The four-wave interactions are the most efficient for the deep-water case, but the resonance wave quartets interact among each other [18, 33]; moreover, quasi-resonance wave interactions obviously have to be taken into account [4, 33] to obtain a reliable result.

Nonresonant interactions lead to the appearance of phase-locked modes in the Fourier spectrum. The free wave components (they may support exact resonant or near resonant wave modes) are considered governing the physics of nonlinear waves. The bound wave components (otherwise the nonlinear Stokes corrections) at multiple frequencies/wavenumbers are naturally observed in experiments. They are an indicator of occurrence of steep waves. The bound waves are coherent waves, but in the spectrum they are far from the spectral peak, and may be accounted by modern models.

The Benjamin-Feir (modulational) instability has been shown to be a nonlinear effect which increases the probability of freak waves in the deep-water narrow-band case. The onset of the Benjamin-Feir nonlinear instability is well controlled by the Benjamin-Feir index, BFI. Initially unstable wave fields tend to a marginally stable state. The self-modulation effect becomes apparent through the occurrence of large-amplitude coherent wave groups.

In this note, we employ the empirically written estimator for the wave coherence, complied with exact model solutions of the nonlinear Schrödinger equation as well as with the general comprehension of the resonance wave quartet nature, see details in [31]. We focus the attention on the coherence in the Fourier space, which corresponds to harmonics, phase-locked due to the nonlinearity. These modes indicate the presence of the mentioned above large-amplitude coherent wave groups (responsible for freak events) rather than individual steep waves.

The estimator, which is actually a kind of an autobicorrelation function, turns out to be efficient to reveal wave coherence in all cases, when the self-modulation effects are significant. The deep-water frequency spectrum is twice narrower than the wavenumber spectrum, and, thus, probably even a stronger wave coherence might be observed in the frequency spectrum. Indeed, a preliminary analysis corroborated the ability of the suggested autobicorrelation estimator to reveal the coherence in laboratory-measured time series of the surface displacement.

We conclude that in the case of modulationally unstable wave fields the wave correlation is quite significant and can be revealed in rather close vicinity to the most energetic spectral area. This part of spectrum cannot be considered as a superposition of independent waves with random phases. The phase coherence should be taken into account when describing the nonlinear waves accurately.

Acknowledgements The author thanks A. Sergeeva for providing with Fig. 6 based on laboratory results, and with the laboratory experiment data displayed in Fig. 1.

The research is supported by grants MK-6734.2010.5 and RFBR 11-02-00483, 11-05-00216, and has received funding from the European Community's Seventh Framework Programme FP7-PEOPLE-2009-IIF under grant agreement No 254389.

References

1. Akhmediev, N., Ankiewicz, A., Soto-Crespob, J.M., Dudley, J.M.: Rogue wave early warning through spectral measurements? Phys. Lett. A **375**, 541–544 (2011).
2. Akhmediev, N., Ankiewicz, A., Taki, M.: Waves that appear from nowhere and disappear without a trace. Phys. Lett. A **373**, 675–678 (2009).
3. Alber, I.E.: The effects of randomness on the stability of two-dimensional wavetrains. Proc. Roy. Soc. Lond. A **363**, 525–546 (1978).
4. Annenkov, S.Yu., Shrira, V.: Role of non-resonant interactions in the evolution of nonlinear random water wave fields. J. Fluid Mech. **561**, 181–207 (2006).
5. Annenkov, S.Y., Shrira, V.I.: Evolution of kurtosis for wind waves. Geophys. Res. Lett. **36**, L13603 (2009).
6. Annenkov, S.Y., Shrira, V.I.: "Fast" nonlinear evolution in wave turbulence. Phys. Rev. Lett. **102**, 024502 (2009).

7. Chalikov, D.: Freak waves: Their occurrence and probability. Phys. Fluids **21**, 076602 (2009).
8. Clamond, D., Francius, M., Grue, J., Kharif, C.: Long time interaction of envelope solitons and freak wave formations. Eur. J. Mech. B / Fluids **25**, 536–553 (2006).
9. Dyachenko, A.I., Zakharov, V.E.: Modulation instability of stokes wave → freak wave. JETP Lett. **81**, 255–259 (2005).
10. Dyachenko, A.I., Zakharov, V.E.: On the formation of freak waves on the surface of deep water. JETP Lett. **88**, 307–310 (2008).
11. Dysthe, K.B.: Note on a modification to the nonlinear Schrödinger equation for application to deep water waves. Proc. Roy. Soc. London A **369**, 105–114 (1979).
12. Dysthe, K., Krogstad, H.E., Muller, P.: Oceanic rogue waves. Annu. Rev. Fluid Mech. **40**, 287–310 (2008).
13. Dysthe, K.B., Trulsen, K.: Note on breather type solutions of the NLS as a model for freak-waves. Physica Scripta T **82**, 48–52 (1999).
14. Dysthe, K.B., Trulsen, K., Krogstad, H.E., Socquet-Juglard, H.: Evolution of a narrow-band spectrum of random surface gravity waves. J. Fluid. Mech. **478**, 1–10 (2003).
15. Henderson, K.L., Peregrine, D.H., Dold, J.W.: Unsteady water wave modulations: fully nonlinear solutions and comparison with the nonlinear Schrodinger equation. Wave Motion **29**, 341–361 (1999).
16. Janssen, P.A.E.M.: Nonlinear four-wave interactions and freak waves. J. Phys. Oceanogr. **33**, 863–884 (2003).
17. Johnson, R.S.: A modern introduction to the mathematical theory of water waves. Cambridge Univ. Press (1997).
18. Kartashova, E., Raab, C., Feurer, Ch., Mayrhofer, G., Schreiner, W.: Symbolic Computation for Nonlinear Wave Resonances. In: Pelinovsky, E., Kharif, C. (eds.) Extreme Waves, pp. 95–126. Springer (2008).
19. Kharif, C., Pelinovsky, E.: Physical mechanisms of the rogue wave phenomenon. Eur. J. Mech. B / Fluids **22**, 603–634 (2003).
20. Kharif, C., Pelinovsky, E., Slunyaev, A.: Rogue Waves in the Ocean. Springer-Verlag, Berlin Heidelberg (2009).
21. Mori, N., Onorato, M., Janssen, P.A.E.M. , Osborne, A.R., Serio, M. : On the extreme statistics of long-crested deep water waves: Theory and experiments. J. Geophys. Res. **112**, C09011 (2007).
22. Onorato, M., Osborne, A.R., Serio, M., Bertone, S.: Freak waves in random oceanic sea states. Phys. Rev. Lett. **86**, 5831–5834 (2001).
23. Onorato, M., Osborne, A.R., Serio, M.: Extreme wave events in directional, random oceanic sea states. Phys. Fluids **14**, L25–L28 (2002).
24. Onorato, M., Osborne, A.R., Serio, M., Cavaleri, L.: Modulational instability and non-Gaussian statistics in experimental random water-wave trains. Phys. Fluids **17**, 078101 (2005).
25. Onorato, M., Waseda, T., Toffoli, A., Cavaleri, L., Gramstad, O., Janssen, P.A., Kinoshita, T., Monbaliu, J., Mori, N., Osborne, A.R., Serio, M., Stansberg, C.T., Tamura, H., Trulsen, K.M.: Statistical properties of directional ocean waves: the role of the modulational instability in the formation of extreme events. Phys. Rev. Lett. **102**, 114502 (2009).
26. Osborne, A.R., Onorato, M., Serio, M.: The nonlinear dynamics of rogue waves and holes in deep water gravity wave trains. Phys. Lett. A **275**, 386–393 (2000).
27. Shemer, L., Sergeeva, A.: An experimental study of spatial evolution of statistical parameters in a unidirectional narrow-banded random wavefield. J. Geophys. Res. **114**, C01015 (2009).
28. Shemer, L., Sergeeva, A., Slunyaev, A.: Applicability of envelope model equations for simulation of narrow-spectrum unidirectional random field evolution: experimental validation. Phys. Fluids **22**, 016601 (2010).
29. Shemer, L., Sergeeva, A., Liberzon, D.: Effect of the initial spectral shape on spatial evolution of the statistics of unidirectional nonlinear random waves. J. Geophys. Res. **115**, C12039 (2010).
30. Slunyaev, A.V.: Numerical simulation of "limiting" envelope solitons of gravity waves on deep water. JETP **109**, 676–686 (2009).

31. Slunyaev, A.V.: Freak wave events and the wave phase coherence. Europ. Phys. J. Special Topics **185**, 67–80 (2010).
32. Socquet-Juglard, H., Dysthe, K.B., Trulsen, K., Krogstad, H.E., Liu, J.: Probability distributions of surface gravity waves during spectral changes. J. Fluid Mech. **542**, 195–216 (2005).
33. Stiassnie, M., Shemer, L.: On the interaction of four water-waves. Wave Motion **41**, 307–328 (2005).
34. Trulsen, K., Kliakhandler, I., Dysthe, K.B., Velarde, M.G.: On weakly nonlinear modulation of waves on deep water. Phys. Fluids **12**, 2432–2437 (2000).
35. Zakharov, V.E., Dyachenko, A.I., Vasilyev, O.A.: New method for numerical simulation of a nonstationary potential flow of incompressible fluid with a free surface. Eur. J. Mech. B/Fluids **21**, 283–291 (2002).
36. Zakharov, V.E., Ostrovsky, L.A.: Modulation instability: the beginning. Physica D **238**, 540–548 (2009).

Quantum Mechanical Treatment of the Lamb Shift Without Taken into Account the Electric Charge

Voicu Dolocan, Andrei Dolocan, and Voicu Octavian Dolocan

1 Introduction

The Lamb shift [7] is a small difference in energy between two energy levels $^2S_{1/2}$ and $^2P_{1/2}$ of the hydrogen atom by an amount now known to be $E/h = 1{,}057.864\,\mathrm{MHz}$. This result is in contradiction with the Dirac and Schrödinger theory which shown that the states with the same n and j quantum numbers but different l quantum numbers ought to be degenerate. The effect is explained by the theory of quantum electrodynamics [1,5,9], in which the electromagnetic interaction itself is quantized. It is assumed that the ground state of the electromagnetic field is not zero, but rather the field undergoes "vacuum fluctuations" that interact with the electron. The contributions to this effect come from the vacuum polarization, electron mass renormalization, and anomalous magnetic moment. Often the "vacuum" is a "refuge" for speculations in science.

By using a Hamiltonian of interaction between fermions based on the coupling through flux lines [3], we have found an equivalent expression for the Coulomb energy of interaction [4], on the form $\alpha\hbar c/R$, where α is the fine structure constant.

In the interaction picture, the effective Hamiltonian is given by [3,4]

$$\begin{aligned}
H_I^{\mathrm{eff}} &= H_{I1}^{\mathrm{eff}} + H_{I2}^{\mathrm{eff}} \\
H_{I1}^{\mathrm{eff}} &= \hbar \sum_{\mathbf{q},\mathbf{q_o},\mathbf{k}} |\mathbf{g}_{\mathbf{q_o}}|^2 \frac{\omega_{\mathbf{q}}}{(\varepsilon_{\mathbf{k}} - \varepsilon_{\mathbf{k}-\mathbf{q}})^2 - \omega_{\mathbf{q}}^2} \times \left(a_{\mathbf{q}} a^+_{\mathbf{q_o}} a_{\mathbf{q}'_o} a^+_{\mathbf{q}'} + a^+_{\mathbf{q}} a_{\mathbf{q_o}} a^+_{\mathbf{q}'_o} a_{\mathbf{q}'} \right) \\
&\quad c^+_{\mathbf{k}-\mathbf{q},\sigma} c^+_{\mathbf{k}',\sigma'} c_{\mathbf{k},\sigma} c_{\mathbf{k}'-\mathbf{q},\sigma'} \\
H_{I2}^{\mathrm{eff}} &= 2\hbar \sum_{\mathbf{q},\mathbf{k}} 2\, |\mathbf{g}_{\mathbf{q}}|^2 \frac{1}{(\varepsilon_{\mathbf{k}} - \varepsilon_{\mathbf{k}-\mathbf{q}}) - \omega_q} a^+_{\mathbf{q}} a_{\mathbf{q}} c^+_{\mathbf{k}-\mathbf{q},\sigma} c_{\mathbf{k}-\mathbf{q},\sigma}
\end{aligned} \tag{1}$$

V. Dolocan (✉)
Faculty of Physics, University of Bucharest, Bucharest, Romania
e-mail: dolocan_voicu@yahoo.com

J.A.T. Machado et al., *Nonlinear and Complex Dynamics: Applications in Physical, Biological, and Financial Systems*, DOI 10.1007/978-1-4614-0231-2_13,

The expectation value of the energy of H_{I1}^{eff} (1) is

$$E_{\text{int}} = \hbar \sum_{\mathbf{q},\mathbf{q}_o,\mathbf{k}} 2\left|g_{\mathbf{q}_o}\right|^2 \frac{\omega_{\mathbf{q}}}{\left(\varepsilon_{\mathbf{k}} - \varepsilon_{\mathbf{k}-\mathbf{q}}\right)^2 - \omega_{\mathbf{q}}^2} \left(n_{\mathbf{q}} + 1\right)\left(n_{\mathbf{q_o}} + 1\right) \times n_{\mathbf{k}-\mathbf{q},\sigma} n_{\mathbf{k},\sigma'} \qquad (2)$$

and the expectation value of the energy of the Hamiltonian H_{I2} (1) is

$$E_{I2} = 4\hbar \sum_{\mathbf{q},\mathbf{q_o},v'} \left|g_{\mathbf{q}_o}\right|^2 \frac{1}{(\varepsilon_{v'} - \varepsilon_v) - \omega_{\mathbf{q}}} (n_{\mathbf{q}} + 1) n_{\mathbf{k}-\mathbf{q}} \qquad (3)$$

where

$$g_{\mathbf{q_o}} = \frac{\hbar D}{8N^2 mR\left(\rho_o + \frac{DR}{c^2}\right)} \frac{\mathbf{q}\mathbf{q}'}{\omega_{\mathbf{q}}\omega_{\mathbf{q_o}}} \sum_n \mathrm{e}^{\mathrm{i}\mathbf{q_o}\mathbf{z_n}} \qquad (4)$$

D is a coupling constant, $\mathbf{q}, \mathbf{q}'$ are the wave vectors associated with the bosons of the connecting field, $\mathbf{q}_o$ is the wave vector associated with the oscillations of the electron, and $\mathbf{k}, \mathbf{k}'$ are the wave vectors of the electrons. $\omega_{\mathbf{q}}, \omega_{\mathbf{q_o}}$ are the classical oscillation frequencies, $a_{\mathbf{q}}^+$ and $a_{\mathbf{q}}$ are the boson creation and annihilation operators, respectively, $c_{\mathbf{k}\sigma}^+$ and $c_{\mathbf{k}\sigma}$ are the creation and annihilation operators for electrons, $\boldsymbol{k}$ is the wave vector of an electron, $n_{\mathbf{q}}$ is the occupation number for bosons and $n_{\mathbf{k}}$ is the occupation number for fermions. We assume $n_{\mathbf{q}},\ n_{\mathbf{q_o}} = 0,\ n_{\mathbf{k}},\ n_{\mathbf{k}-\mathbf{q}} = 1$.

If instead of the Fröhlich fraction

$$\frac{\omega_{\mathbf{q}}}{\left(\varepsilon_{\mathbf{k}} - \varepsilon_{\mathbf{k}-\mathbf{q}}\right)^2 - \omega_{\mathbf{q}}^2} \qquad (5a)$$

we use the fraction

$$\frac{-1}{\left(\varepsilon_{\mathbf{k}} - \varepsilon_{\mathbf{k}-\mathbf{q}}\right) - \omega_{\mathbf{q}}} \qquad (5b)$$

then for $\rho_o = 0$ (mass less interacting field), (2) becomes

$$E_I = \frac{\hbar^3 c^4}{32m^2 R^4} \sum_{\mathbf{q},\mathbf{q_o},\mathbf{k}} \frac{(\mathbf{q}.\mathbf{q_o})^2}{\omega_{\mathbf{q}}^2 \omega_{\mathbf{qo}}^2} \left|\sum_n \mathrm{e}^{\mathrm{i}\mathbf{q_o}.\mathbf{R_n}}\right|^2 \frac{-1}{\varepsilon_{\mathbf{k}} - \varepsilon_{\mathbf{k}-\mathbf{q}} - \omega_{\mathbf{q}}} \qquad (2a)$$

Now we apply this equation to a system of two electrons at $\mathbf{R}_1$ and $\mathbf{R}_2$ acting on the vacuum of the less mass boson field. In this case $\sum_n |\mathrm{e}^{\mathrm{i}\mathbf{q_o}.\mathbf{z_n}}|^2 = 2(1 + \cos(\mathbf{q_o}.\mathbf{R}))$, $\varepsilon_{\mathbf{k}} = \hbar k^2/2m$, and

$$\omega_{\mathbf{q}} = cq, \omega_{\mathbf{q_o}} = \hbar q_o^2/2m,$$

$\mathbf{R} = \mathbf{R}_1 - \mathbf{R}_2$.

Further, if we assume $(\varepsilon_{\mathbf{k}} - \varepsilon_{\mathbf{k-q}}) \ll \omega_{\mathbf{q}}$, we write

$$\sum_{\mathbf{q}} \frac{(\mathbf{q}.\mathbf{q_o})^2}{\omega_{\mathbf{q}}^3 \omega_{\mathbf{q_o}}^2} = \left(\frac{2m}{\hbar}\right)^2 \frac{1}{q_o^2 c^3} \frac{\Omega}{(2\pi)^2} \int_0^{\pi} \cos^2\alpha \sin\alpha d\alpha \int_0^{q_o} q dq = \left(\frac{2m}{\hbar}\right)^2 \frac{R^3}{9\pi c^3}$$

and

$$2\sum_{\mathbf{q_o}} [1 + \cos(\mathbf{q}_o.\mathbf{R})] = 2\sum_{\mathbf{q_o}} 1 + 2\sum_{\mathbf{q_o}} \cos(\mathbf{q_o}.\mathbf{R})$$

$$= 2 + 2\frac{\Omega}{(2\pi)^2} \int_0^{0.76\pi/R} dq_o q_o^2 \int_0^{\pi} d\theta \cos(q_o R\cos\theta) \sin\theta = 3.028$$

We have considered $N = 1$, $\sum_{\mathbf{k}} 1 = 1$ and $\Omega = 4\pi R^3/3$. The upper limit of the integrals over q_o appears from the requirement

$$\frac{4\pi R^3/3}{(2\pi)^3} \times 4\pi \frac{q_m^3}{3} = 1$$

The interaction energy (2a) becomes

$$E_I \approx 2 \times 7.24 \times 10^{-3} \frac{\hbar c}{R} \approx 2 \times \frac{1}{137} \frac{\hbar c}{R} = 2\frac{e^2}{4\pi\varepsilon_o R}$$

The factor 2 appears because we have considered the two nearest neighbours of an electron. The interaction energy between two electrons is

$$E_I = \alpha \frac{\hbar c}{R} \tag{6}$$

where $\alpha = 1/137$ is the fine structure constant. This is an equivalent expression for the Coulomb's law. Expression (6) is obtained from (2a), containing the fraction (5b), which is valid for the interaction between the like charges acting in a mass less boson field. In this case, the flux lines of the two particles do not interfere, the two particles absorb bosons from the ambient space, and move apart from one another. In the case where the two particles have opposite charges, then in a mass less boson field, the interaction energy is given by (2a) where the fraction (5b) is substituted by fraction (5a) and the interaction energy (6) becomes negative (in front off the term from the right hand side is a negative sign). The explanation for this is as follows. Fröhlich obtained the fraction (5a) by dividing fraction (5b) into two parts

$$\frac{1}{2} \frac{1}{\varepsilon_{\mathbf{k}} - \varepsilon_{\mathbf{k-q}} - \omega_{\mathbf{q}}} - \frac{1}{2} \frac{1}{\varepsilon_{\mathbf{k'}} + \omega_{\mathbf{q}} - \varepsilon_{\mathbf{k'+q}}}$$

one part for absorption and the other for emission of a boson. When the two charges have an opposite sign, there is a continuity of the field lines from one particle to the other, one particle absorbs and the other emits a boson, so that an attraction is assured. Also, an attraction may be assured between the like charges, when the connecting field is a massive field; likewise, in this case the interaction energy is given by (2).

When a particle has a charge Q, that is Q/e electronic charges, the term from the right hand side of (6) is multiplied by $Q_1 Q_2/e^2$, because in this case we must define $|\Psi|^2 = Q/e$ (the number of electronic charges per particle).

2 Lamb Shift in the Three-Dimensional Space

From (3) and (4) one obtains

$$E_{I2} = \frac{\hbar^3 D^2}{16m^2R^2\left(\rho_o + \frac{DR}{c^2}\right)^2} \frac{4\pi V}{(2\pi)^3} \int\limits_{mc/\hbar}^{\infty} q_o^2 \mathrm{d}q_o \frac{4\pi V}{(2\pi)^3} \times \int\limits_0^{q_o} q^2 \mathrm{d}q \frac{q^2 q_o^2}{\omega_q^2 \omega_{q_o}^2} \sum_{\nu'} \frac{1}{\varepsilon_{\nu'} - \varepsilon_\nu - \omega_{\mathbf{q}}} \tag{7a}$$

We recognize that there is also a shift for free states. For free electrons (7a) becomes

$$E_{I2f} = \frac{\hbar^3 D^2}{8m^2R^2\left(\rho_o + \frac{DR}{c^2}\right)^2} \frac{4\pi V}{(2\pi)^3} \int\limits_{mc/\hbar}^{\infty} q_o^2 \mathrm{d}q_o \frac{4\pi V}{(2\pi)^3} \times \int\limits_0^{q_o} q^2 \mathrm{d}q \frac{q^2 q_o^2}{\omega_q^2 \omega_{q_o}^2} \frac{-1}{\omega_{\mathbf{q}}} \tag{7b}$$

One gets the physical energy shift by subtracting expression (7b) from (7a). The expression renormalized in this way is

$$\begin{aligned}
\delta E = E_{I2} - E_{I2f} &= \frac{\hbar^3 D^2 V^2}{64\pi^4 m^2 R^2 \left(\rho_o + \frac{DR}{c^2}\right)^2} \int\limits_{mc/\infty}^{\infty} q_o^2 \mathrm{d}q_o \\
&\quad \times \int\limits_0^{q_o} q^2 \mathrm{d}q \frac{q^2 q_o^2}{\omega_{\mathbf{q}}^2 \omega_{\mathbf{q_o}}^2} \sum_{\nu'} \left(\frac{1}{\varepsilon_{\nu'} - \varepsilon_\nu - \omega_{\mathbf{q}}} + \frac{1}{\omega_{\mathbf{q}}} \right) \\
&= \frac{\hbar^3 D^2 \left(4\pi R^3/3\right)^2}{64\pi^4 m^2 R^2 \left(\rho_o + \frac{DR}{c^2}\right)^2} \int\limits_{mc/\hbar}^{\infty} q_o^2 \mathrm{d}q_o \int\limits_0^{q_o} q^2 \mathrm{d}q \\
&\quad \times \frac{q^2 q_o^2}{\omega_{\mathbf{q}}^3 \omega_{\mathbf{q_o}}^2} \sum_{\nu'} \frac{\varepsilon_{\nu'} - \varepsilon_\nu}{\varepsilon_{\nu'} - \varepsilon_\nu - \omega_{\mathbf{q}}}
\end{aligned} \tag{8}$$

Now we assume $\rho_o = 0$ $\omega_{\mathbf{q}} = cq$, $\omega_{\mathbf{q_o}} = \hbar q_o^2/2m_{\mathbf{q}} = \mathrm{cq}$, and $\varepsilon_{v'} - \varepsilon_v \ll \omega_{\mathbf{q}}$. Therefore

$$\delta E = -\frac{\hbar R^2}{9\pi^2} \int\limits_{mc/\hbar}^{\infty} \mathrm{d}q_o \int\limits_{0}^{q_o} \mathrm{d}q \sum_{v'} (\varepsilon_{v'} - \varepsilon_v) \tag{9}$$

The expectation values of the energy $\hbar\varepsilon_v = E_v$ are determined from the Schrödinger equations

$$H_o \Psi_v = \left(-\hbar^2 \frac{\nabla^2}{2m} - \frac{Z\alpha\hbar c}{r}\right)\Psi_v = E_v \Psi_v$$

$$H\Psi_{v'} = \left[-\hbar^2 \frac{\nabla^2}{2m} - Z\alpha\hbar c\left(\frac{1}{r} - \frac{1}{2}(\delta r)^2 \nabla^2 \left(\frac{1}{r}\right)\right)\right]\Psi_{v'} = E_{v'}\Psi_{v'} \tag{10}$$

Ψ_v is the nonrelativistic wave function in the hydrogen like atom. $(\delta r)^2 = s^2$, where s^2 is the mean square value of s_l [3, 4]

$$s^2 = \frac{1}{R^2} \frac{\hbar^2}{4m^2 \omega_{\mathbf{q}_o}^2} (2n_{\mathbf{q}_o} + 1) \tag{11}$$

Even in the lowest state ($n_{qo} = 0$) the oscillator has a finite amplitude with a finite probability. In this case, $s^2 = \left(1/R^2 q_o^4\right)$. By using that

$$\nabla^2 \left(\frac{1}{r}\right) = -4\pi\delta(\mathbf{r})$$

and by using (6) may be written

$$\sum_{v'} (\varepsilon_{v'} - \varepsilon_v) = -2 \times \frac{1}{2} Z\alpha c \frac{1}{R^2 q_o^4} \int \Psi_{vl}^*(\mathbf{r}) \times \nabla^2 \left(\frac{1}{r}\right) \Psi_{vl}(\mathbf{r}) \mathrm{d}\mathbf{r}$$

$$= -\frac{4\pi Z\alpha c}{R^2 q_o^4} |\Psi_{vl}(0)|^2 \delta_{l0} = -\frac{Z\alpha c}{2a_o^5 q_o^4} \delta_{l0} \tag{12}$$

A factor 2 appears because of the two values of the electron spin. For p orbitals, the nonrelativistic wave function vanishes at the origin, so there is no energy shift. But for s orbitals there is some finite value at the origin

$$\Psi_{2s}(0) = \left(\frac{1}{8\pi a_o^3}\right)^{1/2}$$

where we have denoted $R \equiv a_o$, the Bohr radius. By substituting (12) in (9), we have

$$\delta E_{2o} = \frac{Z\alpha\,\hbar c}{18\pi^2 a_o^3} \int_{mc/\hbar}^{\infty} \frac{\mathrm{d}q_o}{q_o^4} \int_0^{q_o} \mathrm{d}q = \frac{Z\alpha\,\hbar^3}{36\pi^2 m^2 c a_o^3} \tag{13}$$

For hydrogen atom, $Z = 1$, this shift is about 6.42×10^{-25} J, which correspond to a frequency of 970 MHz. Further we consider the contribution to the Lamb shift of the interaction terms from the fine-structure Hamiltonian in according to Dirac theory [2]. The first term of interaction is the usual spin-orbit coupling

$$H_2' = \frac{1}{2m^2c^2}\frac{1}{r}\frac{\mathrm{d}V}{\mathrm{d}r}\mathbf{L}.\mathbf{S} = \frac{1}{2m^2c^2}\frac{Z\alpha\hbar c}{r^3} \times \frac{1}{2}\left(\mathbf{J^2} - \mathbf{L^2} - \mathbf{S^2}\right) \tag{14}$$

and the second term of interaction is the Darwin term due to the nonlocalized interaction between the electron and the field

$$H_3' = \frac{\hbar^2}{8m^2c^2}\nabla^2 V = -\frac{\hbar^2}{8m^2c^2}\frac{2Z\alpha\hbar c}{r^3} \tag{15}$$

In the case of the spin-orbit coupling

$$\begin{aligned}
\sum_{v'}(\varepsilon_{v'} - \varepsilon_v) &= 2 \times \frac{1}{2m^2c^2}\frac{Z\alpha\hbar^2 cs^2}{2}\int \Psi^*_{vljm_j}(\mathbf{r}) \times \nabla^2\left(\frac{1}{r^3}\right)\mathbf{LS}\Psi_{vljm_j}\mathrm{d}\mathbf{r} \\
&= \frac{6Z\alpha\hbar^2 s^2}{m^2 c} \times \left[j(j+1) - l(l+1) - \frac{3}{4}\right]\int \Psi^*_{vl}(\mathbf{r})\frac{1}{r^5}\Psi_{vl}(\mathbf{r})\mathrm{d}\mathbf{r} \\
&= \frac{6Z\alpha\hbar^2 s^2}{m^2 c}\left[j(j+1) - l(l+1) - \frac{3}{4}\right]\left\langle\frac{1}{r^5}\right\rangle
\end{aligned} \tag{16}$$

By substituting (16) in (9), one obtains

$$\delta E^{21}_{\mathbf{LS}} = -\frac{Z\hbar^3\alpha}{3\pi^2 m^2 c}\frac{\hbar^2}{m^2c^2}\left\langle\frac{1}{r^5}\right\rangle_{21} \times \begin{cases} 1 \text{ for j} = 1 + 1/2 \\ -2 \text{ for j} = 1 - 1/2 \end{cases} \tag{17}$$

By using the radial wave function of the electron in the hydrogen atom

$$R_{21}(r) = \frac{1}{\sqrt{3}}\left(\frac{Z}{2a_o}\right)^{3/2}\left(\frac{Zr}{a_o}\right)\exp(-Zr/a_o)$$

and

$$\left\langle\frac{1}{r^5}\right\rangle_{21} = \int_{\lambda_C}^{\infty} |R_{21}|^2 \frac{\mathrm{d}r}{r^3}$$

one obtains

$$\left\langle \frac{1}{r^5} \right\rangle_{21} = \frac{0.15}{a_o^5}$$

and $\delta E_{\mathbf{LS}}^{21} = 1.2 \times 10^{-28}\,\text{J}$, which correspond to a frequency of $\sim 0.18\,\text{MHz}$. In the above expression, $\lambda_C = h/mc$ is the Compton wave length. The spin-orbit coupling contribution is zero for s electrons. Next we consider the Darwin term (15). We write

$$\begin{aligned}\hbar \sum_{\nu'} (\varepsilon_{\nu'} - \varepsilon_\nu) &= -2 \times \frac{\hbar^2}{8m^2c^2}\frac{1}{2}\alpha s^2 \hbar c \times \int \Psi_{\nu l}^*(\mathbf{r}) \nabla^4 \left(\frac{1}{r}\right) \Psi_{\nu l}(\mathbf{r}) \\ &= \frac{3\hbar^3\alpha}{m^2 c}\frac{1}{R^2 q_o^4}\left\langle \frac{1}{r^5} \right\rangle \end{aligned} \tag{18}$$

and

$$\delta E_{21}^{\text{Darwin}} = \frac{\hbar^3 \alpha}{3\pi^2 m^2 c}\left\langle \frac{1}{r^5} \right\rangle_{21} \int\limits_{mc/\hbar}^{\infty} \frac{\mathrm{d}q_o}{q_o^5} \int\limits_0^{q_o} \mathrm{d}q = \frac{\hbar^3\alpha}{6\pi^2 m^2 c}\frac{0.15}{a_o^5}\frac{\hbar^2}{m^2c^2} = 2.9 \times 10^{-29}\text{J} \tag{19}$$

The Darwin contribution to the 2s level is

$$\delta E_{20}^{\text{Darwin}} = \frac{\hbar^3 \alpha}{3\pi^2 m^2 c}\frac{\hbar^2}{m^2c^2}\left\langle \frac{1}{r^5} \right\rangle_{20} \tag{20}$$

By using the radial wave function of the electron in the hydrogen atom

$$R_{20} = 2\left(\frac{1}{2a_o}\right)^{3/2}\left(1 - \frac{r}{2a_o}\right)\exp(-r/2a_o)$$

one obtains

$$\left\langle \frac{1}{r^5} \right\rangle_{20} = \int\limits_{h/mc}^{\infty} |R_{20}|^2 \frac{dr}{r^3} = \frac{101.392}{a_o^5}$$

and $\delta E_{20}^{\text{Darwin}} = 2 \times 10^{-26}\,\text{J}$. In the last equations, we have considered $Z = 1$. The contribution of the Darwin term is of the order of 60.2 MHz. By adding the Darwin contribution to the contribution (13), one obtains $\nu = 1{,}030.2\,\text{MHz}$, which is close to the experimental value of the Lamb shift.

3 The Fine Structure Constant in the Two-Dimensional Space

In the two-dimensional space may be written

$$\left|\sum_{l} e^{i\mathbf{q_o}.\mathbf{R}_l}\right|^2 \delta_{R_{12},R_o} = 2\left[1+\frac{1}{2\pi}\int_0^{2\pi} e^{iq_oR_o\cos\theta}d\theta\right] = 2\left[1+J_o(q_oR_o)\right] \tag{21}$$

where $J_o(x)$ is the Bessel function of the first kind, $\mathbf{R}_{12} = \mathbf{R}_1 - \mathbf{R}_2$. Further,

$$\sum_{\mathbf{q}} \frac{(\mathbf{q}.\mathbf{q_o})^2}{\omega_{\mathbf{q}}^2\omega_{\mathbf{q}_o}^2} = \left(\frac{2\mu}{\hbar}\right)^2 \sum_q \frac{q^2q_o^2\cos^2\alpha}{q_o^4q^3c^3} \left(\frac{2\mu}{\hbar}\right)^2 \times \sum_q \frac{\cos^2\alpha}{q_o^2qc^3}$$

$$= \frac{S}{(2\pi)^2}\left(\frac{2\mu}{\hbar}\right)^2 \frac{1}{c^3q_o^2}\int_0^{2\pi}\cos^2\alpha d\alpha \int_0^{q_o} dq = \frac{\pi R^2}{(2\pi)^2}\left(\frac{2\mu}{\hbar}\right)^2 \frac{\pi}{c^3q_o}$$

$$= \left(\frac{2\mu}{\hbar}\right)^2 \frac{R^2}{4c^3q_o} \tag{22a}$$

Next,

$$\sum_{\mathbf{q},\mathbf{q}_o} \frac{(\mathbf{q}.\mathbf{q}_o)^2}{\omega_{\mathbf{q}}^3\omega_{\mathbf{q}_o}^2}[1+J_o(q_oR_o] = \left(\frac{2\mu}{\hbar}\right)^2 \frac{R^2}{4c^3}\frac{S}{(2\pi)^2} \times 2\pi \int_0^{2/R} \frac{1+J_o(q_oR_o)}{q_o} q_o dq_o$$

$$= \left(\frac{2\mu}{\hbar}\right)^2 \frac{R^2}{4c^3}\frac{\pi R^2}{2\pi}\frac{1}{R}\left[2+\int_0^2 J_o(q_oR_o)d(q_oR_o)\right]$$

$$= \left(\frac{2\mu}{\hbar}\right)^2 \frac{R^3}{8c^3}[2+1.426] \tag{22b}$$

The interaction energy in the two-dimensional space becomes

$$E_I^{2D} = -\frac{\hbar^3c^4}{16\mu^2R^4}\frac{4\mu^2}{\hbar^2}\frac{R^3}{8c^3} \times 3.426 = -2\times 0.053\frac{\hbar c}{R} = -2\times\frac{\alpha_{2D}\hbar c}{R} \tag{23}$$

where the fine structure constant in the two-dimensional space is

$$\alpha_{2D} = 0.053 = 7.26\alpha \tag{23a}$$

where α is the fine structure constant in the three-dimensional space. It appears that in the two-dimensional space the Coulomb interaction is approximately 7 times stronger than in the three-dimensional space. Here $1/\mu = 1/M_p + 1/m$, where μ

is the reduced mass of the proton and electron in the hydrogen atom. We consider $\mu \approx m$, the electron mass. The electron radius in the hydrogen atom in the two-dimensional space is

$$a_n^{2D} = \frac{\hbar}{\alpha_{2D}\mu c}\left(n - \frac{1}{2}\right)^2 \tag{24}$$

where n is the principal quantum number. The Bohr radius a_1^{2D} is 29 times larger than the Bohr radius in the three-dimensional space. The increasing of binding energy of the electron in the hydrogen atom is

$$\frac{E_n^{2D}}{E_n^{3D}} = \frac{\alpha_{2D}^2}{\alpha^2}\frac{n^2}{\left(n - \frac{1}{2}\right)^2} \tag{25}$$

For the ground state, the binding energy in the two-dimensional space is 200 times larger than that in the three-dimensional space. It results that if should be a two-dimensional space, in this space the matter should be more condensate than in the three-dimensional space. We specify that the two-dimensional hydrogen atom was studied in the past by many authors [6, 8, 10, 11].

4 Lamb Shift in the Two-Dimensional Space

In the two-dimensional space the expression (8) for the Lamb shift becomes

$$\delta E = \frac{\hbar^3 D^2}{16m^2 R^2\left(\rho_o + \frac{DR}{c^2}\right)^2}\frac{S^2}{(2\pi)^2}\int\limits_{mc/\hbar}^{\infty} q_o \mathrm{d}q_o \times \int\limits_{0}^{q_o} q\mathrm{d}q \frac{q^2 q_o^2}{\omega_{\mathbf{q}}^2 \omega_{\mathbf{q}_o}^2}\sum_{v'}\frac{\varepsilon_{v'} - \varepsilon_v}{(\varepsilon_{v'} - \varepsilon_v) - \omega_{\mathbf{q}}} \tag{26}$$

No, we assume $\rho_o = 0,\ \omega_{\mathbf{q}} = cq,\ \omega_{\mathbf{q}_o} = \hbar q_o^2/2m,\ \varepsilon_{v'} - \varepsilon_v \ll \omega_{\mathbf{q}}$, so that

$$\delta E = -\frac{\hbar}{16}\int\limits_{mc/\hbar}^{\infty}\frac{\mathrm{d}q_o}{q_o}\int\limits_{mc/\hbar}^{q_o}\frac{\mathrm{d}q}{q}\sum_{v'}(\varepsilon_{v'} - \varepsilon_v) \tag{26a}$$

We write

$$\begin{aligned}\sum_{v'}(\varepsilon_{v'} - \varepsilon_v) &= 2 \times \frac{1}{2} Z\alpha_{2D} c \frac{1}{R^2 q_o^4} \times \int \Psi_{vl}^*(\mathbf{r})\nabla^2\left(\frac{1}{r}\right)\Psi_{vl}(\mathbf{r})\,\mathrm{d}\mathbf{r} \\ &= \frac{2Z\alpha_{2D} c}{R^2 q_o^4}\left\langle\frac{1}{r^3}\right\rangle_{vl}\end{aligned}$$

For the 3D case, since $\nabla^2 V(\mathbf{r}) = -4\pi\delta(\mathbf{r})$, δE is nonzero only for the s electron. However, this is no longer true for the 2D hydrogen atom, this term is nonzero for all the electrons. By using that

$$\nabla^2\left(\frac{1}{r}\right) = \frac{2}{r^3}; \quad \Psi_{2o}(\mathbf{r}) = \frac{\beta_2}{\sqrt{3}}\mathrm{e}^{-\beta_2 r/2}(1-\beta_2 r) \times \frac{1}{\sqrt{\pi}}\mathrm{e}^{\mathrm{i}|m|\phi} = R_{2o}(\mathbf{r})\frac{\mathrm{e}^{\mathrm{i}|m|\phi}}{\sqrt{2\pi}}$$
$$\Psi_{21}(\mathbf{r}) = \frac{\beta_2^2 r}{\sqrt{6}}\mathrm{e}^{-\beta_2 r/2}\frac{\mathrm{e}^{\mathrm{i}|m\phi|}}{\sqrt{2\pi}} = R_{21}(\mathbf{r})\frac{\mathrm{e}^{\mathrm{i}|m|\phi}}{\sqrt{2\pi}};$$
$$\beta_2 = \frac{4}{3}\frac{mc}{\hbar}\alpha_{2D} \tag{27}$$

For 2s level, we have

$$\left\langle \frac{1}{r^3} \right\rangle_{20} = \frac{\beta_2^2}{3} \times 0.091\beta_2 = 0.03\beta_2^3$$

and therefore

$$\delta E_{2s}^{2D} = \frac{\hbar}{16}\frac{2Z\alpha_{2D}c}{\left(a_2^{2D}\right)^2} \times 0.03\beta_2^2 \int_{mc/\hbar}^{\infty}\frac{\mathrm{d}q}{q}\int_{mc/\hbar}^{q}\frac{\mathrm{d}q_o}{q_o^5} = \frac{3.75\times 10^{-3}}{4}\frac{\alpha_{2D}\hbar c}{\left(a_2^{2D}\right)^2}\beta_2^3$$
$$\times\left\{\int_{mc/\hbar}^{\infty}\frac{\mathrm{d}q}{q^5} - \left(\frac{\hbar}{mc}\right)^4\int_{mc/\hbar}^{2/R}\frac{\mathrm{d}q}{q}\right\}$$
$$= 4.38\times 10^{-4}\alpha_{2D}^6 mc^2\left\{\frac{1}{4} - \ln\left(\frac{8}{9}\alpha_{2D}\right)\right\} = 1.98\times 10^{-25}\mathrm{J} \tag{28}$$

which correspond to a frequency of 300 MHz. For 2p level

$$\left\langle \frac{1}{r^5} \right\rangle_{21} = \frac{0.193}{6}\beta_2^5$$

and $\delta E_{2p}^{2D} = 5.28\times 10^{-5}$ J which corresponds to a frequency of ~800 MHz.

For the spin-orbit coupling we have

$$\sum_{v'}(\varepsilon_{v'} - \varepsilon_v) = 2\frac{\hbar}{m^2c^2}\frac{\alpha_{2D}\hbar c}{\left(a_2^{2D}\right)^2} \times [j(j+1) - l(l+1) - 3/4)]\left\langle \frac{1}{r^5} \right\rangle_{21} \tag{29}$$

and

$$\delta E_{\mathbf{LS}}^{2p} = 0.0065\alpha_{2D}^8 mc^2 = 3.31\times 10^{-26}\mathrm{J} \tag{30}$$

which corresponds to a frequency of 50 MHz.

For the Darwin term we have

$$\hbar \sum_{v'} (\varepsilon_{v'} - \varepsilon_v) = \frac{3\hbar^3 \alpha_{2D}}{m^2 c} \frac{1}{R^2 q_o^4} \left\langle \frac{1}{r^5} \right\rangle \tag{31}$$

and

$$\delta E_{2D}^{\text{Darwin}} = \frac{\hbar^5 \alpha_{2D}^3}{27 m^4 c^3} \left\langle \frac{1}{r^5} \right\rangle \tag{32}$$

so that

$$\delta E_{2D,2s}^{\text{Darwin}} = 8.27 \times 10^{-4} \alpha_{2D}^8 mc^2 = 4.22 \times 10^{-27} \text{J}$$

which corresponds to a frequency of 6.3 MHz, and

$$\delta E_{2D,2p}^{\text{Darwin}} = 5 \times 10^{-3} \alpha_{2D}^8 mc^2 = 2.55 \times 10^{-26} \text{J}$$

which corresponds to a frequency of 38 MHz. The total shift in the two-dimensional space is of the order of 500 MHz. This value of the shift is some smaller than that in the three-dimensional space.

5 Conclusions

We have presented a theory of the Lamb shift without taking into account the electron charge. This appears as a natural result of the equivalent expression for the Coulomb's interaction energy, $\alpha\hbar c/R$, just derived from our Hamiltonian of interaction. This gives rise to the questions: may be the electron taken off the charge, or the electron is an indestructible charge? What is this the charge? We specify that in 1947 Hans Bethe was the first to explain the Lamb shift in the hydrogen spectrum and thus laid the foundation for the modern development of quantum electrodynamics. Neither the mass nor the charge of the electron or any other charged particle can actually be calculated in QED–they have to be assumed. In our theory, it is not necessary to assume a priori the charge of the electron. Also, we have found that in the two-dimensional space, the matter is more condensate than in the three-dimensional space. In another paper, we extend these results to the interaction between nucleons via mass less bosons and massive particles [12].

References

1. Bethe H. A. (1947), The electromagnetic shift of energy levels, *Phys. Rev.* 72, 339–341.
2. Bransden B. H. and C. J. Joachain (1983), *The Physics of Atoms and Molecules*, Longman Group Ltd.

3. Dolocan A., V. O. Dolocan and V. Dolocan (2005), A new Hamiltonian of interaction for fermions, *Mod. Phys. Lett.* B19, 669–681.
4. Dolocan V., A. Dolocan and V. O. Dolocan (2010), Quantum Mechanical Treatment of the Electron-Boson Interaction Viewed as a Coupling through Flux Lines. Polarons, *Int. J. Mod. Phys.* B24, 479–495; On the Chemical Bindings in Solids, *Rom. J. Phys.* 55, 153–172.
5. Greiner W. and J. Reinhardt (1994), *Quantum Electrodynamics*, Springer-Verlag, Berlin.
6. Guo S. H., X. L. Yang, F.T. Chen, K. W. Wong and W. W. Y. Ching (1991), Analytic Solution of a Two-Dimensional Hydrogen Atom. I. Nonrelativistic Theory. *Phys. Rev.* A43, 1197–1205.
7. Lamb E., Jr. and R. C. Retherford (1947), Fine Structure of the Hydrogen Atom by a Microwave Method, *Phys. Rev.* 72, 241–243.
8. Taut M. (1995), Two-dimensional Hydrogen in a Magnetic Field, *J. Phys.* A28, 2081–2085.
9. Welton T. A. (1948), Some Observable Effects of the Quantum-Mechanical Fluctuations of the Electromagnetic Field, *Phys. Rev.* 74, 1157–1167.
10. Yang X. L., S. H. Guo, F. T. Chan, K. W. Wong an, W. Y. Ching (1991), Analytic Solution of a two Dimensional Hydrogen Atom. II. Relativistic Theory, *Phys. Rev.* A43, 1186–1196.
11. Zaslow B. and M. E. Zandler (1967), Two-Dimensional analog to the Hydrogen Atom, *Am. J. Phys.* 35, 1118–1119.
12. Dolocan V., V. O. Dolocan and A. Dolocan (to be published), On the interaction between fermions via mass less bosons and massive particles.

Determining the Climate Zones of Turkey by Center-Based Clustering Methods

Fidan M. Fahmi, Elçin Kartal, Cem İyigün, Ceylan Yozgatligil, Vilda Purutcuoğlu, İnci Batmaz, Murat Türkeş, and Gülser Köksal

1 Introduction

The increase in atmospheric concentrations of so-called greenhouse gasses such as carbon dioxide (CO_2) and methane (CH_4) has resulted in greenhouse effect and global warming. It is not surprising to see the effect of these changes also in Turkey. Recent analyses of Turkey's climate data reveal that minimum and maximum temperatures are increasing [5]. On the contrary, observed temporal distribution patterns of winter rainfalls are significantly changing while moderate decreases in the total amount of rainfall have also been recorded. As a result of global warming, alterations in the weather patterns and the existence of extreme events can be considered as important indicators of the change in climate zones. To reduce the future vulnerability, it is important to examine the changes and new climate zones (if any) so that appropriate strategies can be developed and precautions can be taken accordingly.

In determining the climate zones, different methods and variables have been used in the literature. The most popular one is the classification method called Koeppen and Thorntwaite [6]. This method is a quantitative one that directly determines the climate types but the rules used in classification are subjectively extracted. Therefore, relatively objective clustering approaches can be applied instead for the same purpose (e.g. [7]). To exemplify, in the study of Ünal et al.'s, Ward's method is proposed to be used as a hierarchical clustering technique.

On the other hand, in literature, temperature and precipitation are treated as the main research variables in this kind of study because of their influence on the distribution of flora and human activities. In our study, temperature measures

---

F.M. Fahmi (✉)
Department of Statistics, Middle East Technical University, Ankara, Turkey
e-mail: fidantelaferli@yahoo.com

J.A.T. Machado et al., *Nonlinear and Complex Dynamics: Applications in Physical, Biological, and Financial Systems*, DOI 10.1007/978-1-4614-0231-2_14,

Table 1 Variables studied and corresponding number of clusters and sample size

Temperature variables	Number of clusters	Sample size
Average	4	61
Minimum	4	65
Maximum	5	65
Average minimum	4	65
Average maximum	4	65
Minimum average	4	65
Maximum Average	5	65

obtained from the Turkish State Meteorological Service stations in the time period 1950–2006 are examined by using two different center-based clustering techniques, which are k-means and fuzzy k-means.

2 Data

The data used in identifying the climate zones of Turkey are temperature measures including minimum and maximum temperature, average temperature, and average of maximum and minimum temperature, minimum and maximum of average temperature. In this data set, the measurements are monthly records belonging to the period 1950–2006. Although 270 Turkish State Meteorological Service stations provide climate data, only 65 stations are used within the 1950–2006 period. The main reason of using such a small number of stations is that the eliminated stations had missing values for more than two successive years. Mean imputation is applied for the missing values recorded for months within a year. For each temperature variable, the number of stations (i.e. sample size) used are listed in Table 1.

3 The Methodology

Clustering is the partitioning of a data set into subsets in such a way that the data in each subset share some common trait—often proximity—according to a distance measure. We assume that our data $D = \{x_{ij}\}$, where $i = 1,2,\ldots,n, j = 1,2,\ldots,p$, consist of p features measured on n independent units. For a given data set, D, and an integer, K, $2 \leq K \leq N$, the clustering problem is to partition D into K disjoint clusters. In other words, $D = C_1 \cup C_2 \cup \ldots \cup C_K$, with $C_j \cap C_k = \phi$ if $j \neq k$ and $C_j \neq \phi$ for $j \in \{1,\ldots,K\}$. Thus, each cluster is consisting of points that are similar in some sense, and points of different clusters are dissimilar. Here, similarity means to be close in the sense of a distance d(**x**, **y**) between any two points **x**, $\mathbf{y} \in R^p$. Note here that the number of clusters denoted by K should be determined beforehand. However, determining the “right” number of clusters is an important issue to be dealt with in clustering. Also note that different distance measures such as Euclidean distance, Manhattan distance, and Mahalanobis distance can be used to measure the similarity between any two points $\mathbf{x}, \mathbf{y} \in \mathrm{R}^{\mathrm{p}}$.

There are many different clustering methods available in the literature [3]. In this study, well-known and easily applied ones, called center-based clustering techniques, were applied as a first attempt.

3.1 Center-Based Clustering Methods

Center-based clustering algorithms construct clusters by using the distances of data points from the cluster centers. The best-known and most commonly used center-based algorithm is the k-means algorithm [2], which minimizes the objective

$$\sum_{k=1}^{K} \sum_{x_i \in C_k} \|x_i - c_k\|^2, \tag{1}$$

where c_k is the centroid of the kth cluster. The steps of the algorithm are as follows:

Step 0: Initialization: Given the data set D and an integer K, $2 \leq K \leq N$, select K initial centers $\{\mathbf{c}_{\mathrm{k}}\}$.

Step 1: Compute the distances $\mathrm{d}(\mathbf{x}_{\mathrm{i}}, \mathbf{c}_{\mathrm{k}})$, $i = 1,\ldots,N$; $k = 1,\ldots,K$, between the data points and cluster centers, and partition the data set by assigning each data point to a cluster whose center is the nearest.

Step 2: Recompute the cluster centers as the cluster mean points.

Step 3: If the centers have not changed, stop; else go to Step 1.

A major challenge in cluster analysis is to determine the correct or natural number of clusters. It may be possible to provide insight into the number of clusters in the data by using the classical "elbow method." Figure 1 shows a typical graph of evaluation measure (*SSE* for our case) for k-means clustering method applied for minimum temperature versus the number of clusters employed. As one can see, the *SSE* shown below in (2) decreases monotonically as the number of clusters increases.

$$SSE = \sum_{k=1}^{K} \sum_{x_i \in C_k} d^2 (c_i, x), \tag{2}$$

where c_i represents the center of the ith cluster.

We can try to find the natural number of clusters in a data set by looking for the number of clusters at which there is a knee in the plot [4]. In Fig. 1, there is a distinct knee in the SSE for k-means clustering applied for the minimum temperature when the number of cluster is equal to four. For the other variables, the optimal number of clusters listed in Table 1 is determined by using the same approach.

Several variants of k-means algorithm have been reported in the literature [1]. Some of them attempt to select a good initial partition so that the algorithm is more

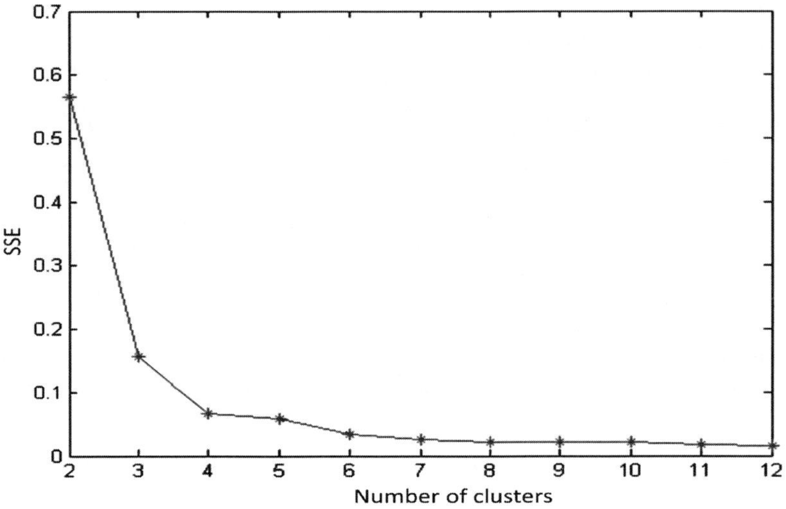

Fig. 1 Plot of SSE versus the number of clusters, K, for the minimum temperature

likely to find the global minimum value. The k-means algorithm can be adapted to soft clustering. A well-known center-based algorithm for soft clustering is the so-called fuzzy k-means algorithm in which the objective function to be minimized is

$$\sum_{i=1}^{N}\sum_{k=1}^{K} u_{\text{ik}}^{m} d_{\text{ik}}^{2} = \sum_{i=1}^{N}\sum_{k=1}^{K} u_{\text{ik}}^{m} \left\| x_i - v_k \right\|^2. \quad (3)$$

Here, u_{ik}^{m} is the membership function of $x_i \in C_k$, and typically satisfy (4); $m > 1$ is a real number known as fuzzifier.

$$\sum_{k=1}^{K} u(x, C_k) = 1, \quad \text{and} \quad u(x, C_k) \geq 0 \quad \text{for all} \quad k = 1, \ldots, K. \quad (4)$$

The same optimal number of clusters which were determined by using elbow approach for the k-means clustering method (in Table 1) is used for fuzzy k-means clustering; therefore, the assignment of the stations to the clusters can be compared for both methods. The clustering results obtained by using both methods are displayed in graphs like in Figs. 2 and 3. By using these graphs, clusters are examined, and the stations that fall into different clusters from their neighbors are determined, and the reasons are discussed. Additionally, all of the graphs obtained for each variable are compared. Clustering methods were also applied to the multivariate data set containing all temperature variables. The optimal number of clusters for this application is found to be five by using the elbow method for the k-means clustering approach and this number of cluster is also used for fuzzy k-means clustering method. As a result, the graphs obtained by using the k-means and the fuzzy k-means clustering methods are displayed in Figs. 2 and 3, respectively.

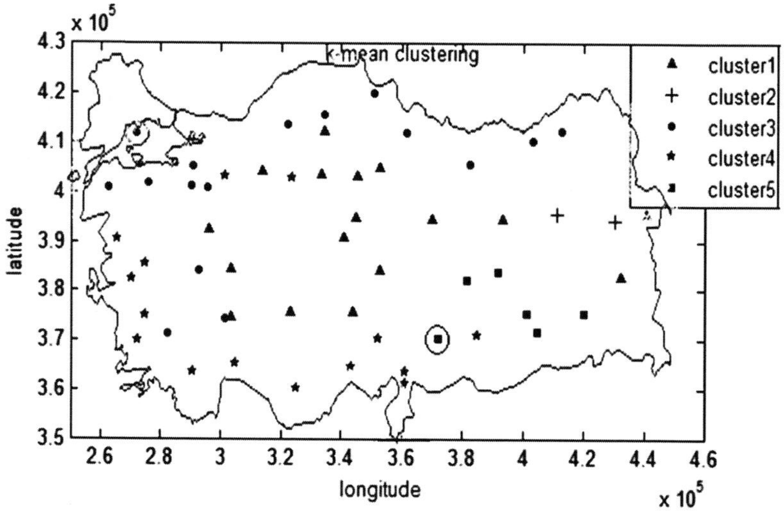

Fig. 2 k-means clustering graph

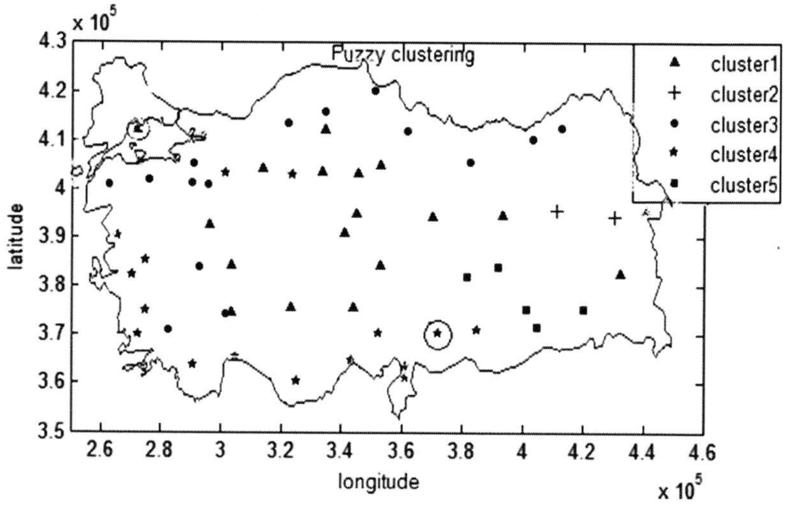

Fig. 3 Fuzzy k-means clustering graph

In both graphs, similar partitioning structures are obtained except two stations indicated by circles. This is not surprising, however, since the membership values of these two stations (i.e. Lüleburgaz and İslahiye) for two clusters are very close to each other.

4 Conclusion and Future Studies

There have been seven recognized climate zones of Turkey for many years. Black sea, Marmara, Aegean, Mediterranean, central Anatolian, eastern Anatolian, and southeastern Anatolian regions. Note here that in addition to climate differences, social and economic variables are also considered in identifying these zones. In this study, however, the climate zones of Turkey are reexamined by using a mathematical methodology of center-based clustering analysis. As a result of applying k-means and fuzzy k-means clustering methods on Turkish data, different clusterings results are obtained for each temperature variable. Moreover, when the multivariate cluster analysis is applied, five clusters are obtained over Turkey, and the partition of regions is found to be similar to the two methods studied, except two stations. As can be seen in Figs. 2 and 3, the Aegean region has the same properties as the Mediterranean and Southeastern Anatolian regions. In addition, in the Marmara region, Black Sea's climate effect become dominant except Trakya station, which is similar to the central Anatolian region. Moreover, two stations have different climate characteristics within the Eastern Anatolian region.

Because the clustering of the temperature variables does not consider the spatial properties of the stations, resulting clusters may not represent the correct partitioning. To increase the efficiency of the method bisecting k-means, which is less susceptible to initialization problems, can be applied. This method is an extension of the k-means algorithm that is based on simple idea: to obtain K clusters, split the set of all points into two clusters, select one of these clusters to split, and so on, until K clusters have been produced [4]. Actually, our main goal is to cluster the stations by considering their spatio and temporal properties. In order to accomplish our aim, as an initial step, bisecting and hierarchical clustering will be applied, and compared with k-means and fuzzy k-means results.

Acknowledgements This study has been supported by Middle East Technical University, Turkey, under the project number BAP-2008-01-09-02. Authors would like to thank all other NINLIL research group members for their support and contribution to this study.

References

1. Anderberg, M.R.: Cluster Analysis for Cluster Applications. Academic Press Inc., New York, NY (1973)
2. Hartigan, J.: Clustering Algorithms. Wiley and Sons Inc., New York, NY (1975)
3. Jain, A.K., Murty, M.M., Flynn, P.J.: Data clustering: a review. ACM Computing Surveys. **31**(3), 264–323 (1999)
4. Tan, P.N., Steinbach, M., Kumar, V.: Introduction to Data Mining. Pearson Education Inc., Boston, MA (2006)
5. Türkeş, M.: Spatial and temporal analysis of annual rainfall variations in Turkey. International Journal of Climatology. **16**, 1057–1076 (1996a)

6. Türkeş, M.: Observed changes in maximum and minimum temperatures in Turkey. International Journal of Climatology. **16**, 463–477 (1996b)
7. Ünal, Y., Kindap, T., Karaca, M.: Redefining the climate zones of Turkey using cluster analysis. International Journal of Climatology. **23**, 1045–1055 (2003)

The Determination of Rainy Clouds Motion Using Optical Flow

O. Raaf and A. Adane

1 Introduction

For several years, rain forecasting in the short time limit has been a topic of meteorologist interest. To account for these phenomena, most meteorological networks are now equipped with automatic measurement systems and remote sensing tools [12]. In these conditions, different geostationary satellites used for meteorological observations, turn around the Earth in the equatorial plane since the last 30 years. A network of rain gauges gives the punctual and direct measures of rain rate on the ground so that it is possible to interpolate in order to reconstitute the complete rainy field. In this case, this operation is not very reliable when rain presents a large spatial variability.

In this case, radars, lidars, and sodars are other remote sensing systems used to collect meteorological data in the large-scale characterizing the air motions and cloud formations in the atmosphere [6, 9]. The radars equipping the meteorological stations are especially designed to detect rainy clouds. The radar equipment represents rain cells with: position, shape, intensity, dimensions, and displacement [9, 13].

Motion estimation from an image sequence is used in several applications. In robotics, it can identify and anticipate changes in the position of objects. In video compression, it allows the fullest possible understanding where the temporal redundancy of the sequence and information describing an image using the surrounding images. In meteorology, it allows the detection and forecasting of rainy clouds including those which are dangerous and predicts their motion.

O. Raaf (✉)
Faculty of Electronics and Computer Science, University of Science and Technology Houari Boumediene (U.S.T.H.B.), PO Box 32, El Alia, Bab Ezzouar 16111, Algiers, Algeria
e-mail: rf_ouarda@yahoo.fr

J.A.T. Machado et al., *Nonlinear and Complex Dynamics: Applications in Physical, Biological, and Financial Systems*, DOI 10.1007/978-1-4614-0231-2_15,

The movement, in a sequence of images is visible through changes in spatial distribution of a photometric variable between two successive images, such as luminance, brightness, or reflectivity. This latter is the variable used in meteorological images. With the use of nonlinear algorithms, signal processing and modern mathematics have been opened to the study of such a random phenomenon.

In this context, we are interested in this chapter for using optical flow in the detection and extraction of the velocity of rainfall from radar images. Specifically, the part "low level" which provides local information on the speed as a field of velocity is treated. Given these objectives, this chapter will be organized as follows: first the database of the images used in this study is presented. Then, a formulation of the optical flow equation used to calculate the velocity fields in artificial images and weather radar images is given. After that, an application of the method on images from the databank is illustrated. Finally, the results of the application are discussed and interpreted.

2 Data Bank

The images to be studied in this chapter were collected with meteorological radar ASWR 81 (Algerian Service Weather Radar) installed in the city of Maghras in Setif (36°11N, 05°25E, altitude 1,730 m) in the north-east of Algeria. This region is mostly known under the name of high plateaus. In Fig. 1, we present an example of a radar image taken in December 26th, 2004 at 17:30, where the radar is in the middle of the picture. In this figure, the fixed echoes reflected from mountains are clearly apparent.

Indeed, at the west of Setif we can discern an orange band due to the presence of regional mountain ranges known as Djurdjura where the highest elevation corresponds to Lalla Khadidja Mountain (2,308 m). At the south and at the north of the radar we can observe a bluish zone corresponding respectively to Bibans and Babor ranges. The resolution of this image is 1km par pixel in PPI (Plan Position Indicator) representation. Its format is 512×512 pixels with 16 reflectivity levels. We can also see in this figure rain echoes represented as a quasi-homogeneous structure, while fixed echoes are scattered in small cells of nonhomogeneous intensity. The use of meteorological radars is delicate because of the earth relief, which produces masque effects that degrades noticeably rain echoes detection and reduces radar performance. It is therefore essential to remove this clutter to be able to follow the evolution of the precipitation echoes [4].

In order to eliminate the ground echoes that constitute a parasitic signal in the forecasting and the quantification of rain cells, a mask method combined with a shape detection is used [11]. To filter the radar images, templates representing the obstacles of the Earth surface are constructed and then employed to hide the ground clutter in these images. An image of templates is therefore obtained by averaging a large number of clear sky images. To improve the filtering method, an algorithm

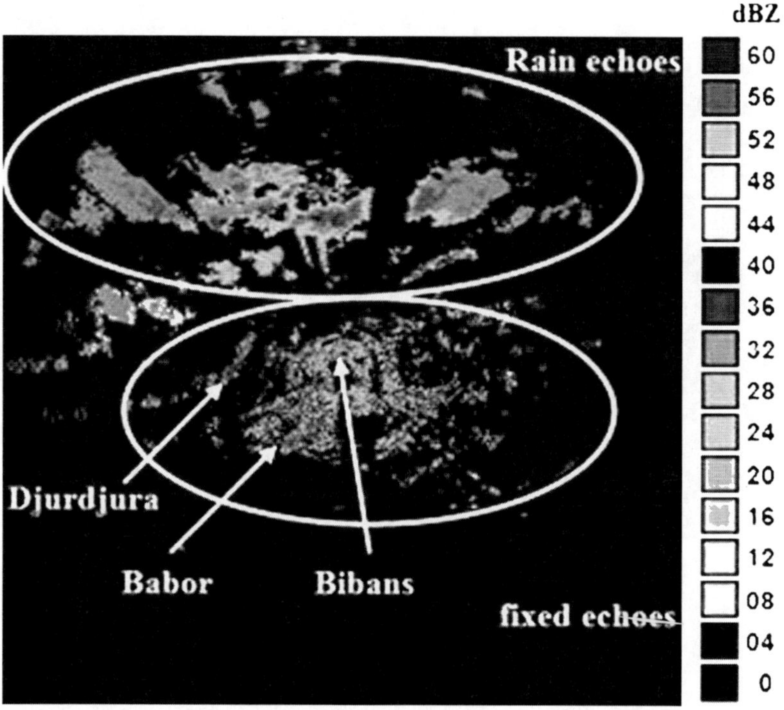

Fig. 1 Radar images collected at Setif during periods of bad weather

of pattern recognition has been developed. It consists in detecting the shape of the radar echoes and recording only those which have their surface greater than 30 km^2. This algorithm operates by exploring the image, pixel by pixel, using an analysis window of 3×3 pixels. Figure 2 exhibits the images of rainfall echoes obtained by applying the templates and pattern recognition filters to the original images given in Fig. 1. The images so filtered show that the rainfall cells are properly reconstituted and all the undesirable ground echoes have practically been eliminated.

3 Identification and Tracking of Rainfall Clouds

3.1 Structure of Rain Clouds

The precipitation fields detected by weather radar are composed of a multitude of water droplets and ice crystals (with diameters lying between 0.1 μm and 5 mm). The resulting clouds are heterogeneous structures classified into two main

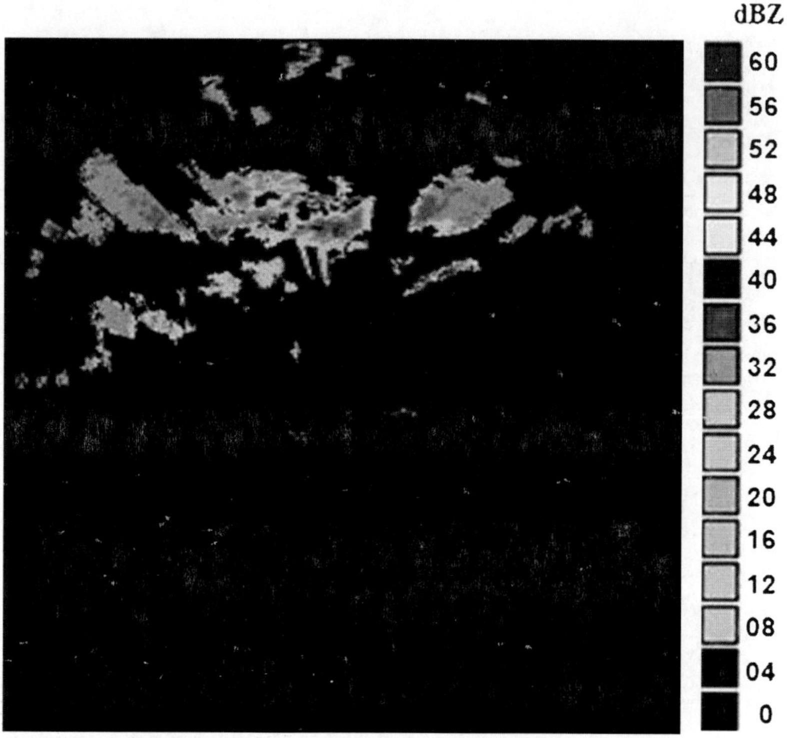

Fig. 2 Images of rainfall echoes for the regions of Setif, filtered using template and pattern recognition based-algorithm

categories: stratified clouds and convective cells. Briefly, stratiform clouds arise from the meeting of masses of warm and dry air with cool and wet atmospheric layers. These clouds essentially caused by advection, spread horizontally over large areas and then, yield rain of low intensity and long duration. Convective clouds are generated by subsidence due to the rise of strong and warm air masses and their condensation during the descent of cold air currents. Owing to the convective motion, these clouds extend mainly in the vertical direction and can even reach the top of troposphere. Most of them are dense and produce strong rainfall during short periods of time. Among the convective clouds, cumulonimbuses are known to be submitted to important atmospheric instabilities. Consequently, they frequently give rise to big thunderstorms and strong showers. Radar images can be processed so as to identify the rainy clouds, follow their motion, and get a good estimate of the intensity of precipitations watering the terrestrial surface.

3.2 Classification of Rain Clouds

The precipitation fields observed on the radar images are made up of several groups of cells cloud resulting from the agglomeration of hydrometeors in varied size and shape. Each cell is cloud-shaped variable and evolves separately for a period of life ranging from one to eight hours. The cells spread rain for most of the roughly 50–1,000 km^2 and are centered on a core of reflectivity to a greater or lesser extent. To characterize a cell and monitor cloud changes from one image to another, the radar images have been previously screened decomposed into elementary images each containing a single cell cloud. The algorithm of pattern recognition described in Sect. 2, enables us to separate a rainfall cell from the rest of the cloud cover, index it in the radar image and calculate its surface, gravity centre reflectivity and its direction.

In order to evaluate the precipitation fields more efficiently for each cell, and to know their evolution in time, a method of processing radar images presented hereafter, is implemented using mainly optical flow.

4 Motion Estimation Using Optical Flow

4.1 Optical Flow Equation

Motion estimation is based on the fact that each point on an image has a fixed intensity. It is therefore possible to estimate the motion of a point by superposing two consecutive images using the method of optical flow [1, 10]. A sequence of images can be represented by its brightness function $I(p, t)$. The hypothesis of brightness conservation says that a physical point of the image sequence does not vary with time, which gives:

$$I(p,t) = I(p + V(p)\,\delta t, t + \delta t), \tag{1}$$

where $p(x, y)$ is one point of the image and $V(p) = V(u, v)$ is the velocity vector at point p and time t, with u and v are, respectively, the velocity along x and y.

Equation (1) leads to the cancelation of the difference between the successive two images. Assuming that $I(x, y, t)$ is a continuous function together with its time derivative, the assumption of conservation of the luminance is given by:

$$\frac{\mathrm{d}I}{\mathrm{d}t} = \frac{\delta I}{\delta t} + \nabla I \cdot V = 0. \tag{2}$$

Where: $\nabla I = \left[\frac{\delta I}{\delta x}, \frac{\delta I}{\delta y}\right]$ is the spatial gradient of I.

Several models have been used to solve the equation of optical flow such as the model of translation, the parametric model, and the nonparametric model. In here we choose to use the nonparametric differential methods to determine the velocity field seen in a sequence of radar images.

4.2 Differential Method

The differential methods belong to the techniques commonly used for the calculation of optical flow in image sequences. Their advantages include the reduction in the complexity of calculations commonly used in the matching methods while increasing the range of measurable displacements. They can be classified into two categories: global methods such as the technique of Horn and Schunck, and local methods as the Lucas–Kanade approach and that based on Gabor filters. These methods are based on the equation of the apparent motion (ECMA) issue to Taylor development of the equation of the intensity *I(x, y, t)* described by:

$$I(x+\mathrm{d}x, y+\mathrm{d}y, t+\mathrm{d}t) = I(x,y,t) + \frac{\delta I}{\delta x}\mathrm{d}x + \frac{\delta I}{\delta y}\mathrm{d}y + \frac{\delta I}{\delta t}\mathrm{d}t + o(\varepsilon), \tag{3}$$

where ε is the term of higher order of the first order Taylor development of I which tends to 0 when dt tends to 0. Considering that the quantities dx, dy, and dt is sufficiently small and that (1) is respected, we have:

$$uI_x + vI_y + I_t = 0, \tag{4}$$

I_x, I_y, I_t: spatiotemporal derivative of intensity at point *p(x, y)* at time t.

To solve this equation by respecting the conditions of validity, two solutions are proposed in this study: the Lucas–Kanade method and Gabor filters.

4.2.1 Lucas and Kanade Method

The method proposed by Lucas and Kanade is a local approach based on the assumption that the motion is uniform over a region of the image or that the optical flow is constant over Ω neighborhood centered on the pixel that we want to calculate the displacement. We are therefore led to minimize the following functional [2, 8]:

$$E = \sum_{\Omega} [\nabla I \cdot V(p_i) + I_t]^2, \tag{5}$$

where Ω is a window around the point at we wish to determine the displacement.

In this method, each calculation is done on a small window in parallel independently of other windows. The solution is to make an estimate in the sense of least squares. In matrix notation the solution is:

$$V = \left(A^{\mathrm{T}}A\right)^{-1}A^{\mathrm{T}}b. \tag{6}$$

With: $A = \begin{bmatrix} I_{x1} & I_{y1} \\ I_{x2} & I_{y2} \\ I_{xn} & I_{xy} \end{bmatrix}, \quad b = \begin{bmatrix} I_{t1} \\ I_{t2} \\ I_{tn} \end{bmatrix}.$

It may happen that the matrix $A^{\mathrm{T}}A$ is ill conditioned (almost zero determinant and therefore not invertible). This is because the pixel is in an area where the luminance intensity is constant with zero gradient (problem opening) leading to the estimate in the sense of least squares of V to become aberrant. We can remedy these problems by using a regularization technique. This is done by using:

$$E = \sum_{\Omega} [\nabla I \cdot V(p_i) + I_t]^2 + \alpha V^2 . \tag{7}$$

With α (adjustable) representing the regularity of the solution. This equation is always linear and can yield to the following:

$$V = \left(A^{\mathrm{T}}A + \alpha J\right)^{-1} A^{\mathrm{T}}b , \tag{8}$$

where J is the identity matrix.

The developed optical flow equations are valid only if the displacement is small, so we have to use the multi-resolution developed in the algorithm of Horn and Schunck to solve them. On the other hand, it is possible to further refine the results at each level of the pyramid minimizing the gap between two successive frames and running again the algorithm after moving one of two images according to the latest calculated field velocity. If the algorithm is stable, the method converges and therefore:

$$V^l = V^{l-1} + \eta^l , \tag{9}$$

with $V^0 = 0$ and $\eta^l = (A^{\mathrm{T}}A + \alpha J)^{-1}A^{\mathrm{T}}b_l$.

If L iterations are necessary to achieve convergence, the final solution is the last calculated vector and is given by:

$$V = \sum_{l=1}^{L} V^l . \tag{10}$$

4.2.2 The Gabor Filters Approach

The technique involves estimating the optic flow locally based on sub-band decomposition of the image sequence. The estimate was made by Gabor filters space by combining information from filter banks positioned at different orientations to robustly estimate the motion. Gabor filters are widely used in computer vision. This success is due both to their spatial and spectral properties. Bruno [3] proposed a novel approach based on the differential method for local estimation of optical flow from Gabor filter banks. It is based on projecting the optical flow equation (21.4) into a pixel image on a bank of N Gabor filters. We obtain an over determined system of N equations that allows us to calculate the two velocity components of the pixel considered.

Assuming that V is constant over the filter support G_i, we can write the optical flow equation as:

$$u\left(\frac{\delta \mathrm{I}}{\delta \mathrm{x}}G_i\right)+v\left(\frac{\delta \mathrm{I}}{\delta \mathrm{y}}G_i\right)+\frac{\delta \mathrm{I}}{\delta \mathrm{t}}G_i=0 \quad i=1\ldots N. \tag{11}$$

The minimization is obtained iteratively by using the Weighted Least Squares algorithm (WLS):

$$V=\arg\min\sum_i w_i\rho(r_i). \tag{12}$$

With $w(x)=\frac{1}{x}\frac{\delta\rho(x)}{\delta x}$ is a weighting function and $\rho(x)$ is a Euclidean norm.

The function $\rho(x)$ can be chosen from a wide range of functions. It must be even defined, positive, and should not have a single minimum at zero. In addition, in order for the function $\rho(x)$ to minimize the influence of erroneous data, it has to be less than the quadratic function. In the case of a local estimate, it is preferable that the aberrant data be deleted. So we chose to use the M-estimator [5].

5 Application

The Lucas and Kanade together with the Gabor algorithm were tested on three different sequence images: a cube, taxi, and a rainy cell (Figs. 2a, 3a, and 4a, b).

- Figure 3a represents the image of an artificial sequence called "cube" consisting of a cube placed on a turntable plateau. The image size is 256×240 pixels.
- Figure 4a represents the image of an artificial sequence called "taxi" in a street intersection. It is composed of three cars and a pedestrian moving. The two bottom vehicles moving in the opposite direction. The taxi turned the corner in the middle of the street. The image size is 256×190 pixels.
- Figure 5a, b represent the sequences of two stratiform cells isolated from images of size 64×64 pixels. It represents one of the rain cells observed in the radar image taken in December 26th, 2004 at 17:30 (Fig. 4a) and 15 min later (Fig. 4b). The first cell has 2,685 km^2 of surface and 19.44 dBZ of reflectivity.

6 Results and Interpretations

6.1 Lucas and Kanade Method

The regularization coefficient alpha (α) is simply to avoid cancellation of the determinant in the least squares estimation. Its value should be close to zero, between 10^{-3} and 10^{-6}. The gradient that give the best results was calculated using the mask $[-1\ 1,\ -1\ 1]$ for a horizontal displacement x, and $[-1\ -1,\ 1\ 1]$ for

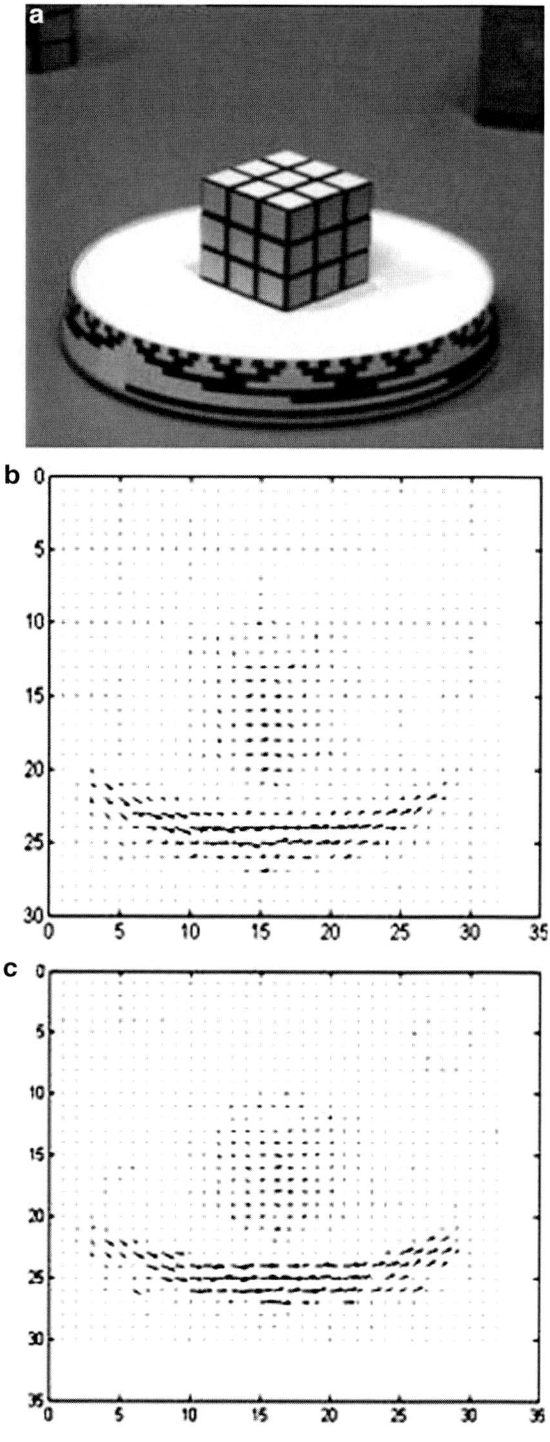

Fig. 3 Optical flow estimated for cube image by patch 7 × 7. (**a**) Cube image; (**b**) field estimated by Lucas–Kanade; (**c**) field estimated by Gabor filters

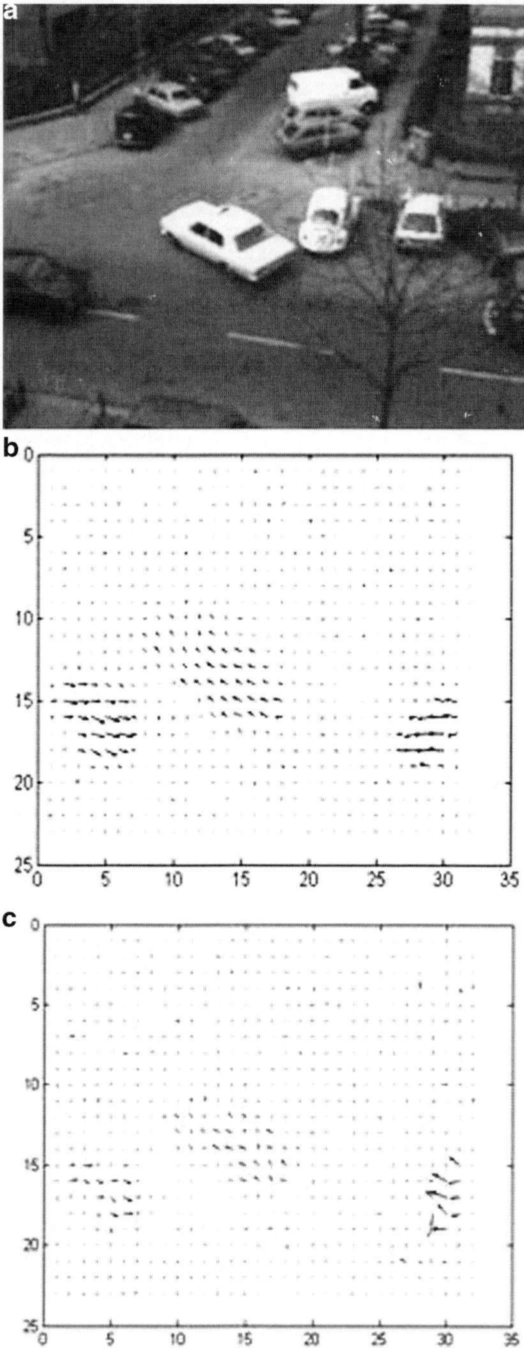

Fig. 4 Optical flow estimated for taxi image by patch 7 × 7. (**a**) Taxi image; (**b**) field estimated by Lucas–Kanade; (**c**) field estimated by Gabor filters

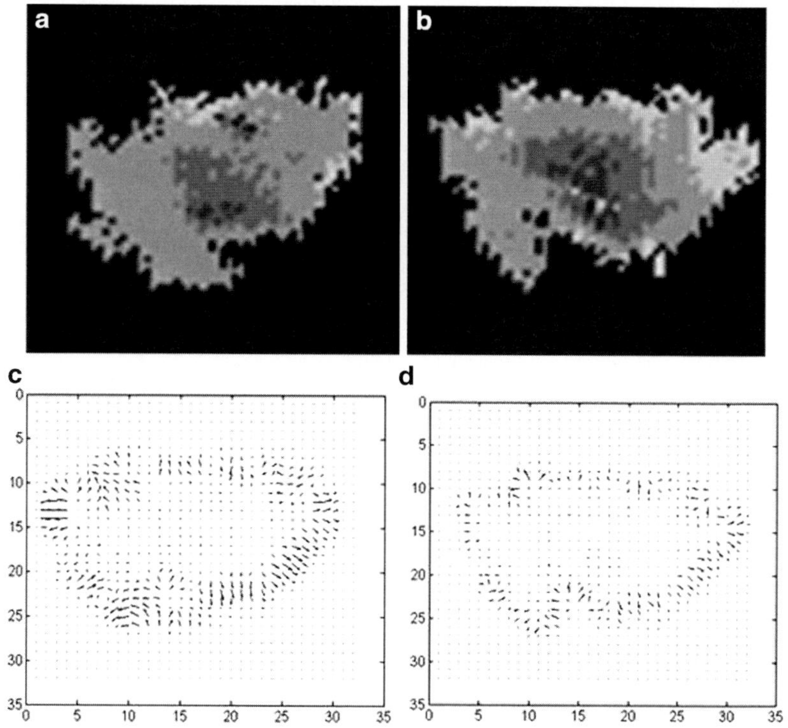

Fig. 5 Optical flow estimated for rainy cell. (**a**) First cell of rain cloud; (**b**) second cell of rain cloud; (**c**) field estimated by Lucas–Kanade method; (**d**) field estimated by Gabor filters

vertical displacement y. After several trial tests, the best results were obtained under the following conditions:

- The cube level of the pyramid 3, patch 7×7 pixels, $\alpha = 10^{-3}$: the velocity field in Fig. 2b shows the rotation of the plateau.
- Taxi: level of the pyramid 3, patch 7×7 pixels, $\alpha = 10^{-3}$: Fig. 3b shows three cars moving and that both gray cars are moving in opposite directions and the white taxi in the center turns on the corner. The movement of the pedestrian is not detected.
- Rainy cell: pyramid level 1 patch of 5×5 pixels, $\alpha = 10^{-3}$: In Fig. 4c appears vectors almost directed to the right indicating the direction of development of the rainy cloud. We also note that the estimated motion is almost zero at the center of the cell and increases gradually as we come near the edges, which is logical because the cloud changes little in 15 min. Outside the rain cell the field is considered zero.

The size of the images used in the sequences "Cube" and "taxi" is important, which justifies the use of third-level pyramid. Concerning the cell extracted from the radar image, first-level pyramid is used directly given the size of the image (5×5 pixels).

6.2 Gabor Filter Method

The estimated flow shows all the objects in motion in the three image sequences. The tuning parameters of this algorithm do not differ greatly in all three cases, thus:

- Cube (Fig. 2c): $K = 3$, $\alpha = 10^{-3}$, $\sigma = 4$, $f_0 = 0.14$.
- Taxi (Fig. 3c): $K = 3$, $\alpha = 10^{-3}$, $\sigma = 4$, $f_0 = 0.18$.
- Rain cell (Fig. 4d): $K = 1$, $\alpha = 10^{-3}$, $\sigma = 4$, $f_0 = 0.14$.

We also note that the density increases with velocity vectors N to reach a high where it does not change, for $N = 6$. However, the increment of N greatly increases the computing time. The density also increases with σ up to a maximum value beyond which we obtain a smooth velocity field, since the assumption of locally constant optical flow is no longer respected.

We also note that variations of the frequency centre are very low. This is due to that the global variations of the considered photometric variable (brightness or reflectivity) are low-frequency type.

On setting the parameters for obtaining good results, there is a trade-off between α and the number of iterations since the higher number of iteration gives a better motion boundaries area.

In these three examples, we have considered a single iteration loop for refinement: i.e., $L = 1$. Increasing L will have little influence on the velocity fields of the moving objects. Tests on the size of the patch showed that whenever the window size is increased, the fields motion appear, which is more and more dense until reaching an optimum value. Thus, if we continue to increase the window size, the optical flow fields became smooth, i.e. the contour areas will begin to disappear and the movement becomes less clear.

7 Conclusions

In this chapter, we have presented two techniques for estimating local motion of rainy cells in which the quality of estimation depends mainly on the right choice of the parameters in each method. In both cases and when applied on three different sequence images, the estimated optical flow detected all the moving objects and their direction. We also found that the method based on Gabor filters gives a less dense optical flow with a lack of precision in the estimated field near the border. This was in contrast with the method of Lucas and Kanade. Moreover, it was found that the computing time required to run the algorithm of the latter is much smaller than that of the Gabor filters. Based on this, we can say that optical flow can be used successfully to determine rainy clouds motion in weather radar images.

References

1. Baron J.L., Fleet DJ. and Beauchemin SS: Performance of optical flow techniques. *In International Journal on Computer Vision*, vol. 12, pp. 43–77 (1994).
2. Bouguet J.: Pyramidal Implementation of the Lucas Kanade Feature tracker, Intel Corporation, Microprocessor Research Labs (2000).
3. Bruno E: De l'estimation locale à l'estimation globale de mouvement dans les séquences d'images. Thèse, Université Joseph Fourier, Grenoble, France (2001).
4. Darricaud, J.(Ed). Physique et theorie du radar, 3ème ed., t.2: principes et performances de base (Physics and theory of the radar, 3rd ed., t.2: Basic principles and performance). Paris: Sodipe (1993).
5. Huber P.: Robust statistics, Wiley(Ed), New York (1981).
6. CRANE, R.K. (Ed): Electromagnetic wave propagation through rain, 269 p. New York: J. Wiley and sons (1996).
7. Keeler, R.J. and Passarelli, R.E.: Signal processing for atmospheric radars. In *Radar probing and measurement of the planetary boundary layer*; D.H. Lenschow (Ed.), pp.199–229.Boston: American Meteorological Society (1986).
8. Mrazat J.: Estimation temps réel du flot optique; Rapport de stage ingénieur; INRIA, Yvelines (2008).
9. Meischner, A.: Weather radar: principles and advanced applications. Springer (2005).
10. Mémin E.: Estimation du flot optique-contribution et panorama des différents approches; Habilitation à diriger des recherches. Université de Rennes 1, Rennes (2003).
11. Raaf, O. and Adane, A.: Image-Filtering Techniques for meteorological radar. *In International Symposium on Industrial Electronics* (IEEE ISIE08), 30 June-02 July 2008, Cambridge, pp. 2561–2565.Sauvageot, H.(1992). Radar Meteorology. Artech House(Ed), Boston (2008).
12. Renaut, D.: Les satellites meteorologiques (Meteorological satellites). *La meteorologie*, **45**, pp. 33–37 (2004).
13. Sauvageot, H.: Radar Meteorology, 366p.Boston: Artech House (1992).

Hydrodynamic Modeling of Port Foster, Deception Island (Antarctica)

Juan Vidal, Manuel Berrocoso, and Bismarck Jigena

1 Introduction

Deception Island is part of the South Shetland Island chain with Smith, Snow and Livingston Islands to the north and west, and King George Island to the north and east. These islands form the northern boundary of the Bransfield Strait and the Antarctic Peninsula forms its southern boundary. Deception Island is situated between latitudes 62.89–63.02° South and longitudes 60.49–60.75° West. Deception Island is an active volcano that has a central flooded caldera, which is open onto Bransfield Strait through a shallow and narrow sill, at Neptune's Bellows [17].

Bransfield Strait tides are a mixture of diurnal and semidurnal frequencies. The M2 component is the most important with an amplitude around 0.40 m while O1, K1 and S2 tidal components, all with amplitudes around 0.28 m, are also significant [13]. Amplitudes of the main tidal constituents are higher in the northwestern Weddell Sea than at the northwestern side of the peninsula [7]. The local circulation in the Bransfield Strait is strongly influenced by tides. Both at the surface and at deeper layer, the flow patterns exhibit a strong diurnal and semidiurnal periodicity in agreement with the barotropic character of tides [12]. The Bransfield Current is a northeastward geostrophic current. Based on geostrophic studies calculated to a reference level, various authors proved that the currents are oriented parallel to the front with speeds from 0.05 to 0.30 m s^{-1}, [14] and [8].

J. Vidal (✉)
Centro Andaluz de Ciencia y Tecnologías Marina, Campus Universitario de Puerto Real, Puerto Real (Cadiz), 11510 Spain
e-mail: juan.vidal@uca.es

J.A.T. Machado et al., *Nonlinear and Complex Dynamics: Applications in Physical, Biological, and Financial Systems*, DOI 10.1007/978-1-4614-0231-2_16,

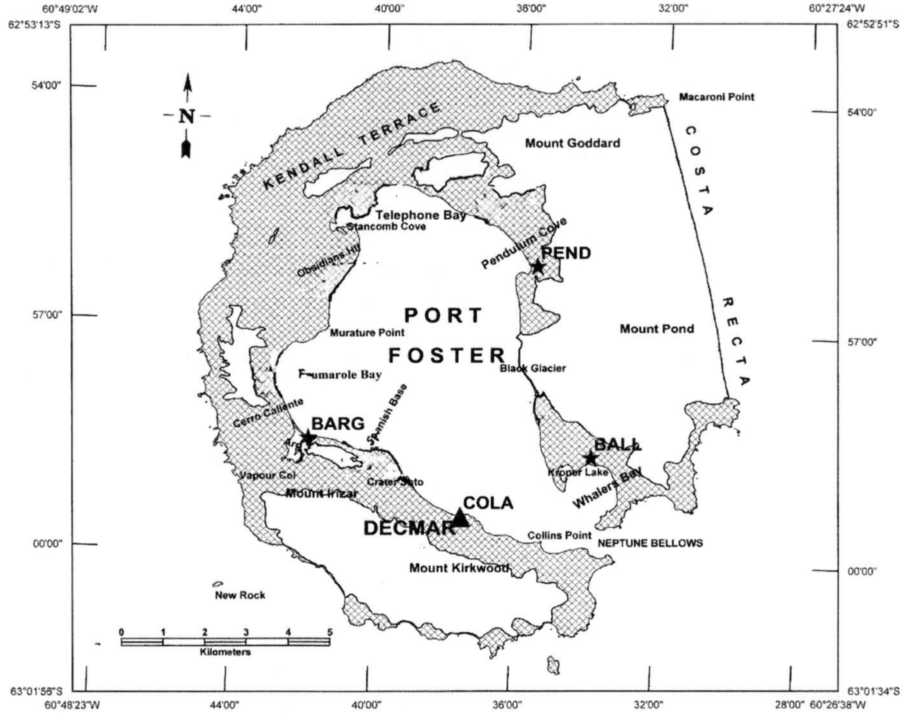

Fig. 1 Tidal Stations in Deception Island: (BALL) Balleneros Station, (COLA) Colatinas Sation and (PEND) Pendulo Station

However, Port Foster water mass characteristics differ significantly from Bransfield Strait water mass characteristics, indicating that the sill at Neptunes Bellow limits cross-sill exchange below 30 m. The small tidal currents in Port Foster, less than 0.1 m s^{-1}, also suggest that the larger Bransfield Strait tidal currents do not direct water into Deception Island [11]. The maximum tidal currents are obtained at Neptune's Bellows, with 0.64 m s^{-1} during spring tides [4].

Several hydrodynamical numeric models, including the Bransfield Strait have been recently implemented by several authors, [7] and [16]. However, these models are larger in scale. The main objectives of this paper are to present a detailed study of the tidal characteristics in Por Foster. The amplitudes and phases obtained through the analysis of the time series generated by the model allowed the attainment of co-tidal maps and ellipses of currents maps; these results are compared with the harmonic constants of several coastal stations, which positions are given in Fig. 1. The tidal simulations are a first step toward the modeling of the hydrodynamic flushing.

2 Model Approach

2.1 Governing Equations

The hydrodynamic model UCA2D here applied has been developed at University of Cadiz. A completed description of the model can be found in [2]. It has already been applied successfully to several coastal environments. See [1, 3, 18].

The model resolves the equations in their formulations with water levels and transports:

$$\frac{\partial U}{\partial t} - fV + gH\frac{\partial \zeta}{\partial x} + RU + X = 0 \tag{1}$$

$$\frac{\partial V}{\partial t} + fU + gH\frac{\partial \zeta}{\partial y2} + RV + Y = 0 \tag{2}$$

$$\frac{\partial \zeta}{\partial t} + \frac{\partial U}{\partial x} + \frac{\partial V}{\partial y} = 0, \tag{3}$$

where f is the Coriolis parameter, ζ is the water level, g is the gravitational acceleration, $H = h + \zeta$ is the total water depth, and U and V are the vertically integrated velocities. R is the friction coefficient which is expressed as:

$$R = C_{\mathrm{b}}\frac{\sqrt{u^2 + v^2}}{H}, \tag{4}$$

where C_{b} is the bottom drag coefficient. The terms X and Y of equations contain the nonlinear terms:

$$X = u\frac{\partial U}{\partial x} + v\frac{\partial U}{\partial y} - A_{\mathrm{H}}\left(\frac{\partial^2 U}{\partial^2 x} + \frac{\partial^2 U}{\partial^2 y}\right) \tag{5}$$

$$Y = u\frac{\partial V}{\partial x} + v\frac{\partial V}{\partial y} - A_{\mathrm{H}}\left(\frac{\partial^2 V}{\partial^2 x} + \frac{\partial^2 V}{\partial^2 y}\right), \tag{6}$$

where the last terms in equations represent the horizontal turbulent diffusion and A_{H} is the horizontal eddy viscosity.

At the open boundaries, the water levels are prescribed in accordance with the Dirichlet condition, while at the closed boundaries the normal velocity is set to zero and the tangencial velocity is a free parameter.

2.2 Residence Times and Water Exchange

The residence time may be defined as the time it takes for a particular water parcel to leave a water body through its inlet. However, if the residence time is interpreted as an average, and does not refer to individual water parcels, it may be expressed as

$T_r = V/Q$, where V is the volume of the water body and Q is the water exchange (sum of the net freshwater supply and exchange with the open sea), [15] and [20].

Hydrodynamic flushing feature in an estuary is a measure of its self-purification capability and it determines the renewal of ocean water. The flushing time is equivalent to the reposition time [5], also known as replacement time. The definition is strictly oriented at the time needed for the volume introduced by the ocean flux to be equal to the fresh water volume in the bay, but says nothing about the processes of replacement.

Conventionally, estuarine flushing time (conversely flushing rate) is computed by simplified approaches, such as the tidal prism or salt balance method [10]. We have carried out a study of its flushing characteristics with a numerical tracer experiment by solving the vertically integrated mass transport equation.

3 Application to Port Foster

3.1 The Model Setup

The numerical computation has been carried out on a spatial domain that represents Port Foster through a finite element grid, which consists of 7,980 rectangular elements with a resolution of 100×100 m (Fig. 2). The bottom drag coefficient has been set to $cd = 3.0 \times 10^{-3}$. In barotropics models, the bottom drag coefficient is usually determined by fitting the modeled M2 tidal elevation and the observations as tide gauge stations [6]. All simulations presented in this work have been carried out using a maximum time step of 5 s.

The open boundary oscillations of the model consider components M2, S2, O1, and K1 that represent more than 90% of the tidal energy in Port Foster [13]. We use the amplitude and phase of the major tidal constituents from Antarctic Tide Gauge Database (http:www.esr.org) at the oceanic border.

For simulations of tidal velocities, all variables, including sea levels and velocities, were initially set to zero and the model was forced at the open boundary by specifying the sea level as above.

Although the seasonal cycle is the dominant signal in the temperature, characterized by the onset of stratification and deepening of the thermocline, using a two-dimensional model, we assumed that the water column is well mixed. This situation occurs during the austral winter [14]. Corresponding wind stress and spatial salinity variation were omitted. For the horizontal turbulent diffusivity, a low value of 3 $m^2\,s^{-1}$ has been chosen.

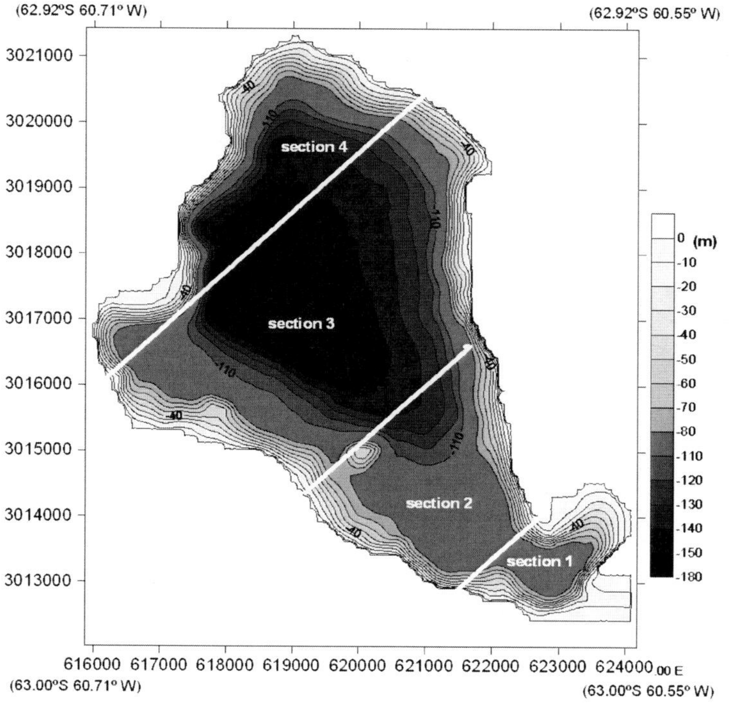

Fig. 2 Grid of Port Foster

3.2 *Tidal and Current Measurements for Model Calibration*

The model has been validated comparing the amplitudes and phases of the major components generated by the model with those derived from harmonic analysis of the tide gauge data within the island. We used the results of the harmonic analysis of series of direct sea levels measurements at Pendulo and Balleneros stations studied by Dragani et al. [7]. Additionally, water measurements were undertaken from December to March 2008, using a recording tide gauges. The tide gauges was deployed at Colatina (DECMAR station) [19]. The accuracy given by the pressure sensor is 1 cm. Harmonic analysis applied to tide gauge data is also shown in Table 1. Current measurements were carried out during four days at Colatinas station. However, for security reasons of the instrument, it was deployed too close to the shore about mid-depth at a total depth of approximately 8 m. Harmonic analysis was carried out also on the current measurements from Colatina.

Table 1 Elevations: harmonic analysis results

	Ball				Pend				Cola			
	A(m)		θ (degrees)		A (m)		θ (degrees)		A(m)		θ (degrees)	
Station	O	M	O	M	O	M	O	M	O	M	O	M
M2	0.46	0.39	280	284	0.44	0.39	281	285	0.40	0.40	281	283
S2	0.28	0.26	x	356	0.29	0.26	x	357	0.26	0.26	351	355
O1	0.29	0.27	48	55	0.29	0.27	55	55	0.27	0.27	53	54
K1	0.26	0.30	66	75	0.26	0.30	73	75	0.30	0.30	74	75
Error		0.03		2		0.03		1		0.00		1

Observed (O) and calculated (M) amplitudes and phase lags at Balleneros (Ball), Pendulo (Pen) and Colatinas (Cola) stations.

For each constituent of each simulation, the error in the amplitudes (e_A) and the error in the phases (e_F) have been calculated by using:

$$e_A = \sqrt{\frac{1}{N}\sum_i (Ao_i - Am_i)^2/2} \tag{7}$$

$$e_F = \frac{1}{N}\sum_i A_{mi}^2(\theta_{oi} - \theta_{mi})/(\Sigma_i A_{mi}^2). \tag{8}$$

The error in the amplitudes was calculated from the difference between the observations (Ao) and the results (Am) of the model, and the error in the phases was calculated from their difference ($\theta_{oi} - \theta_{mi}$), but weighted with the values of the observed amplitudes (Am).

3.3 Modeling of Tidal Flushing

Based on the validated hydrodynamic model, a numerical dye tracer experiment is performed. The entire bay is initially labeled (at high water) at a uniform 100% concentration with a passive tracer at time $t = 0$. The subsequent change in the tracer mass in different parts of the bay is then tracked by solving the 2D mass transport equation. At the open boundary, the inflow tracer dye concentration is 0. The computed flushing time is taken as the time when mass removal falls below 1/e of the original vertically integrated mass.

4 Results

Table 1 presents the comparison of the analysis of model results (M) with harmonic constants (O) of coastal stations (Balleneros, Pendulo, and Colatinas). The tidal amplitudes are in meters and the phases angle are in degrees relative to local time.

Model results show that the amplitudes of the components decrease as they spread into the caldera from the ocean through Neptune's Bellows. The phase lags are also important in this area due to the increased friction. Inside Port Foster, where depths are around 100 m deep, the amplitudes and phase lags do not vary significantly.

One of the most important results obtained by modeling is that the amplitudes of the velocities are very low (less than 5 cm s^{-1}) within Port Foster, excluding the area of Neptune's Bellows. Amplitudes and phase lags for the M_2 constituent are shown in Fig. 3. The velocities near the narrow channel of Neptune's Bellows show that the east–west direction is more important than the north–south, coinciding with the orientation of the narrow passage. For this area, through which it produces all the water exchange between the caldera and the open ocean, the results for simulations

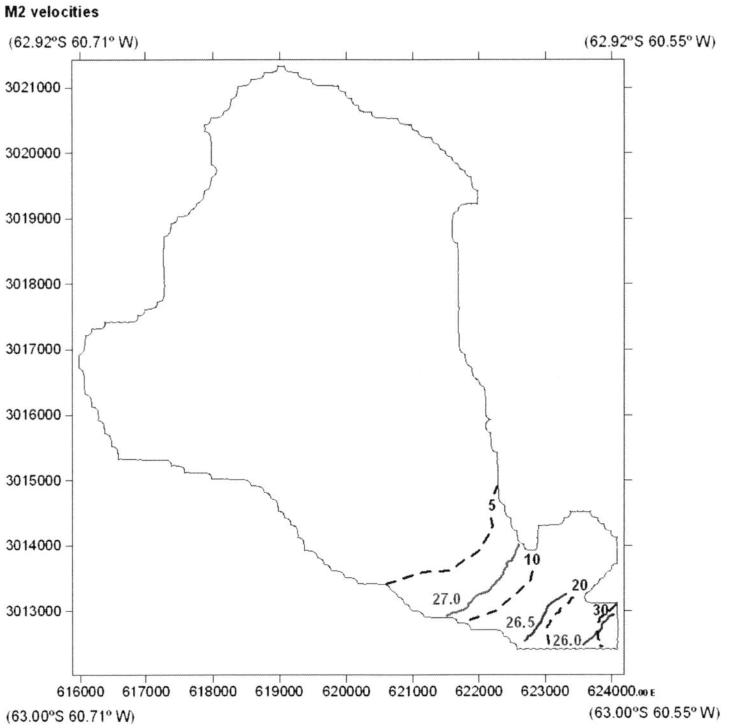

Fig. 3 Tidal velocities: amplitudes (*black dashed line*, in cm s^{-2}) and phases lags (*gray solid line*, in degrees) for M_2 constituent

of tidal velocities show maximum speeds of 0.76 m s^{-1} during spring tides. Table 2 shows the amplitudes of the velocities for the east–west components of the main harmonic constituents in a center point of the narrow and near Cola Station.

The volume of the basin is approximately 3,500 Hm^3 and the cross-section at Neptune's Bellows is about 5,500 m^2. Figure 4 shows tidal elevation data obtained by the model at Neptune's Bellows. A mean tidal level variation is about 1.20 m. Tidal elevation data obtained by the model are shown in Fig. 4. A height of 1.20 m implies that about 1.15% of the total volume of Port Foster is exchanged with Bransfield Strait during the flood and the ebb tide. This implies a exchange of 42 Hm^3 during filling or emptying of the Bay (in this case the predominant semi-diurnal tide cycle of 12.42 h), obtaining a calculated residence time of 82 days. Additionally, the total sea water volume V_d within the bay can be estimated as [9]:

$$V_d = \left(1 - \frac{S_m}{S_o}\right) V_l \qquad (9)$$

where S_m is the mean salinity of the bay, S_o is the sea water salinity entering through the bay and V_l is the bay volume at high water. According to Lenn et al. [11], the

Table 2 Velocities: harmonic analysis results

	Neptuno's Bellows		Cola			
	A (m s^{-1})	θ (degrees)	A (m s^{-1})		θ (degrees)	
Station	M	M	O	M	O	M
M2	0.40	27	0.13	0.10	33	36
S2	0.25	90	x	0.07	x	98
O1	0.13	148	x	0.05	x	156
K1	0.15	166	0.09	0.06	173	179
Error				0.02		2

Observed (O) and calculated (M) amplitudes and phase lags at Neptuno's Bellows and Pendulo Colatinas (Cola) stations.

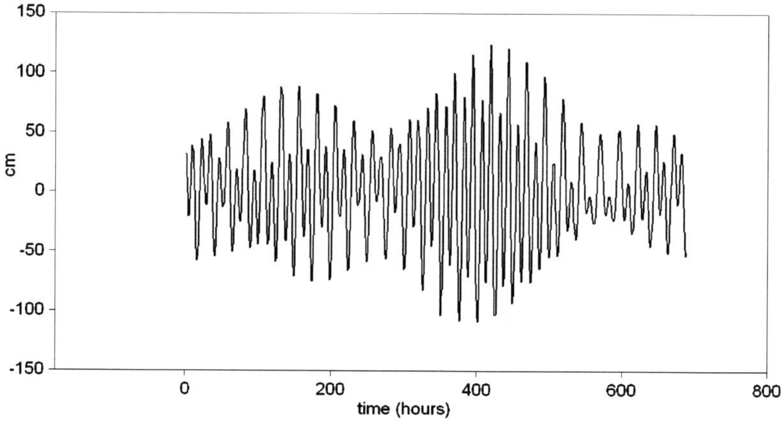

Fig. 4 Tidal elevation data obtained by the model

mean salinity in Port Foster is about 34.04 psu (practical salinity units). The mean salinity in the wester and central basin of the Bransfield strait is 34.4 psu [12]. For salinity to remain in a steady state, the volume of sea water required to this is 40.7 Hm^3, similar to the previous result. The maximum range, defined as the largest difference (at spring) between a maximum and the next minimum, is around 2.20 m. For the case of spring tides, this amount is approximately 2.2% of the total volume. The residence time is 45 days.

To compute the different flushing feature in Deception Bay, the entire bay water was grouped into four segments as it is shown in Fig. 2. Using the typical tidal and associated salinity conditions, the flushing time was calculated as shown in Table 3. A quick renewal occurs in the lower bay, because the water flushed out in a few days. The upper bay is characterized by large flushing time, because the velocities are weaker-less than 0.1 m s^{-1}.

Table 3 Flushing time in deception bay

Section	Days
1	2.7
2	16.1
3	51.2
4	78.0

5 Conclusion

The tide in Port Foster was satisfactorily reproduced for all cases studied in this work. Currents obtained with the model are very small inside, with maximum values of 0.76 $\mathrm{m\,s^{-1}}$ in the area of the Neptune's Bellows. Neptune's Bellows is the region where most of the tidal energy is dissipated. Approximately 90% of the total energy dissipates in this area.

The residual circulation due to tides is very small in the center of the caldera where the residence times are more than 35 days. Port Foster in Deception Island has a rate of poor water exchange (between 1.1% and 2.2% volume exchange over each tidal cycle). The two-dimensional model results presented here are in general consistent with previous three-dimensional model experiments on Deception. However, in a two-dimensional model, the absence of the vertical dispersion of the tracer may lessen the spread of the tracer patch. To implement a realistic simulation of the dye trace experiment, a three-dimensional model must be employed with the real hydrography and topography.

Acknowledgements This work was possible thanks to the project CGL2005-0789-C03-01/ANT INVESTIGACIONES GEODESICAS, GEOFISICAS Y DE TELEDETECCION EN LA ISLA DECEPCION Y SU ENTRONO (PENINSULA ANTARTICA, ISLAS SHETLAND DEL SUR), financed by Spanish Ministry of Science and Technology through the National Program of Antarctic Research of Natural Resources. We also thank the Las Palmas Crew and the members of the Spanish Antarctic Station Gabriel de Castilla for their collaboration during the surveying campaigns.

References

1. Alvarez, O., Tejedor, B., Tejedor, L.: Simulación hidrodinámica en el área de la Bahía de Cádiz: Análisis de las constituyentes principales. IV Jornadas Espaolas de Puertos y Costas, pp. 125–136. Servicio de Publicaciones de la Universidad Politécnica de Valencia, Valencia (1997)
2. Alvarez, O.: Simulación numérica de la dinámica de marea en la Bahía de Cádiz: análisis de las constituyentes principales, interacción marea-brisa e influencia del sediemnto en suspensión. Ph. D. Thesis, Universidad de Cádiz, Cádiz (1999)
3. Alvarez, O., Izquierdo, A., Tejedor, B., Maanes, R., Tejedor, L., Kagan, B. A.: The Influence of Sediment Load on Tidal Dynamics, a Case Study: Cadiz Bay. Estuarine, Coastal and Shelf Science. **48**, 439–450 (1999)
4. Antartic Pilot: Comprising the coasts of Antarctica and all Islands Southward of the Usual Route of Vessels, 4th Edition. Hydrographer of the Navy. Taunton, Somerset (1994)

5. Bolin, B., Rodhe, H.: A note on the concepts of age distribution and transit time in natural reservoirs. Tellus **25**, 58–62 (1973)
6. Crean, P.B., Murty, T. S., Stronach, J. A.: Mathematical Modelling of tides and Estuarine Circulation: the Coastal Seas of Southern British Columbia and Washington State. Coastal Estuarine Stud. **30**, pp.1-47, Washington (1991)
7. Dragani, W. C., D'onofrio, E. E., Fiore, M. E., Speroni, J. O.: Propagación y amplificación de la marea en el sector norte de la Península Antártica. V Simposio Argentino y Latinoamericano sobre Investigaciones Antárticas, Buenos Aires (2004)
8. Garcia, M. A., Castro, C. G., Rios, A. F., Doval, M. D., Roson, G., Gomis, D., Lopez, O.: Water masses and distribution of physico-chemical properties in the western Bransfield Strait and Gerlache Strait during austral summer 1995/96. Deep-Sea Research II. **49**, 585–602 (2002)
9. Ketchum, B.H.: The exchanges of fresh and salt waters in tidal estuaries. L. Mar. Res. **10**, 18–37 (1951)
10. Lee, J. H., Qian, A.: Three-dimensional modeling of hydrodynamic and flushing in deep bay. International Conference on Estuaries and Coasts November 9-11, Hangzhou (2003)
11. Lenn, Y. D., Chereskin, T. K., Glatts, R. C.: Seasonal to tidal variability in currents, stratification and acoustic backscatter in a Antarctic ecosystem at Deception Island. Deep-Sea Research II. **50**, 1665–1683 (2003)
12. Lopez, O., Garcia, M. A., Sanchez-Arcilla, A. S.: Tidal and residual currents in the Bransfield Strait, Antarctica. Ann Geophysicae. **12**, 887–902 (1994)
13. Lopez, O., Garcia, M. A., Gomis, D., Rojas, P., Sospedra, J, Sanchez-Arcilla, A. S.: Hydrographic and hydrodynamica. characteristics of the eastern basin of the Bransfield Strait (Antartica). Deep-Sea Research I. **46**, 1755–1778 (1999)
14. Niiler, P. P., Amos, A., Hu, J. H.: Water masses and 200m relative geostrophic circulation in the western Bransfield Strait region. Deep-See Research A. **38**, 943–959 (1991)
15. Officer, C.B., Kester, D.R.: On estimating the non-advective tidal exchanges and advective gravitational circulation exchanges in an estuary. Estuarine, Coastal and shelf Science. **32**, 99–103 (1991)
16. Padman, L., Fricker, H. A., Coleman, R., Howard, S., Erofeeva, L.: A new Tide Model for the Antarctic Ice and Seas. Annals of Glaciology. **3**, 1–14 (2002)
17. Smith, K. L. Jr., Baldwin, R. J., Kaufmann, R. S., Sturz, A.: Ecosystem studies at Deception Island, Antartica: an overview. Deep-See Research II. **50**, 1595– 1609 (2003)
18. Vidal, J.: Caracterización dinámica de la marea y del sedimento en el canal de Sancti Petri. Ph. D. Thesis, Universidad de Cádiz, Cádiz (2002)
19. Vidal, J., Berrocoso, M., Fernandez, A.: Study of the tide in Deception and Livingston Islands (Antarctica). Antarctic Science. **in press**, (2010)
20. Wijeratne, E.M.S., Rydberg, L.: Modelling and observations of tidal wave propagation, circulation and residence times in Puttalam Lagoon, Sri Lanka. Estuarine Coastal and Shelf Science. **74**, 697–708 (2007)

Part II
Nonlinear and Complex Dynamics: Applications in Biological Systems

Localized Activity States for Neuronal Field Equations of Feature Selectivity in a Stimulus Space with Toroidal Topology

Evan C. Haskell and Vehbi E. Paksoy

1 Introduction

The formation of spatially localized activity profiles or "bumps" have long been proposed as a mechanism for feature selectivity in models of cortex. Conditions for the existence and stability of persistent localized activity were first obtained by Amari [1]. Amari's original results have been extended to networks that are not of lateral inhibition type [9, 10], spiking models [8], planar networks [10, 13, 14] and to sphere topology [4].

We seek to extend the mean field approach of Amari to study the formation of localized activity states in the topology of a torus. This is motivated by recent evidence that the topology of population activity is correlated to the topology of the stimulus space [12]. When the population is tuned to respond preferentially to two circular variables such as orientation and color hue, the underlying topology could be represented by a torus. Indeed, areas of primary visual cortex do demonstrate spatial maps that are clustered for orientation [6] and those clustered for color [3] implying the appropriateness of considering a torus model for areas with an ordered set of two-dimensional feature maps with circular variables. To what extent such a topology is applicable to other cortical areas is unclear. However, throughout cortex cells sharing similar receptive field properties tend to be arranged in a columnar structure [5]. It is hypothesized that such a clustering of cells by receptive field properties in the visual system facilitates the computations required to generate linking features used to associate related parts of an image [2, 7]. Here we explore a model of the computations performed for cells with receptive-fields tuned for two circular stimulus variables.

E.C. Haskell (✉)
Division of Math, Science, and Technology, Farquhar College of Arts and Sciences, Nova Southeastern University, Ft. Lauderdale, FL 33314, USA
e-mail: haskell@nova.edu

J.A.T. Machado et al., *Nonlinear and Complex Dynamics: Applications in Physical, Biological, and Financial Systems*, DOI 10.1007/978-1-4614-0231-2_17,

2 Model

To model feature selectivity with two circular variables, we consider a continuum of neurons distributed on a Torus. The mean membrane potential of neurons located at a position (θ, φ) in the stimulus space with $\theta \in [-\pi, \pi)$ and $\phi \in [-\pi, \pi)$ is designated by $u(\theta, \varphi, t)$. Following Amari, the field equation for the evolution of $u(\theta, \varphi, t)$ for neurons receiving a homogeneous external input h, and an inhomogeneous external input $s(\theta, \varphi, t)$ is taken to be of the form:

$$\tau \frac{\partial u(\theta, \varphi, t)}{\partial t} = -u(\theta, \varphi, t) + \int_T w(\theta, \varphi | \theta', \varphi') f\left[u(\theta', \varphi', t)\right] D(\theta', \varphi') + h + s(\theta, \varphi, t) \quad (1)$$

where T designates the torus, $w(\theta, \varphi | \theta', \varphi')$ is the mean efficacy of synapses from neurons at (θ', φ') to (θ, φ), and $f[u]$ is the mean activity level of the neurons. The integration measure, $D(\theta, \varphi) = \frac{(a+b\cos(\phi))}{4\pi^2 a} d\theta d\phi$, is defined for the torus given in parametric form as:

$$x = (a + b\cos(\phi))\cos(\theta)$$
$$y = (a + b\cos(\phi))\sin(\theta)$$
$$z = b\sin(\phi)$$

with meridian radius a and vertical radius b. For this analysis we will use an indicator function for $f[u]$, which is 1 if $u > 0$ and 0 otherwise.

Given the torus topology it is natural to construct a weight function, $w(\theta, \varphi | \theta', \varphi')$, that is invariant to the orientation preserving isometries of the torus. Although there are two such orientation preserving isometries corresponding to individual angular coordinate rotations and reflections, nontrivial dynamics should occur with a weighting function that is invariant to one of these isometries. We construct a weight function that is invariant to rotations about the z-axis by any angle. This is represented by the rotation group $O(2)$. Such a weighting function can be constructed from the angular separation, σ, between two points (θ, φ) and (θ', φ') on the torus. The angular separation between these points is given by the relation $\cos(\sigma) = \frac{(a+b\cos(\phi))(a+b\cos(\phi'))\cos(\theta-\theta')+b^2\sin(\phi)\sin(\phi')}{\sqrt{(a^2+2ab\cos(\phi)+b^2)}\sqrt{(a^2+2ab\cos(\phi')+b^2)}}$. Thus, we suggest a simple nontrivial form for a local weight distribution that will lead to nontrivial dynamics given by:

$$w(\theta, \phi | \theta', \phi') = W_0 + W_1 \frac{(a+b\cos(\phi))(a+b\cos(\phi'))\cos(\theta-\theta') + b^2\sin(\phi)\sin(\phi')}{\sqrt{(a^2+2ab\cos(\phi)+b^2)}\sqrt{(a^2+2ab\cos(\phi')+b^2)}}, \quad (2)$$

where W_0 and W_1 are parameters of the weighting function. The use of rotational symmetry to construct an $O(2)$ invariant weight function for the ring or an $O(3)$ invariant weight function for the sphere implies that the pattern of connections

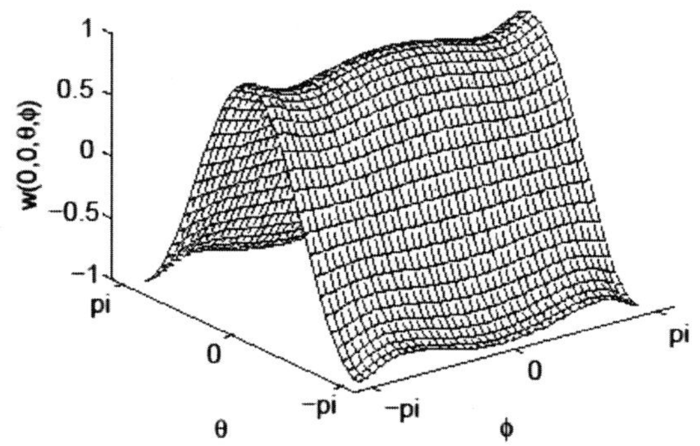

Fig. 1 Surface plot of $w(0,0|\theta,\varphi)$ in the (θ,φ) plane for the ring torus with $a=2$, $b=1$, $W_0=0$ and $W_1=1$

within the network depends only upon the relative distance between the cells [4]. For a spherical topology, the angular separation coincides with the metric distance with respect to the intrinsic geometry of the sphere. The same cannot be said for the torus. Nevertheless, the metric preserving automorphisms and symmetries of the torus can be extended to that of ambient Euclidean space and symmetries preserve the angles. Therefore, one can consider angular separation for weight function.

It should be pointed out that for a horn torus where $a=b$, angular separation is not defined everywhere as the inclusion of the origin as a point on the torus allows for vectors of zero length. However, angular separation can be used to define a weight function for the ring torus where $a>$ b and the spindle torus where $a<$ b. Using angular separation to define a weighting function for the sphere, we can construct a network of lateral inhibition type where neurons that are similar in feature preference are mutually excitatory and those that are dissimilar are mutually inhibitory. This is not the case for the torus. Nevertheless, the $O(2)$ invariant weight function for the torus utilized here provides similar results as that found with the $O(3)$ invariant weighting function for a sphere.

Figure 1 gives an example of the $O(2)$ invariant weight function $w(\theta,\varphi|\theta',\varphi')$ defined by (2) for the ring torus. As we can see this defines a network of lateral inhibition type in the θ variable only. The φ variable provides only a small modulation to the weighting level set by the θ variable. This would be indicative of a network that receives input representing two independent circular variables but is primarily selective to one of the variables. However, as we will see in the analysis the φ variable will have an impact on the radius of a localized activity state.

Figure 2 gives an example of the $O(2)$ invariant weight function $w(\theta,\varphi|\theta',\varphi')$ defined by (2) for the spindle torus. This weighting function is not properly of lateral inhibition type. While the strength of interaction does initially decay from the point of identical feature selectivity (0, 0), at the points of self-intersection of the torus with itself the weighting function becomes positive and returns to a maximum value when both variables are maximally dissimilar, $|\theta-\theta'|=\pi$ and $|\varphi-\varphi'|=\pi$. The spindle torus intersects itself in the two circles $\varphi=\cos^{-1}\left(\frac{-a}{b}\right)$ and $\varphi=\cos^{-1}\left(\frac{a}{b}\right)$,

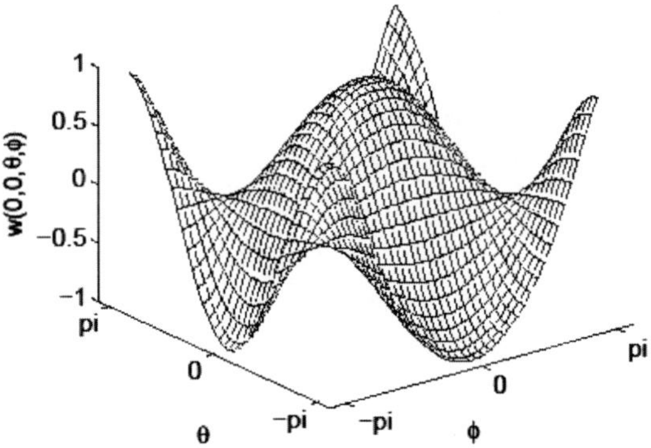

Fig. 2 Surface plot of $w(0,0|\theta,\varphi)$ in the (θ,φ) plane for the spindle torus with $a=2$, $b=4$, $W_0=0$ and $W_1=1$

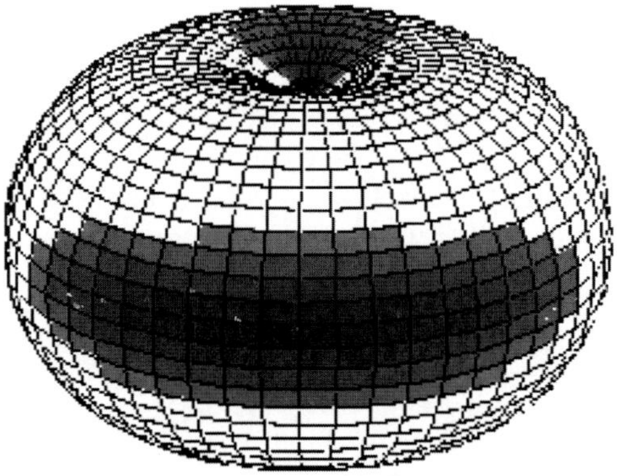

Fig. 3 Example of a mirror bump for the spindle torus with $a=2$, $b=3$, $W_0=-1$, $W_1=4$, and $h=-0.05$

thus, the region where cells with dissimilar preferences for both features excite each other can be restricted by keeping the ratio of a to b close to 1. The spindle torus has the added complication that the integration measure defined above vanishes on these circles. However, by having these regions where cells with dissimilar preferences in each preference excite each other allows for the potential formation of two mirror bumps to the primary localized activity state generated by a transient initial stimulus $s(\theta,\varphi,0)$. An example of such a localized activity state and a corresponding mirror bump is shown in Fig. 3.

It is unclear if there are cortical areas with cells that are mutually excitatory when their feature preferences are both similar and very dissimilar while being mutually inhibitory when one feature preference is similar and the other dissimilar. Also, although multiple localized activity states can arise through long-range interactions between cells with similar feature preferences in different columns, it is unclear whether such phenomena would occur in the scenario described here. As such, we will focus our analysis for conditions on the existence of localized activity states to the ring torus.

3 Equilibrium Solutions for Ring Torus

We follow the approach of Amari [1] to classify and derive conditions for equilibrium solutions of the neural field equation with homogeneous inputs (i.e. $s(\theta,\varphi,t)=0$) since this is applicable to arbitrary choices of $O(2)$ invariant weight distributions. equilibria solutions, $U(\theta,\varphi)$, of (1) in the absence of an inhomogeneous input satisfy

$$U(\theta,\phi)=\int_T w(\theta,\phi|\theta',\phi')f\left[U(\theta',\phi')\right]D(\theta',\phi')+h. \tag{3}$$

Letting $R[U]=\{(\theta,\varphi)|\mathrm{U}(\theta,\varphi)>0\}$ denote the region over which the field is excited, (3) can be rewritten as

$$U(\theta,\phi)=\int_{R[U]} w(\theta,\phi|\theta',\phi')D(\theta',\phi')+h. \tag{4}$$

Following the previous analysis for the sphere topology [4] we distinguish three types of equilibrium solution: $\emptyset-$, $\pi-$, and α-solutions. A $\emptyset$-solution satisfies $U(\theta,\varphi)\leq 0$ for $(\theta,\varphi)\in T$ so that the excited region is empty, $R[U]=\phi$. This is complementary to the π-solution where $R[U]=T$ in which the entire space is excited. α-solutions are localized activity states where a simply-connected proper subset of the torus is excited. Given the homogeneity and $O(2)$ invariance of the weighting function (2), we can choose the center of the α-solution to be at any arbitrary point (Θ,Φ) on the torus. The boundary of the excited region over which $U(\theta,\varphi)=0$ is then given by $\cos(\alpha)=\frac{(a+b\cos(\phi))(a+b\cos(\Phi))\cos(\theta-\Theta)+b^2\sin(\phi)\sin(\Phi)}{\sqrt{(a^2+2ab\cos(\phi)+b^2)}\sqrt{(a^2+2ab\cos(\Phi)+b^2)}}$ with the angular radius $\alpha\in(0,\pi)$ determined self consistently from the equilibrium equation (4). Furthermore, without loss of generality we consider α-solutions centered at $\theta=0$. Given symmetry considerations we have that for α-solutions, $R[U]=\{(\theta,\varphi)|-\alpha<\theta<\alpha,\varphi\in[-\pi,\pi),\alpha\in(0,\pi)\}$. Solutions of (4) then reduce to the form $U(\theta,\varphi)=G(\theta,\varphi,\alpha)+h$ where,

$$G(\theta,\phi,\alpha)=\int_{-\pi}^{\pi}\int_{-\alpha}^{\alpha} w(\theta,\phi|\theta',\phi')D(\theta',\phi')=\frac{W_0}{\pi}\alpha+\frac{W_1}{2\pi^2 a}\frac{a+b\cos(\phi)}{\sqrt{a^2+2ab\cos(\phi)+b^2}}$$
$$\times\left(\int_{-\pi}^{\pi}\frac{(a+b\cos(\phi'))^2}{\sqrt{a^2+2ab\cos(\phi')+b^2}}\mathrm{d}\phi'\right)\cos(\theta)\sin(\alpha). \tag{5}$$

It can be seen for any fixed value of φ that for $W_1>0$, this function decreases as we move away from the center of the bump at $\theta=0$. Hence as in the ring and sphere models, a necessary condition for the existence of an α-solution in which $\mathrm{U}>0$ for $-\alpha<\theta<\alpha$ and $U<0$ for $|\theta|>\alpha$ is $W_1>0$. Further note, that for the ring torus the terms containing φ in (5) are strictly positive contributions to the

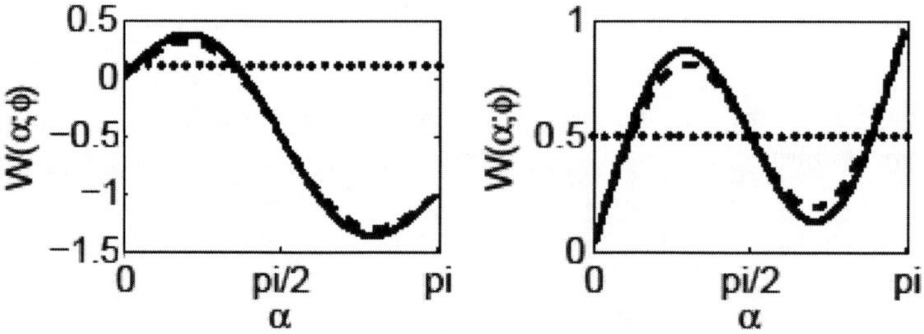

Fig. 4 Plot of $W(\alpha;\varphi)$ as a function of α for $\varphi = 0$ (*solid line*) and $\varphi = \pi/2$ (*dashed-line*). In the left panel $W_0 = -1$ and $W_1 = 4$. In the right panel $W_0 = 1$ and $W_1 = 4$. The *dotted line* represents $-h$. The existence of α-solutions can be found by solving $W(\alpha;\varphi) = -h$ graphically for various values of h. For example, the left panel has two solutions for $h = -0.1$ and the right panel has three solutions for $h = -0.5$. In both panels $a = 2$ and $b = 1$

equation indicating that the radius α of the solution is dependent upon φ. If we define $W(\alpha;\varphi) = G(\alpha,\varphi,\alpha)$, then finding the radius of the α-system becomes a problem of solving $W(\alpha\,;\varphi)+h=0$ for α as a function of φ. This is in contrast to the previously studied cases for the ring and sphere the width of the localized activity state did not have dependence upon a feature variable [4]. Examples of $W(\alpha,\varphi)$ are shown in Fig. 4 which correspond to $W_0 < 0$ and $W_0 > 0$, respectively. Notice that φ has impacts on both the minimum and maximum values of $W(\alpha;\varphi)$ and the location of these critical values. It is important to note that $W(0;\varphi) = 0$, $W(\pi;\varphi) = W_0$, and $W(\alpha;\varphi) = W_0 - W(\pi-\alpha)$.

Theorem 1. *In the absence of an inhomogeneous input:*

(a) There exists a ϕ*-solution if and only if* $h < 0$
(b) There exists a π*-solution if and only if* $W_0 > -h$
(c) There exists an α*-solution if and only if* $W_1 > 0$ *and* $W(\alpha;\varphi) + h = 0$ *with* $\pi > \alpha > 0$

Proof. (a) If there is a ø-solution then $\mathrm{U}(\alpha) = \mathrm{h}$ requiring $\mathrm{h} < 0$. If $\mathrm{h} < 0$ then $\mathrm{U}(\alpha) = \mathrm{h}$ is a ø-solution.
(b) For a π-solution, $u = W(\pi;\phi) + h = W_0 + h > 0$ which means that $W_0 > -h$. If $W_0 > -h$ then $u = W_0 + h$ is a π-solution.
(c) The conditions for the existence of an α-zolution follows from (5) as discussed above. An α-solution is zero on the boundary of R[U] if and only if $W(\alpha,\varphi) = 0$ and is negative outside the domain of R[U] if and only if $W_1 > 0$.

An example of an α-solution for the ring torus is shown in Fig. 5. Notice that the solution is a band around the torus centered at the preferred value of the θ variable that is widest at the preferred value of the φ variable.

We can further partition the parameter space (W_0, W_1) according to the sign of W_0 and the nature of the stationary points $W_{\max} = \overset{\max}{\underset{\alpha}{}} W(\alpha;\phi)$ and $W_{\min} = \overset{\min}{\underset{\alpha}{}} W(\alpha;\phi)$. In the ring and sphere topologies, five cases were identified

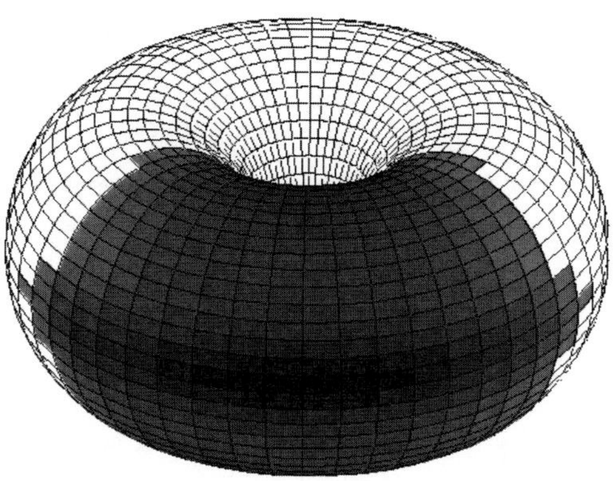

Fig. 5 Numerical example of an α-solution for the ring torus. The solution has been half-wave rectified for ease of presentation. Light and dark regions correspond to active and inactive regions respectively. Here $W_0 = -1$, $W_1 = 5$, $h = -0.1$, $a = 5$, and $b = 4$

in the parameter space that gave a classification of the equilibrium solutions for various values of h in the different portions of the parameter space [4]. We point out that the width of the localized activity state will depend on both the value of φ and the partitioning of the parameter space will depend on the torus radii a and b. The five cases are:

Case Ia: $W_1 > 0$, $W_0 < 0$ and $W_1 + W_0 < 0$: $W(\alpha;\varphi)$ is a monotonically decreasing negative function (no stationary points).

Case Ib: $W_1 > 0$, $W_0 < 0$ and $W_1 + W_0 > 0$: $W(\alpha;\varphi)$ has two stationary points with $W_{\max} > 0 > W_{\min}$ (see left panel of Fig. 4).

Case IIa: $W_0 > 0$, $0 < W_1 < \gamma_1 W_0$: $W(\alpha;\varphi)$ is a monotonically increasing positive function (no stationary points).

Case IIb: $W_0 > 0$, $\gamma_1 W_0 < W_1 < \gamma_2 W_0$: $W(\alpha;\varphi)$ has two stationary points with $W_0 > W_{\max} > W_{\min} > 0$ (see right panel of Fig. 4).

Case IIc: $W_0 > 0$, $W_1 > \gamma_2 W_0$: $W(\alpha;\varphi)$ has two stationary points with $W_{\max} > W_0 > 0 > W_{\min}$.

Case 1 follows from recognizing that $\frac{a+b\cos(\phi)}{\sqrt{a^2+2ab\cos(\phi)+b^2}} \leq 1$ and $\int_{-\pi}^{\pi} \frac{(a+b\cos(\phi))^2}{\sqrt{a^2+2ab\cos(\phi)+b^2}} \mathrm{d}\phi \leq \int_{-\pi}^{\pi} (a+b\cos(\phi))\mathrm{d}\phi = 2\pi a$. The parameters γ_1 and γ_2 satisfying $0 < \gamma_1 < \gamma_2$ depend upon the torus radii a and b and φ. With this notation, the taxonomy of the sets of equilibrium solutions for the various values of h is identical to that of the sphere model presented by Haskell and Bressloff [4] and is reproduced here in Table 1.

Theorem 2. *The possible sets of equilibrium solutions for various values of h is given in Table 1.*

Proof. The existence of $\phi-$, α-, and π-solutions follows from the qualitative properties of the function $W(\alpha;\varphi)$ in each of the five cases. Solving for $W(\alpha;\varphi) + h = 0$ will result in the various types of α-solution.

Table 1 Equilibria solutions for various values of h with ϕ, π, and a denoting the existence of $\phi-, \pi$-, and a-solutions, respectively. Also, $a_1 < a_2 < a_3$ and ă indicating unstable equlibria (analysis not shown here). The bottom row of each table specifies the minimum value of h within a given interval with h increasing from left to right

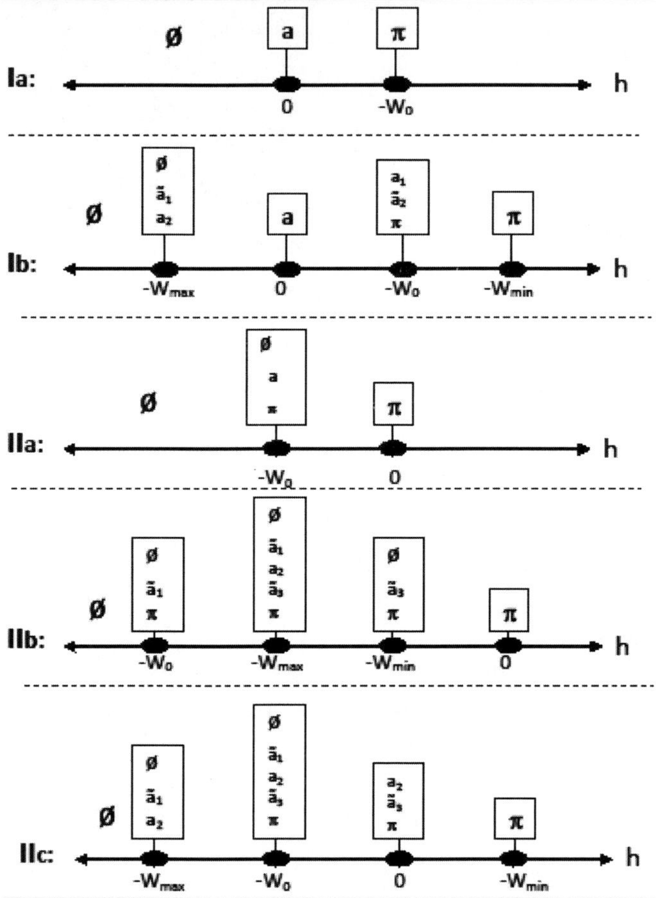

What remains to be determined are the coefficients γ_1 and γ_2. To find γ_1 we note that the condition $W_0 > 0$ and $W_1 = \gamma_1 W_0$ is the critical point when the two stationary points $W_{\max}$ and $W_{\min}$ coincide. By symmetry arguments this occurs when $\alpha = \pi/2$. We have that

$$W'(\alpha;\phi) = \frac{W_0}{\pi} + \frac{W_1}{2\pi^2 a} K(a,b;\phi)\cos(2\alpha), \tag{6}$$

where,

$$K(a,b;\phi) = \frac{a + b\cos(\phi)}{\sqrt{a^2 + 2ab\cos(\phi) + b^2}} \int_{-\pi}^{\pi} \frac{(a + b\cos(\phi'))^2}{\sqrt{a^2 + 2ab\cos(\phi') + b^2}} \mathrm{d}\phi'.$$

Table 2 Values of the parameters γ_1 and γ_2. The value of γ_2 given for the ring model is approximate

	Ring	Sphere	Ring torus
γ_1	1	2	$\frac{2\pi a}{K(a,b;\phi)}$
γ_2	4.6	8	$\frac{1}{\gamma_2}\frac{4\pi a}{K(a,b;\phi)}\alpha+\sin(2\alpha)=0$

From (6) see that $W'(\pi/2) = 0$ *when* $W_l = \gamma_l W_0$ *where* $\gamma_1 = \frac{2\pi a}{K(a,b;\varphi)}$. *The condition* $W_0 > 0$ and $W_l = \gamma_2 W_0$ *corresponds to the case where a pair of zeros of* $W(\alpha;\varphi)$ *coincide such that* $W_{\min} = 0$. *To find the value of* γ_2 *requires solving the transcendental equation* $\frac{W_0}{W_1}\frac{4\pi a}{K(a,b;\phi)}\alpha+\sin(2\alpha)=0$ to find the ratio of W_0 to W_1 that provides a solution for a torus with given radii a and b.

4 Tristability

Tristable states in networks of lateral inhibition type in which a stable α-solution coexists with stable Ø- and π-solutions for noncompact topologies are not found in one-dimension [1] and were first reported for two-dimensional planar networks [14]. For compact topologies, tristable states have been identified for the one-dimensional ring and two-dimensional sphere [4]. However, in contrast to the planar and line networks, the existence of a localized activity state is not dependent upon a presence of global inhibition [4]. In the earlier analysis of the ring model, the value of γ_2 required to have a stable α-solution in the presence of a predominantly excitatory network was misidentified. The correct value comes from finding the value of γ_2 where two roots of the transcendental equation $W(a) = \frac{W_0}{\pi}(a+\gamma_2\sin(2a))$ coalesce so that the minimum value is zero. Numerically this is found to occur approximately when $\gamma_2 = 4.6$.

Table 2 provides a summary of the parameter values γ_1 and γ_2 that partition the (W_0, W_1) parameter space by the different possible behaviors of the network. For the ring torus, these parameters will depend on the radii of the torus and are given in functional form. As with the ring model, finding the values of γ_2 for the ring torus involves finding the solution of a transcendental equation.

5 Discussion

We have analyzed the existence of localized activity states in neuronal network models of feature selectivity with a torus topology. Our worked is based upon a foundation presented by Amari for studying localized activity states for one-dimensional infinite neural fields of lateral inhibition type [1]. Amari showed that such fields are either monostable or bistable. When extended to a two-dimensional planar network, a regime of tristability appears in which a localized activity state coexists with both a quiescient and fully excited homogeneous state [14]. Extension to compact topologies of the ring and sphere showed that these topologies supported the monostable, bistable, and tristable states as well; however, in compact topologies a global inhibition is no longer required for the presence of a localized activity state [4]. In the compact topology of the torus, we are able to repeat these results as we have done in this paper. Experimental evidence continues to grow between the relationship between the structure of stimulus space and neuronal activity [11, 12] increasing the need to further understand the relationship between network topology and the computations permitted by cortical networks.

The feature selectivity properties of neurons is generally not limited to two features and the interactions of those features may be quite complex. As part of a first step toward understanding the computational behavior of neuronal netwoks with higher dimensional feature spaces, we have examined here the common properties found in networks of lateral inhibition type for the compact network topologies of the ring, sphere, and ring torus.

References

1. Amari, S. (1977). Dynamics of pattern formation in lateral-inhibition type neural fields. *Biol Cybernetics*, **27**, 77–87.
2. Barlowe, H. B. (1986). Why have multiple cortical areas? *Vision Res*, **26**, 81–90.
3. Conway, B. R., & Tsao, D. Y. (2009). Color-tuned neurons are spatially clustered according to color preference within alert macaque posterior inferior temporal cortex. *PNAS*, **106**, 18034–18039.
4. Haskell, E. C., & Bressloff, P. C. (2003). On the Formation of Persistent States in Neuronal Network Models of Feature Selectivity. *J Integ Neurosci*, **2**, 103–123.
5. Horton, J. C., & Adams, D. L. (2005). The cortical column: A structure without a function. *Philos Trans R Soc Lond B Biol Sci*, **360**, 837–862.
6. Hubel, D. H., & Wiesel, T. N. (1968). Receptive fields and functional echitecture of monkey striate cortex. *J Physiol*, **195**, 215–243.
7. Hubel, D. H., & Wiesel, T. N. (1974). Sequence regularity and geometry of orientation columns in the monkey striate cortex. *J Comp Neurol*, **158**, 267–293.
8. Laing, C. R., & Chow, C. C. (2001). Stationary bumps in networks of spiking neurons. *Neural Comp*, **13**, 1473–1494.
9. Laing, C. R., Troy, W. C., Gutkin, B., & Ermentrout, G. B. (2002). Multiple bumps in a neuronal model of working memory. *SIAM J on Appl Math*, **63**, 62–97.
10. Potthast, R., & Graben, P. B. (2010). Existence and Properties of solutions for neural field equations. *Math Meth Appl Sci*, **33**, 935–949.
11. Simoncelli, E.P., & Olshausen, B.A. (2001). Natural image statistics and neural representation. *Ann Rev Neurosci*, **24**, 1193–1216.
12. Singh, G., Memoli, F., Ishkhanov, T., Sapiro, G., Carlsson, G., & Ringach, D. L. (2008). Topological analysis of population activity in visual cortex. *J Vis*, **8**, 1–18.
13. Taylor, J. G. (1999). Neural 'bubble' dynamics in two dimensions: foundations. *Biol Cybernetics*, **80**, 393–409.
14. Werner, H., & T. Richter. (2001). Circular stationary solutions in two-dimensional neural fields. *Biol Cybernetics*, **85**, 211–217.

Intrinsic Fractal Dynamics in the Respiratory System by Means of Pressure–Volume Loops

Clara M. Ionescu and J. Tenreiro Machado

1 Introduction

The study of fractional order systems received considerable attention recently [20], due to the fact that many physical systems are well characterized by fractional models [18]. With the success in the synthesis of real noninteger differentiators, the emergence of new electrical elements and the design of fractional controllers [15, 19], fractional calculus (FC) has been applied in a variety of dynamical processes [21]. The importance of fractional order mathematical models is that it can be used to make a more accurate description and to give a deeper insight into the physical processes underlying long range memory behaviors. In previous works, it was demonstrated that the respiratory system has fractal dynamics and constitutes a good test-bed for their study [9, 10].

The pseudo phase space (PPS) is used to analyze signals with nonlinear behavior. For the two-dimensional case, it is called pseudo phase plane (PPP) [14]. In this paper, we propose the use of pressure–volume loops to analyze data from lung function tests in healthy subjects and in patients with respiratory disorders. The novelty of the proposed methodology is to combine the information from pressure–volume loops with the corresponding fractal dimension. In this way, the fractal dynamics of the respiratory system can be assessed and further analyzed [21]. One of the most common features extracted from lung function tests is the air-pressure and the air-flow variation during forced breathing or during breathing at rest. The data used in this application has been acquired in lung function tests during breathing at rest, namely the forced oscillation technique [17]. We aim to analyze three sets of patient data: healthy children, children with asthma and children with cystic fibrosis.

C.M. Ionescu (✉)
Ghent University, Technologiepark 913, B9052 Gent-Zwijnaarde, Belgium
e-mail: ClaraMihaela.Ionescu@UGent.be

J.A.T. Machado et al., *Nonlinear and Complex Dynamics: Applications in Physical, Biological, and Financial Systems*, DOI 10.1007/978-1-4614-0231-2_18,

This paper is organized as follows. Section two introduces the materials and methods employed in the paper. The breathing recordings from patients in time domain is presented for one patient in each group, along with information upon the various sets of data, biometric details of the population sets and measurement protocol. A short description of the pressure–volume (PV) loop is given in the same section. Next, the processing of the information extracted from the PV loops and the calculus of the fractal dimension follows, along with a power law model structure. Section three presents the results of the PV loop analysis and the feature study and section four discusses the geometrical and clinical interpretation of the emerged results. Finally, a conclusion section summarizes the main outcome of this paper.

2 Materials and Methods

2.1 Patient Database

There are three sets of patients available for analysis, whose biometric details are given in Table 1. This data has been previously published in [12].

The measurements on the 33 *healthy children* were performed at the St. Vincentius Basis School in Zwijnaarde, Belgium; the biometric details are given in Table 1. The children had no history of pulmonary disease, and were selected using a specific questionnaire. The questionnaire verified the absence of dyspnea, chronic cough, wheeze in the chest, etc. The measurements performed on healthy children have been verified with predicted values [7], based on their height. All measured data was validated within the 95% confidence interval values.

Asthma denotes a pulmonary disease in which there is obstruction to the flow of air out of the lungs, but the obstruction is usually reversible and, between attacks of asthma, the flow of air through the airways is usually good [5]. Asthma is caused by chronic (ongoing, long-term) inflammation of the airways, making them highly sensitive to various triggers. Asthma can be controlled using specific medication (inhaled steroids). The data for this study was recorded at the University Hospital Antwerp-Belgium from 44 asthmatic children data sets whose corresponding biometric and spirometric values are given in Table 1.

Table 1 Biometric parameters of the investigated children subjects

Children	Healthy $n = 33$	Asthma $n = 44$	CF $n = 38$
Age (years)	9 ± 1	11 ± 4	14 ± 6
Height (m)	1.35 ± 0.05	1.40 ± 0.2	1.49 ± 0.15
Weight (kg)	32 ± 6	36 ± 15	40 ± 11

Values are presented as mean $\pm$ standard deviation
CF cystic fibrosis, *n* the number of tested volunteers

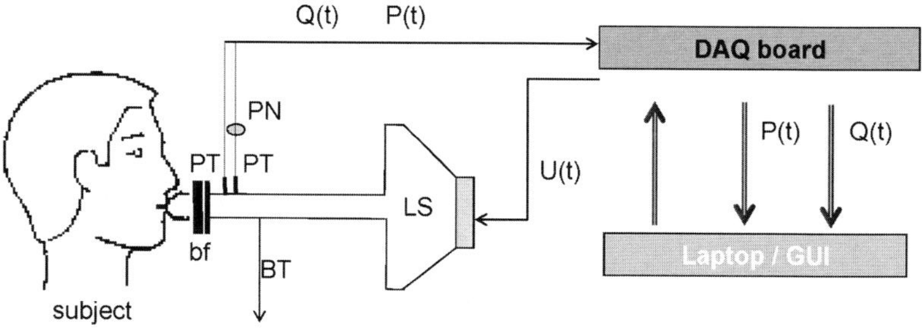

Fig. 1 A schematic overview of the device with patient performing the FOT lung function test. *LS* loudspeaker, *BT* bias-tube, *PN* pneumotachograph, *PT* pressure transducer, *bf* biological filter with mouthpiece, *Q* flow, *P* pressure, $U(t)$ driving signal (test input)

Cystic fibrosis (CF) is one of the most common severe (genetic) diseases, characterized by the production of abnormal secretions, leading to mucus build-up and persistent inflammation in a variety of organs [4, 8]. Inflammation and infection also cause injury and structural changes to the lungs, originating a variety of symptoms and eventually respiratory failure. The data used in this study was recorded at the University Hospital Antwerp from 38 data sets from children diagnosed with cystic fibrosis whose biometric and spirometric values are given in Table 1.

2.2 Lung Function Testing: The Forced Oscillation Technique

The impedance was measured using the Forced Oscillation Technique (FOT) standard setup, commercially available, assessing respiratory mechanics from 4 to 48 Hz in steps of 2 Hz. The subject is connected to the typical setup from Fig. 1 via a mouthpiece, suitably designed to avoid flow leakage at the mouth and dental resistance artifact. The oscillation pressure in most recent FOT devices is generated by a loudspeaker (LS) connected to a chamber, namely the SPH-165KEP MONACOR, with a range from 3 to 1,000 Hz. The LS is driven by a power amplifier fed with the oscillating signal generated by a computer, namely a HP Pavilion dv1000 with a Pentium M processor, 1.5 GHz, 512 MB with 266 MHz SDRAM. The movement of the LS cone generates a pressure oscillation inside the chamber, which is applied to the patient's respiratory system by means of a flexible respiratory tube of 1 m length and 2 cm diameter, connecting the LS chamber and the bacterial filter (bf). A side opening (BT) of the main tubing allows the patient to decrease total dead space re-breathing (i.e., 40 ml). This bias tube exhibits high impedance at the excitation frequencies to avoid the loss of power from the LS pressure chamber [2]. During the measurements, the patient wears a nose clip and keeps the cheeks firmly supported to reduce the artifact of upper airway shunt.

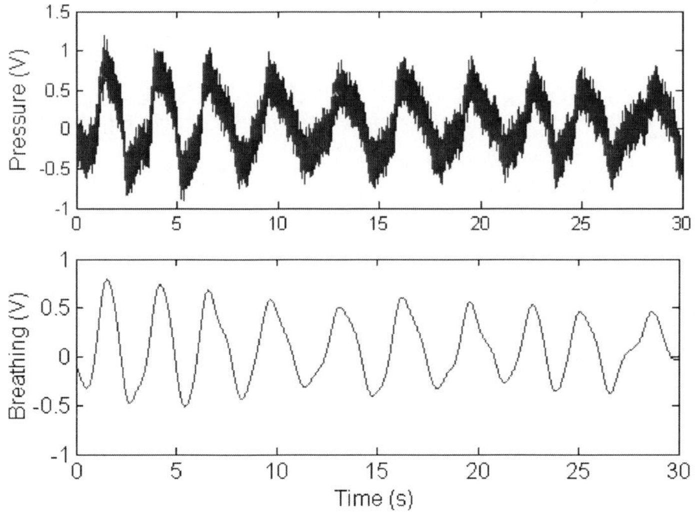

Fig. 2 Typical recorded airway pressure signal from the patient and filtered breathing signal

Pressure and flow are measured at the mouthpiece, respectively, by means of (a) a pressure transducer (PT) and (b) a pneumotachograph (PN) plus a differential pressure transducer (PT). The high precision pressure transducers are BSDX0050D4D, with a bipolar pressure range from 0 to 1 kPa, accuracy of 0.004 kPa and a common mode rejection ratio of 80 dB over the frequency range of interest. The working range is a peak-to-peak size between 0.1 and 0.3 kPa, in order to ensure optimality, patient comfort and stay within a narrow range in order to assume linearity [17]. The flow is measured using a Hans Rudolph pneumotachograph, covering a range from 0 to 400 l/min, (6.6 l/s), 4830B series, with a dead space volume between 0 and 6.66 ml.

Averaged measurements from 3 to 5 technically acceptable tests were taken into consideration for further signal processing (i.e., coherence function above 0.8). The signals were acquired using a PCMCIA DAQ card 4026E series, 12-bit from National Instruments, sampled every 1 ms. Typical time-records are depicted in Fig. 2. All patients were tested in the sitting position, with cheeks firmly supported and elbows resting on the table.

The output of the system is the airway pressure, where both the low-frequency component (i.e., breathing) and the excitation signal (i.e., multisine) are superimposed. Since the breathing frequency is one decade below the first excited harmonic, a low-pass filter can be applied and the breathing signal $b(t)$ can be extracted, as given in Fig. 2 (right). For the purpose of this investigation, we applied a Butterworth filter of order four and cut-off frequency at 1 Hz. From the flow measurement, volume has been obtained by integration.

2.3 The Pressure–Volume Curve

In clinical terms, the air-pressure and air-volume variations in one breathing cycle plotted against each other form a closed loop known as the PV loop [16]. The area inside this loop, and the slope of the axis of the minimal-to-maximal points in the PV loop are used to evaluate the respiratory mechanics of the patient. The interpretation of the PV loop is then made with respect to inspiratory and expiratory parameters, such as air-flow resistance and work of breathing.

The PV loops are defined by

$$\text{Area} = \int_0^T V(t)\mathrm{d}P(t) = \int_0^T P(t)\mathrm{d}V(t) \tag{1}$$

with $P(t)$ the pressure and $V(t)$ the volume at time instants t. The air-flow is related to the air-volume by $Q(t) = \mathrm{d}V(t)/\mathrm{d}t$, and using this in (1) we obtain that the area is the integral of the power:

$$\text{Area} = \int_0^T P(t)Q(t)\mathrm{d}t, \tag{2}$$

which is by definition the work (energy) of breathing to perform the cycle over the period T. With disease, the work of breathing is increased and the ratio between peak pressure and peak volume is altered. However, for the purpose of this study, the area within the PV loop is of interest to us within a geometrical context.

2.4 Fractal Dimension and Power Law Model

The fractal dimension F_d is a quantity that gives an indication of how completely a spatial representation appears to fill the space. There are many specific methods to compute the fractal dimension. The most popular and simple methods are the Hausdorff dimension and box-counting dimension [1]. In this paper, the box-counting dimension method is used due to its simplicity of implementation and is defined as:

$$F_\mathrm{d} = \frac{ln\,[N(\varepsilon)] - ln(C)}{ln(1/\varepsilon)}, \tag{3}$$

where C is a constant related to the total area, $N(\varepsilon)$ represents the minimal number of covering cells (e.g., boxes) of size ε required to cover the PV graph. For each box size value ε, follows a corresponding number of boxes $N(\varepsilon)$. The slope for various values of box-sizes provides an estimate of the fractal dimension F_d:

$$N(\varepsilon) = C(1/\varepsilon)^{F_\mathrm{d}} \tag{4}$$

Further on, given for each patient we have a PV plot, we shall obtain for each patient a value for C and a value for F_{d}. If one represents these values in a log–log plot, a power law model can be fitted to each group of patients, allowing deriving a model for each pathology:

$$C = A \times F_{\mathrm{d}}^{B} \tag{5}$$

with A and B identified constants.

3 Results

For each measured set of signals, we have extracted the breathing signal and the pressure–volume curves. An illustrative example is given in Fig. 3.

Based on the PV plots, the box-counting method can be applied to obtain the fractal dimension F_{d} and the constant C for each set of data. The models are identified for each set of measurements from volunteer using the least-squares algorithm [13]. Consequently, the box-counting values $N(\varepsilon)$ and the box-sizes $1/\varepsilon$ values are obtained for each patient, in each data set. The result is depicted for each group in Figs. 4–6, respectively. From the F_{d} and C data obtained, one may fit a trend-line for each data set, following an identified model parameters from (5) valid for each respiratory disease. The values of the fitted trend-lines are given in Table 2, with the corresponding plots in Fig. 7.

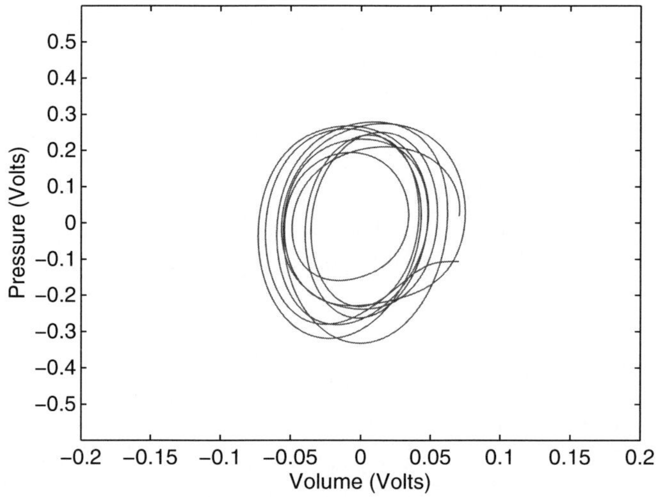

Fig. 3 Illustrative example of the PV loop for a healthy child

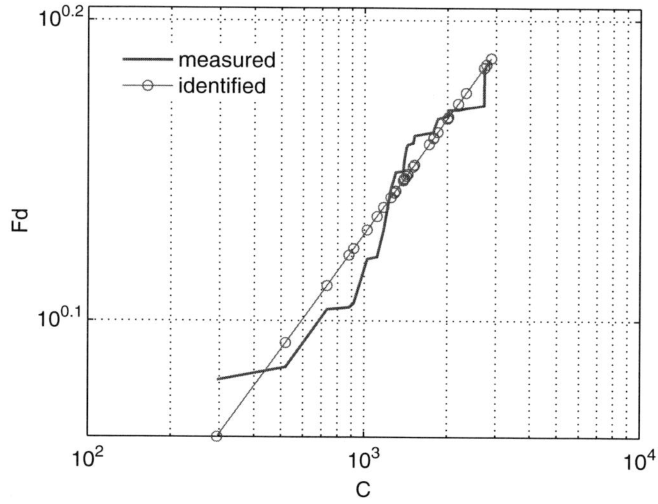

Fig. 4 Healthy children: the information extracted for each patient in terms of C and F_d, and the result of the identification

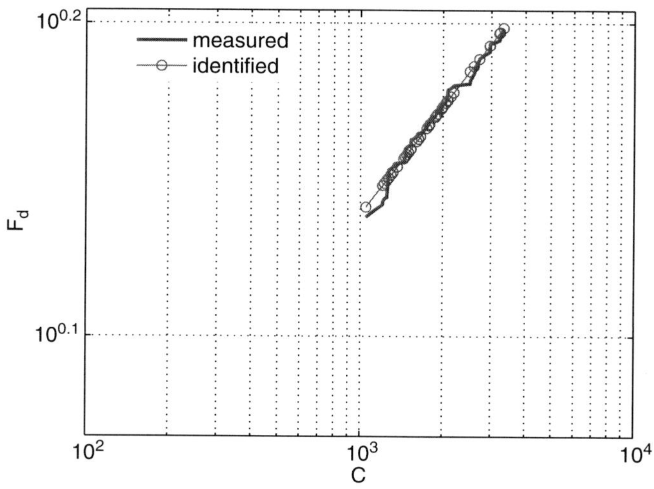

Fig. 5 Asthma children: the information extracted for each patient in terms of C and F_d, and the result of the identification

4 Discussion

In its most simple representation, the respiratory system can be represented as a series connection of a resistance R_e and a compliance C_e. It assumes patient's respiratory muscles inactive and the external equipment is driving the flow into the

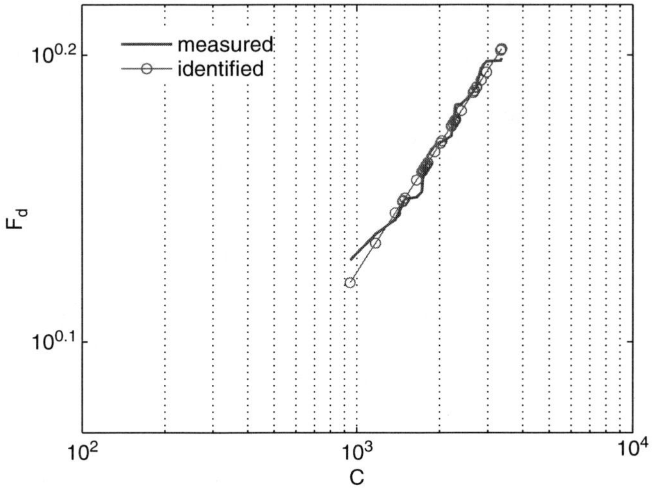

Fig. 6 Cystic fibrosis children: the information extracted for each patient in terms of C and F_{d}, and the result of the identification

Table 2 Fitted power-law models (5) for each group of data

Data	A	B
Healthy children	0.5579	0.1275
Asthma	0.6245	0.1145
CF	0.4786	0.1475

CF cystic fibrosis

lungs [3]. The driving pressure $P(t)$ generates flow $Q(t)$ across the resistance and the volume $V(t)$ changes in the compliance. If $P_r(t)$ and $P_e(t)$ are the resistive and elastic pressure drops respectively, we have that:

$$R_r = \frac{P_r(t)}{Q(t)};\ C_r = \frac{V(t)}{P_e(t)} \text{ and } P(t) = P_e(t) + P_r(t). \tag{6}$$

It results that:

$$P(t) = R_r \times Q(t) + \frac{V(t)}{C_r} \tag{7}$$

This represents the first-order equation in the motion-equation for a single compartment model of the respiratory system: a single balloon with compliance C_r on a pipeline with a resistance R_r. This system can be studied using the exponential decay of volume $V(t)$ as resulting from a step input V_0:

$$V(t) = V_0 \times \mathrm{e}^{-\frac{t}{\tau}}, \tag{8}$$

where t is time and τ is the time constant which characterizes the system, denoted by the product of R_rC_r [9, 10, 17].

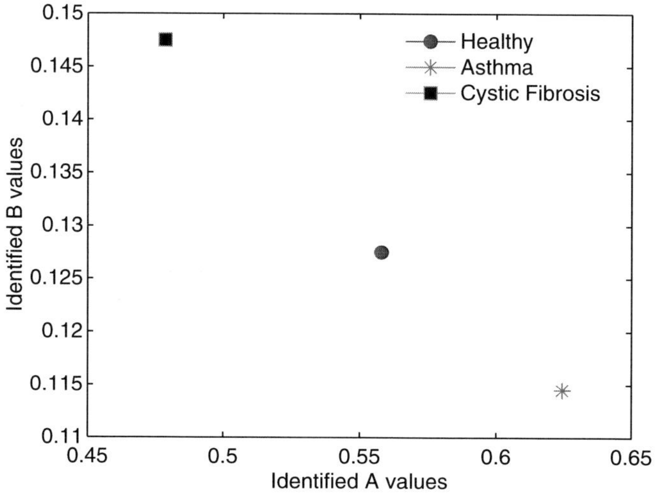

Fig. 7 The plot of the identified A and B values from model (5) for each set of data

In the representation of the PV plots, we have the breathing signal expressed as pressure and the volume. From (7) can be observed that there exists a relation between pressure and flow ($Q(t) = \mathrm{d}V/\mathrm{d}t$). In clinical terms, the pressure–volume loop during one breathing period is able to tell the clinician something about the dynamic compliance of the respiratory system and its work. The area enclosed by the PV loop is called the physiologic work of breathing, denoting the resistive work performed by the patient to overcome the resistance present in the airways.

During cycling loading, the stress that develops in the viscoelastic body (respiratory tissue) displays:

- A component in phase with strain, which is the elastic stress contributing to the storage modulus E_S (elastance).
- A component out of phase with strain, corresponding to the viscous dissipation and contributing to the loss of modulus E_D (damping).

In [6, 11] was shown that the respiratory system can be indeed modeled as a combination of series RC elements in a cascade arrangement of consecutive airways, by using their mechanical analogue representation, springs K_{rs} and dashpots B_{rs}. In this mechanical model, it follows that the pressure–volume relationship equivalent to the stress-strain relationship is given by:

$$P(t) = \frac{K_{\mathrm{rs}}\ell_{\mathrm{rs}}}{A_{\mathrm{rs}}}V(t) + \frac{B_{\mathrm{rs}}\ell_{\mathrm{rs}}}{A_{\mathrm{rs}}}\frac{\mathrm{d}V(t)}{\mathrm{d}t}, \tag{9}$$

with P the air-pressure, V the air-volume, ℓ_{rs} and A_{rs} the changes in length and area of airways during the breathing cycle, and $K_{\mathrm{rs}}, B_{\mathrm{rs}}$ the constants of the spring and dashpot, respectively [6]. This relation suggests that the PV loop is indeed related to position and speed of air during breathing.

In summary, it can be concluded that the PV plots are able to provide information upon the intrinsic fractal dynamics of the breathing which changes with disease. Using the derived parameters extracted from the information delivered by the PV plots, i.e., models (4) and (5), we were able to obtain a classification between all three sets of measurements. It is interesting to notice that changes in the respiratory dynamics implies changes in the fractal dimension, as suggested by the values given in Table 2. This is indeed in agreement with the fact that airways and tissue structure are changing with disease, leading to changes in the fractal dimension of the respiratory tree [9, 10].

5 Conclusion

In this paper, a pressure–volume analysis has been performed on respiratory data from three groups of children: healthy, asthmatic, and cystic fibrosis. The results suggest the usefulness of using pressure–volume loops for mapping purposes, describing (dis)similarities within the breathing dynamics in groups of patients and between groups of patients.

Acknowledgments C. Ionescu gratefully acknowledges the children from primary school in Zwijnaarde who volunteered to perform lung function testing in our laboratory. We also acknowledge the technical assistance provided at University Hospital Antwerp, Belgium, to measure asthma and CF diagnosed children. J. Tenreiro Machado would like to acknowledge FCT, FEDER, POCTI, POSI, POCI, POSC, POTDC, and COMPETE for their support to R&D Projects and GECAD Unit.

References

1. Baker G.L., Gollub J.B., (1996) Chaotic Dynamics: An Introduction, 2nd ed. Cambridge, England: Cambridge University Press.
2. Birch M., MacLeod D., Levine M., (2001) An analogue instrument for the measurement of respiratory impedance using the forced oscillation technique, Phys Meas 22:323-339
3. Blom J., (2004) Monitoring of respiration and circulation, CRC Press.
4. Brennan S., Hall G., Horak F., Moeller A., Pitrez P., Franzamann A., Turner S., de Clerck N., Franklin P., Winfield K., Balding E., Stick S., Sly P., (2005) Correlation of forced oscillation technique in preschool children with cystic fibrosis with pulmonary inflammation, Thorax 60:159-163
5. Busse W., Lemanske R., (2001) Asthma, New Engl J Med, 344(5): 350-362
6. De Geeter N., Ionescu C., De Keyser R., (2009) A mechanical model of soft biological tissue - an application to lung parenchyma, In: Proceedings of the 31st Annual Int Conf of the IEEE Engineering in Medicine and Biology Society, Minneapolis, USA, 2-6 September, ISBN 978-1-4244-3296-7, 2863-2866
7. Duiverman E., Clement J., Van de Woestijne K., Neijens H., van den Bergh A., Kerrebijn K., (1985) Forced oscillation technique: reference values for resistance and reactance over a frequency spectrum of 2-26 Hz in healthy children aged 2.3-12.5 years, Clinical Resp Physiol 21:171-178

8. Elizur A., Cannon C., Ferkol T., (2008) Airway inflammation in cystic fibrosis, Chest 133(2):489-495
9. Ionescu C., Segers P., De Keyser R., (2009) Mechanical properties of the respiratory system derived from morphologic insight, IEEE Trans Biomed Eng 56(4):949-959
10. Ionescu C. , Muntean I., Machado J.T., De Keyser R., Abrudean M., (2010) A theoretical study on modelling the respiratory tract with ladder networks by means of intrinsic fractal geometry, IEEE Trans Biomed Eng 57(2):246-253
11. Ionescu C., Kosinsky W., De Keyser R., (2010) Viscoelasticity and fractal structure in a model of human lungs, Archives of Mechanics 62(1): 21-48
12. Ionescu C., (2009) Fractional-order models for the respiratory system, Doctoral Thesis, ISBN 978-90-8578-318-3
13. Ljung L., (1999) System identification: theory for the user, Prentice Hall
14. Moon F.C., (1987) Chaotic Vibration, New York: John Wiley,
15. Monje A. , Chen Y., Vinagre B., Xue D., Feliu V., (2010) Fractional order systems and controls, Springer-Verlag
16. Northrop R., (2002) Non-invasive instrumentation and measurement in medical diagnosis, CRC Press
17. Oostveen E.,Macleod D. , Lorino H., Farré R., Hantos Z., Desager K., Marchal F., (2003) The forced oscillation technique in clinical practice: methodology, recommendations and future developments, Eur Respir J 22:1026-1041
18. Podlubny I., (2002) Geometrical and physical interpretation of fractional integration and fractional differentiation, Journal of Fractional Calculus and Applied Analysis 5(4):357-366
19. Tenreiro Machado J. A., (1997) Analysis and Design of Fractional-Order Digital Control Systems, Journal Systems Analysis-Modelling-Simulation, Gordon and Breach Science Publishers 27:107-122
20. Tenreiro Machado J. A., (2003) A Probabilistic Interpretation of the Fractional-Order Differentiation, Journal of Fractional Calculus and Applied Analysis 6(1):73-80
21. West B., (2010) Fractal physiology and the fractional calculus: a perspective, Frontiers in Fractal Physiology 1(12):1-17 (open source: www.frontiersin.org)

Part III
Nonlinear and Complex Dynamics: Applications in Financial Systems

Forecasting Project Costs by Using Fuzzy Logic

M. Bouabaz, M. Belachia, M. Mordjaoui, and B. Boudjema

1 Introduction

The concept of fuzzy logic is derived from the theory of fuzzy sets. The theory of fuzzy sets was developed by Zadeh [1]. It is ranked among the methods of artificial intelligence. The method of fuzzy logic provides a way for coping with problems arising from unexpected situations. It is a means for solving hard problems, by determining a mathematical model that describes the system behavior, known as an unsupervised learning method.

The proposed model consists of two parts: First, optimization is the process to determine the number of clusters from data input–output, which respectively serves for subsequent use. The second part involves the extraction of fuzzy rules.

In this paper, we evaluate the use of the expectation maximization clustering algorithm in modeling and estimating projects costs.

1.1 The Fuzzy Clustering Algorithm

The Takagi–Sugeno [2] was the first model developed in 1985. This model can effectively represent complex nonlinear systems using fuzzy sets. The clustering technique is an essential method in data analysis and pattern recognition. Fuzzy clustering allows natural grouping of data in a large data set and provides a basis for constructing a rule-based fuzzy model. It is a partitioning method of data intosubsets

M. Bouabaz (✉)
Civil Engineering Department, University of 20August, Skikda, Algeria
e-mail: mbouabaz@hotmail.fr

J.A.T. Machado et al., *Nonlinear and Complex Dynamics: Applications in Physical, Biological, and Financial Systems*, DOI 10.1007/978-1-4614-0231-2_19,

or groups based on similarities between the data [2]. The representation of fuzzy rules for the Takagi–Sugeno model takes the form:

$$\begin{aligned} R_i : \text{If } x_1 \text{ is } A_{i,1} \text{ and} \ldots \text{and } x_p \text{ is } A_{i,p} \\ \text{Then } y_I = a_{i0} + a_{i1}x_1 + \ldots a_{ip} \end{aligned} \tag{1}$$

where, R_i is the rule number, x_j is the j-th input variable, A_{ij} is the fuzzy set of the j-th input variable in the i-th rule, y_i is the output of the i-th fuzzy rule.

1.2 The Fuzzy C-Means Algorithm

Fuzzy C-means algorithm (FCM) is a fuzzy clustering technique that is different from C-means that uses hard partitioning. The fuzzy c-means uses fuzzy partitioning in which a data point can belong to all clusters with different grades between 0 and 1. The FCM is an iterative algorithm that aims to find cluster centers that minimize the objective function.

$$J_{\text{FCM}}(Z;\, \Phi,\, V) = \sum_{i=1}^{c} \sum_{k=1}^{N} (\mu_{ik})^m \; D_{ikA}^2 \tag{2}$$

where $V = [v_1,\, v_2,\, \ldots,\, v_c]$, v_{i} are the clusters centers to be determined. $\phi = \{\mu_{ik}\}$ is a fuzzy partition matrix; (μ_{ik}) is a membership degree between the ith cluster and kth data which is subject to conditions (3).

$$\mu_{ik} \in [0,1], \quad 0 < \sum_{k=1}^{N} \mu_{ik} < N, \quad \sum_{i=1}^{c} \mu_{ik} = 1 \tag{3}$$

1.3 The Expectation Maximization Algorithm

The EM algorithm was proposed by Abonyi [3]. It is an extension of the algorithm of Gath and Geva [4], with a covariance matrix has nonzero diagonal elements, which creates an error in the projection of these elements. The methodology of the used algorithm is described as follows:

The partition matrix is expressed by:

$$\mu_{ik} = \frac{1}{\sum_{j=1}^{c} \left(D_{ikA}^2 / D_{jkA}^2\right)^{1/(m-1)}} \qquad 1 \le i \le c, \quad 1 \le k \le N \tag{4}$$

The prototypes (center) are:

$$V_i^x = \frac{\sum_{k=1}^{N} \mu_{i,k}^{(l-1)} X_k}{\sum_{k=1}^{N} \mu_{i,k}^{(l-1)}} \tag{5}$$

The standard deviation of the Gaussian membership functions is:

$$\sigma_{i,j}^2 = \frac{\sum_{k=1}^{N} \mu_{i,k}^{(l-1)} (x_{j,k} - v_{j,k})^2}{\sum_{k=1}^{N} \mu_{i,k}^{(l-1)}} \tag{6}$$

The local model parameters are extracted as follows:

$$\theta_i = \left(X_e^T \Phi_i X_e\right)^{-1} X_e^T \Phi_i y \tag{7}$$

where Φ is the weights matrix having the membership degrees defined by:

$$\Phi_i = \begin{bmatrix} \mu_{i,1} & 0 & \cdots & 0 \\ 0 & & \cdots & 0 \\ \vdots & \vdots & \ddots & \vdots \\ 0 & 0 & \cdots & \mu_{i,N} \end{bmatrix} \quad X = \begin{bmatrix} X_1^T \\ \cdot \\ \cdot \\ X_N^T \end{bmatrix} \quad y = \begin{bmatrix} y_1 \\ \cdot \\ \cdot \\ y_N \end{bmatrix} \tag{8}$$

The extended matrix X_e is given by; $\mathtt{X_e} = [\mathtt{X} \quad \mathtt{1}]$
And the prior probability is expressed by:

$$\alpha_i = \frac{1}{N} \sum_{k=1}^{N} \mu_{i,k} \tag{9}$$

The weights on rules are expressed as follows:

$$w_i = \prod_{j=1}^{n} \frac{\alpha_i}{\sqrt{2\pi\sigma_{i,j}^2}} \tag{10}$$

with a distance norm given by the following expression:

$$\frac{1}{D_{i,k}^2} = w_i \times \exp\left(-\frac{1}{2} \frac{(x_{j,k} - v_{i,j})^2}{\sigma_{i,j}^2}\right).$$

$$\frac{1}{\sqrt{2\pi\sigma_i^2}} \times \exp\left(\frac{(y_k - f_i(x_k,\ \theta_i))^T (y_k - f_i(x_k,\ \theta_i))}{2\sigma_i^2}\right) \tag{11}$$

where $f_i(\mathrm{x}_k,\ \theta_i)$ is the model consequents.

The proposed algorithm is summarized as follows:

Let $Z = \{z_{k1}, z_{k2}, \ldots, z_{kn}\}^T$
Select the number of clusters $c > 1$
Select the weithning exponent ($\mathbf{m} = 2$)
and the termination criterion ($\varepsilon > 0$)
Initialize the partition matrix such as (**3**)

Start

1: Compute the clusters centers using (**5**)
2: Compute fuzzy covariance matrix by (**6**)
3: Compute (7), (9), and (10)
4: Compute the distances using (**11**)
5: Update the partition matrix using (**4**)

If stopping criterion $|\boldsymbol{U}^{(l)} - \boldsymbol{U}^{(l-1)}| \leq \varepsilon$ satisfied then stop
Otherwise $l \leftarrow l + 1$ and go to step 2
End

2 Validation

2.1 Indices

It is important to determine the number of clusters for use in simulation. For this, different indices for validation have been proposed by Bezdek [5] in data clustering. This can be done by the partition coefficient (PC) and the partition entropy (PE).

The PC measures the amount of overlapping between clusters.

$$PC\ (c) = \frac{1}{N} \sum_{i=1}^{c} \sum_{k=1}^{N} (\mu_{ik})^2 \tag{12}$$

The PE measures the fuzzyness of the cluster

$$PE(c) = -\frac{1}{N} \sum_{i=1}^{c} \sum_{k=1}^{N} \mu_{ik} \log(\mu_{ik}) \tag{13}$$

where $PC(c) \in [1/c,\ 1]$, $PE(c) \in [0,\ \log_a]$, with an increase of c, the values of PC and PE are decreased/increased, respectively. The above mentioned cluster validity indices are sensitive to fuzzy coefficient m. When $\mathtt{m} \to \mathtt{1}$, the indices give the same value for all c. When $\mathtt{m} \to \infty$, both PC and PE exhibit significant knee at $c = 2$. The number corresponding to a significant knee is selected as the optimal number of clusters.

2.2 The Performance Error

To evaluate the performance of the Fuzzy model, we use the following criteria proposed by Bezdek [6]. They are used for evaluating the output of the model.

The root mean squared error is expressed as follows:

$$\text{RMSE} = \sqrt{\frac{1}{N}\sum_{k=1}^{N}(y_k - \hat{y}_k)^2} \tag{14}$$

where y is the output of the process and y' is the output of the model.

The variance accounted for is expressed by:

$$\text{VAF} = 100\% \left[1 - \frac{\text{var}(y-\hat{y})}{\text{var}(y)}\right] \tag{15}$$

Where n is the number of projects and i is the project number ($i = 1\ldots n$), and y' is the actual output

3 Application of Fuzzy Logic to Cost Estimation

The production of an accurate estimate for estimating projects costs is a challenging task for the estimator at the early stage of a project.

In an attempt to overcome the problem, a soft computing method has been used to solve the problem of uncertainties in estimating and constructing an accurate model for forecasting the final cost of a project.

3.1 The Model Development

The present model has been developed in three phases. First, it consists of the determination of the number of cluster. Second, the learning phase, and finally the testing phase.

3.2 The Data

A wide range of project contracts made on cost–significance work packages were used for modelling purposes [7].

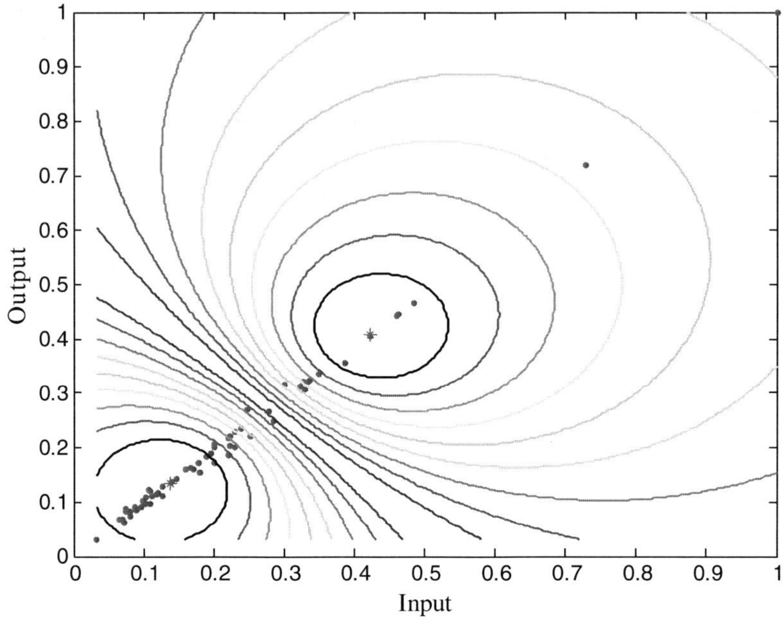

Fig. 1 The clustering map using expectation maximisation algorithm

3.3 The Clustering

The clustering consists of the selection of the number of clusters, which depends on the PC and the classification entropy.

The clustering map by the expectation maximisation algorithm is presented in Fig. 1.

We can deduce that the clustering number is equal to 2 clusters according to PC and CE from Fig. 2.

The values for PC and PE at optimization stage are given in Tables 1 and 2.

3.4 The Model Simulation

The proposed model was developed utilizing a set of projects. The developed model was generated from 68 data samples using Matlab toolbox [3,8] in a microcomputer. It has 1 input and 1 output. The training was stopped when the variance accounted-for (VAF) reached the maximum percentage value of 99.5304 in an elapsed time of 1.335000 s. The termination tolerance of the clustering algorithm was 0.01. The training error (RMSE) is 0.0107. The simulation model at learning phase is shown graphically on Fig. 3.

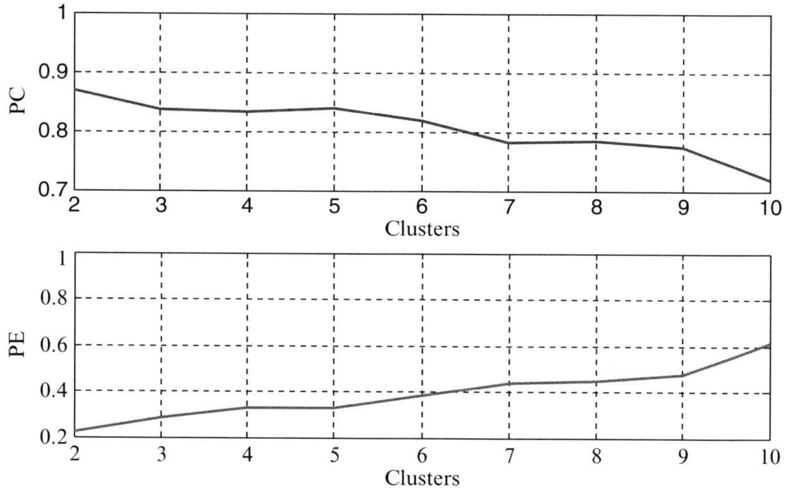

Fig. 2 Clusters validity indices for PC and PE

Table 1 Summarize the clustering results

Indices					Clusters			
2	3	4	5	6	7	8	9	10
PC = 0.869	0.837	0.832	0.839	0.818	0.781	0.783	0.774	0.719
CE = 0.224	0.284	0.328	0.327	0.385	0.436	0.449	0.475	0.611

Table 2 The values for the clusters centres

Rules	$y(k-1)$	U
R1	1.61×10^{-1}	1.69×10^{-1}
R2	4.20×10^{-1}	4.28×10^{-1}

In order to illustrate sense of membership functions of the actual data versus simulated data obtained by the projection of the clusters by the expectation maximization algorithm. Fig. 4 shows fuzzy sets of the model with their rules presented in Table 3.

3.5 *The Testing Model*

A testing phase was investigated on the adopted rules in order to determine the accuracy of the model. Some flattering results are shown in Table 4.

Figure 5 shows the plot of the results given in Table 4 from the obtained rules at testing phase.

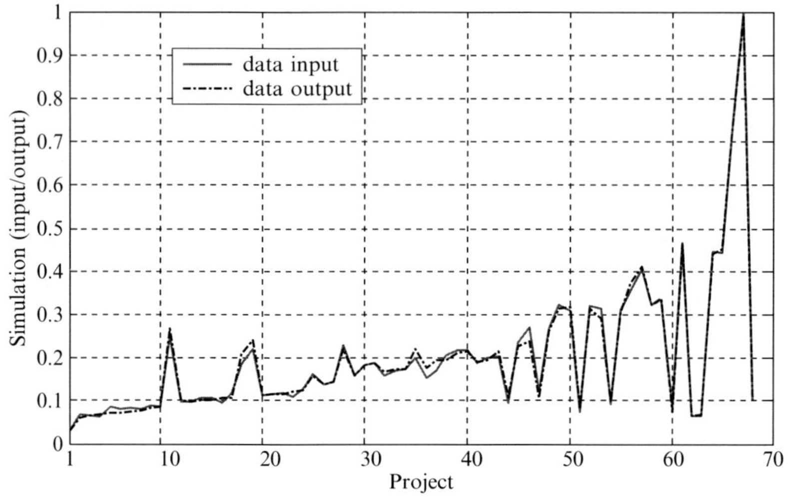

Fig. 3 Predicted model at learning phase

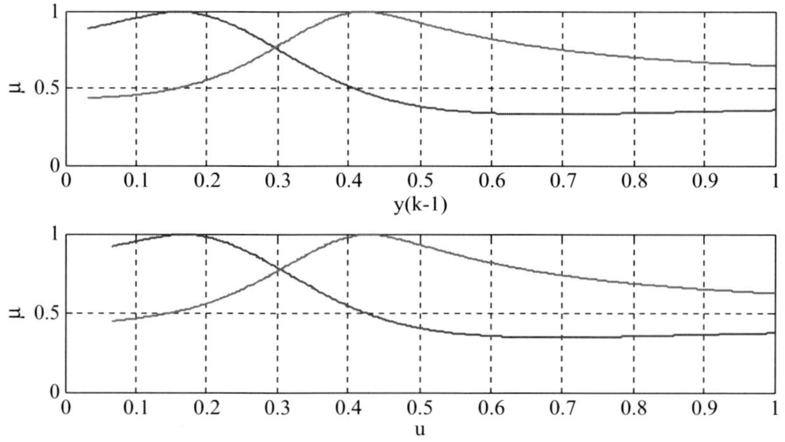

Fig. 4 The membership functions of the model

Table 3 Fuzzy rules obtained by expectation maximization algorithm

Rule 1
If $y(k-1)$ is A_{11} and u is A_{12} Then $y(k) = 4.50 \times 10^{-2} y(k-1) + 9.23 \times 10^{-1} u + 1.04 \times 10^{-3}$
Rule 2
If $y(k-1)$ is A_{21} and u is A_{22} Then $y(k) = 5.27 \times 10^{-3} y(k-1) + 1.04 \times 10^{0} \mathrm{u} + 270 \times 10^{-2}$

Table 4 Results of testing rules

Project	Value of cswp's	Simulated fuzzy model	Actual bill value	Cpe
N°	(£)	(£)	(£)	(%)
1	21,402	26,662	26,753	0.000
2	47,158	57,931	57,775	−0.267
3	37,520	47,145	47,451	0.650
4	29,668	37,086	37,794	1.909
5	44,500	54,884	55,152	0.489
6	39,400	50,407	49,663	−1.475
7	57,898	72,373	71,702	−0.925
8	91,023	112,374	115,301	2.605
9	62,890	80,481	78,833	−2.046
10	110,264	137,241	140,700	2.520
Mean error				0.345
Standard deviation				1.617

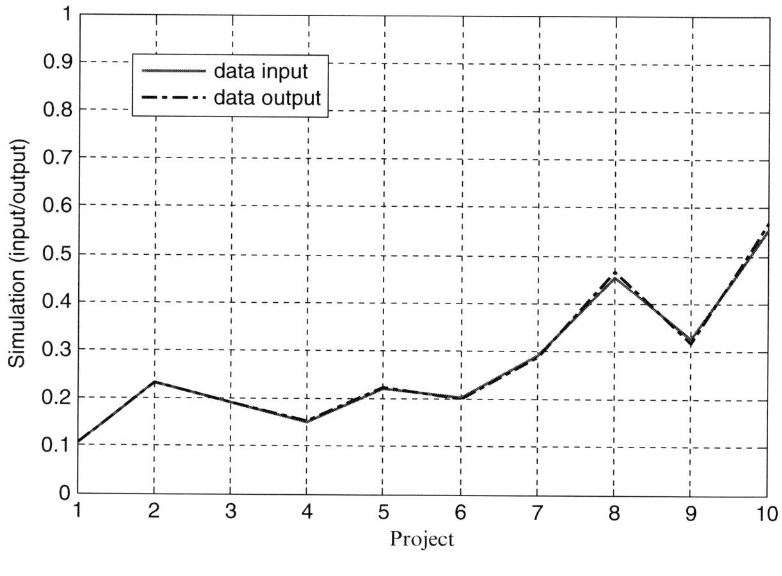

Fig. 5 Plot of simulated values on actual at testing phase

4 Conclusion

A modified fuzzy clustering algorithm based on the expectation maximization algorithm was used to construct a fuzzy model. Based on the results of the study, it can be concluded that the objectives of the research outlined has been achieved and the accuracy is significantly improved by means of fuzzy logic. It is also encouraged that fuzzy logic could be used in estimating and financing projects, were bills of quantities made on cost-significance work packages for civil engineering projects contracts models are ready for use. As a conclusion fuzzy clustering approach based on the expectation–maximization algorithm seems to reveal promising results in modeling and forecasting projects costs in terms of accuracy.

References

1. Zadeh, L. A. Fuzzy sets. Information and Control. **8**, pp 338–353 (1965)
2. Takagi T. and M. Sugeno Fuzzy Identification of Systems and its Applications to Modeling and Control, IEEE Transactions on Systems, Man, and Cybernetics **15**(1), pp 116–132 (1985)
3. Abonyi J (2003). Fuzzy Modeling Identification for Control. Birkhäuser, Boston (2003)
4. Gath I and Geva, A.B (1989). Unsupervised Optimal Fuzzy Clustering. IEEE Trans Pattern and Machine Intell, vol. **11** (7), pp 773–781 (1989)
5. Bezdeck J C.; Ehrlich, R,; Full, W. FCM: Fuzzy C-means Algorithm. Computers and Geoscience, 10(2–3):pp 191–203 (1984)
6. Bezdeck, J C. Cluster Validity With Fuzzy Sets, Cybernetics **3**(3), pp 58–73 (1975)
7. Bouabaz M. and Horner RMW. Modeling and predicting bridge repair and maintenance costs, Proc. Int. Conf. on Bridge Management, London pp 187–197 (1990)
8. Abonyi J Balasko B, and Balazs L Fuzzy Clustering and Data Analysis Toolbox for Use With Matlab (2005)

Why You Should Consider Nature-Inspired Optimization Methods in Financial Mathematics

A. Egemen Yilmaz and Gerhard-Wilhelm Weber

1 Introduction

Financial Mathematics is a flourishing area of modern science. The subject has developed rapidly into a substantial body of knowledge since the days of pioneering people of this discipline such as Black, Scholes, and Merton. As of today, numerous applications of financial mathematics have become vital for the financial institutions, especially as regards to trading, asset management, and risk control of complicated financial positions.

The basic mathematics that underlies the subject is probability theory, with its strong connections to partial differential equations and numerical analysis. On the finance side, the main topics of importance are the pricing of derivatives, the evaluation of risk, and the management of portfolios. In fact, in today's world, many aspects of capital markets management are becoming more quantitatively and computationally sophisticated; however, it is still a valid argument to say that everything began with derivatives.

As complicated as the problems in consideration get, or as complicated as the approaches/models for handling of these problems get, conventional tools or methodologies become insufficient. In this paper, we will try to focus on the portfolio optimization problem, which is one of the main topics in financial mathematics. We will try to identify why and when conventional methods become insufficient, and metaheuristics (especially nature-inspired optimization methods) might constitute a remedy.

The organization of this chapter is as follows. After this introductory section, we will try to revisit the definition of the portfolio optimization problem with existing models in the literature. In Sect. 3, we will give brief descriptions of

A.E. Yilmaz (✉)
Ankara University Department of Electronics Engineering, 06100 Tandoğan, Ankara/Turkey
e-mail: aeyilmaz@eng.ankara.edu.tr

J.A.T. Machado et al., *Nonlinear and Complex Dynamics: Applications in Physical, Biological, and Financial Systems*, DOI 10.1007/978-1-4614-0231-2_20,

some popular nature-inspired optimization algorithms. Section 4 is nothing but a condensed literature review about the application of the nature-inspired methods for the solution of the portfolio optimization problem. In Sect. 5, we will try to give our concluding remarks.

2 Portfolio Optimization

Portfolio is nothing but the allocation of wealth (or resources in hand) among several assets. Portfolio optimization, which addresses the ideal assignment of resources to existing assets, have been one of the important research fields in modern risk management, or more generally financial management.

A fundamental answer to this problem was given by Markowitz [53, 54], who proposed the mean-variance model, which is now considered as the basis of modern portfolio theory. In Markowitz's approach, the problem was formulated as an optimization problem with two criteria:

- The profit (sometimes also referred to as reward or return) of a portfolio (measured by the mean) that should be maximized.
- The risk of the portfolio (measured by the variance of return) that should be minimized.

In the presence of two criteria, there is not a single optimal solution to the problem (i.e., a single optimal portfolio), but a set of optimal portfolios. Certainly, there is a trade-off between risk and return.

Since the mean-variance theory of Markowitz, research has been performed about extending or modifying the basic model in three directions [1]:

1. The simplification of the type and amount of input data.
2. The introduction of alternative measures of risk.
3. The incorporation of additional criteria and/or constraints.

In the following sections, we will try to summarize the basic models extending that of Markowitz while trying to identify the differences.

2.1 Mean–Variance Model

Markowitz's mean–variance model, in which the variance or the standard deviation is considered as a measure of risk, has been regarded as a quadratic programming problem. In spite of its popularity during the past, the mean–variance model is based on the assumptions that an investor is risk averse and that either (1) the distribution of the rate of return is multivariate normal or (2) the utility of the investor is a quadratic function of the rate of return [6].

However, neither (1) nor (2) holds in practice unfortunately. It is now widely recognized that the real world portfolios do not follow a multivariate normal distribution. Many researchers suggested that one cannot blindly depend on mean–variance model. That is why various risk measures such as semi-variance model, mean absolute deviation model, and variance with skewness model have been proposed.

2.2 Semi–Variance Model

Standard mean–variance model is based on the following assumptions:

- An investor is risk averse.
- The distribution of the rate of return is multivariate normal.

With this model, eventually the variance component of the Markowitz's quadratic objective function can be replaced by other risk functions such as semi-variance. With an asymmetric return distribution, the mean–variance approach leads to an unsatisfactory prediction of portfolio behavior. Indeed, Markowitz himself suggested that a model based on semi-variance would be preferable.

2.3 Mean Absolute Deviation Model

Konno and Yamazaki [44] were the ones who first proposed a mean absolute deviation portfolio optimization model as an alternative to the Markowitz mean–variance portfolio selection model, with the advantage of the portfolio selection problem to be formulated and solved via linear programming.

It has been shown that this model yields similar results to the mean–variance model. Moreover, due to its simplicity, computational it outperforms to the mean–variance model [43, 45].

2.4 Variance with Skewness

Samuelson [68] was the one who first noticed the importance of the third order moment in portfolio optimization. A portfolio return may not be a symmetric distribution. The distribution of individual asset returns tends to exhibit a higher probability of extreme values than is consistent with normality.

In order to capture the characteristics of the return distribution and to provide further decision-making information to investors, this model includes skewness into the mean–variance model. Although the existence of skewness in portfolios has been demonstrated many times, only a few studies to date have proposed incorporating skewness into the portfolio optimization problem [6]. Konno and Yamamoto [46] showed that a mean–variance skewness portfolio optimization model can be solved exactly in a fast manner by using the integer programing approach.

3 Nature-Inspired Optimization Methods

Nature-inspired optimization methods fall into the class of metaheuristics. These are nothing but some methods influenced by the existing behaviors/phenomena for the solution of an optimization-like problem in nature. A very simple example of inspiration is the behavior of a colony or a swarm while searching for the best food source.

In computer science, the term metaheuristic is used to describe a computational method which optimizes a problem by iteratively trying to improve a candidate solution with regard to a given measure of quality. In other words, such methods are nothing but systematical trial-and-error approaches. Metaheuristics (sometimes also referred to as derivative free, direct search, black box, or indeed just heuristic methods) make few or no assumptions about the problem (such as modality or dimension) being optimized and can search very large spaces of candidate solutions. Moreover, most of these algorithms by definition are easily adaptable to parallel computing, which makes them applicable in very large scale problems.

However, it should be noted that metaheuristics do not guarantee that an optimal solution is ever found. On the other hand, for each algorithm, numerous studies (most of which are empirical) have been carried out in order to understand how the algorithm parameters should be adjusted for increasing the success probability.

Originally, metaheuristics were proposed for combinatorial optimization in which the optimal solution is sought over a discrete search space. An example is the traveling salesman problem, where the search space of candidate solutions grows exponentially as the size of the problem increases, which makes an exhaustive search for the optimal solution infeasible. Popular metaheuristics for combinatorial problems include simulated annealing [42], tabu search [32, 33], genetic algorithms [35], and ant colony optimization [21–24].

Later, metaheuristics for problems over real-valued search-spaces were also proposed. In such problems, the conventional approach was to derive the gradient of the function to be optimized, and then to employ gradient descent or a quasi-Newton method. Metaheuristics do not use the gradient or Hessian matrix; hence their advantage is that the function to be optimized need not be continuous or differentiable; moreover, it can also have constraints. Popular metaheuristic optimizers for real-valued search spaces include particle swarm optimization [41], differential evolution [75] and evolution strategies [66, 72].

All algorithms of this sort were initially proposed for single-objective problems. However, throughout the years, multiobjective extensions of these algorithms have been proposed. One of the early attempts was the extension of simulated annealing to multiobjective problems by Czyzak and Jaszkiewicz [16]. Another one was that of

Hansen [34] for extension of tabu search. In a review article, Ehrgott and Gandibleux [26] listed a bibliography of multiobjective optimization approaches for combinatorial (not only considering metaheuristics, but also the conventional approaches). In another review article, Coello [12] identified the historical development of the research studies about extending these algorithms to multiobjective problems. This review was not limited to the combinatorial problems; but only the studies regarding population-based metaheuristics were considered.

In this paper, we will focus on the nature-inspired metaheuristics and their applications to the portfolio optimization problem. In the upcoming sections, we will briefly summariz the most popular and well-known ones.

3.1 Genetic Algorithm

Influenced from the "survival of the fittest" principle in the evolution theory, Holland [35] proposed genetic algorithms for the solution of combinatorial optimisation problems. The method simply relies on representation of the solution candidates by means of chromosomes, via which the relevant objective function is evaluated. The solution candidates constitute a population, which will be evolved throughout the generations (with the programmer's perspective, the generations correspond to the populations in succeeding iterations). Performing these evaluations and considering the fitness of each solution candidate (i.e., the value of the objective function corresponding to that candidate), it is decided which candidates deserve to survive and to be transferred to the next generation. Certainly, as in the evolution process, diversity is added by means of some operators such as crossover (yielding the hybridization of high-quality solution candidates) and mutation. The main flow and the basic idea of the method are illustrated in Fig. 1.

Even though genetic algorithm was originally proposed by Holland for single-objective combinatorial problems, later it has been extended to real-valued optimization problems; even to multiobjective optimization problems (such as Schaffer [70], Corne et al. [13], Zitzler and Thiele [84], Zitzler et al. [85], Deb [17], Deb et al. [18]). A review by Coello [11] lists and identifies the genetic algorithm based multiobjective optimization techniques.

3.2 Genetic Programming

Inspired from the genetic algorithm, Koza [47] proposed the genetic programming approach in order to achieve a software program with a desired capability defined in terms of numerous input–output pairs. The approach is quite similar to the genetic algorithms; but this time, the genes inside the chromosomes are the building blocks (i.e., the functions, components, or modules) of the sought computer program [73].

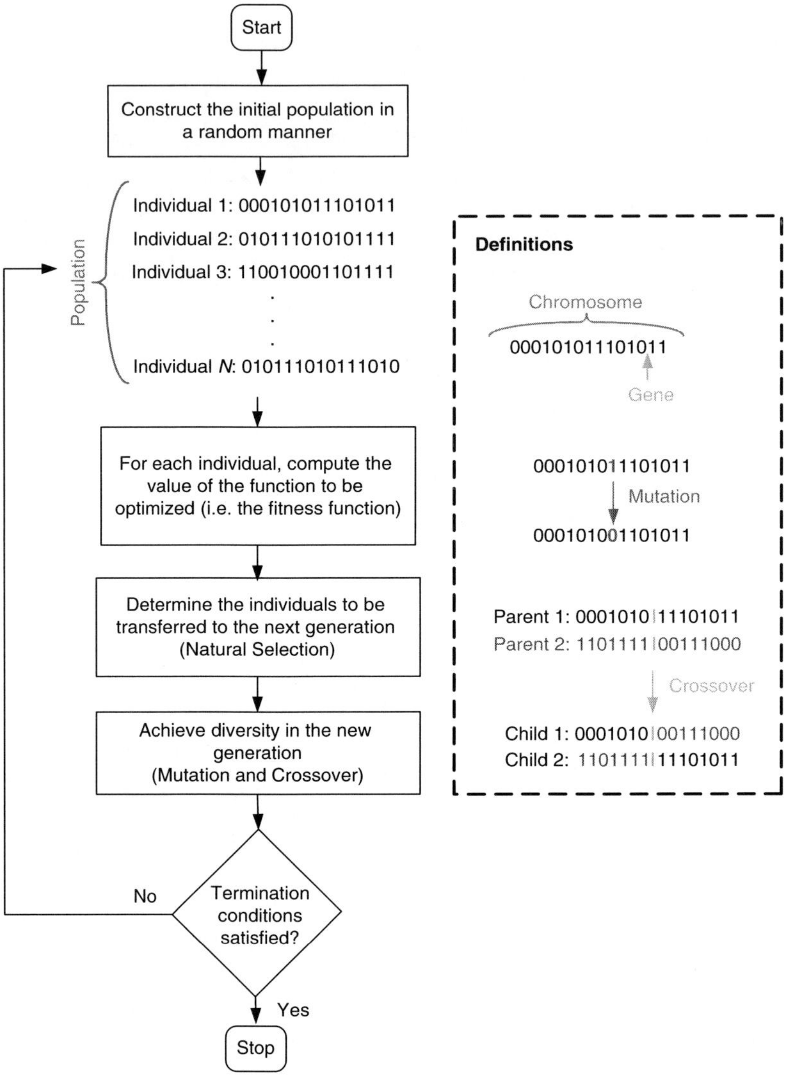

Fig. 1 The terminology and the main flow of the genetic algorithm

3.3 Differential Evolution

Extended from the genetic algorithm, the differential evolution is a recent meta-heuristic originally proposed by Storn and Price [75] for single-objective continuous problems. Again, the method relies on evolutionary operators' crossover and mutation, in addition to the concept of so-called differential weight. Despite its simplicity, the method has so far proven itself in numerous occasions with benchmark problems

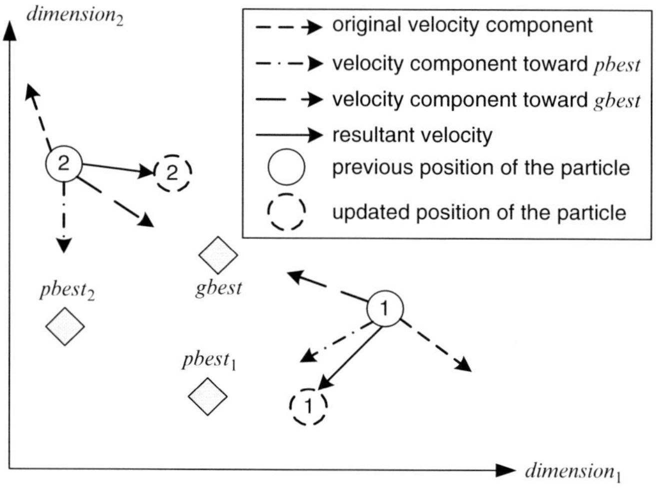

Fig. 2 The pictorial description of the particle tendencies in particle swarm optimization for a 2-D optimization problem

and outperformed many other optimization algorithms. Later, many differential evolution variants for other purposes (i.e., for the solution of the continuous and multiobjective problems, etc.) have been proposed. Unlike other nature-inspired optimization methods, it is possible to guarantee the trial-error procedure imposed by the algorithm to converge.

3.4 Particle Swarm Optimization

Particle swarm optimization is a method proposed by Kennedy and Eberhart [41] after getting inspired by the behaviors of the animal colonies/swarms. Similar to such swarms searching for the best place for nutrition in 3-dimensional space, this method relies on the motions of the swarm members (so-called particles) searching for the global best in an N-dimensional continuous space. This time, the position of each particle is a candidate solution of the problem in hand. As seen in Fig. 2, each member of the swarm has:

- A cognitive behavior (i.e., having tendency to return positions related with good memories); as well as
- A social behavior (i.e., having tendency to go where the majority of the swarm members are located); in addition to
- An exploration capability (i.e., the tendency for random search throughout the domain).

The balance among these three tendencies is the key to the success and the power of the method. So far, the method has been successfully applied to various multidimensional continuous and discontinuous problems. In fact, the results of a similar analysis were recently reported by Poli [63] in a review article (More detailed version of this review is also available on the web [62]). The power of the method is its simplicity allowing implementation in almost every platform and every programming language as well as its ease of parallelization.

Even though particle swarm optimization was originally proposed for single-objective continuous problems, later its discrete variants have also been published. Also, so far more than 30 versions of multiobjective particle swarm optimization extensions have been proposed, most of which have been reviewed by Reyes-Sierra and Coello [67].

3.5 Ant Colony Optimization

Ant Colony Optimization is another algorithm originally proposed by Dorigo [21] and later by Dorigo et al. [22–24] for the solution of combinatorial problems such as the traveling salesman or the shortest path.

Dorigo was inspired from the behaviors of ants while transporting food to their nests. The algorithm depends on the following principles: Initially, ants have random movements; but upon finding food they lay down pheromone trails returning home. Other ants have a tendency to follow these pheromones instead of keeping their random behavior. By this time, all pheromone trails start to evaporate and reduce their attractiveness. However, since pheromones over shorter paths are traced faster, and new pheromones are laid over the same path; new pheromone laying-out rate overcomes the evaporation rate. Due to this positive feedback mechanism, the popularity of shorter paths (i.e., pheromone density) increases in an accelerated manner as seen in Fig. 3. This is the key to the success of the ant colony optimization for the solution of relevant problems.

Similarly, continuous and multiobjective variants of ant colony optimization have later been published.

3.6 Other Nature-Inspired Optimization Methods

There are some other recent nature-inspired optimization algorithms with self-descriptive names such as:

- Bees algorithm [61]
- Invasive weed optimization [55]
- Artificial bee colony algorithm [37]
- Saplings growing-up algorithm [38]

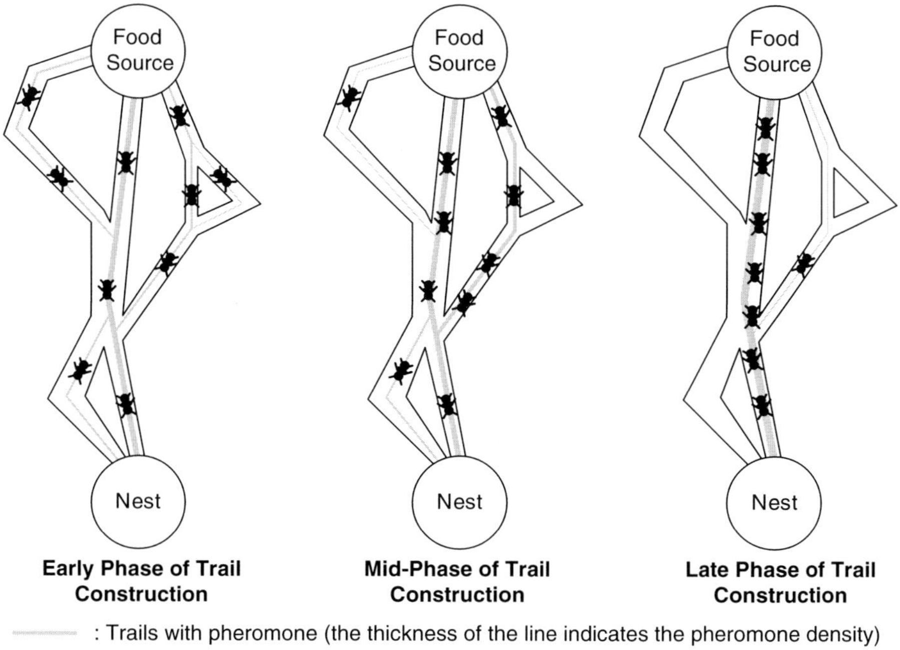

Fig. 3 Pictorial description of the basic idea underlying the ant colony optimization

- Monkey search [56]
- Central force optimization [29–31]
- Viral systems [14]
- Intelligent water drops algorithm [74]
- Gravitational search algorithm [65]
- Glowworm swarm optimization [49]
- Cuckoo search [82, 83]
- Firefly algorithm [80, 81]

which are still waiting for further research and promotion.

4 Application of Metaheuristics to the Portfolio Optimization Problem: A Literature Review

As stated before, portfolio optimization problem is actually a constrained multiobjective optimization problem. But in the early decades (between 1950s and early 1990s), due to lack of powerful methods and computational power, the problem used to be handled with oversimplifying assumptions. With the diffusion of metaheuristics to all disciplines, researchers started to apply them to problems of their own branches. Eventually, portfolio optimization took its share. As of 2001, as pointed by Chen and Kuo [7], there have been about 400 publications regarding the application of metaheuristics to problems in economy and finance. Certainly, publications related to portfolio optimization constituted the majority.

Early attempts were applications of metaheuristics to the single-objective portfolio optimization problem with no constraints. Dueck and Winker [25] used a local-search based heuristic for the solution. Arnone et al. [4] were the ones who first applied genetic algorithm to the portfolio selection problem.

Later research can be considered in two main directions:

1. Incorporation of constraints in the problem model
2. Handling the problem as a multiobjective one

Regarding the studies about incorporation of the constraints in the problem: Jobst et al. [36] and Fieldsend et al. [28] discussed the computational aspects in the presence of the discrete asset choice constraints. Chang et al. [5] applied tabu search, simulated annealing and genetic algorithm to the portfolio optimization problem considering the cardinality constraint; afterwards Schaerf [69] and Kellerer et al. [39] applied local search algorithms; Streichert et al. [76] and Diosan [19] applied various evolutionary algorithms for the same purpose.

One of the very early multiobjective solution approaches in portfolio optimization problem was by Lin et al. [52], who applied genetic algorithm for this purpose [51]. Crama and Schyns [15] discussed how to apply simulated annealing in complex portfolio optimization problems. Ong et al. [59] applied a multiobjective evolutionary algorithm, whereas Armananzas and Lozano [2] applied multiobjective greedy-search, simulated annealing and ant colony optimization by considering the portfolio optimization problem as a triobjective one. Doerner et al. [20] applied ant colony optimization; Subbu et al. [77], Diosan [19], and Chiam et al. [10] applied various evolutionary algorithms; Kendall and Su [40] applied particle swarm optimization; Yang [79] applied the genetic algorithm. Application of differential evolution to the field is brand new. As of today, Ardia et al. [3] and Krink et al. [48] are the ones who have so far applied differential evolution to the portfolio optimization problem.

For deeper and broader surveys of the literature, interested readers can take a look at Schlottmann and Seese [71] or Tapia and Coello [78] the on application of multiobjective evolutionary algorithms in economics and finance in general.

Application of metaheuristics, or more specifically nature-inspired methods to similar problems such as index fund management, credit portfolio construction, etc. is also possible. Orito et al. [60], Kyong et al. [50], Oh et al. [57, 58] used genetic algorithms for the index fund management problem.

Another issue in portfolio management is the cost of transactions, which is usually neglected during the modelling of the problems. It is possible to incorporate this factor while applying metaheuristics. For example, Chen and Zhang [8] applied a particle swarm optimization variant to the portfolio optimization problem considering the transaction costs.

Meanwhile, genetic programming also found application in finance. Potvin et al. [64] applied genetic programming for generating trading rules in stock markets; more recently, Etemadi et al. [27] used it for bankruptcy prediction, meanwhile Chen et al. [9] used a time-adaptive version of the technique to portfolio optimization.

5 Conclusions

Metaheuristics, more specifically nature-inspired optimization algorithms, constitute powerful means for the solution of existing problems in economy and finance. The main factors promoting the usage of such algorithms can be summarized as follows:

- The algorithms make no assumptions (or require no a priori information) about the objective function.
- They do not require the objective function to be continuous or differentiable.
- They can handle complicated models with constraints.
- Almost all of them have variants for handling continuous and combinatorial problems.
- Almost all of them have extensions to multiobjective problems.
- Almost all of them support parallelization, which yields the solution of very large-scale problems.

Eventually, the literature is full of a plethora of publications about successful applications of such algorithms to problems, which could not have been considered and handled previously with conventional approaches. By observing the rate of increase in such publications, it can be easily forecasted that more and more applications will occur in the near future.

References

1. Anagnostopoulos, K.P. and G. Mamanis (2010). A portfolio optimization model with three objectives and discrete variables. *Computers & Operations Research*, **37**, 1285–1297.
2. Armananzas, R. and J.A. Lozano (2005). A multiobjective approach to the portfolio optimization problem. In: *Proceedings of IEEE Congress on Evolutionary Computation*, vol. 2, p. 1388–1395.
3. Ardia, D., K. Boudt, P. Carl, K.M. Mullen and B.G. Peterson (2010). Differential evolution (DEoptim) for non-convex portfolio optimization. *Munich Personal Repec Archive*, MPRA Paper No. 22135, [online], April 2010, Available at: http://mpra.ub.uni-muenchen.de/22135/1/MPRA_paper_22135.pdf.
4. Arnone, S., A. Loraschi and A. Tettamanzi (1993). A genetic approach to portfolio selection. *Neural Network World*, **6**, 597–604.
5. Chang, T.J., N. Meade, J.E. Beasley and Y.M. Sharaiha (2000). Heuristics for cardinality constrained portfolio optimization. *Computers & Operations Research*, **27**, 1271–1302.

6. Chang, T.-J., S.-C. Yang and K.-J. Chang (2009). Portfolio optimization problems in different risk measures using genetic algorithm. *Expert Systems with Applications*, **36**, 10529–10537.
7. Chen, S.-H. and T.-W. Kuo (2002). Evolutionary computation in economics and finance: A bibliography. In *Evolutionary Computation in Economics and Finance* (S.-H. Chen, (Ed.)), 419–455, Physica-Verlag, Heidelberg.
8. Chen, W. and W.-G. Zhang (2010). The admissible portfolio selection problem with transaction costs and an improved PSO algorithm. *Physica A*, **389**, 2070–2076.
9. Chen, Y., S. Mabub and K. Hirasawa (2010). A model of portfolio optimization using time adapting genetic network programming. *Computers & Operations Research*, **37**, 1697–1707.
10. Chiam, S.C., K.C. Tan and A.A.L. Mamum (2008). Evolutionary multi-objective portfolio optimization in practical context. *International Journal of Automation and Computing*, **5**(1), 67–80.
11. Coello, C.A.C. (2002). An updated survey of GA-based multiobjective optimization techniques. *ACM Computing Surveys*, **32**(2), 109–143.
12. Coello, C.A.C. (2006). Evolutionary multi-objective optimization: a historical view of the field. *IEEE Computational Intelligence Magazine*, **1**(1), 28–36.
13. Corne, D.W., J.D. Knowles and M.J. Oates (2000). The Pareto envelope-based selection algorithm for multiobjective optimization. In: *Proceedings of the Parallel Problem Solving from Nature VI Conference - Lecture Notes in Computer Science*, vol. 1917, p. 839–848.
14. Cortés, P., J.M. García, J. Muñuzuri, and L. Onieva (2008). Viral systems: a new bio-inspired optimisation approach. *Computers & Operations Research*, **35**(9), 2840–2860.
15. Crama, Y. and M. Schyns (2003). Simulated annealing for complex portfolio selection problems. *European Journal of Operational Research*, **150**, 546–571.
16. Czyzak, P. and A. Jaszkiewicz (1998). Pareto simulated annealing: a metaheuristic technique for multiple-objective combinatorial optimization. *Journal of Multi-Criteria Decision Analysis*, **7**, 34–47.
17. Deb, K. (2001). *Multi-Objective Optimization Using Evolutionary Algorithms*. John Wiley & Sons, Chichester.
18. Deb, K., A. Ptratap, S. Agarwal and T. Meyarivan (2002). A fast and elitist multiobjective genetic algorithm: NSGA-II. *IEEE Transactions on Evolutionary Computation*, **6**(2), 182–197.
19. Diosan, L. (2005). A multi-objective evolutionary approach to the portfolio optimization problem. In: *Proceedings of the International Conference on Computational Intelligence for Modelling, Control and Automation*, p. 183–188.
20. Doerner, K., W.J. Gutjahr, R.F. Hartl, C. Strauss and C. Stummer (2001). Ant colony optimization in multiobjective portfolio selection. In: *Proceedings of the 4th Metaheuristics International Conference*, p. 243–248.
21. Dorigo, M. (1992). *Optimisation, Learning, and Natural Algorithms*, Ph.D. Thesis, Politecnico diMilano, Italy.
22. Dorigo, M., V. Maniezzo, and A. Colorni. (1996). The ant system: optimization by a colony of cooperating agents. *IEEE Transactions on Systems, Man and Cybernetics*, **26**, 29–41.
23. Dorigo, M. and L. Gambardella (1997). Ant colony system: a cooperative learning approach to the travelling salesman problem. *IEEE Transactions on Evolutionary Computation*, **1**, 53–66.
24. Dorigo, M. and G. Di Caro (1999). The ant colony optimization meta-heuristic. In *New Ideas in Optimization* (D. Corne, M. Dorigo and F. Glover (Eds.)), p. 11–32, McGraw-Hill, London.
25. Dueck, G. and P. Winker (1992). New concepts and algorithms for portfolio choice. *Applied Stochastic Models and Data Analysis*, **8**, 159–178.
26. Ehrgott, M. and X. Gandibleux (2000). A survey and annotated bibliography of multiobjective combinatorial optimization. *OR Spektrum*, **22**, 425–460.
27. Etemadi, H., A.A.A. Rostamy and H.F. Dehkordi (2009). A genetic programming model for bankruptcy prediction: empirical evidence from Iran. *Expert Systems with Applications*, **36**(2), 3199–3207.
28. Fieldsend, J.E., J. Matatko and M. Peng (2004). Cardinality constrained portfolio optimisation. In: *Lecture Notes in Computer Science (including Subseries Lecture Notes in Artificial Intelligence and Lecture Notes in Bioinformatics)*, vol. 3177, p. 788–793.

29. Formato, R.A. (2007). Central force optimization: a new nature inspired computational framework for multidimensional search and optimization. In: *Proceedings of Nature Inspired Cooperative Strategies for Optimization (NICSO 2007) – Studies in Computational Intelligence*, vol. 129, p. 221–238.
30. Formato, R.A. (2009a). Central force optimization: a new deterministic gradient-like optimization metaheuristic. *OPSEARCH*, **46**(1), 25–51.
31. Formato, R.A. (2009b). Central force optimisation: a new gradient-like metaheuristic for multidimensional search and optimisation. *International Journal of Bio-Inspired Computation*, **1**(4), 217–238.
32. Glover, F. (1989). Tabu search - Part I. *ORSA Journal on Computing*, **1**(3), 190–206.
33. Glover, F. (1990). Tabu search - Part II. *ORSA Journal on Computing*, **2**(1), 4–32.
34. Hansen, M. (2000). Tabu search for multiobjective combinatorial optimization: TAMOCO. *Control and Cybernetics*, **29**, 799–818.
35. Holland, J.H. (1975). *Adaptation in Natural and Artificial Systems: An Introductory Analysis with Applications to Biology, Control and Artificial Intelligence*. University of Michigan Press, Michigan.
36. Jobst, N.J., M.D. Horniman, C.A. Lucas and G. Mitra (2001). Computational aspects of alternative portfolio selection models in the presence of discrete asset choice constraints. *Quantitative Finance*, **1**, 1–13.
37. Karaboga, D. and B. Basturk (2007). A powerful and efficient algorithm for numerical function optimization: artificial bee colony (ABC) algorithm, *Journal of Global Optimization*, **39**(3), 459–471.
38. Karci, A. (2007). Theory of saplings growing up algorithm. In: *Proceedings of the International Conference on Adaptive and Natural Computing Algorithms*, vol. 4431, p. 450–460.
39. Kellerer, H. and D.G. Maringer (2003). Optimization of cardinality constrained portfolios with a hybrid local search algorithm. *OR Spectrum*, 25(4), 481–495.
40. Kendall, G. and Y. Su (2005). A particle swarm optimisation approach in the construction of optimal risky portfolios. In: *Proceedings of Artificial Intelligence and Applications*, vol. 453, p. 140–145.
41. Kennedy, J. and R.C. Eberhart (1995). Particle swarm optimization. In: *Proceedings of IEEE International Conference on Neural Networks*, vol. 4, p. 1942–1948.
42. Kirkpatrick, S., C.D. Gelatt and M.P. Vecchi (1983). Optimization by simulated annealing, *Science - New Series*, **220**(4598), 671–680.
43. Konno, H. (2003). Portfolio optimization of small fund using mean-absolute deviation model. *International Journal of Theoretical and Applied Finance*, **6**(4), 403–418.
44. Konno, H. and H. Yamazaki (1991). Mean-absolute deviation portfolio in optimization model and its application to Tokyo stock market. *Management Science*, **37**, 519–531.
45. Konno, H., and T. Koshizuka (2005). Mean-absolute deviation model. *IIE Transactions*, **37**, 893–900.
46. Konno, H. and R. Yamamoto (2005). A mean–variance–skewness model: Algorithm and applications. *International Journal of Theoretical and Applied Finance*, **8**(4), 409–423.
47. Koza, J.R. (1992). *Genetic Programming, on the Programming of Computers by means of Natural Selection*. MIT Press, Cambridge (MA).
48. Krink, T. and S. Paterlini (2010). Multiobjective optimization using differential evolution for real-world portfolio optimization. *Computational Management Science*, doi: 10.1007/s10287-009-0107-6 (*in press*).
49. Krishnanand, K.N. and D. Ghose 2009. Glowworm swarm optimisation: a new method for optimising multi-modal functions. *International Journal of Computational Intelligence Studies*, **1**(1), 93–119.
50. Kyong, J.O., Y.K. Tae and M. Sungky (2005). Using genetic algorithm to support portfolio optimization for index fund management. *Expert Systems with Applications*, **28**, 371–379.
51. Lin, C.C. and Y.T. Liu (2008). Genetic algorithms for portfolio selection problems with minimum transaction lots. *European Journal of Operational Research*, **185**(1), 393–404.

52. Lin, D., S. Wang and H. Yan (2001). A multiobjective genetic algorithm for portfolio selection problem. In: *Proceedings of the ICOTA 2001*, Hong Kong.
53. Markowitz, H. (1952). Portfolio selection. *Journal of Finance*, **7**, 77–91.
54. Markowitz, H. (1959). *Portfolio Selection: Efficient Diversification of Investments*. John Wiley & Sons, New York.
55. Mehrabian, A. R. and C. Lucas (2006). A novel numerical optimization algorithm inspired from weed colonization. *Ecological Informatics*, **1**, 355–366.
56. Mucherino, A and O. Seref (2007). Monkey search: a novel meta-heuristic search for global optimization. In: *Proceedings of the Conference Data Mining, System Analysis and Optimization in Biomedicine (AIP Conference Proceedings 953)*, p. 162–173.
57. Oh, K.J., T.Y. Kim, S.H. Min (2005). Using genetic algorithm to support portfolio optimization for index fund management. *Expert Systems with Applications*, **28**, 371–379.
58. Oh, K.J., T.Y. Kim, S.H. Min and H.Y. Lee (2006). Portfolio algorithm based on portfolio beta using genetic algorithm. *Expert Systems with Applications*, **30**(3), 527–534.
59. Ong, C.S., J.J. Huang and G.H. Tzeng (2005). A novel hybrid model for portfolio selection. *Applied Mathematics and Computation*, **169**, 1195–1210.
60. Orito, Y., H. Yamamoto and G. Yamazaki (2003). Index fund selections with genetic algorithms and heuristic classifications. *Computers and Industrial Engineering*, **45**, 97–109.
61. Pham D.T., A. Ghanbarzadeh, E. Koç, S. Otri, S. Rahim and M. Zaidi (2006). The bees algorithm – a novel tool for complex optimisation problems, In: *Proceedings of IPROMS 2006 Conference*, p. 454–461.
62. Poli, R. (2007). An analysis of publications on particle swarm optimisation applications, *Technical Report- University of Essex Computer Science and Electrical Engineering CSM-469*, [online], May 2007 (rev. Nov. 2007), Available at: http://www.essex.ac.uk/dces/research/publications/technicalreports/2007/tr-csm469-revised.pdf.
63. Poli, R. (2008). Analysis of the publications on the applications of particle swarm optimisation. *Journal of Artificial Evolution and Applications*, Article ID: 685175, doi:10.1155/2008/685175.
64. Potvin, J., P. Soriano and M. Vallee (2004). Generating trading rules on the stock markets with genetic programming. *Computers & Operations Research*, **31**(7), 1033–1047.
65. Rashedi, E., H. Nezamabadipour, and S. Saryazdi (2009). GSA: A gravitational search algorithm. *Information Sciences*, 179(13), 2232–2248.
66. Rechenberg, I. (1971). *Evolutionsstrategie – Optimierung Technischer Systeme nach Prinzipien der Biologischen Evolution*, Ph.D. Thesis, Germany.
67. Reyes-Sierra, M. and C.A.C. Coello (2006). Multi-objective particle swarm optimizers: a survey of the state-of-the-art. *International Journal of Computational Intelligence Research*, **2**(3), 287–308.
68. Samuelson, P. (1958). The fundamental approximation theorem of portfolio analysis in terms of means variances and higher moments. *Review of Economic Studies*, **25**, 65–86.
69. Schaerf, A (2002). Local search techniques for constrained portfolio selection problems. *Computational Economics*, **20**(3), 177–190.
70. Schaffer, J.D. (1985). Multiple objective optimization with vector evaluated genetic algorithm. In: *Proceedings of the First International Conference on Genetic Algorithms and their Applications*, p. 93–100.
71. Schlottmann, F. and D. Seese (2004). Modern heuristics for finance problems: A survey of selected methods and applications. In *Handbook of Computational and Numerical Methods in Finance* (S. Rachev (Ed.)), p. 331–360, Birkhauser, Berlin.
72. Schwefel, H.-P. (1974). *Numerische Optimierung von Computer-Modellen*, Ph.D. Thesis, Germany.
73. Sette, S. and L. Boullart (2001). Genetic programming: principles and applications. *Engineering Applications of Artificial Intelligence*, **14**(6), 727–736.
74. Shah-Hosseini, H. (2009). The intelligent water drops algorithm: a nature-inspired swarm-based optimization algorithm. *International Journal of Bio-Inspired Computation*, **1**(1/2), 71–79.

75. Storn, R. and K. Price (1997). Differential evolution - a simple and efficient heuristic for global optimization over continuous spaces. *Journal of Global Optimization*, **11**, 341–359.
76. Streichert, F., H. Ulmer and A. Zell (2003). Evolutionary algorithms and the cardinality constrained portfolio optimization problem. In: *Selected Papers of the International Conference on Operations Research*, p. 253–260.
77. Subbu, R., B. Bonissone, N. Eklund, S. Bollapragada and K. Chalermkraivuth (2005). Multiobjective financial portfolio design: a hybrid evolutionary approach. In: *Proceedings of IEEE Congress on Evolutionary Computation*, vol. 2, p. 1722–1729.
78. Tapia, M.G.C. and C.A.C. Coello (2007). Applications of multi-objective evolutionary algorithms in economics and finance. In: *Proceedings of IEEE Congress on Evolutionary Computation*, p. 532–539.
79. Yang, X. (2006). Improving portfolio efficiency: A genetic algorithm approach. *Computational Economics*, **28**, 1–14.
80. Yang X.-S. (2009). Firefly algorithms for multimodal optimization, In: *Stochastic Algorithms: Foundations and Applications (Lecture Notes in Computer Sciences 5792)*, p. 169–178.
81. Yang, X. S., (2010). Firefly algorithm, stochastic test functions and design optimisation. *International Journal of Bio-Inspired Computation*, **2**(2), 78–84.
82. Yang, X.-S. and S. Deb (2009). Cuckoo search via Lévy flights. In: *Proceedings of World Congress on Nature & Biologically Inspired Computing (NaBIC 2009)*, p. 210–214.
83. Yang, X.-S. and S. Deb (2010). Engineering optimisation by cuckoo search. *International Journal of Mathematical Modelling and Numerical Optimisation*, **1**(4), 330–343.
84. Zitzler, E. and L. Thiele (1999). Multiobjective evolutionary algorithms: a comparative case study and the strength Pareto approach. *IEEE Transactions on Evolutionary Computation*, **3**(4), 257–271.
85. Zitzler, E., M. Laumanns and L. Thiele (2001). SPEA2: improving the strength of Pareto evolutionary algorithm. *TIK-103*, Department of Electrical Engineering, Swiss Federal Institute of Technology, Zurich, Switzerland.

Desirable Role in a Revenue-Maximizing Tariff Model with Uncertainty

Fernanda A. Ferreira and Flávio Ferreira

1 Introduction

It is well known that the outcome of a duopoly market varies when competition takes different forms such as Cournot or Stackelberg. Most of the economic literature assumes Cournot competition with simultaneous play as the natural order of moves in a quantity-setting game. However, recent advances in game theory argue that the assumed order of play should also be consistent with the players preferences over the time of actions. There are several comparative studies of these market structures. Notice that, in the typical models, compared to Cournot competition the Stackelberg leader can never be worse off: The leader can choose the Cournot quantity and then for the follower it is optimal to produce the same amount and we have Cournot outcome; so, leader either chooses Cournot quantity or if it chooses a different level of production it must be better off otherwise it would not do that. Here, we examine optimal trade in a market with demand uncertainty, in a duopoly in which a home firm competes with a foreign firm. Furthermore, we also assume that in that international market the home government imposes a tariff on the imported goods. In this case, we will see that the results can be different than the ones refereed above.

Tariff revenue may be an important source of government revenue for developing countries that do not have an efficient tax system. So, the government may use the maximum-revenue tariff. Brander and Spencer [1] have shown that a tariff has a profit-shifting effect in addition to its effect on tariff revenue. Larue and Gervais [8] studied the effect of maximum-revenue tariff in a Cournot duopoly. Ferreira and Ferreira [4] examined the maximum-revenue tariff under international Bertrand competition with differentiated products when rivals' production costs are unknown.

F.A. Ferreira (✉)
ESEIG - Polytechnic Institute of Porto and CMUP, Rua D. Sancho I, 981,
4480-876 Vila do Conde, Portugal
e-mail: fernandaamelia@eu.ipp.pt

J.A.T. Machado et al., *Nonlinear and Complex Dynamics: Applications in Physical, Biological, and Financial Systems*, DOI 10.1007/978-1-4614-0231-2_21,

Clarke and Collie [2] studied a similar question, when there is no uncertainty on the production costs. The propose of this paper is to study the maximum-revenue tariff under international quantity competition with demand uncertainty, with different possible timings of decisions.

We consider a two-country, two-good model where a domestic and a foreign good are produced by a home and a foreign monopolist, respectively. Since we assume that the two countries are perfectly symmetric, it is sufficient to describe only the domestic economy. We should mention that issues related to those of this paper have been studied by Ferreira et al. [5, 6], Ferreira and Pinto [7] and Spulber [9].

2 The Benchmark Model

There are two countries, home and foreign. Each country has one firm, firm F_1 (home firm) and firm F_2 (foreign firm) that produces homogeneous goods. Consider the home market, where the two firms compete in quantities (see [10]). We consider that the domestic government imposes an import tariff t per unit of imports from the foreign firm.

The inverse demand function is given by $p = A - q_i - q_j$, where $i, j \in \{1, 2\}$ with $i \neq j$ and q_i stands for quantity. In this section, we assume that the intercept A is commonly known since the begin of the game.

The model consists in the following two-stage game:

- In the first stage, the domestic government chooses the import tariff t per unit of imports from the foreign firm.
- In the second stage, both firms choose output levels.

Firms' profits, π_1 and π_2, are, respectively, given by

$$\pi_1 = (A - q_1 - q_2)q_1 \quad \text{and} \quad \pi_2 = (A - q_1 - q_2 - t)q_2.$$

2.1 Simultaneous Decision

In this section, we suppose that, in the second stage of the game, both home and foreign firms play a Cournot-type game, i.e., each firm F_i independently chooses q_i. Let the superscript C denote the equilibrium outcome of the Cournot-type game.

We determine the subgame perfect Nash equilibrium by backwards induction. Suppose that the domestic government has chosen the import tariff t per unit of imports in the first stage.

Maximizing simultaneously both firms' profits, we get the following output levels:

$$q_1 = \frac{A+t}{3} \quad \text{and} \quad q_2 = \frac{A-2t}{3}.$$

Now, we can use the above results to derive the maximum-revenue tariff. The maximum-revenue tariff is the tariff rate that maximizes the tariff revenue collected by the government in the home country:

$$R = t\frac{A-2t}{3}.$$

Theorem 1. *In the Cournot-type game, the maximum-revenue tariff is given by*

$$t^C = \frac{A}{4}.$$

So, we get the following result.

Theorem 2. *In the case of Cournot competition, the output levels at equilibrium are given by*

$$q_1^C = \frac{5A}{12} \quad \text{and} \quad q_2^C = \frac{A}{6}.$$

Thus, the aggregate quantity in the market is given by $Q^C = 7A/12$ and the price is given by $p^C = 5A/12$. The following results are also obtained straightforwardly.

Theorem 3. *In the case of Cournot competition, home firm's profit π_1^C and foreign firm's profit π_2^C are, respectively, given by*

$$\pi_1^C = \frac{25A^2}{144} \quad \text{and} \quad \pi_2^C = \frac{A^2}{36}.$$

Corollary 1. *In the case of Cournot competition, home firm profits more than foreign firm.*

2.2 Home Firm Is the Leader

In this section, we suppose that, in the second stage of the game, the home firm is the leader. Home firm F_1 chooses q_1, and foreign firm F_2 chooses q_2 after observing q_1. Let the superscript L denote the equilibrium outcome of the game where the home firm F_1 is the leader.

We determine the subgame perfect Nash equilibrium by backwards induction. Suppose that the domestic government has chosen the import tariff t per unit of imports in the first stage. Also, suppose that, the leader home firm F_1 produces q_1. Then, maximizing foreign firm profit's $\pi_2 = (A - q_1 - q_2 - t)q_2$, we get

$$q_2 = \frac{A-t-q_1}{2}. \tag{1}$$

Now, maximizing home firm profit's $\pi_1 = (A - q_1 - q_2)q_1$, knowing the above quantity q_2, we get

$$q_1 = \frac{A+t}{2}. \tag{2}$$

Putting (2) into (1), we get

$$q_2 = \frac{A-3t}{4}.$$

Now, we can use the above results to derive the maximum-revenue tariff. The maximum-revenue tariff is the tariff rate that maximizes the tariff revenue collected by the government in the home country. From the above results, the tariff revenue is

$$R = t\frac{A-3t}{4},$$

which leads us to the following result.

Theorem 4. *In the case of Stackelberg competition, with the home firm as the leader, the maximum-revenue tariff is given by*

$$t^L = \frac{A}{6}.$$

So, the output levels at equilibrium are as follows.

Theorem 5. *In the case of Stackelberg competition, with the home firm as the leader, home and foreign firms produce, respectively,*

$$q_1^L = \frac{7A}{12} \quad \text{and} \quad q_2^L = \frac{A}{8}.$$

Thus, the aggregate quantity in the market is given by $Q^L = 17A/24$ and the price is given by $p^L = 7A/24$. The following results are also obtained straightforwardly.

Theorem 6. *In the case of Stackelberg competition, with the home firm as the leader, home firm's profit π_1^L and foreign firm's profit π_2^L are, respectively, given by*

$$\pi_1^L = \frac{49A^2}{288} \quad \text{and} \quad \pi_2^L = \frac{A^2}{64}.$$

2.3 Home Firm Is the Follower

In this section, we suppose that, in the second stage of the game, the home firm is the follower. Foreign firm F_2 chooses q_2, and home firm F_1 chooses q_1 after observing q_2. Let the superscript F denote the equilibrium outcome of the game where the home firm F_1 is the follower.

We determine the subgame perfect Nash equilibrium by backwards induction. Suppose that the domestic government has chosen the import tariff t per unit of imports in the first stage. Also, suppose that, the leader foreign firm F_2 produces q_2. Then, maximizing home firm profit's $\pi_1 = (A - q_1 - q_2)q_1$, we get

$$q_1 = \frac{A - q_2}{2}. \tag{3}$$

Now, maximizing foreign firm profit's $\pi_2 = (A - q_1 - q_2 - t)q_2$, knowing the above quantity q_1, we get

$$q_2 = \frac{A - 2t}{2}. \tag{4}$$

Putting (4) into (3), we get

$$q_1 = \frac{A + 2t}{4}.$$

Now, we can use the above results to derive the maximum-revenue tariff. The maximum-revenue tariff is the tariff rate that maximizes the tariff revenue collected by the government in the home country. From the above results, the tariff revenue is

$$R = t\frac{A - 2t}{4},$$

which leads us to the following result.

Theorem 7. *In the case of Stackelberg competition, with the home firm as a follower, the maximum-revenue tariff is given by*

$$t^F = \frac{A}{4}.$$

So, the output levels at equilibrium are as follows.

Theorem 8. *In the case of Stackelberg competition, with the home firm as a follower, home and foreign firms produce, respectively,*

$$q_1^F = \frac{3A}{8} \quad \text{and} \quad q_2^F = \frac{A}{4}.$$

Thus, the aggregate quantity in the market is given by $Q^F = 5A/8$ and the price is given by $p^F = 3A/8$. The following result is also obtained straightforwardly.

Theorem 9. *In the case of Stackelberg competition, with the home firm as a follower, home firm's profit π_1^F and foreign firm's profit π_2^F are, respectively, given by*

$$\pi_1^F = \left(\frac{3A}{8}\right)^2 \quad \text{and} \quad \pi_2^F = \frac{A^2}{32}.$$

Corollary 2. *In the case of Stackelberg competition, with the home firm as a follower, the home firm profits more than the foreign leader firm.*

2.4 Comparisons

In this section, we are going to compare the results obtained in each model. First, we observe that, independently of the role, the sales of the home firm are increasing in the tariff, and the sales of the foreign firm are decreasing in the tariff. Furthermore, the total sales in the home country are decreasing in the tariff. Next corollary states that the domestic government imposes a lower tariff in the game where the home firm is the leader; and the tariffs are equal in the Cournot-type game and in the game where the home firm is the follower.

Corollary 3. *The tariffs in the different games are related as follows:*

$$t^L < t^F = t^C.$$

The total sales in the home market are higher in the game where the home firm is the leader; and they are lower in the Cournot-type game, as stated in the following corollary.

Corollary 4. *The total sales in the home market are related as follows:*

$$Q^C < Q^F < Q^L.$$

In the next corollary, we compare the profits of the firms obtained in each game. We note that the home firm profits more when it plays a Cournot-type game than in any other game. This result is in contrast to the well-known result that a Stackelberg leader firm profits more than a Cournot firm. Furthermore, the foreign firm prefers to be a Stackelberg leader firm, and the worse situation is to be a Stackelberg follower firm.

Corollary 5. *Home firm's profits are related as follows:*

$$\pi_1^F < \pi_1^L < \pi_1^C.$$

Foreign firm's profits are related as follows:

$$\pi_2^L < \pi_2^C < \pi_2^F.$$

3 The Model with Demand Uncertainty

In this section, we consider that the inverse demand function is given by

$$p = A - q_i - q_j,$$

where the intercept A is ex-ante unobservable, although its prior cumulative function $F(a)$ is commonly known to domestic government and both firms, with strictly positive finite mean $E(A)$ and variance $V(A)$. We assume that the exact realization of this intercept A becomes observable after the decision on the import tariff t fixed by the domestic government, and before both firms decide their output levels.

The model consists in the following two-stage game:

- In the first stage, the domestic government chooses the import tariff t per unit of imports from the foreign firm, without knowing the demand realization.
- In the second stage, both firms choose output levels, knowing the exact realization a of the demand,

Firms' profits, π_1 and π_2, are given by

$$\pi_1 = (a - q_1 - q_2)q_1 \quad \text{and} \quad \pi_2 = (a - q_1 - q_2 - t)q_2.$$

3.1 *Simultaneous Decision*

In this section, we suppose that, in the second stage of the game, both home and foreign firms play a Cournot-type game, i.e., each firm F_i independently chooses q_i. Let the superscript C denote the equilibrium outcome of the Cournot-type game.

We determine the subgame perfect Nash equilibrium by backwards induction. Suppose that the domestic government has chosen the import tariff t per unit of imports in the first stage.

Maximizing simultaneously both firms profits, we get the following output levels:

$$q_1 = \frac{a+t}{3} \quad \text{and} \quad q_2 = \frac{a-2t}{3}.$$

Now, we can use the above results to derive the maximum-revenue tariff. The maximum-revenue tariff is the tariff rate that maximizes the tariff revenue collected by the government in the home country. Since the government does not know the exact demand, it will use the expected demand to compute that tariff. The tariff is t per unit of imports and the expected demand is $E(A)$; so, expected tariff revenue is

$$E(R) = t\frac{E(A) - 2t}{3}.$$

Theorem 10. *The maximum-revenue tariff is given by*

$$t^C = \frac{E(A)}{4}.$$

So, we get the following result.

Theorem 11. *In the case of Cournot competition, the output levels at equilibrium are given by*

$$q_1^C = \frac{4a + E(A)}{12} \quad \text{and} \quad q_2^C = \frac{2a - E(A)}{6}.$$

Thus, the aggregate quantity in the market is given by $Q^C = (8a - E(A))/12$ and the price is given by $p^C = (4a + E(A))/12$. The following result is also obtained straightforwardly.

Theorem 12. *Expected ex-ante profits of the two firms are given by*

$$E(\pi_1^C) = \left(\frac{5E(A)}{12}\right)^2 + \frac{V(A)}{9} \quad \text{and} \quad E(\pi_2^C) = \left(\frac{E(A)}{6}\right)^2 + \frac{V(A)}{9}.$$

3.2 Home Firm Is the Leader

In this section, we suppose that, in the second stage of the game, the home firm is the leader. Home firm F_1 chooses q_1, and foreign firm F_2 chooses q_2 after observing q_1. Let the superscript L denote the equilibrium outcome of the game where the home firm F_1 is the leader.

We determine the subgame perfect Nash equilibrium by backwards induction. Suppose that the domestic government has chosen the import tariff t per unit of imports in the first stage. Also, suppose that, the leader home firm F_1 produces q_1. Then, maximizing foreign firm profit $\pi_2 = (a - q_1 - q_2 - t)q_2$, we get

$$q_2 = \frac{a - t - q_1}{2}. \tag{5}$$

Now, maximizing home firm profit $\pi_1 = (a - q_1 - q_2)q_1$, knowing the above quantity q_2, we get

$$q_1 = \frac{a + t}{2}. \tag{6}$$

Putting (6) into (5), we get

$$q_2 = \frac{a - 3t}{4}.$$

Now, we can use the above results to derive the maximum-revenue tariff. The maximum-revenue tariff is the tariff rate that maximizes the expected tariff revenue collected by the government in the home country. From the above results, the expected tariff revenue is

$$E(R) = t\frac{E(A) - 3t}{4},$$

which leads us to the following result.

Theorem 13. *In the case of Stackelberg competition, with the home firm as a leader, the maximum-revenue tariff is given by*

$$t^L = \frac{E(A)}{6}.$$

So, the output levels at equilibrium are as follows.

Theorem 14. *In the case of Stackelberg competition, with the home firm as a leader, home and foreign firms produce, respectively,*

$$q_1^L = \frac{6a + E(A)}{12} \quad \text{and} \quad q_2^L = \frac{2a - E(A)}{8}.$$

Thus, the aggregate quantity in the market is given by $Q^L = (18a - E(A))/24$ and the price is given by $p^L = (6a + E(A))/24$. The following result is also obtained straightforwardly.

Theorem 15. *In the case of Stackelberg competition, with the home firm as the leader, home and foreign firms' ex-ante expected profits are, respectively, given by*

$$E(\pi_1^L) = \frac{49(E(A))^2}{288} + \frac{36V(A)}{288} \quad \text{and} \quad E(\pi_2^L) = \frac{(E(A))^2}{64} + \frac{V(A)}{16}.$$

3.3 *Home Firm Is the Follower*

In this section, we suppose that, in the second stage of the game, the home firm is the follower. Foreign firm F_2 chooses q_2, and home firm F_1 chooses q_1 after observing q_2. Let the superscript F denote the equilibrium outcome of the game where the home firm F_1 is the follower.

We determine the subgame perfect Nash equilibrium by backwards induction. Suppose that the domestic government has chosen the import tariff t per unit of imports in the first stage. Also, suppose that, the leader foreign firm F_2 produces q_2. Then, maximizing home firm profit

$$\pi_1 = (a - q_1 - q_2)q_1,$$

we get

$$q_1 = \frac{a - q_2}{2}. \tag{7}$$

Now, maximizing foreign firm profit $\pi_2 = (a - q_1 - q_2 - t)q_2$, knowing the above quantity q_1, we get

$$q_2 = \frac{a - 2t}{2}. \tag{8}$$

Putting (8) into (7), we get

$$q_1 = \frac{a + 2t}{4}.$$

Now, we can use the above results to derive the maximum-revenue tariff. The maximum-revenue tariff is the tariff rate that maximizes the expected tariff revenue collected by the government in the home country. From the above results, the expected tariff revenue is

$$E(R) = t\frac{E(A) - 2t}{4},$$

which leads us to the following result.

Theorem 16. *In the case of Stackelberg competition, with the home firm as the follower, the maximum-revenue tariff is given by*

$$t^F = \frac{E(A)}{4}.$$

So, the output levels at equilibrium are as follows.

Theorem 17. *In the case of Stackelberg competition, with the home firm as the follower, home and foreign firms produce, respectively,*

$$q_1^F = \frac{2a + E(A)}{8} \quad \text{and} \quad q_2^F = \frac{2a - E(A)}{4}.$$

Thus, the aggregate quantity in the market is given by $Q^F = (6a - E(A))/8$ and the price is given by $p^F = (2a + E(A))/8$. The following result is also obtained straightforwardly.

Theorem 18. *In the case of Stackelberg competition, with the home firm as the follower, home and foreign firms' ex-ante expected profits are, respectively, given by*

$$E(\pi_1^F) = \left(\frac{3E(A)}{8}\right)^2 + \frac{V(A)}{16} \quad \text{and} \quad E(\pi_2^F) = \frac{(E(A))^2}{32} + \frac{V(A)}{8}.$$

3.4 Comparisons

In this section, we are going to compare the results obtained in each way of moving. First, we observe that, independently of the role, the sales of the home firm are increasing in the tariff, and the sales of the foreign firm are decreasing in the tariff. Furthermore, the total sales in the home country are decreasing in the tariff. Next corollary states that the domestic government imposes a lower tariff in the game where the home firm is the leader; and the tariffs are equal in the Cournot-type game and in the game where the home firm is the follower.

Corollary 6. *The tariffs in the different games are related as follows:*

$$t^L < t^F = t^C.$$

The total sales in the home market are higher in the game where the home firm is the leader; but, they can be lower either in the Cournot-type game or in the Stackelberg-type game where the foreign firm is the leader, depending upon the value a of the realized demand, as stated in the corollary below. This result is in contrast to the benchmark model, in which the total sales are always lower in the Cournot-type game.

Corollary 7. *The total sales in the home market are related as follows:*

1. $Q^L > Q^C$ *and* $Q^L > Q^F$
2. $Q^C < Q^F$ *if, and only if,* $a > E(A)/2$

In the next corollary, we compare the ex-ante expected profits of the firms in each game. We note that, in the ex-ante analysis, the worse situation for the home firm is to play a Stackelberg-type game being a follower firm; and the best situation is to play a Cournot-type game, if the variance of the demand parameter is low, and to be leader in a Stackelberg-type game, if the variance is high. This result is in contrast to the benchmark model, in which the home firm always prefers to play a Cournot-type game (see [3]). Furthermore, in the ex-ante analysis, the foreign firm prefers to be a Stackelberg leader firm, and its worse situation is to be a Stackelberg follower firm.

Corollary 8. *Home firm's ex-ante expected profits are related as follows:*

1. $E\left(\pi_1^F\right) < E\left(\pi_1^L\right)$ *and* $E\left(\pi_1^F\right) < E\left(\pi_1^C\right)$
2. $E\left(\pi_1^C\right) > E\left(\pi_1^L\right)$ *if, and only if,* $V(A) < (E(A))^2/4$

Foreign firm's ex-ante expected profits are related as follows: $E\left(\pi_2^L\right) < E\left(\pi_2^C\right) < E\left(\pi_2^F\right)$.

4 Conclusions

In this paper, we studied an international quantity competition with demand uncertainty, where each firm is a Cournot competitor or a Stackelberg leader. We computed the maximum-revenue tariff, the quantities, the prices and the profits in each role of the model.

Acknowledgements We are grateful to Alberto A. Pinto for a number of very fruitful and useful discussions on this work and for his friendship and encouragement. We thank ESEIG – Instituto Politécnico do Porto, Centro de Matemática da Universidade do Porto and the Programs POCTI and POCI by FCT and Ministério da Ciência, Tecnologia e do Ensino Superior for their financial support.

References

1. Brander JA, Spencer BJ (1984) Tariff protection and imperfect competition. In: Kierzkowski H (ed) Monopolistic Competition and International Trade: 194–207, Oxford University Press, Oxford
2. Clarke R, Collie DR (2006) Optimum-welfare and maximum-revenue tariffs under Bertrand duopoly. Scottish Journal of Political Economy 53: 398–408
3. Ferreira FA, Ferreira F (2010) Simultaneous decisions or leadership in an international competition. In: Simos TE et al. (eds) Numerical Analysis and Applied Mathematics, AIP Conference Proceedings 1281: 804–807, American Institute of Physics, New York
4. Ferreira FA, Ferreira F (2009) Maximum-revenue tariff under Bertrand duopoly with unknown costs. Communications in Nonlinear Science and Numerical Simulation 14: 3498–3502
5. Ferreira FA, Ferreira F, Pinto AA (2007) Bayesian price leadership. In: Kenan Tas et al. (eds) Mathematical Methods in Engineering: 371–379, Springer, Dordrecht
6. Ferreira FA, Ferreira F, Pinto AA (2007) Unknown costs in a duopoly with differentiated products. In: Kenan Tas et al. (eds) Mathematical Methods in Engineering: 359–369, Springer, Dordrecht
7. Ferreira FA, Pinto AA (2011) Uncertainty on a Bertrand duopoly with product differentiation. In: Tenreiro Machado JA et al. (eds) Nonlinear Science and Complexity: 389–395 Springer, Dordrecht
8. B. Larue B, Gervais J-P (2002) Welfare-maximizing and revenue-maximizing tariffs with few domestic firms. Canadian Journal of Economics 35: 786–804
9. Spulber D (1995) Bertrand competition when rivals' costs are unknown. Journal of Industrial Economics 43: 1–11
10. Tirole J (1994) The Theory of Industrial Organization. MIT Press, Cambridge, Mass.

Can Term Structure of Interest Rate Predict Inflation and Real Economic Activity: Nonlinear Evidence from Turkey?

Tolga Omay

1 Introduction

Policymakers may use the information provided by financial indicators as a complementary to the information provided by monetary aggregates to make monetary policy decisions. Therefore, it is crucial for policymakers to know whether these financial indicators can predict future real economic activity and inflation (i.e., [1]). There are numerous studies, for instance, on whether the interest rate spread (i.e., the difference between the long-term interest rate and the short-term interest rate) can predict future economic activity [1–9]. However, most of the previous literature applied a linear approach to tackle the problem. More recently, Venetis et al. [10] has suggested that Smooth Transition Regression (STR) models are better suited to investigate spread-output relationships as they can contain regime switching-type behavior and time-varying parameters. On the other hand, there are numerous studies on whether the interest rate spread can predict future inflation [11–15]. More recently, Telatar et al. [16] has also suggested that nonlinear models (Markov Switching) are useful and better suited to investigate spread-inflation relationship as they can accommodate to regime switching-type behavior and time-varying parameters.

In addition to the bivariate single equation investigations of the predictability of term structure, there are also some "equation system" studies in order to identify the predictability relationship. In the equation-system approach class, we see that the most influential study is Estrella's [1] 1997 paper. In this paper, he has derived reduced form relationships of output growth and spread, inflation and spread and finally, interest rate and spread from a simple structural model of the

T. Omay (✉)
Department of Economics, Çankaya University, Ogretmenler Cad. No:14, YY., Balgat, Ankara, Turkey
e-mail: omayt@cankaya.edu.tr

J.A.T. Machado et al., *Nonlinear and Complex Dynamics: Applications in Physical, Biological, and Financial Systems*, DOI 10.1007/978-1-4614-0231-2_22,

economy. Estrella [1] has found that empirical relationships are not structural, and alternative monetary policy regimes could lead to very different outcomes. Estrella's [1] approach in this study is a unification of the theories which uses the term structure of interest rates for predicting different important economic variables. In the empirical part of his paper, he has employed a theoretical impulse response analysis to this unified model and found out that it can imitate a real economic system. Moreover, Jardet [17] has used this methodology in her study to analyze US and Canadian economies and employed a VAR-VECM model. She has found out that the predictability of the term structure of interest rate is reduced by the structural breaks, related to monetary policy changes. Furthermore, Estrella [1] has also found similar results for the US economy in his 2004 research.

In this paper, we have applied a nonlinear approach to an emerging market, Turkey, concerning a particular period, a period that was influenced by high inflation, high budget deficits, and political instability. Besides these handicaps, Turkey has been encompassed by other disadvantages of testing the predictability issue like not having wide and deep financial markets for the same periods.[1] Because of these features, all possible methods must be applied only very carefully to the Turkish economy. In particular, linear models are not sufficient to analyze these types of economies. For example, for predictability of inflation, Mishkin [11] has stated that "when the relationship between the term structure and inflation is unstable, it would not be possible to provide correct information about the inflationary pressures in the economy by using the estimated parameters of the linear model," and therefore the term structure of interest rates are no longer a guide for future monetary policies aiming at ensuring price stability. Following this advice and other reasons, we have employed a nonlinear model. Unfortunately, all papers mentioned above have investigated the stability issue with a structural breaks analysis, in linear bivariate single equation models except [10] to investigate spread-output relationships and [16] for spread-inflation relationship. In our analysis, we have employed STR-type nonlinearity to both the relationships like [10], but our analysis is the generalization of their analysis applied in multivariate case. By using this analysis, the stability and the causes of the predictability issue among these nexus will further be investigated. For this purpose, we are employing a Logistic Smooth Transition Vector Autoregressive (LSTVAR) model too obtain the Generalized Impulse Response Function (GIRF). GIRF analysis has several advantages over the traditional counterparts in analyzing these kinds of systems. Traditional Impulse Response Function (TIRF) analysis has its limitations and confines researchers to give one impulse response for one specific situation in the estimation period. For example, giving an impulse to interest rate variable has one response by inflation variable during the estimation period in a linear VAR model. But in a LSTVAR model, one can apply an impulse response to specific histories (events) in the estimation period. This feature enables researchers to employ impulse response analysis to specific variables both before and after an important event.

[1]For the detailed information about the Turkish economy, [16] can be used.

Therefore, we have incorporated these advantages into our research to investigate the highlighted issues in our study. We are also investigating, the Expectation Theory, Interest Transmission Channel and Fisher Effect with assistance of the GIRF for Turkey through the estimation period. This analysis sheds light on the theoretical connections of predictability of term structure for Turkey. Moreover, we have mentioned above that the empirical relationships are not structural, and alternative monetary policy regimes could lead to varying outcomes. This result has the obvious but interesting empirical implication that shifts in the relationship between real output changes (or inflation) and the spread may provide evidence of shifts in the relative weights policy makers give to deviations of inflation or output from their targets (see [18]). In order to analyze the stability of these relationships, researchers can estimate the monetary policy reaction function and test whether the parameters of the monetary policy reaction function have changed or not. This feature can be cross-checked by a parameter constancy test which was established by Lin and Terasvirta [19]. In their specification test, they are testing linearity against multiple structural changes. On the other hand, there are other structural change tests like Bai-Perron, Chow, and Andrews. Therefore, after estimating a linear model of monetary policy reaction function, a researcher can apply these tests in addition to parameter constancy test. In our analysis, we are estimating a linear monetary policy reaction function in order to robustify the results of the stability of the relationship between real output changes (or inflation) and the spread. Therefore, this investigation may be seen as further proof of Estrella's [1] claim or not, in the case of Turkey.

The remainder of the paper is organized as follows: Sect. 2 reviews the previous literature and model. Section 3 discusses the data and methodology used. Section 4 presents the empirical results. Section 5 estimates the policy reaction function to establish the connection with monetary policy. Finally, Sect. 6 offers conclusions drawn from the study undertaken.

2 Previous Literature and Model

Harvey [2] is the pioneer study that investigates the relationship between real economic activity and spread. Harvey [2] shows that there is useful information in the real yield spread about future consumption growth. Estrella and Hardouvelis [4] argue that the nominal spread is a useful predictor of future growth in output, consumption, and investment. Plosser and Rounwenhorst [5] study shows that the term structure has a significant predictive power for long-term economic growth. Haubrich and Dombrosky [6] contend that the yield spread is an excellent predictor of four quarter economic growth but its predictive content has changed over time. Dotsey [7] has thoroughly investigated the forecasting properties of the yield spread for economic activity. To our knowledge, the only other study that has investigated the yield spread-future economic activity relationship in an emerging market is by

Kim and Limpaphayom [20]. Their results show that the term structure of interest rates has significant power in forecasting Indian real economic activity.

Mishkin [11] is the first study that has investigated the relationship between inflation and spread. In his study, he uses the difference between an n-months interest rate and an m-months interest rate to predict the difference between average inflation rates over n-months and m-months concerning the future, where $n,\ m \leq 12$. He has examined the spread for US Treasury bills. Although, he has found that for maturities of 6 months or less, the term structure of nominal interest rates have provided so little information about the future path of inflation, he has concluded, nevertheless, that for longer maturities of 9 and 12 months, the spread of nominal interest rates indeed provides information about the future path of inflation. Mishkin [12] obtains even better results with longer maturities, i.e., from 1 to 5 years. In both of these papers, the researchers have developed the theoretical background from the application of both the Fisher equation and the Expectations Hypothesis of the term structure of interest rates. On the other hand, Mishkin [11] has also examined data obtained from ten OECD countries and has found out that shorter maturities have provided almost no information about the future path of inflation. Estrella and Mishkin [15] study shows that the term structure has a significant predictive power for subsequent economic activities and the inflation rate by using Europe and the US data. Finally, Frankel [14] shows that a different measure of spread has a better predictive power by using US data.

All of these empirical studies above use linear models, which have limitations as they cannot control the possibility of asymmetric effects or structural shifts. Moreover, it is common knowledge that a number of economic variables exhibit a nonlinear behavior. Following this common knowledge, a limited number of studies have been motivated to employ nonlinear approaches like [10], and [16]. Venetis et al. [10] uses a TV-STAR model in order to investigate the relationship between real economic activity and the spread, for three developed countries. Telatar et al. [16] uses a time-varying parameter model with Markov-switching heteroskedastic disturbances in order to investigate the relationship between inflation and the spread for Turkey. Besides [1], there are two other studies that have investigated the Turkish economy, namely [21], and [22], which have employed linear models to test the predictability relationship between inflation and term structure of interest rate. Şahinbeyoğlu and Yalçın [21] have concluded that the term structure and inflation have a negative relationship between each other. They claim that the high volatility of real interest rates relative to expected inflation, as well as the negative correlation between these two variables, have produced significant and negative term structure coefficients as mentioned in [11] and [12].

Hence, we aim to contribute to the existing literature on this issue by employing a LSTVAR model to an emerging market and investigating the sources of the negative relationship between these nexus.

In this paper, we have used the model which is given below. This model was developed by Estrella [1]:

1. Phillips curve: $\pi_{t+1} = \pi_t + ay_t + e_{t+1}$
2. IS curve: $y_{t+1} = b_1 y_t - b_2 \rho_t + \eta_{t+1}$

3. Monetary policy reaction function:
   $r_{t+1} = g_1 r_t + g_2 \pi_{t+1} + g_3 y_{t+1} + (1 - g_1 - g_2) z_{t+1}$
4. Monetary shock: $z_{t+1} = z_t + v_{t+1}$
5. Fisher equation: $R_t = \rho + \frac{1}{2}\left(E_t \pi_{t+1} + E_t \pi_{t+2}\right)$
6. Expectation hypothesis: $R_t = \frac{1}{2}\left(r_t + E_t r_{t+1}\right)$

Parameter restrictions in the equation system:

$$0 \prec a \prec 1,\ 0 \prec b_1 \prec 1,\ 0 \leq b_2 \leq 1,\ 0 \leq g_1 \leq 1,\ g_2,\ g_3 \geq 0$$

where π_t is the inflation rate in period t, y_t is the output gap in period t, r_t is the short term (1 period) nominal interest rate, R_t is the long-term (2 period) nominal interest rate in period t, ρ_t is the long-term (2 period) real interest rate in period t, Z_t is the inflation target in period t, E_t is the rational expectations operator based on period t information and ε_t, $v_t \eta_t$: i.i.d. random variables.

Estrella [1] derives reduced form equation for the relationship between inflation, changes in real output, and the spread with a simple structural model of an economy given as above. Estrella's [1] model is based on the model used by Fuhrer and Moore [23] for empirical purposes. But, Estrella [1] has found a theoretical, implicit solution for this model. The advantage of Estrella's [1] model is the flexible choice of a monetary policy reaction function. The monetary policy reaction function can be simplified by the choice of parameters to either the [24] reaction function $(g_1 = 0)$ or Fuhrer–Moore $(g_1 = 1)$ version of the Taylor rule.

Estrella [1] and researches derived the reduced forms of this model by forward and backward methods and show that the coefficient linking expected future output, inflation, and short-term interest rates to spread, depending on the parameters in the monetary reaction function. For example, if the central bank only reacts to output deviations where $g_2 = 0$, the coefficient linking expected future output and spread in the reduced form is given as $(2/g_3)$. As a conclusion of Estrella's [1] work, monetary policy is a key determinant of the precise relationship between the term structure of interest rates and macroeconomic variables such as real output and inflation. With respect to Estrella's [1] rational expectation's model, he notes that with alternative monetary policy rules, one or more of the following can occur: the term structure spread is the optimal predictor of real output; the term structure is the optimal predictor of changes in inflation; the short-term nominal rate is the best predictor of real output; the short-term nominal rate is perfectly correlated with the long-term real rate; interest rates are not informative with regard to future output and inflation. However, Estrella [1] has found that the term structure of interest rates has at least some predictive power for both real output and inflation under almost all circumstances. From this discussion, it is clear that the empirical relationships are not structural, and alternative monetary policy regimes could lead to very different outcomes. This issue can be further analyzed by the method that we mentioned in the introduction.

3 Data and Methodology

3.1 Data

Forthe paper, we are using industrial production index and the consumer price index (CPI) downloaded from the website of the Republic of Turkey's Central Bank. In order to calculate the term structure of interest rate (spread), 1-month and 3-month government security interest rates are taken from the secondary market. These data are extracted from the Istanbul Stock Exchange (ISE) monthly journal.[2] Therefore, the spread variable sp is obtained by subtracting 1-month interest rate from 3-month interest rate.

3.2 Basic Regression of Predictability: Real Economic Activity and Inflation

In the paper, we consider the annual growth rate of the monthly industrial production index as "cumulative," whereas some other papers use marginal growth rates:

$$\Delta^k y_t = \frac{12}{k}(y_t - y_{t-k}) \tag{1}$$

where y_t is the logarithm of the industrial production index at time t. We compute the slope of the nominal yield curve as the difference between the long-term bond yield LR i_t and the short-term yield SR i_t as (LR i_t-SR i_t) at time t.

The following basic regression is a way to describe the relationship between the spread and future activity:

$$\Delta^3 y_t = \zeta + \beta sp_{t-1} + \varepsilon_t \tag{2}$$

where k is the forecast horizon and ε_t forecast error. Depending on the previous literature, the best forecasting horizon is a 3-month period for Turkey. Therefore, we only consider the forecast horizon of 3 months in this study. Moreover, Turkey is a high inflationary country thus, a long horizon can be at most 3 months with regard to long term economic decision. The fact that we are working with monthly data creates some temporal correlation between the successive error terms. In order

[2]The daily interest rates obtained from ISE are weighted with the transaction volumes to calculate the interest rates for 1-month and 3-month maturities. The 1-month interest rate is defined to have a maturity in the range of 20–40 days and the 3-month interest rate is defined to have a maturity in the range of 80–100 days.

to remedy this serial correlation problem, we have used [25] for standard error terms (see [17]). The estimated equation is given as below:

$$\Delta^3 y_t = \underset{(0.039)}{0.018} - \underset{(0.0001)}{0.0004} sp_{t-1} + \varepsilon_t$$

$$\text{r} - \text{square} = 0.006, \hat{\sigma}_e = 0.337, \text{SK} = 0.313\,(0.197)\,,\ \text{EK} = -0.043\,(0.930)\,,$$

$$\text{JB} = 1.715\,(0.424)\,\text{DW} = 1.254, \text{ARCH}\,(1) = 3.284\,(0.069) \qquad (3)$$

where heteroscedasticity consistent (hcc) standard errors are given in the parentheses below the parameter estimates, ε_t denotes the regression residual at time t, $\hat{\sigma}_e$ is the residual standard deviation, SK is skewness, EK excess kurtosis, JB the Jarque–Bera test of normality of the residuals, and ARCH is the LM test of no Autoregressive Conditional Heteroscedasticity (ARCH). Normality of residual is not rejected with Jarque–Bera test. SK and EK are also rejected. But, Durbin–Watson test shows that there is a significant autocorrelation problem. Moreover, the LM test for ARCH has significant value which can be the indication of neglected nonlinearity (see [26]). This final conjecture is investigated further by applying the LM-type linearity test of [27]. From (3), we can observe that the relationship between real economic activity and the spread is significant and negative like the previous findings. This linear model suggested that the spread has information content about real economic activity for Turkey; but this relationship is negative, whereas the relationship has been found to be positive in developed countries. This issue has one of the main questions posed in this paper and will be analyzed further in Sect. 4. On the other hand, the other main question is the stability of this relationship. For investigating the stability issue, we have employed two different approaches; one of them is a recursive Chow structural break test with which we can date the structural breaks; and the other one is parameter stability test, which is developed by Lin and Terasvirta [19]. A recursive Chow structural test for (3) shows three consecutive structural change points between 1999 and 2000. These structural points are 1999:3 = 2.901(0.038), 1999:4 = 2.885(0.039), and 1995:5 = 2.930(0.037). The first value is Chow test and the probabilities are given in parenthesis. The parameter stability test LM = 2.283(0.081) suggests that the equation can be modeled by a TV-STAR model which shows that the parameter of the model is not constant. Hence, stability issue is not satisfied for the Turkish case by using linear model.

For the inflation case, we consider the marginal growth rate of monthly inflation rate as:

$$\Delta^k \pi_t = (\pi_t - \pi_{t-k}) \qquad (4)$$

where π_t is the logarithm of the inflation rate, which is obtained by the CPI at time t.

The following basic regression shows the relationship between spread and future activity:

$$\Delta^k \pi_t = \zeta + \beta sp_{t-1} + \varepsilon_t \qquad (5)$$

where k is the forecast horizon and ε_t forecast error. In this study we are only dealing with the forecast horizon of 3 months, because we are not dealing with the best predictability horizon. On the other hand, other papers, dealing with the predictability issue, have found that the best forecasting horizon is 3-month period for Turkey. This is theoretically sensible as well, because Turkey is a high inflationary country and a long horizon can be at most 3 months to decide to make sound long term investments.

$$\Delta^3 \pi_t = \underset{(31.173)}{-40.063} - \underset{(0.905)}{6.870} sp_{t-1} + e_t$$

$$\text{r} - \text{square} = 0.580\, \hat{\sigma}_e = 0.054 \text{SK} = 0.001\,(0.197)\, \text{EK} = -0.564\,(0.930)$$

$$\text{JB} = 1.529\,(0.424)\, \text{DW} = 0.707 \text{ARCH}\,(1) = 60.096\,(0.000) \qquad (6)$$

All the misspecification test results and their explanations are the same. However, this time linear model has better information content than the real economic activity case, which we can decide as looking at the r-square criteria.[3] Again we have investigated the stability issue; we employed two different approaches, which are mentioned above. The recursive Chow structural test for (2) shows that there are many structural breaks at 1% significance level, starting from 96:3 = 5.815(0.000) until the end of the sample period. The most significant one is 99:8 = 29.771 (0.000). The parameter stability test suggests that the equation can be modeled by a TV-STAR model, which shows that the parameter of the model is not constant and the test statistics here LM = 6.835(0.000). From these analyses, we have seen that the parameters of the linear models are not constant and the stability issues are not satisfied with the Turkish data.

3.3 *Specification of Smooth Transition Vector Autoregressive Model and Linearity*

3.3.1 Tests

The specification and estimation of multivariate STAR models are discussed in further details in [26] (Chap. 5). In the specification of a smooth transition vector auto-regression model (STVAR), we follow [26]. Let $x_t = (x_{1t}, \ldots x_{kt})'$ be a $(k \times 1)$ vector time series. We have $x_t = (y_t, p_t, i_t, sp_t)'$ where y_t is the log of industrial production index, i_t the log of nominal interest, p_t the log of the **CPI**, and sp_t the log of spread values. A k-dimensional STVAR then can be formulated as

[3] In both of the regression equations, we are using different dependent variables; hence, there is no direct comparison of R-squares.

$$\Delta x_t = \left\{ \phi_1 + \sum_{i=1}^{p-1} \varphi_{1i} \Delta x_{t-i} \right\} . (1 - F(s_t;\gamma,c)) + \left\{ \phi_2 + \sum_{i=1}^{p-1} \varphi_{2i} \Delta x_{t-i} \right\} . (F(s_t;\gamma,c)) + \varepsilon_t \tag{7}$$

where $\phi_j j = 1,2$ are $(k \times 1)$ vectors, $\varphi_{j,i}, j = 1,2,\ I = 1,\ldots,p-1$ are $(k \times k)$ matrices, and $\varepsilon_t = (\varepsilon_{1t},\ldots,\varepsilon_{kt})$ is a k-dimensional vector white-noise process with mean zero and $(k \times k)$ covariance matrix Σ. The transition function $F(s_t;\gamma,\ c)$ is assumed to be a continuous function between zero and one, with parameters γ and c determining the smoothness and location of the change in the value of $F(s_t;\gamma,\ c)$, respectively. Here we have the following form logistic function

$$F(s_t;\gamma,c) = \frac{1}{1+\exp\{-\gamma[s_t - c]\}}\ ,\ \gamma > 0 \tag{8}$$

The specific-to-general approach for specifying univariate STAR models put forward by Terasvirta [28] can be adapted to the multivariate case. This procedure starts with specifying a vector autoregressive model for $x_t = (y_t, p_t, i_t, sp_t)'$, that is,

$$\Delta x_t = \phi_1 + \sum_{i=1}^{p-1} \varphi_i \Delta x_{t-i} . \varepsilon_t \tag{9}$$

where the order p should be such that residuals $\hat{\varepsilon}_t$ have zero autocorrelations at all lags. The choice of p is based on applying both the SIC and AIC in a linear VAR model for x_t with a deterministic linear trend. Both information criteria select $p = 1$ as the appropriate lag order.

The next step of the specification procedure is testing linearity against STAR-type nonlinearity as given in (1), with (2). However, the testing problem is complicated by the presence of unidentified nuisance parameters under the null hypothesis. This can be understood by noting that the null hypothesis can be expressed in multiple ways, either as $H_0 : \phi_1 = \phi_2$ and $\varphi_{1,i} = \varphi_{2,i}$ for $I = 1,\ldots,p-1$ or as $H_0' : \gamma = 0$. In order to overcome this problem, following the approach of [27], we replace the transition function $F(s_t;\gamma,\ c)$ with a suitable Taylor approximation. For example, a first-order Taylor approximation of the transition function results in the following auxiliary regression

$$\Delta x_t = A_0 + \sum_{i=1}^{p-1} B_i \Delta x_{t-i} + A_1 s_t + \sum_{i=1}^{p-1} B_{1,i} \Delta x_{t-i} s_t + e_t \tag{10}$$

where e_t comprises the original shocks ε_t as well as the error arising from the Taylor approximation. In (10) it is assumed that the transition variable s_t is not one of the elements of $\Delta x_{t-i},\ i = 1,\ldots,p-1$ or their linear combination. If this is not the case, the term $A_1 s_t$ must be dropped from the auxiliary regression. The parameters in A_j and $B_{j,i} j = 0,1,\ I = 1,\ldots,p-1$, are functions of the parameters in the original STVAR. In this case, the null hypothesis of linearity can be expressed as $H_0'' : A_1 = B_{1,i} = 0,\ I = 1,\ldots,p-1$, that is, the parameters associated with the auxiliary regressors are all zeros. This null hypothesis can be tested by a standard variable addition test in a straightforward manner. The test statistics, to be denoted

Table 1 LM-type test against linearity

Candidate transition variables	Equation Δp_t	Δi_t	Δy_t	sp_t	System-wide test
Δp_{t-6}	3.880 (0.000)	1.135 (0.344)	1.809 (0.089)	2.415 (0.022)	65.664 (0.000)
$\Delta^3 p_{t-6}$	1.937 (0.067)	0.519 (0.818)	1.316 (0.246)	1.803 (0.091)	67.824 (0.000)
Δy_{t-1}	3.014 (0.005)	0.569 (0.779)	4.373 (0.000)	3.573 (0.001)	104.128 (0.000)
Δy_{t-6}	0.898 (0.509)	0.613 (0.744)	0.473 (0.852)	3.217 (0.003)	87.663 (0.000)
sp_{t-1}	6.205 (0.000)	4.385 (0.000)	4.585 (0.000)	23.774 (0.000)	110.094 (0.000)
Δsp_{t-1}	5.143 (0.000)	1.537 (0.159)	2.947 (0.006)	17.922 (0.000)	101.927 (0.000)
Δi_{t-2}	3.760 (0.000)	1.463 (0.184)	2.218 (0.036)	2.164 (0.040)	49.827 (0.006)
Δi_{t-2}	1.430 (0.197)	0.745 (0.634)	1.602 (0.139)	0.762 (0.619)	47.709 (0.011)
t	6.412 (0.000)	3.650 (0.001)	3.287 (0.002)	3.285 (0.002)	42.891 (0.035)

The most suitable candidate variables are given. Besides, the other test statistics are available upon request

as LM_1, have an asymptotic χ^2 distribution with $k(q+1)+(p-1)k^2$ degrees of freedom under the null of linearity. Since the LM_1 statistic does not test the original null hypothesis $H_0' : \gamma = 0$ but rather the auxiliary null hypothesis $H_0'' : A_1 = B_{1,i} = 0$, this statistic is usually referred to as an LM-type statistic.

As noted by Luukkonnen et al. [27], the LM_1 statistic has no power in situations where only the intercept in the VAR varies across regimes, that is, when $\phi_1 = \phi_2$ but $\varphi_{1,i} = \varphi_{2,i}$ for $I = 0, 1, \ldots, p-1$, in STVAR given in (1). Luukkonnen et al. [27] suggest to remedy this problem by replacing the transition function $F(s_t; \gamma,\ c)$ by a third-order Taylor approximation, which results in the auxiliary regression (4) with, s_t^j and $\Delta x_{t-i} s_t^i$, $j = 2,\ 3,\ i = 1, \ldots, p-1$ as additional auxiliary regressors. The test statistic computed from the augmented auxiliary regression is LM_2 statistic. Since only the parameters corresponding to s_t^2 and s_t^3 are functions of ϕ_1 and ϕ_2, a parsimonious, or "economy" version of the LM_2 statistic can be obtained by augmenting (10) with additional regressors s_t^2 and s_t^3. The resultant statistic is the LM_3 statistic.

To identify an appropriate transition variable s_t, the LM-type statistic can be computed for several candidates, and the one for which the associated p-value of the test statistic is smallest can be selected. In this paper, we consider 54 different candidate transition variables; lagged growth rates in output (Δy_{t-i}), lagged growth rates in interest (Δi_{t-i}), lagged inflation rates (Δp_{t-i}), and lagged rate of change of the spread rates (Δsp_{t-i}). By being a linear combination of real and monetary variables, such a transition variable can capture both nominal and real rigidities. Since monthly time series exhibit considerable short-run fluctuations which do not necessarily represent changes in regimes, we also consider quarterly changes in the above-mentioned variables (i.e., $\Delta_3 \theta_{t-i} = \theta_{t-i} - \theta_{t-i-3}$ where θ_t is one of the elements in the time series vector x_t) We set $i = 1, \ldots, 12$ in all cases.

The results of the linearity tests are shown in Table 1, which reports only the candidate transition variables for which the null of linearity of the output equation is rejected.

The results of both system-wide linearity tests as well as a test of linearity of the output equation suggest Δp_{t-i}, Δy_{t-i}, Δi_{t-i} and the spread as appropriate transition

Table 2 Estimation results of LSTVAR

Parameters	DCPI	DGSMH	SP	DSRI
Constant	−0.059 (0.017)	0.194 (0.056)	−2.345 (32.931)	−2.352 (0.475)
$L-\Delta p_{t-1}$	0.983 (0.015)	0.072 (0.049)	−9.827 (28.813)	−1.519 (0.424)
$L-\Delta y_{t-1}$	−0.057 (0.024)	0.585 (0.077)	−60.994 (45.551)	−1.271 (0.651)
$L-sp_{t-1}$	−0.0002 (0.000)	0.0002 (0.000)	0.174 (0.068)	−0.003 (0.000)
$L-\Delta i_{t-1}$	0.015 (0.005)	−0.051 (0.016)	0.731 (9.430)	0.751 (0.138)
$U-\Delta p_{t-1}$	0.002 (0.001)	−0.015 (0.005)	−5.310 (3.070)	0.145 (0.044)
$U-\Delta y_{t-1}$	0.081 (0.042)	0.062 (0.136)	14.422 (79.991)	−1.701 (1.1436)
$U-sp_{t-1}$	0.0002 (0.000)	−0.0004 (0.000)	1.407 (0.139)	−0.006 (0.001)
$U-\Delta i_{t-1}$	−0.005 (0.004)	0.030 (0.013)	13.072 (7.713)	−0.236 (0.111)
R^2	0.985	0.521	0.634	0.487
γ	5.119 (8.550)			
C	10.736 (0.394)			

variables. Since the null of linearity is rejected more strongly for Δsp_{t-1}, in price, output, and short run interest rate equations, we use this variable as the system-wide transition variable in estimation of the STVAR.

Given the choice of the transition variable s_t, estimation of the parameters in the STVAR (1) is a relatively straightforward application of nonlinear least squares, which is equivalent to quasi-maximum likelihood based on a normal distribution. Under certain (weak) regularity conditions, which are discussed by White and Domowitz [29], and Pötscher and Prucha [30], among others, the NLS estimates are consistent and asymptotically normal.

The estimation can be performed using any conventional nonlinear optimization procedure. The burden on the optimization algorithm can be alleviated by using good starting values. For fixed values of the parameters in the transition function, γ and c, the STVAR model is linear in the parameters ϕ_j, $\varphi_{j,i}$, $j=1,2$, $I=1,\ldots,p-1$, and therefore, can be estimated by OLS. Hence, a convenient way to obtain sensible starting values for the nonlinear optimization algorithm is to perform a two-dimensional grid search over γ and c. Furthermore, the objective function (the log of the determinant of the residual covariance matrix) can be concentrated with respect to ϕ_j, $\varphi_{j,i}$, $j=1,2$, $I=1,\ldots,p-1$. This considerably reduces the dimensionality of the NLS estimation problem, as the objective function needs to be minimized with respect to the two parameters γ and c only. Parameter estimates of LSTVAR model are given in Table 2.

After estimating the LSTVAR model, we test whether the estimated model adequately captures the nonlinear features of the time series under examination. Specifically, we test the model for remaining nonlinearity and for parameter constancy. The testing procedure is outlined in Appendix D in [31], who generalize the procedure developed by Eitrheim and Terasvirta [32] to the multivariate context. In addition to the candidate transition variables used for testing linearity of the baseline model, we also use semiannual changes in the same variables (Table 3).

Table 3 Test against multiple regime LSTVAR

Candidate transition variables	Additive	Multiplicative	Additive and multiplicative together
Δp_{t-6}	57.106 (0.172)	56.313 (0.191)	34.420 (0.999)
$\Delta^3 p_{t-6}$	28.056 (0.990)	46.493 (0.534)	30.028 (0.999)
Δy_{t-1}	29.912 (0.981)	40.417 (0.773)	26.116 (1.000)
Δy_{t-6}	27.175 (0.993)	36.434 (0.889)	19.379 (1.000)
sp_{t-1}	49.731 (0.404)	51.182 (0.349)	39.359 (0.999)
Δsp_{t-1}	30.595 (0.976)	40.728 (0.762)	37.753 (0.999)
Δi_{t-2}	38.831 (0.824)	51.920 (0.323)	36.046 (0.999)
Δi_{t-4}	40.525 (0.769)	38.643 (0.830)	26.623 (1.000)
T	40.366 (0.775)	47.586 (0.489)	33.781 (0.999)

Table contains the smallest p value test statistics. Other test statistics are available upon request

4 Investigations of Expectation Theory Interest Transmission Mechanism Fisher Effect and Stability of Predictability: Generalized Impulse Response Analysis

In this section, we have computed a generalized impulse response analysis in order to detect the effects of structural changes and crises in predicting future output and inflation for Turkey. From this analysis, we will detect whether the predictability relationship variables mentioned are stable in the estimation period or not. Generalised Impulse Response Functions have advantages on their linear counter parts per [33]. Hence, we use GIRF as an indicator tool for visualizing the effects of structural changes and crises.

The TRIF has characteristic properties in case the model is linear. First, the TRIF then is symmetric. In the sense that a shock of $-\delta$ has exactly the opposite effect as a shock of size $+\delta$. Furthermore, it might be called linear, as the impulse response is proportional to the size of shock. Finally, the impulse response is history independent as it does not depend on the particular history w_{t-1} [26]. These properties do not carry over to nonlinear models. In nonlinear models, the impact of a shock depends on the sign and the size of the shock, as well as on the history of the process. The GRIF, introduced by Koop et al. [33] provides a natural solution to the problems involved in defining impulse responses in nonlinear models. The GIRF is a function of δ (shock) and w_{t-1} (history), which are the realizations of the random variables ε_t and Ω_{t-1}. Koop et al. [33] stress this; hence the GIRF is a realization of a random variable. Using this interpretation of the GIRF as a random variable, various conditional versions can be defined, which are of potential interest. In our case, to analyze the stability and source of the relationship, we give a shock to the estimated model for every specific year in the sample period.

During this analysis, we have considered whether "Expectations Theory" and "Interest Transmission Mechanism" operate simultaneously in the same direction or not, in order to investigate the source of the predictability relationship between spread and real economic activity. In almost all empirical studies, these mechanisms are assumed to be the theoretical structure of real economic activity and the spread relationship. For example, if the Central Bank decides to exercise an expansionary monetary policy, the first effect of this policy will be a fall in the interest rate of all maturities. According to the Expectations Theory of the interest rate's term structure, long-term interest rates will be less than short-term interest rates. The simultaneously falling interest rates result in an increase in investment and, successively, an increase in real economic activity, whereas the higher fall in short-term interest rates relative to long-term interest rates causes an increase in the spread. As it can be seen from the above discussion, the mechanism is very simple. The spread and real economic activity move in the same direction with expansionary monetary policy. The move in the same direction can be demonstrated as the theoretical reason why the spread is an indicator of future real economic activity. On the other hand, Expectation Theory with Fisher Hypothesis is the key theoretical underpinning for the predictability of inflation. Moreover, there are two important assumptions; rational expectation and constancy of ex-ante real rates must be satisfied. Hence, we are also inspecting the spread and inflation relationships with a GIRF analysis. The second important relationship for the predictability of inflation, Fisher Hypothesis is analyzed by dynamic correlation analysis, which is also obtained by GIRF analysis. High volatility of the real interest rates relative to expected inflation, as well as the negative correlation between these two variables, produce significant and negative term structure coefficients as mentioned in [11] and [12]. Both of these arguments have revealed the predictability relationship of inflation for Turkey while we are investigating the relationship.

Among many authors who have investigated the relationship between monetary policy and the spread are [34]. Cook and Hahn [34], first firmly established the positive empirical relationships between target rates and long-term rates, and interpret their findings as supportive of the Expectations Theory of the term structure. The expectations theory says that a long-term interest rate should be equal to the sum of short-term interest rates over the same period of time plus a term premium; thus an increase in the first couple of short-term rates should drive up the long-term rate too, but by less. However, Romer [35] produced the opposite results to [34]. To them, positive movement in the long-term rate is inconsistent with standard monetary theory; this is, in short, a puzzle. According to received theory, they claim, an increase in short-term rates should reduce inflation, and hence reduce the level of long-term rates sufficiently. Romer [35] suggests that the puzzle can be resolved if the central bank has access to private information about economic fundamentals, but they do not develop their argument formally. In short, we can say that some authors argue that long-term rates should increase as monetary policy is tightened, mainly via the expectations hypothesis of the term structure. Others support the hypothesis that a monetary tightening should increase short rates but decrease long-term rates as inflation expectations fall.

Ellingsen and Söderström [36] expand Romer's [35] idea. According to [36], if monetary policy reveals information about economic developments interest rates of all maturities move in the same direction in response to policy innovation as if monetary policy reveals the central bank's policy preferences interest rates short-term and long-term rates move in the opposite direction. So the first proposition of [36] supports the Expectation Hypothesis and they called this type of "policy endogenous". For both of the theoretical explanations, an increase in short-term interest rate (tight monetary policy) leads to a decrease in the spread variable. From Estrella's [1] argument, we can derive a new relationship which is revealing the Expectation Hypothesis; if the short-term rate increases (decreases) then the spread will decrease (increase), because short-term rates should drive up the long rate too, but by less; when monetary policy reveals the central bank's policy preferences then short rate and long rate move in opposite direction, which again leads to a decrease. Therefore, we can deduce Expectation Hypothesis by analyzing impulse response between the short-term interest rate (which can be the monetary policy tool) and the spread variable; giving positive one standard shock to the short-term interest rate leads to a negative response of the spread variable in impulse response analysis, which indicates that the Expectation Hypothesis is valid for that period for any type of theoretical explanation. In an opposite case, we will conclude that the Expectation Hypothesis is not valid.

We have given GIRF to nine different periods covering the sample period. From the impulse response given to extract the relationship of short-term interest rates and the spread, we have found out that six of them are negative, which indicates that the Expectation Hypothesis is valid for those periods. Besides, we have found two significant positive relationships covering the periods through 1998:5 to 2000:8 and one insignificant positive relationship covering the period through 2000:5 to 2001:8. These periods are important for the Turkish economy, which experienced heavy banking crises. The structural break analysis which is applied to univaried predictability equations shows the same dates as significant structural breaks. From these results, we can conclude that the GIRF findings are consistent with the earlier research in this study. The results of GIRF analysis can be seen from the Fig. 1 given below.

The relationships between monetary policy (or the short-term interest rate) and real economic activity can also be analyzed employing impulse response analysis in the estimation period. This analysis shows us whether the interest rate transmission channel of Turkey is working or not. When an increase (decrease) occurred in the short-term interest rate, this causes a decrease (increase) in real economic activity, e.g., by stimulated investments; this mechanism is called as "Interest Rate Transmission Mechanism" (see [37]). Hence, we can analyze this Transmission Channel by giving positive one standard impulse to short-term rates and tracing the response of this impulse on real economic activity. From the above argument, the short-term interest rate has to have a negative relationship with real economic activity; hence when we observe a positive response in terms of real economic activity, which indicates that the Transmission Channel is not working for that period.

We have given a GIRF to nine different periods covering the sample period. From the impulse response given to extract the relationships of short-term interest rate and real economic activity, we have found that six of them were positive, which indicates that the Interest Rate Transmission Channel was not working for these

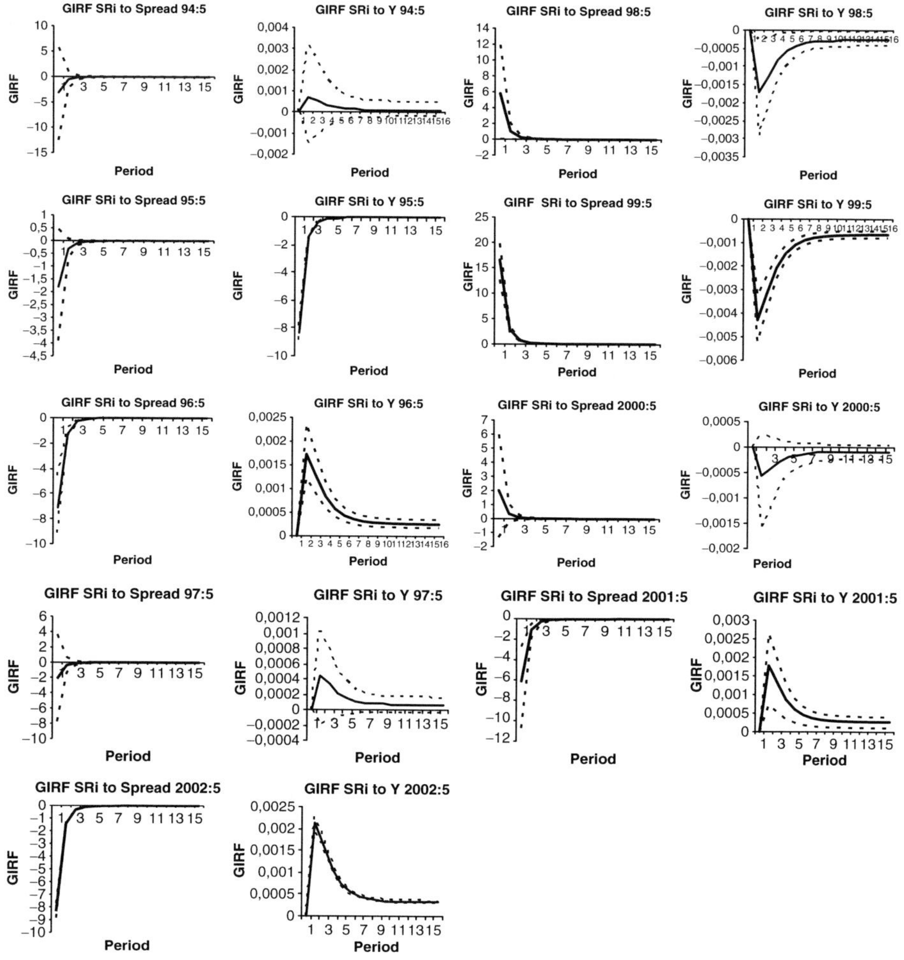

Fig. 1 Investigation of expectation theory and transmission channel: GIRF analysis

periods. Besides, we have found two significant negative relationships which cover the period through 1998:5 to 2000:8; one insignificant negative relationship which covers the period through 2000:5 to 2001:8. These periods are important periods in which the Turkish economy experienced banking crises, and which we mentioned above for the Expectation hypothesis. In the sample period, Turkey has a very high real interest rate which prevents domestic investors from making investments in real sector. Hence, this evidence shows the main obstacle of the Interest Transmission Channel to work properly for the sample period.[4] From Fig. 1, we can trace the GIRF analysis of Interest Transmission Channel through the sample period.

[4]Conversely, during the crises the Interest Transmission Channel worked. What could be the cause of this abrupt phenomenon? This question is a subject of a further study.

One can easily recognize that generalized impulse response analysis shows an asymmetric relationship between Expectation Hypothesis and Interest Transmission Channel. When the Expectation Hypothesis is valid for Turkey, the Interest Rate Transmission Channel is not working and vice versa, along the sample period. On the other hand, this systematic and reverse symmetric movement explains the negative relationships between the spread and real economic activity. This systematic reverse (symmetric) movement explains why the spread predicts real economic activity negatively. For example, again if the Central Bank decides to exercise an expansionary monetary policy, the first effect of this policy will be a fall in the interest rate of all maturities. According to the Expectations Theory of the term structure interest rate, long-term interest rates will fall less than short-term interest rates. Unfortunately, the simultaneously falling interest rates do not result in an increase in investment and, successively, an increase in real economic activity, because of high real interest rates. As can be seen from this discussion, the mechanism does not lead to a move in the same direction between spread and real economic activity. Thus, we have obtained a negative predictability relationship between these variables. Alternatively, this result can be obtained by discussing an opposite argument.

Fama's [38] seminal paper found that real interest rates are constant over time, with fluctuations in nominal rates mainly reflecting changes in expected inflation (the so-called Fisher Effect). We have two possible methods to investigate the Fisher Hypothesis. First one is a direct approach of employing the GIRF analysis. In the first approach, the expected inflation must be specified which is an important component of Fisher equation:

$$i_t = \rho + \beta \pi_t^e, \tag{11}$$

i_t, nominal interest rate, ρ is constant, including constant real interest rate, and π_t^e is expected inflation rate. Barro and Gordon [39] and Lucas [40] suggest that the expected inflation rate, which is unobservable, may be systematically related to past rates of inflation. Hence, using the lag of past rates of inflation as a proxy for the expected rate of inflation, we can employ GIRF analysis to inspect Fisher Effect. The second approach is dynamic correlation approach, which is one of the methods used for investigating the long-run relationships between the nominal interest rate and inflation. In order to obtain dynamic correlation, a VAR analysis must be held in which the variables are filtered from structural breaks in level (see [41]). In our case, we are not in need of filtering data because of using nonlinear VAR technique, which considers the structural breaks while we are dealing with the data. The dynamic correlation is obtained by giving orthogonal shock in the inflation rate, which is taken as exogenous, and by tracing the effects of this shock in the interest rate and

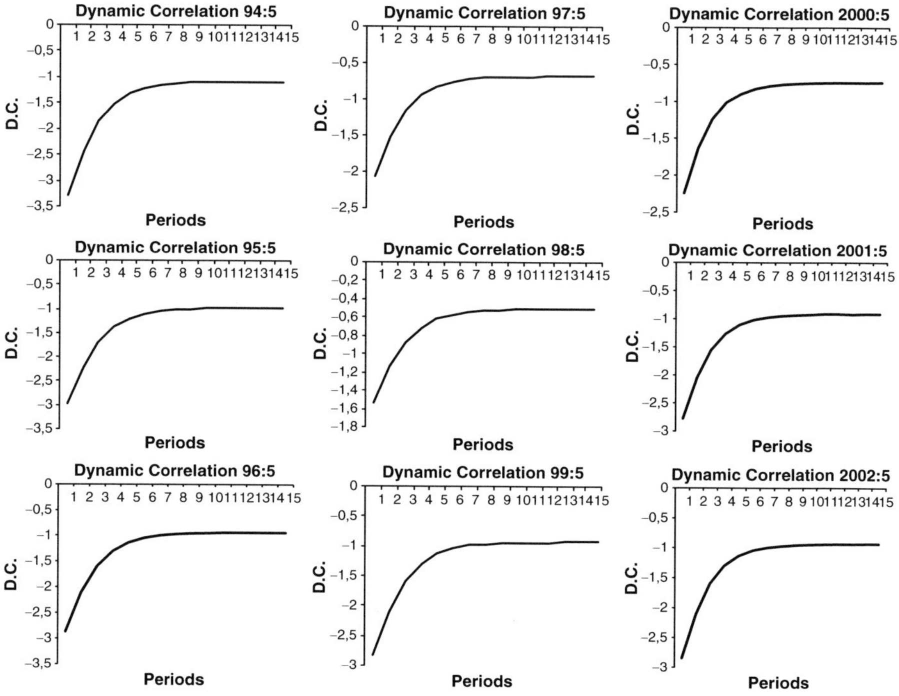

Fig. 2 Dynamic correlation analysis for long run Fisher effect

the inflation rate. The dynamic correlation is the ratio of the cumulative response function for the interest rate to that the inflation rate, calculated as:

$$\rho_s = \frac{\sum\limits_{i=0}^{s} C_{11,i}}{\sum\limits_{i=0}^{s} C_{21,i}}, \quad s = 1, 2, \ldots, S. \tag{12}$$

If the long-run Fisher Effect is true, then the dynamic correlation has converged to 1.0 for large s (s is period for our GIRF analysis which is taken as 15 months[5]). Both analyses give the same results except for the 98:5, 99:5, and 2000:5 periods which are also sensitive for Expectation Theory and Transmission Mechanism. The first analysis is changing the sign negative to positive in these periods, but dynamic correlation is the same negative sign all through the sample period. Also dynamic correlation has changed in these periods as decreasing effect of the correlation. These periods are the significant structural breaks in the linear model so again these analyses are consistent with the other analyses, which we held before. The dynamic correlation analyses are given in below Fig. 2.

[5]Fifteen months is a very long period for an emerging market which has a high and volatile inflation rate.

Except 97:5, 98:5, and 2000:5 the dynamic correlation is converging to −1.0. These negative correlations cannot be explained as inverted Fisher Effect. In inverted Fisher Effect the relationship is obtained by a real interest rate where as we are using a nominal interest rate. Hence, we can conclude that the relationships between a nominal interest rate and expected inflation is negative for the first analysis that we designed except the 98:5, 99:5, and 2000:5 periods and for long-run relationships it is negative all over the sample period. As we have mentioned before, the results are traced to a combination of the Fisher equation and the Expectation Hypothesis of the term structure of interest rate. The strength of the results, however, is seen to hinge on the nature of the variability of the unobserved real interest rates and on their relationships with the observed variables [1]. Both Expectation Hypothesis and Fisher equation are negative in our sample period. Negative results have been attributed to three factors: time-varying risk premia, the complexities of movement in the real rate, and the influences of monetary policy [1]. If we think about the predictability of the inflation equation (5), it is obtained by taking the difference of the Fisher equation for two interest rates with maturities m and n : $i_t^m = r_t^m + E_t \pi_{t+m}$, and $i_t^n = r_t^n + E_t \pi_{t+n}$, respectively. The results become an (5) if ex-ante real interest rates are constant and expectations are rational. Therefore, we can easily conclude that the negative predictability of inflation is caused by negative Fisher effect and Expectation Hypothesis.

From the linear bivariate single-equation analysis of inflation and real economic activity, we have found significant structural breaks with the recursive Chow test. From the GIRF analysis, we have found the same pattern in these relationships. As we have stated before, the GIRF analysis has its own advantages while analyzing data. Hence, we use these advantages in order to deduce stability relationship from the estimated model. Both generalized impulse responses of the spread to inflation and also real economic activity have significant and negative impacts. For the spread to inflation GIRF, the periods 94:5 to 95:8, 95:5 to 96:8, 96:5 to 97:8, 97:5 to 98:8, 2001:5 to 2002:8, and 2002:5 to 2003:8 have a significant negative impact of between −0.005 and −0.01, whereas for periods 98:5 to 99:8 and 2000:5 to 2001:8 there is a significant negative impact around −0.01. For the period 99:5 to 2000:8, there is a significant negative impact greater than −0.01. These impulse response results are consistent with the recursive Chow test results. Again the most significant Chow test result date is the biggest impact period for the GIRF analysis. GIRF analysis shows us a more detailed picture of the structural break points without losing any degree of freedom. Moreover, we can follow the conditions of the relationships at every point of sample. For the spread to real economic activity GIRF, the periods 94:5 to 95:8, 95:5 to 96:8, 96:5 to 97:8, 97:5 to 98:8, 2001:5 to 2002:8, and 2002:5 to 2003:8 have a significant negative impact of between −0.006 and −0.008, whereas for the periods 98:5 to 99:8 and 2000:5 to 2001:8 there was a significant negative impact around −0.01 and for period 99:5 to 2000:8 there was a significant negative impact greater than −0.01. These impulse response results are consistent with the recursive Chow test results. Again the most significant Chow test result date is the biggest impact period for the GIRF analysis. From these results, we can conclude that the GIRF analysis can be used as a helpful tool for finding the

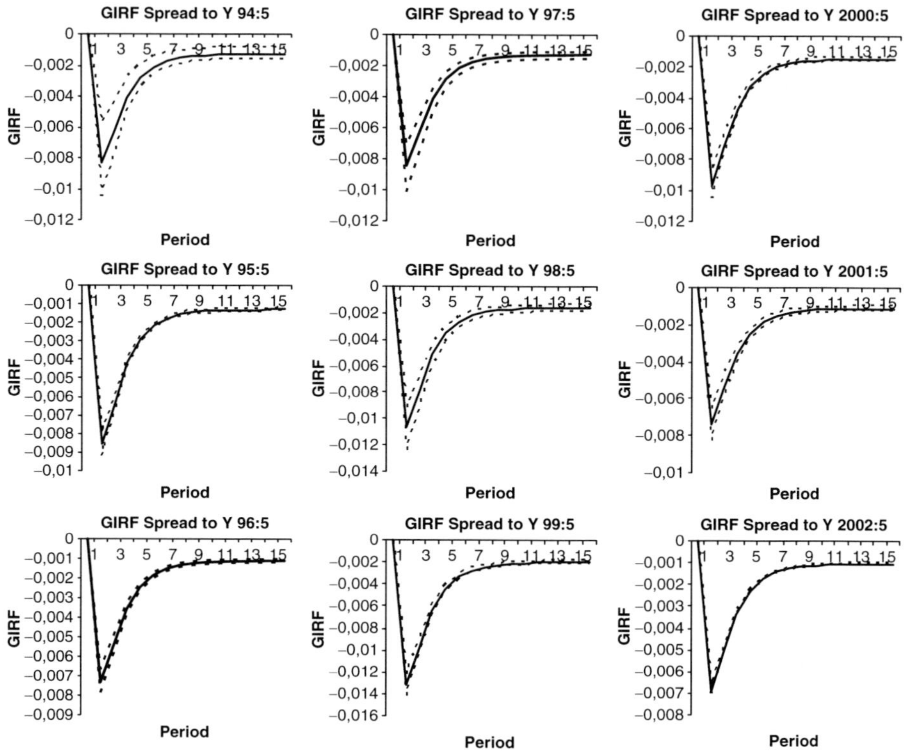

Fig. 3 Stability analysis of real economic activity predictability: GIRF analysis

break points of a relationship. On the other hand, there are numerous structural break points in the sample period, which cannot be analyzed by linear models, because of dividing them into subperiods. Most probably, subperiods have a very small sample size that cannot be analyzed by conventional statistical techniques. Especially, there are consecutive break points in the inflation predictability equation which leads to only one observation point for each subsample. Therefore, analyzing the changing structure of predictability can only be done by the GIRF method, which we have suggested in the beginning of this section (Fig. 3).

From the GIRF analysis, we have recognized that the negative predictability power increases in the periods of crises, when the Transmission Channel is working, providing that the Expectation Theory is not valid. In our opinion, this point needs further investigation. From this analysis, we can conclude that the Interest Rate Transmission Channel has more influential effects than the Expectation Hypothesis on predictability of the real economic activity. This point indicates that the information content of spread is more derived from monetary policy than Expectation Theory, which is the rival theory in the predictability literature. In a sense, then, this result supports Estrella's [1] conclusions, which we will analyze in the next section in order to prove our assessment (Fig. 4).

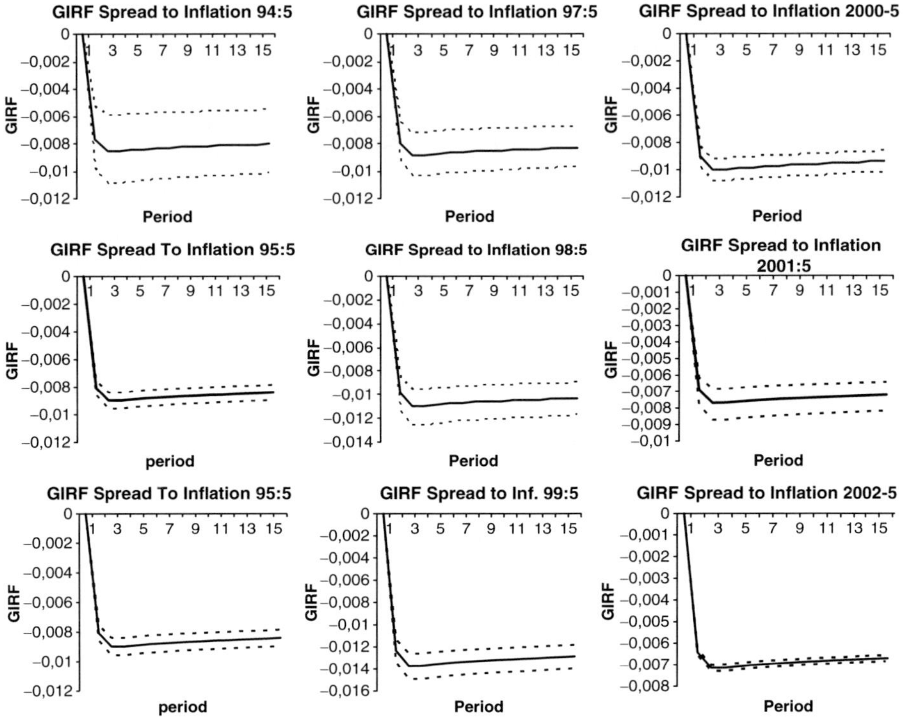

Fig. 4 Stability analysis of inflation predictability: GIRF analysis

5 Investigation of Effects of Monetary Policy Reaction Functions to Stability of Term Structure of Interest Rate

From Sect. 4, we have investigated the source of the negative relationship between real economic activity (and inflation) and the spread. Sections 3.2 and 4 has shed light on the stability of prediction relationships of variables mentioned for Turkey. On the other hand, from the Estrella's [1] discussion, it is clear that empirical relationships are not structural, and alternative monetary policy regimes could lead to very different outcomes. This result has the obvious but interesting empirical implication that shifts in the relationships between real output changes (or inflation) and the spread may provide evidence of shifts in the relative weights policy makers give to deviations of inflation or output from their targets (see [18]). In order to analyze the stability of these relationships then, a researcher should estimate a monetary policy reaction function and test whether the parameters of the monetary policy reaction function change or not. This feature can be examined by a parameter constancy test which is developed by Lin and Terasvirta [19]. In their specification test, they test linearity against multiple structural changes with STAR non-linearity.

Table 4 Estimation of policy reaction function

Model 1 $r = \beta + \beta r_{t-1} + \beta r_{t-2} + \beta \pi^* + \beta ygap$
Model 2 $r = \beta + \beta r_{t-1} + \beta r_{t-2} + \beta \pi^* + \beta ygap + \beta \pi_{t-1}$

Coefficient	Constant	$r(1)$	$r(2)$	π_{t+3}	$ygap$	$\pi(-1)$
Estimation	0.476	0.475 (0.195)	0.398	−0.564	0.811	–
Model 1	(0.638)		(0.225)	(0.534)	(0.475)	
Statistics	$R^2 = 0.336$	ARCH(4) = 0.691		$\hat{\sigma}_e = 0.297$	J.test = 0.999	
Estimation	0.859	0.427	0.320	−0.991	0.419	2.570
Model 2	(0.430)	(0.197)	(0.206)	(0.374)	(0.443)	(1.523)
Statistics	$R^2 = 0.349$	ARCH(4) = 0.691		$\hat{\sigma}_e = 0.293$	J.test = 0.999	

The values which are given under the parameter estimates are heteroscedastic consistent standard errors which are obtained by Newey-West (1987) procedure.

Berüment and Malatyalı [42] and Kalkan et al. [43], show that the inter-bank interest rate is the best variable indicating the policy stance of the Turkish Central Bank. Consequently, in this paper, we are using the inter-bank interest rate as the dependent variable of the policy reaction function estimation. Independent variables are forward looking inflation rates ($\pi_{t+3} = \log p_{t+3} - \log p_t$) and the output gap.[6] The data before have been obtained again from the same source as before. The variables in the question are selected according to two criteria. First the consistency of the criteria in the investigation per previous literature and the second is Turkey's specific economic environment. Although the most influential paper about the monetary policy reaction estimation is [44], suggesting the use of a forward-looking policy reaction function; we are using the backward-looking type of this function as well. We are using the forward-looking version in model 1 in order to be consistent with the previous studies, and we are using the backward-looking version in model 2 in order to be consistent with the Turkish economy's special conditions. Turkey is a high inflationary and politically unstable country; hence these conditions let Central Bank use backward version of the function (see [42]). Besides these issues, the estimation is made by GMM technique, which is recommended by Clarida et al. [44] and Berüment and Malatyalı [42]. Estimated models of policy reaction functions are given below[7] (Table 4).

With respect to model 1, the output gap has a significant coefficient estimate, whereas forward-looking inflation has an insignificant coefficient estimate. From these estimates, we can conclude that the Turkish Central Bank has targeted the output gap instead of the forward-looking inflation rate. When the lagged value of inflation was included into model 2, the output gap coefficient has an insignificant

[6]Output gap is obtained from seasonally adjusted industrial production index which is the deviation from the trend taht is calculated with Hodrick Prescott filter.

[7]The nonlinear estimation of the same sample periods of monetary reaction function can be found in [45] and [46].

statistic, whereas the lag value of inflation has a significant coefficient estimate. This result indicates that the Turkish Central Bank has targeted the backward-looking inflation rate. From now on we are only using model 2, because, this model describes the Turkish monetary authority behavior. In order to analyze the stability of this relationship, we have estimated a monetary policy reaction function and tested whether the parameters have changed or not. The test statistic for model 1 is $LM = 0.852(0.599)$, and model 2 is $LM = 2.743(0.001)$. From model 2, we have seen that the parameter estimates of the model are not constant. The recursive Chow structural test for model 2 shows that there are many structural breaks at 1% significance level, starting from $96 : 1 = 7.739(0.000)$ until the end of the sample period. The most significant one is $99 : 7 = 13.933(0.000)$. The first number is Chow test and the probabilities are given in parenthesis. The linear model of inflation predictability's structural dates and the structural dates of policy reaction function have a similar pattern; this shows there is a close relationship between these variables (inflation and monetary policy, which are shown by the short run interest rate) for Turkey. For the real economic activity, the same pattern cannot be observed; so we cannot make the same kind of argument with respect to these variables (real economic activity and monetary policy, which are shown by the short run interest rate). Estrella's [1] conclusion can be taken as a theoretical result for developed countries, but this conclusion may not be a valid result for some developing countries, especially, for a developing country which has multiple structural breaks in their data. Multiple structural breaks do not permit a researcher to come to a conclusion that the spread variable has information on monetary policy nor can it enable a prediction of real economic activity or inflation. In the case of inflation in Turkey, the above evidence is relatively weak. But for the real economic activity, the above structural break analysis is a contradiction for information content of spread. Besides, Peel and Ioannidis [18] have concluded that the ideal data for analyzing the theoretical prediction of Estrella's analysis are those of the United States and Canada. For the US and Canada data, one significant structural break has been found and this leads us to understand the predictability content of term structure. Hence, multiple structural breaks and the policy preference of Turkish money authority are not appropriate for analyzing Estrella's [1] theoretical prediction.

6 Concluding Remarks

This study investigates whether the term structure of interest rates contains useful information about future real economic activity and inflation for Turkey covering the period 1991–2004. We employ a recursive Chow Structural Break Test to the linear models and from this recursive test procedure, we have seen that the spread-real economic activity and spread-inflation relationships are not stable. This conclusion is also reaffirmed by the linearity test against STR nonlinearity, which is given as a parameter stability test. On the other hand, we have employed GIRF analysis to LSTVAR model in order to understand the source of the negative relationship between these variables. This analysis shows us that the negative relationship between spread and real economic activity has occurred because of the negative symmetric relationship between Expectation Hypothesis and Interest Rate Transmission Channel. And, also this analysis shows us that the negative relationship between spread and inflation has occurred because of the negative Expectation Hypothesis and Fisher Effect. Furthermore, we have employed a GIRF analysis in order to see whether the predictability relationships are stable for Turkey. This analysis shows us that the stability relationship of these variables is not established for the period that we have analyzed. From this analysis, we have seen that the GIRF analysis is consistent with the recursive Chow test and parameter stability test at the same time. The GIRF analysis has its own advantages for analyzing the data which we have mentioned before. Especially, we have seen that the predictability power of these relationships can be analyzed by the GIRF without losing any degrees of freedom which, otherwise, cannot be obtained by linear bivariate equation estimation of subsamples as we have pointed out in Sect. 4. Finally, use of GIRF enables us to make a computation about the Expectation Hypothesis, Interest Rate Transmission Channel, and Fisher Effect, which cannot be obtained by TIRF analysis.

From the GIRF analysis, we have recognized that the negative predictability power is increased during periods of crisis, when the Interest Rate Transmission Channel is working, whereas the Expectation Theory is not valid. In order to further investigate this issue, we have estimated the policy reaction function of Turkey, covering the estimation period of the LSTVAR analysis. We have found out that the policy reaction function has multiple structural breaks that can prevent sound conclusions on the above issues. Hence Estrella's [1] theoretical prediction, "empirical relationships are not structural, and alternative monetary policy regimes could lead to very different outcomes", cannot beproven by the Turkish data [47].

Appendix A

GIRFs are obtained by making bootstrapping. Hence we have to construct their confidence band again by designing a bootstrapping instead of Monte-Carlo design. For GIRF, we handle 2 hundred of impulse responses in order to get one specific histories' impulse response. We have obtained these 2 hundred impulse responses in order to average the effect of intermediate shocks. For their confidence band again we design a bootstrap and handle the confidence band from this computation. For this purpose, we run 1,000 impulse responses which are the averaged from 200 impulse responses and create 10% confidence band for every histories impulse response.

Appendix B

Figure 5

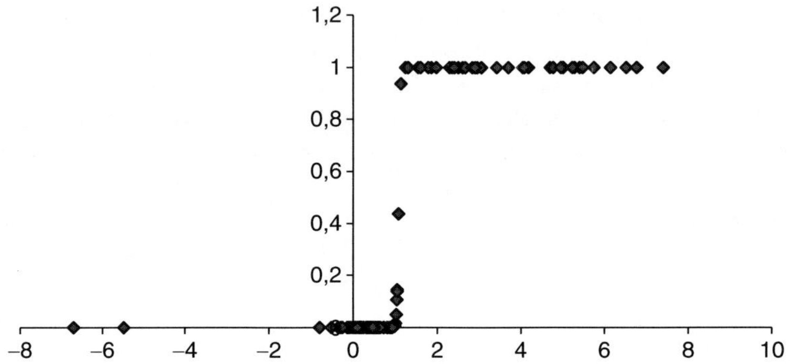

Fig. 5 Transition function

References

1. Estrella, A. (1997), "Why Do Interest Rates Predict Macro Outcomes? A Unified Theory of Inflation, Output, Interest, and Policy," *Federal Reserve Bank of New York Research Paper* No. 9717.
2. Harvey, C. R. (1988) "The Real Term Structure and Consumption Growth." *Journal of Financial Economics*, 22, 305–333.
3. Chen, N. (1991), "Financial Investment Opportunities and the Macroeconomy," *Journal of Finance* 46, 529–554.
4. Estrella, A., ve G. A. Hardouvelis. (1991),"The Term Structure as a Predictor of Real Economic Activity." *Journal of Finance*, 46, 555–576.
5. Plosser, C., ve K. G. Rounwenhorst. (1994), "International Term Structure and Real Economic Growth." *Journal of Monetary Economics*, 33, 133–155.
6. Haubrich, J. G. and A. M. Dombrosky. (1996), "Predicting Real Growth Using the Yield Curve." *Federal Reserve Bank of Cleveland, Economic Review*, 32, 26–35.
7. Dotsey, M. (1998), "The Predictive Content of Interest Rate Term Spread for Future Economic Growth." *Federal Reserve Bank of Richmond Economic Quarterly*, 84, Summer 31–51.
8. Moody, A., ve M.P. Taylor (2000), "The High Yield Spread as a Predictor of real economic Activity: Evidence of a Financial Accelerator for United States." *IMF Stuff Paper*.
9. Hamilton, J. D. and D. H. Kim. (2002), "A Re-examination of the Predictability of Economic Activity Using the Spread." *Journal of Money, Credit, and Banking*, 34, 2–10.
10. Venetis, I. A., I. Paya and D. A. Peel (2003), "Reexamination of the Predictability of Economic Activity Using the Yield Spread: A nonlinear approach" *International Review of Economics and Finance*, 2,
11. Mishkin, F. (1990a),"What Does the Term Structure Tell Us about Future Inflation?" *Journal of Monetary Economics*, 25, 77–95.

12. Mishkin, F. (1990b), "The Information in the Longer Maturity Term Structure About Inflation." *Quarterly Journal of Economics*, 55, 815–828.
13. Mishkin, F. (1991), "A Multi-Country Study of the Information in the Shorter Term Structure About Future Inflation." *Journal of International Money and Finance*, 10, 2–22.
14. Frankel, J. A. ve C. S. Lown (1994), "An Indicator of Future Inflation Extracted from the Steepness of the Interest Rate Yield Curve Along its Entire Length." *The Quarterly Journal of Economics*, 517–530.
15. Estrella, A. and F. S. Mishkin. (1995), "The Term Structure of Interest Rates and Its Role in Monetary Policy for the European Central Bank." *National Bureau of Economic Research Working Paper*, 5279.
16. Telatar, E., F. Telatar and R.A. Ratti (2003), "On the Predictive Power of the Term Structure of Interest Rates for Future Inflation Changes in the Presence of Political Instability: The Turkish Economy." *Journal of Policy Modeling*, 25, 931–946.
17. Jardet, C., (2004), "Why Did the Term Structure of Interest Rates Lose its Predictive Power?" *Economic Modeling*. 21: 509–524.
18. Peel, D. A. ve C. Ioannidis (2003), "Empirical Evidence on the Relationship Between the Term Structure of Interest and Future Real Output Changes When There are Changes in Policy Regime." Economic Letters, 78, 147–152.
19. Lin, H. and T. Terasvirta (1994), "Testing the Constancy of Regression Parameters against Continuous Structure Change" *Journal of Econometrics*, 62, 211–228.
20. Kim K. A., ve P. Limpaphayom (1997), "The Effect of Economic Regimes on the Relation between the Term Structure and Real Activity in Japan" *Journal of Economics and Business*, 49, 379–392.
21. Şahinbeyoğlu, G. and C. Yalçın (2000), "The Term Structure of Interest Rates: Does it tell about inflation?" *TCMB Discussion Paper* 2000:2.
22. Akyıldız, K., (2003), "Getiri Farkı Ekonomik Aktivitenin Tahmininde Öncü Gösterge İşlevi Görebilir Mi? Türkiye Örneği" *Hazine Dergisi*, Sayı: 16, 1–20.
23. Fuhrer, J.C. ve G.R. Moore (1995) "Monetary Policy Trade offs and the Correlation Between nominal interest rates and real output", *American Economic Review* 85, 219–239.
24. Taylor, J. B. (1993), "Discretion Versus Policy Rules in Practice," *Carnegie-Rochester Conference Series On Public Policy* 39:195–214.
25. Newey, W. ve K. West (1987), "A Simple, Positive Definite, Heteroskedasticty and Autocorrelation Consistent Variance Matrix" *Econometrica*, 55, 703–708.
26. van Dijk D. (1999), *STR Models Extensions and Outlier Robust Inference*. Tinbergen Institute Research Series no. 200.
27. Luukkonnen, R., P. Saikkonen and T. Terasvirta (1988), Testing Linearity against STAR Models." *Biometrika*, 75, 491–499.
28. Terasvirta, T. (1994), "Specification, Estimation and Evaluation of STAR Models." *Journal of American Statistical Association*, 89, 208–218.
29. White, H. and Domowitz, I. (1984) "Nonlinear regression with dependent observations", Econometrica, 52, pp. 143–161
30. Pötscher, B.M. and Prucha, I.V. (1997) Dynamic Nonlinear Econometric Models – Asymptotic Theory, Berlin, Springer-Verlag
31. Anderson, M. H., and Vahid, F. (1998), "Testing multiple equation systems for common nonlinear component", *Journal of Econometric,s* Volume 84, Issue 1, May 1998, Pages 1–36
32. Eitrheim, Q. ve T. Terasvirta (1996), "Testing the Adequacy of STAR Models." *Journal of Econometrics*, 74, 59–76.
33. Koop, G., M.H. Pesaran and S.M. Porter (1996) "Impulse response analysis in nonlinear multivariate models", *Journal of Econometrics*, 74, 1 119–147.
34. Cook, T. and Hahn, T., (1989), "The effect of changes in the federal funds rate target on market interest rates in the 1970's" *Journal of Monetary Economics*,19, 31–72.
35. Romer, D. (1996), *Advanced Macroeconomics*, New York: McGraw-Hill.
36. T. Ellingsen and U. Söderström "Why are Long Rates Sensitive to Monetary Policy?" Riksbank Working Paper No:5

37. Mishkin, F. (1995), "Symposium on the Monetary Transmission Mechanism." *Journal of Economic Perspective*, 9, 3–10.
38. Fama, E. F. (1975),"Short-term Interest Rates as a Predictors of Inflation." *The American Economic Review*, 65, 269–282.
39. Barro R.J. and Gordon D.B.(1981) "Rules, Discreation and Reputation in a Model of Monetary Policy." *Journal Of Monetary Economics*, 12:101–121
40. Lucas R.E. J. (1972) "Expectations and the Neutrality of Money." *Journal of Economics Theory*, 4: 103–124.
41. Chen, N. (1991), "Financial Investment Opportunities and the Macroeconomy," *Journal of Finance* 46, 529–554.
42. Berüment H. and Malatyalı K., (2000), "The implicit reaction function of th Central Bank of Repuclic of Turkey", *Applied Economics Letters*, 7, 425–430.
43. Kalkan, M., Kipici, A. and Peker, A., (1997), "Monetary policy and leading indicators of inflation in Turkey", *Irvin Fisher Committee Bulletin*, 1.
44. Clarida, R., J. Gali ve M. Gertler, (2000), "Monetary Policy Rules and Macroeconomic stability: Evidence and Some Theory", *Quarterly Journal of Economics*, 115, 147–180
45. Omay T. and Hasanov M. (2006) "Türkiye için Reaksiyon Fonksiyonunun Doğrusal Olmayan Modelle Tahmin Edilmesi" MPRA Paper 20154.
46. Hasanov, M. and Omay, T. (2008), "Monetary Policy Rules in Practice: Re-examining the Case for Turkey. *Physica A Statistical Mechanics and Its Applications*. 387(17), p. 4309–4318
47. Omay T. (2006) Türkiyeìde Fazin Vade Yapısı ile Reel Ekonomik Aktivite Arasındaki İlişki. İktisadi Araştırmalar Vakfı Yayınları.

Licensing in an International Competition with Differentiated Goods

Fernanda A. Ferreira

1 Introduction

The effect of entry on social welfare has been studied by Collie [1], Cordella [2], and Klemperer [8]. In a closed economy, Klemperer [8] shows that entry reduces social welfare if the cost of the entrant is sufficiently higher than that of the incumbent (see also Lahiri and Ono [9]). Collie [1] examines this issue in an open economy and shows that entry of a foreign firm reduces domestic welfare unless the cost of the foreign firm is sufficiently lower than that of the incumbent. Cordella [2] also considers this issue in an open economy, showing the effects of the number of firms. Technological difference is an important reason for cost differences between firms, which may encourage them to share their technological information through licensing. Faul-Oller and Sandonis [3] show that higher welfare under entry is more likely in the presence of licensing by the technologically efficient incumbent compared with no licensing. Their results suggest that entry always increases welfare if there is licensing with output royalty but licensing with fixed fee only increases the likelihood of higher welfare under entry rather than eliminating the possibility of lower welfare under entry. While they have considered the situation of a closed economy, Mukherjee and Mukherjee [10] show the welfare implications of entry in the presence of technology licensing in an open economy. If either the entrant or the incumbent has a relatively superior technology,[1] it creates the possibility of technology licensing. Mukherjee and Mukherjee [10] show that if there is licensing with an upfront fixed fee, entry of a foreign firm not only increases

[1]In our analysis, technology is defined by the marginal cost of production. Lower marginal cost implies better technology.

F.A. Ferreira (✉)
ESEIG - Polytechnic Institute of Porto and CMUP, Rua D. Sancho I, 981,
4480-876 Vila do Conde, Portugal
e-mail: fernandaamelia@eu.ipp.pt

J.A.T. Machado et al., *Nonlinear and Complex Dynamics: Applications in Physical, Biological, and Financial Systems*, DOI 10.1007/978-1-4614-0231-2_23,

domestic welfare when the foreign firm is sufficiently technologically superior to the domestic firm, it also increases domestic welfare if the foreign firm's technological inferiority is neither very small nor very large. However, if there is licensing with output royalty, foreign entry increases domestic welfare when the foreign firm is either sufficiently technologically superior or sufficiently technologically inferior to the domestic firm. So, the presence of technology licensing significantly affects the result of Collie [1], which considers the welfare effect of foreign entry without licensing. In this paper, we follow Mukherjee and Mukherjee [10], by doing a similar study for differentiated goods, in the case of a technologically superior entrant. Since differentiation of the goods reduces competition between firms, it increases the possibility of licensing. These results have important implications for competition policies and show that the policymakers need to be concerned about the technological efficiency of the foreign firm and the type of licensing contract (i.e. fixed-fee or royalty licensing) available to the firms.

We should mention that issues related to those of this paper have been studied by Ferreira [4, 5], Ferreira and Ferreira [6], and Ferreira et al. [7].

2 The Model and the Results

Consider a country, called the domestic country, in which there is a monopolist, called incumbent. To study the implications of entry, we will consider the following two situations in our analysis: (a) where the incumbent is a monopolist in the domestic country; and (b) where a foreign firm, called entrant, enters the market and competes with the incumbent. We suppose that the entrant is technologically superior to the incumbent. This situation may be consistent for trade between the developed countries with technological leapfrogging by the technologically lagging country.

2.1 *The Case of a Monopoly*

Let us first consider the situation where the incumbent is a monopolist in the domestic country, where the inverse market demand function is given by $p = a - q$, where p is the price, q the quantity in the market, and $a > 0$ the demand intercept. We assume that the incumbent can produce a good with the constant marginal production cost c_1. For simplicity, we assume that there is no other production cost. The incumbent maximizes the following objective function to maximize its profit: $\max_q (a - q - c_1)q$. Optimal output of the incumbent is $q^* = (a - c_1)/2$ and its profit and consumer surplus are, respectively,

$$\pi^* = \frac{(a - c_1)^2}{4} \quad \text{and} \quad CS = \frac{(a - c_1)^2}{8}.$$

Therefore, in the monopoly case, welfare W^m of the domestic country, which is the summation of consumer surplus and profit of the incumbent, is given by

$$W^m = \frac{3(a-c_1)^2}{8}.$$

2.2 *Entry Without Licensing*

To show the implications of licensing, let us first consider entry of a foreign firm without licensing. Assume that there is a foreign firm, called entrant, who can produce the good at the constant marginal production cost c_2 less than the marginal cost c_1 of the domestic firm. So, we consider the following assumption:

Assumption 1. $c_2 < c_1$.

Assumption 1 can be interpreted as the foreign firm being technologically superior to the domestic firm. We also assume that there is no other production cost of the entrant, namely we assume that there is no transportation costs and/or tariff. In our stylized framework, we assume that the entrant exports its product to the domestic country and the firms (the incumbent and the entrant) compete like Cournot duopolists with differentiated products. The inverse demands are given by

$$p_i = a - q_i - \gamma q_j,$$

with $\alpha > 0$ and $0 < \gamma \leq 1$, where p_i is the price and q_i the amount produced of good i, for $i, j \in \{1,2\}$. We note that the two products are substitutes, and, since $\gamma \leq 1$, "cross effects" are dominated by "own effect." The value of γ expresses the degree of product differentiation. When γ is equal to one, the goods are homogeneous, and when γ tends to zero, we are close to independent goods. In what follows, we restrict the parameters of the model to satisfy the following assumption:

Assumption 2.

$$\max\left\{0, \frac{2c_1-(2-\gamma)a}{\gamma}\right\} < c_2 < \frac{(2-\gamma)a+\gamma c_1}{2}.$$

This assumption requires that, in case of entry, both firms always produce positive outputs.

The incumbent and the entrant choose their outputs to maximize, respectively, their profits, i.e.,

$$\max_{q_1}(a-q_1-\gamma q_2-c_1)q_1 \;\text{ and }\; \max_{q_2}(a-\gamma q_1-q_2-c_2)q_2,$$

where q_1 and q_2 are the outputs of the incumbent and the entrant, respectively. Optimal outputs of the incumbent and the entrant are, respectively,

$$q_1^* = \frac{(2-\gamma)a - 2c_1 + \gamma c_2}{4-\gamma^2}$$

and

$$q_2^* = \frac{(2-\gamma)a - 2c_2 + \gamma c_1}{4-\gamma^2}.$$

Profits of the incumbent, the entrant, and consumer surplus are, respectively,

$$\pi_1^* = \frac{((2-\gamma)a - 2c_1 + \gamma c_2)^2}{(4-\gamma^2)^2},$$

$$\pi_2^* = \frac{((2-\gamma)a - 2c_2 + \gamma c_1)^2}{(4-\gamma^2)^2}$$

and

$$CS = \frac{2(2-\gamma)\left(2+\gamma-\gamma^2\right)a^2 - 2\left(4-3\gamma^2+\gamma^3\right)(c_1+c_2)a}{2\left(4-\gamma^2\right)^2} + \frac{(4-3\gamma^2)c_1^2 + \left(2\gamma^3 c_1 + \left(4-3\gamma^2\right)c_2\right)c_2}{2\left(4-\gamma^2\right)^2}.$$

So, in the case of entry without licensing, domestic welfare W_{nl}^e is given by

$$W_{\mathrm{nl}}^e = \frac{2(2-\gamma)a^2 - 2((3-\gamma)c_1 + (1-\gamma)c_2)a + 3c_1^2 - (2\gamma c_1 - c_2)c_2}{2\left(4-\gamma^2\right)}. \tag{1}$$

Theorem 1. *Assume that there is no possibility of technology licensing between the firms. If*

$$c_1 > \frac{(3\gamma-2)a}{3\gamma} \quad \text{and} \quad c_2 < \frac{3\gamma c_1 - (3\gamma-2)\mathrm{a}}{2},$$

entry increases domestic welfare.

It reduces welfare otherwise.

Proof. The inequality $W_{\mathrm{nl}}^e > W^m$ is equivalent to $c_1 > (3\gamma-2)a/(3\gamma)$ and $c_2 < (3\gamma c_1 - (3\gamma-2)a)/2$. Noting that $(3\gamma c_1 - (3\gamma-2)a)/2 < ((2-\gamma)a + \gamma c_1)/2$, we get the result. □

2.3 *Entry with Licensing*

Now, we are going to analyze the case of the entry under licensing. We consider two important types of licensing contracts (see, for example, Wang [11]): (a) fixed-fee licensing, where the licenser charges an upfront fixed fee for its technology; and (b) licensing with output royalty, where the licenser charges royalty per unit of output.

We consider the following game under entry. At stage 1, the technologically efficient entrant decides whether to license its technology to the incumbent, and the incumbent accepts the licensing contract, if it is not worse off under licensing compared with no licensing. At stage 2, the firms compete *à la* Cournot.

2.3.1 Fixed-Fee Licensing

We have seen above that profits of the incumbent and the entrant under no licensing are, respectively,

$$\pi^*_{1,\text{nl}} = \frac{((2-\gamma)a - 2c_1 + \gamma c_2)^2}{(4-\gamma^2)^2}$$

and

$$\pi^*_{2,\text{nl}} = \frac{((2-\gamma)a - 2c_2 + \gamma c_1)^2}{(4-\gamma^2)^2}.$$

Now, consider the situation under licensing. If licensing occurs, both firms produce with c_2, since the entrant charges an upfront fixed fee for its technology. Profits of the incumbent and the entrant are, respectively,

$$\frac{(a-c_2)^2}{(2-\gamma)^2} - F \quad \text{and} \quad \frac{(a-c_2)^2}{(2-\gamma)^2} + F,$$

where F is the optimal licensing fee charged by the entrant. So, licensing is profitable, if the following two conditions are satisfied for the incumbent and the entrant, respectively (with at least one strict inequality):

$$\frac{(a-c_2)^2}{(2+\gamma)^2} - F \geq \frac{((2-\gamma)a - 2c_1 + \gamma c_2)^2}{(4-\gamma^2)^2}$$

and

$$\frac{(a-c_2)^2}{(2+\gamma)^2} + F \geq \frac{((2-\gamma)a - 2c_2 + \gamma c_1)^2}{(4-\gamma^2)^2}.$$

Since the entrant gives a take-it-or-leave-it offer to the incumbent, the fixed fee makes the incumbent indifferent between licensing and no licensing, i.e.

$$F = \frac{(a-c_2)^2}{(2+\gamma)^2} - \frac{((2-\gamma)a-2c_1+\gamma c_2)^2}{(4-\gamma^2)^2}.$$

So, licensing occurs if, and only if,

$$\frac{2(a-c_2)^2}{(2+\gamma)^2} > \frac{((2-\gamma)a+\gamma c_1-2c_2)^2+((2-\gamma)a-2c_1+\gamma c_2)^2}{(4-\gamma^2)^2},$$

which is satisfied for $c_2 > \frac{2(2-\gamma)^2 a-(4+\gamma^2)c_1}{\gamma^2-8\gamma+4}$ with $\gamma \neq 4-2\sqrt{3}$. Thus, we have proved the following lemma.

Lemma 1. *Licensing occurs if, and only if,*

$$c_2 > \frac{2(2-\gamma)^2 a-(4+\gamma^2)\,c_1}{\gamma^2-8\gamma+4} \quad \text{with} \quad \gamma \neq 4-2\sqrt{3}.$$

Under fixed-fee licensing, the profit of the incumbent and consumer surplus are, respectively,

$$\pi^*_{1,\mathrm{lf}} = \frac{((2-\gamma)a-2c_1+\gamma c_2)^2}{(4-\gamma^2)^2}$$

and

$$CS_{\mathrm{lf}} = \frac{(1+\gamma)(a-c_2)^2}{(2+\gamma)^2},$$

if $c_2 > \frac{2(2-\gamma)^2 a-(4+\gamma^2)c_1}{\gamma^2-8\gamma+4}$. So, domestic welfare W^e_{lf} under fixed-fee licensing is given by

$$W^e_{\mathrm{lf}} = \frac{((2-\gamma)a-2c_1+\gamma c_2)^2}{(4-\gamma^2)^2} + \frac{(1+\gamma)(a-c_2)^2}{(2+\gamma)^2}.$$

Theorem 2. *Consider the possibility of fixed-fee licensing. Entry increases domestic welfare in the following situations:*

(i) For $0<\gamma<2/3$, *if*

$$c_2 < \frac{\left(8\gamma+(2-\gamma)\sqrt{\Theta}\right)c_1}{4\,(\gamma^3-2\gamma^2+4)} - \frac{(2-\gamma)\left(\sqrt{\Theta}-4\,(2-\gamma^2)\right)a}{4\,(\gamma^3-2\gamma^2+4)};$$

(ii) For $2/3 \leq \gamma \leq 1$, if

$$c_1 > \frac{(2-\gamma)\left(\sqrt{\Theta} - 4\,(2-\gamma^2)\right)a}{8\gamma + (2-\gamma)\sqrt{\Theta}}$$

and

$$c_2 < \frac{\left(8\gamma + (2-\gamma)\sqrt{\Theta}\right)c_1}{4\,(\gamma^3 - 2\gamma^2 + 4)} - \frac{(2-\gamma)\left(\sqrt{\Theta} - 4\,(2-\gamma^2)\right)a}{4\,(\gamma^3 - 2\gamma^2 + 4)},$$

where $\Theta = 6\gamma^5 + 12\gamma^4 - 24\gamma^3 - 24\gamma^2 + 32\gamma + 32$.

It reduces welfare otherwise.

Proof. We have that $W^e_{\mathrm{lf}} > W^m$ is equivalent to

$$c_2 < \frac{\left(8\gamma + (2-\gamma)\sqrt{36\gamma^5 + 12\gamma^4 - 24\gamma^3 - 24\gamma^2 + 32\gamma + 32}\right)c_1}{4\,(\gamma^3 - 2\gamma^2 + 4)} - \frac{(2-\gamma)\left(\sqrt{36\gamma^5 + 12\gamma^4 - 24\gamma^3 - 24\gamma^2 + 32\gamma + 32} - 4\,(2-\gamma^2)\right)a}{4\,(\gamma^3 - 2\gamma^2 + 4)} \qquad (2)$$

or

$$c_2 > \frac{\left(8\gamma + (2-\gamma)\sqrt{36\gamma^5 + 12\gamma^4 - 24\gamma^3 - 24\gamma^2 + 32\gamma + 32}\right)c_1}{4\,(\gamma^3 - 2\gamma^2 + 4)} + \frac{(2-\gamma)\left(\sqrt{36\gamma^5 + 12\gamma^4 - 24\gamma^3 - 24\gamma^2 + 32\gamma + 32} - 4\,(2-\gamma^2)\right)a}{4\,(\gamma^3 - 2\gamma^2 + 4)}. \qquad (3)$$

Since inequality (3) contradicts Assumption 1, we only have inequality (2). The result follows by noting that

(i) For $0 < \gamma \leq 2/3$, the expression

$$\left(8\gamma + (2-\gamma)\sqrt{36\gamma^5 + 12\gamma^4 - 24\gamma^3 - 24\gamma^2 + 32\gamma + 32}\right)c_1 - (2-\gamma)\left(\sqrt{36\gamma^5 + 12\gamma^4 - 24\gamma^3 - 24\gamma^2 + 32\gamma + 32} - 4\,(2-\gamma^2)\right)a$$

is always positive.

(ii) For $2/3 \leq \gamma \leq 1$, we have that

$$\left(8\gamma + (2-\gamma)\sqrt{36\gamma^5 + 12\gamma^4 - 24\gamma^3 - 24\gamma^2 + 32\gamma + 32}\right) c_1$$

$$- (2-\gamma)\left(\sqrt{36\gamma^5 + 12\gamma^4 - 24\gamma^3 - 24\gamma^2 + 32\gamma + 32} - 4\left(2-\gamma^2\right)\right) a > 0$$

if, and only if,

$$c_1 > \frac{(2-\gamma)\left(\sqrt{36\gamma^5 + 12\gamma^4 - 24\gamma^3 - 24\gamma^2 + 32\gamma + 32} - 4\left(2-\gamma^2\right)\right) a}{8\gamma + (2-\gamma)\sqrt{36\gamma^5 + 12\gamma^4 - 24\gamma^3 - 24\gamma^2 + 32\gamma + 32}}.$$

□

2.3.2 Licensing with Output Royalty

Now, consider licensing with per-unit output royalty, where the entrant charges a per-unit output royalty for its technology. In that case of licensing, the effective marginal cost of the incumbent is $c_2 + r$, where r is the optimal per-unit output royalty. The optimal outputs of the incumbent and the entrant are, respectively,

$$q^*_{1,\mathrm{lr}} = \frac{(2-\gamma)a - (2-\gamma)c_2 - 2r}{4-\gamma^2}$$

and

$$q^*_{2,\mathrm{lr}} = \frac{(2-\gamma)a - (2-\gamma)c_2 + \gamma r}{4-\gamma^2}.$$

So, their profits are, respectively,

$$\pi^*_{1,\mathrm{lr}} = \frac{((2-\gamma)a - (2-\gamma)c_2 - 2r)^2}{(4-\gamma^2)^2}$$

and

$$\pi^*_{2,\mathrm{lr}} = \frac{(2-\gamma)^2(a-c_2)^2 + (2-\gamma)\left(4+2\gamma-\gamma^2\right)(a-c_2)r - \left(8-3\gamma^2\right)r^2}{\left(4-\gamma^2\right)^2}.$$

The entrant solves the problem

$$\max_r \pi^*_{2,\mathrm{lr}} \tag{4}$$

subject to the constraint $r \leq c_1 - c_2$, to determine the optimal royalty rate.

Lemma 2. *The optimal output royalty r^* is as follows:*

(i) If

$$c_2 \geq \frac{2(8-3\gamma^2)c_1 - (2-\gamma)(4+2\gamma-\gamma^2)a}{8-2\gamma^2-\gamma^3},$$

then

$$r^* = c_1 - c_2;$$

(ii) If

$$c_2 < \frac{2\left(8-3\gamma^2\right)c_1 - (2-\gamma)\left(4+2\gamma-\gamma^2\right)a}{8-2\gamma^2-\gamma^3},$$

then

$$r^* = \frac{(2-\gamma)\left(4+2\gamma-\gamma^2\right)(a-c_2)}{2\left(8-3\gamma^2\right)}. \tag{5}$$

Proof. The solution of (4) is $r = (2-\gamma)\left(4+2\gamma-\gamma^2\right)(a-c_2)/(2\left(8-3\gamma^2\right))$. The result follows, since we have that $(2-\gamma)\left(4+2\gamma-\gamma^2\right)(a-c_2)/\left(2\left(8-3\gamma^2\right)\right) \geq c_1 - c_2$, for $c_2 \geq \left(2\left(8-3\gamma^2\right)c_1 - (2-\gamma)\left(4+2\gamma-\gamma^2\right)a\right)/\left(8-2\gamma^2-\gamma^3\right)$, and we note that, for all $0 < \gamma \leq 1$, we have that

$$\frac{2c_1-(2-\gamma)a}{\gamma} \leq \frac{2\left(8-3\gamma^2\right)c_1 - (2-\gamma)\left(4+2\gamma-\gamma^2\right)a}{8-2\gamma^2-\gamma^3} \leq \frac{(2-\gamma)a+\gamma c_1}{2}.$$

□

Theorem 3. *Suppose that there is royalty licensing.*

(i) If

$$c_2 \geq \frac{2\left(8-3\gamma^2\right)c_1 - (2-\gamma)\left(4+2\gamma-\gamma^2\right)a}{8-2\gamma^2-\gamma^3},$$

welfare implications of entry remain the same under licensing and no licensing.

(ii) If

$$c_2 < \frac{2\left(8-3\gamma^2\right)c_1 - (2-\gamma)\left(4+2\gamma-\gamma^2\right)a}{8-2\gamma^2-\gamma^3}, \tag{6}$$

then entry always increases welfare.

Proof. In case (i), by Lemma 2, the optimal per-unit output royalty is $r^* = c_1 - c_2$. Therefore, under entry, the incumbent's profit and consumer surplus remain the same under licensing and no licensing. Hence, domestic welfare under entry with royalty licensing is given by (1), and so we get the result.

In case (ii), by Lemma 2, the optimal per-unit output royalty is given by (5). Therefore, under entry, the incumbent's profit, consumer surplus and domestic welfare are, respectively, given by

$$\pi^*_{1,\text{lr}} = \frac{4(1-\gamma)^2(a-c_2)^2}{(8-3\gamma^2)^2},$$

$$CS_{\text{lr}} = \frac{\left(80-76\gamma^2+12\gamma^3+9\gamma^4\right)(a-c_2)^2}{8\,(8-3\gamma^2)^2}$$

and

$$W^e_{\text{lr}} = \frac{\left(112-64\gamma-44\gamma^2+12\gamma^3+9\gamma^4\right)(a-c_2)^2}{8\,(8-3\gamma^2)^2}.$$

The inequality $W^e_{\text{lr}} > W^m$ is equivalent to

$$c_2 < a - \frac{\sqrt{3}\left(8-3\gamma^2\right)(a-c_1)}{\sqrt{112-64\gamma-44\gamma^2+12\gamma^3+9\gamma^4}}$$

$$\vee\; c_2 < a + \frac{\sqrt{3}\left(8-3\gamma^2\right)(a-c_1)}{\sqrt{112-64\gamma-44\gamma^2+12\gamma^3+9\gamma^4}}.$$

This, together with condition (6), implies the result. □

3 Conclusions

We showed the effects of foreign entry on social welfare in the presence of licensing, when the firms produce differentiated goods. We considered both fixed-fee licensing and royalty licensing, in the case where the foreign firm (the entrant) is technologically superior to the domestic firm (the incumbent). We found that the welfare implications of entry depend upon the type of licensing contract, and since differentiation of the goods reduces competition between firms, it increases the possibility of licensing.

Acknowledgments We are grateful to Alberto A. Pinto for a number of very fruitful and useful discussions on this work and for his friendship and encouragement. We thank ESEIG – Instituto Politécnico do Porto, Centro de Matemática da Universidade do Porto and the Programs POCTI and POCI by FCT and Ministério da Ciência, Tecnologia e do Ensino Superior for their financial support.

References

1. Collie D (1996) Gains and losses from unilateral free trade under oligopoly. Recherches Economiques de Louvain 62: 191–202
2. Cordella T (1993) Trade liberalization and oligopolistic industries: a welfare appraisal. Recherches Economiques de Louvain 59: 355–363

3. Faul-Oller R, Sandonis J (2003). To merge or to license: implications for competition policy. International Journal of Industrial Organization 21: 655–672
4. Ferreira FA (2009) Privatization and entry of a foreign firm with demand uncertainty. In: Simos TE et al. (eds) Numerical Analysis and Applied Mathematics, AIP Conference Proceedings 1168: 971–974, American Institute of Physics, New York
5. Ferreira FA (2008) Licensing in an international market. In: Simos TE et al. (eds) Numerical Analysis and Applied Mathematics, AIP Conference Proceedings 1048: 201–204, American Institute of Physics, New York
6. Ferreira FA, Ferreira F (2008) Welfare effects of entry into international markets with licensing. In: Todorov D (ed) Applications of Mathematics in Engineering and Economics, AIP Conference Proceedings 1067: 321–327, American Institute of Physics, New York
7. Ferreira FA, Ferreira F, Ferreira M, Pinto AA (2008) Quantity competition in a differentiated duopoly. In: Tenreiro JA et al. (eds) Intelligent Engineering Systems and Computational Cybernetics 365–374. Springer Science+Business Media B.V., New York
8. Klemperer P (1988) Welfare effects of entry into markets with switching costs. Journal of Industrial Economics 37: 159–165
9. Lahiri S, Ono T (1988) Helping minor firms reduces welfare. Economic Journal 98: 1199–1202
10. Mukherjee A, Mukherjee S (2005) Foreign competition with licensing. The Manchester School 73: 653–663
11. Wang XH (1988) Fee versus royalty licensing in a Cournot duopoly model. Economics Letters 60: 55–62

Multidimensional Scaling Analysis of Stock Market Indexes

Gonçalo M. Duarte, J. Tenreiro Machado, and Fernando B. Duarte

1 Introduction

Economical indexes measure the performance of segments of the stock market and are normally used to benchmark the performance of stock portfolios. This chapter proposes a descriptive method that analyzes possible correlations in international stock markets. The study of the correlation of international stock markets may have different motivations. Economic motivations to identify the main factors that affect the behavior of stock markets across different exchanges and countries. Statistical motivations to visualize correlations in order to suggest some potentially plausible parameter relations and restrictions. The understanding of such correlations would be helpful to the design good portfolios [1–4].

Bearing these ideas in mind, the outline of our chapter is as follows. In Sect. 2, we give the fundamentals of the multidimensional scaling (MDS) technique, which is the core of our method, and we discuss the details that are relevant for our specific application. In Sect. 3, we apply our method for daily data on 25 stock markets, including major American, Asian/Pacific, and European stock markets. In Sect. 4, we conclude the chapter with some final remarks and potential topics for further research.

2 Fundamental Concepts

MDS is a set of data analysis techniques for analysis of similarity or dissimilarity data. It is used to represent (dis)similarity data between objects by a variety of distance models.

J.T. Machado (✉)
Department of Electrical Engineering of Engineering, Porto, Portugal
e-mail: jtm@isep.ip.pt

J.A.T. Machado et al., *Nonlinear and Complex Dynamics: Applications in Physical, Biological, and Financial Systems*, DOI 10.1007/978-1-4614-0231-2_24,

The term similarity is used to indicate the degree of "likeness" between two objects, while dissimilarity indicates the degree of "unlikeness". MDS represents a set of objects as points in a multidimensional space in such a way that the points corresponding to similar objects are located close together, while those corresponding to dissimilar objects are located far apart. The researcher then attempts to "make sense" of the derived object configuration by identifying meaningful regions and/or directions in the space.

In this article, we introduce the basic concepts and methods of MDS. We then discuss a variety of (dis)similarity measures and the kinds of techniques to be used. The main objective of MDS is to represent these dissimilarities as distances between points in a low dimensional space such that the distances correspond as closely as possible to the dissimilarities.

Let n be the number of different objects and let the dissimilarity for objects i and j be given by δ_{ij}. The coordinates are gathered in an $n \times p$ matrix $\mathbf{X}$, where p is the dimensionality of the solution to be specified in advance by the user. Therefore, row i from $\mathbf{X}$ gives the coordinates for object i. Let d_{ij} be the Euclidean distance between rows i and j of $\mathbf{X}$ defined as

$$d_{ij} = \sqrt{\sum_{s=1}^{p} (x_{is} - x_{js})^2} \tag{1}$$

that is, the length of the shortest line connecting points i and j. The objective of MDS is to find a matrix $\mathbf{X}$ such that d_{ij} matches δ_{ij} as closely as possible. This objective can be formulated in a variety of ways but here we use the definition of raw-Stress σ^2, that is,

$$\sigma^2 = \sum_{i=2}^{n} \sum_{j=1}^{i-1} w_{ij} \left(\delta_{ij} - d_{ij}\right)^2 \tag{2}$$

by Kruskal [5] who was the first one to propose a formal measure for doing MDS. This measure is also referred to as the least-squares MDS model. Note that due to the symmetry of the dissimilarities and the distances, the summation only involves the pairs i, j where $i > j$. Here, w_{ij} is a user defined weight that must be nonnegative. The minimization of σ^2 is a complex problem. Therefore, MDS programs use iterative numerical algorithms to find a matrix $\mathbf{X}$ for which σ^2 is a minimum. In addition to the raw stress measure there exist other measures for doing stress. One of them is normalized raw stress, which is simply raw stress divided by the sum of squared dissimilarities. The advantage of this measure over raw stress is that its value is independent of the scale and the number of dissimilarities. The second measure is Kruskal's stress-1 which is equal to the square root of raw stress divided by the sum of squared distances. A third measure is Kruskal's stress-2, which is similar to stress-1 except that the denominator is based on the variance of the distances instead of the sum of squares. Another measure that seems reasonably popular is called S-stress and it measures the sum of squared error between squared distances and squared dissimilarities.

Because Euclidean distances do not change under rotation, translation, and reflection, these operations may be freely applied to MDS solution without affecting the raw-stress. Many MDS programs use this indeterminacy to center the coordinates so that they sum to zero dimension wise. The freedom of rotation is often exploited to put the solution in so-called principal axis orientation. That is, the axis are rotated in such a way that the variance of $\mathbf{X}$ is maximal along the first dimension, the second dimension is uncorrelated to the first and has again maximal variance, and so on.

In order to assess the quality of the MDS solution we can study the differences between the MDS solution and the data. One convenient way to do this is by inspecting the so-called Shepard diagram. A Shepard diagram shows both the transformation and the error. Let p_{ij} denote the proximity between objects i and j. Then, a Shepard diagram plots simultaneously the pairs (p_{ij}, d_{ij}) and (p_{ij}, δ_{ij}). By connecting the (p_{ij}, δ_{ij}) points a line is obtained representing the relationship between the proximities and the disparities. The vertical distances between the (p_{ij}, δ_{ij}) points and (p_{ij}, d_{ij}) are equal to $\delta_{ij} - d_{ij}$, that is, they give the errors of representation for each pair of objects. Hence, the Shepard diagram can be used to inspect both the residuals of the MDS solution and the transformation. Outliers can be detected as well as possible systematic deviations.

Measuring and predicting human judgment is an extremely complex and problematic task. There have been many techniques developed to deal with such type of problems. These techniques fall under a generic category called Multidimensional Scaling (MDS). Generally speaking, MDS techniques develop spatial representations of psychological stimuli or other complex objects about which people make judgments (e.g., preference, relatedness), that is, they represent each object as a point in an n-dimensional space. What distinguishes MDS from other similar techniques (e.g., factor analysis) is that in MDS there are no preconceptions about which factors might drive each dimension. Therefore, the only data needed are measures for the similarity between each possible pair of objects under study. The result is the transformation of the data into similarity measures that can be represented by Euclidean distances in a space of unknown dimensions [6]. The greater the similarity of two objects, the closer they are in the n-dimensional space. After having the distances between all the objects, the MDS techniques analyze how well they can be fitted by spaces of different dimensions. The analysis is normally made by gradually increasing the number of dimensions until the quality of fit (measured for example by the correlation between the data and the distance) is little improved with the addition of a new dimension. In practice, a good result is normally reached well before the number of dimensions theoretically needed to a perfectly fit is reached (i.e., $N-1$ dimensions for N objects) [7–10].

In the MDS method, a small distance between two points corresponds to a high correlation between two stock markets and a large distance corresponds to low or even negative correlation [11, 12]. A correlation of one should lead to zero distance between the points representing perfectly correlated stock markets. MDS tries to estimate the distances for all pairs of stock markets to match the correlations as close as possible. MDS may thus be seen as an exploratory technique without any distributional assumptions on the data. The distances between the points in the MDS maps are generally not difficult to interpret and thus may be used to formulate more specific models or hypotheses. Also, the distance between two points should be interpreted as being the distance conditional on all the other distances. One possibility to obtain such an approximate solution is given by minimizing the stress function. The obtained representation of points is not unique in the sense that any rotation or translation of the points retains the distances [13].

3 Dynamics of Financial Indexes

In this section, we study numerically the 25 selected stock markets, including seven American, eleven European, and seven Asian/Pacific markets.

Our data consist of the n daily close values of $S = 25$ stock markets,listed in Table 1, from January 2, 2000, up to December 31, 2009, to be denoted as $x_i(t), 1 \leq t \leq n, i = 1, \ldots, S$.

The data are obtained from data provided by Yahoo Finance web site [14] and [15] , and they measure indexes in local currencies.

Figure 1 depicts the time evolution, of daily, closing price of the 25 stock markets versus year with the well-known noisy and "chaotic-like" characteristics.

Table 1 Twenty-five stock markets

i	Stock market index	Abbrev.	Country
1	Dutch Euronext Amsterdam	aex	Netherlands
2	Index of the Vienna Bourse	atx	Austria
3	EURONEXT BEL-20	bfx	Belgium
4	Bombay Stock Exchange Index	bse	India
5	So Paulo (Brazil) Stock	bvsp	Brazil
6	Budapest Stock Exchange	bux	Hungary
7	Dow Jones Industrial	dji	USA
8	Cotation Assiste en Continu	cac	France
9	Footsie	ftse	United Kingdom
10	Deutscher Aktien Index	dax	Germany
11	Standard & Poor's	sp500	USA
12	Toronto Stock Exchange	tsx	Canada
13	Stock Market Index in Hong Kong	hsi	Hong Kong
14	Iberia Index	ibex	Spain
15	Jakarta Stock Exchange	jkse	Indonesia
16	Stock Market Index of South Korea	ks11	South Korea
17	Italian Bourse	mibtel	Italy
18	Bolsa Mexicana de Valores	mxx	Mexico
19	Tokyo Stock Exchange	nikkei	Japan
20	NASDAQ	ndx	USA
21	New York Stock Exchange	nya	USA
22	Stock Exchange of Portugal	psi20	Portugal
23	Shanghai Stock Exchange	ssec	China
24	Swiss Market Index	ssmi	Switzerland
25	Straits Times Index	sti	Singapore

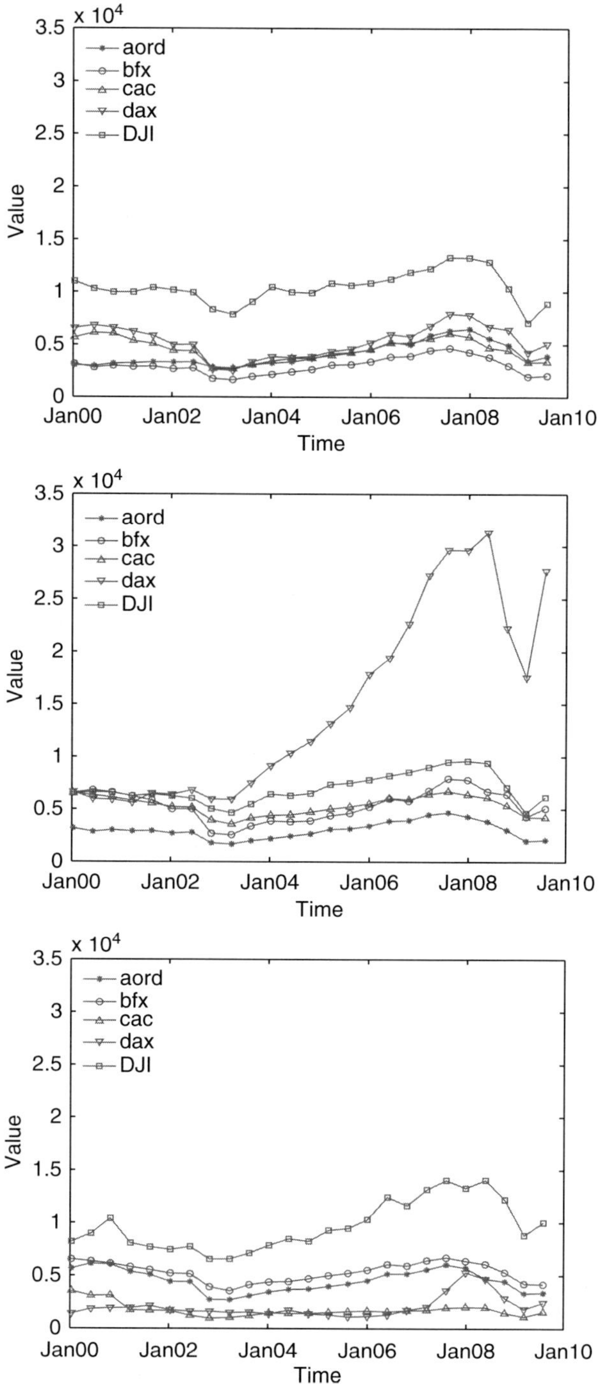

Fig. 1 Time series for the 25 indexes from January 2000, up to December 2009

The section is organized into three subsections, the first adopts an analysis based on a Fourier transform (FT) “distance measure”, the second adopts an analysis based on the correlation of the time evolution, and the third adopts a metrics based on histogram distances.

3.1 MDS Analysis Based on Fourier Transform

We calculate the Fourier transform (FT) for each one of the indexes, leading to the values $\mathrm{Re}\left[\mathscr{F}\left\{x_k(t)\right\}\right] + \mathrm{j}\ \mathrm{Im}\left[\mathscr{F}\left\{x_k(t)\right\}\right]$, $k = \{1, \ldots, S\}$. It is adopted the “distance measure” defined as $(i, j = 1, \ldots, S)$:

$$s(i,j) = \frac{\sum\limits_{\Omega}\left(|\mathrm{Re}_i - \mathrm{Re}_j|^2 + |\mathrm{Im}_i - \mathrm{Im}_j|^2\right)}{\sum\limits_{\Omega}\left(|\mathrm{Re}_i|^2 + |\mathrm{Im}_i|^2\right) \cdot \sum\limits_{\Omega}\left(|\mathrm{Re}_j|^2 + |\mathrm{Im}_j|^2\right)}, \tag{3}$$

where Ω is the set of sampling frequencies for the FT calculation, and Re_i, Im_i, Re_j and Im_j are the values of real and imaginary parts of the FT. We get the matrix **M**, where each cell represents the distance between a pair of FTs.

In order to reveal possible relationships between the signals the MDS [7] technique is used and several distance criteria are tested. The Sammon criterion [16, 17], that tries to optimize a cost function that describes how well the pairwise distances in a data set are preserved, revealed good results and is adopted in this work.

Figures 2 and 3 show the 2D and 3D MDS locus of index positioning in the perspective of the expressions (3), respectively. Figure 4 depicts the stress as function of the dimension of the representation space, revealing that, as usual, a high dimensional space would probably describe slightly better the “map” of the 25 indexes. However, the three dimensional representations were adopted because the graphical representation is easier to analyze while yielding a reasonable accuracy. Moreover, the resulting Sheppard plot, represented in Fig. 5 shows that a good distribution of points around the 45° line is obtained.

Analyzing Figs. 2 and 3 is visible that the MDS indexes forms an arc shapes in all the stock market indexes under analysis. How can we interpret these regularities? This is an interesting question for further research and a matter of decision of the stock market handlers.

3.2 MDS Analysis Based on Time Correlation

In this section, we apply the MDS method to the time correlation of all the stock markets.

For the 25 markets, we consider the time correlations between the daily close values. We first compute the correlations among the 25 stock markets obtained a $S \times S$ matrix and then apply MDS. In this representation, points represent the stock markets.

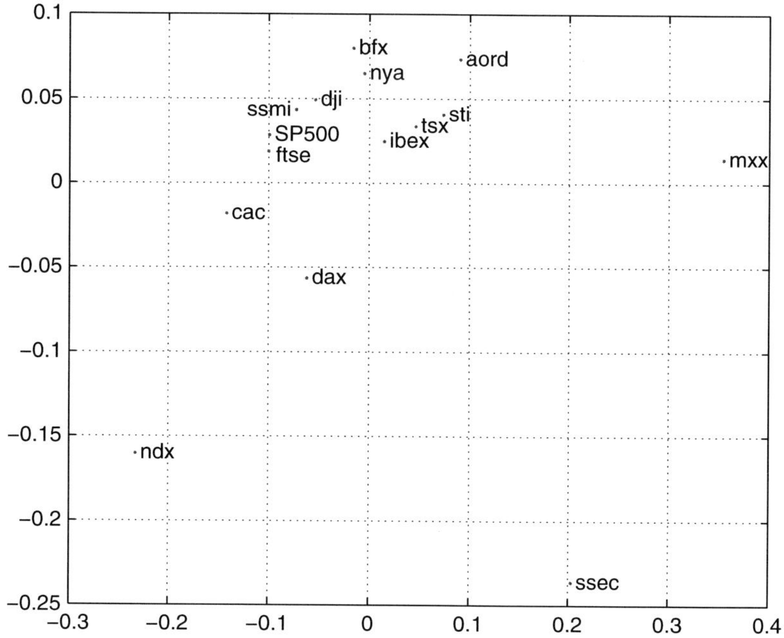

Fig. 2 Two-dimensional MDS graph for the 25 indexes based on the FT distance (3)

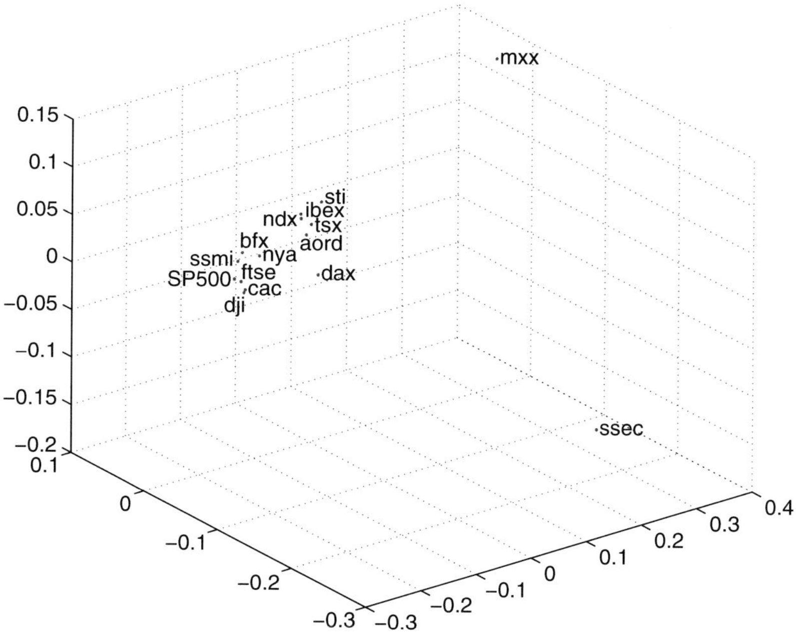

Fig. 3 Three-dimensional MDS graph for the 25 indexes based on the FT distance (3)

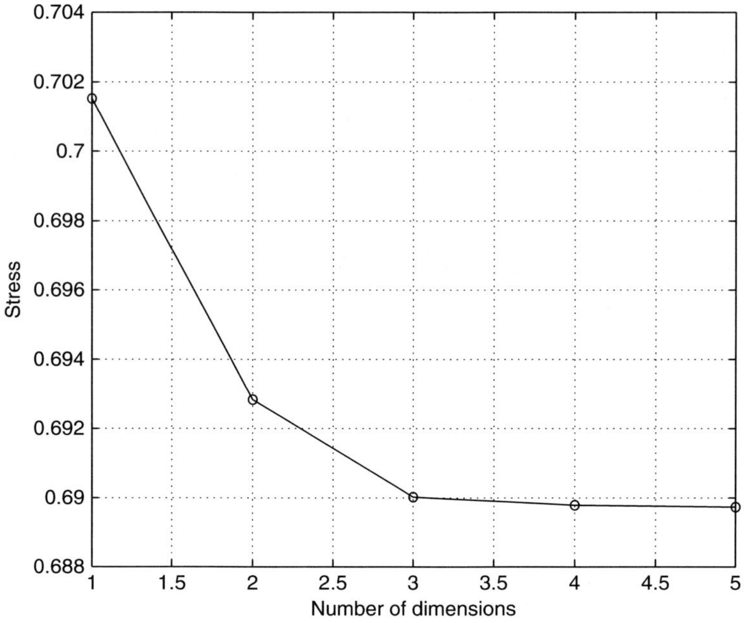

Fig. 4 Stress plot of the MDS representation, for the 25 indexes, *vs* number of dimensions, using the FT distance (3)

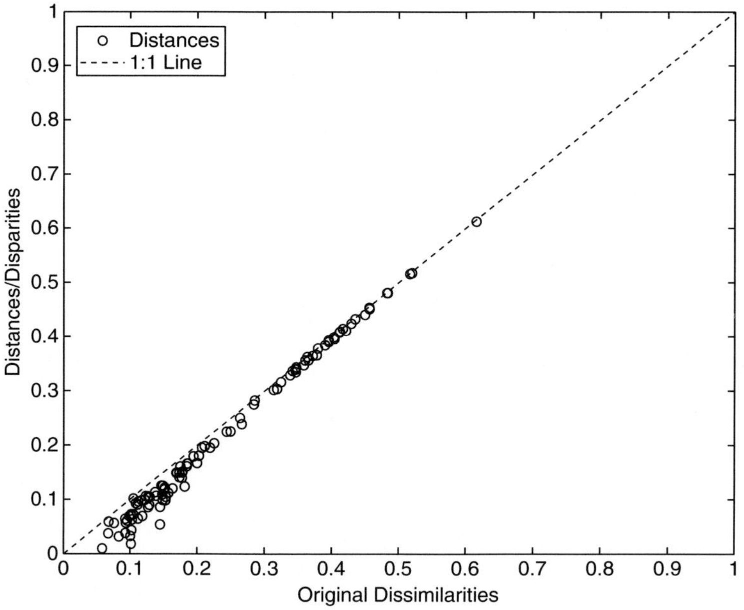

Fig. 5 Shepard plot of the three-dimensional MDS representation, for the 25 indexes,using the FT distance (3)

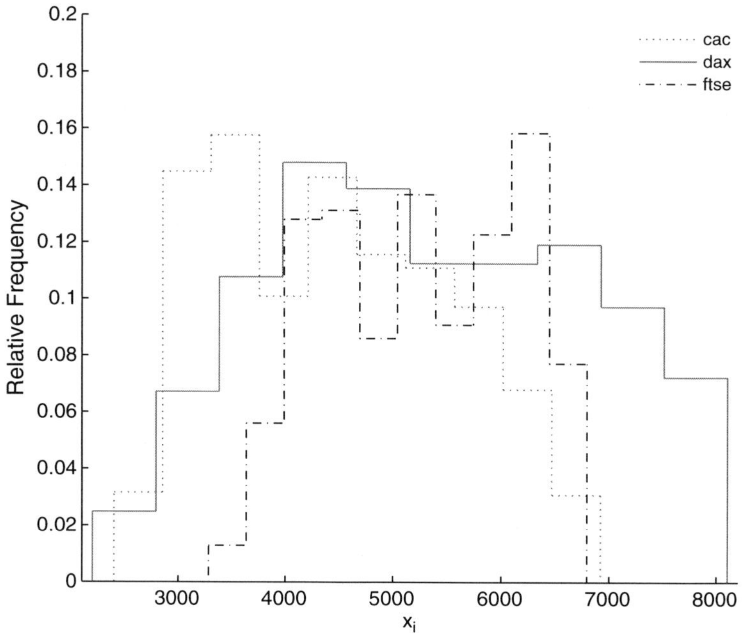

Fig. 6 Two-dimensional MDS graph for the 25 indexes using time correlation, according (4)

In order to reveal possible relationships between the market stocks index, the MDS technique is used. In this perspective, several MDS criteria are tested. The Sammon criterion revealed good results and is adopted in this work [18]. For this purpose, we calculate 25×25 matrix **M** based on a correlation coefficient $c(i, j)$, that provides a measurement of the similarity between two indexes and is defined in (4). In matrix **M**, each cell represents the time correlation between a pair of indexes, $i, j = 1, \ldots, S$.

$$c(i,j) = \left(\frac{\frac{1}{n} \sum_{t=1}^{n} x_i(t) \cdot x_j(t)}{\sqrt{\frac{1}{n} \sum_{t=1}^{n} (x_i(t))^2 \cdot \frac{1}{n} \sum_{t=1}^{n} (x_j(t))^2}} \right)^2 \tag{4}$$

Figures 6 and 7 show the 2D and 3D locus of each index positioning in the perspective of expression (4), respectively. Figure 8 depicts the stress as function of the dimension of the representation space, revealing that a three-dimensional space describe a with reasonable accuracy the "map" of the 25 signal indexes. Moreover, the resulting Sheppard plot, represented in Fig. 9, shows that a good distribution of points around the 45° line is obtained.

There are several empirical conclusions one can draw from the graphs in Figs. 6 and 7, and we will mention just a few here. We can clearly observe that there seem

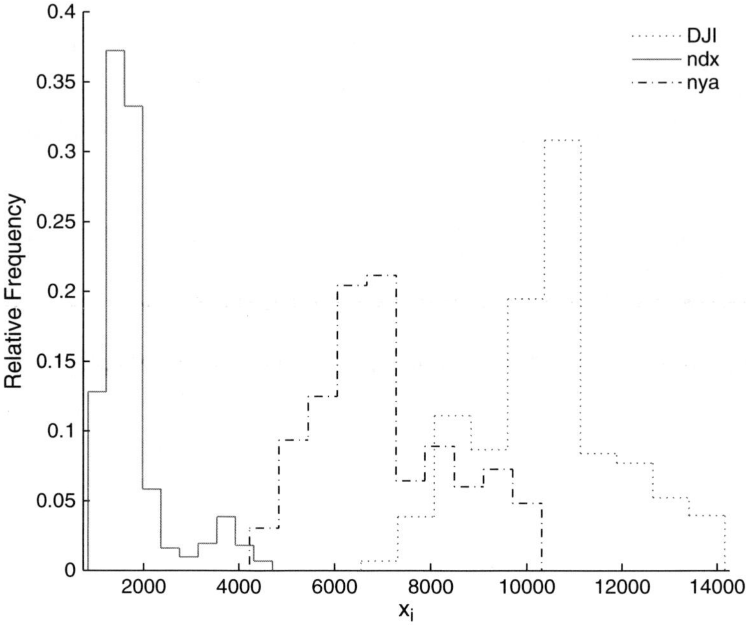

Fig. 7 Three-dimensional MDS graph for the 25 indexes using time correlation, according (4)

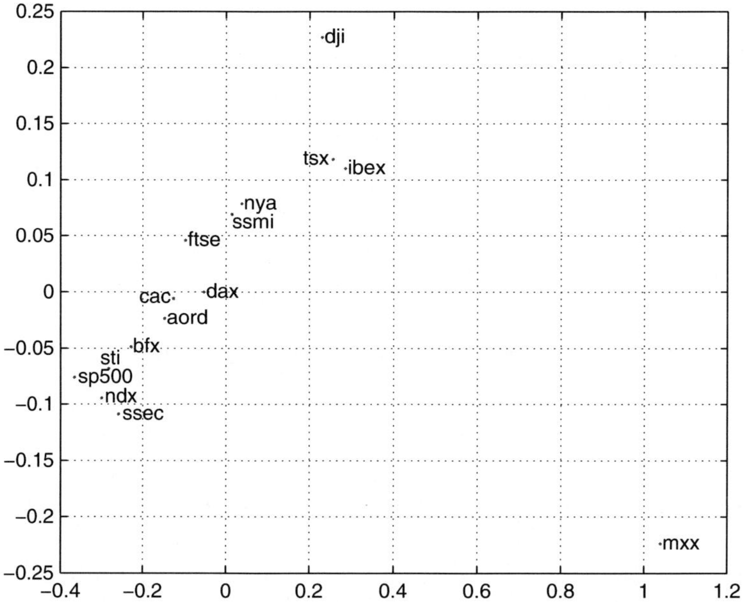

Fig. 8 Stress plot of MDS representation of the 25 indexes *vs* number of dimension using time correlation, according (4)

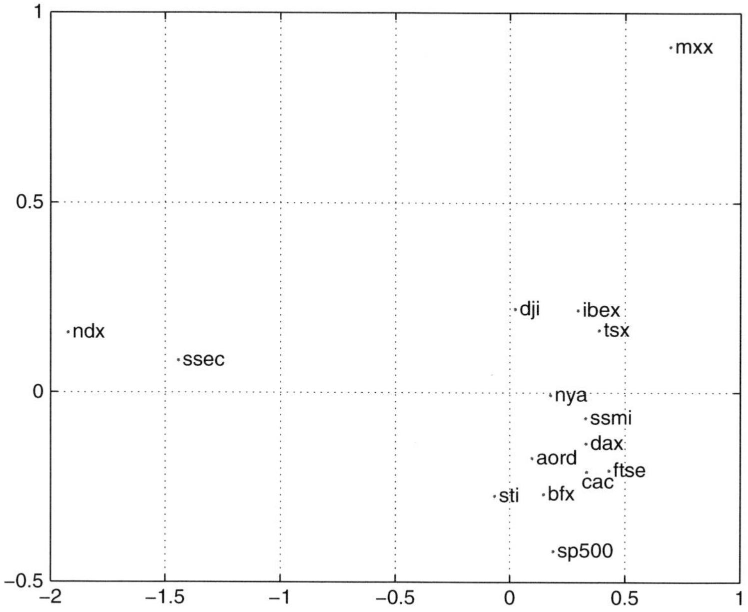

Fig. 9 Shepard plot for MDS with a three-dimensional representation of the 25 indexes using time correlation, according (4)

to emerge clusters, which show similar behavior [19]. Hence, there does not seem to be a single world market, but perhaps there are several important regional markets. This last observation would match with standard financial theory which tells us that higher (lower) volatility corresponds with higher (lower) returns. Indeed, if this would be the case, one would expect to see similar patterns over time across returns and volatility.

3.3 MDS Analysis Based on Histograms

For each of the 25 indexes, we draw the corresponding histogram of relative frequency and we calculate statistical descriptive parameters like the arithmetic mean (μ_i), the standard deviation (σ_i) and the Pearson's Kurtosis coefficient γ_i.

For all the 25 indexes, we calculate the "histogram's distance" [20, 21], d_{h}, using (5):

$$d_{\mathrm{h}}(i,j) = \sqrt{\frac{(\mu_i - \mu_j)^2}{A} + \frac{(\sigma_i - \sigma_j)^2}{B} + \frac{(\gamma_i - \gamma_j)^2}{C}}, \tag{5}$$

where $i, j = 1, \ldots, S$, $A = [\max(\mu_i - \mu_j)]^2$, $B = [\max(\sigma_i - \sigma_j)]^2$, $C = [\max(\gamma_i - \gamma_j)]^2$.

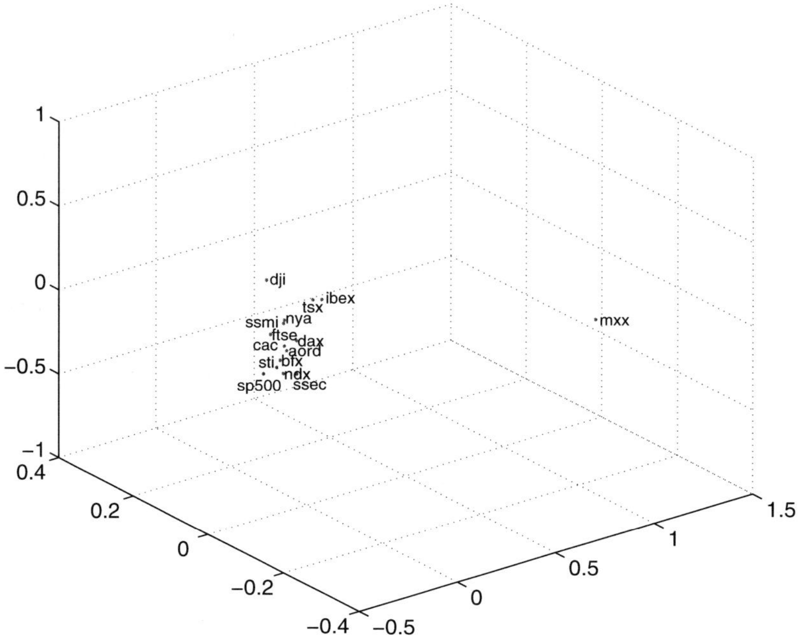

Fig. 10 Two-dimensional MDS graph for the 25 indexes using histogram's distance d_h

We get the matrix **M**, where each cell represents the histogram's distance between a pair of indexes.

Figures 10 and 11 show, respectively, the 2D and 3D locus of each index positioning in the perspective of the expression (5), demonstrating differences between the corresponding MDS plots.

Figure 12 depicts the stress *vs* the dimension of the representation space, for d_h, revealing that a three-dimensional space describes with reasonable accuracy the "map" of the 25 indexes. The resulting Sheppard plots, represented in Fig. 13, shows that a good distribution of points around the 45° line is obtained for the indices [22].

Curiously in the chart corresponding to the MDS based on correlation (Fig. 6), we can see a parabolic shape with the S&P500 index at the vertex, and the NDX, and ATX at the top. However, in the chart corresponding to the MDS based on the histogram distances (Fig. 10) and in the chart corresponding to the MDS based on the FT distance (Fig. 2), such parabolic shape form cannot be found.

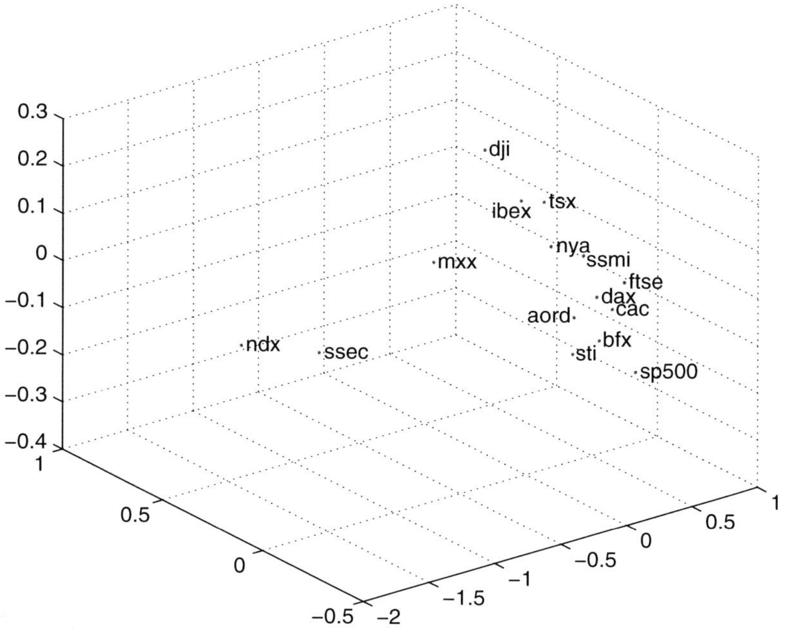

Fig. 11 Three-dimensional MDS graph for the 25 indexes using histogram's distance d_{h}

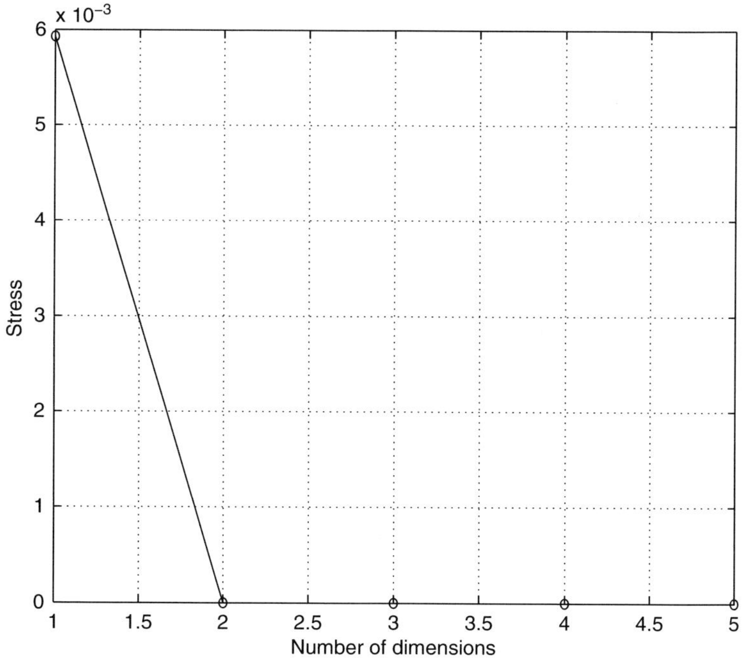

Fig. 12 Stress plot of MDS representation of the 25 indexes *vs* number of dimension using histogram's distance d_{h}

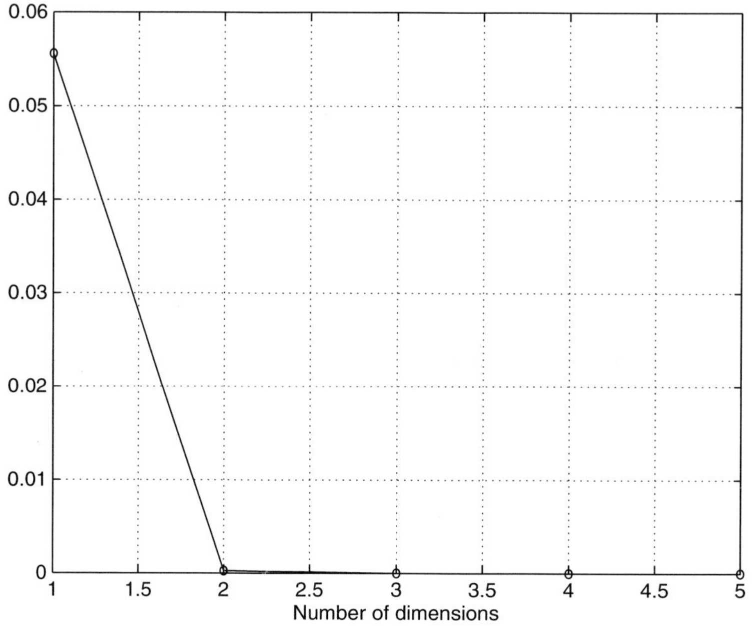

Fig. 13 Shepard plot for MDS with a three-dimensional representation of the 25 indexes, using histogram's distance d_{h}

4 Conclusion

In this chapter, we proposed simple graphical tools to visualize time-varying correlations between stock market behavior. We illustrated our MDS-based method daily close values of 25 stock markets.There are several issues relevant for further research. A first issue concerns applying our method to alternative data sets, with perhaps different sampling frequencies or returns and absolute returns, to see how informative the method can be in other cases. A second issue concerns taking the graphical evidence seriously and incorporating it in an econometric time series model to see if it can improve empirical specification strategies.

References

1. R. Nigmatullin, Communications in Nonlinear Science and Numerical Simulation **15**(3), 637 (2010)
2. R.V. Mendes, M.J. Oliveira, Economics: The Open-Access, Open-Assessment E-Journal **2**(2008-22) (2008). URL http://www.economics-ejournal.org/economics/discussionpapers/2008-22
3. R.V. Mendes, Nonlinear Dynamics **55**(4), 395 (2009)
4. R.B.A.L.S.H. Plerou V., Gopikrishnan P., Physica A **279**, 443 (2000)

5. J. Kruskal, Psychometrika **29**(1), 1 (1964). URL http://dx.doi.org/10.1007/BF02289565
6. I. Borg, P. Groenen, *Modern Multidimensional Scaling:Theory and Applications* (Springer, New York, 2005)
7. T. Cox, M. Cox, (Chapman & HallCrc, New York, 2001)
8. J. Kruskal, M. Wish, *Multidimensional Scaling* (Sage Publications, Inc., Newbury Park, CA, 1978)
9. J. Woelfel, G.A. Barnett, Quality and Quantity **16**(6), 469 (1982)
10. J.O. Ramsay, Psychometrika **45**(1), 139 (1980)
11. S. Nirenberg, P.E. Latham, Proc. National Academy of Sciences **100**(12), 7348 (2003)
12. F.B. Duarte, J.T. Machado, G.M. Duarte, Nonlinear Dynamics **61**(4), 691 (2010)
13. A. Buja, D. Swayne, M. Littman, N. Dean, H.Hofmann, L.Chen, Journal of Computational and Graphical Statistics **17**(85), 444 (2008)
14. http://finance.yahoo.com
15. http://www.bse.hu
16. J.W. Sammon, IEEE Trans. Comput. **18**(5), 401 (1969)
17. G.A.F. Seber, *Multivariate Observations* (J. Wiley & Sons, New York, 1986)
18. B. Ahrens, Hydrology and Earth System Sciences (10), 197 (2006)
19. J.T. Machado, G.M. Duarte, F.B. Duarte, Nonlinear Dynamics
20. F. Serratosa, A. Sanroma, G.and Sanfeliu, Lecture Notes in Computer Science **4756**(1), 115 (2008)
21. B. Sierra, E. Lazkano, E. Jauregi, I. Irigoien, Decis. Support Syst. **48**(1), 180 (2009). DOI http://dx.doi.org/10.1016/j.dss.2009.07.010
22. J.T. Machado, F.B. Duarte, G.M. Duarte, in *Proceedings of 3rd Conference on Nonlinear Science and Complexity*, ed. by D. Baleanu (Cankaya University, 2010)

Index

J.A.T. Machado et al., *Nonlinear and Complex Dynamics: Applications in Physical, Biological, and Financial Systems*, DOI 10.1007/978-1-4614-0231-2,

Printed by Books on Demand, Germany